GATE PREP SERIES

GATE

Graduate Aptitude Test in Engineering

2022

15 Years

Section-Wise Solved Papers

(2007-2021)

Aerospace Engineering

G. K. PUBLICATIONS (P) LTD.

Title : GATE 2022 - Aerospace Engineering - 15 Years' Section-wise Solved Papers 2007-21

Language : English

Editor's Name : Vinit Garg

Typeset & Published by :

Career Launcher Infrastructure (P) Ltd.

A-45, Mohan Cooperative Industrial Area, Near Mohan Estate Metro Station, New Delhi - 110044

Marketed by :

G.K. Publications (P) Ltd.

Plot No. 9A, Sector-27A, Mathura Road, Faridabad, Haryana-121003

ISBN : **978-93-91061-32-6**

Printer's Details :

For product information :

Visit ***www.gkpublications.com*** or email to ***gkp@gkpublications.com***

About Authors

BIPLAB SADHUKHAN

Biplab Sadhukhan received his B.Tech degree in Aerospace Engineering in 2013. He completed his M.Tech from Indian Institute of Technology Kharagpur, in 2017 with an excellence of his class. He is currently working toward the Ph.D. degree at Indian Institute of Technology Kharagpur and working on several projects for Government. of India.

IQBAL SINGH

Iqbal Singh comes from a humble background, born and brought up in the village Lahal of Chamba district of Himachal Pradesh. He did his high and senior secondary school at his native place. He completed B.Tech in Aeronautical engineering from IK Gujral Punjab technical university, Punjab in 2015.After working for 1 year at DRDO Chandigarh, he decided to pursue masters in Aerospace Engineering from Indian Institute of Science, Bangalore. In 2018, he graduated from IISc Bangalore with masters and joined Tata Advanced Systems, Bangalore. Apart from aerospace engineering, he has great passion for music, dance and writing.

He dedicates this book to his parents (Mr. Purushotam Kumar and Sumna Devi) for their love, support and motivation. He would like to thank Sohan Mal, Sc. Vikrant Satya and Dr. S.B.Kandagal for sowing the first seeds of his research career.

PRABHAKAR KUMAR

Prabhakar Kumar has completed B.Tech in Aerospace Engineering with first class degree with distinction in 2013. After completing his bachelors degree, he worked at Hindustan Aeronautics Limited, Aeronautical Development Agency, Bangalore, and several project of Ministry of Defence Government of India. He has completed his M.E in Aerospace Engineering from Birla Institute of Technology, Mesra, Ranchi in 2018.

RANJAY KR SINGH

Ranjay kr Singh has completed B.E. from aeronautical society of India in aero-mechanical stream. Then M.E. from Birla Institute of Technology, Mesra, Ranchi in space engineering and rocketry with specialisation in aerodynamics. He has more than two year teaching experience in aeronautical stream, currently he is employed as a assistant professor in school of aeronautics, Neemrana Rajasthan.

GATE
Aerospace Engineering

Contents

GATE

Graduate Aptitude Test in Engineering

IIT Institutes

GATE 2021 conducted by
Indian Institute of Technology, Mumbai

GATE 2022 will be conducted by
Indian Institute of Technology, Kharagpur

Indian Institute of Technology, Delhi

Indian Institute of Technology, Chennai

Indian Institute of Technology, Guwahati

Indian Institute of Technology, Roorkee

Indian Institute of Technology, Kanpur

Preface

The Graduate Aptitude Test in Engineering (GATE) is an online exam conducted by the IITs for admissions to PG courses in IITs, IISc Bangalore, NITs and many state run universities as well as private universities. Also there are more than 37 PSUs that use GATE score for recruitments. A large number of corporates are also using GATE score as a tool to screen students for placements.

GK Publications is well known as the "publisher of choice" to students preparing for GATE and other technical examinations in the country. We published the first set of books in 1994 when GATE exam, both objective and conventional, was conducted in the paper and pencil environment, and used as a check point for entry to post graduate courses in IITs and IISCs. At that time, students had little access to technology and relied mainly on instructor led learning followed by practice with books available for these examinations.

A lot has changed since then!

Today, GATE is conducted in an online only mode with multiple choice and numerical based questions. The score is valid for three years and is used not only for post graduate courses but is also used by major PSUs for recruitment. Today's students have easy access to technology and the concept of a monologue within the classroom has changed to dialogue where students come prepared with concepts and then discuss topics. They learn a lot of things on the go with their mobile devices and practice for mock tests online.

We, as a leading publisher of GATE books, have also embraced change. Today, our books are no more guides and papers only but come with a fully supported mobile app and a web portal. The mobile app provides access to video lectures, short tests and regular updates about the exam. The web portal in additional to what is available on the app provides full length mock tests to mimic the actual exam and help you gauge your level of preparedness. The combination of practice content in print, video lectures, and short and full length tests on mobile and web makes this product a complete courseware for GATE preparation.

This book includes previous years GATE questions along with detailed solution of each question for better understanding. It will help the GATE aspirants to know an idea about the pattern of questions asked in GATE examination.

We also know that improvement is a never ending process and hence we welcome your suggestions and feedback or spelling and technical errors if any. Please write to us at gkp@gkpublications.com

We hope that our small effort will help you prepare well for the examination.

We wish you all the best!

GK Publications Pvt. Ltd.

About GATE

The Graduate Aptitude Test in Engineering (GATE) conducted by IISc and IITs has emerged as one of the bench mark tests for engineering and science aptitude in facilitating admissions for higher education (M.Tech./Ph.D.) in IITs, IISc and various other Institutes/Universities/Laboratories in India. With the standard and high quality of the GATE examination in 27 disciplines of engineering and science, Humanities and Social Sciences subjects, it identifies the candidate's understanding of a subject and aptitude and eligibility for higher studies. During the last few years, GATE score is also being used as one of the criteria for recruitment in Government Organizations such as Cabinet Secretariat, and National/State Public Sector Undertakings in India. Because of the importance of the GATE examination, the number of candidates taking up GATE exams has increased tremendously. GATE exams are conducted by the IITs and IISc as a computer based test having multiple choice questions and numerical answer type questions. The questions are mostly fundamental, concept based and thought provoking. From 2017 onwards GATE Exam is being held in Bangladesh, Ethiopia, Nepal, Singapore, Sri Lanka and United Arab Emirates. An Institute with various nationalities in its campus widens the horizons of an academic environment. A foreign student brings with him/her a great diversity, culture and wisdom to share. Many GATE qualified candidates are paid scholarships/assistantship, especially funded by Ministry of Human Resources Development, Government of India and by other Ministries. IIT, Kharagpur is the Organizing Institute for GATE 2022.

Why GATE?

Admission to Post Graduate and Doctoral Programmes

Admission to postgraduate programmes with MHRD and some other government scholarships/ assistantships in engineering colleges/institutes is open to those who qualify through GATE. GATE qualified candidates with Bachelor's degree in Engineering/Technology/Architecture or Master's degree in any branch of Science/Mathematics/Statistics/Computer Applications are eligible for admission to Master/Doctoral programmes in Engineering/Technology/Architecture as well as for Doctoral programmes in relevant branches of Science with MHRD or other government scholarships/ assistantships. Candidates with Master's degree in Engineering/Technology/Architecture may seek admission to relevant Ph.D programmes with scholarship/assistantship without appearing in the GATE examination.

Financial Assistance

A valid GATE score is essential for obtaining financial assistance during Master's programs and direct Doctoral programs in Engineering/Technology/Architecture, and Doctoral programs in relevant branches of Science in Institutes supported by the MHRD or other Government agencies. As per the directives of the MHRD, the following procedure is to be adopted for admission to the post-graduate programs (Master's and Doctoral) with MHRD scholarship/assistantship. Depending upon the norms adopted by a specific institute or department of the Institute, a candidate may be admitted directly into a course based on his/her performance in GATE only **or** based on his/her performance in GATE **and** an admission test/interview conducted by the department to which he/she has applied **and/or** the candidate's academic record. If the candidate is to be selected through test/interview for post-graduate programs, a minimum of 70% weightage will be given to the performance in GATE and the remaining 30% weightage will be given to the candidate's performance in test/interview and/or academic record, as per MHRD guidelines. The admitting institutes could however prescribe a minimum passing percentage of marks in the test/interview. Some colleges/institutes specify GATE qualification as the mandatory requirement even for admission without MHRD scholarship/assistantship.

To avail of the financial assistance (scholarship), the candidate must first secure admission to a program in these Institutes, by a procedure that could vary from institute to institute. Qualification in GATE is also a minimum requirement to apply for various fellowships awarded by many Government organizations. Candidates are advised to seek complete details of admission procedures and availability of MHRD scholarship/assistantship from the concerned admitting institution. The criteria for postgraduate admission with scholarship/assistantship could be different for different institutions. The management of the post-graduate scholarship/assistantship is also the responsibility of the admitting institution. Similarly, reservation of seats under different categories is as per the policies and norms prevailing at the admitting institution and Government of India rules. *GATE offices will usually not entertain any enquiry about admission, reservation of seats and / or award of scholarship / assistantship.*

PSU Recruitments

As many as 37 PSUs are using GATE score for recruitment. It is likely that more number of PSUs may start doing so by next year. Below is the list of PSUs:

MDL, BPCL, GAIL, NLC LTD, CEL, Indian Oil, HPCL, NBPC, NECC, BHEL, WBSEDCL, NTPC, ONGC, Oil India, Power Grid, Cabinet Secretariat, Govt. of India, BAARC, NFL, IPR, PSPCL, PSTCL, DRDO, OPGC Ltd., THDC India Ltd., BBNL, RITES, IRCON, GHECL, NHAI, KRIBHCO, Mumbai Railway Vikas Corporation Ltd. (MRVC Ltd.), National Textile Corporation, Coal India Ltd., BNPM, AAI, NALCO, EdCIL India.

Important :

1. Admissions in IITs/IISc or other Institutes for M.Tech./Ph.D. through GATE scores shall be advertised separately by the Institutes and GATE does not take the responsibility of admissions.
2. Cabinet Secretariat has decided to recruit officers for the post of Senior Field Officer (Tele) (From GATE papers of EC, CS, PH), Senior Research Officer (Crypto) (From GATE papers of EC, CS, MA), Senior Research Officer (S&T) (From GATE papers EC, CS, CY, PH, AE, BT) in the Telecommunication Cadre, Cryptographic Cadre and Science & Technology Unit respectively of Cabinet Secretariat. The details of the scheme of recruitment shall be published in National Newspaper/Employment News by the concerned authority.
3. Some PSUs in India have expressed their interest to utilize GATE scores for their recruitment purpose. The Organizations who intend to utilize GATE scores shall make separate advertisement for this purpose in Newspapers etc.

Who Can Appear for GATE?

Eligibility for GATE

Before starting the application process, the candidate must ensure that he/she meets the eligibility criteria of GATE given in Table.

Eligibility Criteria for GATE 2022

Degree/Program	Qualifying Degree/Examination	Description of Eligible Candidates	Expected Year of Completion
B.E. / B.Tech. / B.Pharm.	Bachelor's degree in Engineering / Technology (4 years after 10+2 or 3 years after B.Sc. / Diploma in Engineering / Technology)	Currently in the 3rd year or higher grade or already completed	2022
B. Arch.	Bachelor's degree of Architecture (5- year course) / Naval Architecture (4- year course) / Planning (4- year course)	Currently in the 3rd year or higher grade or already completed	2023 (for 5-year program), 2022 (for 4-year program)
B.Sc. (Research) / B.S.	Bachelor's degree in Science (Post-Diploma/4 years after 10+2)	Currently in the 3rd year or higher grade or already completed	2022
Pharm. D. (after 10+2)	6 years degree program, consisting of internship or residency training, during third year onwards	Currently in the 3rd/4th/5th/6th year or already completed	2024
M.B.B.S.	Degree holders of M.B.B.S. and those who are in the 5th/6th/7th semester or higher semester of such programme.	5th, 6th, 7th or higher semester or already completed	2022
M. Sc. / M.A. / MCA or equivalent	Master's degree in any branch of Arts/Science/Mathematics/Statistics/ Computer Applications or equivalent	Currently in the first year or higher or already Completed	2022
Int. M.E./ M.Tech. (Post-B.Sc.)	Post-B.Sc Integrated Master's degree programs in Engineering/ Technology (4-year program)	Currently in the 1st / 2nd/3rd/4th year or already completed	2024
Int. M.E./ M.Tech. or Dual Degree (after Diploma or 10+2)	Integrated Master's degree program or Dual Degree program in Engineering/Technology (5-year program)	Currently in the 3rd /4th/5th year or alreadycompleted	2023
B.Sc. / B.A. / B.Com.	Bachelor degree in any branch of Science / Arts / Commerce (3 years program)	Currently in the 3rd year or already completed	2021
Int. M.Sc. / Int. B.S. / M.S.	Integrated M.Sc. or 5-year integrated B.S.-M.S. program	Currently in the 3rd year or higher or already completed	2022
Professional Society Examinations (equivalent to B.E. / B.Tech. / B.Arch.)	B.E./B.Tech./B.Arch. equivalent examinations of Professional Societies, recognized by MoE/UPSC/AICTE (e.g. AMIE by Institution of Engineers-India, AMICE by the Institute of Civil Engineers-India and so on)	Completed Section A or equivalent of such professional courses	NA

In case a candidate has passed one of the qualifying examinations as mentioned above in 2020 or earlier, the candidate has to submit the degree certificate / provisional certificate / course completion certificate / professional certificate / membership certificate issued by the society or institute. In case, the candidate is expected to complete one of the qualifying criteria in 2022 or later as mentioned above, he/she has to submit a certificate from Principal or a copy of marks card for section A of AMIE.

Certificate From Principal

Candidates who have to submit a certificate from their college Principal have to obtain one from his/ her institution beforehand and upload the same during the online submission of the application form.

Candidates With Backlogs

Candidates, who have appeared in the final semester/year exam in 2022, but with a backlog (arrears/ failed subjects) in any of the papers in their qualifying degree should upload a copy of any one of the mark sheets of the final year,

OR

obtain a declaration from their Principal along with the signature and seal beforehand and upload the same during the online submission of the application form.

GATE Structure

GATE 2022 Papers

GATE 2022 will be conducted on 29 subjects (papers). Table below shows the list of papers and paper codes for GATE 2022. A candidate is allowed to appear in ANY ONE or UP TO TWO papers of the GATE examination. However, note that the combination of TWO papers in which a candidate can appear MUST be from the pre-defined list as given in Table. Also note that for a paper running in multiple sessions, a candidate will be mapped to appear for the examination in one of the sessions ONLY.

List of GATE 2022 Papers and Corresponding Codes

GATE Paper	Code	GATE Paper	Code
Aerospace Engineering	AE	Instrumentation Engineering	IN
Agricultural Engineering	AG	Mathematics	MA
Architecture and Planning	AR	Mechanical Engineering	ME
Bio-medical Engineering	BM	Mining Engineering	MN
Biotechnology	BT	Metallurgical Engineering	MT
Civil Engineering	CE	Naval Architecture and Marine Engineering	NM
Chemical Engineering	CH	Petroleum Engineering	PE
Computer Science and Information Technology	CS	Physics	PH
Chemistry	CY	Production and Industrial Engineering	PI
Electronics and Communication Engineering	EC	Statistics	ST
Electrical Engineering	EE	Textile Engineering and Fibre Science	TF
Environmental Science & Engineering	ES	Engineering Sciences	XE*
Ecology and Evolution	EY	Humanities & Social Sciences	XH**
Geomatics Engineering	GE	Life Sciences	XL***
Geology and Geophysics	GG		

Note: *"Geomatics Engineering (GE)" and "Naval Architecture and Marine Engineering (NM)" are two new papers introduced in GATE 2022.*

*XE Paper Sections	Code	**XH Paper Sections	Code	***XL Paper Sections	Code
Engineering Mathematics (Compulsory) (15 marks)	A	Reasoning and Comprehension (Compulsory) (25 marks)	B1	Chemistry (Compulsory) (25 marks)	P
Any TWO optional Sections (2x35 = 70 marks)		Any ONE optional Section (60 marks)		Any TWO optional Sections (2x30 = 60 marks)	
Fluid Mechanics	B	Economics	C1	Biochemistry	Q
Materials Science	C	English	C2	Botany	R
Solid Mechanics	D	Linguistics	C3	Microbiology	S
Thermodynamics	E	Philosophy	C4	Zoology	T
Polymer Science and Engineering	F	Psychology	C5	Food Technology	U
Food Technology	G	Sociology	C6		
Atmospheric and Oceanic Sciences	H				

*XE (Engineering Sciences), **XH (Humanities & Social Sciences), ***XL (Life Sciences), papers are of general nature and will be comprised of Sections listed in the above table

Note: *Each subject/paper is of total 100 marks. General Aptitude (GA) section of 15 marks is common for all papers. Hence remaining 85 marks are for the respective subject/paper code.*

Combination of Two Papers Allowed to Appear in GATE 2022 (subject to availability of infrastructure and schedule)

Code of The First (Primary) Paper	Codes of Papers Allowed as The Second Paper
AE	XE
AG	ES
AR	CE
BM	BT / XL
BT	BM / XL
CE	AR / ES
CH	CY / PE / XE
CS	MA
CY	CH / XL
EC	IN / PH
EE	IN
ES	AG / CE
EY	XL
GG	MN / PE / PH
IN	EC / EE / PH
MA	CS / PH / ST
ME	XE
MN	GG / XE
MT	PH / XE
PE	CH / GG / XE
PH	EC / GG / IN / MA / MT / ST
PI	XE
ST	MA / PH
TF	XE
XE	AE / CH / ME / MN / MT / PE / PI / TF
XH	----
XL	BM / BT / CY / EY

General Aptitude Questions

All the papers will have a few questions that test the General Aptitude (Language and Analytical Skills), apart from the core subject of the paper.

Duration and Examination Type

All the papers of the GATE 2022 examination will be for 3 hours duration and they consist of 65 questions for a total of 100 marks. Since the examination is an ONLINE computer based test (CBT), at the end of the stipulated time (3-hours), the computer screen will automatically close the examination inhibiting any further action.

Candidates will be permitted to occupy their allotted seats 40 minutes before the scheduled start of the examination. Candidates can login and start reading the instructions 20 minutes before the start of examination. The late login time (if any) recorded by the computer system MUST NOT be beyond 30 minutes from the actual starting time of the examination. Under NO circumstances, will a candidate be permitted to login after 30 minutes from the actual examination starting time. Candidates will NOT be permitted to leave the examination hall before the end of the examination.

Pattern of Question Papers

GATE 2022 may contain questions of THREE different types in all the papers:

(i) **Multiple Choice Questions (MCQ)** carrying 1 or 2 marks each, in all the papers and sections. These questions are objective in nature, and each will have choice of four answers, out of which ONLY ONE choice is correct.

Negative Marking for Wrong Answers: For a wrong answer chosen in a MCQ, there will be negative marking. For 1-mark MCQ, 1/3 mark will be deducted for a wrong answer. Likewise, for 2-mark MCQ, 2/3 mark will be deducted for a wrong answer.

(ii) **Multiple Select Questions (MSQ)** carrying 1 or 2 marks each in all the papers and sections. These questions are objective in nature, and each will have choice of four answers, out of which ONE or MORE than ONE choice(s) is / are correct.

Note: There is NO negative marking for a wrong answer in MSQ questions. However, there is NO partial credit for choosing partially correct combinations of choices or any single wrong choice.

(iii) **Numerical Answer Type (NAT)** Questions carrying 1 or 2 marks each in most of the papers and sections. For these questions, the answer is a signed real number, which needs to be entered by the candidate using the virtual numeric keypad on the monitor (keyboard of the computer will be disabled). No choices will be shown for these types of questions. The answer can be a number such as 10 or –10 (an integer only). The answer may be in decimals as well, for example, 10.1 (one decimal) or 10.01 (two decimals) or –10.001 (three decimals). These questions will be mentioned with, up to which decimal places, the candidates need to present the answer. Also, for some NAT type problems an appropriate range will be considered while evaluating these questions so that the candidate is not unduly penalized due to the usual round-off errors. Candidates are advised to do the rounding off at the end of the calculation (not in between steps). Wherever required and possible, it is better to give NAT answer up to a maximum of three decimal places.

Note: *There is NO negative marking for a wrong answer in NAT questions.*
Also, there is NO partial credit in NAT questions.

Marking Scheme – Distribution of Marks and Questions

General Aptitude (GA) Questions

In all papers, GA questions carry a total of 15 marks. The GA section includes 5 questions carrying 1-mark each (sub-total 5 marks) and 5 questions carrying 2-marks each (sub-total 10 marks).

Question Papers other than GG, XE, XH and XL

These papers would contain 25 questions carrying 1-mark each (sub-total 25 marks) and 30 questions carrying 2-marks each (sub-total 60 marks) consisting of some MCQ type questions, while the remaining may be MSQ and / or NAT questions.

GG (Geology and Geophysics) Paper

Apart from the General Aptitude (GA) section, the GG question paper consists of two parts: Part A and Part B. Part A is compulsory for all the candidates. Part B contains two sections: Section 1 (Geology) and Section 2 (Geophysics). Candidates will have to attempt questions in Part A and questions in either Section 1 or Section 2 of Part B. The choice of Section 1 OR Section 2 of Part B has to be made at the time of online registration in GOAPS. At the examination hall, candidate cannot request for change of section.

Part A consists of 25 questions carrying 1-mark each (sub-total 25 marks and some of these will be MSQ and/or numerical answer type questions while remaining questions will be MCQ type). Either section of Part B (Section 1 and Section 2) consists of 30 questions carrying 2-marks each (sub-total 60 marks and some of these will be MSQ and/or numerical answer type questions while remaining questions will be MCQ type).

XE Paper (Engineering Sciences)

A candidate appearing in the XE paper has to answer the following:

- GA – General Aptitude carrying a total of 15 marks.
- Section A – Engineering Mathematics (Compulsory): This section contains 11 questions carrying a total of 15 marks: 7 questions carrying 1-mark each (sub-total 7 marks), and 4 questions carrying 2-marks each (sub-total 8 marks). Some questions will be of numerical answer type while remaining questions will be MCQ type.
- Any two of XE Sections B to H: The choice of two sections from B to H can be made during the examination after viewing the questions. Only TWO optional sections can be answered at a time. A candidate wishing to change midway of the examination to another optional section must first choose to deselect one of the previously chosen optional sections (B to H). Each of the optional sections of the XE paper (Sections B through H) contains 22 questions carrying a total of 35 marks: 9 questions carrying 1-mark each (sub-total 9 marks) and 13 questions carrying 2-marks each (sub-total 26 marks). Some questions will be of MSQ and/or numerical answer type while remaining questions will be MCQ type.

XH Paper (Humanities and Social Sciences)

A candidate appearing in the XH paper has to answer the following:

- **GA – General Aptitude** carrying a total of 15 marks.
- **Section B1 – Reasoning and Comprehension (Compulsory):** This section contains 15 questions carrying a total of 25 marks: 5 questions carrying 1-mark each (sub-total 5 marks) and 10 questions carrying 2-marks each (sub-total 20 marks). Some questions will be of MSQ and/or numerical answer type while remaining questions will be MCQ type.
- **Any ONE of XH Sections C1 to C6:** The ONE choice of section from C1 to C6 has to be made at the time of online registration in GOAPS. At the examination hall, candidate cannot request for change of section. Each of the optional sections of the XH paper (Sections C1 through C6) contains 40 questions carrying a total of 60 marks: 20 questions carrying 1-mark each (sub-total 20 marks) and 20 questions carrying 2-marks each (sub-total 40 marks). Some questions will be of MSQ and/or numerical answer type while remaining questions will be MCQ type.

XL Paper (Life Sciences)

A candidate appearing in the XL paper has to answer the following:

- **GA – General Aptitude** carrying a total of 15 marks.
- **Section P–Chemistry (Compulsory):** This section contains 15 questions carrying a total of 25 marks: 5 questions carrying 1-mark each (sub-total 5 marks) and 10 questions carrying 2-marks each (sub-total 20 marks). Some questions will be of MSQ and/or numerical answer type while remaining questions will be MCQ type.
- **Any two of XL Sections Q to U:** The choice of two sections from Q to U can be made during the examination after viewing the questions. Only TWO optional sections can be answered at a time. A candidate wishing to change midway of the examination to another optional section must first choose to deselect one of the previously chosen optional sections (Q to U). Each of the optional sections of the XL paper (Sections Q through U) contains 20 questions carrying a total of 30 marks: 10 questions carrying 1-mark each (sub-total 10 marks) and 10 questions carrying 2-marks each (sub-total 20 marks). Some questions will be of MSQ and/or numerical answer type while remaining questions will be MCQ type.

GATE Score

After the evaluation of the answers, the actual (raw) marks obtained by a candidate will be considered for computing the GATE Score. For multi-session papers (subjects), raw marks obtained by the candidates in different sessions will be converted to Normalized marks for that particular subject. Thus, raw marks (for single session papers) or normalized marks (for multi-session papers) will be used for computing the GATE Score, based on the qualifying marks.

Calculation of Normalized Marks for Multi-Session Papers

In GATE 2022 examination, some papers may be conducted in multi-sessions. Hence, for these papers, a suitable normalization is applied to take into account any variation in the difficulty levels of the question papers across different sessions. The normalization is done based on the fundamental assumption that "in all multi-session GATE papers, the distribution of abilities of candidates is the same across all the sessions". This assumption is justified since the number of candidates appearing in multi-session papers in GATE 2022 is large and the procedure for allocation of session to candidates is random. Further, it is also ensured that for the same multi-session paper, the number of candidates allotted in each session is of the same order of magnitude.

Based on the above, and considering various normalization methods, the committee arrived at the following formula for calculating the normalized marks for the multi-session papers.

Nsormalization mark of jth candidate in the ith session $\widehat{M_{ij}}$ is given by

$$\widehat{M_{ij}} = \frac{\bar{M}_t^g - M_q^g}{\bar{M}_{ti} - M_{iq}}(M_{ij} - M_{iq}) + M_q^g$$

where

M_{ij} : is the actual marks obtained by the jth candidate in i^{th} session

$\bar{M}_t^g$: is the average marks of the top 0.1% of the candidates considering all sessions

M_q^g : is the sum of mean and standard deviation marks of the candidates in the paper considering all sessions

$\bar{M}_{ti}$: is the average marks of the top 0.1% of the candidates in the ith session

M_{iq} : is the sum of the mean marks and standard deviation of the ith session

Calculation of GATE Score for All Papers

For all papers for which there is only one session, actual marks obtained by the candidates will be used for calculating the GATE 2022 Score. For papers in multi-sessions, normalized marks will be calculated corresponding to the raw marks obtained by a candidate and the GATE 2022 Score will be calculated based on the normalized marks.

The GATE 2022 score will be computed using the formula given below.

$$\text{GATE Score} = S_q + (S_t - S_q)\frac{(M - M_q)}{(\bar{M}_t - M_q)}$$

where

M : marks obtained by the candidate (actual marks for single session papers and normalized marks for multi-session papers)

M_q : is the qualifying marks for general category candidate in the paper

$\bar{M}_t$: is the mean of marks of top 0.1% or top 10 (whichever is larger) of the candidates who appeared in the paper (in case of multi-session papers including all sessions)

S_q : 350, is the score assigned to M_q

S_t : 900, is the score assigned to $\bar{M}_t$

In the GATE 2022 the qualifying marks (M_q) for general category student in each subject will be 25 marks (out of 100) or $\mu + \sigma$, whichever is larger. Here μ is the mean and σ is the standard deviation of marks of all the candidates who appeared in the paper.

After the declaration of results, GATE Scorecards can be downloaded by the GATE qualified candidates ONLY.

The GATE 2022 Committee has the authority to decide the qualifying mark/score for each GATE paper. In case of any claim or dispute with respect to GATE 2022 examination or score, the Courts and Tribunals in Mumbai alone will have the exclusive jurisdiction to entertain and settle them.

GATE Syllabus

AE: Aerospace Engineering

Important Note for Candidates: *In each of the following subjects the topics have been divided into two categories – Core Topics and Special Topics. The corresponding sections of the question paper will contain 90% of their questions on Core Topics and the remaining 10% on Special Topics.*

Section1: Engineering Mathematics

Core Topics:

Linear Algebra: Vector algebra, Matrix algebra, systems of linear equations, rank of a matrix, eigenvalues and eigenvectors.

Calculus: Functions of single variable, limits, continuity and differentiability, mean value theorem, chain rule, partial derivatives, maxima and minima, gradient, divergence and curl, directional derivatives. Integration, Line, surface and volume integrals. Theorems of Stokes, Gauss and Green.

Differential Equations: First order linear and nonlinear differential equations, higher order linear ODEs with constant coefficients. Partial differential equations and separation of variables methods.

Special Topics: Fourier Series, Laplace Transforms, Numerical methods for linear and nonlinear algebraic equations, Numerical integration and differentiation. Complex analysis. Probability and statistics.

Section 2: Flight Mechanics

Core Topics:

Basics

Atmosphere: Properties, standard atmosphere. Classification of aircraft. Airplane (fixed wing aircraft) configuration and various parts. Pressure altitude; equivalent, calibrated, indicated air speeds; Primary flight instruments: Altimeter, ASI, VSI, Turn-bank indicator. Angle of attack, sideslip; Roll, pitch & yaw controls. Aerodynamic forces and moments.

Airplane performance: Drag polar; takeoff and landing; steady climb & descent; absolute and service ceiling; range and endurance, load factor, turning flight, V-n diagram. Winds: head, tail & cross winds.

Static stability: Stability & control derivatives; longitudinal stick fixed & free stability; horizontal tail position and size; directional stability, vertical tail position and size; lateral stability. Wing dihedral, sweep & position; hinge moments, stick forces.

Special Topics: Dynamic stability: Euler angles; Equations of motion; Decoupling of longitudinal and lateral-directional dynamics; longitudinal modes; lateral-directional modes.

Section 3: Space Dynamics

Core Topics:

Central force motion, determination of trajectory and orbital period in simple cases. Kepler's laws; escape velocity.

No Special Topics:

Section 4: Aerodynamics

Core Topics:

Basic Fluid Mechanics: Conservation laws: Mass, momentum and energy (Integral and differential form); Dimensional analysis and dynamic similarity;

Potential flow theory: sources, sinks, doublets, line vortex and their superposition. Elementary ideas of viscous flows including boundary layers.

Airfoils and wings: Airfoil nomenclature; Aerodynamic coefficients: lift, drag and moment; Kutta-Joukoswki theorem; Thin airfoil theory, Kutta condition, starting vortex; Finite wing theory: Induced drag, Prandtl lifting line theory; Critical and drag divergence Mach number.

Compressible Flows: Basic concepts of compressibility, One-dimensional compressible flows, Isentropic flows, Fanno flow, Rayleigh flow; Normal and oblique shocks, Prandtl-Meyer flow; Flow through nozzles and diffusers.

Special Topics: Wind Tunnel Testing: Measurement and visualization techniques. Shock - boundary layer interaction.

Section 5: Structures

Core Topics:

Strength of Materials: Stress and strain: Three-dimensional transformations, Mohr's circle, principal stresses, Three-dimensional Hooke's law, Plane stress and strain. Failure theories: Maximum stress, Tresca von Mises. Strain energy. Castigliano's principles. Statically determinate and indeterminate trusses and beams. Elastic flexural buckling of columns.

Flight vehicle structures: Characteristics of aircraft structures and materials. Torsion, bending and shear of thin-walled sections. Loads on aircraft.

Structural Dynamics: Free and forced vibrations of undamped and damped SDOF systems. Free vibrations of undamped 2-DOF systems.

Special Topics: Vibration of beams. Theory of elasticity: Equilibrium and compatibility equations, Airy's stress function.

Section 6: Propulsion

Core Topics:

Basics: Thermodynamics, boundary layers, heat transfer, combustion and thermochemistry.

Aerothermodynamics of aircraft engines: Thrust, efficiency, range. Brayton cycle.

Engine performance: ramjet, turbojet, turbofan, turboprop and turboshaft engines. Afterburners.

Turbomachinery: Axial compressors: Angular momentum, work and compression, characteristic performance of a single axial compressor stage, efficiency of the compressor and degree of reaction, multi-staging.

Centrifugal compressor: Stage dynamics, inducer, impeller and diffuser.

Axial turbines: Stage performance.

Rockets: Thrust equation and specific impulse, rocket performance. Multi-staging. Chemical rockets.

Performance of solid and liquid propellant rockets.

Special Topics: Aerothermodynamics of non-rotating propulsion components such as intakes, combustor and nozzle. Turbine blade cooling. Compressor-turbine matching, Surge and stall.

GATE SOLVED PAPERS
(Section-Wise)

1
CHAPTER

ENGINEERING MATHEMATICS

1. If $f(\theta)=\begin{bmatrix}\cos\theta & \sin\theta\\ -\sin\theta & \cos\theta\end{bmatrix}$, then $f(\alpha)f(\beta)=$

(*a*) $f(\alpha|\beta)$ (*b*) $f(\alpha+\beta)$
(*c*) $f(\alpha-\beta)$ (*d*) 2×2 zero matrix

[GATE 2007, 1 M]

2. The Euler iteration formula for numerically integrating a first order nonlinear differential equation of the form $\dot{x}=f=(x)$, with a constant step size of Δt is

(*a*) $x_{k+1}=x_k-\Delta t\times f(x_k)$
(*b*) $x_{k+1}=x_k+(\Delta t^2/2)\times f(x_k)$
(*c*) $x_{k+1}=x_k-(1/\Delta t)\times f(x_k)$
(*d*) $x_{k+1}=x_k+\Delta t\times f(x_k)$

[GATE 2007, 1 M]

3. The minimum value of $J(x)=x^2-7x+30$ occurs at

(*a*) $x=7/2$ (*b*) $x=7/30$
(*c*) $x=30/7$ (*d*) $x=30$

[GATE 2007, 1 M]

4. Let P and Q be two square matrices of same size. Consider the following statements

(i) $PQ=0$ implies $P=0$ or $Q=0$ or both
(ii) $PQ=l^2$ implies $P=Q^{-1}$
(iii) $(P+Q)^2=P^2+2PQ+Q^2$
(iv) $(P-Q)^2=P^2-2PQ+Q^2$

where I is the identity matrix. Which of the following statements is correct?

(*a*) (i), (ii) and (iii) are false, but (iv) is true
(*b*) (i), (ii) and (iv) are false, but (iii) is true
(*c*) (ii), (iii) and (iv) are false, but (i) is true
(*d*) (i), (iii) and (iv) are false, but (ii) is true

[GATE 2007, 2 M]

5. The eigenvalues of the matrix, $A=\begin{bmatrix}2 & 1\\ 0 & 3\end{bmatrix}$ are

(*a*) 1 and 2 (*b*) 1 and 2
(*c*) 2 and 3 (*d*) 2 and 4

[GATE 2007, 2 M]

6. The eigenvalues of the matrix A^{-1},

where $A=\begin{bmatrix}2 & 1\\ 0 & 3\end{bmatrix}$, are

(*a*) 1 and 1/2 (*b*) 1 and 1/3
(*c*) 2 and 3 (*d*) 1/2 and 1/3

[GATE 2007, 2 M]

7. Let a system of linear equations be as follows:

$x-y+2z=0$
$2x+3y-z=0$
$2x-2y+4z=0$

This system of equations has

(*a*) No non-trivial solution
(*b*) Infinite number of non-trivial solutions
(*c*) An unique non-trivial solution
(*d*) Two non-trivial solutions

[GATE 2007, 2 M]

8. At a stationary point of a multi-variable function, which of the following is true?

(*a*) Curl of the function becomes unity
(*b*) Gradient of the function vanishes
(*c*) Divergence of the function vanishes
(*d*) Gradient of the function is maximum

[GATE 2007, 2 M]

9. Numerical value of the integral

$J=\int_0^l \frac{1}{1+x^2}\,dx$, if evaluted numerically using the Trapezoidal rule with $dx=0.2$ would be

(*a*) 1 (*b*) $\frac{\pi}{4}$
(*c*) 0.7837 (*d*) 0.2536

[GATE 2007, 2 M]

10. The Newton-Raphson iteration formula to find a cube root of a positive number c is

(*a*) $x_{k+1}=\frac{2x_k^3+\sqrt[3]{c}}{3x_k^2}$ (*b*) $x_{k+1}=\frac{2x_k^3-\sqrt[3]{c}}{-3x_k^2}$

(*c*) $x_{k+1}=\frac{2x_k^3+c}{3x_k^2}$ (*d*) $x_{k+1}=\frac{x_k^3+c}{3x_k^2}$

[GATE 2007, 2 M]

11. $\lim_{x\to\infty} \frac{\sin x}{e^x x} =$

(*a*) 10 (*b*) 0

(*c*) 1 (*d*) ∞

[GATE 2007, 2 M]

12. Let $F(s) = \frac{(s+10)}{(s+2)(s+20)}$

The partial fraction expansion of $F(s)$ is

(*a*) $\frac{1}{s+2} + \frac{1}{s+20}$ (*b*) $\frac{5}{s+2} + \frac{2}{s+20}$

(*c*) $\frac{2}{s+2} + \frac{20}{s+20}$ (*d*) $\frac{4/9}{s+2} + \frac{5/9}{s+20}$

[GATE 2007, 2 M]

13. Let $F(s) = \frac{(s+10)}{(s+2)(s+20)}$

The inverse Laplace transform of $F(s)$ is

(*a*) $2e^{-2t} + 20e^{-20t}$ (*b*) $\frac{4}{9}e^{-2t} + \frac{5}{9}e^{-20t}$

(*c*) $5e^{-2t} + 2e^{-20t}$ (*d*) $\frac{9}{4}e^{-2t} + \frac{9}{5}e^{-20t}$

[GATE 2007, 2 M]

14. The function defined by

$f(x) = \sin x, \; x < 0$

$= 0, \quad x = 0$

$= 3x^2, \quad x > 0$

(*a*) is neither continuous nor differentiable at $x = 0$

(*b*) is continuous and differentiable at $x = 0$

(*c*) is differentiable but not continuous at $x = 0$

(*d*) is continuous but not differentiable at $x = 0$

[GATE 2008, 1 M]

15. The product of the eigenvalues of the matrix

$$\begin{bmatrix} 1 & 0 & 1 \\ 0 & 2 & 1 \\ 1 & 1 & -3 \end{bmatrix}$$

(*a*) 4 (*b*) 0

(*c*) –6 (*d*) –9

[GATE 2008, 1 M]

16. Which of the following equations is a LINEAR ordinary differential equation?

(*a*) $\frac{d^2y}{dx^2} + \frac{dy}{dx} + 2y^2 = 0$

(*b*) $\frac{d^2y}{dx^2} + y\frac{dy}{dx} + 2y = 0$

(*c*) $\frac{d^2y}{dx^2} + x\frac{dy}{dx} + 2y = 0$

(*d*) $\left(\frac{dy}{dx}\right)^2 + \frac{dy}{dx} + 2y = 0$

[GATE 2008, 1 M]

17. The function $f(x,y,z) = \frac{1}{2}x^2y^2z^2$ satisfies

(*a*) grad f = 0 (*b*) div (grad f) = 0

(*c*) curl (grad f) = 0 (*d*) grad (div(grad f)) = 0

[GATE 2008, 2 M]

18. Which of the following is true for all choices of vectors $\vec{p}, \vec{q}, \vec{r}$?

(*a*) $\vec{p}, \times \vec{q} + \vec{q}, \times \vec{r} + \vec{r}, \times \vec{p} = 0$

(*b*) $(\vec{p}.\vec{q})\vec{r} + (\vec{q}.\vec{r})\vec{p} + (\vec{r}.\vec{p})\vec{q} = 0$

(*c*) $\vec{p}.(\vec{q} \times \vec{r}) + \vec{q}.(\vec{r} \times \vec{p}) + \vec{r}.(\vec{p} \times \vec{q}) = 0$

(*d*) $\vec{p} \times (\vec{q} \times \vec{r}) + \vec{q} \times (\vec{r} \times \vec{p}) + \vec{r} \times (\vec{p} \times \vec{q}) = 0$

[GATE 2008, 2 M]

19. The value of the line integral $\frac{1}{2\pi}\oint(x\,dy - y\,dx)$ taken anticlockwise along a circle of unit radius is

(*a*) 0.5 (*b*) 1

(*c*) 2 (*d*) π

[GATE 2008, 2 M]

20. Which of the following is a solution of

$\frac{d^2y}{dx^2} + 2\frac{dy}{x} + y = 0$?

(*a*) $e^{-x} + xe^{-x}$ (*b*) $e^{x} + xe^{-x}$

(*c*) $e^{x} + e^{-x}$ (*d*) $e^{-x} + xe^{x}$

[GATE 2008, 2 M]

21. Suppose the non-constant function F(x) and G(t) satisfy

$$\frac{d^2F}{dx^2} + p^2F = 0, \frac{dG}{dt} + c^2p^2G = 0,$$

where p and c are constants.

Then the function u(x, t) = F(x) G(t) definitely satisfies

(a) $\frac{\partial^2 u}{\partial t^2} = c^2 \frac{\partial^2 u}{\partial x}$ (b) $\frac{\partial u}{\partial^2 u} = c^2 \frac{\partial^2 u}{\partial x^2}$

(c) $\nabla^2 u = 0$ (d) $\frac{\partial^2 u}{\partial t^2} + c^2u^2 = 0$

[GATE 2008, 2 M]

22. The following set of equations

$$\begin{bmatrix} 1 & 1 & 2 \\ 1 & 0 & 1 \\ 0 & 1 & 1 \end{bmatrix} \begin{bmatrix} x_1 \\ x_2 \\ x_3 \end{bmatrix} = \begin{bmatrix} 1 \\ -1 \\ 0 \end{bmatrix} \text{ has}$$

(a) no solution (b) a unique solution

(c) two solutions (d) infinite solutions

[GATE 2008, 2 M]

23. The function $f(x) = x^2 - 5x + 6$

(a) has its maximum value at x = 2.0

(b) has its maximum value at x = 2.5

(c) is increasing on the interval (2.0, 2.5)

(d) is increasing on the interval (2.5, 3.0)

[GATE 2008, 2 M]

24. Let Y(s) denote the Laplace transform L(y(t)) of the function y(t) = cosh (at) sin (at). Then

(a) $L\left(\frac{dy}{dt}\right) = \frac{dy}{ds}, L(ty(t)) = sY(s)$

(b) $L\left(\frac{dy}{dt}\right) = sY(s), L(ty(t)) = -\frac{dy}{ds}$

(c) $L\left(\frac{dy}{dt}\right) = \frac{dy}{ds}, L(ty(t)) = Y(s-1)$

(d) $L\left(\frac{dy}{dt}\right) = sY(s), L(ty(t)) = e^{as}Y(s)$

[GATE 2008, 2 M]

Statement for Linked Answer Questions 76 and 77: The following two questions relate to Simpon's rule for approximating the integral $\int_a^b f(x)\,dx$ on the interval [a, b].

25. Which of the following gives the correct formula for Simpson's rule?

(a) $\frac{(b-a)}{2}\left[f(b) + f\left(\frac{a+b}{2}\right)\right]$

(b) $\frac{(b-a)}{2}\left[\frac{f(a)+f(b)}{2} + \frac{4}{3}f\left(\frac{a+b}{2}\right)\right]$

(c) $\frac{(b-a)}{2}\left[\frac{f(a)+f(b)}{3} + \frac{4}{3}f\left(\frac{a+b}{2}\right)\right]$

(d) $\frac{(b-a)}{2}\left[\frac{f(a)+f(b)}{3} + \frac{4}{3}f\left(\frac{a+b}{3}\right)\right]$

[GATE 2008, 2 M]

26. The percentage error (with respect to the exact solution) in estimation of the integral $\int_0^1 x^3\,dx$ using Simpson's rule is

(a) 5.3 (b) 3.5

(c) 2.8 (d) 0

[GATE 2008, 2 M]

27. The ordinary differential equation $\frac{d^2y}{dx^2} + ky = 0$ where k is real and positive

(a) is non-linear

(b) has a characteristic equation with one real and one complex root

(c) has a characteristic equation with two real roots

(d) has a complementary function that is simple harmonic

[GATE 2009, 1 M]

28. A non-trivial solution to the (n xn) system of equations [A] {x} = {0}, where {0} is the null vector

(a) can never be found

(b) may be found only if [A] is not singular

(c) may be found only if [A] is an orthogonal matrix

(d) may be found only if [A] has at least one eigenvalue equal to zero

[GATE 2009, 1 M]

29. The value of the integral $\int_0^{\pi} \frac{dx}{1+x+\sin x}$ evaluated using the trapezoidal rule with two equal intervals is approximately

(a) 1.27 (b) 1.81

(c) 1.41 (d) 0.71

[GATE 2009, 2 M]

30. The product of the eigenvalues of the matrix

$$\begin{bmatrix} 2 & 1 & 1 \\ 1 & 3 & 1 \\ 1 & 1 & 4 \end{bmatrix} \text{ is}$$

(a) 20 (b) 24

(c) 9 (d) 17

[GATE 2009, 2 M]

31. In the interval $1 \le x \le 2$, the function $f(x) = e^{\pi x} + \sin \pi x$ is

(a) maximum at x = 1

(b) maximum at x = 2

(c) maximum at x = 1.5

(d) monotonically decreasing

[GATE 2009, 2 M]

32. The inverse Laplace transform of

$$F(s) = \frac{(s+1)}{(s+4)(s-3)} \text{ is}$$

(a) $\frac{3}{7}e^{4t} + \frac{4}{7}e^{-3t}$ (b) $\frac{3}{7}e^{-4t} + \frac{4}{7}e^{3t}$

(c) $\frac{5}{7}e^{-4t} + \frac{6}{7}e^{3t}$ (d) $\frac{5}{7}e^{4t} + \frac{6}{7}e^{-3t}$

[GATE 2009, 2 M]

33. The linear system of equations Ax = b where

$$A = \begin{bmatrix} 1 & 2 \\ 2 & 4 \end{bmatrix} \text{ and } b = \begin{Bmatrix} 3 \\ 3 \end{Bmatrix} \text{ has}$$

(a) no solution

(b) infinitely many solutions

(c) a unique solution $x = \begin{Bmatrix} 1 \\ 1 \end{Bmatrix}$

(d) a unique solution $x = \begin{Bmatrix} 0.5 \\ 0.5 \end{Bmatrix}$

[GATE 2009, 2 M]

34. The correct iterative scheme for finding the square root of a positive real number R using the Newton Raphson method is

(a) $x_{n+1} = \sqrt{R}$

(b) $x_{n+1} = \frac{1}{2}\left(x_n + \frac{R}{x_n}\right)$

(c) $x_{n+1} = \frac{1}{2}\left(\sqrt{x_n} + \sqrt{x_{n-1}}\right)$

(d) $x_{n+1} = \frac{1}{2}\left(\sqrt{R} + x_n\right)$

[GATE 2009, 2 M]

35. Two position vectors are indicated by $\bar{V}_1 = \begin{Bmatrix} x_1 \\ y_1 \end{Bmatrix}$ and $\bar{V}_2 = \begin{Bmatrix} x_2 \\ y_2 \end{Bmatrix}$. If $a^2 + b^2 = 1$, then the operation $\bar{V}_2 = \begin{bmatrix} a & -b \\ b & a \end{bmatrix} \bar{V}_1$ amounts to obtaining the position vector $\bar{V}_2$ from $\bar{V}_1$ by

(a) translation

(b) rotation

(c) magnification

(d) combination of translation, rotation and magnification

[GATE 2010, 1 M]

36. The linear second order partial differential equation $5\frac{\partial^2 \phi}{dx^2} + 3\frac{\partial^2 \phi}{dxdy} + 2\frac{d^2 \phi}{dy^2} + 9 = 0$ is

(a) Parabolic

(b) Hyperbolic

(c) Elliptic

(d) None of the above

[GATE 2010, 1 M]

37. The eigen-values of a real symmetric matrix are always

(a) positive

(b) imaginary

(c) real

(d) complex conjugate pairs

[GATE 2010, 1 M]

38. The concentration x of a certain chemical species at time t in a chemical reaction is described by the differential equation $\frac{dx}{dt} + kx = 0$, with $x(t = 0) = x_0$. Given that e is the base of the natural logarithms, the concerntration x at $t = \frac{1}{k}$

(*a*) falls to the value $0.5\ x_0$
(*b*) rises to the value $2x_0$
(*c*) falls to the value $\frac{x_0}{e}$
(*d*) rises to the value ex_0

[GATE 2010, 1 M]

39. The definite integral $\int_{-1}^{+1} \frac{dx}{x^2}$

(*a*) does not exist (*b*) is equal to 2
(*c*) is equal to 0 (*d*) is equal to –2

[GATE 2010, 1 M]

40. Given that the Laplace transform of $y(t) = e^{-1}(2\cos 2t - \sin 2t)$ is $y(s) = \frac{2s}{(s+1)^2 + 4}$, the Laplace transform of $y_t(t) = e^t(2\cos 2t - \sin 2t)$ is

(*a*) $\frac{2(s-2)}{(s-1)^2+4}$ (*b*) $\frac{2(s+2)}{(s+3)^2+4}$

(*c*) $\frac{2(s+2)}{(s+1)^2+4}$ (*d*) $\frac{2(s-1)}{(s-1)^2+4}$

[GATE 2010, 2 M]

41. In a certain region a hill is described by the shape $z(x, y) = \frac{1}{50}x^4 + y^2 - xy - 3y$, where the axes x and y are in the horizontal plane and axis z points vertically upward. If $\hat{i}, \hat{j}$ and $\hat{k}$ are unit vectors along x, y and z, respectively, then at the point x = 5, y = 10 the unit vector in the direction of the sleepest slope of the hill will be:

(*a*) $\hat{i}$ (*b*) $\hat{j}$

(*c*) $\hat{k}$ (*d*) $\hat{i} + \hat{j} + \hat{k}$

[GATE 2010, 2 M]

42. The function $f(x, y) = x^2 + y^2 - xy - 3y$ has an extremum at the point

(*a*) (1, 2) (*b*) (3, 0)
(*c*) (2, 2) (*d*) (1, 1)

[GATE 2010, 2 M]

43. Consider x, y, z to be right-handed Cartesian coordinates. A vector function is defined in this coordinate system as $v = 3xi + 3xyj - yz^2k$, where i. j and k are the unit vectors along x, y and z axes, respectively. The curl of v is given by

(*a*) $z^2i - 3yk$
(*b*) $z^2j + 3yk$
(*c*) $z^2i + 3yj$
(*d*) $-z^2i + 3yk$

[GATE 2011, 1 M]

44. Which of the following functions is periodic?

(*a*) $f(x) = x^2$ (*b*) $f(x) = \log x$
(*c*) $f(x) = e^x$ (*d*) f(x) = const.

[GATE 2011, 1 M]

45. The function $f(x_1, x_2, x_3) = x_1^2 + x_2^2 + x_3^2 - 2x_1 - 4x_2 - 6x_3 + 14$ has its minimum value at

(*a*) (1, 2, 3)
(*b*) (0, 0, 0)
(*c*) (3, 2, 1)
(*d*) (1, 1, 3)

[GATE 2011, 1 M]

46. Consider the function $f(x_1, x_2) = x_1^2 + x_2^2 + e^{-x_1-x_2}$. The vector pointing in the direction of maximum increases of the function at the point (1, –1) is

(*a*) $\begin{pmatrix} 2 \\ -5 \end{pmatrix}$ (*b*) $\begin{pmatrix} 1 \\ -5 \end{pmatrix}$

(*c*) $\begin{pmatrix} -0.73 \\ -6.73 \end{pmatrix}$ (*d*) $\begin{pmatrix} 2 \\ -4 \end{pmatrix}$

[GATE 2011, 1 M]

47. Two simultaneous equations given by $y = \pi + x$ and $y = \pi - x$ have

(*a*) a unique solution
(*b*) infinitely many solutions
(*c*) no solution
(*d*) a finite number of multiple solutions

[GATE 2011, 1 M]

48. Consider the function f(x) = x − sin (x). The Newton-Raphson iteration formula to find the root of the function starting from an initial guess $x^{(0)}$ at iteration k is

(a) $x^{(k+1)} = \dfrac{\sin x^{(k)} - x^{(k)} \cos x^{k}}{1 - \cos x^{(k)}}$

(b) $x^{(k+1)} = \dfrac{\sin x^{(k)} - x^{(k)} \cos x^{k}}{1 + \cos x^{(k)}}$

(c) $x^{(k+1)} = \dfrac{\sin x^{(k)} + x^{(k)} \cos x^{k}}{1 - \cos x^{(k)}}$

(d) $x^{(k+1)} = \dfrac{\sin x^{(k)} + x^{(k)} \cos x^{k}}{1 + \cos x^{(k)}}$

[GATE 2011, 2 M]

49. Consider the matrix $\begin{bmatrix} 2 & a \\ b & 2 \end{bmatrix}$ where a and b are real numbers. The two eigenvalues of this matrix λ_1 and λ_1 are real and distinct ($\lambda_1 \neq \lambda_2$) when

(a) a < 0 and b > 0 (b) a > 0 and b < 0

(c) a < 0 and b < 0 (d) a = 0 and b < = 0

[GATE 2011, 2 M]

50. The solution of $\dfrac{dy}{dt} = y^3 e^t t^2$ with initial condition y(0) = 1 is given by

(a) $\dfrac{1}{9} e^t (t+3)^2$

(b) $\sqrt{\dfrac{9}{5 + 2e^t (t^2 - 2t + 2)}}$

(c) $\dfrac{4e^t}{(t+2)^2}$

(d) $\sqrt{\dfrac{1}{5 - 2e^t (t^2 - 2t + 2)}}$

[GATE 2011, 2 M]

51. The constraint $A^2 = A$ on any square matrix A is satisfied for

(a) the identity matrix only.

(b) the null matrix only.

(c) both the identity matrix and the null matrix.

(d) no square matrix A.

[GATE 2012, 1 M]

52. The general solution of the differential equation

$\dfrac{d^2y}{dt^2} + \dfrac{dy}{dt} - 2y = 0$ is

(a) $Ae^{-t} + Be^{2t}$ (b) $Ae^{-2t} + Be^{-t}$

(c) $Ae^{-2t} + Be^{t}$ (d) $Ae^{t} + Be^{2t}$

[GATE 2012, 1 M]

53. The value of k for which the system of equations x + 2y + kz =1; 2x + ky +8z = 3 has no solution is

(a) 0 (b) 2

(c) 4 (d) 8

[GATE 2012, 1 M]

54. If u(t) is a unit step function, the solution of the differential equation $m\dfrac{d^2x}{dt^2} + kx = u(t)$ in Laplace domain is

(a) $\dfrac{1}{s(ms^2 + k)}$ (b) $\dfrac{1}{ms^2 + k}$

(c) $\dfrac{s}{ms^2 + k}$ (d) $\dfrac{1}{s^2(ms^2 + k)}$

[GATE 2012, 1 M]

55. The general solution of the differential equation

$\dfrac{dy}{dx} - 2\sqrt{y} = 0$ is

(a) $y - \sqrt{x} + C = 0$ (b) $y - x + C = 0$

(c) $\sqrt{y} - \sqrt{x} + C = 0$ (d) $\sqrt{y} - x + C = 0$

[GATE 2012, 1 M]

Question Q.56 is numerical answer type. The answer of this questions is either a positive whole number, or a positive real number with maximum of 2 decimal places.

56. The integration $\int_0^1 x^3\, dx$ computed using trapezoidal rule with n = 4 intervals is ____.

[GATE 2012, 2 M]

57. The nth derivative of the function $y = \frac{1}{x+3}$ is

(a) $\frac{(-1)^n n!}{(x+3)^{n+1}}$ (b) $\frac{(-1)^{n+1} n!}{(x+3)^{n+1}}$

(c) $\frac{(-1)^n (n+1)!}{(x+3)^n}$ (d) $\frac{(-1)^n n!}{(x+3)^n}$

[GATE 2012, 2 M]

58. The volume of a solid generated by rotating the region between semi-circle $y = 1 - \sqrt{1-x^2}$ and straight line y = 1, about x axis, is

(a) $\pi^2 - \frac{4}{3}\pi$ (b) $4\pi^2 - \frac{1}{3}\pi$

(c) $\pi^2 - \frac{1}{3}\pi$ (d) $\frac{3}{4}\pi^2 - \pi$

[GATE 2012, 2 M]

59. One eigenvalue of the matrix $A = \begin{bmatrix} 2 & 7 & 10 \\ 5 & 2 & 25 \\ 1 & 6 & 5 \end{bmatrix}$ is – 9.33. One of the other eigenvalues is

(a) 18.33 (b) –18.33

(c) 18.33 – 9.33i (d) 18.33 + 9.33i

[GATE 2012, 2 M]

60. The directional derivative of the function $f(x,y) = \frac{x^2 + xy^2}{\sqrt{5}}$ in the direction $\vec{a} = 2\hat{i} = 2\hat{i} - 4\hat{j}$ at (x, y) = (1, 1) is

(a) $-\frac{1}{\sqrt{5}}$ (b) $-\frac{2}{\sqrt{5}}$

(c) 0 (d) $-\frac{1}{5}$

[GATE 2013, 1 M]

61. The value of $\int_4^5 \frac{x+2}{x^2+4x-21} dx$ is

(a) $\ln\sqrt{\frac{24}{11}}$ (b) $\ln\sqrt{\frac{12}{11}}$

(c) $\ln\sqrt{2}$ (d) $\ln\left(\frac{12}{11}\right)$

[GATE 2013, 1 M]

62. At x = 0 the function y = |x| is

(a) continuous but not differentiable

(b) continuous and differentiable

(c) not continuous but differentiable

(d) not continuous and not differentiable

[GATE 2013, 1 M]

63. One of the eigenvectors of the matrix

$$A = \begin{pmatrix} 1 & -1 & 0 \\ 0 & 1 & -1 \\ -1 & 0 & 1 \end{pmatrix} \text{ is } v = \begin{pmatrix} 1 \\ 1 \\ 1 \end{pmatrix}$$

The corresponding eigenvalue is ___________

[GATE 2013, 1 M]

64. Let $\iint_S \left(y^2 z\hat{i} + z^2 x\hat{j} + x^2 y\hat{k}\right)\left(x\hat{i} + y\hat{j} + z\hat{k}\right) ds$, where S denotes the surface of the sphere of unit radius centered at the origin. Here $\hat{i}$, $\hat{j}$ and $\hat{k}$ enote three orthogonal unit vectors. The value of I is _________

[GATE 2013, 2 M]

65. Given that the Laplace transform, $L\left(e^{at}\right) = \frac{1}{s-a}$ then $L\left(3e^{5t} \sin h\, 5t\right) =$

(a) $\frac{3s}{s^2 - 10s}$ (b) $\frac{15}{s^2 - 10s}$

(c) $\frac{3s}{s^2 + 10s}$ (d) $\frac{15}{s^2 + 10s}$

[GATE 2013, 2 M]

66. Values of a, b and c, which render the matrix

$$Q = \begin{bmatrix} \frac{1}{\sqrt{3}} & \frac{1}{\sqrt{2}} & a \\ \frac{1}{\sqrt{3}} & 0 & b \\ \frac{1}{\sqrt{3}} & -\frac{1}{\sqrt{2}} & c \end{bmatrix}$$

orthonormal are, respectively

(a) $\frac{1}{\sqrt{2}}, \frac{1}{\sqrt{2}}, 0$ (b) $\frac{1}{\sqrt{6}}, -\frac{2}{\sqrt{6}}, \frac{1}{\sqrt{6}}$

(c) $-\frac{1}{\sqrt{3}}, -\frac{1}{\sqrt{3}}, \frac{1}{\sqrt{3}}$ (d) $-\frac{1}{\sqrt{6}}, \frac{2}{\sqrt{6}}, -\frac{1}{\sqrt{6}}$

[GATE 2013, 2 M]

67. A function y(t) satisfies the differential equation $\frac{d^2y}{dt^2} - 2\frac{dy}{dt} + y = 0$ and is subject to the initial conditions $y(t = 0)$ and $\frac{du}{dt}(t = 0) = 1$. The value of y(t = 1) is

(a) e (b) 0

(c) 1 (d) –1

68. For a real symmetric matrix [A], which of the following statements is true:

(a) The matrix is always diagonalizable and invertible

(b) The matrix is always invertible but not necessarily diagonalizable.

(c) The matrix is always diagonalizable but not necessarily invertible.

(d) The matrix is always neither diagonalizable nor invertible.

[GATE 2014, 1 M]

69. The series

$$s = \sum_{m=1}^{\infty} m^2 \left(\frac{x-2}{3}\right)^m$$

converges for all x with $|x - 2| \le R$ given by

(a) R = 0 (b) R = 3

(c) R = ∞ (d) R = 1/3

[GATE 2014, 1 M]

70. The function given by

$$f(x) = \begin{cases} \sin(1/x), & x \neq 0 \\ 0, & x = 0 \end{cases}$$

(a) Unbounded everywhere

(b) Bounded and continuous everywhere

(c) Bounded but not continuous at x = 0

(d) Continuous and differentiable everywhere

[GATE 2014, 1 M]

71. The value of $I = \int_0^1 1000x^4 dx$, obtained by using Simpson's rule with 2 equally spaced intervals is,

(a) 200 (b) 400

(c) 180 (d) 208

[GATE 2014, 1 M]

72. If $[A] = \begin{bmatrix} 3 & -3 \\ -3 & 4 \end{bmatrix}$. Then det(– [A]² + 7[A] – 3[I]) is

(a) 0 (b) –324

(c) 324 (d) 6

[GATE 2014, 2 M]

73. For the periodic function given by

$$f(x) = \begin{cases} -2, & -\pi < x < 0 \\ 2, & 0 < x < \pi \end{cases}$$

With $f(x + 2\pi) = f(x)$, using Fourier series, the sum $s = 1 - \frac{1}{3} + \frac{1}{5} - \frac{1}{7} + ...$ converges to

(a) 1 (b) 3π

(c) 4π (d) 5π

[GATE 2014, 2 M]

74. Let Γ be the boundary of the closed circular region A given by $x^2 + y^2 = 1$. Then $I = \int_\Gamma (3x^3 - 9xy^2) ds$ (where ds means integration along the bounding curve) is

(a) π (b) –π

(c) 1 (d) 0

[GATE 2014, 2 M]

75. Solution to the boundary-value problem

$$-9\frac{d^2u}{dx^2} + u = 5x, 0 < x < 3 \text{ with } u(0) = 0, u'(3) = 0$$

(a) $u(x) = \frac{15e}{1+e^2}\left(\frac{x}{e^3} - \frac{x}{e^3}\right) + 5x$

(b) $u(x) = \frac{15e}{1+e^2}\left(\frac{x}{e^3} + \frac{x}{e^3}\right) + 5x$

(c) $u(x) = \frac{15\sin(x/3)}{\cos(1)} + 5x$

(d) $u(x) = \frac{15\sin(x/3)}{\cos(1)} - \frac{5}{54}x^3$

[GATE 2014, 2 M]

76. The Laplace transform $L(u(t)) = U(s)$, for the solution u(t) of the problem $\frac{d^2u}{dt^2} + \frac{2du}{dt} + u = 1$, t < 0 with initial conditions u(0) = 0, u'(0) = 5, is given by:

(a) $\frac{6}{s(s+1)^2}$ (b) $\frac{1+5s}{s(s+1)^2}$

(c) $\frac{1-5s}{s(s+1)^2}$ (d) $\frac{1+5s^2}{s(s+1)^2}$

[GATE 2014, 2 M]

77. The partial differential equation is :

$$\frac{\partial u}{\partial t} + \frac{\partial\left(\frac{u^2}{2}\right)}{\partial x} = 0$$

(a) linear and first order

(b) linear and second order

(c) non-linear and first order

(d) non-linear and second order

[GATE 2015, 1 M]

78. The system of equations for the variable x and y

$ax + by = e$

$cs = dy = f$

has a unique solution only if

(a) $ab - dc \neq 0$ (b) $ac - bd \neq 0$

(c) $a + c \neq b + d$ (d) $a - c \neq b - d$

[GATE 2015, 1 M]

79. In the solution of $\frac{d^2y}{dx^2} - 2\frac{dy}{dx} + y = 0$, if the values of the integration constants are identical and one of the initial conditions is pecified as y(0) = 1, the other initial condition y'(0) = _______.

[GATE 2015, 2 M]

80. For x > 0, the general solution of the differential equation $\frac{dy}{dx} = 1 - 2y$ asymptotically approaches _______.

[GATE 2015, 2 M]

81. For a parabola defined by $y = ax^2 + bx + c$, $a \neq 0$ the coordinates (x, y) of the extremum are

(a) $\left(\frac{-b}{2a} + \frac{\sqrt{b^2 - 4ac}}{2a}, 0\right)$

(b) $\left(\frac{-b}{2a}, \frac{-b^2 + 4ac}{2a}\right)$

(c) $\left(\frac{-b}{2a}, \frac{-b^2 + 4ac}{4a}\right)$

(d) (0, c)

[GATE 2015, 2 M]

82. If all the eigenvalues of a matrix are real and equal, then

(a) the matrix is diagonalizable

(b) its eigenvectors are not necessarity linearly independent

(c) its eigenvectors are linearly independent

(d) its determinant is necessarily zero

[GATE 2015, 2 M]

83. The value of the integral $\int_1^2 \left(4x^3 + 3x^2 + 2x + 1\right)dx$ evaluated numerically using simpson's rule with one step is

(a) 26.5 (b) 26

(c) 25.5 (d) 25.3

[GATE 2015, 2 M]

84. Consider an eigenvalue problem given by $AX = \lambda_i X$. If λ_i represent the eigenavalues of the non-singular square matrix A, then what will be the eigenvalues of matrix A^2?

(a) λ_i^4 (b) λ_i^2

(c) $\lambda_i^{1/2}$ (d) $\lambda_i^{1/4}$

[GATE 2016, 1 M]

85. If A and B are both non-singular n × n matrices, then which of the following statement is NOT TRUE.

(a) det(AB) = det(A) det(B)

(b) det(A + B) = det(A) + det(B)

(c) $\det(AA^{-1}) = 1$

(d) $\det(A^T) = \det(A)$

[GATE 2016, 1 M]

86. Let x be positive real number. The function $f(x) = x^2 + \frac{1}{x^2}$ has its minima at x = _______.

[GATE 2016, 1 M]

87. The vector $\vec{u}$ is defined as $\vec{u} = y\hat{e}_x - x\hat{e}_y$, where $\hat{e}_x$ and $\hat{e}_y$ are the unit vector along x any y directions, respectively. If the vector $\vec{\omega}$ is defined as $\vec{\omega} = \vec{\nabla} \times \vec{u}$, then $\left|(\vec{\omega} \cdot \vec{\nabla})\vec{u}\right|$ _______.

[GATE 2016, 1 M]

88. The partial differential equation

$$\frac{\partial u}{\partial t} = \alpha \frac{\partial^2 u}{\partial x^2}$$

where α is a positive constant, is

(a) circular (b) elliptic

(c) hyperbolic (d) parabolic

[GATE 2016, 1 M]

89. Consider a second order linear differential equation $\frac{d^2y}{dx^2} - 4\frac{dy}{dx} + 4y = 0$, with the boundary conditions $y(0) = 1; \left.\frac{dy}{dx}\right|_{x=0} = 1$, the value of y at x = 1 is :

(a) 0 (b) 1

(c) e (d) e^2

[GATE 2016, 2 M]

90. Consider the following system of linear equations :

$2x - y + z = 1$

$3x - 3y + 4z = 6$

$x - 2y + 3z = 4$

This system of linear equations has

(a) no solution (b) one solution

(c) two solutions (d) three solutions

[GATE 2016, 2 M]

91. The value of definite integral $\int_0^{\pi} x \sin x \, dx$ is ______.

[GATE 2016, 2 M]

92. Use Newton-Raphson method to solve the equation :

$xe^x = 1$. Begin with the intial guess $x_0 = 0.5$. The solution after one step is x = ______.

[GATE 2016, 2 M]

93. Given the vectors $\vec{v}_1 = \hat{l} + 3\hat{j}, \vec{v}_2 = 2\hat{l} - 4\hat{j} + 3\hat{k}$, the vector perpendicular to both $\vec{v}_1, \vec{v}_2$ is given by

(a) $\vec{v}_3 = \vec{v}_1 - (\vec{v}_1 \cdot \vec{v}_2)\frac{\vec{v}_2}{|\vec{v}_2|}$

(b) $\vec{v}_3 = \hat{k}$

(c) $\vec{v}_3 = \vec{v}_2 - (\vec{v}_1 \cdot \vec{v}_2)\frac{\vec{v}_1}{|\vec{v}_1|}$

(d) $\vec{v}_3 = \frac{\vec{v}_1 \times \vec{v}_2}{|\vec{v}_1 \times \vec{v}_2|}$

[GATE 2017, 1 M]

94. The value of the integral $I = \int_c ((x - y)dx + x^2 dy)$, with C the boundary of the square $0 \le x \le 2$; $0 \le y \le 2$ is ______.

[GATE 2017, 1 M]

95. Let $\overline{v(t)}$ be a unit vector that is a function of the parameter f; Then $\vec{v} \cdot \frac{d\vec{v}}{dt} =$ ______.

[GATE 2017, 1 M]

96. The eigenvalues λ_n and eigenfunctions $u_n(x)$ of the sturm-Liouville problem.

$$\frac{d^2y}{dx^2} + k^2\lambda y = 0,\ 0 < x < 1;\ y(0), y(1) = 0$$

are given by

(a) $\lambda_n = n^2\pi^2$; un(x) = $\sin\lambda_n x$, n = 0, ±1, ±2 ... ∞

(b) $\lambda_n = \frac{n^2\pi^2}{k^2}; u_n(x) = \sin kn\pi x, n = 0, \pm 1, \pm 2 \ldots \infty$

(c) $\lambda_n = \frac{n^2\pi^2}{k^2}; u_n(x) = \sin n\pi x, n = 0, \pm 1, \pm 2 \ldots \infty$

(d) $\lambda_n = n^2\pi^2; u_n(x) = \sin n\pi x, n = 0, \pm 1, \pm 2 \ldots \infty$

[GATE 2017, 1 M]

97. 3 point Guassion integration formula is given by

$$\int_{-1}^{1} f(x)dx \approx \sum_{j=1}^{3} A_j f(x_j), \text{ with}$$

$$x_1 = 0, x_2 = -x_3 = -\sqrt{\frac{3}{5}};\ A_1 = \frac{8}{9}, A_2 = \frac{5}{9}$$

This formula exactly integrates

(a) $f(x) = 5 - x^7$ (b) $f(x) = 2 + 3x^3 + 6x^4$

(c) $f(x) = 13 + 6x^3 + x^6$ (d) $f(x) = e^{-x2}$

[GATE 2017, 1 M]

98. Matrix $[A] = \begin{pmatrix} 2 & 0 & 2 \\ 3 & 2 & 7 \\ 3 & 1 & 5 \end{pmatrix}$ and vector $\{b\} = \begin{Bmatrix} 4 \\ 4 \\ 5 \end{Bmatrix}$ are given. If vector {x} is the solution to the system of equation [A]{x} = {b}. Which of the following is true for (x).

(a) Solution does not exist

(b) Infinite solution exist

(c) Unique solution exists

(d) five possible solution exists

[GATE 2017, 2 M]

99. Consider the initial value problem

$$\frac{d^2y}{dt^2} + 4\frac{dy}{dt} + 6y = f(t);\ y(0) = 2,\ \left(\frac{dy}{dt}\right)_{t=0} = 1$$

If $Y(s) = \int_0^{\infty} y(t)e^{-5t}dt$ and $F(s) = \int_0^{\infty} f(t)e^{-5t}dt$ are the laplace transforms of y(t) and f(t) respectively, then Y(s) is given by

(a) $\frac{F(s)}{(s^2 + 4s + 6)}$ (b) $\frac{F(s) + 2s + 9}{(s^2 + 4s + 6)}$

(c) $\frac{F(s)}{(-s^2 + 4s + 6)}$ (d) $\frac{F(s) - 2s + 9}{s^2 + 4s + 6}$

[GATE 2017, 2 M]

100. Let u(x, t) denote the displacement of a point on a rod. The displacement satisfies the following equation of motion :

$$\frac{\partial^2 u}{\partial t^2} - 25\frac{\partial^2 u}{\partial x^2} = 0,\ 0 < x < 1$$

with u(x, 0) = 0.01 sin(10πx), $\frac{\partial u}{\partial t}(x,0) = 0$; u(0, t) = 0, u(1, t) = 0. The value of u(0.25, 1) is ______. (in three decimal places).

[GATE 2017, 2 M]

101. The equation $x^2\frac{d^2y}{dx^2} + 5x\frac{dy}{dx} + 4y = 0$ has a solution y(x) that is :

(a) A polynomial in x

(b) finite series in terms of non-integer fractional power of x

(c) Consists of negative integer powers of x and logarithmic function of x

(d) consists of exponential function of x.

[GATE 2017, 2 M]

102. Let matrix $[A] = \begin{pmatrix} 2 & -6 \\ 0 & 2 \end{pmatrix}$. Then for any non-trivial vector $\{x\} = \begin{Bmatrix} x_1 \\ x_2 \end{Bmatrix}$, which of the following is true for $K = \{x\}^+ [A]\{x\}$.

(a) K is always less than zero

(b) K is always greater than zero

(c) K is non-negative

(d) K can be anything **[GATE 2017, 2 M]**

103. Let $\vec{a}, \vec{b}$ be two distinct vectors that are not parallel. The vector $\vec{c} = \vec{a} \times \vec{b}$ is :

(a) zero

(b) orthogonal to $\vec{a}$ alone

(c) orthogonal to $\vec{a} + \vec{b}$

(d) orthogenal to $\vec{b}$ alone

[GATE 2018, 1 M]

104. Consider the function $f(x,y) = \frac{x^2}{2} + \frac{y^2}{3} - 5$. All the roots of this function

(a) from a finite set of points

(b) lie on an elliptical curve

(c) lie on thr surface of a sphere

(d) lie on a hyperbolic curve

[GATE 2018, 1 M]

105. Consider a vector field given by $x\hat{i} + y\hat{j} + z\hat{k}$. Thus vecot field is :

(a) divergence-free and curl-free.

(b) curl-free but not divergence-free.

(c) divergence-free but not curl-free.

(d) neither divergence-free nor curl-free.

[GATE 2018, 1 M]

106. The determinant of the matrix $\begin{bmatrix} 1 & 2 & -1 \\ 2 & 1 & 0 \\ 3 & 1 & 1 \end{bmatrix}$ is ______ (accurate to one decimal place).

[GATE 2018, 1 M]

107. The solution of the differential equation $\frac{d^2y}{dx^2} + 3\frac{dy}{dx} = 0$, given they y = 0 and $\frac{dy}{dx} = 1$ at x = 0 is :

(a) $x(1 - e^{-3x})$ (b) $\frac{1}{3}(1 - e^{-3x})$

(c) $\frac{1}{3}(1 + e^{-3x})$ (d) $\frac{1}{3}xe^{-\frac{3x}{2}}$

[GATE 2018, 2 M]

108. Consider the vector field $\vec{v} = \frac{-y}{r^2}\hat{i} + \frac{x}{r^2}\hat{j}$, where $r = \sqrt{x^2 + y^2}$. The contour integral $\oint \vec{v} \cdot d\vec{S}$, where $d\vec{S}$ is tangent to the contour that encloses the origin is ______. (two decimal places).

[GATE 2018, 2 M]

109. The magnitude of the x-component of a unit vector at the point (1, 1) that is normal to equipotential lines of the potential function $\phi(r) = \frac{1}{r^2 + y}$, where $r = \sqrt{x^2 + y^2}$, is ______.

[GATE 2018, 2 M]

110. The maximum of the function $f(x) = xe^{-x}$ (where x is real) is

(a) $\frac{1}{e}$ (b) $\frac{2}{e^2}$

(c) $\frac{\left(\frac{-1}{e^2}\right)}{2}$ (d) ∞

[GATE 2019]

111. Vector $\vec{b}$ is obtained by rotating $\vec{a} = \hat{i} + \hat{j}$ by 90° about $\hat{k}$, where $\hat{i}, \hat{j}$ and $\hat{k}$ are unit vectors along the x, y and z axes, respectively. $\vec{b}$ is given by

(a) $\hat{i} - \hat{j}$ (b) $-\hat{i} + \hat{j}$

(c) $\hat{i} + \hat{j}$ (d) $-\hat{i} - \hat{j}$

[GATE 2019]

112. A scalar function is given by $f(x, y) = x^2 + y^2$. Take $\hat{i}$ and $\hat{j}$ as unit vectors along the x and y axes, respectively. At (x, y) = (3, 4), the direction along which f increases the fastest is

(a) $\frac{1}{5}(4\hat{i} - 3\hat{j})$ (b) $\frac{1}{5}(3\hat{i} - 4\hat{j})$

(c) $\frac{1}{5}(3\hat{i} + 4\hat{j})$ (d) $\frac{1}{5}(4\hat{i} + 3\hat{j})$

[GATE 2019]

113. A function f(x) is defined $f(x) = \frac{1}{2}(x + |x|)$. The value of

$$\int_{-1}^{1} f(x)\,dx$$

is _______ (round off to 1 decimal place).

[GATE 2019]

114. The value of the following limit is _______ (round off to 2 decimal places).

$$\lim_{\theta \to 0} \frac{\theta - \sin\theta}{\theta^3}$$

[GATE 2019]

115. The following system of equations

$$2x - y - z = 0,$$
$$-x + 2y - z = 0,$$
$$-x - y + 2z = 0$$

(a) has no solution.

(b) has a unique solution.

(c) has three solutions.

(d) has an infinite number of solutions.

[GATE 2019]

116. For real x, the number of points of intersection between the curves y = x and y = cos x is ______.

[GATE 2019]

117. One of the eigenvalues of the following matrix is 1.

$$\begin{pmatrix} x & 2 \\ -1 & 3 \end{pmatrix}$$

The other eigenvalue is _______.

[GATE 2019]

118. The curve y = f(x) is such that its slope is equal to y^2 for all real x. If the curve passes through (1, –1), the value of y at x = –2 is _______ (round off to 1 decimal place).

[GATE 2019]

119. For f(x) = |x|, with $\frac{df}{dx}$ denoting the derivative, the mean value theorem is not applicable because

(*a*) f(x) is not continuous at x = 0

(*b*) f(x) = 0 at x = 0

(*c*) $\frac{df}{dx}$ is not defined at x = 0

(*d*) $\frac{df}{dx} = 0$ at x = 0

[GATE 2020]

120. For the function $f(x) = \frac{e^{-\lambda}}{\sigma\sqrt{2\pi}}$, where $\lambda = \frac{1}{2\sigma^2}(x-\mu)^2$, and σ and μ are constants, the maximum occurs at

(*a*) $x = \sigma$ (*b*) $x = \sigma\sqrt{2\pi}$

(*c*) $x = 2\sigma^2$ (*d*) $x = \mu$

[GATE 2020]

121. $y = Ae^{mx} + Be^{-mx}$, where A, B and m are constants, is a solution of

(*a*) $\frac{d^2y}{dx^2} - m^2y = 0$ (*b*) $A\frac{d^2y}{dx^2} + m^2y = 0$

(*c*) $B\frac{d^2y}{dx^2} + Ay = 0$ (*d*) $\frac{d^2y}{dx^2} + my = m^2$

[GATE 2020]

122. Given $A = \begin{pmatrix} \sin\theta & \tan\theta \\ 0 & \cos\theta \end{pmatrix}$, the sum of squares of eigen values of A is

(*a*) $\tan^2\theta$ (*b*) 1

(*c*) $\sin^2\theta$ (*d*) $\cos^2\theta$

[GATE 2020]

123. The similarity solution for the diffusion equation, $\frac{\partial u}{\partial t} = \alpha\frac{\partial^2 u}{\partial x^2}$ is $u(x, t) = u(\eta)$, where similarity variable, $\eta = \frac{x}{\sqrt{\alpha t}}$. If $u(x,0) = e^{-x^2}$ the ratio, $\frac{u(0,1)}{u(0,4)} =$ _______ (*round off to one decimal place*).

[GATE 2020]

124. The equation $x\frac{dx}{dy}+y=c$, where c is a constant, represents a family of

(*a*) exponential curves (*b*) parabolas

(*c*) circles (*d*) hyperbolas

[GATE 2020]

125. A closed curve is expressed in parametric form as $x = a\cos\theta$ and $y = b\sin\theta$, where $a = 7$ m and $b = 5$m. Approximating $\pi = \frac{22}{7}$, which of the following is the area enclosed by the curve?

(*a*) $110\,m^2$ (*b*) $74\,m^2$

(*c*) $35\,m^2$ (*d*) $144\,m^2$

[GATE 2020]

126. In the equation $AX = B$, $A=\begin{bmatrix}\frac{1}{\sqrt{2}} & 0 & \frac{1}{\sqrt{2}}\\ 0 & 1 & 0\\ \frac{1}{\sqrt{2}} & 0 & \frac{-1}{\sqrt{2}}\end{bmatrix}, X=\begin{bmatrix}x\\y\\z\end{bmatrix}$

$B=\begin{bmatrix}0\\1\\-\sqrt{2}\end{bmatrix}$, where A is an orthogonal matrix, the sum of the unknowns, $x + y + z$ = _____ (*round off to one decimal place*).

[GATE 2020]

127. If $\int_0^1 (x^2-2x+1)dx$ is evaluated numerically using trapezoidal rule with four intervals, the difference between the numerically evaluated value and the analytical value of the integral is equal to ________ (*round off to three decimal places*).

[GATE 2020]

128. Consider the differential equation $\frac{d^2y}{dx^2}+8\frac{dy}{dx}+16y=0$ and the boundary conditions y(0) = 1 and $\frac{dy}{dx}(0)=0$. The solution to this equation is:

(a) $y = (1 + 2x)e^{-4x}$ (b) $y = (1 - 4x)e^{-4x}$

(c) $y = (1 + 8x)e^{-4x}$ (d) $y = (1 + 4x)e^{-4x}$

[GATE 2021]

129. *u(x, y)* is governed by the following equation

$$\frac{\partial^2 u}{\partial x^2}-4\frac{\partial^2 u}{\partial x\partial y}+6\frac{\partial^2 u}{\partial y^2}=x+2y.$$

The nature of this equation is:

(a) linear (b) elliptic

(c) hyperbolic (d) parabolic

[GATE 2021]

130. $\lim_{x\to 0}\left(\frac{1}{\sin x}-\frac{1}{x}\right)$ = ______ (*round off to nearest integer*).

[GATE 2021]

131. Given that ς is the unit circle in the counter-clockwise direction with its center at origin, the integral $\oint_{\varsigma}\frac{z^3}{4z-i}dz$ = ______ (*round off to three decimal place*).

[GATE 2021]

132. Which of the following statement(s) is/are true about the function defined as $f(x) = e^{-x}|\cos x|$ for $x > 0$?

(a) Differentiable at $x = \frac{\pi}{2}$

(b) Differentiable at $x = \pi$

(c) Differentiable at $x = \frac{3\pi}{2}$

(d) Continuous at $x = 2\pi$

[GATE 2021]

133. The ratio of the product of eigenvalues to the sum of the eigenvalues of the given matrix

$\begin{bmatrix}3 & 1 & 2\\ 2 & -3 & -1\\ 1 & 2 & 1\end{bmatrix}$ is ______ (*round off to nearest integer*).

[GATE 2021]

134. The definite Integral $\int_1^5 x^2 dx$ is evaluated using four equal intervals by two methods-first by the trapezoidal rule and then by the Simpson's one - third rule. The absolute value of the difference between the two calculations is ______ (*round off to the decimal places*).

[GATE 2021]

ANSWER KEY

1. (*b*)	**2.** (*d*)	**3.** (*a*)	**4.** (*d*)	**5.** (*c*)	**6.** (*d*)	**7.** (*b*)	**8.** (*b*)	**9.** (*c*)	**10.** (*c*)
11. (*c*)	**12.** (*d*)	**13.** (*b*)	**14.** (*d*)	**15.** (*b*)	**16.** (*c*)	**17.** (*c*)	**18.** (*d*)	**19.** (*b*)	**20.** (*a*)
21. (*b*)	**22.** (*a*)	**23.** (*d*)	**24.** (*b*)	**25.** (*c*)	**26.** (*d*)	**27.** (*d*)	**28.** (*d*)	**29.** (*c*)	**30.** (*d*)
31. (*b*)	**32.** (*b*)	**33.** (*a*)	**34.** (*b*)	**35.** (*b*)	**36.** (*c*)	**37.** (*c*)	**38.** (*c*)	**39.** (*a*)	**40.** (*a*)
41. (*c*)	**42.** (*a*)	**43.** (*d*)	**44.** (*d*)	**45.** (*a*)	**46.** (*b*)	**47.** (*c*)	**48.** (*a*)	**49.** (*c*)	**50.** (*d*)
51. (*c*)	**52.** (*c*)	**53.** (*c*)	**54.** (*a*)	**55.** (*d*)	**56.** 0.265625	**57.** (*a*)	**58.** (*a*)	**59.** (*a*)	**60.** (*d*)
61. (*a*)	**62.** (*a*)	**63.** (*c*)	**64.** 0	**65.** (*b*)	**66.** (*b*)	**67.** (*a*)	**68.** (*c*)	**69.** (*b*)	**70.** (*c*)
71. (*d*)	**72.** (*a*)	**73.** (*c*)	**74.** (*d*)	**75.** (*a*)	**76.** (*b*)	**77.** (*c*)	**78.** (*a*)	**79.** (*b*)	**80.** 0.5
81. (*c*)	**82.** (*b*)	**83.** (*b*)	**84.** (*b*)	**85.** (*b*)	**86.** (*a*)	**87.** 0	**88.** (*d*)	**89.** (*a*)	**90.** (*a*)
91. 6.28	**92.** 0.57	**93.** (*d*)	**94.** 12	**95.** 0	**96.** (*c*)	**97.** (*b*)	**98.** (*b*)	**99.** (*b*)	
100. .008 to .012		**101.** (*c*)	**102.** (*d*)	**103.** (*c*)	**104.** (*b*)	**105.** (*b*)	**106.** 0.0	**107.** (*b*)	**108.** 6.28
109. –0.707	**110.** (a)	**111.** (b)	**112.** (c)	**113.** (0.5)	**114.** (0.166)		**115.** (d)	**116.** (1)	**117.** (2)
118. (0.5)	**119.** (*c*)	**120.** (*d*)	**121.** (*a*)	**122.** (b)	**123.** (1.9 to 2.1)		**124.** (*c*)	**125.** (a)	
126. (0.9 to 1.1)		**127.** (0.010 to 0.012)		**128.** (d)	**129.** (b)		**130.** (0 to 0)	**131.** (0.02 to 0.03)	
132. (b; d)	**133.** (8 to 8)			**134.** (0.66 to 0.68)					

EXPLANATIONS

1. $f(\alpha+\beta)$

$$f(\theta) = \begin{bmatrix} \cos\theta & \sin\theta \\ -\sin\theta & \cos\theta \end{bmatrix}$$

$$f(\alpha).f(\beta) = \begin{bmatrix} \cos\alpha & \sin\alpha \\ -\sin\alpha & \cos\alpha \end{bmatrix}\begin{bmatrix} \cos\beta & \sin\beta \\ -\sin\beta & \cos\beta \end{bmatrix}$$

$$\begin{bmatrix} \cos\alpha\cos\beta - \sin\alpha\sin\beta & \cos\alpha\cos\beta + \sin\alpha\sin\beta \\ -\sin\alpha\cos\beta - \cos\alpha\sin\beta & -\sin\alpha\cos\beta + \cos\alpha\sin\beta \end{bmatrix}$$

$$f(\alpha).f(\beta) = \begin{bmatrix} \cos(\alpha+\beta) & \sin(\alpha+\beta) \\ -\sin(\alpha+\beta) & \cos(\alpha+\beta) \end{bmatrix}$$

$$= f(\alpha+\beta)$$

2. Taylor's expansion of x(t) at $t = t_{k+1}$

$$x(t_{k+1}) = x(t_k) + \Delta t.\frac{dx(t)}{dt}_{(t=t_k)} + \frac{(\Delta t)^2}{2!}\frac{dx(t)}{dt}_{(t=t_k)} + \ldots$$

In Euler iteration formula, we neglect higher terms other than first order term and in Taylor expansion.

$$\Rightarrow x(t_{k+1}) = x(t_k) + \Delta t.\frac{dx(t)}{dt}_{(t=t_k)}$$

∵ From given differential equation,

$$\frac{dx}{dt} = f(x)$$

Putting this in Euler iteration formula,

$$x(t_{k+1}) = x(t_k) + \Delta t.fx(t_k)$$

represents

$$x(t_{k+1}) = x_{k+1}$$

$$x(t_k) = x_k$$

$$\Rightarrow x_{k+1} = x_k + \Delta t \times f(x_k)$$

3. $J(x) = x^2 - 7x + 30$

$$J'(x) = 2x - 7$$

$$J''(x) = x$$

At $J(x)_{min}$,

$J'(x) = 0$ and $J''(x)$ = Positive

$$\therefore x = \frac{7}{2}$$

4. (3) and (4) will be true only if PQ = QP and so they are false.

(1) is false, (2) is true.

$PQ = I^2 \Rightarrow PQ = I$

$P = Q^{-1}$ = true

5. Given, $A = \begin{bmatrix} 2 & 1 \\ 0 & 3 \end{bmatrix}$

To get the eigen values of A, solve the characteristic equation $|A - \lambda I| = 0$

$$\therefore \begin{vmatrix} 2-\lambda & 1 \\ 0 & 3-\lambda \end{vmatrix} = 0$$

$(2 - \lambda)(3 - \lambda) - 0 = 0,$

$\lambda = 2, 3$

If eigen value of matrix A is λ, eigen value of A^{-1} will be $\frac{1}{\lambda}$.

6. $A = \begin{bmatrix} 2 & 1 \\ 0 & 3 \end{bmatrix}$

$$A^{-1} = \frac{\text{adj}(A)}{|A|} = \frac{1}{6}\begin{bmatrix} 3 & 0 \\ -1 & 2 \end{bmatrix}$$

$$= \begin{bmatrix} 1/2 & 0 \\ -1/6 & 1/3 \end{bmatrix}$$

Eien values of A^{-1} is

Now let A^{-1} = B.

Characteristic equation for B = (λI – B) = 0 and roots of this equation will be eigen values of b = A^{-1}.

$\Rightarrow |B - \lambda I| = 0$

$$\Rightarrow (B - \lambda I) = \begin{bmatrix} 1/2 - \lambda & 0 \\ -1/6 & 1/3 - \lambda \end{bmatrix}$$

$$\Rightarrow \left(\frac{1}{2} - \lambda\right)\left(\frac{1}{3} - \lambda\right) - 0 = 0$$

Roots $\lambda = \frac{1}{2}, \frac{1}{3}$ = eigen values

7. System of equations in matrix form AX = B

$$\begin{bmatrix} 1 & -1 & 2 \\ 2 & 3 & -1 \\ 2 & -2 & 4 \end{bmatrix}\begin{bmatrix} X \\ Y \\ Z \end{bmatrix} = \begin{bmatrix} 0 \\ 0 \\ 0 \end{bmatrix}$$

AX = 0 has either a trivial solution or infinite many solutions.

Check for |A|

Determinant of A = 0 in our case, so this system will have infinite number of non-trivial solutions.

8. Here, the function is not said to be a vector function, so we will take first it as a multivariable scalar function and then multivariable vector function.

First assume F as scalar function.

$F : D \subseteq R^n \to R$

At Stationary point,

$$\frac{\partial F}{\partial x_i} = 0$$

where, x_i is the coordinate of R^n.

In this case, curl and divergence are not defined, only gradient is defined.

$$\Rightarrow \nabla F = \frac{\partial F}{\partial x_i}\hat{e}_i = 0$$

($\hat{e}_i$ is the unit vector of base vector of R^n)

Now, take F as a vector function.

$F : D \subseteq R^n \to R^n$

At stationary point,

$$\frac{\partial F_k}{\partial n_j} = 0 \qquad k, j = 1 \text{ to } n$$

Curl of $\vec{F}$

$$\left(\vec{\nabla} \times \vec{F}\right)_i = \xi_{ijk}\frac{\partial F_k}{\partial n_j};\ i, j, k = 1 \text{ to } n$$

$$\left(\vec{\nabla} \times \vec{F}\right)_i = 0\ \forall\ i$$

divergent of $\vec{F}$

$$\Rightarrow (\vec{\nabla}.\vec{F})_i = \frac{\partial f_j}{\partial x_j};\ i, j = 1 \text{ to } n$$

$$\Rightarrow (\vec{\nabla}.\vec{F})_i = 0$$

Gradient of $\vec{F}$

$$\Rightarrow (\vec{\nabla}.\vec{F})_{ij} = \frac{\partial f_i}{\partial x_j};\ i, j = 1 \text{ to } n$$

= 0

So, the gradient of $\vec{F}$ vanishes in both cases.

9. Taking

$$\Delta n = 0.2 \quad y = \frac{1}{1+n^2},$$

Then for $\Delta n = 0.2$

x	0.0	0.2	0.4	0.6	0.8	1.0
y	1	0.9615	0.862	0.7353	0.6097	0.5

By using trapezoidal rule

$$J = \int_0^1 \frac{1}{1+x^2} = \int_0^1 y\,dx$$

$$J = \frac{\Delta x}{2}[y(x = 0.0) + 2y(x = 0.2) + 2y(x = 0.4) +$$

$$2y(x = 0.6) + 2y(x = 0.8) + y(x = 1.0)]$$

$$= \frac{0.2}{2}(1 + 2 \times 0.9615 + 2 \times 0.862 +$$

$$2 \times 0.7353 + 2 \times 0.6097 + 0.5)$$

$$= 0.7837$$

10. Taking a function

$$f(x) = x^3 - c$$

The roots of this polynomial function gives cube root of c.

$$\because f'(x) = 3x^2$$

In Newton-Raphson iteration formula,

$$x_{k+1} = x_k - \frac{f(x_k)}{f'(x_k)}$$

$$= x_k - \frac{x_k^3 - c}{3x_k^2} = \frac{3x_k^3 - x_k^3 + c}{3x_k^2}$$

$$= \frac{2x_k^3 + c}{3x_k^2}$$

11. $$\lim_{x\to 0} \frac{\sin x}{e^x.x} = \left(\lim_{x\to 0} \frac{\sin x}{x}\right)\left(\lim_{x\to 0} \frac{1}{e^x}\right) = 1 \times 1 = 1$$

12. $$F(s) = \frac{(s+10)}{(s+2)(s+20)}$$

$$= \frac{A}{s+2} + \frac{B}{s+20}$$

$$s + 10 = s(A + B) + 20A + 2B$$

$$A + B = 1$$

$$20A + 2B = 10$$

$$A = \frac{8}{18}, B = \frac{10}{18}$$

Thus partial fraction of F(s) is

$$F(s) = \frac{4/9}{s+2} + \frac{5/9}{s+20}.$$

13. Inverse Laplace transform of F(s)

$$L[F(s)] = \frac{4}{9}L\left(\frac{1}{s+2}\right) + \frac{5}{9}L\left(\frac{1}{s+20}\right)$$

$$L[F(s)] = \frac{4}{9}e^{-2t} + \frac{5}{9}e^{-20t}$$

14. Given,

$$f(x) = \begin{Bmatrix} \sin x, & \text{for } x < 0 \\ 0 & , \text{for } x = 0 \\ 3x^3 & , \text{for } x > 0 \end{Bmatrix}$$

$$\lim_{x\to 0^+} f(x) = \lim_{x\to 0} 3x^3 = 0$$

$$\lim_{x\to 0^-} f(x) = \lim_{x\to 0} \sin x = 0$$

$$f(0) = 0$$

$$\because \lim_{x\to 0^+} f(x) = \lim_{x\to 0^-} f(x) = f(0)$$

Therefore, f(x) is continuous at x = 0

LHD (Left Hand Derivative)

$$= \lim_{x\to 0^-} \frac{f(x) - f(0)}{x - 0}$$

$$= \lim_{x\to 0^-} \frac{\sin x - 0}{x - 0} = 1$$

RHD (Right Hand Derivative)

$$= \lim_{x\to 0^+} \frac{f(x) - f(0)}{x - 0}$$

$$= \lim_{x\to 0^+} \frac{3x^3 - 0}{x - 0} = \lim_{x\to 0} 3x^2 = 0$$

$\therefore$ LHD $\neq$ RHD

15. Given,

$$A = \begin{pmatrix} 1 & 0 & 1 \\ 0 & 2 & 1 \\ 1 & 1 & -3 \end{pmatrix}$$

the characteristic equation is given by

$$|A - \lambda I| = 0$$

$$\Rightarrow \begin{vmatrix} 1-\lambda & 0 & 1 \\ 0 & 2-\lambda & 1 \\ 1 & 1 & -3-\lambda \end{vmatrix} = 0$$

$$\Rightarrow (1-\lambda)[-(2-\lambda)(3+\lambda)-(1)(1)] - 0 + 1[0 - 1(2-\lambda)] = 0$$

$$-(\lambda-1)(\lambda^2+3\lambda-2\lambda-6-1)+\lambda-2=0$$

$$\Rightarrow (\lambda-1)(\lambda^2+\lambda-7)-(\lambda-2)=0$$

$$\Rightarrow \lambda^3+\lambda^2-7\lambda-\lambda^2+7-\lambda+2=0$$

$$\Rightarrow \lambda^3-9\lambda+9=0$$

the roots of this equation are eigen values.

Product of roots $\lambda_1\lambda_2\lambda_3 = \frac{0}{1} = 0$

16. In a linear ordinary differential equation, none of the terms containing unknown variable or its derivatives are multiplied by themselves or each other.

In option (a) y^2 term is present, in option (b) $y\frac{dy}{dx}$ term makes it nonlinear, option (d) also consists of nonlinear term $\left(\frac{dy}{dx}\right)^2$

So, the right answer is option (c) which is an ordinary differential equation.

17. For any differentiable function, f

$$\vec{\nabla} \times \vec{\nabla} f = \vec{0}$$

18. $\vec{p} \times (\vec{q} \times \vec{r}) + \vec{q} \times (\vec{r} \times \vec{p}) + \vec{r} \times (\vec{p} \times \vec{q})$

$= [(\vec{p}.\vec{r})\vec{q} - (\vec{p}.\vec{q})\vec{r}] + [(\vec{q}.\vec{p})\vec{r} - (\vec{q}.\vec{r})\vec{p}]$
$+[(\vec{r}.\vec{q})\vec{p} - (\vec{r}.\vec{p})\vec{q}]$

$= 0$

For other cases,

let, $\vec{A} = \vec{p} \times \vec{q} + \vec{q} \times \vec{r} + \vec{r} \times \vec{p}$

$\vec{A}.\vec{p} = \vec{p}(\vec{p} \times \vec{q}) + \vec{p}(\vec{q} \times \vec{r}) + \vec{p}(\vec{r} \times \vec{p})$

$= 0 + [\vec{p}\vec{q}\vec{r}] + 0$

$\Rightarrow \vec{A}.\vec{p} \neq 0$ in general

$\Rightarrow$ A can be non-zero.

Let $\vec{B} = (\vec{p}.\vec{q})\vec{r} + (\vec{q}.\vec{r})\vec{p} + (\vec{r}.\vec{p})\vec{q}$

$\vec{p} \times \vec{B} = (\vec{p}.\vec{q})(\vec{p} \times \vec{r}) + 0 + (\vec{r}.\vec{p})(\vec{p} \times \vec{q})$

$\neq 0$in general

$\vec{C} = \vec{p}.(\vec{q} \times \vec{r}) + \vec{q}.(\vec{r} \times \vec{p}) + \vec{r}(\vec{p} \times \vec{q})$

$= [\vec{p}\vec{q}\vec{r}] + [\vec{q}\vec{r}\vec{p}] + [\vec{r}\vec{p}\vec{q}]$

$= 3[\vec{p}\vec{q}\vec{r}] \neq 0$ for general vectors.

19. $x = \cos\theta$

$dx = -\sin\theta d\theta$

$y = \sin\theta$

$dy = \cos\theta d\theta$

$$\frac{1}{2\pi}(xdy - ydx)$$

$$= \frac{1}{2\pi}\left(\int_0^{2\pi} \cos\theta(\cos\theta)d\theta - \sin\theta(-\sin\theta)d\theta\right)$$

$$= \frac{1}{2\pi}\int_0^{2\pi}(\cos^2\theta + \sin^2\theta)d\theta$$

$$= \frac{1}{2\pi}\int_0^{2\pi} 1 d\theta = \frac{2\pi}{2\pi} = 1$$

20. $\frac{d^2y}{dx^2} + 2\frac{dy}{dx} + y = 0$

Let $y = e^{kx}$

$$\frac{dy}{dx} = ke^{kx} \; ; \; \frac{d^2y}{dx^2} = k^2e^{kx}$$

$$\Rightarrow k^2e^{kx} + 2ke^{kx} + e^{kx} = 0$$

$$\Rightarrow e^{kx}(k^2 + 2k + 1) = 0$$

$^{kx} \neq 0, (k+1)^2 = 0$

Repeated root k = –1

Therefore, $y = e^{-x}$ and $y = xe^{-x}$ are solutions of this differential equation.

General solution is

$y = ae^{-x} + bxe^{-x}$; for $a = b = 1$

21. $\frac{d^2F}{dx^2} = -p^2F$

$$\frac{dG}{dt} = -c^2p^2G$$

$u = F(x)\,G(t)$

$$\frac{\partial u}{\partial t} = F(x)\frac{dG}{dt} = -(c^2p^2G)F$$

$$\frac{\partial^2 u}{\partial t^2} = \frac{d^2F}{dx^2}G(t) = -p^2FG$$

$$\therefore \frac{\partial u}{\partial t} = c^2\frac{\partial^2 u}{\partial x^2}$$

22. $A\vec{x} = \vec{b}$

$$\text{Det } A = \begin{vmatrix} 1 & 1 & 2 \\ 1 & 0 & 1 \\ 0 & 1 & 1 \end{vmatrix}$$

$$= 1(0\times1 - 1\times1) - 1(1\times1 - 2\times1) + 0$$

$= -1 + 1 = 0$

$\Rightarrow$ The system does not have unique solution

$$D_2 = \begin{vmatrix} 1 & 1 & 2 \\ 1 & -1 & 1 \\ 0 & 0 & 1 \end{vmatrix}$$

$$= 0 - 0 + 1[1\times(-1) - 1\times1]$$

$= -2 \neq 0$

$\Rightarrow$ System has no solution.

23. $f(x) = x^2 - 5x + 6 = (x-2)(x-3)$

$$= \left(x - \frac{5}{2}\right)^2 - \frac{1}{4}$$

f(x) has its minima a at x = 2.5.

24. For any function y,

$$L\left(\frac{dy}{dt}\right) = sY(s) - y(0)$$

$$y(0) = \cosh(0)\sin 0 = 0$$

$$\therefore L\left(\frac{dy}{dx}\right) = sY(s) \text{ for } y = \cosh(at)\sin(at)$$

$L[t\ y(t)] = -y'(s)$ for any function y.

25. Simpson's rule,

$$\int_a^b f(x)dx \approx \frac{b-a}{6}\left[f(a) + 4f\left(\frac{a+b}{2}\right) + f(b)\right]$$

$$= \frac{b-a}{2}\left[\frac{f(a)+f(b)}{3} + \frac{4}{3}f\left(\frac{a+b}{2}\right)\right]$$

26. $\int_0^1 x^3\,dx = \left[\frac{x^4}{4}\right]_0^1 = \frac{1^4 - 0}{4} = \frac{1}{4} = 0.25$

Using Simpson's rule,

$$\int_0^1 x^3\,dx \approx \frac{1-0}{6}\left[0^3 + 1^3 + 4\times\left(\frac{0+1}{2}\right)^3\right]$$

$$= \frac{1}{6}\left(1 + 4\times\frac{1}{8}\right) = \frac{1}{6}\times\frac{3}{2} = \frac{1}{4} = 0.25$$

$$\%\text{ error} = \frac{0.25 - 0.25}{0.25}\times 100 = 0\%$$

27. Given, $\frac{d^2y}{dx^2} + ky = 0$

$y'' = -ky$

Where, k > 0 and k is real

Therefore, Characteristic equation is

$D^2 = k \times (-1) = -k$

$$\Rightarrow D = \pm ik^{\frac{1}{2}}$$

$$y = A\sin\omega x + B\cos\omega x$$

where, $\omega = k^{\frac{1}{2}}$

28. $[A] = \{x\} = \{0\}$

Non-trivial solution can be found only when $|A| = 0$. Now, if [A] is diagonalised, $|A|$ remains invariant and the diagonal matrix obtained is

$$= \begin{pmatrix} \lambda_1 & 0 & \cdots \\ 0 & \lambda_2 & \cdots \\ \vdots & \vdots & \lambda_n \end{pmatrix}$$

Where, λ_i's are eigen values

$$|A| = \prod_{i=1}^{n} \lambda_i$$

$$\Rightarrow \lambda_i = 0 \text{ for any } i = |A| = 0$$

29. Given,

$$\int_0^\pi \frac{dx}{1 + x + \sin x}$$

Using trapezoidal rule, using two equal intervals

$\left(0,\frac{\pi}{2}\right)$ and $\left(\frac{\pi}{2},\pi\right)$

$$I=\frac{\pi}{2}\left(\frac{f\left(\frac{\pi}{2}\right)+f(0)}{2}\right)+\frac{\pi}{2}\left(\frac{f\left(\frac{\pi}{2}\right)+f(n)}{2}\right)$$

$$f(x)=\frac{1}{1+x+\sin x}$$

$$f(0)=1, f\left(\frac{\pi}{2}\right)=0.28,$$

$$f(\pi)=0.3183$$

$$\therefore I=\frac{\pi}{2}\times 0.9369=1.47$$

30. Characteristic equation of

$$\begin{pmatrix}2 & 1 & 1\\ 1 & 3 & 1\\ 1 & 1 & 4\end{pmatrix}$$

$\Rightarrow$ $(2-s)(3-s)(4-s)-1)(4-s-1)+1(1-3+s)$ $=0$

Collecting the constant terms,

$(2\times 3\times 4)-2-3+1-3=17$

The product of the eigen values is the constant term.

31. $f(x)=e^{\pi x}+\sin \pi x$

$$f'(x)=\pi e^{\pi x}+\pi\cos \pi x$$

$$f'(1)=75.835, f'(2)=1685.488$$

f'(x) fas no zero is (1, 2) as there is only one maximum/minimum of $\pi\cos\pi x$ in (1, 2).

hence, the maximum is at x = 2.

32. $F(s)=\frac{(s+1)}{(s+4)(s-3)}$

$$=\frac{s+1}{s^2+s-12}$$

$$=\frac{s+\frac{1}{2}+\frac{1}{2}}{\left(s+\frac{1}{2}\right)^2-\frac{49}{4}}$$

$$=\frac{s+\frac{1}{2}}{\left(s+\frac{1}{2}\right)^2-\left(\frac{7}{2}\right)^2}+\frac{1}{7}\times\frac{\frac{7}{2}}{\left(s+\frac{1}{2}\right)^2-\left(\frac{7}{2}\right)^2}$$

$$\therefore L^{-1}(F(s))=e^{-\frac{1}{2}}\sinh\left(\frac{7}{2}t\right)+1$$

$$\therefore L^{-1}(F(s))=e^{-\frac{1}{2}}\left\{\cosh\left(\frac{7}{2}t\right)+\frac{1}{7}\sinh\left(\frac{7}{2}t\right)\right\}$$

$$=e^{-\frac{t}{2}}\left\{\frac{e^{\frac{7}{2t}}+e^{-\frac{7}{2t}}}{2}+\frac{e^{\frac{7}{2t}}+e^{-\frac{7}{2t}}}{14}\right\}$$

$$=\left\{\frac{7e^{3t}+7e^{-4t}+e^{3t}-e^{-4t}}{14}\right\}$$

$$=\frac{4}{7}e^{3t}+\frac{3}{7}e^{-4t}$$

33. $A=\begin{pmatrix}1 & 2\\ 2 & 4\end{pmatrix}; |A|=0$

and $B=\left\{\begin{pmatrix}3\\3\end{pmatrix}\right\}\neq\left\{\begin{pmatrix}0\\0\end{pmatrix}\right\}$

$\therefore$ No solution can exist

34. Since, $\sqrt{R}$ is to be obtained, option (a) and (d) are trivial.

In option (B), as

$$n\to\infty, x_{n+1}=x_n=x$$

$$\Rightarrow 2x=x+\frac{R}{x}$$

$$\Rightarrow 2x^2=x^2+R$$

$$\Rightarrow x^2=R$$

hence, $x=\sqrt{R}$ upon iterative convergence.

Newton-Raphson's method is given by

$$x_{n+1}=x_n-\frac{f(x_n)}{f'(x_n)}$$

where, $f(x)=0$

$$f(x)=x^2-R$$

$$f'(x)=2x$$

$$\therefore\ x_{n+1}=x_n-\frac{x_n^2-R}{2x_n}$$

$$\therefore\ x_{n+1} = x_n - \frac{x_n}{2} + \frac{R}{2x_n}$$

$$= \frac{x_n}{2} + \frac{R}{2x_n}$$

$$\therefore\ x_{n+1} = \frac{1}{2}\left\{x_n + \frac{R}{x_n}\right\}$$

Hence, obtained.

35. Given,

$a^2 + b^2 = 1$

Assume

$a = \cos\theta$

$b = \sin\theta$

$$\text{also, } |\vec{V}_2| = \begin{pmatrix} a & -b \\ b & a \end{pmatrix} |\vec{V}_1| = |\vec{V}_1|$$

$\therefore$ No magnification

$$\text{Morever, } V_2 = \begin{pmatrix} \cos\theta & -\sin\theta \\ \sin\theta & \cos\theta \end{pmatrix} V_1$$

which is rotation matrix for rotation by angle θ.

36. Given,

$$5\frac{\partial^2\phi}{\partial x^2} + 3\frac{\partial^2\phi}{\partial x\partial y} + 2\frac{\partial^2\phi}{\partial y^2} + 9 = 0$$

As per formula

$D = b^2 - 4ac$

$D = 3 \times 3 - 4 \times 5 \times 2 = -31$

For, $D < 0$; elliptical

$D = 0$; parabolic

$D > 0$; hyperbolic

37. As per Cauchy's principle

Real symmetric matrices have real eigen value, hence applicable for stress tensor also. If real skew symmetric matrix, then imaginary eigen value.

38. Given,

$$\frac{dx}{dt} = -kx,\ x(0) = x_o$$

$$\Rightarrow \int \frac{dx}{x} = -\int k dt$$

$$\Rightarrow x = Ae^{-kt}$$

At $t = 0, x = x_o$

$$\Rightarrow x_o = Ae^{-k(0)} = A$$

$$\therefore x = x_o e^{-kt}$$

At $t = \frac{1}{k}, x = \frac{x_o}{e}$

39. Given,

$$\int_{-1}^{+1} \frac{dx}{x^2}$$

$\frac{1}{x^2}$ is not defined in [-1,1] (since, discontinuity at $x = 0$)

Therefore, its integral does not exist.

40. $y_1(t) = e^{2t} y(t)$

$$Y_1(s) = \int e^{2t} y(t) e^{-st} dt$$

$$= \int y(t) e^{-(s-2)} dt = Y(s-2)$$

41. Given,

$$z(x,y) = \frac{x^2}{50} + y^2 - xy - 3y$$

$$\nabla z = \left(\frac{2x}{50} - y\right)\hat{i} + (2y - x - 3)\hat{j} + 0\hat{k}$$

$$\nabla z_{(5,10)} = \left(\frac{1}{5} - 10\right)\hat{i} + 12\hat{j} + 0\hat{k}$$

Thus, vector is perpendicular to the plane. We want to find the direction of steepest slope of the hill. Therefore, parallel to k vector will be that slope.

42. Given,

$$f(x,y) = x^2 + y^2 - xy - 3y$$

$$\nabla f = (2x - y)\hat{i} + (2y - x - 3)\hat{j}$$

Now, for extremum,

$\nabla f = 0$

$\therefore 2x = y$

also, $2y = x + 3$

This gives $x = 1, y = 2$

43. $\vec{V} = 3x\hat{i} + 3xy\hat{j} - yz^2\hat{k}$

Curl of $\vec{V} = \vec{\nabla} \times \vec{V}$

$$= \begin{vmatrix} \hat{i} & \hat{j} & \hat{k} \\ \frac{\partial}{\partial x} & \frac{\partial}{\partial y} & \frac{\partial}{\partial z} \\ 3x & 3xy & -yz^2 \end{vmatrix}$$

$$= \hat{i}\left[\frac{\partial}{\partial y}\left(-yz^2\right) - \frac{\partial}{\partial z}(3xy)\right]$$

$$\hat{j}\left[\frac{\partial}{\partial x}\left(-yz^2\right) - \frac{\partial}{\partial z}(3x)\right]$$

$$+\hat{k}\left[\frac{\partial}{\partial x}(3xy) - \frac{\partial}{\partial y}(3x)\right]$$

$$= \left(-z^2\right)\hat{i} - (0)\hat{j} + (3y)\hat{k}$$

$$= -z^2\hat{i} + 3y\hat{k}$$

44. Function f(x) is said to be periodic if there exist a positive real number T such that

f (x + T) = f(x)

This condition is satisfied by only

f(x) = constant

say constant = c

Let w take real no. T

f(x) = C ... (i)

f(x + T) = C ... (ii)

say constant = C

For any value T, function will always be equal to constant

∴ f (x +T) = f(x)

if we take any other option

e.g. $f(x) = x^2$

take any real no. T

$f(x + T) = (x + T)^2$

$= x^2 + T^2 + 2xT$

for function to be periodic

f (x) = f (x + T)

$$\Rightarrow \quad \cancel{x^2} = \cancel{x^2} + T^2 + 2xT$$

⇒ T (T + 2x) = 0

⇒ T = 0, T= –2x

Here T itself is a function of x

Thus x^2 can not be periodic function T should be a real number.

45. $f(x_1, x_2, x_3) = x_1^2 + x_2^2 + x_3^2 - 2x_1 - 4x_2 - 6x_3 + 14$

For getting minimum value,

$$\frac{\partial f}{\partial x_1} = 0, \frac{\partial f}{\partial x_2} = 0, \frac{\partial f}{\partial x_3} = 0$$

$$\frac{\partial f}{\partial x_1} = 2x_1 - 2 = 0$$

$\Rightarrow x_1 = 1$

$$\frac{\partial f}{\partial x_2} = 2x_2 - 4 = 0$$

$\Rightarrow x_2 = 2$

$$\frac{\partial f}{\partial x_3} = 2x_3 - 6 = 0$$

$\Rightarrow x_3 = 3$

Second order partial derivatives

$$\nabla^2 = \begin{bmatrix} \frac{\partial^2 f}{\partial x_1^2} & \frac{\partial^2 f}{\partial x_1 \partial x_2} & \frac{\partial^2 f}{\partial x_1 \partial x_3} \\ \frac{\partial^2 f}{\partial x_1 \partial x_2} & \frac{\partial^2 f}{\partial x_2^2} & \frac{\partial^2 f}{\partial x_2 \partial x_3} \\ \frac{\partial^2 f}{\partial x_1 \partial x_3} & \frac{\partial^2 f}{\partial x_2 \partial x_3} & \frac{\partial f}{\partial x_3^2} \end{bmatrix}$$

$$= \begin{bmatrix} 2 & 0 & 0 \\ 0 & 2 & 0 \\ 0 & 0 & 2 \end{bmatrix}$$

In above matrix all the diagonal value are positive non–zero values

∴ It is a positive definite matrix

If $x_1 = 1$

$x_2 = 2$

$x_3 = 3$

will give minimum value for the function

46. $f(x_1, x_2) = x_1^2 + 2x_2^2 + e^{-x_1 - x_2}$

Gradient of operator scalar function will give a vector pointing in the direction of maximum increase of function

$$\vec{\nabla} f = \begin{bmatrix} 2x_1 - & e^{-x_1 - x_2} \\ 4x_2 - & e^{-x_1 - x_2} \end{bmatrix}$$

at point (1, –1)

$$\vec{\nabla} f = \begin{bmatrix} 1 \\ -5 \end{bmatrix}$$

47. $y = \pi + x = x + \pi$ ...(A)

$y = x - \pi$ ...(B)

Rearranging equation

$y - x = \pi$

$y - x = -\pi$

- such a system of simultaneous equation will never have any solution we can see clearly from (i) and (ii) that subtraction of sum variables is giving the constants and – ve constant
- Other interpretation : Both equation (A) and (B) represents two parallel lines which never intersect each other thus this kind of system will not have solution.

48. $f(x) = x - \sin(x)$

initial guess $x^{(0)}$

According to Newton – Raphson iteration formula

$$x_1 = x^{(0)} - \frac{f(x^0)}{f^1\left(x^{(0)}\right)}$$

$$x^{(k+1)} = x^{(k)} - \frac{f\left(x^{(k)}\right)}{f^1\left(k^{(k)}\right)} \quad \text{...(i)}$$

$$\left.\begin{aligned} f(x^{(k)}) &= x^{(k)} - \sin(x^{(k)}) \\ f^1(x^{(k)}) &= 1 - \cos(x^{(k)}) \end{aligned}\right\} \quad \text{... (ii)}$$

∴ from (i) and (ii)

$$x^{(k+1)} = x^{(k)} - \frac{\left(x^{(k)} - \sin\left(x^{(k)}\right)\right)}{1 - \cos\left(x^k\right)}$$

$$= \frac{x^{(k)} - x^{(k)}\cos\left(x^{(k)}\right) - x^{(k)} + \sin\left(x^{(k)}\right)}{1 - \cos\left(x^{(k)}\right)}$$

$$= \frac{\sin\left(x^{(k)}\right) - x^{(k)}\cos\left(x^{(k)}\right)}{1 - \cos\left(x^{(k)}\right)}$$

49. Let $A = \begin{bmatrix} 2 & a \\ b & 2 \end{bmatrix}$

Feigen value analysis:

$$\begin{bmatrix} 2-\lambda & a \\ b & 2-\lambda \end{bmatrix} = 0$$

$(2 - \lambda)^2 - ab = 0$

$4 + \lambda^2 - 4\lambda - ab = 0$

$\lambda^2 - 4\lambda + 4 - ab = 0$

For roots of this equation to be real and distinct

$(-4)^2 - 4(1)(4 - ab) > 0$

$\Rightarrow 16 - 16 + 4\,ab > 0$

$\Rightarrow ab > 0$

which is only possible either

$a, b > 0$

$a, b < 0$

50. $\dfrac{dy}{dt} = y^3e^tt^2$

$$\frac{dy}{y^3} = e^tt^2dt$$

Integrating both sides, we get

$$\int\frac{dy}{y^3} = \int e^tt^2dt$$

$$\frac{-1}{2y^2} = t^2e^t - \int 2t \cdot e^tdt$$

$= t^2\ e^t - 2t\ e^t + 2e^t + c$ ... (1)

using initial condition

$y(0) = 1$ in eq (1)

$$\frac{-1}{2x1} = 2 + C$$

$$\Rightarrow C = \frac{-1}{2} - 2 = \frac{-5}{2}$$

$$\frac{-1}{2y^2} = t^2e^t - 2te^t + 2e^t - \frac{5}{2}$$

$$\Rightarrow -\frac{1}{y^2} = 2t^2e^t - 4te^t + 4e^t - 5$$

$$\Rightarrow y = \sqrt{\frac{-1}{2t^2e^t - 4te^t + 4e^t - 5}}$$

$$= \sqrt{\frac{-1}{-\left[-2t^2e^t - 4e^{t+5}\right]}}$$

$$\boxed{y = \sqrt{\frac{1}{5 - 2e^t\left[t^2 - 2t + 2\right]}}}$$

51. $A^2 = A$ is satisfied by both identity matrix and null matrix.

Prove: $A^2 = A$

$\Rightarrow A^2 - A = 0$

$\Rightarrow A(A - I) = 0$

$\Rightarrow A = 0$ (Null Matrix) or $A = I$ (Identity matrix)

52. $\frac{d^2y}{dt^2} + \frac{dy}{dt} - 2y = 0$

General solution = Complimentary (C.F) function + Particular Integral (P.I)

For this equation (P.I) = 0

C.F. :

$D^2 + D - 2 = 0$

$D^2 + 2D - D - 2 = 0$

$D(D + 2) - 1(D + 2) = 0$

$(D - 1)(D + 2) = 0$

$\Rightarrow D = 1, D = -2$

$\therefore$ Complimentary function C.F. = $Ae^{-2t} + Be^{t}$

where A and B are constants

53. $x + 2y + kz = 1$ ... (i)

$2x + ky + 8z = 3$ ... (ii)

this system will not have a solution if

$$\left[\frac{a_2}{a_1} = \frac{b_2}{b_1} = \frac{c_2}{c_1} \neq \frac{d_2}{d_1}\right]$$

if k = 4

$$\frac{2}{1} = \frac{k}{2} = \frac{8}{k} \neq \frac{3}{1}$$

$$\Rightarrow \frac{2}{1} = \frac{4}{2} = \frac{8}{4} \neq \frac{3}{1}$$

$\Rightarrow 2 = 2 = 2 \neq 3$

54. $m\frac{d^2x}{dt^2} + kx = u(t)$

u(t) is unit step function

Taking laplace of equation

$$m\left[s^2 \times (s) - s \times (0) - x^1(0)\right] + kx(s) = \frac{1}{s}$$

say, initial conditions are zero

$x(0) = 0, x^1(0) = 0$

$$\left(ms^2 + k\right)x(s) = \frac{1}{s}$$

$$\boxed{x(s) = \frac{1}{s\left(ms^2 + k\right)}}$$

$$L\left[u(t)\right] = \frac{1}{s}$$

$$L\left[\frac{d^2x}{dt^2}\right] = s^2 \times (s) - s\,x(0) - x'(0)$$

55. $\frac{dy}{dx} - 2\sqrt{y} = 0$

$$\Rightarrow \frac{dy}{2\sqrt{y}} = +dx$$

Integrating both sides, we get

$$\int \frac{dy}{2\sqrt{y}} = \int dx$$

$$\frac{1}{2} \times \frac{y^{\frac{-1}{2}+1}}{\left(\frac{-1}{2}+1\right)} + c = x$$

$$\boxed{\sqrt{y} - x + c = 0}$$

56. $\int_0^1 x^3 dx$

According to Trapezoidal Rule : If integration partial [a, b] is divided into n number of intervals

then $\int_a^b f(x)dx$ can be approximated as:

$$\int_a^b f(x)dx = \frac{\Delta x}{2}\left[\begin{array}{r} f(x_0) + 2f(x_1) + 2f(x_2) + 2f(x_3) \\ + \ldots\ldots.. + 2f(x_{n-1}) + f(x_n) \end{array}\right]$$

here, $\Delta x = \frac{b-a}{n}$

$b = 1, a = 0, n = 4$

$\therefore \quad \Delta x = \frac{1}{4} = 0.25$

$f(x) = x^3$

$x_0 = 0$	$f(x_0) = 0$
$x_1 = x_0 + \Delta x = 0.25$	$f(x_1) = 0.015625$
$x_2 = 0.5$	$f(x_2) = 0.125$
$x_3 = 0.75$	$f(x_3) = 0.421875$
$x_4 = 1$	$f(x_4) = 1$

$$\therefore \int_0^1 x^3 dx = \frac{0.25}{2}\left[\begin{array}{r} 0 + 2(0.015625) + 2(0.125) \\ + 2(0.421875) + 1 \end{array}\right]$$

$= 0.265625$

$\simeq 0.27$

57. $y = \frac{1}{x+3} = (x+3)^{-1}$

1st derivative $y' = -1 \times (x+3)^{-2}$

2nd derivative $y'' = (-1) \times (-2) \times (x+3)^{-3}$

nth derivative $y^n = (-1)^n n!(x+3)^{-(n+1)}$

58. $y = 1-\sqrt{1-x^2}$ – circle

$y = 1$ – line

Rotating area dA by an angle dθ about x-axis will produce volume $dV = y.dAd\theta$ integrating it from $\theta = 0$ to 2π

$$V = (2\pi)\int_{-1}^{1}\int_{1-\sqrt{1-x^2}}^{1} ydydx$$

$y = 1-\sqrt{1-x^2}$ semi circle

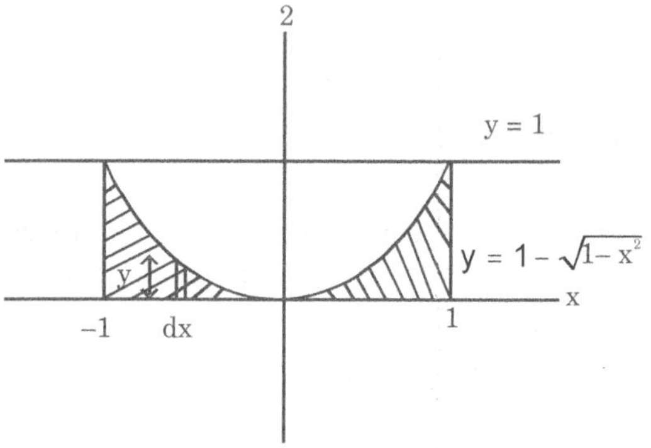

y = 1 straight line

volume of redution by circle about x-axis

$V = \pi r^2 h$ – volume generated by shaded postion

here r = 1, h = 2r = 2

$\boxed{V = 2\pi - V_1}$

$$V_1 = \int_{-1}^{1} \pi y^2 dx$$

$$= \int_{-1}^{1} \pi\left(1-\sqrt{1-x^2}\right)^2 dx$$

$$= \pi\int_{-1}^{1}\left[1+\left(1-x^2\right)-2\sqrt{1-x^2}\right]dx$$

$$= \pi\left[\left(2x-\frac{x^3}{3}\right)_1^{-1} \int_{-1}^{1} 2\sqrt{1-x^2}dx\right]$$

$$= \pi\left[\left(2-\frac{1}{3}+2-\frac{1}{3}\right)-I\right]$$

$$= \pi\left[4-\frac{2}{3}-I\right]$$

$$= \pi\left[\frac{10}{3}-I\right]$$

$$I = \int_{\frac{-\pi}{2}}^{\frac{\pi}{2}} 2\sqrt{1-x^2}\theta x$$

$x = \sin\theta$

$dx = \cos\theta \, d\theta$

$$= \int_{\frac{-\pi}{2}}^{\frac{\pi}{2}} 2\sqrt{1-\sin^2\theta}.\cos\theta \, d\theta$$

$$= \int_{\frac{-\pi}{2}}^{\frac{\pi}{2}} 2\cos^2\theta \, d\theta = \int_{\frac{-\pi}{2}}^{\frac{\pi}{2}} 2\left[\frac{1+\cos^2\theta}{2}\right]d\theta$$

$$= \theta + \frac{\sin^2\theta}{2}\Bigg|_{\frac{-\pi}{2}}^{\frac{\pi}{2}} = \frac{\pi}{2}+0+\frac{\pi}{2}+0 = \pi$$

$$V_1 = \pi\left[\frac{10}{3}-\pi\right] = \left[\frac{10}{3}\pi-\pi^2\right]$$

$V = 2\pi - V_1$

$$= 2\pi - \frac{10}{3}\pi + \pi^2$$

$$\boxed{V = \pi^2 - \frac{4}{3}\pi}$$

59. $|A-\lambda I| = 0$

$$\begin{vmatrix} 2-\lambda & 7 & 10 \\ 5 & 2-\lambda & 25 \\ 1 & 6 & 5-\lambda \end{vmatrix} = 0$$

$(2-\lambda)[(2-\lambda)(5-\lambda)-150]-7[5(5-\lambda)-25]+10[30-(2-\lambda)] = 0$

$(2-\lambda)[\lambda^2-7\lambda+10-150]-7[25-5\lambda-25]+10[30-2+\lambda] = 0$

$(2-\lambda)[\lambda^2-7\lambda-140]+7\times5\lambda+280+10\lambda=0$

$-\lambda^3+9\lambda^2+171\lambda=0$

By solving above characteristic equation, we get

$\lambda_1=-9.33$

$\lambda_2=18.33$

$\lambda_3=0$

60. $f(x,y)=\frac{x^2+xy^2}{\sqrt{5}}$

$\vec{a}=2\hat{i}-4\hat{j}$

$(x, y)=(1, 1)$

Unit vector of $\vec{a}=\hat{a}=\frac{\vec{a}}{|a|}$

$$\frac{2\hat{i}-4\hat{j}}{\sqrt{2^2+(-4)^2}}=\frac{2\hat{i}-4\hat{j}}{\sqrt{20}}$$

$$=\frac{i-2\hat{j}}{\sqrt{5}}$$

Gradient of function f is given by:

$$\nabla=\frac{\partial f}{\partial x}\hat{i}+\frac{\partial f}{\partial y}\hat{j}$$

$$=\frac{2x+y^2}{\sqrt{5}}\hat{i}+\frac{2xy}{\sqrt{5}}\hat{j}$$

Directional derivative of function f in direction of vector a is defined as:

$$\nabla.f.\hat{a}=\left[\frac{2x+y^2}{\sqrt{5}}\hat{i}+\frac{2xy}{\sqrt{5}}\hat{j}\right]\cdot\left[\frac{i-2j}{\sqrt{5}}\right]$$

$$=\frac{1}{5}\left[2x+y^2-4xy\right]$$

$|\nabla f.\hat{a}|_{(1,1)}$

$$=\frac{1}{5}\left[2\times1+(1)^2-4\times1\times1\right]$$

$$=\frac{-1}{5}$$

61. $\int_4^5\frac{x+2}{x^2+4x-21}dx$

$$=\int_4^5\frac{\frac{1}{1}\times2(x+2)}{x^2+4x-21}$$

$$=\frac{1}{2}\int_4^5\frac{2x+4}{x^2+4x-21}$$

if $\frac{d}{dx}$(Denominator) = Numerator

$\frac{d}{dx}\left[x^2+4x-21\right]=2x+4$ which is equal to numerator

$$\therefore=\frac{1}{2}\left[\ell n\left[x^2+4x-21\right]\right]_4^5$$

$$=\frac{1}{2}\left[\ell n\left[5^2+4\times5-21\right]-\ell n\left(4^2+4\times4-21\right)\right]$$

$$=\frac{1}{2}\left[\ell n(24)-\ell n(11)\right]$$

$$=\frac{1}{2}\ell n\left[\frac{24}{11}\right]$$

$$=\ell n\left[\frac{24}{11}\right]$$

62. At x = 0

Function y = |x| is continuous but not differentiable. if we plot y = |x|

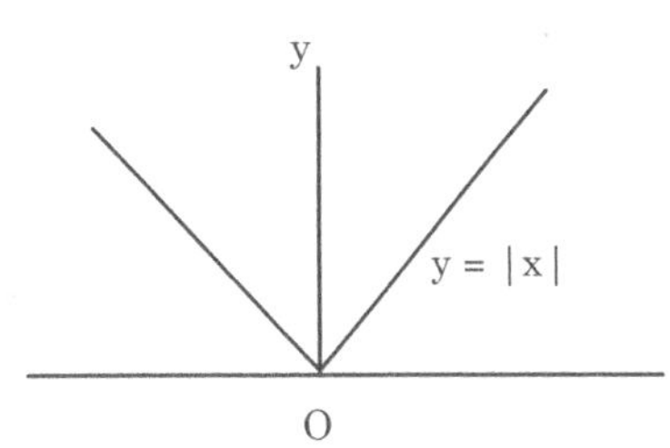

at x = 0

y has a sharp turn but continuous.

63. $A=\begin{bmatrix}1&-1&0\\0&1&-1\\-1&0&1\end{bmatrix}$

eigen vector of matrix A

$$V=\begin{bmatrix}1\\1\\1\end{bmatrix}$$

Eigne value analysis:

$|A-\lambda I|=0$

$$\Rightarrow \begin{vmatrix} 1-\lambda & -1 & 0 \\ 0 & 1-\lambda & -1 \\ -1 & 0 & 1-\lambda \end{vmatrix} = 0$$

$$\Rightarrow (1-\lambda)(1-\lambda)^2 - 2 = 0$$

$$\Rightarrow (1-\lambda)^3 = 1$$

$$\Rightarrow 1-\lambda = 1^{\frac{1}{3}}$$

$\boxed{\lambda = 1-1 = 0}$ $\boxed{\therefore \quad \lambda_1 = \lambda_2 = \lambda_3 = 0}$

It can also be interpreted in other ways directly if an eigen vector have same value in all rows like

$V = \begin{bmatrix} 1 \\ 1 \\ 1 \end{bmatrix}$, corresponding eigen value is always zero.

64. $I = \int_1 \left(y^2 z\hat{i} + 2^2 x\hat{j} + x^2 y\hat{k}\right)\left(x\hat{i} + y\hat{j} + z\hat{k}\right) ds$

$$= \int_s \left(y^2 zx + z^2 xy + x^2 yz\right) ds$$

$$= \int xyz(x+y+z) dx$$

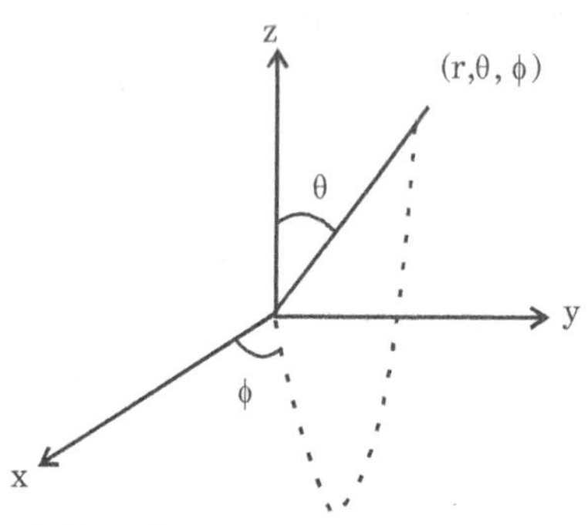

$x = r \sin\theta \cos\phi$

$y = r \sin\theta \sin\phi$

$z = r \cos\theta$

$r = 1$ for unit sphere

$s = d\theta\, d\phi$

$$I = \int_0^{2\pi}\int_0^{2\pi} \sin^2\theta \cos\theta \cos\phi \sin\phi$$

$$[\sin\theta\cos\phi + \sin\theta\sin\phi + \cos\theta] 2\theta d\phi$$

$$I = \int_0^{2\pi}\int_0^{2\pi} \begin{bmatrix} \sin^3\theta\cos\theta\cos^2\phi\sin\phi + \sin^3 \\ \theta\cos\theta + \cos\phi\sin^2\phi + \sin^2 \\ \theta\cos^2\theta\cos\phi\sin\phi \end{bmatrix} d\theta d\phi$$

$$I = \int_0^{2\pi}\int_0^{2\pi} \begin{bmatrix} \sin^3\theta\cos\theta\cos\phi\sin\phi[\cos\phi + \sin\phi] \\ + \sin^2\theta + \cos^2\theta\cos\phi\sin\phi \end{bmatrix} d\theta d\phi$$

$$= \int \left[\left\{\left[\frac{\sin^4\theta}{4}\cos\phi\sin\phi[\cos\phi + \sin\phi]\right]_0^{2\pi}\right\}\right.$$

$$\left. + \left\{\int_0^{2\pi} \left[\frac{\sin^2 2\theta}{4}\cos\phi\sin\phi\right] d\phi\right\}\right]$$

$$= \int_0^{2\pi} [I_1 + I_2]$$

$\boxed{I_1 = 0}$

$$I_2 = \int_0^{2\pi}\left[\int_0^{2\pi}\left[\frac{1-\cos 4\theta}{8}\right]\right] \cos\phi \sin\phi\, d\phi$$

$$= \int_0^{2\pi} \left.\frac{\theta}{8} - \frac{\sin 4\theta}{32}\right|_0^{2\pi} \cos\phi \sin\phi\, d\phi$$

$$= \int_0^{2\pi} \frac{\pi}{4}\cos\phi \sin\phi\, d\phi = \frac{\pi}{8}\left[\frac{-\cos^2\phi}{2}\right]_0^{2\pi}$$

$$= \frac{\pi}{8}\left[-\frac{1}{2} + \frac{1}{2}\right] = 0$$

$\therefore \boxed{I = 0}$

65. $L\left(e^{at}\right) = \dfrac{1}{s-a}$

$$\alpha\left(2e^{st}\sin h\, st\right) = \alpha\left[3e^{st}\left[\frac{e^{st} - e^{-st}}{2}\right]\right]$$

$$= \alpha\left[\frac{3}{2}\left[e^{10t} - e^{0t}\right]\right]$$

$$= \frac{3}{2}\left[\frac{1}{s-10} - \frac{1}{s-0}\right]$$

$$= \frac{3}{2}\left[\frac{s-s+10}{s(s-10)}\right]$$

$$= \frac{15}{s^2 - 10^5}$$

66. $Q = \begin{bmatrix} \frac{1}{\sqrt{3}} & \frac{1}{\sqrt{2}} & a \\ \frac{1}{\sqrt{3}} & 0 & b \\ \frac{1}{\sqrt{3}} & -\frac{1}{\sqrt{2}} & c \end{bmatrix}$

for orthonormal matrix

$Q^T = Q^T Q = I$

$$Q = \begin{bmatrix} \frac{1}{\sqrt{3}} & \frac{1}{\sqrt{2}} & a \\ \frac{1}{\sqrt{3}} & 0 & b \\ \frac{1}{\sqrt{3}} & -\frac{1}{\sqrt{2}} & c \end{bmatrix} \begin{bmatrix} \frac{1}{\sqrt{3}} & \frac{1}{\sqrt{3}} & \frac{1}{\sqrt{3}} \\ \frac{1}{\sqrt{2}} & 0 & -\frac{1}{\sqrt{2}} \\ a & b & c \end{bmatrix} = I$$

$$\Rightarrow \begin{bmatrix} \frac{1}{3}+\frac{1}{2}+a^2 & \frac{1}{3}+0+ab & \frac{1}{3}-\frac{1}{2}+ac \\ \frac{1}{3}+0+ab & \frac{1}{3}+0+b^2 & \frac{1}{3}+0+bc \\ \frac{1}{3}-\frac{1}{2}+ac & \frac{1}{3}+0+bc & \frac{1}{3}+\frac{1}{2}+c^2 \end{bmatrix} = I$$

$$\Rightarrow \begin{bmatrix} \frac{5}{6}+a^2 & \frac{1}{3}+ab & -\frac{1}{6}+ac \\ \frac{1}{3}+ab & \frac{1}{3}+b^2 & \frac{1}{3}+bc \\ ac-\frac{1}{6} & \frac{1}{3}+bc & \frac{5}{6}+c^2 \end{bmatrix} \begin{bmatrix} 1 & 0 & 0 \\ 0 & 1 & 0 \\ 0 & 0 & 1 \end{bmatrix}$$

Thus, $\frac{5}{6}+a^2 = 1$ ... (i)

$\frac{1}{3}+b^2 = 1$ ... (ii)

$\frac{5}{6}+c^2 = 1$ ... (iii)

using (i), (ii), (iii)

$$\boxed{c = \sqrt{1-\frac{5}{6}} = \pm\sqrt{\frac{1}{6}}}$$

$$\boxed{b = \sqrt{1-\frac{1}{3}} = \sqrt{\frac{2}{3}} = \sqrt{\frac{2\times 2}{3\times 2}} = \pm\frac{2}{\sqrt{6}}}$$

67. $\frac{d^2y}{dt} - 2\frac{dy}{dt} + y = 0$

$y(0) = 0.\ y(0) = 1$

C.F. $\Rightarrow D^2 - 20 + 1 = 0$

$D_1 = 1, \qquad D_2 = 1$

$\therefore$ Solution $y(t) = A e^t + Bt e^t$

using initial conditions:

$y(0) = 0$ ... (i)

$\boxed{A = 0}$

$\dot{y}(0) = 0$ ... (ii)

$\dot{y}(t) = Ae^t + B[te^t + e^t]$

using (ii)

$Be^{(0)}[1 + 0] = 1$

$\Rightarrow \boxed{B = 1}$

$\therefore\ y(t) = te^t$

$\boxed{y(1) = e}$

68. Real matrix is always diagonalisable but not necessarily invertible.

69. $s = \sum_{m=1}^{\infty} \frac{m^2}{3^m}(x-2)^m$ converges for all x with

$|x-2| \le R$

If wes put

$x - 2 = R$

$$s = \sum_{m=1}^{\infty} \frac{m^2}{3^m}(R)^m$$

$$= \sum_{m=1}^{\infty} m^2 \left[\frac{R}{3}\right]^m$$

If R is greater than 3 value $\left(\frac{R}{3}\right)^m$ will shoot up if

it 3 or loss than 3 than $\left(\frac{R}{3}\right)^m$ will never

increase more than 1, it will always approach zero f(x) will converge if R = 3

$|x-2| \le 3$

70. $f(x) = \begin{cases} \sin\left(\frac{1}{x}\right), & x \ne 0 \\ 0, & x = 0 \end{cases}$

function is bounded as for all value of x f (x) is finite, But it is not continous at x = 0, as it fails in existance of limit.

$\lim_{x\to\infty} f(x)$ doesn't exit

Thus not continous

71. $I = \int_0^1 1000\, x^4 dx$

$= 1000\int_0^1 x^4 dx$

$= 1000\int_0^1 f(x)\,dx$

Simpson's rule :

$\int_a^b f(x)\,dx = \frac{\Delta x}{3}$

$\left[\begin{array}{l} f(x_0)+4f(x_1)+2f(x_2) \\ \quad +....+4f(x_{n-1})+f(x_n) \end{array}\right]$

$\Delta x = \frac{b-a}{2}, a = 0, b = 1,$

$\Delta x = \frac{1-0}{2} = \frac{1}{2}$

$f(x) = x^4$

$f(x_0) = f(a) = f(0) = 0$

$4f(x_1) = 4f\left(\frac{1}{2}\right) = 0.25$

$f(x_2) = f(b) = f(1) = 1$

$\therefore \int_0^1 f(x)\,dx = \frac{1}{6}[0+0.25+1] = 0.2084$

$\therefore I = 1000 \times \int_0^1 f(x)\,dx$

$= 1000 \times 0.2084 = 208.4 \approx 208$

72. $[A] = \begin{bmatrix} 3 & -3 \\ -3 & 4 \end{bmatrix}$

$[A]^2 = \begin{bmatrix} 3 & -3 \\ -3 & 4 \end{bmatrix}\begin{bmatrix} 3 & -3 \\ -3 & 4 \end{bmatrix}$

$= \begin{bmatrix} 9+9 & -9-12 \\ -9-12 & +9+16 \end{bmatrix}$

$= \begin{bmatrix} 18 & -21 \\ -21 & 25 \end{bmatrix}$

$-[A]^2 = \begin{bmatrix} -18 & 21 \\ 21 & -25 \end{bmatrix}$

$7[A] = \begin{bmatrix} 21 & -21 \\ -21 & 28 \end{bmatrix}$

$[A]^2 + 7[A] - 3[I] = \begin{bmatrix} -18 & 21 \\ 21 & -25 \end{bmatrix} + \begin{bmatrix} 21 & -21 \\ -21 & 28 \end{bmatrix} - \begin{bmatrix} 3 & 0 \\ 0 & 3 \end{bmatrix}$

$= \begin{bmatrix} 0 & 0 \\ 0 & 0 \end{bmatrix}$

$\therefore \det\left[-[A]^2 + 7[A] - 3[I]\right] = 0$

73. $f(x) = \begin{bmatrix} -2 & -\pi < x < 0 \\ 2 & 0 < x < \bar{u} \end{bmatrix}$

$f(x + \Delta x) = 2\pi$

$\therefore T = 2\pi$

$w = \frac{2\pi}{T} = 1$

$a_0 = \frac{2}{T}\int_0^T f(x)\,dx = \frac{1}{\pi}\left[\int_{-\bar{u}}^0 -2dx + \int_0^\pi 2dx\right]$

$= \frac{1}{\pi}\left[2 - [x]_{-\pi}^0 + 2[x]_0^{-\pi}\right]$

$= \frac{1}{\pi}[-2\pi + 2\pi] = 0$

$a_k = \frac{2}{T}\int_0^T f(x)\cos(kwx)\,dx$

$= \frac{1}{\pi}\left[\int_{-\pi}^0 -2\cos kx\, dx + \int_0^\pi 2\cos kx\, dx\right]$

$= \frac{2}{\pi}\left[\left.\frac{-\sin kx}{k}\right|_{-\pi}^0 + \left.\frac{\sin kx}{k}\right|_0^{-\pi}\right]$

$= 0$

$b_k = \frac{2}{T}\int_0^T f(x)\sin(kwx)\,dx$

$$= \frac{2}{\pi}\left[\int_{-\pi}^{0} -\sin(kx)dx + \int_{0}^{\pi}\sin(kx)dx\right]$$

$$= \frac{2}{\pi}\left[\frac{\cos kx}{k}\Big|_{-\pi}^{0} - \frac{\cos kx}{k}\Big|_{0}^{\pi}\right]$$

$$= \frac{2}{\pi}\left[\frac{1}{k} - \frac{\cos kx}{k} - \frac{\cos k\pi}{k} + \frac{1}{k}\right]$$

$$= \frac{2}{\pi}\left[\frac{2}{k} - \frac{2\cos k\pi}{k}\right]$$

$$f(x) = \sum_{k=1}^{\infty} b_k \sin(kwx)$$

$$= \sum_{k=1}^{\infty} \frac{2}{\pi}\left[\frac{2}{k} - \frac{2\cos k\pi}{k}\right]\sin(kwx)$$

$$= \frac{4}{\pi}\sum_{k=1}^{\infty}\left[\frac{\sin kx}{k} - \frac{\cos k\pi \sin kx}{k}\right]$$

Now, from (1)

For $x = \frac{\pi}{2}$ $f(x) = 2$

$$2 = \frac{4}{\pi}\sum_{k=1}^{\infty} \frac{\sin k\frac{\pi}{2}}{k} - \frac{\cos k\frac{\pi}{2}\sin k\frac{\pi}{2}}{k}$$

For all over k $\sin k\frac{\pi}{2} = 0$

$$\Rightarrow \frac{\sin k\frac{\pi}{2}}{k} = 0 \;\&\; \frac{\cos k\frac{\pi}{2}}{k} = 0$$

$\therefore$ (A) becomes

$$2 = \frac{4}{\pi}\sum_{\substack{k=1\\ k\text{ odd}}}^{\infty} \frac{\sin k\frac{\pi}{2}}{k} - \frac{\cos k\pi \sin k\frac{\pi}{2}}{k}$$

For all odd k $\cos k\pi = -1$

$$2 = \frac{4}{\pi}\sum_{k=1}^{\infty} \frac{\sin k\frac{\pi}{2}}{k} + \frac{\sin k\frac{\pi}{2}}{k}$$

$$2 = \frac{4}{\pi}\sum_{k=1}^{\infty} \frac{2\sin k\frac{\pi}{2}}{k}$$

$$\frac{2}{4\times 2\times \pi} = \sum_{k=1}^{\infty}\frac{\sin k\frac{\pi}{2}}{k}$$

expanding R.H.S.

$$\frac{1}{4\pi} = \frac{\sin\frac{\pi}{2}}{1} + \frac{\sin 3\frac{\pi}{2}}{3} + \frac{\sin 5\frac{\pi}{2}}{5} + \frac{\sin 7\frac{\pi}{2}}{7} = \ldots\ldots$$

$$\Rightarrow \boxed{1 - \frac{1}{3} + \frac{1}{5} - \frac{1}{7}\ldots\ldots = \frac{1}{4\pi}}$$

74. $I = \int_{1}\left(3\times 3^3 - 9xy^2\right)ds$

Circular region $x^2 + y^2 \le 1$

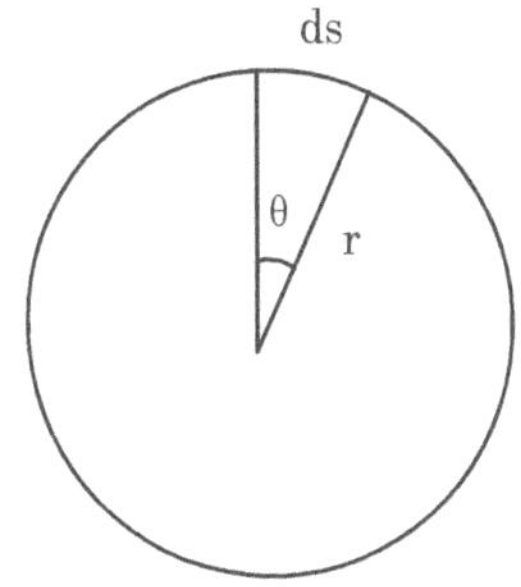

radius r = 1

$x = r\cos\theta = \cos\theta$

$y = r\sin\theta = \sin\theta$

ds = rds

$$I = \int_{0}^{2\pi}\left[3\cos^3\theta - 9\cos\theta\sin^2\theta\right]d\theta$$

$$= \int_{0}^{2\pi}\left[3\left(1-\sin^2\theta\right)\cos\theta - 9\cos\theta\sin^2\theta\right]d\theta$$

$$= \int_{0}^{2\pi}\left[3\cos\theta - 12\cos\theta\sin^2\theta\right]d\theta$$

$$= 3\sin\theta\Big|_{0}^{2\pi} - I_1$$

$$= 0 - I_1 = I_1$$

$$I_1 = \int_{0}^{2\pi} 12\cos\theta\sin^2\theta\, d\theta$$

$$\boxed{\frac{d}{d\theta}\left(\sin^3\theta\right) = 3\sin^2\theta\cos\theta}$$

$\therefore \quad I_1 = 4\sin^3\theta\Big|_0^{2\pi} = 0$

$\therefore \quad \boxed{I = 0}$

75. $-9\dfrac{d^2u}{dx^2} + u = 5x, \quad 0 < x < 3$

$u(0) = 0, \left.\dfrac{du}{dx}\right|_{x=0} = 0$

u (x) = C.F. + P.I.

C.F. Auxilary equations

$-9D^2 + 1 = 0$

$-D + \dfrac{1}{9} = 0$

$D = \sqrt{\dfrac{1}{9}} = \pm\dfrac{1}{3}$

$\text{C.F. } u(x) = C_1 e^{\frac{x}{3}} + C_2 e^{\frac{-x}{3}}$

P.I. say u (x) = $\propto x$

$-9(0) + \propto x = 5x$

$\boxed{\propto = 5}$

$\therefore$ P.I. = 5x

$\therefore \quad u(x) = C_1 e^{\frac{x}{3}} + C_2 e^{\frac{-x}{3}} + 5x$

$\dfrac{du}{dx} = \dfrac{C_1}{3} e^{\frac{x}{3}} - \dfrac{C_2}{3} e^{\frac{-x}{3}} + 5$

u (0) = 0

$C_1 + C_2 = \Rightarrow \boxed{C_1 = -C_2}$

$\dfrac{du}{dx}(3) = 0$

$\dfrac{C_1}{3} e - \dfrac{C_2}{3} e^{-1} + 5 = 0$

$\dfrac{-C_2}{3}\left[e + \dfrac{1}{e}\right] + 5 = 0$

$$\boxed{C_2 = -\left[\dfrac{-5\times 3}{\left(\dfrac{e^2+1}{e}\right)}\right] = \dfrac{15}{\left(\dfrac{e^2+1}{e}\right)}}$$

$$\boxed{C_1 = \dfrac{-15}{\dfrac{e^2+1}{C}}}$$

$$4(x) = \dfrac{-15}{\left(\dfrac{e^2+1}{e}\right)} e^{\frac{x}{3}} + \dfrac{15}{\left(\dfrac{e^2+1}{e}\right)} e^{\frac{-x}{3}} + 5x$$

$$= \dfrac{15e}{1+e^2}\left[e^{\frac{-x}{3}} - e^{\frac{x}{3}}\right] + 5x$$

76. $\dfrac{d^2u}{dt^2} + 2\dfrac{du}{dt} + u = 1$ ...(1)

u (0) = 0 $\dfrac{du(0)}{dt} = u'(0) = 5$

Laplace transform of eq. (1)

$S^2U(s) - Su(0) - u'(0) + 2\,SU(s) - 2u(0)$

$+U(s) = \dfrac{1}{S}$

$S^2U(s) - 5 + 2SU(s) + U(s) = \dfrac{1}{s}$

$\left[s^2 + 2S + 1\right]U(s) = \dfrac{1}{s} + 5$

$U(s) = \dfrac{5s+1}{s\left(s^2 + 2s + 1\right)}$

$= \dfrac{5s+1}{5(s+1)^2}$

77. $\dfrac{\partial u}{\partial t} \rightarrow$ First order

$\dfrac{\partial\left(\dfrac{u^2}{2}\right)}{\partial x} \rightarrow$ Non – linear

79. $\dfrac{d^2y}{dx^2} - 2\dfrac{dy}{dx} + y = 0$

Auxiliary equation, $m^2 - 2m + 1 = 0$

$(m-1)^2 = 0$

Roots are real and equal

c.f. = $(c_1 + xc_2)e^{mx}$

$y = (c_1 + xc_2)e^x$

Given y(0) = 1

$1 = c_1$

Since constants are identical, so $c_1 = c_2 = 1$

$y = (1 + x)e^x$

$y' = (1 + x)e^x + e^x$

$y'(0) = (1 + 0)e^0 + e^0 = 1 + 1 = 2$

80. $\frac{dy}{dx} = 0$

$1 - 2y = 0$

$2y = 1$

$y = 0.5$

81. $y = ax^2 + bx + c$

$$\frac{d^2y}{dx^2} - 2\frac{dy}{dx} + y = 0$$

for extremum, $\frac{dy}{dx} = 0$

$2ax + b = 0$

$$\Rightarrow x = -\frac{-b}{2a}$$

$$y = ax\left(\frac{-b}{2a}\right)^2 + b \times \left(\frac{-b}{2a}\right) + c$$

$$= \frac{ab^2}{4a^2} - \frac{b^2}{2a} + c$$

$$= \frac{ab^2}{4a^2} - \frac{b^2}{2a} + c$$

$$y = \frac{-b^2 + 4ac}{4a}$$

$$\left[\frac{-b}{2a}, \frac{-b^2 + 4ac}{4a}\right]$$

82. By matrix properties.

83. $I = \int_1^2 (4x^3 + 3x^2 + 2x + 1)dx$

x	1	1.5	2
f(x)	10	24.25	49

$f(x) = 4x^3 + 3x^2 + 2x + 1$

From simpson's rule :

$$\int_{x_0}^{x_n} f(x)dx = \frac{h}{3}[(y_0 + y_n) + 2(y_2 + y_4 + ... + y_{n-2}) + 4(y_1 + y_3 + ... + y_{n-1})]$$

$= h = 1.5 - 1 = 0.5$

$$= \frac{0.5}{3}[(10 + 49) + 4 \times 24.25]$$

$= 26$

84. By matrix property

85. By the property of determinant option (*a*), (*c*) and (*d*) are correct.

86. 1

$$f(x) = x^2 + \frac{1}{x^2}$$

$$f'(x) = 2x - \frac{2}{x^3}$$

$$f'(x) = 0$$

$$\Rightarrow 2x - \frac{2}{x^3} = 0$$

$2x^4 - 2 = 0$

$x = 1, 1, 1, 1$

$$f''(x) = 2 + \frac{6}{x^4}$$

$$f''(x)\big|_{x=1} = 8 \text{ i.e., } > 0$$

So at $x = 1$ the given function has minimum value.

87. (0)

$$\vec{u} = y\hat{e}_x - x\hat{e}_y$$

$$\vec{w} = \vec{\nabla} \times \vec{u} = \begin{vmatrix} \hat{e}_x & \hat{e}_y & \hat{e}_z \\ \frac{\partial}{\partial x} & \frac{\partial}{\partial y} & \frac{\partial}{\partial z} \\ y & -x & 0 \end{vmatrix}$$

$$= \hat{e}_x(0) - \hat{e}_y(0) + \hat{e}_z(-1-1)$$

$$\vec{w} = 0\,\hat{e}_x - 0\,\hat{e}_y - 2\hat{e}_z$$

$$\vec{w} \cdot \vec{\nabla} = (0\,\hat{e}_x - 0\,\hat{e}_y - 2\hat{e}_z) \cdot \left(e_x\frac{\partial}{\partial x} + e_y\frac{\partial}{\partial y} + e_z\frac{\partial}{\partial z}\right)$$

$$= -\frac{\partial}{\partial z}(2) = 0$$

88. General form of partial differential equation is

$$A\frac{\partial^2 u}{\partial x^2} + B\frac{\partial^2 u}{\partial x \partial y} + c\frac{\partial^2 u}{\partial y^2} + D\frac{\partial u}{\partial x} + E\frac{\partial u}{\partial y} + fu = G$$

If $B^2 - 4AC > 0 \rightarrow$ hyperbola

$B^2 - 4AC = 0 \rightarrow$ parabola

$B^2 - 4AC < 0 \rightarrow$ ellipse

Here, $A = \alpha$, $B = 0$, $C = 0$

$B^2 - 4AC = (0^2 - 4 \times \alpha \times 0 = 0)$

So, $B^2 - 4AC = 0 \rightarrow$ parabolic

89. $\frac{d^2y}{dx^2} - 4\frac{dy}{dx} + 4y = 0$

A.E. $m^2 - 4m + 4 = 0$

$(m - 2)^2 = 0$

$m = 2, 2$

Roots are real and equal

$y = (c_1 + xc_2)e^{2x}$

$y(0) = 1$

$\Rightarrow 1 = c_1 \times e^0$

$\Rightarrow c_1 = 1$

$y = (1 + xc_2)c^{2x}$

$$\frac{dy}{dx} = (1 + xc_2)e^{2x} \times 2 + e^{2x} \times c_2$$

$$\left.\frac{dy}{dx}\right|_{x=0} = 1 = 2 + c_2$$

$\Rightarrow c_2 = -1$

$y = (1 - x)e^{2x}$

$y|_{x=1} = 0$

90. The given equation in matrix form, with interchanging the position of equation (1) and (3)

$$\begin{bmatrix} 1 & -2 & 3 \\ 3 & -3 & 4 \\ 2 & -1 & 1 \end{bmatrix} \begin{bmatrix} x \\ y \\ z \end{bmatrix} = \begin{bmatrix} 4 \\ 6 \\ 1 \end{bmatrix}$$

$R_2 \to R_2 - 3R_1,\ R_3 \to R_3 - 2R_1$

$$\begin{bmatrix} 1 & -2 & 3 \\ 0 & 3 & -5 \\ 0 & 3 & -5 \end{bmatrix} \begin{bmatrix} x \\ y \\ z \end{bmatrix} = \begin{bmatrix} 4 \\ -6 \\ -7 \end{bmatrix}$$

Rows 2nd and third are identical, it means these two lines are parallel, so no solution exists between these lines.

91. 6.28

$$\int_0^\pi x \sin x\, dx = \left[x\int_0^\pi \sin x\, dx - \int_0^\pi \frac{d}{dx}(x)\int \sin x\, dx\right]dx \Bigg]_0^\pi$$

$$= \left[x \times (-\cos x)_0^\pi + \int_0^\pi \cos x\, dx \right]_0^\pi$$

$$= \left[2x + (\sin x)_0^\pi\right]_0^\pi$$

$$= [2x]_0^\pi = 2\pi = 2 \times 3.14 = 6.28$$

92. 0.57

$f(x) = xe^x - 1$

$f(0.5) = -0.1756$

$f'(x) = xe^x + e^x$

$f'(0.5) = 2.4731$

$$x_1 = x_0 - \frac{f(x_0)}{f'(x_0)}$$

$$= 0.5 + \frac{0.1756}{2.4731}$$

$x_1 = 0.571$

93. The cross product of two vector is perpendicular to the third vector.

94. 12

Along line AB, $y = 0$

$dy = 0$

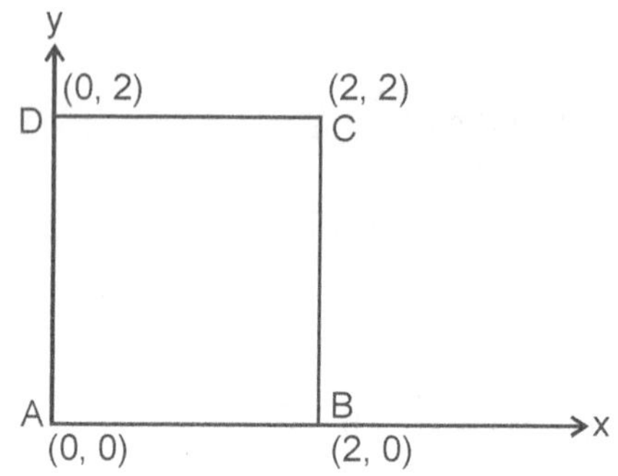

$$I_1 = \int_0^2 x\, dx = 2$$

Along line BC, $x = 2$

$$I_2 = \int_0^2 4\, dy = 8$$

Along line, CD, $y = 2$

$dy = 0$

$$I_3 = \int_2^0 (x - 2)\, dx = 2$$

Along line DA, $x = 0$

$dx = 0$

$I_y = 0$

$I = I_1 + I_2 + I_3 + I_7 = 2 + 8 + 2 + 0 = 12$

95. 0

96. $\frac{d^2y}{dx^2} + k^2\lambda y = 0$

A.E., $m^2 + k^2\lambda = 0$

$m^2 = -k^2\lambda = k^2\lambda i^2$

$m = \pm\sqrt{\lambda}\ ki$

Roots are complex

$y = e^{\alpha x}(c_1 \cos \beta x + c_2 \sin \beta_x x)$

$m = \pm\sqrt{\lambda}\ ki$

$$y = c_1 \cos\sqrt{\lambda}\,kx + c_2 \sin\sqrt{\lambda}\,kx$$

Given, $y(0) = 0$

$0 = c_1$

$$y = c_2 \sin\sqrt{\lambda}\,kx$$

$y(1) = 0$

$$\Rightarrow\ 0 = c_2 \sin\sqrt{\lambda}\,k$$

$$\sin\sqrt{\lambda}\,k = 0$$

$$\sin\sqrt{\lambda}\,k = \sin n\pi$$

$$\sqrt{\lambda}\,k = n\pi$$

$$\lambda = \frac{n^2\pi^2}{k^2}$$

$$y = \sin\cdot\frac{n\pi}{k}kx = \sin n\pi x$$

97. $\int_{-1}^{1} f(x)dx = A_1 f(x_1) + A_2 f(x_2) + A_3 f(x_3)$

Option (b) satisfy this equation.

98. $\begin{bmatrix} 2 & 0 & 2 \\ 3 & 2 & 7 \\ 3 & 1 & 5 \end{bmatrix}\begin{bmatrix} x_1 \\ x_2 \\ x_3 \end{bmatrix} = \begin{bmatrix} 4 \\ 4 \\ 5 \end{bmatrix}$

$R_1 \leftrightarrow R_3$

$$\begin{bmatrix} 3 & 1 & 5 \\ 3 & 2 & 7 \\ 2 & 0 & 2 \end{bmatrix}\begin{bmatrix} x_1 \\ x_2 \\ x_3 \end{bmatrix} = \begin{bmatrix} 5 \\ 4 \\ 4 \end{bmatrix}$$

$$R_2 \to R_2 - R_1,\ R_3 \to R_3 - \frac{2}{3}R_1$$

$$\begin{bmatrix} 3 & 1 & 5 \\ 0 & 1 & 2 \\ 0 & -\frac{2}{3} & -\frac{4}{3} \end{bmatrix}\begin{bmatrix} x_1 \\ x_2 \\ x_3 \end{bmatrix} = \begin{bmatrix} 5 \\ -1 \\ -\frac{2}{3} \end{bmatrix}$$

$$-\frac{2}{3}\begin{bmatrix} 3 & 1 & 5 \\ 0 & 1 & 2 \\ 0 & 1 & 2 \end{bmatrix}\begin{bmatrix} x_1 \\ x_2 \\ x_3 \end{bmatrix} = -\frac{2}{3}\begin{bmatrix} 5 \\ -1 \\ -1 \end{bmatrix}$$

Hence rows 2nd and 3rd are identical, means these are the coincident lines, hence infinite solution exist.

99. $\frac{d^2y}{dt^2} + 4\frac{dy}{dt} + 6y = f(t)$

$$\Rightarrow\ s^2y(s) - sy(0) - y'(0) + 4[sy(s) - y(0)] + 6y(s) = F(s)$$

$$\Rightarrow\ s^2y(s) - 2s - 1 + 4sy(s) - 4 \times 2 + 6y(x) = F(s)$$

$$y(s)[s^2 + 4s + 6] - 2s - 9 = F(s)$$

$$y(s) = \frac{f(s) + 2s + 9}{s^2 + 4s + 6}$$

100. The general solution of the PDE of the form

$u_{tt} - C^2u_{xx} = 0$ is given by d'Alembert's formula.

where,

$u(x, t) = \alpha(x)$

$u_t(x, 0) = \beta(x)$

Given equation is

$u_{tt} - 25u_{xx} = 0,$

$u(x, 0) = 0.1\sin(10\pi x),\ u_t(x, 0) = \beta(x)$

$C = 5$

$u(x, t)$

$$= \frac{0.01\sin\left(10\pi(x + 5t) + 0.1\sin(10\pi(x - 5t) + 0\right)}{2}$$

$$\Rightarrow u(0.25,\ 1) = \frac{0.01\left[\sin(10\pi \times 5.25) + \sin(10\pi \times 4.75)\right]}{2}$$

$u(0.25, 1) = 0.01$

101. Let $x = e^z$

$$x\frac{dy}{dx} = Dy,\ \text{where } D = \frac{d}{dz}$$

$$x^2\frac{d^2y}{dx^2} = D(D - 1)y$$

$$\therefore x^2\frac{d^2y}{dx^2} + 5x\frac{dy}{dx} + 4y = 0$$

$$D(D - 1)y + 5Dy + 4y = 0$$

$$\left(D^2 - D + 5D + 4\right)y = 0$$

$\therefore$ A.E,

$(D^2 + 4D + 4) = 0$

$D = -2, -2$

$\therefore\ y = (C_1 + 2C_2)e^{-2Z}$

$$= \left(C_1 + \left(\log_e x\right)C_2\right)e^{-2\log_e x}$$

$$= \left(C_1 + (\log_e x)C_2\right)e^{\log_e x^{-2}}$$

$$= \left(\frac{C_1}{x^2} + \frac{\log_e x}{x^2}C_2\right) = \left(C_1 + \log_e xC_2\right)\frac{x_1}{x^2}$$

102. The value of 'k' depends on matrix {x}

103. Vector $\vec{c}$ is cross product of $\vec{a}$ and $\vec{b}$, so it will orthogonal to $\vec{a}$ and $\vec{b}$ and $\vec{a}$, is perpendicular to $\vec{b}$.

104. Putting f(x, y) = 0

$$\frac{x^2}{2} + \frac{y^2}{3} - 5 = 0$$

$$\frac{x^2}{2} + \frac{y^2}{3} = 5$$

This is the equation of ellipse, so all of its roots lies on ellipse.

105. $\vec{A} = x\hat{i} + y\hat{j} + z\hat{k}$

$$\vec{\nabla}\cdot\vec{A} = \left(\vec{i}\frac{\partial}{\partial x} + \vec{j}\frac{\partial}{\partial y} + \vec{k}\frac{\partial}{\partial z}\right)\cdot\left(x\hat{i} + y\hat{j} + z\hat{k}\right)$$

$$= \hat{i} + \hat{j} + \hat{k}$$

$$\vec{\nabla}\cdot\vec{A} = \begin{vmatrix} i & j & k \\ \frac{\partial}{\partial x} & \frac{\partial}{\partial y} & \frac{\partial}{\partial z} \\ x & y & z \end{vmatrix}$$

$$= \hat{i}(0-0) - \hat{j}(0-0) + \hat{k}(0-0) = 0$$

106. 0.0

$$\begin{vmatrix} 1 & 1 & -1 \\ 2 & 1 & 0 \\ 3 & 1 & 1 \end{vmatrix}$$

$= 1(1-0) - 1(2-0) + (-1)(2-3)$
$= 1 - 2 + 1 = 0$

107. $\dfrac{d^2y}{dx^2} + 3\dfrac{dy}{dx} = 0$

A.E. $m^2 + 3m = 0$
$m(m + 3) = 0$
$m = 0, -3$
Roots are real and unequal

$$y = c_1e^{m_1x} + c_2e^{m_2x}$$

$$y = c_1e^0 + c_2e^{-3x}c_1 + c_2e^{-3x}$$

at x = 0, y = 0

$0 = c_1 + c_2$ ... (i)

$$\frac{dy}{dx} = -3c_2e^{-3x}$$

$$\left.\frac{dy}{dx}\right|_{x=0} == 1 = -3\times c_2e^0$$

$$\Rightarrow c_2 = -\frac{1}{3}$$

$$c_1 = \frac{1}{3}$$

$$y = \left(\frac{1}{3} - \frac{1}{3}e^{-3x}\right) = \frac{1}{3}(1 - e^{-3x})$$

108. 6.28

$$\oint \vec{v}\cdot\overrightarrow{dx} = \oint\left(\frac{-y}{r^2}\hat{i} + \frac{x}{r^2}\hat{j}\right)(dx\hat{i} + dy\hat{j})$$

$$= \oint \frac{-y}{r^2}dx + \frac{x}{r^2}dy$$

r = r cos θ   y = r sin θ
dx = –r sin θ dθ   dy = r cos θ dθ

$$= \int_0^{2\pi} \frac{(-r\sin\theta)}{r^2}\times(-r\sin\theta)d\theta + \frac{r\cos\theta\times r\cos\theta d\theta}{r^2}$$

$$= \int_0^{2\pi}(\sin^2\theta + \cos^2\theta)d\theta$$

$$= \int_0^{2\pi} d\theta = [\theta]_0^{2\pi} = 2\pi = 2\times 3.14$$

109. –0.707

$$\phi = \frac{1}{x^2 + y^2 + 4}$$

$$\frac{\partial\phi}{\partial x} = \frac{-2x}{(x^2 + y^2 + 4)^2}, \frac{\partial\phi}{\partial y} = \frac{-2y}{(x^2 + y^2 + 4)^2}$$

Normal vector = $\nabla\phi = \left(\hat{i}\dfrac{\partial}{\partial x} + \hat{j}\dfrac{\partial}{\partial y}\right)\phi$

$$= \hat{i}\frac{\partial\phi}{\partial x} + \hat{j}\frac{\partial\phi}{\partial y}$$

$$= \frac{-2x}{(x^2 + y^2 + 4)^2}\hat{i} - \frac{2y}{(x^2 + y^2 + 4)^2}\hat{j}$$

Noramal vector at $(1,1) = \frac{-2}{36}\hat{i} - \frac{2}{36}\hat{j}$

$$= -\frac{1}{18}\hat{i} - \frac{1}{18}\hat{j}$$

Unit normal vector = $\dfrac{-\frac{1}{18}(\hat{i}+\hat{j})}{\sqrt{\left(-\frac{1}{18}\right)^2 + \left(-\frac{1}{18}\right)^2}}$

$$= \frac{-\frac{1}{18}(\hat{i}+\hat{j})}{\frac{\sqrt{2}}{18}}$$

$$= -\frac{1}{\sqrt{2}}\hat{i} - \frac{1}{\sqrt{2}}\hat{j}$$

x-compon ent of unit vector $= -\frac{1}{\sqrt{2}} = -0.707$

110. Given $f(x) = xe^{-x}$

Now, $f'(x) = x(-e^{-x}) + e^{-x}$

$= e^{-x}(1-x)$

For maxima of minima

$f'(x) = 0$

$\Rightarrow \quad e^{-x}(1-x) = 0$

$\Rightarrow \quad e^{-x} = 0$

$\Rightarrow \quad x = \infty$

and $\quad 1 - x = 0$

$x = 1$

$\therefore \quad f''(x) = e^{-x}(-1)+(1-x)(-e^{-x})$

$= e^{-x}(x-2)$

$f'(x-1) = e^{-1}(1-2) = -e^{-1}$

$= \frac{-1}{e}(-ve) \Leftrightarrow \max$

$\text{At}, f(x) = f(1) = \frac{1}{e}$

111. If any vector is rotated by angle θ than new vector is obtained by matrix between rotation and given vector

$$B = \begin{bmatrix} \cos\theta & -\sin\theta \\ \sin\theta & \cos\theta \end{bmatrix} A$$

where B = new vector

A = old vector

Let, $A = \begin{bmatrix} p \\ q \end{bmatrix}$ and $B = \begin{bmatrix} p' \\ q' \end{bmatrix}$

$$\therefore \begin{bmatrix} p' \\ q' \end{bmatrix} = \begin{bmatrix} \cos\theta & -\sin\theta \\ \sin\theta & \cos\theta \end{bmatrix}\begin{bmatrix} p \\ q \end{bmatrix}$$

Given $\vec{a} = \hat{i} + \hat{j}$

$\therefore \begin{bmatrix} p \\ q \end{bmatrix} = \begin{bmatrix} 1 \\ 1 \end{bmatrix}$ is rotated by $\theta = 90°$

to obtain $\vec{b} = \begin{bmatrix} p' \\ q' \end{bmatrix}$, we get

$$\begin{bmatrix} p' \\ q' \end{bmatrix} = \begin{bmatrix} \cos 90 & -\sin 90 \\ \sin 90 & \cos 90 \end{bmatrix}\begin{bmatrix} 1 \\ 1 \end{bmatrix} = \begin{bmatrix} 0 & -1 \\ 1 & 0 \end{bmatrix}\begin{bmatrix} 1 \\ 1 \end{bmatrix}$$

$$\therefore \begin{bmatrix} p' \\ q' \end{bmatrix} = \begin{bmatrix} 1 \\ 1 \end{bmatrix} = \hat{i} + \hat{j}$$

112. Given function

$f(x,y) = x^2 + y^2$

$\nabla f = 2x\hat{i} + 2y\hat{j}$

At $(x,y) = (3,4)$

$\nabla f = 6\hat{i} + 8\hat{j}$

$$\therefore \text{unit vector} = \frac{6\hat{i}+8\hat{j}}{\sqrt{6^2+8^2}} = \frac{6\hat{i}+8\hat{j}}{\sqrt{100}} = \frac{6}{10}\hat{i} + \frac{8}{10}\hat{j}$$

$$= \frac{3}{5}\hat{i} + \frac{4}{5}\hat{j}$$

113. $f(x) = \frac{1}{2}(x + |x|)$

$$|x| = \begin{cases} -x & x < 0 \\ +x & x < 0 \end{cases}$$

$$\int_{-1}^{1} \frac{1}{2}(x+|x|)dx = \int_{-1}^{0} \frac{1}{2}(x-x)dx + \int_{0}^{1} \frac{1}{2}(x+x)dx$$

$$= \frac{1}{2} \times 2\left|\frac{x^2}{2}\right|_0^1 = \frac{1}{2}\left|1^2\right| = \frac{1}{2} = 0.5$$

114. $\lim\limits_{\theta\to 0} \dfrac{\theta - \sin\theta}{\theta^3}$

Now, applying L–Hospital rule

$$\lim_{\theta\to 0} \frac{\theta - \cos\theta}{3\theta^3} = \lim_{\theta\to 0} \frac{2\sin^2 \theta/2}{2\theta^2}$$

$$\lim_{\theta\to 0} \frac{2}{3}\left(\frac{\sin \theta/2}{\theta/2}\right)^2 \times \frac{1}{4} = \frac{1}{6} = 0.166$$

115. Applying determinant method to solve the systems of equations

of $D = 0$ $D_1, D_2, D_3, \neq 0$ the system has no solution of $D = 0$, $D_1 = D_2 = D_3 = 0$ the system has infinite numbers of solution

of $D \neq 0$, $D_1 = D_2 = D_3 = 0$ has only one trival solutions.

$2x - y - z = 0$

$-x + 2y - z = 0$

$-x - y + 2z = 0$

$$\Delta = \begin{vmatrix} 2 & -1 & -1 \\ -1 & 2 & -1 \\ -1 & -1 & 2 \end{vmatrix}$$

$$= 2(4-1) + (-1)(-2-1) - 1(1+2)$$
$$= 6 - 3 - 3 = 0$$

$$\Delta_1 = \begin{vmatrix} 2 & 0 & -1 \\ -1 & 0 & -1 \\ -1 & 0 & 2 \end{vmatrix}$$

$\Delta_1 = 0$

∴ Infinite no. of solution.

116. $y = x$

$y = \cos x$

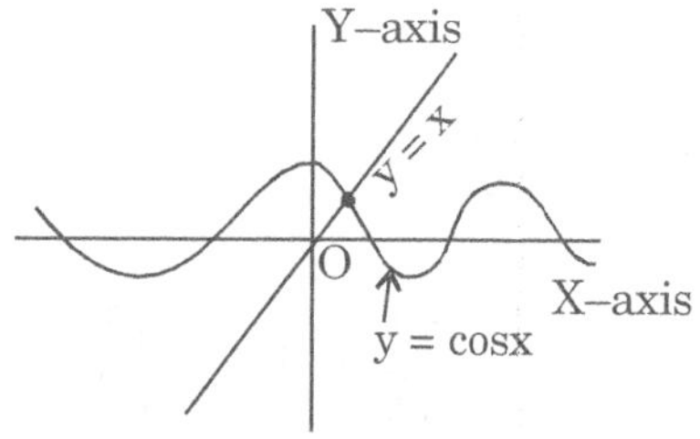

Hence no. of intersection point is one.

117. $\begin{pmatrix} x & 2 \\ -1 & 3 \end{pmatrix}$

Let λ is the eigen value of the matrix.

$$\begin{pmatrix} x-\lambda & 2 \\ -1 & 2-\lambda \end{pmatrix} = 0$$

$(x - \lambda)(3 - \lambda) + 2 = 0$

$3x - 3\lambda - \lambda x + \lambda^2 + 2 = 0$

$\lambda^2 - (3 + x)\lambda + 3x + 2 = 0$ ...(1)

Let λ_1, λ_2 be the root of eqn, Then

$$\lambda_1 + \lambda_2 = \frac{+(3+x)}{+1} = 3 + x \quad ...(2)$$

$\lambda_1 \lambda_2 = 3x + 2$ ...(3)

given $\lambda_1 = 1$

$1 + \lambda_2 = 3 + x$ ...(4)

$1, \lambda_2 = 3x + 2$

$\lambda_2 = 3x + 2$ ...(5)

putting value of x from eq (4) in equations we get

$\lambda_2 = 3(\lambda_2 - 2) + 2$

$\Rightarrow \lambda_2 = 2\lambda_2 - 6 + 2$

$\Rightarrow +2\lambda_2 = +4$

$\boxed{\lambda_2 = 2}$

118. Given

$y = f(x)$

$$\frac{dy}{dx} = y^2$$

$$\frac{dy}{y^2} = dx$$

$$\Rightarrow \frac{y^{-2+1}}{-2+1} = x + C$$

$$\Rightarrow \frac{-1}{y} = x + C$$

$$\Rightarrow y = -\frac{1}{x+C} \quad ...(A)$$

At (1, –1)

$$+1 = +\frac{1}{1+C}$$

$\Rightarrow 1 + C = 1$

$\boxed{C = 0}$

substituting this values in equation (A) we get

$$\frac{-1}{x+0} = y$$

$\Rightarrow xy = -1$

$xy + 1 = 0$ ....(B)

At $x = -2$

$$\frac{1}{x+0} =$$

$$y = -\frac{1}{x+C} = -\frac{1}{-2+0} = \frac{-1}{-2} = \frac{1}{2}$$

$\boxed{y = 0.5}$

119.

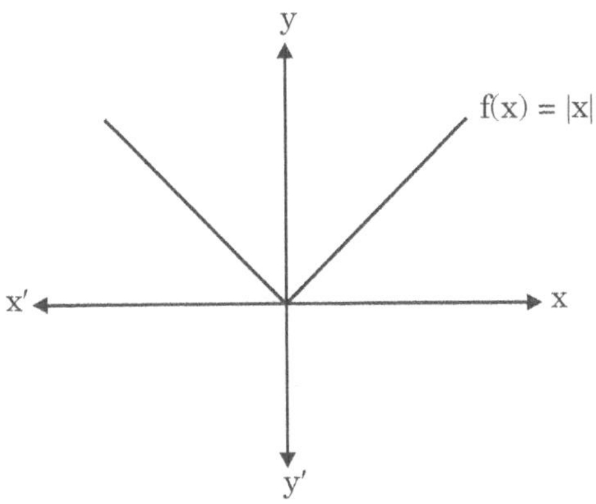

$$f(x) = |x|$$

$$|x| = \begin{Bmatrix} x, x \geq 0 \\ -x, x < 0 \end{Bmatrix}$$

$$f'(0) = \lim_{x\to 0} \frac{f(x) - f(0)}{x - 0}$$

$$= \lim_{x\to 0} \frac{|x| - |0|}{x - 0}$$

$$= \lim_{x\to 0} \frac{|x|}{0}$$

Check : $f(x) = |x|$

$$\lim_{x\to 0^+} \frac{|x|}{x} = \lim_{x\to 0^+} \frac{x}{x}$$

$$= \lim_{x\to 0^+}(1) = 1 \quad ...(1)$$

$$\lim_{x\to 0^-} \frac{|x|}{x} = \lim_{x\to 0^-} \frac{-x}{x}$$

$$= \lim_{x\to 0^-}(-1) = -1 \quad ...(2)$$

$$① \neq ②$$

$$f'(x) = \begin{Bmatrix} 1 \text{ if } x > 0 \\ -1 \text{ if } x < 0 \end{Bmatrix}$$

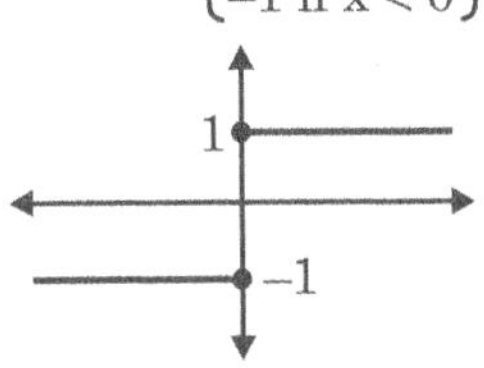

$\frac{\partial f}{\partial x}$ is not defined at x = 0 because derivative does not exist at the point

120. Differentiate f(x) with respect to x and solve for x to find the maximum. Because our f(x) contains the natural exponential function, however, it is easier to take the derivative of the natural log of f(x) with respect to x and solve for x to find the maximum. [The maximum of f(x) is the same as the maximum of the natural log of f(x), because $\log_e(x)$ is an increasing function of x. That is $x_1 < x_2$ implies that $\log_e(x_1) < \log_e(x_2)$. Therefore, $f(x_1) < f(x_2)$ implies $\text{loge}(f(x_1) < \log_e(f(x_2)))$.] That said, taking the natural log of f(x), we get :

$$\text{loge}(f(x)) = \log\left(\frac{1}{\sigma\sqrt{2\pi}}\right) - \frac{1}{2\sigma^2}(x-\mu)^2$$

Taking the derivative of $\log_e(f(x))$ with respect to x, we get :

$$\frac{d \log f(x)}{dx} = -\frac{1}{2\sigma^2}.2(x-\mu)$$

Now, setting the derivative of loge(f(x)) to 0 :

$$\frac{d \log f(x)}{dx} = -\frac{1}{2\sigma^2}.2(x-\mu) \text{ skr } 0$$

and solving for x, we get that x = μ. Taking the second derivative of $\log_e(f(x))$ with respect to x, we get :

$$\frac{d^2 \log f(x)}{dx^2} = \frac{1}{\sigma^2}$$

Because the second derivative of $\log_e(f(x))$ is negative (for all x, in fact), the point x = μ is deemed a local maximum.

121.

$$y = Ae^{mx} + Be^{-mx}$$

$$y' = mAe^{mx} + (-m)Be^{-mx}$$

$$y'' = m^2Ae^{mx} + (-m \times -m)Be^{-mx}$$

$$y'' = m^2(Ae^{mx} + Be^{-mx})$$

$$y'' = m^2y$$

$$y'' - m^2y = 0$$

$$\frac{d^2y}{dx^2} - m^2y = 0$$

122.

$$A = \begin{pmatrix} \sin\theta & \tan\theta \\ 0 & \cos\theta \end{pmatrix}$$

In case of diagonal matrix, diagonal elements are eigen values (roots)

$$\lambda_1 = \sin\theta, \lambda_2 = \cos\theta$$

$$\therefore \quad \lambda_1^2 + \lambda_2^2 = \sin^2\theta + \cos^2\theta = 1$$

123. Given,

$$u(x, 0) = e^{-x^2}$$

$$\eta = \frac{x}{\sqrt{\alpha t}}$$

$$u(x, t) = u(\eta)$$

Now, $u(0, t) = e^0 = 1$

Therefore, $\frac{u(0,1)}{u(0,4)} = \frac{1}{\sqrt{\alpha}} \times \frac{\sqrt{\alpha 4}}{1} = 2$

124. $x\frac{dx}{dy}+y=c$

$\Rightarrow xdx+ydy=cy$

Integrating we get,

$\Rightarrow \frac{x^2}{2}+\frac{y^2}{2}=cy+k$

$\Rightarrow x^2+y^2-2cy+2k$

$\Rightarrow x^2+y^2-2cy=2k$

$\Rightarrow x^2+y^2-2cy+c^2=2k+c^2$

$\Rightarrow x^2+(y-c)^2=2k+c^2 \ldots\ldots\ldots\ldots(i)$

Thus, euation 1 represent the equation of circle.

125. Given,

x = a cos θ

y = b sin θ

a = 7, b = 4

$$\frac{x}{a} = \cos\theta$$

$$\frac{y}{b} = \sin\theta$$

$$\frac{x^2}{a^2}+\frac{y^2}{b^2} = \cos^2\theta+\sin^2\theta = 1$$

$$\frac{x^2}{a^2}+\frac{y^2}{b^2} = 1$$

It is equation of ellipse, hence the area enclosed by ellipse = πab = π × 7 × 5 = 110 m^2

126.

$$A = \begin{bmatrix} \frac{1}{\sqrt{2}} & 0 & \frac{1}{\sqrt{2}} \\ 0 & 1 & 0 \\ \frac{1}{\sqrt{2}} & 0 & \frac{-1}{\sqrt{2}} \end{bmatrix}, X = \begin{bmatrix} x \\ y \\ z \end{bmatrix}$$

$$B = \begin{bmatrix} 0 \\ 1 \\ -\sqrt{2} \end{bmatrix}$$

Now, AX = B

$$\begin{bmatrix} \frac{1}{\sqrt{2}} & 0 & \frac{1}{\sqrt{2}} \\ 0 & 1 & 0 \\ \frac{1}{\sqrt{2}} & 0 & \frac{-1}{\sqrt{2}} \end{bmatrix}\begin{bmatrix} x \\ y \\ z \end{bmatrix} = \begin{bmatrix} 0 \\ 1 \\ -\sqrt{2} \end{bmatrix}$$

$$\begin{bmatrix} \frac{x}{\sqrt{2}}+\frac{z}{\sqrt{2}} \\ y \\ \frac{x}{\sqrt{2}}-\frac{z}{\sqrt{2}} \end{bmatrix} = \begin{bmatrix} 0 \\ 1 \\ -\sqrt{2} \end{bmatrix}$$

$$\frac{x}{\sqrt{2}}+\frac{z}{\sqrt{2}} = 0 \quad \ldots(i)$$

$$y = 1 \quad \ldots(ii)$$

$$\frac{x}{\sqrt{2}}-\frac{z}{\sqrt{2}} = -\sqrt{2} \quad \ldots(iii)$$

From equation (i) and (ii), we get

x = – 1 and by then putting x = – 1 in equation (i), we get

z = 1

∴ x + y + z = – 1 + 1 + 1 = 1

127. $\int_0^1 (x^2-2x+1)dx$

As given to evaluate with four intervals

$f(x) = x^2 - 2x + 1$

x	0	0.25	0.5	0.75	1
f(x)	1	0.562	0.25	0.062	0

Trapezoidal rule in given by

$$\int_a^b f(x)dx = \frac{b-a}{2n}\left[f(x_0)+f(x_n)+2[f(x_1)+f(x_2)]+\ldots\ldots+f(x_{n-1})\right]$$

For given Interval

$$\int_a^b f(x)dx = \frac{0.25}{2}\left[f(0)+f(1)+2\left[f(0.25)+f(0.5)+f(0.75)\right]\right]$$

= 0.3435

For analytical value

$$\int_0^1 (x^2-2x+1)dx = \int_0^1 x^2dx - 2\int_0^1 xdx + \int_0^1 dx$$

$$= \left|\frac{x^3}{3}\right|_0^1 - 2\left|\frac{x^2}{2}\right|_0^1 + |x|_0^1$$

$$= \frac{1}{3}-1+1$$

$$= \frac{1}{3}$$

$$\text{Difference} = \left|0.3435-\frac{1}{3}\right| = 0.0102$$

128. Given,

$$\frac{d^2y}{dx^2}+8\frac{dy}{dx}+16y=0$$

The auxillary equation is

$(D^2 + 8D + 16)y = 0$

$(D + 4)(D + 4) = 0$

D = –4, –4

Now, The C.F. is

$y = (c_1 + c_2x)e^{-4x}$

Differentiating we get,

$$\frac{dy}{dx} = (c_1 + c_2x) - 4e^{-4x} + c_2e^{-4x}$$

Now for, $x = 0$, $y = 1$

$\therefore\ 1 = (c_1 + 0)e^0$

$\Rightarrow c_1 = 1$

For $\frac{dy}{dx} = 0, x = 0$

$\Rightarrow 0 = (c_1 + 0) \times - 4e^0 + c_2e^0$

$\Rightarrow c_2 = 4c_1$

$\therefore\ c_2 = 4$

$y = (1 + 4x)e^{-4x}$

129. Given,

$$\frac{\partial^2 u}{\partial x^2} - 4\frac{\partial^2 u}{\partial x \partial y} + \frac{\partial^2 u}{\partial y^2} = x + 2y$$

The standard form is given as

$$A\frac{\partial^2 u}{\partial x^2} + B\frac{\partial^2 u}{\partial x \partial y} + C\frac{\partial^2 u}{\partial y^2} + D\frac{\partial y}{\partial x} + E\frac{\partial x}{\partial y} + F = 0$$

Comparing with the standard form

we get, A = 1, B = –4, C = 6

Now, the determinant

$B^2 - 4AC = 16 - 4 \times 1 \times 6 = -8$

As the determinant is negative.

Thus, the function elliptic in nature.

130. $\lim_{x\to 0}\left(\frac{1}{\sin x} - \frac{1}{x}\right)$

$$\lim_{x\to 0}\left(\frac{x - \sin x}{x \sin x}\right)$$

Substituting $x = 0$, we get $\frac{0}{0}$

Thus, applying L. Hospital's rule

$$\lim_{x\to 0}\left(\frac{1-\cos x}{x\cos x + \sin x}\right), \text{again } \frac{0}{0} \text{ from}$$

Thus, $\lim_{x\to 0}\left(\frac{\sin x}{-x\sin x + \cos x + \cos x}\right) = \frac{0}{0+1+1} = \frac{0}{2} = 0$

131. Given,

$$\oint \frac{z^3}{4z - i} dz$$

$\xi = 1$

Now, $f(z) = \frac{z^3}{4z - i}$

For Poles

$4z - i = 0,$

$\Rightarrow z = \frac{i}{4}$

Using residue theorem

I = $2\pi i$ × sum of residues

$$= 2\pi i \times \lim_{z\to \frac{i}{4}}\left(z - \frac{i}{4}\right) \times \frac{z^3}{4z - i}$$

$$= 2\pi i \times \frac{\left(\frac{i}{4}\right)^3}{4} = \frac{2\pi \times 1}{256} = 0.0245 \approx 0.025$$

132. Given,

$f(x) = e^{-x}|\cos x|$ for $x > 0$

we can write the equation in two expressions

$$f(x) = \begin{cases} -e^{-x}\cos x & \cos x < 0 \\ e^{-x}\cos x & \cos x > 0 \end{cases}$$

Differentiating we get,

$$f'(x) = \begin{cases} e^{-x}\cos x + e^{-x}\sin x & \cos x < 0 \\ -e^{-x}\cos x - e^{-x}\sin x & \cos x > 0 \end{cases}$$

Taking option (A)

At $x = \frac{\pi}{2}, \cos\frac{\pi}{2} = 0$

$\therefore$ L.H.S = $e^{-\frac{\pi}{2}} \times 0 + e^{-\frac{\pi}{2}} \times 1 = e^{-\frac{\pi}{2}}$

R.H.S = $-e^{-\frac{\pi}{2}} \times 0 - e^{-\frac{\pi}{2}} \times 1 = -e^{-\frac{\pi}{2}}$

$\therefore$ L.H.S ≠ R.H.S

Thus, the function is not differentiable at $x = \frac{\pi}{2}$

Taking option (B)

For $x = \pi$, $\cos \pi = -1 < 0$

The above criteria is valid for only one expression

Thus, the function is differentiable at $x = \pi$

Taking option (C)

$$x = \frac{3\pi}{2}, \cos\frac{3\pi}{2} = 0$$

$\therefore$ L.H.S = $e^{-\frac{3\pi}{2}} \times 0 + e^{-\frac{3\pi}{2}} \times 1 = e^{-\frac{3\pi}{2}}$

R.H.S = $-e^{-\frac{3\pi}{2}} \times 0 - e^{-\frac{3\pi}{2}} \times 1 = -e^{-\frac{3\pi}{2}}$

$\therefore$ L.H.S ≠ R.H.S

Thus, the function is not differentiable at $x = \frac{3\pi}{2}$

Taking option (D)

For $x = 2\pi$

$\cos 2\pi = 1 > 0$

The above criteria is valid for only one expression

Thus, the function is continuous at $x = 2\pi$

133. Given,

$$\text{Let A} = \begin{bmatrix} 3 & 1 & 2 \\ 2 & -3 & -1 \\ 1 & 2 & 1 \end{bmatrix}$$

As we know

Product of eigen values = Determinant of matrix

and Sum of eigen values = Trace of matrix

∴ Determinant of matrix A is

$$\begin{vmatrix} 3 & 1 & 2 \\ 2 & -3 & -1 \\ 1 & 2 & 1 \end{vmatrix} = 3(-3+2)-1(2+1)+2(4+3)$$

$= -3 - 3 + 14 = 8$

Trace of matrix A is

$3 - 3 + 1 = 1$

$$\therefore \frac{\text{Product of eigenvalues}}{\text{Sum of eigen values}} = \frac{8}{1} = 8$$

134. Given,

$$\int_1^5 x^2 dx$$

evaluated at four equal intervals

$$h = \frac{b-a}{n} = \frac{5-1}{4} = 1$$

$y = x^2$

x	1	2	3	4	5
y	1	4	9	16	25

Using trapezoidal method

$$I_t = \frac{h}{2}[(1+25)+2(4+9+16)] = \frac{1}{2}(26+58) = 42$$

Using Simphon's one-third rule

$$I_s = \frac{h}{3}[(1+25)+4(4+16)+2(9)]$$

$$= \frac{1}{3}(26+80+18) = 41.33$$

Difference

$I_t - I_s = 42 - 41.33 = 0.67$

2
CHAPTER
FLIGHT MECHANICS

1. For maximum range of a glider, which of the following conditions is true?
   (*a*) lift to drag ratio is maximum
   (*b*) rate of descent is minimum
   (*c*) descent angle is maximum
   (*d*) lift to weight ratio is maximum
   [GATE 2007, 1 M]

2. An airplane with a larger wing as compared to a smaller wing will necessarily have
   (*a*) more longitudinal static stability
   (*b*) less longitudinal static stability
   (*c*) same longitudinal static stability
   (*d*) more longitudinal static stability for an aft tail airplane if aerodynamic center of the larger wing is behind the center of gravity of the airplane
   [GATE 2007, 1 M]

3. Two airplanes are identical except for the location of the wing. The longitudinal static stability of the airplane with low wing configuration will be
   (*a*) more than the airplane with high wing configuration
   (*b*) less than the airplane with high wing configuration
   (*c*) same as the airplane with high wing configuration
   (*d*) more if elevator is deflected
   [GATE 2007, 1 M]

4. For a fixed center of gravity location of an airplane, when the propeller is mounted on the nose of the nose of the fuselage
   (*a*) longitudinal static stability increase
   (*b*) longitudinal static stability decrease
   (*c*) longitudinal static stability remains same
   (*d*) longitudinal static stability is maximum
   [GATE 2007, 1 M]

5. Let an airplane in a steady level flight be trimmed at a certain speed. A level and steady flight at a higher speed could be achieved by changing
   (*a*) engine throttle only
   (*b*) elevator only
   (*c*) throttle and elevator together
   (*d*) rudder only
   [GATE 2007, 1 M]

6. Wing A has a constant chord c and span b. Wing B is identical but has a span 4b. When both wings are operating at the same geometric angle of attack at subsonic speed, then:
   (*a*) wings A and B produce the same lift coefficient
   (*b*) wing A produces a smaller lift coefficient than wing B
   (*c*) wing A produces a greater lift coefficient than wing B
   (*d*) the free stream Mach number decides which wing produces the greater lift coefficient
   [GATE 2007, 2 M]

7. A piston-prop airplane having propeller efficiency, η_p = 0.8 and weighing 73108 N could achieve maximum climb rate of 15 m/s at flight speed of 50 m/s.The excess Brake Power (BP) at the above flight condition will be
   (*a*) 1700 kW (*b*) 2100k W
   (*c*) 1371 kW (*d*) 6125 kW
   [GATE 2007, 2 M]

8. An airplane model with a symmetric airfoil was tested in a wind tunnel. C_{m0} (C_m at angle of attack, $\alpha = 0$) was estimated to be 0.08 and 0 respectively for elevator settings (δe) of 5 degrees up and 5 degrees down. The estimated value of the elevator control power $\left(\frac{\partial C_m}{\partial \delta e}\right)$ of the model will be
   (*a*) 0.07 per deg (*b*) –1.065 per deg
   (*c*) –0.008 per deg (*d*) –0.762 per deg
   [GATE 2007, 2 M]

9. The lateral-directional characteristic equation for an airplane gave the following set of roots: $\lambda_1 = -0.6$, $\lambda_2 = -0.002$, $\lambda_{3,4} = -0.06 \pm j1.5$, where $j = \sqrt{-1}$. The damping ratio corresponding to the Dutch-roll mode will be
   (*a*) 0.04 (*b*) 0.66
   (*c*) 0.35 (*d*) 0.18
   [GATE 2007, 2 M]

10. An airplane is flying at an altitude of 10 km above the sea level. Outside air temperature and density at 10km altitude are 233 K and 0.413kg/m^3 respectively. The airspeed indicator of the airplane

indicates a speed of 60 m/s Density of air at sea level is 1.255 kg/m^3 and value of the gas constant R is 288 J/kg/K. The stagnation pressure (P_0) measured by the Pitot tube mounted on the wing tip of the airplane will be magnitude.

(*a*) 3.5×10^4 Nm/^{2} (*b*) 2.0×10^4 Nm/^{2}

(*c*) 2.87×10^4 Nm/^{2} (*d*) 0.6×10^4 Nm/^{2}

[GATE 2007, 2 M]

11. If the center of gravity of an airplane is moved forward towards the nose of the airplane, the $C_{L\,max}$ (maximum value of the lift coefficient) value for which the airplane can be trimmed ($C_m = 0$) will

(*a*) decrease

(*b*) increase

(*c*) remain same

(*d*) depend upon ruder deflection

[GATE 2007, 2 M]

12. If the contribution of only the horizontal tail of an airplane was considered for estimating $\frac{\partial c_m}{\partial \alpha}$, and if the tail moment arm l_1 was doubled, then how many times the original value the new $\frac{\partial c_m}{\partial \alpha}$ become?

(*a*) two times (*b*) three times

(*c*) 1.414 times (*d*) 1.732 times

[GATE 2007, 2 M]

13. If the vertical tail of an airplane is inverted and put below the horizontal tail, then the contribution to roll derivative, $\frac{\partial c_l}{\partial \beta}$, will be

(*a*) negative (*b*) positive

(*c*) zero (*d*) imaginary

[GATE 2007, 2 M]

14. If horizontal tail area in increased while the elevator to horizontal tail area ratio is kept same, then

(*a*) both longitudinal static stability and elevator control power will increase

(*b*) only longitudinal static stability will increase

(*c*) only elevator control power will increase

(d) neither stability nor control power changes

[GATE 2007, 2 M]

15. An airfoil section is known to generate lift when placed in a uniform stream of speed U¥ at an incidence α. A biplane consisting of two such sections of identical chord c, separated by a distance h is shown in the following figure:

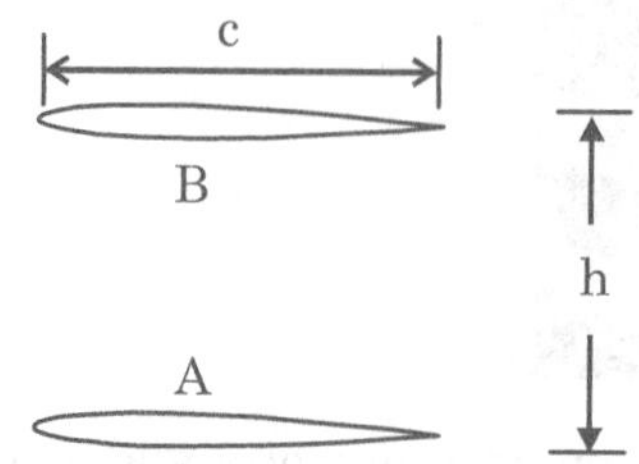

With regard to this biplane, which of the following statements is true?

(*a*) Both the airfoils experience an upwash and an increased approach velocity

(*b*) Both the airfoils experience a downwash and a decreased approach velocity

(*c*) Both the airfoils experience an upwash and airfoil A experiences a decreased approach velocity while airfoil B experiences an increased approach velocity

(*d*) The innocence for the individual sections of the biplane is not altered

[GATE 2007, 2 M]

16. The span-wise lift distribution for three wings is shown in the following figure:

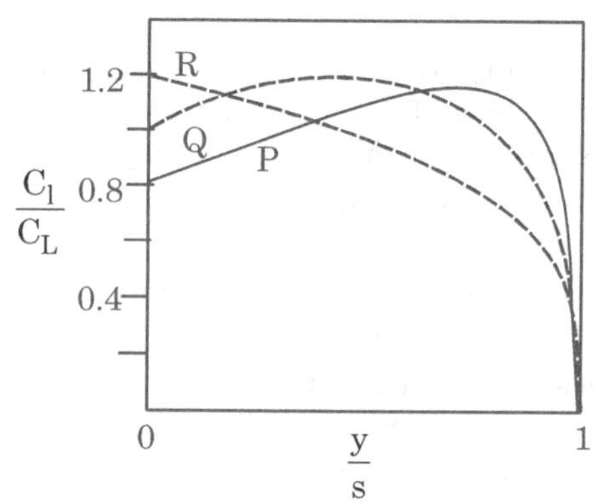

In the above figure, c_l refers to the section lift coefficient, C_L refers to the lift coefficient of the wing, y is the coordinates along the span and s is the span of the wing. A designer prefers to use a wing for which the stall begins at the root. Which of the wings will be choose?

(*a*) P (*b*) Q

(*c*) R (*d*) None

[GATE 2007, 2 M]

Common Data for Questions 17, 18 & 19: An airplane designer wants to keep longitudinal static stability margin (SM) within 5% to 15% of mean aerodynamic chord. A wind tunnel test of the model showed that for $\overline{X}_{CG} = 0.2$, $\frac{dC_m}{dC_L} = -0.1$. Note that the distance from the wing leading edge to the centre gravity (X_{CG}) has been non-dimensionalized by dividing it with mean aerodynamic chord, $\bar{c}$, such that $\bar{X}_{CG} = \frac{X_{CG}}{\bar{c}}$. Note also that the relation $\frac{dC_m}{dC_L} = -SM$ holds true for this airplane.

17. The most forward location of the airplane centre of gravity permitted to fulfill the designer's requirement on longitudinal static stability margin is

(*a*) $0.35\ \bar{c}$ (*b*) $0.25\ \bar{c}$

(*c*) $0.15\ \bar{c}$ (*d*) $0.52\ \bar{c}$

[GATE 2007, 2 M]

18. The most aft location of the airplane centre of gravity permitted to fulfill designer's requirement on longitudinal static stability is

(*a*) $0.35\ \bar{c}$ (*b*) $0.45\ \bar{c}$

(*c*) $0.52\ \bar{c}$ (*d*) $0.67\ \bar{c}$

[GATE 2007, 2 M]

19. The center of gravity location to have

$\frac{d\delta_e}{dC_L} = 0$ is

(*a*) $0.35\ \bar{c}$ (*b*) $0.45\ \bar{c}$

(*c*) $0.5\ \bar{c}$ (*d*) $0.4\ \bar{c}$

Statement for Linked answer questions 20 & 21: For a piston propeller airplane weighing 20000 N, the flight testing at 5 km pressure altitude in standard atmosphere gave the variation of power required versus true air speed as shown in figure below. The student forgot to label the air speed axis. The maximum climb rate at sea level was calculated to be 4 m/s. Assume shaft power available to the independent of speed of flight. For piston propeller airplane, it can be assumed that the shaft power available is proportional to ambient density. Values of air density at sea level and at 5 pressure altitude are 1.225 kg/m^3 and 0.74 kg/m^3, respectively.

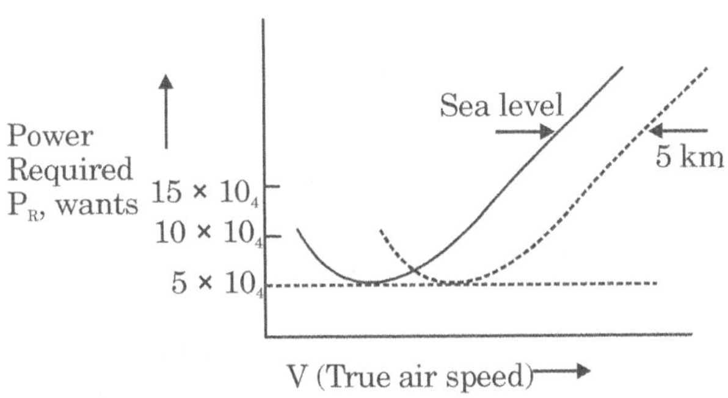

20. The maximum rate of climb achievable by this airplane at 5 km altitude will be

(*a*) 1.65 m/s (*b*) 0.51 m/s

(*c*) 1.43 m/s (*d*) 3.65 m/s

[GATE 2007, 2 M]

21. If during the maximum rate of climb at 5 km altitude, the airplane was flying at an angle of attack of 4 degrees and attitude (pitch) angle of 5 degrees, what was equivalent airspeed of the airplane?

(*a*) 40.2 m/s (*b*) 63.7 m/s

(*c*) 130.3 m/s (*d*) 20.2 m/s

[GATE 2007, 2 M]

22. The service ceiling of a transport aircraft is defined as the altitude

(a) that is halfway between sea-level and absolute ceiling

(b) at which it can cruise with one engine operational

(c) at which its maximum rate of climb is zero

(d) at which its maximum rate of climb is 0.508 m/s

[GATE 2008, 1 M]

23. The drag of an aircraft in steady climbing flight at a given forward speed is

(*a*) inversely proportional to climb angle

(*b*) higher than drag in steady level flight at the same forward speed

(*c*) lower than drag in steady level flight at the same forward speed

(*d*) independent of climb angle

[GATE 2008, 1 M]

24. In steady, level turning flight of an aircraft at a load factor 'n', the ratio of the horizontal component of lift and aircraft weight is

(*a*) $\sqrt{n-1}$ (*b*) $\sqrt{n+1}$

(*c*) $\sqrt{n^2-1}$ (*d*) $\sqrt{n^2+1}$

[GATE 2008, 1 M]

25. The parameters that remain constant in a cruise-climb of an aircraft are

(*a*) equivalent airspeed and lift coefficient

(*b*) altitude and lift coefficient

(*c*) equivalent airspeed and altitude

(*d*) lift coefficient and aircraft mass

[GATE 2008, 1 M]

26. The figure below shows the variation of C_m versus α for an aircraft for three combinations of elevator deflections and locations of centre of gravity. In the figure, lines P and Q are parallel, while lines Q and R have the same intercept on the C_m axis.

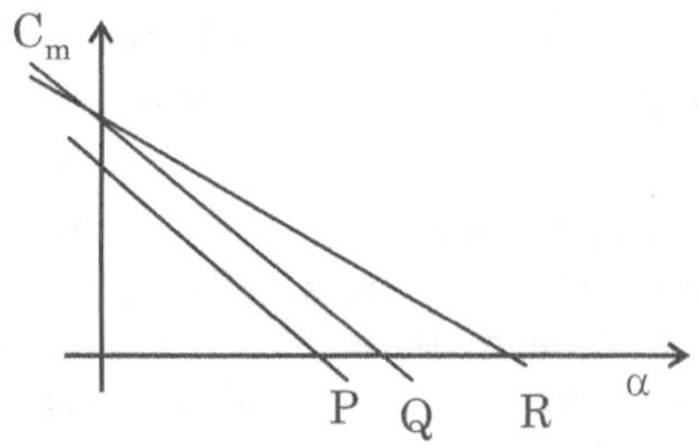

Which of the following statements is true?

(a) Lines P and Q correspond to the same centre of gravity location.

(b) Lines Q and R correspond to the same centre of gravity location.

(c) Lines P and Q correspond to the same elevator deflection.

(d) Lines P and R correspond to the same centre of gravity location.

[GATE 2008, 2 M]

27. Which of the following statements is TRUE as the altitude increase in stratosphere of International Standard Atmosphere?

(a) Temperature increase and dynamic viscosity decreases.

(b) Temperature remains constant and pressure increases.

(c) Temperature decrease and sound speed decreases.

(d) Temperature remains constant and density decreases.

[GATE 2008, 2 M]

28. Which of the following statements is TRUE?

(a) Wing dihedral reduce roll stability while a low wing increase roll stability.

(b) Wing dihedral increases roll stability while a low wing reduces roll stability.

(c) Wing dihedral, as well as low wing reduces roll stability.

(d) Wing dihedral, as well as low wing increases roll stability.

[GATE 2008, 2 M]

29. An aircraft has a level flight stalling speed of 60 m/s EAS (equivalent air speed). As per the V-n diagram, what is the minimum speed at which it should be designed to withstand the maximum vertical load factor of 9?

(a) 20 m/s (b) 60 m/s

(c) 120 m/s (d) 180 m/s

[GATE 2008, 2 M]

30. Match each mode of aircraft motion listed in Group I to its corresponding property from Group II.

Group I: Aircraft mode	Group II: Property
P. Short period mode	1. Coupled roll-yaw oscillations
Q. Wing rock	2. Angle of attack remains constant
R. Phugoid mode	3. Roll oscillations
S. Dutch roll	4. Speed remains constant

(a) P-2, Q-1, R-4, S-3 (b) P-4, Q-3, R-2, S-1

(c) P-4, Q-1, R-2, S-3 (d) P-2, Q-3, R-4, S-1

[GATE 2008, 2 M]

31. An aircraft is cruising at a true air speed (TAS) of 100 m/s under ISA conditions, at an altitude at which the density of free stream is 0.526 kg/m^3. What will be the equivalent air speed (EAS)?

(a) 65.5 m/s (b) 72.5 m/s

(c) 110.5 m/s (d) 152.7 m/s

[GATE 2008, 2 M]

32. In the definition of the aircraft Euler angles φ (roll), θ(pitch), and ψ(yaw), the correct sequence of rotations required to make the inertial frame coincide with the aircraft body frame is

(a) first ψ about z axis, second θ about y axis, third φ about x axis

(b) first θ about y axis, second φ about x axis, third ψ about z axis

(c) first φ about x axis, second θ about y axis, third ψ about z axis

(d) first ψ about z axis, second φ about x axis, third θ about y axis

[GATE 2008, 2 M]

33. To maximize range of jet engine aircraft, it should be flown at a velocity that maximizes

(a) $\frac{C_L}{C_D}$ (b) $\frac{C_L^{0.5}}{C_D}$

(c) $\frac{C_L^{1.5}}{C_D}$ (d) $\frac{C_L^{2}}{C_D}$

[GATE 2008, 2 M]

34. The primary function of the fin in the vertical tail of an aircraft is to provide

(a) yaw control (b) yaw stability

(c) roll damping (d) roll stability

[GATE 2008, 2 M]

35. An aircraft requires the trailing edge of the elevator to the deflected upwards from its initial position to lower the trim speed. Which of the following statements about the static stick-fixed stability of this aircraft is true?

(*a*) The aircraft is unstable.

(*b*) The aircraft is neutrally stable.

(*c*) The aircraft is stable.

(*d*) The stability of the aircraft cannot be determined from the given information.

[GATE 2008, 2 M]

36. Which of the following statements is true for an aircraft flying at a low angle of attack?

(*a*) Yawing motion generates yawing moment and pitching moment.

(*b*) Rolling motion generates rolling moment and pitching moment.

(*c*) Yawing motion generates yawing moment and rolling moment.

(*d*) Pitching motion generates yawing moment and rolling moment.

[GATE 2008, 2 M]

37. Which one of the following high lift device results in higher stalling angle?

(*a*) split flap (*b*) Fowler flap

(*c*) plain flap (*d*) leading edge flap

[GATE 2008, 2 M]

Statement for Linked Answer Questions 38 and 39: An aircraft has a zero-lift drag coefficient C_{DO} = 0.0223, wing aspect ratio AR_w = 10.0, and Oswald's efficiency factor e = 0.7

38. The thrust required for steady level flight will be minimum when the aircraft operates at a lift coefficient of

(*a*) 0.65 (*b*) 0.70

(*c*) 0.75 (*d*) 0.80

[GATE 2008, 2 M]

39. The glide angle that results in maximum range in a power-off glide is

(*a*) 1.82 degrees (*b*) 2.68 degrees

(*c*) 3.64 degrees (*d*) 5.01 degrees

[GATE 2008, 2 M]

40. For two-dimensional irrotational and incompressible flows

(*a*) Both potential and stream functions satisfy the Laplace equation.

(*b*) Potential function must satisfy the Laplace equation but the stream function need not.

(*c*) Stream function must satisfy the Laplace equation but the potential function need not.

(*d*) Neither the stream function nor the potential function need to satisfy the Laplace equation.

[GATE 2009, 1 M]

41. A conventional altimeter is a

(*a*) Pressure transducer

(*b*) Temperature transducer

(*c*) Density transducer

(*d*) Velocity transducer

[GATE 2009, 1 M]

42. The relation between an airplane's true airspeed V_{TAS} and equivalent airspeed V_{EAS} in terms of the density ratio ($\sigma = \frac{\rho}{\rho_0}$), where ρ_0 is the air density at sea-level and p is the air density at the altitude at which the airplane is flying, is given by the formula:

(*a*) $\frac{V_{EAS}}{V_{TAS}} = \sigma$ (*b*) $\frac{V_{EAS}}{V_{TAS}} = \sigma^2$

(*c*) $\frac{V_{EAS}}{V_{TAS}} = \sqrt{\sigma}$ (*d*) $\frac{V_{EAS}}{V_{TAS}} = \frac{1}{\sqrt{\sigma}}$

[GATE 2009, 1 M]

43. An unswept fixed-winged aircraft has a large roll stability if the wing is placed

(*a*) low on the fuselage and has negative dihedral angle

(*b*) low on the fuselage and has positive dihedral angle

(*c*) high on the fuselage and has negative dihedral angle

(*d*) high on the fuselage and has positive dihedral angle

[GATE 2009, 1 M]

44. If $C_{m_{CG}}$ is the pitching moment coefficient about the center of gravity of an aircraft, and α is the angle of attack, then $\frac{dC_{m_{CG}}}{d\alpha}$ is

(*a*) a stability derivative which represents stiffness in pitch

(*b*) a stability derivative which represents stiffness in pitch

(*c*) a control derivative in pitch

(*d*) positive for an aircraft that is stable in pitch

[GATE 2009, 1 M]

45. An airplane flying at 100 m/s is pitching at the rate of 0.2 deg/s. Due to this pitching, the horizontal tail surface located 4 metres behind the centre-of-mass of the airplane will experience a change in angle of attack. which is

(*a*) 0.01 deg (*b*) 0.008 deg

(*c*) 0.04 deg (*d*) 0.004 deg

[GATE 2009, 2 M]

46. The contribution of the horizontal tail to the pitching moment coefficient about the center of gravity $\left(C_{m_{CG}}\right)$ of an aircraft is given by $C_{m_{tail}}$ $= 0.2 - 0.0215\ \alpha$, where α is the angle of attack of the aircraft. The contribution of the tail to the aircraft longitudinal stability

(*a*) is stabilizing

(*b*) is destabilizing

(*c*) is nil

(*d*) cannot be determined from the given information

[GATE 2009, 2 M]

47. The linearized dynamics of an aircraft (which has no large rotating components) in straight and level flight is governed by the equations

$$\frac{d\vec{x}}{dt} = \begin{bmatrix} [A] & [B] \\ [C] & [D] \end{bmatrix} \vec{x}$$

where

$\vec{x} = [u \quad w \quad q \quad \theta \quad v \quad p \quad r \quad \phi]^T, [\ ]^T$ represents the transpose of a matrix, [A], [B], [C] and [D] are 4×4 matrices and [0] is the 4×4 null matrix. Which of the following is true?

(*a*) $[A] \neq [0]; [B] \neq [0]; [C] = [0]; [D] \neq [0]$

(*b*) $[A] = [0]; [B] \neq [0]; [C] \neq [0]; [D] = [0]$

(*c*) $[A] \neq [0]; [B] = [0]; [C] = [0]; [D] \neq [0]$

(*d*) $[A] \neq [0]; [B] = [0]; [C] \neq [0]; [D] = [0]$

[GATE 2009, 2 M]

48. The velocity vector of an aircraft along its body-fixed axis is given by $\vec{V} = \begin{Bmatrix} u \\ v \\ w \end{Bmatrix}$. If V is the magnitude $\vec{V}, \alpha$ is the angle of attack and β is the angle of sideslip, which of the following set of relations is correct?

(*a*) $u = V \sin\beta \cos\alpha$; $v = V \sin\beta$; $w = V \cos\beta \sin\alpha$

(*b*) $u = V \cos\beta \cos\alpha$; $v = V \cos\beta$; $w = V \cos\beta \sin\alpha$

(*c*) $u = V\cos\beta \cos\alpha$; $v = V \sin\beta$; $w = V \sin\beta \sin\alpha$

(*d*) $u = V \cos\beta \cos\alpha$; $v = V \sin\beta$; $w = V \cos\beta \sin\alpha$

[GATE 2009, 2 M]

49. An aircraft of mass 2500 kg in straight and level flight at a constant speed of 100 mls has available excess power of 1.0 x 10^6 W. The steady rate of climb it can attain at that speed is

(*a*) 100 m/s (*b*) 60 m/s

(*c*) 40 m/s (*d*) 20 m/s

[GATE 2009, 2 M]

Common Data for Questions 51 and 52:

The roots of the characteristic equation for the longitudinal dynamics of a certain aircraft are:

$\lambda_1 = -0.02 + 0.2i$; $\lambda_2 = -0.02 - 0.2i$; $\lambda_3 = -2.5 + 2.6i$; $\lambda_4 = -2.5 - 2.6i$, where $i = \sqrt{-1}$.

50. The pair of eigenvalues that represent the phugoid mode is

(*a*) λ_1 and λ_3 (*b*) λ_2 and λ_4

(*c*) λ_3 and λ_4 (*d*) λ_1 and λ_2

[GATE 2009, 2 M]

51. The short period damped frequency is

(*a*) 2.6 rad/s (*b*) 0.2 rad/s

(*c*) 2.5 rad/s (*d*) 0.02 rad/s

[GATE 2009, 2 M]

52. An aircraft is climbing at a constant speed in a straight line at a steep angle of climb. The load factor it sustains during the climb is:

(*a*) equal to 0.1

(*b*) greater than 1.0

(*c*) positive but less than 1.0

(*d*) dependant on the weight of the aircraft

[GATE 2010, 1 M]

53. All other factors remaining constant, if the weight of an aircraft increases by 30% then tile takeoff distance increases by approximately:

(*a*) 15% (*b*) 30%

(*c*) 70% (*d*) 105%

[GATE 2010, 1 M]

54. An aircraft stalls at a speed of 40 m/s in straight and level flight. The slowest speed at which this aircraft can execute a level tum at a bank angle of 60 degrees is:

(*a*) 28.3 m/s (*b*) 40.0 m/s

(*c*) 56.6 m/s (*d*) 80.0 m/s

[GATE 2010, 1 M]

55. The absolute ceiling of an aircraft is the altitude above which it:

(*a*) can never reach

(*b*) cannot sustain level flight at a constant speed

(*c*) can perform accelerated flight as well as straight and level flight at a constant speed

(*d*) can perform strnight and level flight at a constant speed only

[GATE 2010, 1 M]

56. A propeller powered aicraft, trimmed to attain maximum range and flying in a straight line, travels a distance R from its take-off point when it has consumed a weight of fuel equal to 20% of its take-off weight. If the aircraft continues to fly and consumes a total weight of fuel equal to 50% of its take-off weight, the distance between it and its take-off point becomes:

(*a*) 2.5 R (*b*) 3.1 R

(*c*) 2.1 R (*d*) 3.9 R

[GATE 2010, 2 M]

57. The trim curves of an aircraft are of the form $C_m = (0.05 - 0.2\delta)$ the elevator deflection angle. δ is in radians. The static margin of the aircraft is:

(*a*) 0.5 (*b*) 0.2

(*c*) 0.1 (*d*) 0.05

[GATE 2010, 2 M]

58. The trim curves of an airc:raft are of tile form $C_m = (0.05 - 0.2\delta) - 0.I\ C_L$ where tile elevator deflection angle, δ is in radians. The change in elevator deflection needed to increase tile lift coefficient from 0.4 to 0.9 is:

(*a*) –0.5 radians (*b*) –0.25 radians

(*c*) 0.25 radians (*d*) 0.5 radians

[GATE 2010, 2 M]

Statement for Linked Answer Questions 54 and 55:

An aircraft is in straight and level flight at a constant speed v. It is disturbed by a symmetric vertical gust, resulting in a phugoid oscillation of time period T.

59. Assuming that g is the acceleration due to gravity, T is given approximately by:

(*a*) $\frac{v}{\pi g}$ (*b*) $\frac{\pi v}{g}$

(*c*) $\frac{v}{\sqrt{2}\pi g}$ (*d*) $\frac{\sqrt{2}\pi v}{g}$

[GATE 2010, 2 M]

60. If v = 200 m/s then the wavelength of the phugoid oscillation, assuming $g = 9.81\ m/s^2$, is, approximately:

(*a*) 1.28×10^4 m (*b*) 1.30×10^3 m

(*c*) 1.81×10^{-4} m (*d*) 918 m

[GATE 2010, 2 M]

61. In an un-powered glide of an aircraft having weight W, lift L and drag D, the equilibrium glide angle is defined as

(*a*) $\tan^{-1}\left(\frac{L}{D}\right)$ (*b*) $\tan^{-1}\left(\frac{D}{L}\right)$

(*c*) $\tan^{-1}\left(\frac{L}{W}\right)$ (*d*) $\tan^{-1}\left(\frac{W}{L}\right)$

[GATE 2011, 1 M]

62. Lift on an aircraft climbing vertically up is

(*a*) equal to its weight (*b*) zero

(*c*) equal to the drag (*d*) equal to the thrust

[GATE 2011, 1 M]

63. If an aircraft is performing a positive yawing manoeuvre, the side slip angle

(*a*) is always zero (*b*) is never zero

(*c*) is always negative (*d*) could be any value

[GATE 2011, 1 M]

64. For an airplane to be statically stable, its centre of gravity must always be

(*a*) ahead of wing aerodynamic centre

(*b*) aft of the wing aerodynamic centre

(*c*) ahead of neutral point

(*d*) aft of neutral point

[GATE 2011, 1 M]

65. Winglets are used on wings to minimize

(*a*) skin friction drag (*b*) profile drag

(*c*) wave drag (*d*) induced drag

[GATE 2011, 1 M]

66. An aircraft is performing a coordinated turn manoeuvre at a bank angle of 30° and forward speed of 100 m/s. Assume $g = 9.81\ ms^{-2}$. The load factor and turn radius respectively are

(*a*) $\left(\frac{2}{\sqrt{3}}\right)$ and 1.76 km (*b*) $\sqrt{3}$ and 17.6 km

(*c*) 2 and 0.18 km (*d*) $\left(\frac{2}{\sqrt{3}}\right)$ and 17.6 km

[GATE 2011, 2 M]

67. An aircraft in a steady level flight at forward speed of 50 m/s suddenly rolls by 180° and becomes inverted. If no other changes are made to the configuration or controls of the aircraft, the nature of the subsequent flight path taken by the aircraft and its characteristic parameter(s) (assume $g = 9.81ms^{-2}$) are

(*a*) straight line path with a speed of 50 m/s

(*b*) upward circular path with a speed of 50 m/s and radius of 127.4 m

(*c*) downward circular path with a speed of 50 m/s and radius of 127.4 m/s

(*d*) downward circular path with a speed of 25 m/s and radius of 254.8 m/s

[GATE 2011, 2 M]

68. An aircraft with a mass of 5000 kg takes off from sea level with a forward speed of 50 m/s and starts to climb with a climb angle of 15°. The rate of climb and excess thrust available at the start of the climb respectively (assume $g = 9.81\ ms^{-2}$) are

(*a*) 13.40 m/s and 13146.0 N

(*b*) 12.94 m/s and 12694.1 N

(*c*) 13.40 m/s and 12694.1 N

(*d*) 12.94 m/s and 13146.0 N

[GATE 2011, 2 M]

69. A glider having a mass of 500 kg is taken to an altitude of 1000 m with a jeep moving on ground at 54 kmph. Upon reaching the required altitude in 50 s, the glider is released and starts its descent. Under the assumption of equilibrium glide, the range and endurance of the glider for a constant lift-to-drag ratio of 15 are

(*a*) 15.0 km and 1002.2 s respectively

(*b*) 15.0 km and 601.3 s respectively

(*c*) 1.0 km and 601.3 s respectively

(*d*) 1.0 km and 50 s respectively

[GATE 2011, 2 M]

70. An aircraft in level flight encounters a vertical gust, which excites the phugoid mode. The phugoid motion completes 10 cycles in 50 s and its amplitude reduces to half of its maximum value in 25 s. The eigenvalues of the phugoid mode are

(*a*) $-0.05 \pm 0.02i$ (c) $-0.5 \pm 0.2i$

(*c*) $-0.028 \pm 1.26i$ (*d*) $0.028 \pm 1.26i$

[GATE 2011, 2 M]

71. An aircraft in trimmed condition has zero pitching moment at

(*a*) its aerodynamic centre.

(*b*) its centre of gravity

(*c*) 25% of its mean aerodynamic chord.

(*d*) 50% of its wing root chord.

[GATE 2012, 1 M]

72. In an aircraft, constant roll rate can be produced using ailerons by applying

(*a*) a step input. (*b*) a ramp input

(*c*) a sinusoidal input (*d*) an impulse input.

[GATE 2012, 1 M]

73. During the ground roll manoeuvre of an aircraft, the force(s) acting on it parallel to the direction of motion

(*a*) is thrust alone

(*b*) is drag alone.

(*c*) are both thrust and drag

(*d*) are thrust, drag and a part of both weight and lift.

[GATE 2012, 1 M]

74. An aircraft in a steady climb suddenly experiences a 10% drop in thrust. After a new equilibrium is reached at the same speed, the new rate of climb is

(*a*) lower by exactly 10%.

(*b*) lower by more than 10%.

(*c*) lower by less than 10%.

(*d*) an unpredictable quantity.

[GATE 2012, 1 M]

75. In an aircraft, the dive manoeuvre can be initiated by

(*a*) reducing the engine thrust alone.

(*b*) reducing the angle of attack alone.

(*c*) generating a nose down pitch rate.

(*d*) increasing the engine thrust alone.

[GATE 2012, 1 M]

76. In an aircraft, elevator control effectiveness determines

(*a*) turn radius.

(*b*) rate of climb.

(*c*) forward-most location of the centre of gravity.

(*d*) aft-most location of the centre of gravity.

[GATE 2012, 1 M]

77. For a wing of aspect ratio AR, having an elliptical lift distribution, the induced drag coefficient is (where C L is the lift coefficient)

(*a*) $\dfrac{C_L}{\pi AR}$ (*b*) $\dfrac{C_L^2}{\pi\ AR}$

(*c*) $\dfrac{C_L}{2\pi\ AR}$ (*d*) $\dfrac{C_L^2}{\pi\ AR^2}$

[GATE 2012, 1 M]

78. An aircraft has a steady rate of climb of 300 m/s at sea level and 150 m/s at 2500 m altitude. The time taken (in sec) for this aircraft to climb from 500 m altitude to 3000 m altitude is ____.

[GATE 2012, 2 M]

79. If an aircraft takes off with 10% less fuel in comparison to its standard configuration, its range is

(*a*) lower by exactly 10%.

(*b*) lower by more than 10%.

(*c*) lower by less than 10%.

(*d*) an unpredictable quantity.

[GATE 2012, 2 M]

80. An aircraft has an approach speed of 144 kmph with a descent angle of 6.6°. If the aircraft load factor is 1.2 and constant deceleration at touch down is 0.25g ($g = 9.81\ m/s^2$), its total landing distance approximately over a 15 m high obstacle is

(*a*) 1830 m (*b*) 1380 m

(*c*) 830 m (*d*) 380 m

[GATE 2012, 2 M]

81. An aircraft is trimmed straight and level at true air speed (TAS) of 100 m/s at standard sea level (SSL). Further, pull of 5 N holds the speed at 90 m/s without re-trimming at SSL (air density = 1.22 kg/m^3). To fly at 3000 m altitude (air density = 0.91 kg/m^3) and 120 m/s TAS without re-trimming, the aircraft needs

(*a*) 1.95 N upward force.

(*b*) 1.95 N downward force.

(*c*) 1.85 N upward force.

(*d*) 1.75 N downward force.

[GATE 2012, 2 M]

Common Data for Questions 82 and 83:

A wing and tail are geometrically similar, while tail area is one-third of the wing area and distance between two aerodynamic centres is equal to wing semi-span $\left(\frac{b}{2}\right)$. In addition, following data is applicable:

$$\epsilon_a = 0.3, C_L = 1.0, C_{l_w} = 0.008/\text{deg}$$

$$\bar{c} = 2.5\text{m}, b = 30\text{m}, C_{M_\alpha} = 0, \eta_t = 1.$$

The symbols have their usual aerodynamic interpretation.

82. The maximum distance that the centre of gravity can be behind aerodynamic centre without destabilizing the wing-tail combination is

(*a*) 0.4 m (*b*) 1.4 m

(*c*) 2.4 m (*d*) 3.4 m

[GATE 2012, 2 M]

83. The angle of incidence of tail to trim the wing-tail combination for a 5% static margin is

(*a*) –1.4° (*b*) –0.4°

(*c*) 0.4° (*d*) 1.4°

[GATE 2012, 2 M]

84. Which one of the following is the most stable configuration of an airplane in roll?

(*a*) Sweep back, anhedral and low wing

(*b*) Sweep forward, dihedral and low wing

(*c*) Sweep forward, anhedral and high wing

(*d*) Sweep back, dihedral and high wing

[GATE 2013, 1 M]

85. Which one of the following flight instruments is used on an aircraft to determine its attitude in flight?

(*a*) Vertical speed indicator

(*b*) Altimeter

(*c*) Artificial Horizon

(*d*) Turn-bank indicator

[GATE 2013, 1 M]

86. A supersonic airplane is expected to fly at both subsonic and supersonic speeds during its whole flight course. Which one of the following statements is TRUE?

(*a*) Airplane will experience less stability in pitch at supersonic speeds than at subsonic speeds

(*b*) Airplane will feel no change in pitch stability

(*c*) Airplane will experience more stability in pitch at supersonic speeds than at subsonic speeds

(*d*) Pitch stability cannot be inferred from the information given

[GATE 2013, 1 M]

87. Which one of the following is favorable for an airplane operation?

(*a*) Tail wind in cruise and head wind in landing

(*b*) Tail wind both in cruise and landing

(*c*) Head wind both in cruise and landing

(*d*) Head wind in cruise and tail wind in landing

[GATE 2013, 1 M]

88. Which one of the following is TRUE with respect to Phugoid mode of an aircraft?

(*a*) Frequency is directly proportional to flight speed

(*b*) Frequency is inversely proportional to flight speed

(*c*) Frequency is directly proportional to the square root of flight speed

(*d*) Frequency is inversely proportional to the square root of flight speed

[GATE 2013, 1 M]

89. A glider is launched from a 500m high hilltop. Following data is available for the glider: Zero lift drag coefficient $C_{D0} = 0.02$, aspect ratio AR = 10 and Oswald efficiency factor e = 0.95. The maximum range of the glider in km is ________

[GATE 2013, 2 M]

90. Which one of the following criteria leads to maximum turn rate and minimum radius in a level turn flight?

(*a*) Highest possible load factor and highest possible velocity

(*b*) Lowest possible load factor and lowest possible velocity (*c*) Highest possible load factor and lowest possible velocity

(*d*) Lowest possible load factor and highest possible velocity

[GATE 2013, 2 M]

91. Consider an airplane with rectangular straight wing at dihedral angle $\Gamma = 10^0$. Lift curve slope of wing airfoil section (constant over the whole span of the wing) is $C_{la} = 5.4/\text{rad}$. The roll stability derivative, $C_{l\beta}$ in per radian is ________

[GATE 2013, 2 M]

Common Data for Questions 92 and 93:

Data for an airplane are given as follows: weight W= 30 kN, thrust available at sea-level T_0 = 4000N, wing planform area S = 30m², maximum lift coefficient C_{Lmax} = 1.4, and drag coefficient $C_D = 0.015 + 0.024$

Assume air density at sea-level $31.22\, C_L^2$. Assume air density at sea-level $\rho_\infty = 1.22\text{kg/m}^3$.

92. Stall speed of the airplane in m/s is

(*a*) 17.36 (*b*) 34.22

(*c*) 45.52 (*d*) 119.46

[GATE 2013, 2 M]

93. Minimum and maximum speeds of the airplane in level flight condition at sea-level in m/s are respectively

(*a*) 17.36 and 180 (*b*) 17.36 and 34.22

(*c*) 34.22 and 119.46 (*d*) 17.36 and 119.46

[GATE 2013, 2 M]

94. For a NACA 5-digit airfoil of chord c, the designed lift coefficient and location of maximum camber along the chord from the leading edge are denote by C_L and X_m respectively. For NACA12018 airfoil, which combination of C_L and X_m given below are correct?

(*a*) $C_L = 0.15$ and $X_m = 0.1c$

(*b*) $C_L = 0.12$ and $X_m = 0.2c$

(*c*) $C_L = 0.12$ and $X_m = 0.18c$

(*d*) $C_L = 0.15$ and $X_m = 0.2c$

[GATE 2014, 1 M]

95. The vertical ground load factor on a stationary aircraft parked in its hangar is :

(*a*) 0 (*b*) –1

(*c*) Not defined (*d*) 1

[GATE 2014, 1 M]

96. Under what conditions a glider should be operated to ensure minimum sink rate?

(*a*) Maximum C_L/C_D (*b*) Minimum C_L/C_D

(*c*) Maximum $C_D/C_L^{3/2}$ (*d*) Minimum $C_D/C_L^{3/2}$

[GATE 2014, 1 M]

97. In most airplanes, the Dutch roll mode can be excited by applying

(*a*) a step input to the elevators

(*b*) a step input to the rudder

(*c*) a sinusoidal input to the aileron

(*d*) an impulse input to the elevators

[GATE 2014, 1 M]

98. For a given airplan with a given wing loading executing a turn in the vertical plane, under what conditions will the turn radius be minimum and the turn rate be maximum?

(*a*) Highest possible C_L and lowest possible load factor

(*b*) Lowest possible C_L and lowest possible load factor

(*c*) Lowest possible C_L and highest possible load factor

(*d*) Highest possible C_L and highest possible load factor

[GATE 2014, 2 M]

99. Lift-off distance for a given aircraft of weight W is S_{LO}. If the take-off weight is reduced by 10%, then the magnitude of percentage change in the lift-off distance (assume all other parameters to remain constant) is ________.

[GATE 2014, 2 M]

100. Which of the following design parameters influence the maximum rate-of-climb for a jet-propelled airplane?

P. Wing loading

Q. Maximum thrust-to-weight ratio

R. Zero-lift drag coefficient

S. Maximum lift-to-drag ratio

(*a*) P and Q alone (*b*) P, Q, R and S

(*c*) P, Q and S alone (*d*) Q, R and S alone

[GATE 2014, 2 M]

101. Consider the following four statements regarding aircraft longitudinal stability:

P. $C_{M,cg}$ at zero lift must be positive

Q. $\partial CM,cg/\partial\alpha_a$ must be negative (α_a is absolute angle of attack)

R. $C_{M,cg}$ at zero lift must be negative

S. Slope of C_L versus α_a must be negative

Which of the following combination is the necessary criterion for stick fixed longitudinal balance and static stability?

(*a*) Q and R only (*b*) Q, R and S only

(*c*) P and Q only (*d*) Q and S only

[GATE 2014, 2 M]

102. Data for a light, single-engine, propeller driven aircraft in steady level flight at sea-level is as follows: velocity V_∞ = 40m/s, weight W = 13000N, lift coefficient C_L = 0.65, drag coefficient $C_D = 0.25 + 0.04\, C_L^2$ and power available P_{av} = 100,000J/s. The rate of climb possible for this aircraft under the given conditions (in m/s) is

(*a*) 7.20 (*b*) 5.11

(*c*) 6.32 (*d*) 4.23

[GATE 2014, 2 M]

103. Consider the density and altitude at the base of an isothermal layer in the standard atmosphere to be ρ_1 and h_1, respectively. The density variation with altitude (ρ versus h) in that layer is governed by (R: specific gas constant, T: temperature, g_0 : acceleration due to gravity at sea level)

(*a*) $\frac{\rho}{\rho_1} = e^{-\left[\frac{g_0}{RT}\right](h-h_1)}$ (*b*) $\frac{\rho}{\rho_1} = e^{-\left[\frac{g_0}{RT}\right](h_1-h)}$

(*c*) $\frac{\rho}{\rho_1} = e^{-\left[\frac{RT}{g_0}\right](h-h_1)}$ (*d*) $\frac{\rho}{\rho_1} = e^{-\left[\frac{RT}{g_0}\right](h_1-h_1)}$

[GATE 2015, 1 M]

104. For constant free stream velocity and density, a change in lift for a large aspect ratio straight wing, with thin cambered airfoil section at small angles of attack, leads to

(*a*) a shift of the aerodynamic center and no shift of the center of pressure

(*b*) a shift of the center of pressure and no shift of the aerodynamic center

(*c*) shift of both the aerodynamic center and the center of pressure

(*d*) no shift either of the aerodynamic center or of the center of pressure

[GATE 2015, 1 M]

105. Which one of the following modes of a stable aircraft has non-oscillatory response characteristics?

(*a*) Short period (*b*) Phugoid

(*c*) Dutch roll (*d*) Spiral

[GATE 2015, 1 M]

106. As a candidate for a vertical tail, which one of the following airfoil sections is appropriate?

(*a*) NACA 0012 (*b*) NACA 2312

(*c*) NACA 23012 (*d*) Clarke Y profile

[GATE 2015, 1 M]

107. An aircraft in level and unaccelerated flight with a velocity of v_∞ = 300 m/s requires a power of 9×10^6 W. If the aircraft weighs 1.5×10^5 N, the lift-to-drag ratio

[GATE 2015, 1 M]

108. The percentage change in the lift-off distance for a 20 % increase in aircraft weight is _____.

[GATE 2015, 1 M]

109. Consider a monoplane wing and a biplane wing with identical airfoil sections, wingspans and incidence angles in identical conditions in a wind tunnel. As compared to the monoplane, the biplane experiences

(*a*) a higher lift and a higher drag

(*b*) a higher lift and a lower drag

(*c*) a lower lift and a lower drag

(*d*) a lower lift and a higher drag

[GATE 2015, 1 M]

110. A statically stable trimmed aircraft experiences a gust and the angle of attack reduces momentarily. As a result, the center of pressure of the aircraft

(*a*) shifts forward

(*b*) shifts rearward

(*c*) does not shift

(*d*) coincides with the neutral point

[GATE 2015, 1 M]

111. An aircraft, with a wing loading $\frac{W}{S} = 500\ N/m^2$, is gliding at $\left(\frac{L}{D}\right)_{max} = 10$ and $C_L = 0.69$ Considering the free stream density ρ_∞ = 0.9 kg/m³, the equilibrium glide speed (in m/s) is _____.

[GATE 2015, 2 M]

112. For a level flight at cruise altitude, C_D = 0.018 with drag coefficient at zero lift, $C_{D.0}$ = 0.015. For a 30° climb at the same altitude and speed, C_D = _____ $\times 10^{-3}$.

[GATE 2015, 2 M]

113. An aircraft is flying with inertial ground and wind speeds of $\vec{v}_g^{\,b} = (100, 5, 5)\, m/s$ and $\vec{v}_W^{\,b}$ = (0, –5, –10) m/s, respectively, as expressed in the body frame. The corresponding sideslip angle (in degrees) is

(*a*) 0 (*b*) 5.65

(*c*) 8.49 (*d*) 9.54

[GATE 2015, 2 M]

114. The damping ratio in phugoid motion for gliders is usually less compared to powered aircraft because

(*a*) gliders are unpowered.

(*b*) gliders are light.

(*c*) lift to drag ratio is higher for gliders.

(*d*) gliders fly at low speed.

[GATE 2016, 1 M]

115. During an aircraft cruising flight, the altitude above the ground is usually measured using

(*a*) dynamic pressure. (*b*) static pressure.

(*c*) radar. (*d*) laser range finder.

[GATE 2016, 1 M]

116. Indicated airspeed is used by a pilot during

(*a*) take-off.

(*b*) navigation.

(*c*) setting the engine RPM.

(*d*) setting the elevator angle.

[GATE 2016, 1 M]

117. The pitch angle and the angle of attack for a fixed wing aircraft are equal during

(*a*) wings level constant altitude flight.

(*b*) unaccelerated climb.

(*c*) unaccelerated descent.

(*d*) landing.

[GATE 2016, 1 M]

118. The load factor of an aircraft turning at a constant altitudeis 2. The coefficient of lift required for turning flight as compared to level flight at the same speed will be

(*a*) same (*b*) half

(*c*) double (*d*) four times

[GATE 2016, 1 M]

119. The rate of change of moment coefficient with respect to the angle of attack, $\frac{dC_m}{d\alpha}$, at half chord point of a thin airfoil, as per approximations from the thin airfoil theory is

(*a*) $\frac{\pi}{4}$ radian^{-1} (*b*) $\frac{\pi}{2}$ radian^{-1}

(*c*) π radian^{-1} (*d*) 2π radian^{-1}

[GATE 2016, 2 M]

120. An aircraft is flying level in the North direction at a velocity of 55 m/s under cross wind from East to West of 5 m/s. For the given aircraft $C_{n\beta}$ = 0.012/deg and $C_{n\delta r}$ = –0.0072/deg, where dr is the rudder deflection and ß is the side slip angle. The rudder deflection exerted by pilot is ________ degrees.

[GATE 2016, 2 M]

121. An aircraft weighing 10000 N is flying level at 100 m/s and it is powered by a jet engine. The thrust required for level flight is 1000 N. The maximum possible thrust produced by the jet engine is 5000 N. The minimum time required to climb 1000 m, when flight speed is 100 m/s, is ________s.

[GATE 2016, 2 M]

122. The aircraft velocity (m/s) components in body axes are given as [u, v, w] = [100, 10, 10]. The air velocity (m/s), angle of attack (deg) and sideslip angle (deg) in that order are

(*a*) [120, 0.1, 0.1]

(*b*) [100, 0.1, 0.1]

(*c*) [100.995, 0.1, 5.73]

(*d*) [100.995, 5.71, 5.68]

[GATE 2016, 2 M]

123. The Dutch roll motion of the aircraft is described by following relationship

$$\begin{bmatrix} \Delta\dot{\beta} \\ \Delta\dot{r} \end{bmatrix} = \begin{bmatrix} -0.26 & -1 \\ 4.49 & -0.76 \end{bmatrix} \begin{bmatrix} \Delta\beta \\ \Delta r \end{bmatrix}$$

The undamped natural frequency (rad/s) and damping ratio for the Dutch roll motion in that order are:

(*a*) 4.68, 1.02 (*b*) 4.49, 1.02

(*c*) 2.165, 0.235 (*d*) 2.165, 1.02

[GATE 2016, 2 M]

124. A glider weighing 3300 N is flying at 1000 m above sea level. The wing area is 14.1 m^2 and the air density is 1.23 kg/m^3 . Under zero wind conditions, the velocity for maximum range is ________ m/s.

α(deg)	C_L	C_D	$\frac{C_L}{C_D}$
11	1.46	0.0865	16.9
9	1.36	0.0675	20.1
7	1.23	0.0535	22.9
5	1.08	0.0440	24.5
3	0.90	0.0350	25.7
1	0.70	0.0275	25.4
–1	0.49	0.0220	22.2
–3	0.25	0.0180	13.8

[GATE 2016, 2 M]

125. Which one of the following statements is NOT true

(*a*) The pitching moment of any airfoil at any angle of attack is always zero at the center of pressure

(*b*) The pitching moment of any airfoil at any angle of attack is always zero at the aerodynamic center

(*c*) The center of pressure and aerodynamic center coincide for a symmetric airfoil

(*d*) The pitching moment about the aerodynamic center, for any airfoil, does not vary with angle of attack

[GATE 2017, 1 M]

126. Let $\overline{N}_m$ and $\overline{N}_0$ be respectively the non-dimensional locations of the stick-fixed maneuver point and stick-fixed neutral point of a low speed conventional aircraft. These distance are measured with respect to the nose of the fuselage.

The numerical value of $\overline{N}_m - \overline{N}_0$

(*a*) will always be negative

(*b*) will always be positive

(*c*) will always be zero

(*d*) can have any value depending on the location of the center of gravity of the aircraft

[GATE 2017, 1 M]

127. The phenomenon of rudder lock in conventional low speed aircraft is primarily due to

(*a*) large value of directional derivative, $C_{n\beta}$

(*b*) the sidewash due to fuselage on the vertical stabilizer

(*c*) the tendency of rudder to float rapidly at high angles of side-slip

(*d*) the sidewash due to wing on the vertical stabilizer

[GATE 2017, 1 M]

128. An aircraft model was tested in a low speed wind-tunnel (Reynolds number = 2×10^6 based on wing mean chord). The variation of pitching moment coefficient (C_m) with angle of attack (α) for two elevator deflections (δ_e) as recorded during this test is presented below.

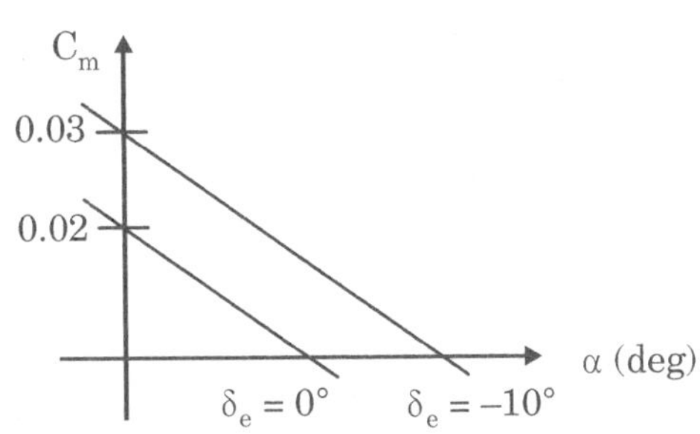

Based on the result presented in the figure above, the value of elevator control power ($C_{m\delta e}$) in per radian will be ________ (in three decimal place).

[GATE 2017, 2 M]

129. A pilot was flying a single engine propeller aircraft and maintaining a steady flight at a lift coefficient, C_L = 0.5 at an altitude of 500 m. Due to some emergency, at the same altitude (500 m), the pilot had to fully deploy the landing gear. If the pilot wants to maintain steady level flights at the same C_L = 0.5 and at the same altitude, which of the following control actions should the pilot undertake:

(*a*) move the elevator up, and decrease the throttle

(*b*) move the elevator up, and increase the throttle

(*c*) move the elevator down, and decrease the throttle

(*d*) move the elevator down, and increase the throttle

[GATE 2017, 2 M]

130. A conventional low speed aircraft had the following aerodynamic characteristics:

C_{D0} - 0.20, e = 1.0 (Oswald efficiency).

The aircraft was flown to maintain a steady level flight and for minimum thrust required, at a lift coefficient of C_L = 0.8. The numerical value of the aspect ratio of the wing is ______ (in three decimal places).

[GATE 2017, 2 M]

131. The roots obtained by solving longitudinal characteristic equations of motion for a statically stable aircraft are given below:

$\lambda_{1.2} = -0.02 \pm 0.30i$, $\lambda_{3.4} = -2.00 \pm 2.50\ i$, where $i = \sqrt{-1}$

The undamped short-period longitudinal natural frequency (radians/sec) and damping ratio, in that order, are close to

(*a*) 3.40, 0.73 (*b*) 3.36, 0.65

(*c*) 3.83, 0.56 (*d*) 3.20, 0.63

[GATE 2017, 2 M]

132. An aircraft is to be designed to ensure that it has enough excess power to achieve steady climb at flight path angle, γ = 10 degrees, maintaining The numerical value of the thrust to weight ratio of the complete aircraft to meet the above requirement under standard atmospheric condition will be _________ (in three decimal place).

[GATE 2017, 2 M]

133. A conventional aircraft was analyzed to estimate the contribution of wing towards C_{m0} (C_m at $\alpha = 0$) of the whole aircraft. The wing was installed with zero setting angle along the fuselage reference line. Further, the wing was laid such that $\overline{X}_{ac,w} = 0.3$, and $\overline{X}_{cg,aircrft} = 0.4$. $\overline{X}_{ac,w}$ and $\overline{X}_{cg,aircraft}$ are the non-dimensional distance, from the leading edge of the wing, of the aerodynamic center of the having and center of gravity of the aircraft respectively. The wing had the following aerodynamic characteristics:

C_{L0} = 0.20, and $C_{mac.wing}$ = –0.02.

The numerical value of $C_{m0,w}$ (contribution of the wing to C_{m0}) about the CG of the aircraft is ______ in (two decimal places).

[GATE 2017, 2 M]

134. A jet aircraft is initially flying steady and level at its maximum endurance condition. For the aircraft to fly steady and level, but faster at the same altitude, the pilot should

(*a*) increase thrust alone.

(*b*) increase thrust and increase angle of attack

(*c*) increase thrust and reduce angle of attack.

(*d*) reduce angle of attack alone.

[GATE 2018, 1 M]

135. The pilot of a conventional airplane that is flying steady and level at some altitude, deflects the port side aileron up and the starboard aileron down. The aircraft will then

(*a*) pitch, nose up.

(*b*) roll with the starboard wing up.

(*c*) pitch, nose down.

(*d*) roll with the port wing up.

[GATE 2018, 1 M]

136. A NACA 0012 airfoil has a trailing edge flap. The airfoil is operating at an angle of attack of 5 degrees with un-deflected flap. If the flap is now deflected by 5 degrees downwards, the C_L versus α curve

(*a*) shifts right and slope increases

(*b*) shifts left and slope increases

(*c*) shifts left and slope stays the same.

(*d*) shifts right and slope stays the same.

[GATE 2018, 1 M]

137. An airplane requires a longer ground roll to lift-off on hot summer days because

(*a*) the thrust is directly proportional to free-stream density.

(*b*) the thrust is directly proportional to weight of the aircraft.

(*c*) the lift-off distance is directly proportional to free-stream density.

(*d*) the runway friction is high on hot summer days.

[GATE 2018, 1 M]

138. The highest limit load factor experienced by a civil transport aircraft is in the range

(*a*) 0.0 – 2.0 (*b*) 2.0 – 5.0

(*c*) 5.0 – 8.0 (*d*) 8.0 – 10.0

[GATE 2018, 1 M]

139. Assuming ISA standard sea level conditions (288.16 K, density of 1.225 kg/m^3, g = 9.81 m/s^2, R = 287 J/(kg-K)), the density (in kg/m^3) of air at Leh, which is at an altitude of 3500 m above mean sea level is ________ (accurate to two decimal places).

[GATE 2018, 2 M]

140. An aircraft with mass of 400,000 kg cruises at 240 m/s at an altitude of 10 km. Its lift to drag ratio at cruise is 15. Assuming g as 9.81 m/s^2, the power (in MW) needed for it to cruise is _______ (accurate to two decimal places).

[GATE 2018, 2 M]

141. A statically-stable aircraft has a $C_{L_\alpha} = 5$ (where the angle of attack, a, is measured in radians). The coefficient of moment of the aircraft about the center of gravity is given as $C_{M,c,g} = 0.05 - 4\alpha$. The mean aerodynamic chord of the aircraft wing is 1 m. The location (positive towards the nose) of the neutral point of the aircraft from the center of gravity is ______ (in m, accurate to two decimal places).

[GATE 2018, 2 M]

142. An aircraft with a gross weight of 2000 kg, has a speed of 130 m/s at sea level, where the conditions are: 1 atmosphere (pressure), 288 K(temperature), and 1.23 kg/m^3 (density). The speed (in m/s) required by the aircraft at an altitude of 9000 m, where the conditions are: 0.31 atmosphere, 230 K, 0.47 kg/m 3, to maintain a steady, level flight is ________ (accurate to two decimal places).

[GATE 2018, 2 M]

143. A pitot probe on an aircraft in a steady, level flight records a pressure of 55,000 N/m^2. The static pressure and density are 45,280 N/m^2 and 0.6 kg/m^3, respectively. The wing area and the lift coefficient are 16 m^2 and 2, respectively. The wing loading (in N/m^2) on this aircraft is ________ (accurate to one decimal place).

[GATE 2018, 2 M]

144. Gross weight of an airplane is 7000 N, wing area is 16 m^2, and the maximum lift coefficient is 2.0. Assuming density at the altitude as 1.23 kg/m^3, the stall speed (in m/s) of the aircraft is ________ (accurate to two decimal places).

[GATE 2018, 2 M]

145. An aircraft wind tunnel model, having a pitch axis mass moment of inertia (I_{yy}) of 0.014 kg-m^2, is mounted in such a manner that it has pure pitching motion about its centre of gravity, where it is supported through a frictionless hinge. If the pitching moment (M) derivative with respect to angle of attack (a), denoted by '$M\alpha$', is –0.504 N-m/rad and the pitching moment (M) derivative with respect to pitch rate (q), denoted by 'M_q', is –0.0336 N-m/(rad/s), the damping ratio of the resulting motion due to an initial disturbance in pitch angle is approximately ________ (accurate to three decimal places).

[GATE 2018, 2 M]

146. When a propeller airplane in ground-roll during take-off experiences headwind, which of the following statements is FALSE?

(a) The drag on the airplane increases.

(b) The thrust from the propellers decreases.

(c) The wing lift increases.

(d) The ground-roll distance increases.

[GATE 2019]

147. Which of the following graphs represents the response of a dynamically unstable airplane?

(a) 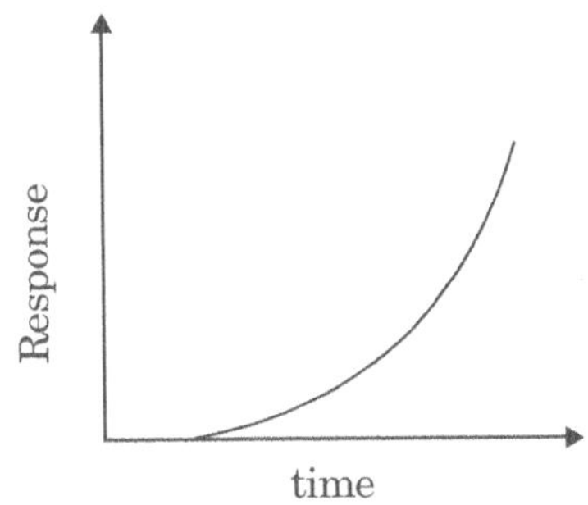

(b) 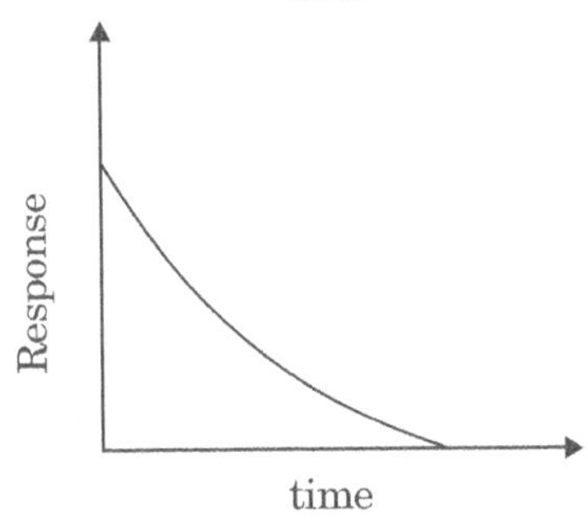

(c) 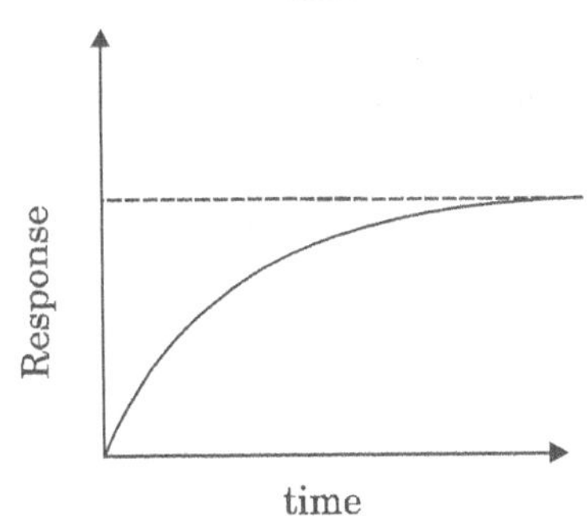

(d) 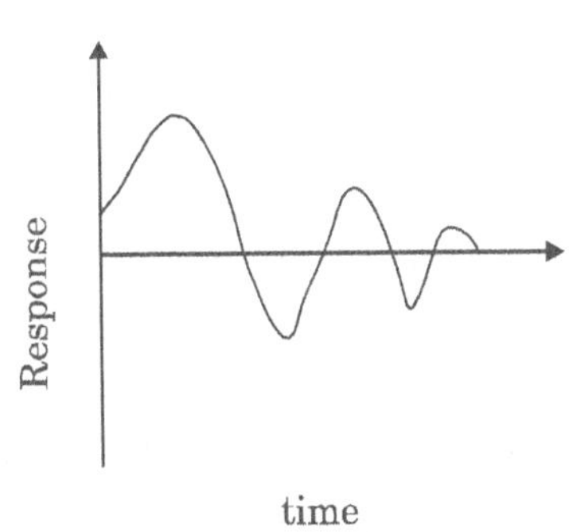

[GATE 2019]

148. The power-off glide range for an airplane with a maximum Lift to Drag ratio of 18, when the glide starts at an altitude of 4 km, is ______ km (round off to the nearest integer).

[GATE 2019]

149. For an airplane flying in a vertical plane, the angle of attack is 3°, the horizontal and vertical components of velocity in wind axis are 300 km/h and 15.72 km/h, respectively.

The pitch attitude of the airplane is ______ degrees (round off to 2 decimal places)

[GATE 2019]

150. An airplane is in steady level flight with a true air speed of 50 m/s. The ambient air density and ambient pressure at the flight altitude are 0.91 kg/m^3 and 7×104 N/m^2, respectively. At sea level, ir density is 1.225 kg/m^3 and ambient pressure is 1.01×10^5 N/m^2. The equivalent or indicated air speed of the airplane is_______ m/s (round off to 2 decimal places).

[GATE 2019]

151. Which of the following statements about adverse yaw of an airplane is/are correct?

P. It is caused by flow separation resulting from large rudder deflection.

Q. It is caused by dissimilar drag forces acting on the two halves of the wing resulting from aileron deflections of same magnitude.

R. It can be eliminated by ensuring that the upward deflection of one aileron is greater than the downward deflection of the opposite aileron.

(a) P only (b) Q only

(c) P and R (d) Q and R

[GATE 2019]

152. The airplane shown in figure starts executing a symmetric pull-up maneuver from steady level attitude with a constant nose-up pitch acceleration of 20 deg/s^2. The vertical load factor measured at this instant at the centre of gravity (CG) is 2. Given that the acceleration due to gravity is 9.81 m/s^2, the vertical load factor measured at point P on the nose of the airplane, which is 2 m ahead of the CG, is _____ (round off to 2 decimal places).

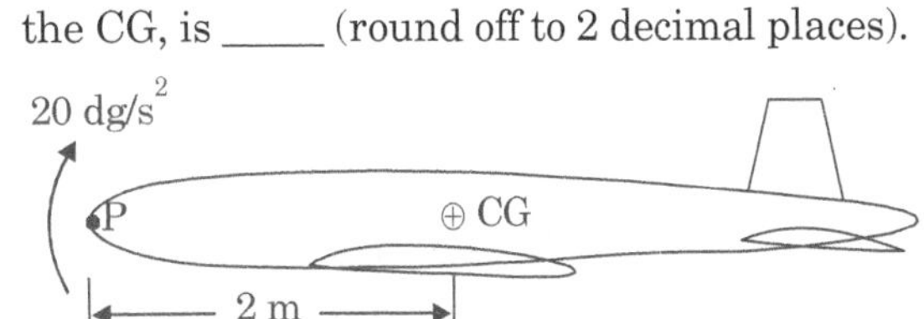

[GATE 2019]

153. Consider an airplane with a weight of 8000 N, wing area of 16 m^2, wing zero-lift drag coefficient of 0.02, Oswald's efficiency factor of 0.8, and wing aspect ratio of 6, in steady level flight with wing lift coefficient of 0.375. Considering the same flight speed and ambient density, the ratio of the induced drag coefficient during steady level flight to that during a 30° climb is _______ (round off to 2 decimal places).

[GATE 2019]

154. A propeller driven airplane has a gross take-off weight of 4905 N with a wing area of 6.84m^2. Assume that the wings are operating at the maximum $\frac{C_L^{\frac{3}{2}}}{C_D}$ of 13, the propeller efficiency is 0.9 and the specific fuel consumption of the engine is 0.76 kg/kW-hr. Given that the density of air at sea level is 1.225 kg/m^3 and the acceleration due to gravity is 9.81m/s^2, the weight of the fuel required for an endurance of 18 hours at sea level is _______N (round off to the nearest integer).

[GATE 2019]

155. The design of an airplane is modified to increase the vertical tail area by 20% and decrease the moment arm from the aerodynamic centre of the vertical tail to the airplane centre of gravity by 20%. Assuming all other factors remain unchanged, the ratio of the modified to the original directional static stability (C_{N_β} due to tail fin) is _______ (round off to 2 decimal places).

[GATE 2019]

156. For conventional airplanes, which one of the following is true regarding roll control derivative $\left(C_{l\delta_r} = \frac{\partial C_l}{\partial \delta_r}\right)$ and yaw control derivative $\left(C_{n\delta_r} = \frac{\partial C_n}{\partial \delta_r}\right)$, where δ_r is rudder deflection?

(*a*) $C_{l\delta_r} > 0$ and $C_{n\delta_r} < 0$ (*b*) $C_{l\delta_r} < 0$ and $C_{n\delta_r} > 0$

(*c*) $C_{l\delta_r} < 0$ and $C_{n\delta_r} < 0$ (*d*) $C_{l\delta_r} > 0$ and $C_{n\delta_r} > 0$

[GATE 2020]

157. Velocity of an airplane in the body fixed axes is given as [100 – 10 20] m/s. The sideslip angle is ________ degress (*round off to two decimal places*).

[GATE 2020]

158. For three different airplanes A, B and C, the yawing moment coefficient (C_n) was measured in a wind-tunnel for three settings of sideslip angle β and tabulated as

	Airplane A	Airplane B	Airplane C
$\beta = -5$ deg	–0.030	–0.025	0.040
$\beta = 0$ deg	0	0	0
$\beta = 5$ deg	0.030	0.025	–0.040

Which one of the following statements is true regarding directional static stability of the airplanes A, B and C?

(*a*) All three airplanes A, B and C are stable.

(*b*) Only airplane C is stable, while both A and B are unstable.

(*c*) Airplane C is unstable, A and B are stable with A being more stable than B.

(*d*) Airplane C is unstable, A and B are both stable with A less stable than B.

[GATE 2020]

159. The eigenvalues for phugoid mode of a general aviation airplane at a stable cruise flight condition at low angle of attack are $\lambda_{1,2} = -0.02 \pm i\,0.25$. If the acceleration due to gravity is 9.8 m/s^2, the equilibrium speed of the airplane is _______ m/s (*round off to two decimal places*).

[GATE 2020]

160. For a general aviation airplane with tail efficiency $\eta = 0.95$, horizontal tail volume ratio $V_H = 0.453$, downwash angle slope $\frac{d\varepsilon}{d\alpha} = 0.35$, wing lift curve slope $C^w_{L\alpha} = 4.8$ rad^{-1}, horizontal tail lift curve slope $C^t_{L\alpha} = 4.4$ rad^{-1}, shift in neutral point location as a percentage of mean aerodynamic chord is ______ (*round off to two decimal places*).

[GATE 2020]

161. A single engine, propeller driven, general aviation airplane is flying in cruise at sea-level condition (density of air at sea-level is 1.225 kg/m^3) with speed to cover maximum range. For drag coefficient $C_D = 0.025 + 0.049\ C_L^2$ and wing loading W/S = 9844 N/m^2, the speed of the airplane is _______ m/s (*round off to one decimal place*).

[GATE 2020]

162. An airplane of mass 4000 kg and wing reference area 25 m^2 flying at sea level has a maximum lift coefficient of 1.65. Assume density of air as 1.225 kg/m^3 and acceleration due to gravity as 9.8 m/s^2. Using a factor of safety of 1.25 to account for additional unsteady lift during a sudden pull-up, the speed at which the airplane reaches a load factor of 3.2 is ________ m/s (*rounde off to two decimal places*).

[GATE 2020]

163. A Pitot tube mounted on the wing tip of an airplane flying at an altitude of 3 km measures a pressure of 0.72 bar, and the outside air temperature is 268.66 K. Take the sea level conditions as, pressure = 1.01 bar, temperature = 288.16 K, and density = 1.225 kg/m^3. The acceleration due to gravity is 9.8 m/s^2 and the gas constant is 287 J/(kg K). Assuming standard atmosphere, the equivalent airspeed for this airplane is ___ m/s (round off to two decimal place).

[GATE 2020]

164. Let V_{TAS} be the true airspeed of an aircraft flying at a certain altitude where the density of air is ρ, and V_{EAS} be the equivalent airspeed. If ρ_0 is the density of air at sea-level, what is the ratio $\frac{V_{TAS}}{V_{EAS}}$ equal to?

(a) $\frac{\rho}{\rho_0}$ (b) $\frac{\rho_0}{\rho}$

(c) $\sqrt{\frac{\rho_0}{\rho}}$ (d) $\sqrt{\frac{\rho}{\rho_0}}$

[GATE 2021]

165. $C_m - \alpha$ variation for a certain aircraft is shown in the figure. Which one of the following statements is true for this aircraft?

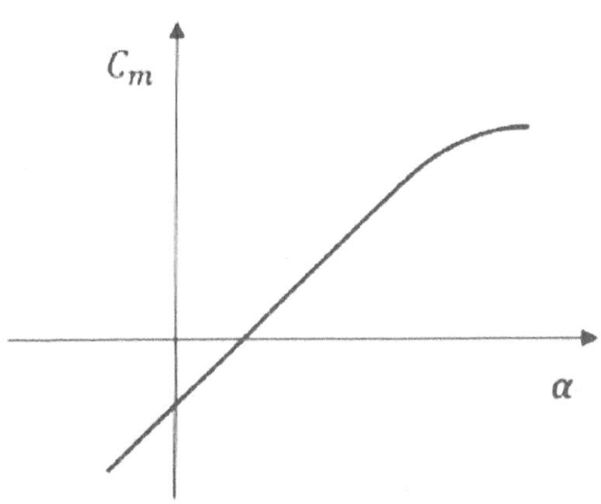

(a) The aircraft can trim at a positive α and it is stable.
(b) The aircraft can trim at a positive α, but it is unstable
(c) The aircraft can trim at a negative α and it is stable.
(d) The aircraft can trim at a negative α, but it is unstable.

[GATE 2021]

166. A jet aircraft has the following specifications: wing loading = 1800 N/m^2, wing area = 30 m^2, drag polar $C_D = 0.02 + 0.04 C_L^2$, and $C_{L,max} = 1.6$. Take density of air at sea level as 1.225 kg/m^3. The speed at which the aircraft achieves maximum endurance in a steady and level flight at sea level is ______ m/s (*round off to two decimal places*).

[GATE 2021]

167. An aircraft with twin jet engines has the following specifications:

Thrust produced (per engine) = 8000 N

Spanwise distance between the two engines = 10 m

Wing area = 50 m^2, Wing span = 10 m

Rudder effectiveness, $C_{n_{\delta r}} = -0.002$ /deg

Density of air at sea level = 1.225 kg/m^3

The rudder deflection, in degrees, required to maintain zero sideslip at 100 m/s in steady and level flight at sea level with a non-functional right engine is ______ (*round off to two decimal places*).

[GATE 2021]

168. For a conventional fixed-wing aircraft in a 360° inverted vertical loop maneuver, what is the load factor (n) at the topmost point of the loop?

Assume the flight to be steady at the topmost point.

(a) n = 1 (b) n < 1
(c) n = – 1 (d) n > – 1

[GATE 2021]

169. For a conventional fixed-wing aircraft, which of the following statements are true?

(a) Making C_{m_α} more negative leads to an increase in the frequency of its short-period mode.

(b) Making C_{m_q} more negative leads to a decreased damping of the short-period mode.

(c) The primary contribution towards C_{l_p} is from the aircraft wing.

(d) Increasing the size of the vertical fin leads to a higher yaw damping,

[GATE 2021]

170. Which of the following statement(s) is/are true?

(a) Service ceiling is higher than absolute ceiling for a piston-propeller aircraft.

(b) For a given aircraft, the stall speed increases with increase in altitude.

(c) Everything else remaining the same, a tailwind increases the range of an aircraft.

(d) For a jet aircraft constrained to fly at constant altitude, there exists an altitude where its range is maximum.

[GATE 2021]

171. A conventional fixed-wing aircraft, with a horizontal tail and vertical fin, in steady and level flight is subjected to small perturbations. Which of the following statement(s) is/are true?

(a) Vertical fin has a stabilizing effect on the lateral stability of the aircraft.

(b) Vertical fin has a destabilizing effect on the directional stability of the aircraft.

(c) Presence of wing anhedral increases the lateral stability of the aircraft.

(d) Horizontal tail has a stabilizing effect on the longitudinal static stability of the aircraft.

[GATE 2021]

172. An airplane weighing 5500 kg is in a steady level flight with a speed of 225 m/s. The pilot initiates a steady pull-up maneuver with a radius of curvature of 775 m. The location of center of gravity (CG), center of pressure on wing (CP) and point of action (T) of tail force are marked in the figure. Use g = 9.81 m/s^2. Neglect drag on the tail and assume that tail force is vertical. Assuming the engine thrust and drag to be equal, opposite and collinear, the tail force is ______ kN (*round off to one decimal place*).

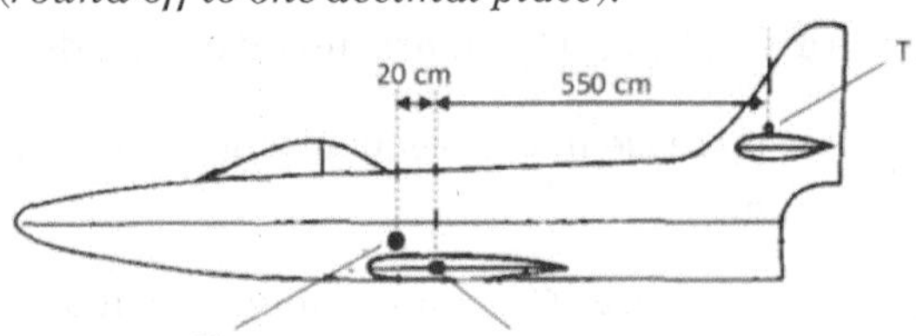

[GATE 2021]

173. A jet aircraft weighing 10,000 kg has an elliptic wing with a span of 10 m and area 30 m^2. The C_{D0}, for the aircraft is 0.025. The maximum speed of the aircraft in steady and level flight at sea level is 100 m/s. The density of air at sea level is 1.225 kg/m^3, and take g = 10 m/s^2. The maximum thrust developed by the engine at sea level is ______ N (*round off to two decimal places*).

[GATE 2021]

174. Consider a jet transport airplane with the following specifications:

Lift curve slope for wing-body $\dfrac{\partial C_{L_{wb}}}{\partial \alpha_{wb}} = 0.1$ /deg

Lift curve slope for tail $\dfrac{\partial C_{L_t}}{\partial \alpha_t} = 0.068$ /deg

Tail area $S_t = 80$ m^2

Wing area S = 350 m^2

Distance between mean aerodynamic centers of tail and wing-body

$\bar{l}_t = 28$ m

Mean aerodynamic chord $\bar{c} = 9m$

Downwash $\varepsilon = 0.4\alpha$

Axial location of the wing-body mean aerodynamic center $x_{cg} / \bar{c} = 0.25$

Axial location of the center of gravity $x_{cg} / \bar{c} = 0.3$

All axial locations are with respect to the leading edge of the root chord and along the body x-axis. Ignore propulsive effects.

The pitching-moment-coefficient curve slope $(C_{m\alpha})$ is ______ /deg (*round off to three decimal places*).

[GATE 2021]

ANSWER KEY

1. (a) **2.** (d) **3.** (c) **4.** *(b)* **5.** *(c)* **7.** *(*)* **8.** *(c)* **9.** *(a)* **10.** *(c)* **11.** *(c)*

12. *(a)* **13.** *(c)* **14.** *(a)* **15.** *(c)* **16.** *(a)* **17.** *(b)* **18.** *(a)* **20.** *(c)* **22.** *(d)* **23.** *(c)*

24. *(c)* **25.** *(a)* **26.** *(a)* **27.** *(d)* **31.** *(b)* **33.** *(a)* **34.** *(b)* **35.** *(b)* **36.** *(a)* **37.** *(c)*

38. *(a)* **39.** *(b)* **40.** *(a)* **41.** *(a)* **42.** *(c)* **43.** *(b)* **44.** *(a)* **45.** *(b)* **46.** *(a)* **48.** *(d)*

50. *(d)* **51.** *(a)* **52.** *(c)* **53.** *(c)* **54.** *(c)* **55.** *(b)* **56.** *(b)* **57.** *(c)* **58.** *(c)* **59.** *(d)*

60. *(c)* **61.** *(b)* **62.** *(b)* **63.** *(c)* **64.** *(a)* **65.** *(d)* **66.** *(a)* **67.** *(c)* **68.** *(b)* **69.** *(a)*

70. *(c)* **71.** *(b)* **72.** *(d)* **73.** *(d)* **74.** *(b)* **75.** *(c)* **76.** *(c)* **77.** *(b)* **79.** *(c)* **80.** *(d)*

81. *(d)* **82.** *(a)* **83.** *(d)* **84.** *(d)* **85.** *(b)* **86.** *(c)* **87.** *(a)* **88.** *(b)* **92.** *(b)* **93.** *(c)*

94. *(a)* **95.** *(d)* **96.** *(d)* **97.** *(b)* **98.** *(d)* **100.** *(b)* **101.** (c) **102.** (b) **103.** *(a)* **104.** (b)

105. *(d)* **106.** (b) **109.** (a) **110.** (a) **113.** (b) **114.** (c) **115.** (b) **116.** (a) **117.** (a) **118.** (c)

119. (b) **122.** (d) **123.** (c) **125.** (b) **126.** (b) **127.** (c) **129.** (b) **134.** (c) **135.** (b) **136.** (c)

137. (a) **138.** (b) **146.** (d) **147.** (a) **148.** (72) **149.** (6°) **150.** (43.09) **151.** (d) **152.** (2.021) **153.** (1.33)

154. (1477.27 N) **155.** (0.96) **156.** *(a)* **157.** (–5.62 to –5.57) **158.** (c) **159.** (55.20 to 55.33)

160. (25.60 to 25.70) **161.** (149.0 to 151.0) **162.** (62.95 to 63.08) **163.** (57.10 to 60.00) **164.** (c) **165.** (b)

166. (64.30 to 64.60) **167.** (6.50 to 6.60) **168.** (d) **169.** (a; c; d) **170.** (b; c; d)

171. (a; d) **172.** (14.8 to 15.2) **173.** (9787 to 9795) **174.** (–0.025 to –0.023)

EXPLANATIONS

2. An Aeroplane having larger wing will have greater lift force and so leads to greater pitching moment with respect to smaller wing. Hence is case of larger wing the longitudinal statice stability will be more if centre of gravity is ahead to Aerodynmaic centre of wing.

3. High wing or low wing configuration doesnot affect the longitudinal static stability, because the pitching moment generated in both the cases are same lateral static stability is affected by high wing or low wing configuration, lateral stability is higher for high wing configuration.

lateral stability (Roll stability) is higher.

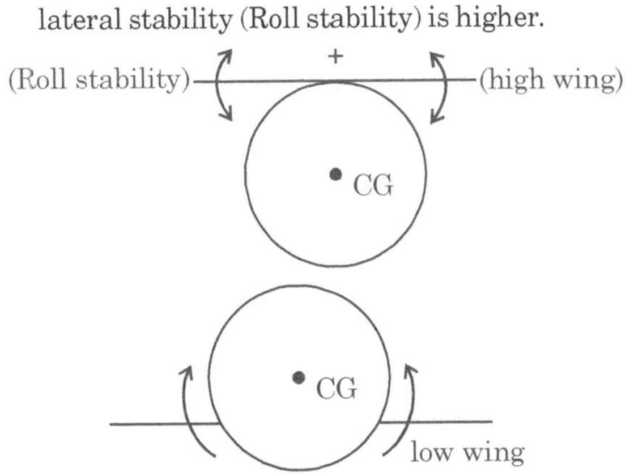

4. N_p = Normal force

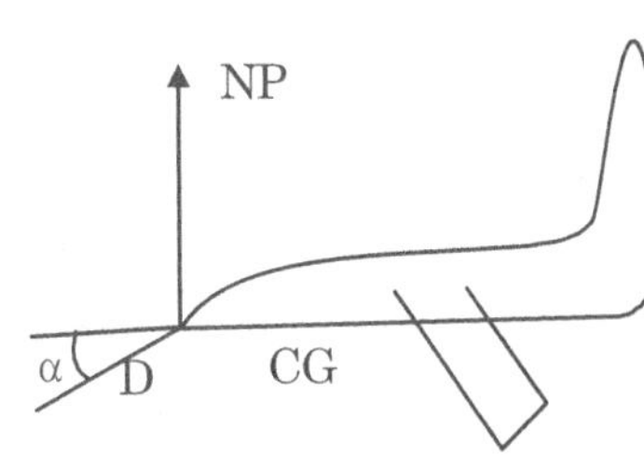

For a fixed C.G aircraft when propeller is mounted on nose of fuselage, a normal force in generated when propeller is at an angle α with the (free stream) velocity, it creats pitching moments which has destabilizing effect when propeller is ahead to centre of gravity.

At the angle of attack increases the moment created by N_p increases which causes more destabilizing effect.

5.

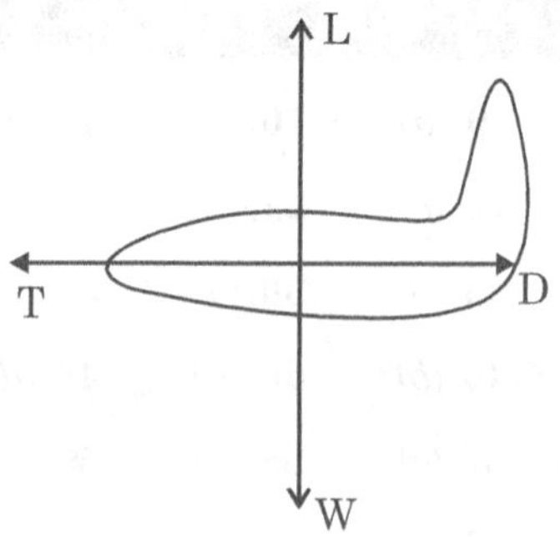

At trimmed condition

$$\boxed{\begin{matrix} T = D \\ L = W \end{matrix}}$$

$L = W$

$$C_L \times \frac{1}{2}\rho A V^2 = W$$

$$V^2 = \frac{2W}{C_L \times \rho \times A}$$

$$V = \sqrt{\frac{d^2 W}{C_L \times \rho \times A}}$$

$$V = \sqrt{\frac{2W}{C_L \times e \times A}}$$

$$V \times \frac{1}{\sqrt{C_L}}$$

for increasing velocity C_L should decrease. C_L can be decreased by using elevator setting.

In level steady flight, higher velocity could be achieved by throttle settings also.

6. Wing A

Area = b × C

$$L_A = C_L \times \frac{1}{2}\rho A V^2$$

$$= C_L \times \frac{1}{2}\rho (bc) V^2$$

Wing B

Area = 4b × C

$$L_B = C_L \times \frac{1}{2}\rho A V^2$$

$$= C_L \times \frac{1}{2}\rho \times (4b \times c) V^2$$

$L_B > L_A$.

Wing B will generate more lift because of more wing area.

List Coefficient C_L is geometrical property of wing and remains same for same angle of attack.

7.

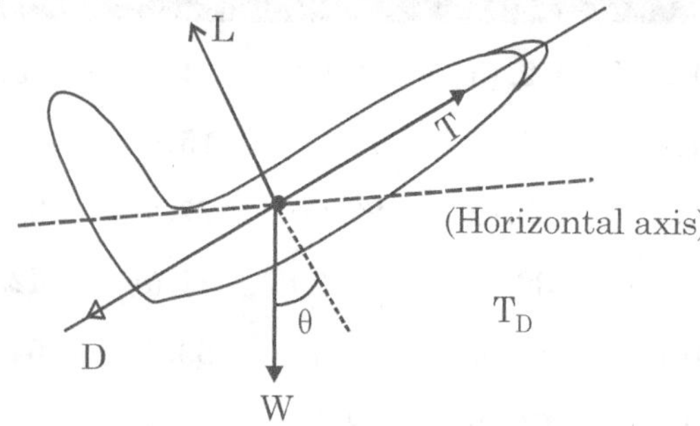

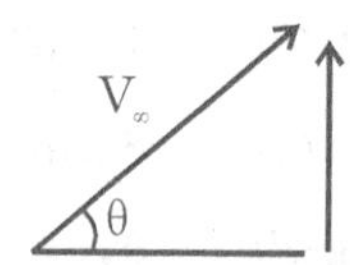

$$\frac{R}{C} = V_\infty \sin\theta$$

Summing force parallel to flight path
Perpendicular to flight path

$T = D + W \sin\theta$ ...(1)

$L = W \cos\theta$ ....(2)

Multiplying eqn (1) and V_∞, we get

$$TV_\infty = DV_\infty + WV_\infty \sin\theta$$

$$\frac{TV_\infty = DV_\infty}{W} = V_\infty \sin\theta$$

$V_\infty \sin\theta$ is Airplane vertical velocity the vertical velocity is called rate of climb $\left(\frac{R}{C}\right)$.

$$\left(\frac{R}{C}\right) = \frac{(TV_\infty - DV_\infty)}{W}$$

$$= \frac{P_A - P_R}{W}$$

$$= \frac{\text{excess power}}{\text{weight}}$$

for a propeller driver airplane

$$\left(\frac{R}{C}\right)_{max} = \frac{(P_A - P_R)n_p}{W}$$

$$\left(\frac{R}{C}\right)_{max} = \frac{(\text{excess break power})\eta_p}{W}$$

Given

$n_p = 0.8$

$W = 73108$ N

$$\left(\frac{R}{C}\right)_{max} = \frac{15m}{sec} \text{ at } v = 15 \text{ m/sec}$$

$$\text{excess break power} = \frac{\left(\frac{R}{C}\right)_{max} \times W}{\eta_p}$$

$$73108 \times 15 \times \frac{1}{0.8}$$

$= 1371KW$

8. Given

δ_e	C_{m_0}
$-5°$	0.08
$+5°$	0

$$\frac{\partial C_m}{\partial \delta_e} = \frac{\Delta Cm}{\Delta \delta_e}$$

$$= \frac{0.08}{-5-5} = \frac{0.08}{-10} = -0.008 \text{ per degree}$$

9. Roots of lateral directional characteristics equation of an airplane gave following set of roots

$\lambda_1 = -0.6$, $\lambda_2 = -0.021$ $\lambda_{3,4} = -0.06 \pm i\,1.5$

$\lambda_1 = -0.6$ (heavily damped, roll mode)

$\lambda_2 = -0.021$ (lightly damped, spiral mode)

$\lambda_{3,4} = -0.06 \pm i\,1.5$ (oscillatony motion, heavily damped dutch roll mode)

$\lambda_{3,4} = -0.06 \pm i\,1.5$

$= \xi\omega_n \pm \omega_n\sqrt{1-\xi^2}$

$\xi\omega_n = 0.06$ ....(1)

$\omega_n = \sqrt{1-\xi^2} = 1.5$ ....(2)

dividing equations (2), by (1) we get

$$\frac{\sqrt{1-\xi^2}}{\xi} = \frac{1.5}{0.06}$$

$3 = 0.04$

10. $\xi = 10$ Km

$T_a = 223$ K

$\rho_s = 0.4/3$ kg/m²

$\rho_s = 1.225$

$R = 288$ J/Kg/K

$V_i = 60$m/sec

Relation between indicated and true airspeed

$$V_i = V_{True}\sqrt{\frac{\rho_a}{\rho_5}}$$

$$V_{True} = V_i\sqrt{\frac{\rho_5}{\rho_a}}$$

$$= 60\sqrt{\frac{1.225}{0.413}}$$

$= 103.33$ m/sec

$\boxed{V_{True} = 103 \text{ m/sec}}$

stagnantion pressure

$$P_{oa} = P_a + \frac{1}{2}\rho_a V_{True}^2$$

$$= ROH_a RT_a + \frac{1}{2} \times ROH_a V_{True}^2$$

$$= 0.413 \times 287 \times 233 + \frac{1}{2} \times 0.43 \times 103^2$$

$$= 2.98 \times 10^4 \text{ N/m}^2$$

11.

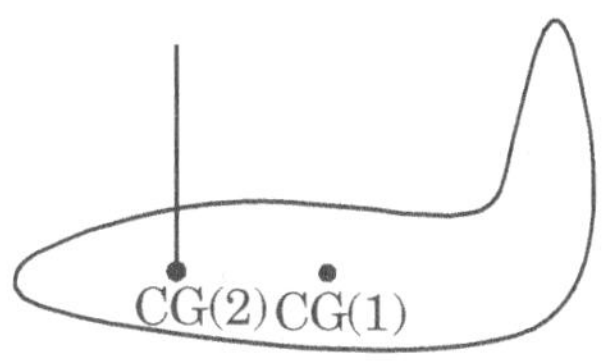

$C_{L_{max}}$ of the wing doesnot depend upon the location of CG of Airplane. Hence it will remains same.

12. $\frac{\partial C_m}{\partial \alpha}$ is one parameter which measure the longitudinal static stability. It is affected by tail moment are doubling tail moment arm will increase the pitching moment due to tail for its double but it will happen for every α.

$$\left(\frac{\partial C_m}{\partial \alpha}\right)_{original} = \frac{C_{m_1} - C_{m_2}}{4\alpha}$$

$$\left(\frac{\partial C_m}{\partial \alpha}\right)_{current} = \frac{2C_{m_1} - 2C_{m_2}}{4\alpha}$$

2 times of original

$\left(\frac{\partial C_m}{\partial \alpha}\right)$ will double.

13.

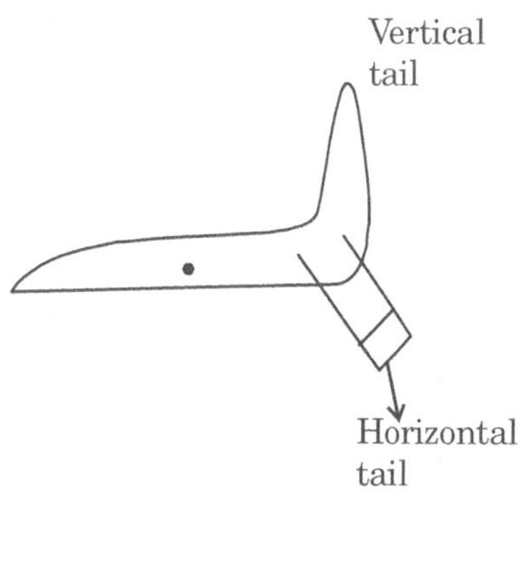

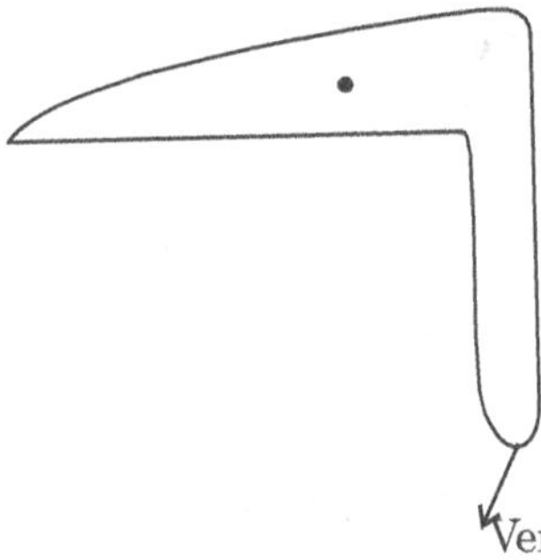

The purpose of vertical tail is to provide directional stability inverting it doesnot effect the roll stability roll stability of the airplane can be achieved by differential deflection of small flats on wing called ailerons. Which is located outward of the wings or by the use of spoilers.

14. If horizontal tail area is increased stabilities of aircraft will more and when the elevator to horizontal area ratio kept same it means elevator area will increase to kept the ratio of areas to be same. So more moment in generated for same elevator deflection. Hence, longitudinal static stability will increase.

Elevator control power is defied as $\frac{\partial C_m}{\partial \delta_e}$ or

$(C_{m\delta_e})$ due to more elevator area ∂C_m is more, hence elevator control power also increased.

15. In this problem the designer as to prefer to use the wing for which the stall begins at the root in the given option R in correct because wing R shows drop of the lift at near to root.

Stalling → stalling of wing in due to break down of flow because of that sudden drop of lift coefficient.

17. $\bar{X}_{CG} = 0.3$

$$\frac{dC_m}{dC_L} = -0.1 = -SM$$

Static stability Margin (SM) of Mean Aerodyhamic chord = 5% – 15%

= 0.05C – 0.15 C

We have

$$\frac{dC_m}{dC_L} = x_n - x_{cg}$$

$$x_n = \frac{dC_m}{dC_L} + x_{cg}$$

= 0.3 + 0.1 = 0.4

$$\boxed{x_n = 0.4}$$

For most forward location of centre of gravity (CG)

$x_n - x_{cg} = 0.15$

$x_{cg} = (x_n - 0.15)$

$= (0.4 - 0.15)$

$$\boxed{x_{cg} = 0.25C}$$

18. For most aft location of centre of gravity

$x_n - x_{cg} = 0.05$

$x_{cg} = x_n - 0.05$

$$\boxed{x_{cg} = 0.35C}$$

19. For $\frac{d\delta_e}{dC_L} = 0$

$\frac{d\delta_e}{dC_L}$ can be written as

$$\frac{d\delta_e}{dC_m} \times \frac{dC_m}{dC_L}$$

$$\frac{d\delta_e}{dC_L} \times \frac{1}{C_{m\delta_e}}$$

Were $C_{m\delta_e}$ is eleuator control power which will not zero.

$$\therefore \frac{dC_m}{dC_L} = 0$$

$$\bar{X}_n - \bar{X}_{cg} = 0$$

$$\bar{X}_{cg} = \bar{X}_n$$

$$\boxed{\bar{X}_{cg} = 0.4C}$$

20. Given data

W = 20000 N

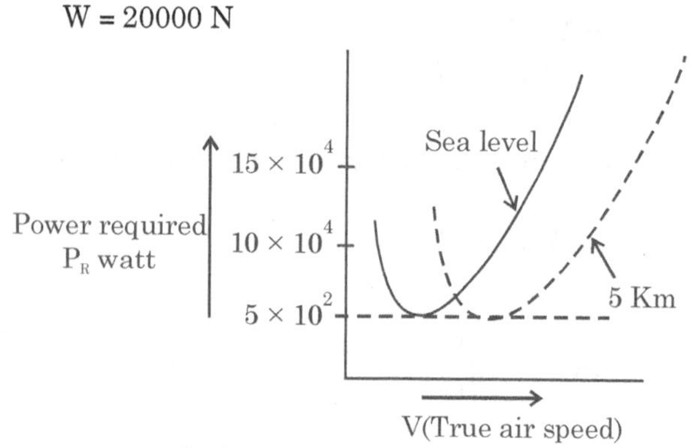

$\rho_a = 0.74$ kg/m³

$\rho_s = 1.225$ kg/m³

$$\left(\frac{R}{C}\right)_{\text{sea level}} = 4\text{m/sec}$$

Maximum rate of climb $\left(\frac{R}{C}\right)_{max}$.

$$= \frac{\text{Maximum excess power}}{\text{Weight}}$$

$$\left(\frac{R}{C}\right)_{max} = \frac{P_{A_0} - P_{Ro}}{W}$$

$$4 = \frac{P_{A_0} - 5 \times 10^4}{2 \times 10^4}$$

$$P_{a_0} = 4 \times 2 \times 10^4 + 5 \times 10^4$$

$$P_{a_0} = 13 \times 10^4 W$$

At 5 Km altitude

$$\frac{P_{A_a}}{P_{A_s}} = \frac{\rho_a}{\rho_s}$$

$$P_{A_a} = P_{A_s} \times \frac{\rho_a}{\rho_s}$$

$$P_a = 13 \times 10^4 \times \frac{0.74}{1.225}$$

$$\boxed{P_{Aa} = 7.853 \times 10^4 W}$$

At Altitude of 5 Km

$$\left(R/C\right)_{max} = \frac{(P_A - P_R)}{W}$$

$$= \frac{7.853 \times 10^4 - 5 \times 10^4}{2 \times 10^4}$$

$$\left(R/C\right)_{max} = 1.43\text{m/sec}$$

21. None of these

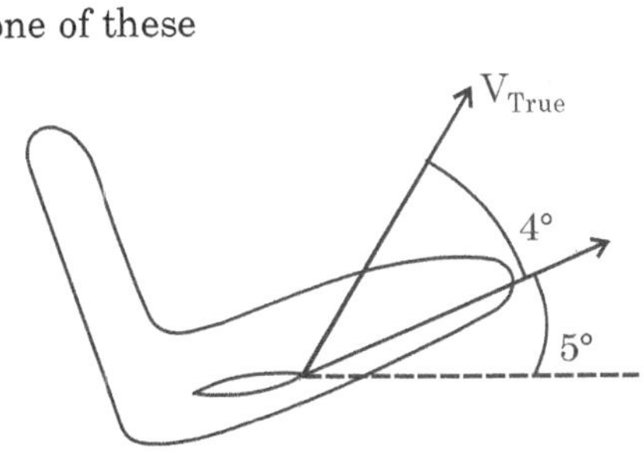

$$\left(R/C\right)_{max} = 1.43$$

$$V_{True} \sin\theta = 1.43$$

$$V_T = \frac{1.43}{\sin 5^0}$$

= 16.41 m/sec

Now relation between true airspeed and efficient airspeed is given by

$$V_{eq} = \sqrt{\frac{\rho_a}{\rho_s}} V_T$$

$$= \sqrt{\frac{0.74}{1.225}} \times 16.41$$

$$\boxed{V_{eq} = 12.75\text{m/sec}}$$

22. The service ceiling of the transport aircraft is defined as the altitude at which maximum rate of climb is 100 ft/min or 0.508 m/sec.

23. The drag of an aircraft in steady climbing flight at a given forward speed is smaller than drag is steady level light at same forward speed.

In steady level flight (flight with no side slip)

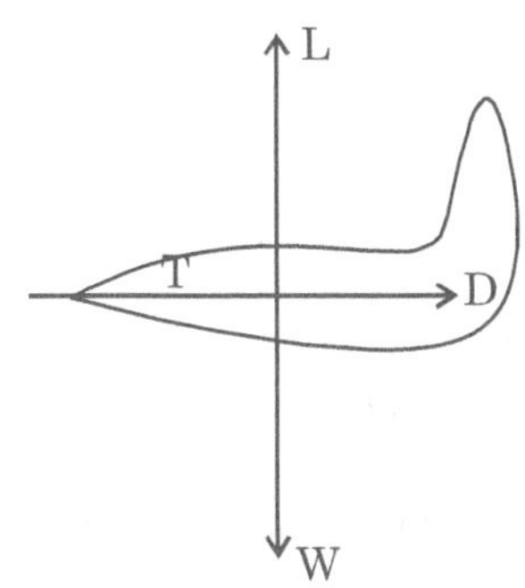

$$\boxed{\begin{array}{l} L = W \\ T = D \end{array}}$$

In Steady climbing flight

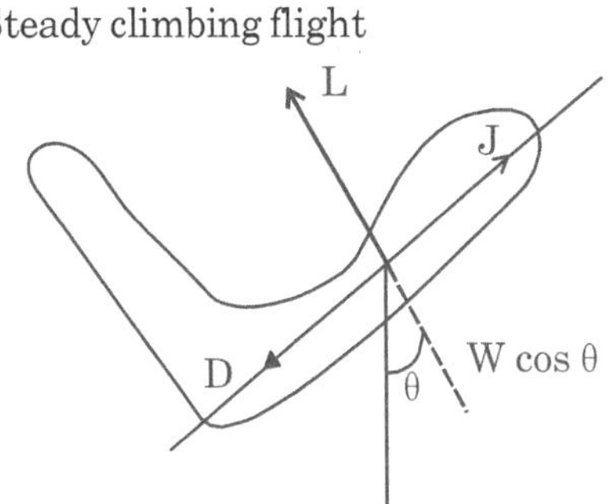

Summing forces parallel and perpendicular to flight path

$T = D + W \sin\theta$ ....(1)

$L = W \cos\theta$ ....(2)

From equation (1) and (2) we get

$$\boxed{\begin{aligned} D &= T - W\sin\theta \\ L &= W\cos\theta \end{aligned}}$$

$(D)_{climb} < (D)_{level}$

$(L)_{climb} < (L)_{level}$

Hence, the lift and drag in steady climbing flight is less than the steady level flight.

24. Level turn manuvers

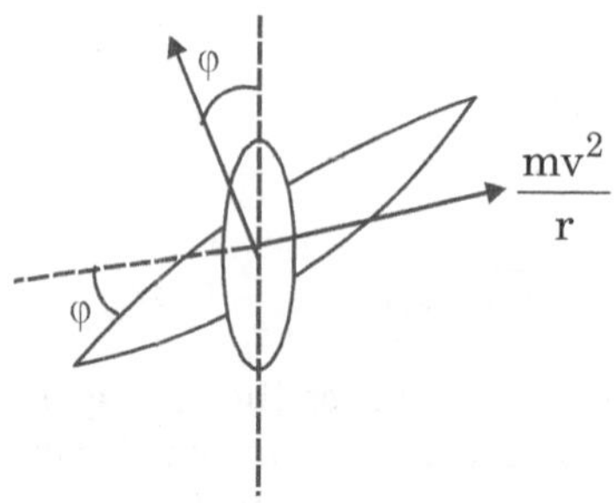

r = radius of turn

φ = roll angle

resolving forces along y and α-axis we get

$W = L\cos\varphi$ ....(1)

$\frac{mv^2}{r} = L\sin\varphi$ ....(2)

from equation (1)

$$\cos\varphi = \frac{1}{L/W} = \frac{1}{\eta} = \frac{W}{L}$$

Horizontal component of lift force = L sin φ.

$$\therefore \frac{L_{Horizontal}}{W} = \frac{L\sin\varphi}{W}$$

$$= n\sqrt{1-\cos^2\varphi} = n\sqrt{1-\frac{1}{n^2}}$$

$$= \frac{n\sqrt{n^2-1}}{n} = \sqrt{n^2-1}$$

$$\boxed{\frac{L_{Horizontal}}{W} = \sqrt{n^2-1}}$$

25. During cruise climb weight of aircraft decrease due to fuel consumptions and during climb altitude increases bases the equivalent airspeed and lift coefficient remains constant.

26. Moment coefficient for aircraft can be expressed as

$$C_m = C_{m_0} + C_{m_\alpha + C_{m\delta e}}\,\delta_e$$

[for α and δ_e]

$$\Rightarrow C_{m_\alpha} = C_{m_\alpha}\left(x_{cg} - x_{ac}\right) - \eta_T V_T C_{m_\alpha T}$$

$$\left(1-\frac{\partial\xi}{\partial\alpha}\right)\left[0 + C_{m_\alpha}\right]_{fuse}$$

C_{m_α} depends apon the centre of gravity location. In the given $C_m - \alpha$ plot the P and Q have same C_m VS α slope. They cores ponds to same centre of gravity location. P and Q have different intercepts on the y-axis so corresponds to different elevator deflection. For Q and R as shown is figure have different slope so different centre of gravity location

27.

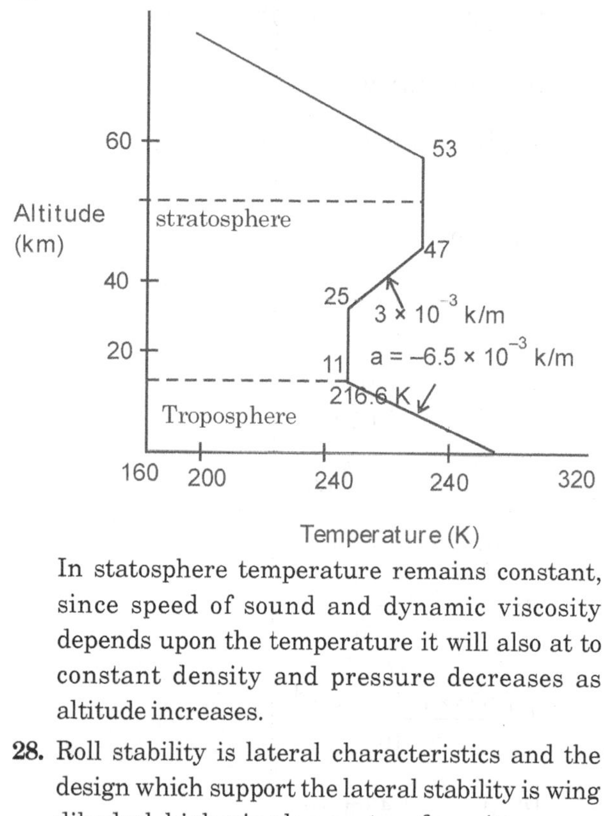

In statosphere temperature remains constant, since speed of sound and dynamic viscosity depends upon the temperature it will also at to constant density and pressure decreases as altitude increases.

28. Roll stability is lateral characteristics and the design which support the lateral stability is wing dihedral, high wing low centre of gravity.

29. V_{stall} = 60 m/All

30. V_{stall} = 60 m/sec.

In case of level flight

L = W, n = 1

$$W = C_L \times \frac{1}{2} e\, A\, V^2 \quad ...(1)$$

When

n = 9

$$\frac{L}{W} = 9$$

$$L = 9 \times C_L \times \frac{1}{2} eAV_{stall}^2 \quad ...(2)$$

[From equation (1)]

$$\Rightarrow 9 \times C_L \times \frac{1}{2} \rho AV_{stall}^2$$

$$\Rightarrow 9 \times C_L \times \frac{1}{2} eAV_{stall}^2$$

$V_{design} = 3V_{stall} = 3 \times 60 = 180$ m/sec

$= 3 \times 60$

31. Short period mode - speed remains constant
Wing rock - Roll oscillation
phugoid mode - Angle of attack remains constant
Dutch roll - Couple roll-yaw oscillation

32. $V_{True} = 100$ m/sec
$e_a = 0.526$ Kg/m³.
$V_{eq} = ?$

$$V_{eq} = \sqrt{\frac{\rho_a}{\rho_s}} \times V_T = \sqrt{\frac{0.526}{1.225}} \times 100$$

$= 65.5$ m/sec

33.

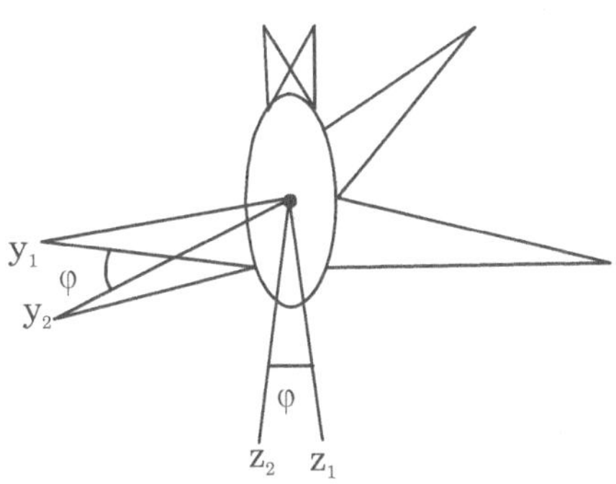

the orientation of the aircraft relative to the earth can be described in terms of what called eulers angles. Three eulers angle φ, θ and ψ are respectively called back angle elevation angle azimeeth angle or heading these three angles are sometimes incorrectly called as roll. Pitch and yaw.

Ψ about z-axis. θ about Y-axis, φ about x-axis

36. When Trailing edge of the elevator deflects upwards from the initial position it causes decrease in C_L. To lower the trim speed the elevator should deflects downward from the initial position to increase the C_L. Hence

$$\frac{d\delta_e}{dC_L} < 0 \; C_{L\delta_e} < 0$$

⇒ Aircraft is was unstable.

37. When angle of attack is low, rolling motion generate yawing and in turn rolling moment, pitching motion generates onlypitching moment due to symmetry, yawing generate yawing as well as rolling.

39. $C_{DO} = 0.0223$
AR = 10
e = 0.7

$$K = \frac{1}{\pi e AR}$$

$$= \frac{1}{3.14 \times 0.7 \times 10}$$

for minimum thrust required

$$C_{DO} = C_{Di}$$

$$= KC_L^2$$

$$C_L = \sqrt{\frac{C_{DO}}{K}}$$

$$= \sqrt{\frac{0.0223}{0.0955}} = 0.7$$

$$\boxed{C_L = 0.7}$$

gliding angle

$$\tan\theta = \frac{1}{\frac{C_L}{C_D}}$$

$$C_{DO} = C_{DO} + KC_L^2$$

$= 2C_{DO}$
$= 2 \times 0.0223 = 0.0446$
As $[C_{DO} = C_{Di})$

For maximum range glider should fly at $\left(\frac{L}{D}\right)_{max}$

$$\tan\theta = \frac{1}{\left(\frac{C_L}{C_D}\right)_m} = \frac{1}{\frac{0.7}{0.0446}}$$

$\theta = \tan^{-1}(0.089)$

$$\boxed{\theta = 3.6°}$$

41. Conventional Altimeter is a pressure transducer. Which is used to measure the static pressure. Static pressure measured is used for obtaining the altitude from international standard atmosphere.

43. A wing placed low on fuselage with dihedral angle positive gives highest resting roll moment against sideslip.

44. $\frac{\partial C_{m}, CG}{\partial \alpha}$ is stability derivative. Hence stability for aircraft.

For longitudinal static stability of the aircraft

$$\frac{\partial C_{m}, CG}{\partial \alpha} < 0, C_{m\alpha} < 0$$

for longitudinally balance

$C_{mo} > 0$

It is associated wit α so, it is stiffness is pitch.

45. Given V_{∞} = 100 m/sec

pitiching rate (q) = 0.2 deg/sec

L = 4 m

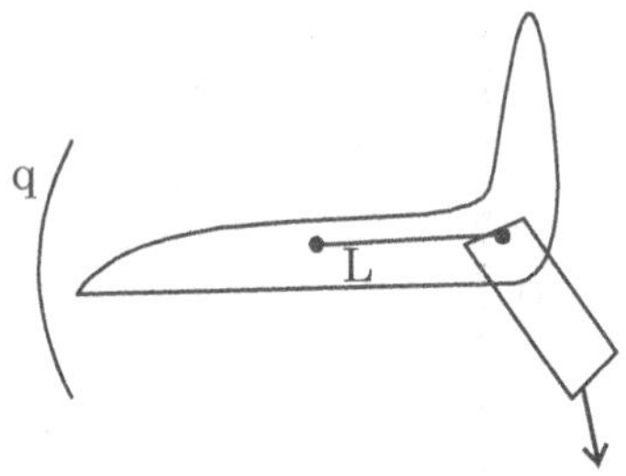

change is angle of attck (Δα) = ?

change in angle of attack

$$(\Delta\alpha) = \left(\frac{qL}{V_{\infty}}\right) = \frac{0.2 \times 9}{100}$$

Δα = 0.008 degree

46. $C_{m.tail} = 0.2 - 0.0215\alpha$

$$\frac{\partial C_{m}}{\partial \alpha} = -0.0215$$

$C_{m\alpha} < 0$

Hence, Aircraft is longitudinal stable.

47. $$\frac{d\bar{x}}{dt} = \begin{bmatrix} [A] & [B] \\ [C] & [D] \end{bmatrix} \vec{\alpha}$$

$$\bar{X} = [u\ w\ q\ \theta\ v\ p\ r\ Q]^{T}$$

separating longitudinal and lateral and directional variables we get

(u, w, q, θ) = longitudinal variables

(v, p, r. Q) = lateral/directional variables these are independent from each other it follows

[B] = [C] = 0

[B] dependent of $\dot{v}$, w, q, θ or v, p, r, φ and [c] dependent on $\dot{v}$, $\dot{r}$, $\dot{q}$ or u, w, q, θ

48. From defiation of sideslip

V = V sin β

w = V cos β sin α

u = V cos β cos α.

49. Rate of climb

$$\left(R/C\right) = \frac{P_A - P_R}{w}$$

$$= \frac{\text{available excess paths power}}{\text{weight}}$$

$$= \frac{1 \times 10^{6}}{2500 \times 9.8} = 40.81 \text{ m/sec}$$

closed option in (L).

Common sets for questions 51, 52

Roots of longitudinal dynamics

$\lambda_1 = -0.02 + 0.2i$

$\lambda_2 = -0.02 - 0.2i$

$\lambda_3 = -2.5 - 2.6\,i$

$\lambda_4 = -2.5 - 2.6i$

50. By definition we know that phugoid mode has low frequency and low damping, peak the roots corresponds to phugoid mode is

–0.02 ± 0.2i

51. Short period mode is highly damped hence the pair of eigen values corresponds to this mode is – 2.5 ± 2.6i

$$\lambda_{3,4} = -2.5 \pm 2.6i$$

$$= -\xi\omega_n \pm i\omega_n\sqrt{1-\xi^2}$$

equating real and imaginary part

$-\xi\omega_n = -2.5$

$\xi\omega_n = 2.5$ ...(1)

$\omega_n\sqrt{1-\xi^2} = 2.6$ ...(2)

ω_d = 2.6 rad/sec

Dividing equation (2) by equation (1), we get

$$\frac{\omega_n\sqrt{1-\xi^2}}{\xi\omega_n} = \frac{2.6}{2.5}$$

$$\frac{\sqrt{1-\xi^2}}{\xi} = 1.04$$

$\xi^2 = 0.98$

$\boxed{\xi = 0.69}$

damping ratio = 0.69

52.

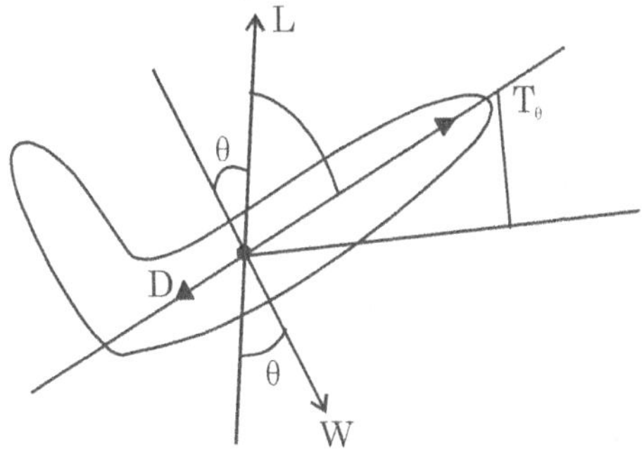

summation of forces parallel to flight path and perpendicular to flight path = 0

from L = W cos θ ...(1)

T = D + W sin θ ...(2)

$\frac{L}{W} = \cos\theta$ $\frac{L}{W} = n = \text{load factor}$

n = cos θ.

the value of cos θ.

maximum value of 1

for step turn θ = 90°.

n = 0

maximum of cos θ = 1

0 < n < 1

53. Lift off distance or take off distance in given by

$$S_{L_o} = \frac{1.44\ w^2}{g\ e_\infty S\ C_{L\,max} T}$$

[ignoring drag OR compound to T]

where T α ρ_∞.

taking other parameter constant te relation between S_{Lo} and ω is

S_{Lo} α w^2

where w is increased by 50%

$(S_{L_o})_{new}\ \alpha\,(w + .3w)2$

$\alpha\,(1.3w)^2$

= K 1.69 w²

K = proposonality constant.

% increase is take off distance

$$= \frac{1.69w^2 - w^2}{w^2}$$

$$= \frac{.69w^2}{w^2} = .69$$

= 69%

closest value is 70%

54. Bank angle = 60°

V_{Stall} = 60 m/sec

(straight level flight)

V_{stall} = ?(at 60° bank)

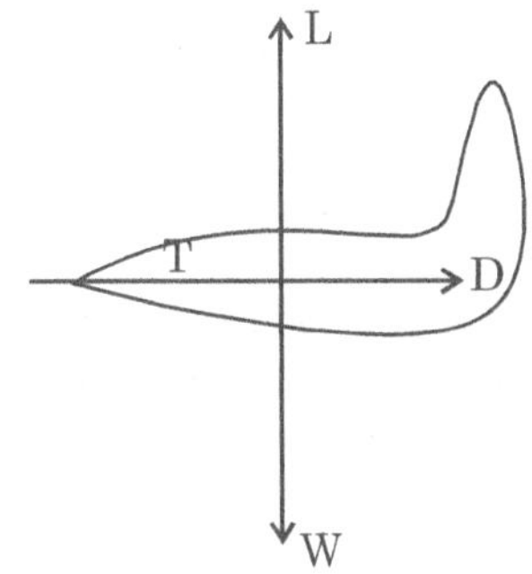

straight and level flight

T = D ...(1)

L = W ...(2)

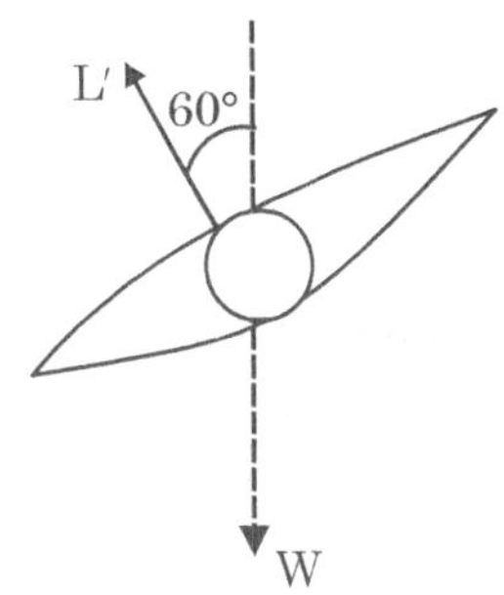

for 60° bank flight

L cos φ = w ...(3)

from equation (2)

$$V = \sqrt{\frac{2\omega}{\rho C_L S}}$$

$$V_{Stall/level} = \sqrt{\frac{2\omega}{\rho_\infty C_{L_{max}} S}} \quad ...(4)$$

from equation (4)

$$V_{Stall/bank} = \sqrt{\frac{2\omega}{\rho_\infty C_{L_{max}} S} \cos 60°}$$

$$= \frac{V_{stall}}{\sqrt{2}} = \frac{40}{\sqrt{2}} = 28.28 \text{ m/sec}$$

55.

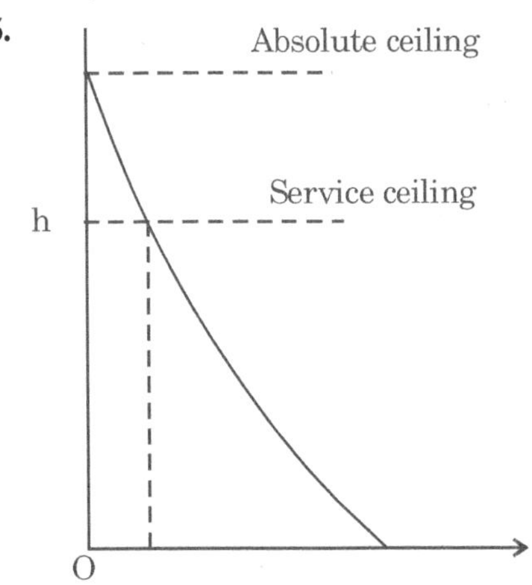

Absolute ceiling is a altitude when $\left(R/C\right)_{max} = 0$

Service ceiling is a altitude where

$\left(R/C\right)_{max} = 100ft/mia$

= 0.5 m/sec

service ceiling represents the practical maximum limit appear for steady, level flight for propeller driven aircraft service ceiling is a altitude were

$\left(R/C\right)_{max} = 5ft/min$

= 0.075 m/sec

56. Brequest range formula for propeller driven aircraft

$$R = \frac{\eta}{C}\left(\frac{L}{D}\right)\ln\frac{W_i}{W_f}$$

$w_f = w_o - w_p$.

w_p = weight of propellant

w_o = gross weight of Aircraft.

According to question when consumptions of weight

$w_f = 0.2w_i$ fuel is 20% of take off weight

$w_f = w_i - 0.2\ w_i$

$= 0.8\ w_i$

$$R = \frac{n}{c}\left(\frac{L}{B}\right)\ln\frac{W_i}{W_f}$$

$$= \frac{\eta}{C}\left(\frac{L}{D}\right)\ln\left(\frac{W_i}{0.8\ W_i}\right)$$

$$R = k\ln\left(\frac{1}{0.8}\right)$$

$= k \ln 12.5$ (A)

When consumption weight of fuel is 50% of take off weight then

$w_p = 0.5\ w_i$

range formula given by

$$R_1 = \frac{n}{c}\frac{C_L}{C_D}\ln\frac{RW_i}{0.5\ Wi}$$

$$= \frac{n}{C}\left(\frac{C_L}{C_D}\right)\ln\left(\frac{1}{0.5}\right)$$

$$= \frac{n}{C}\left(\frac{C_L}{C_D}\right)\ln 2$$

$R_1 = K \ln 2$ (B)

dividing equation (A) by equation (B) we get

$$\frac{R}{R_1} = \frac{K \ln 1.25}{K \ln 2}$$

$$R_1 = \frac{R\ \ell n\ 2}{\ell a\ 1.25} = 3.1R$$

$$\boxed{R_1 = 3.1R}$$

57. Trim curve of aircraft

$$C_{m_{cg}} = 0.05 - 0.2\ \delta_e - 0.14\ C_L$$

$$\text{Static margin} = \frac{-dC_m}{d\ C_L}$$

$= -(-0.1) = 0.1$

$$\boxed{\text{Static marg} = 0.1}$$

58. Given

$$C_{m,\ cg} = 0.05 - 0.8\ \delta_e - 0.1\ C_L \quad ...(1)$$

at trim (condition), $C_{m,\ cg} = 0$

$$\therefore 0.05 - 0.2\ \delta_{C_1} - 0.1C_{L_1} = 0$$

$$0.05 - 0.2\ \delta_{C_2} - 0.1C_{L_2} = 0$$

$$0.2\delta_{e_1} - 0.1\ \delta_{C_1} = 0.05 \quad ...(2)$$

$$0.2\delta_{e_2} - 0.1\ \delta_{C_2} = 0.05 \quad ....(3)$$

from eqn. (2) and eqn. (3), we get

$$0.2\delta_{e_1} - 0.1\ C_{L_1} = 0.2\ \delta_{e_2} + 0.1\ C_{L_2}$$

$$\Rightarrow 0.2\delta_{e_2} = 0.2\ \delta_{e_1} + 0.1\left(C_{L_1} - C_{L_2}\right)$$

$$0.2\delta_{e_2} = 0.2\ \delta_{e_1} + 0.1\ (0.4 - 0.9)$$

$$\delta_{e_2} = \delta_{e_1} + \frac{0.1}{0.2}(-0.5)$$

$$\delta_{e_2} = \delta_{e_1} - 0.25$$

$$\boxed{\delta_{e_2} = \delta_{e_1} - 0.25\ \text{rad}}$$

$$\boxed{\delta_{e_1} = \delta_{e_2} - 0.25\ \text{rad}}$$

59. Time period of period of phugoid oscillation is given by

$$\boxed{T = \sqrt{2}\pi\ \frac{V}{g}}$$

60. V = 200 m/sec. g = 9.8 m/sec^2.

$$T = \sqrt{2}\,\pi \frac{V}{g}$$

$$= \sqrt{2}\pi \times \frac{200}{9.8} = 90.62 \text{ sec}$$

relation between time period and wave length is given by

X = VT

$= 1.81 \times 10^4$ m

61. Equilibrium gliding angle

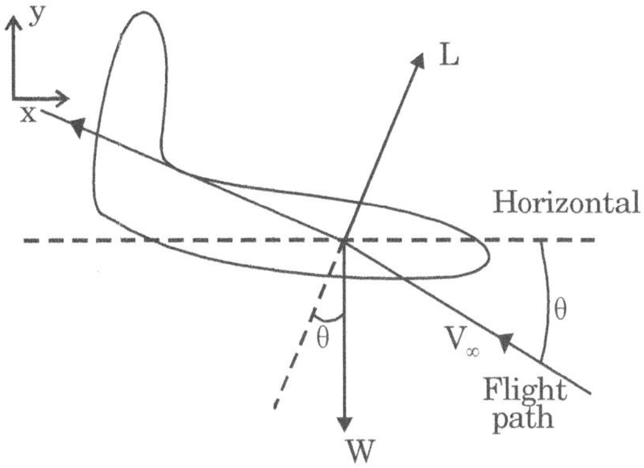

revolving component of forces along x and y direction respectively, we get

$L = W \cos\theta$ ...(1)

$D = W \sin\theta$ ...(2)

Dividing equation (2) by equation (1), we get

$$\frac{\sin\theta}{\cos\theta} = \frac{D}{L}$$

$$\tan\theta = \frac{D}{L} = \frac{1}{\left(\frac{L}{D}\right)}$$

$$\theta = \tan^{-1}\left(\frac{D}{L}\right)$$

$$\tan(\theta)_{min} = \frac{1}{\left(\frac{L}{D}\right)_{max}}$$

62.

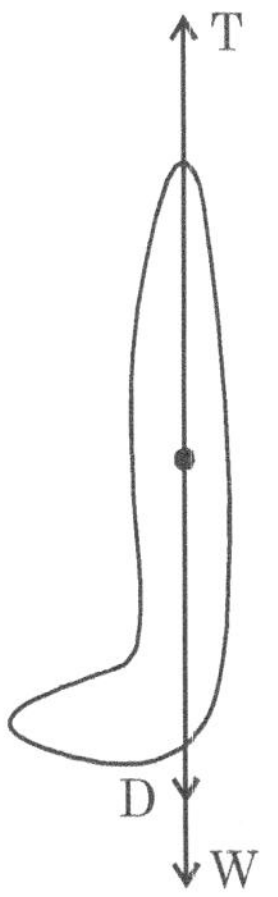

When aircraft climbing vertically up lift force will be zero force acts an the aircraft will be T, D, W.

63. When aircraft is performing positive yawing moment the side slip angle will always negative.

64. for static stable $\frac{\partial C_{m_{cg}}}{\partial \alpha} < 0$

$$C_{m_{cg}} = C_{m_{ac}} + C_L\left(h - h_{ac}\right)$$

$$\frac{\partial C_{m_{cg}}}{\partial \alpha} = 0 + C_{L\alpha}\left(h - h_{ac}\right)$$

$$\frac{\partial C_{m_{cg}}}{\partial \alpha} < 0$$

$C_{L\alpha}\left(h - h_{ac}\right) < 0$ for an aircraft

$\therefore h - h_{ac} < 0,\ C_{L\alpha} > 0$

Hence, centre of gravity should ahead to wing aerodynamic centre.

65. Winglets of wings is used to minimize the induce drag. Induce drag is an aerodynmaic drag force that occurs when even a moving objects redirects the airflow coming at it.

66.

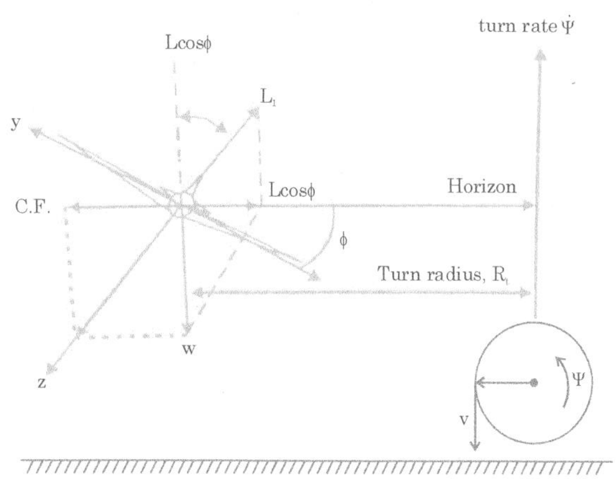

Coordinated turn is defined as the one is which the velocity is the y-axis is constant acceleration is y-direction is zero.

$L \cos\varphi = W$ ....(1)

$L \sin\varphi = \frac{mv^2}{R}$ ....(2)

From equation (1)

$$\frac{L}{W} = \frac{1}{\cos\varphi}$$

$\frac{L}{W}$ = load factor = η

$$n = \frac{1}{\cos\varphi} = \frac{1}{\cos 30°} = \frac{1}{\frac{\sqrt{3}}{2}} = \frac{2}{\sqrt{3}}$$

component of lift force which will provide necessary force of turn maneuvering.

$$L\sin\varphi = \frac{mv^2}{R}$$

putting value of L is equation (2) from equation 1 we get

$$w\tan\varphi = \frac{mv^2}{R}$$

$$R = \frac{mv^2}{w\tan\varphi}$$

$$R = \frac{v^2}{g\tan\varphi} = \frac{100^2}{9.8\times\tan 30°} = 1.76\text{ Km}$$

$$\boxed{R = 1.76\text{ Km}}$$

67. When aircraft takes roll over by 180° from steady level flight, so in the initial; situation is in steady level flight. There will be no change took place so, after rolling velocity will be same angle of attack will be same.

(i)

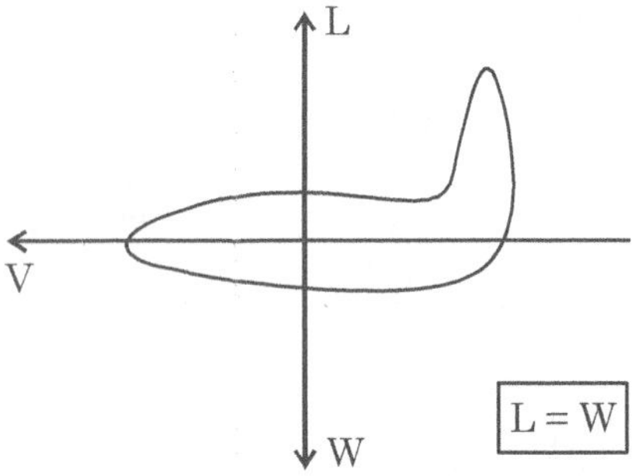

L = mg

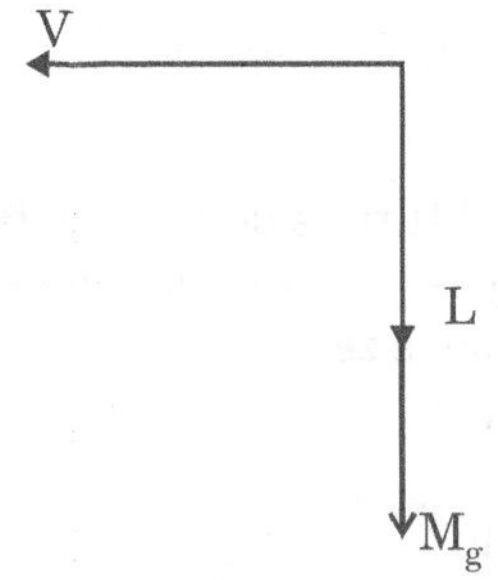

Now angle of attack will be negative to previous condition, so direction of lift will be changed both lift and weight will acts in same downward direction, and so this will be a circular motion pull down manevers.

Hence, Lift and weight forces balance the centrifugal force

$$L + W = \frac{mv^2}{R}$$

$$L + mg = \frac{mv^2}{R}$$

$$2mg = \frac{mv^2}{R}$$

$$\Rightarrow \frac{v^2}{R} = 2g$$

$$\Rightarrow \boxed{R = \frac{v^2}{2g}}$$

$$= \frac{50^2}{2\times 9.81}$$

= 127.4 m.

68.

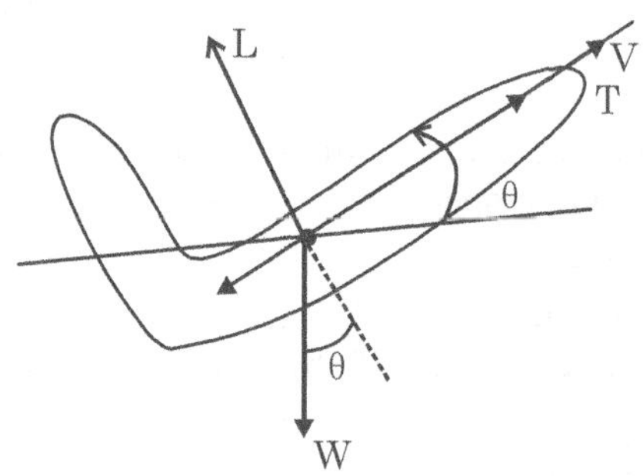

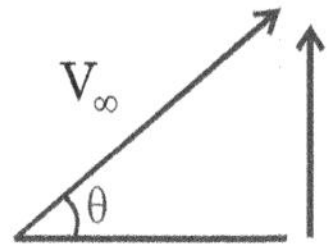

Rate of climb (R/C)

$$(R/C) = V_\infty \sin\theta$$

= 50 sin 50

= 12.94 m/sec

Expression of rate of climb is terms of thrust and drag

$$R/C = \frac{(T - D)v_\infty}{W}$$

$$= \frac{(\text{excess thrust available})v_\infty}{W}$$

⇒ excess thrust available

$$= \frac{(R/C)W}{v_\infty} = \frac{12.94\times 5000\times 10}{50} = 12940\text{ N}$$

69.

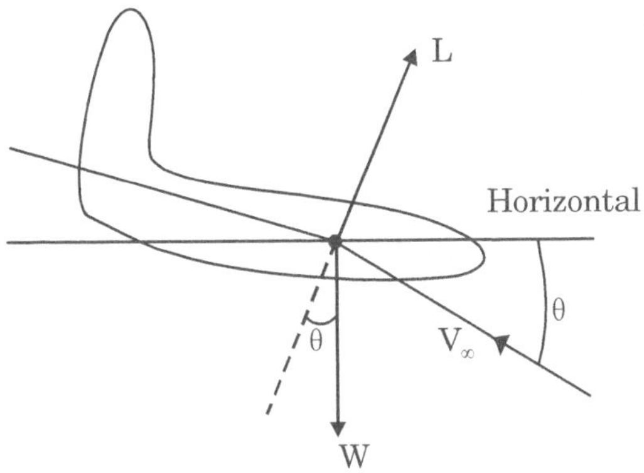

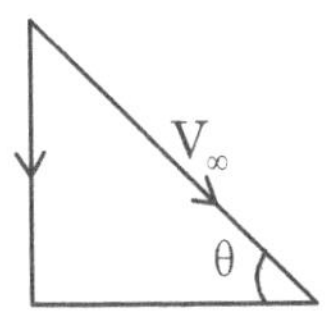

$$\sum f_x = 0$$

$$\sum f_y = 0$$

$L = W \cos\theta$ ....(1)

$D = W \sin\theta$ ....(2)

dividing equation (2) equation (1), we get

$$\frac{D}{L} = \tan\theta$$

$$\tan\theta = \frac{1}{\left(\frac{L}{D}\right)}$$

$$\theta = \tan^{-1}\left(\frac{1}{\frac{L}{D}}\right)$$

$$\tan^{-1}\left(\frac{1}{15}\right) = 3.8^{6}.$$

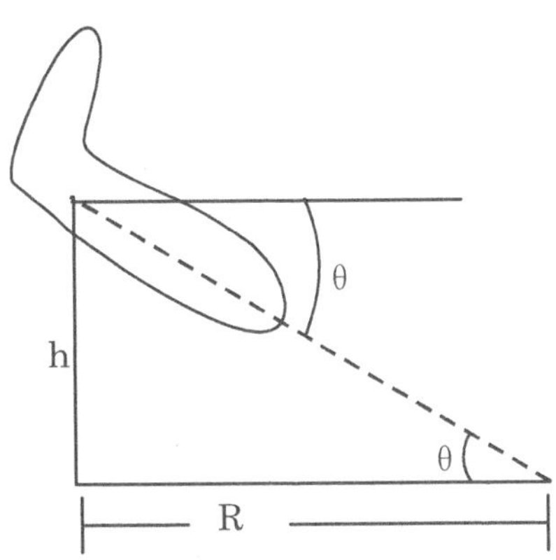

$$\frac{1}{\left(\frac{L}{D}\right)} = \frac{h}{R}$$

$$\Rightarrow r = H\left(\frac{L}{D}\right)$$

$= 1000 \times 15$

$= 15000 \text{ m} = 15\text{Km}$

Endurance

$$h = (v_\infty \sin\theta) xt$$

$$t = \frac{h}{v_\infty \sin\theta}$$

$$= \frac{1000}{54 \times \frac{5}{18} \times \sin 3.8}$$

$= 1005.92$ sec

70. Time period of oscillation of phugoid motion

$$T = \frac{50}{10} = 5\,\text{sec}$$

In 25 sec it will complete its 5 time periods.

$$x_5 = \frac{x_0}{2}$$

$$\frac{x_0}{5} = \frac{x_0}{x_1}\frac{x_1}{x_2}\frac{x_2}{x_3}\frac{x_3}{x_4}\frac{x_4}{x_5}$$

When x_i is amplitude after completing its i time periods.

In damping the ratio of amplitude after completing i time periods]

In damping, the ratio of amplitude in consecutive steps is same.

$$\frac{x_0}{x_1} = \frac{x_1}{x_2} = \frac{x_2}{x_3} = \frac{x_3}{x_4} = \frac{x_4}{x_5}$$

$$\ln\left(\frac{x_0}{x_4}\right) = 5\ln\left(\frac{x_0}{x_1}\right)$$

$$\ln 2 = 58$$

$$\Rightarrow 8\frac{\ln^2}{5}$$

$$\frac{2\pi\xi}{\sqrt{1-\xi^2}} = \frac{\ln 2}{5}$$

$$= \frac{\xi}{\sqrt{1-\xi^2}} = \frac{\ln 2}{10\pi}$$

$$\Rightarrow \frac{\xi^2}{1-\xi^2} = \left(\frac{\ln 2}{10\,\pi}\right)^2$$

$\Rightarrow \xi = 0.022$

So, the eigen values are

$$\lambda_{1,2} = \left(-\xi \pm \sqrt{1-\xi^2}\right) \times \frac{2\pi}{T}$$

$$\boxed{\lambda_{1,2} = -0.028 \pm 1.26i}$$

72. Lateral controls in aircraft are rate producing devices.

Since $i_L = \frac{L_{\delta_a}\,\delta_a}{Ixx}$ an impulse input in the aileraq only changes ϕ and P, the integral quantities. But since the input is impulse type, P returns to 0 after a new value of p is attained. Therefore constant roll rate is produced by an impulse input in the airleroas.

73. During ground roll maneuver of aircraft the forces acting on the it parallel to direction of both weight and lift part of lift and weight bot acts because the line of motion of aircraft is lower at an angle than the aircraft fixed longitudinal axis.

74. $R/C = \frac{(T-D)v_\infty}{W}$

R/C is proportional to excess power is steady climb

Hence, $\frac{R}{C} \propto (T-D)$ for constant w, v_∞.

There is 10 percent is thrust, so T′ = 0.4T

Therefore.

$$\frac{\text{Change in } \frac{R}{C}}{\frac{R}{C}}$$

$$\frac{T-D(0.9T-D)}{T-D} = \frac{0.1T}{(T-D)}$$

$T - D < T$

So $\frac{0.1T}{T-D} > \frac{0.1T}{T} > 0.1$

rate of climb is lower more than 10 percent.

75. For nose dive, a nose down pitch rate is achieved using an eleuator input.

76. Eleuator control effectiveness is defined as

$$\frac{d\,Cm}{d\delta_e} = \frac{\eta\,\delta t}{C\,\delta}\frac{dC_L t}{d\delta_e}\ell$$

Whose ℓ is the distance of eleuator from the CG. elevator control effeteness restricts ℓ and hence the forward most position of the centre of gravity.

77. For elliptical lift distribution

$$C_{D_i} = \frac{C_L^2}{\pi AR}$$

where general formula is

$$C_{D_i} = \frac{C_L^2}{\pi e AR}$$

e = oswald efficiency factor for elliptical loading e = 1

$$\boxed{C_{D_i} = \frac{C_L^2}{\pi AR}}$$

78. 14 sec

Rate of climb an aircraft can be assumed to be decreases with altitude. There for expression for rate of climb

R = a + bh {assembled}

a rate of climb at sea level

$$b = \frac{(R)_{altitude} - (R)_{sea}}{h_{altitude} - h_{sea}}$$

$$= \frac{150-300}{2500-0} = -0.06$$

Now we have the expenses for rate of climb

$$R(h) = \frac{dh}{dt}$$

$$dt = \frac{dh}{R(h)}$$

$$\int_{t_1}^{t_2} dt = \int_{R_1}^{R_2} \frac{dh}{R(h)}$$

$$t_2 - t_1 = \int_{h_1}^{h_2} \frac{dh}{a+bh}$$

$$\boxed{\Delta t = \frac{1}{b}\ln\frac{a+bh_1}{a+bh_2}}$$

$$\Delta t = \frac{1}{0.06}\ln\frac{300-0.06\times 300}{300-0.06\times 500}$$

$= \frac{1}{0.06}\ln\left(\frac{9}{4}\right)$

$= \frac{\ln 2.5}{0.06} = 13.51 \text{ sec}$

= 14 sec

79. W = weight of structure

W_f = weight of fuel

for get aircraft

$$R = 2\sqrt{\frac{2}{\rho_\infty S}} \times \frac{1}{C_t}\frac{C_L^{\frac{1}{2}}}{C_D}\left(W_o^{\frac{1}{2}} - W_1^{\frac{1}{2}}\right)$$

$$R \propto \sqrt{W_o} - \sqrt{W}$$

$$\propto \sqrt{W_o + W_f} - \sqrt{W}$$

fraction reduction is range $\left(\frac{\Delta R}{R}\right)$

$$= \frac{\sqrt{W + Wf} - \sqrt{W + 0.9Wf}}{\sqrt{W + W_f} - \sqrt{W}}$$

tany extreme practical values of Wf,

case – 1 $^{Wf}/_{W} > 1$ we have

$$\frac{\Delta R}{R} = \frac{\sqrt{1 + {}^{Wf}/_{W}} - \sqrt{1 + 0.9}\ {}^{Wf}/_{W}}{\sqrt{1 - {}^{Wf}/_{W}} - \sqrt{1}}$$

$$\frac{4R}{R} = \frac{\sqrt{1+1} - \sqrt{1+0.9}}{\sqrt{2} - \sqrt{1}}$$

$$= \frac{\sqrt{2} - \sqrt{1.9}}{\sqrt{2} - 1} = 8.6\%$$

case - 2

$\frac{Wf}{W} << 1$, using by binomial expansion of root.

We find that $\frac{\Delta R}{R}$ changes to 10%.

for propeller driven Aircraft

$$R = \frac{n}{c}\frac{C_L}{C_D}\ell n\frac{w_0}{w_1}$$

$$R \propto \ell a\frac{N_1}{w}$$

$$\propto \ell a\left(\frac{w + wf}{w}\right)$$

$$\propto \ell a\left(1 + {}^{wf}/_{w}\right)$$

∴ fraction of reduction in range

$$= \frac{\Delta R}{R} = \frac{\ell a\left(1 + {}^{wf}/_{w}\right) - \ell a\left(1 + \frac{(.9wf)}{w}\right)}{\ell a\left(1 + {}^{wf}/_{w}\right)}$$

$$= \frac{\ell a\left(\frac{w + wf}{w + 0.9wf}\right)}{\ell a\left(\frac{w + wf}{w}\right)}$$

for Case – 1

$\frac{wf}{w} \approx 1$

$\frac{\Delta R}{R} = 7.4\%$

Case – 2

$^{wf}/_{w} << 10$ reduction in range

$\frac{4R}{\Delta} = 10\%$

Hence is all the possible cases the reduction is range is less than 10%.

80.

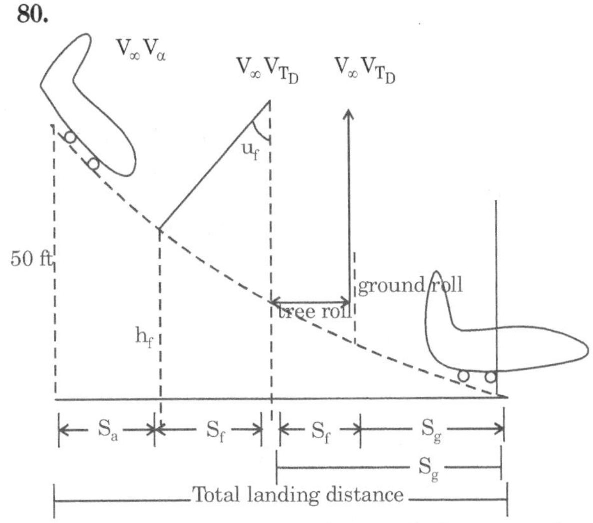

Total landing distance = Approach distance + plane distance + ground roll distance.

$= S_a + S_f + S_g.$

$$= \frac{50 - h_f}{\tan\theta_a} + R\sin\theta_a + \frac{V^2}{2a}$$

where $R_f = R(1 - \cos\theta_a)$

$$R = \frac{V^2}{(n-1)g}$$

n = Load factor

$$= \frac{50 - R(1-\cos\theta_a)}{\tan\theta_a} + R\sin\theta_a + \frac{V^2}{2a}$$

$$= \frac{50 - \frac{V^2}{(n-1)g}(1-\cos\theta_a)}{\tan\theta_a} + \frac{V2}{(n-1)g}\sin\theta_a + \frac{V^2}{2a}$$

In the above problem height of obstacle is 15 m JM

$$\therefore\ S = \frac{15 - \frac{V^2}{(n-1)g}(1-\cos\theta)}{\tan\theta_a} + \frac{V^2}{(n-1)g}\sin\theta_a + \frac{V^2}{2a}$$

Given V = 90 m/sec, θ_a = 6.6°, n = 1.2, 9 - 0.25g

S = 82.9 + 93.7 + 326.2 m

$\boxed{S = 502m}$.

81. V_{True} = 100 m/sec

L = 5N at V = 90 m/sec

According to Questions

$$L = C_L \times \frac{1}{2}\rho_s A V_{100}^A$$

$$\Rightarrow C_L \times \frac{1}{2}\rho_s A V_{100}^2$$

$$\Rightarrow Ke_5 V_{100}^2 = Ke_5 V_{90}^2 + 5 \quad \left(\text{where } k = \frac{G}{2}A\right)$$

$$\Rightarrow K\left(1.25 \times 100^2 - 0.91 \times 90^2\right) = 5$$

$$\Rightarrow K = \frac{5}{\left(1.25 \times 100^2 - 0.91 \times 90^2\right)}$$

$\boxed{K = 0.0021052}$

For avoiding re-trimming at = 300 m. Let the upward force acting on the aircraft be x.

$$L = x + K\rho_{3000} V_{120}^2$$

$$\Rightarrow x = \left(L - K\rho_{3000} V_{120}^2\right)$$

$$= \left(K\ \rho_s V_{1100}^2 - K\ \rho_{3000} V_{120}^2\right)$$

$$= K\left(\rho_s V_{100}^2 - \rho_{3000} V_{120}^2\right)$$

$$x = 0.00216\left(1.225 \times 100^2 - 0.91 \times 120^2\right)$$

= 0.00216 (–804)

= –1.736 N

the force taken as upward, here direction is downward.

Common data question

given data

$\varepsilon_a = 0.3,\ C_L = 1$

$C_{L\alpha} = 0.08/\text{deg}\ C = 2.5$ m

Wing and tail are geometrical similar,

$\therefore\ \alpha_t = \alpha_w$

$S_t = Sw/3$

82. $$C_{m_{cg}} = C_{m_{ac}} + C_{LW}\left(h - h_{ac} - V_H \eta_t \frac{a_t}{a}\left(1 - \frac{\partial\varepsilon}{\partial\alpha}\right) + V_H a_t (i_t + \varepsilon_0)\right)$$

for trimmed condition

$C_{M_{cg}} = 0$

We know static margin = $h_n - h$

S.M = $h_n - h_{ac} - h - h_{ac}$.

0.05 = 1.4 – h + h_{ac}.

h – h_{ac} = 1.4 – 0.05

= 1.35

$\ell_t = 15 - 1.35 \times 2.5 = 11.625$m

$$V_H = \frac{S_t \ell_t}{SC} = \frac{1}{3} \times \frac{11.625}{2.5}$$

=1.55

putting all these values in main equation

$$0 = 0 + 1\left(1.35 - 1.55 \times 1(1 - 0.3) + 1.55 \times 1 \times 0.08(j_t + 0)\right)$$

$\boxed{i_t = -2.13°}$

(d) should be taken.

83. Now static margin

$$h_n = h_{ac} + V_H n_t \frac{a_t}{a}(1 - \varepsilon_\alpha)$$

$$h - h_{ac} = \frac{\frac{1}{3} \times \frac{30}{2}}{2.5} \times 1(1 - 0.3)$$

= 1.4

$h_n - h_{ac} = 1.4$

1.4 is non dimensional

with chord so need to multiply with 2.5

$h_n - h_{ac} = 1.4 \times 2.5$

$= 3.4$ m

84. Sweep back, dihedral, high wing, low C.G support lateral stability roll stability is type of lateral stability it also supports roll stability.

86. At supersonic speed velocity is high hence lift requirement is low.

87. V_g = velocity of Aircraft relative to ground

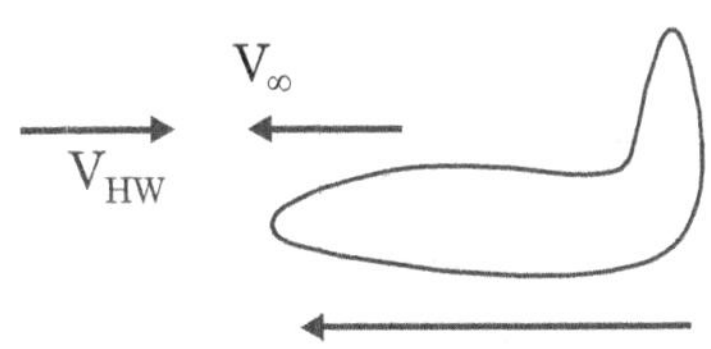

$V_g = V_\infty - V_{HW}$

Head Wind

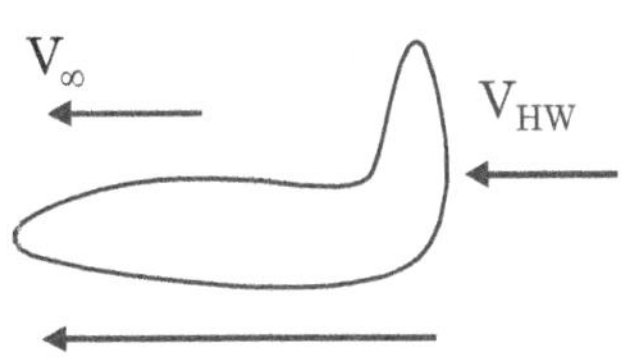

$V_g = V_\infty + V_{HW}$

Tail Wind

Head wind favourable condition for landing, because it reduce the leading distance.

At the time of cruise, tail wind increase the ground speed hence it is favorable in cruise.

88. For phugoid mode

$$T = \frac{\sqrt{2}\pi V_0}{g_0}$$

$$T = 1/f$$

$$f = \frac{1}{T} = \frac{g_0}{\sqrt{2}\pi V_0}$$

$$f \propto \frac{1}{V_0}$$

$$w = 2\pi f$$

$$w \propto f$$

$$\propto \frac{1}{V_0}$$

$$\boxed{W \propto \frac{1}{V_0}}$$

89. $C_{DO} = 0.02$

AR = 10

e = 0.95

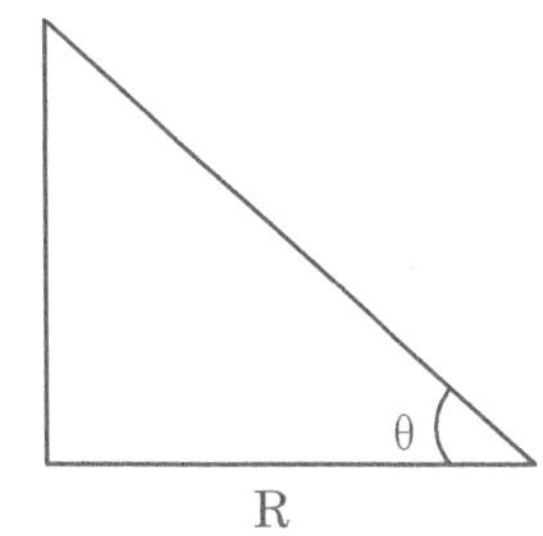

gliding angle θ is given by

$$\tan\theta = \frac{1}{\left(\frac{L}{D}\right)}$$

for θ to be minimum $\left(\frac{L}{D}\right)$ should maximum

$$\left(\frac{L}{D}\right)_{max} = \frac{1}{\sqrt{4C_{D_0}K}}$$

$$= \frac{1}{\sqrt{4 \times 0.02} \times \sqrt{\frac{1}{3.14 \times 0.95 \times 10}}}$$

$= 18.08$

$$\tan\theta = \frac{1}{18.08}$$

$R = h \times 18.08$

$= 500 \times 18008$

$= 9050$

$$\boxed{R = 9 \text{ Km}}$$

90. for level turn maximum

$$\text{turn rate, } W = \frac{g_0\sqrt{n^2 - 1}}{V}$$

$$\text{turn radius } R = \frac{V^2}{g\sqrt{n^2 - 1}}$$

W – larger

n – larger

R – larger

V – larger

R – smaller

V – smaller

91. $\Gamma = 10°$

$$C_{\ell\alpha} = 5.4/_{rad} \quad C_{\ell\beta} = \frac{C_{\ell\alpha} \times \alpha}{4}$$

$$= \frac{5.4 \times \frac{10 \times \pi}{180}}{4} = -0.2358 / \text{rad}$$

92. W = 30 kN
$T_o = 4$ kN
S = 30 m².
$C_{L\,max} = 1.4$

$$C_D = 0.015 + 0.024\ C_L^2$$

$$e_\infty = 1.22\ \text{Kg/m}^3$$

$$V_{stall} = \sqrt{\frac{2W}{\rho_\infty S\ C_{L\,max}}}$$

$$= \sqrt{\frac{2 \times 30 \times 10}{1.25 \times 30 \times 1.4}} = 34.12 \text{ m/sec.}$$

$$\boxed{V_{stall} = 34 \text{ m/sec}}$$

93. $$V_{max},\ _{min} = \left\{\frac{\frac{T}{W} \pm \frac{W}{S}\sqrt{\left(T/W\right) - 4C_{Do}14}}{2\,\rho_\infty\,C_{Do}}\right\}^{\frac{1}{2}}$$

$$V_{max} = \left\{\frac{\frac{4000}{30 \times 10} + \frac{30 \times 10}{30}\sqrt{\left(\frac{4000}{30 + 10^3}\right)^2 - 4 \times 0.015 \times 0.024}}{2 \times 1.225 \times 0.015}\right\}$$

= 119.46 m/sec

$$V_{min} = \left\{\frac{\frac{4000}{30 \times 10} - \frac{30 \times 10}{30}\sqrt{\left(\frac{4000}{30 + 10^3}\right)^2 - 4 \times 0.015 \times 0.024}}{2 \times 1.225 \times 0.015}\right\}$$

94. NACA 12018

95. When aircraft par ked in larger

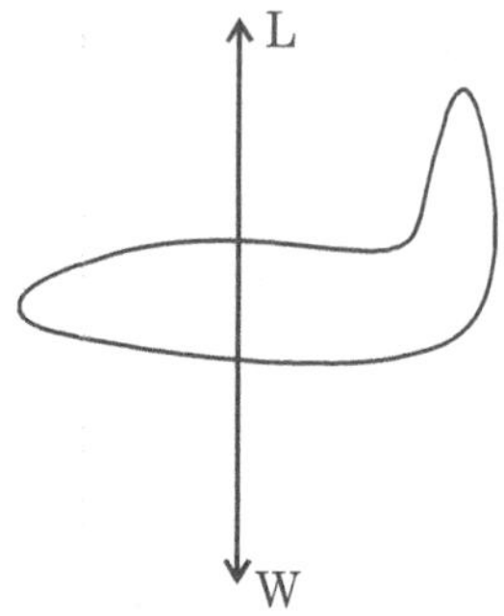

$$L = W$$

$$n = \frac{L}{W}$$

$$= 1$$

$$\boxed{n = 1}$$

96. for minimum sink rate

Aircraft should fly at $\frac{C_L^{3/2}}{C_D}$ maximum or velocity for minimum power required.

98. $$L + W = \frac{mv^2}{R}$$

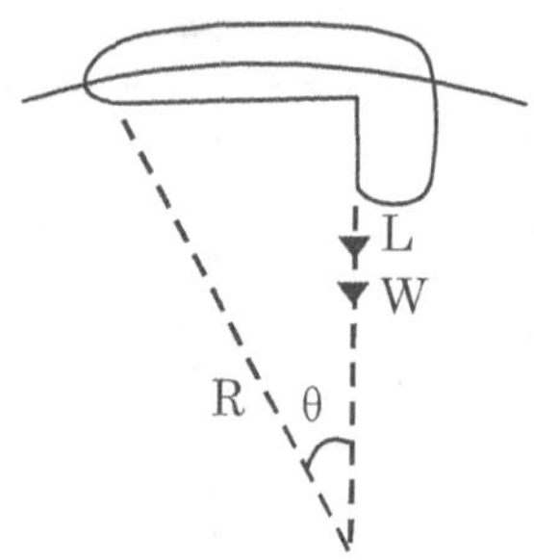

$$R = \frac{mv^2}{(L + W)}$$

for steady level condition (L = W)

$$R = \frac{v^2}{g} \times \frac{1}{(n + 1)}$$

we have, turn rate

$$\omega = \frac{V}{R}$$

$$= \frac{g(n + 1)}{V}$$

for turn rate ω to be maximum load faster n will be highest

$$W \propto n$$

99. Simplified expression for lift off distance is gives by

$$S_{L_o} = \frac{1.44W^2}{\rho_\infty g\ SC_{L\,max}\ T}$$

$$\delta_S \propto W^2$$

When W is reduced by 10%

W − 0.1W = 0.9 W

$$\delta_S \propto (0.9W)^2$$

magnitude of % change of lift off distance

$$= \frac{W^2 - (0.9W)^2}{W^2}$$

$$= \frac{W^2(1-0.9^2)}{W^2}$$

$= 1 - 0.9^2 = 0.19$

$= 19\%$

100. $\left(\frac{R}{C}\right)_{max}$ for jet aircraft in given by

$$\left(\frac{R}{C}\right)_{max} = \left(\frac{\left(\frac{W}{S}\right)Z}{3\rho_\infty C_{D,0}}\right)^{\frac{1}{2}}\left(\frac{T}{W}\right)^{\frac{3}{2}}$$

$$\left[1-\frac{Z}{6}-\frac{3}{2}\frac{1}{\left(\frac{T}{W}\right)^2\left(\frac{L}{D}\right)^2_{max} Z}\right]$$

Where

$$Z = 1+\sqrt{1+\frac{3}{\left(L/D\right)^2_{max}\left(T/W\right)^2}}$$

101. For longitudinal stability of aircraft

$\frac{\partial C_{M,cg}}{\partial \alpha_a}$ must be negative

$C_{M_{cg}}$ at zero lift must be positive.

102. $V_\infty = 40$ m/sec

W = 13000 N

$C_L = 0.65$

$C_D = 0.025 + 0.04\ C_L^2$.

$P_{av} = 100000$ J/sec

Now,

$$P_R = T_R V_\infty$$

$$= \frac{W}{\frac{C_L}{C_D}}V_\infty = \frac{W}{\frac{C_L}{C_{D_0} \times KC_L^2}}V_\infty$$

$$= \frac{13000}{\frac{0.65}{0.025+0.14\times 0.65^2}}\times 40$$

$$= \frac{13000}{15.51}\times 40$$

$= 33250$ Watt

$= 33520$ J/sec

$$\frac{R}{C} = \frac{P_A - P_R}{W}$$

$$= \frac{100000 - 33520}{13000}$$

$= 5.11$ m/sec

103. Using hydrostatic equation we can write

$dP = -\rho g_0 dh$

$$\frac{dP}{P} - \frac{\rho g_0 dh}{\rho RT} = -\frac{g_0 dh}{RT}$$

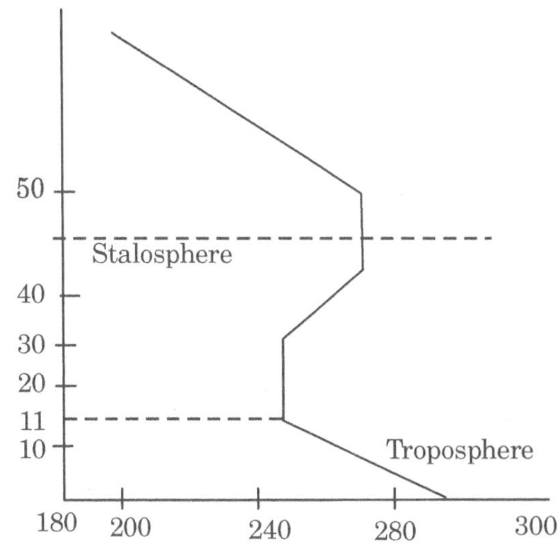

Let temperature pressure density at starting altitude of 80 mal isothermal region are devoted by T_1, P_1, ρ_1

$$\int_{P_1}^{P}\frac{dp}{P} = -\frac{g_o}{RT}\int_{\mu}^{h}dh$$

$$\ln\frac{P}{P_1} = -\frac{g_0}{R-1}(h-h_1)$$

$$\frac{P}{P_1} = -\rho^{\left[-\frac{g_0}{RT}(h-h_1)\right]}$$

using equation of state we get

$$\frac{P}{P_1} = \frac{\rho T}{\rho_1 T_1} = \frac{\rho}{\rho_1}$$

$$\frac{\rho}{\rho_1} = \rho^{-}\frac{g_0}{R-1}(h-h_1)$$

104. Shift of centre of pressure and no shift on aerodynamic centre by definition.

105. Spiral mode has nor oscillatory response characteristics because the roots of spiral mode is real te oscillatory characteristics is given by complex roots.

S = a non - oscillatory

S = a ± ib oscillatory motion

If the sign of real part in – VL then stable system if positive then unstable.

107. 5, Given $P_R = 9 \times 10^6$ KW

$$V_\infty = 300 \text{ m/sec}$$

$$W = 1.5 \times 10^6 \text{N}$$

We have

$$P_R = T_R \times V_\infty$$

$$T_R = \frac{P_R}{V_\infty}$$

$$= \frac{9 \times 10^6}{300} = 30{,}000 = 3 \times 10^4 \text{ N}$$

$$T_R = \frac{W}{\frac{C_L}{C_D}}$$

$$\frac{C_L}{C_D} = \frac{W}{T_R} = \frac{1.5 \times 10^6}{3 \times 10^4} = 5$$

$$\boxed{\frac{C_L}{C_D} = 5}$$

108. 44%

lift off distance $S_{L_0} \propto w^2$

When weight is increased by 20%

$$S_{L_0} \propto (w + 0.200)^2$$

$$S_{L_0} \propto (1.2w)^2$$

$$S_{L_0} \propto 1.44w^2$$

% change in lift off distance

$$= \frac{1.44w^2 - w^2}{w^2} = 44 = 44\%$$

109. For same wing plane form area biplane wing will create less lift due to downwash and drag will be more because of more surface area facing the flow.

110. Since angle of attack decreases centre of pressure of aircraft shifts forward.

111. 40 metre equilibrium gliding velocity

$$V_\infty = \frac{2W}{e_\infty C_{L_{max}} S} = \sqrt{\frac{2 \times 500}{0.9 \times 0.69}}$$

= 40 m/sec

112. $C_D = 0.018$

$C_{D,0} = 0.015$

$C_{Di} = C_D - C_{D0}$

$= 0.018 - 0.015 = 0.003$

$$C_{Di} = \frac{C_L^2}{\pi \rho AR} = KC_L^2$$

$$C_L = \sqrt{\frac{C_{Di}}{K}} = \sqrt{\frac{0.003}{K}}$$

for 30° climb

$L = W\cos\theta = W\cos 30°$

$$C_L = \frac{2W \cos 30°}{\rho_\infty V_\infty^2 S} \qquad \ldots (1)$$

For level flight

$L = W$

$$V_\infty^2 = \frac{2W}{\rho_\infty S C_L} = \frac{2W}{\rho_\infty S \times \sqrt{\frac{0.003}{K}}} \qquad \ldots (2)$$

From equation (1) and (2), we get

$$C_L = \frac{2W \cos 30°}{\rho_\infty S \times \frac{2W}{\rho_\infty S \times \sqrt{\frac{0.003}{K}}}} = \frac{K \cos 30°}{\sqrt{0.003}}$$

$$\therefore \text{ New } C_D = C_{D_0} + KC_L^2$$

$$= 0.015 + K \times \frac{0.022}{K}$$

$$= 0.015 + 0.022 = 17.2 \times 10^{-3}$$

113. Inertial ground and wind speed

$$\overline{V}_g^b = (100, 5, 6) \text{ m/sec},$$

$$\overline{V}_w^b = (0, -5, -10) \text{ m/sec},$$

$$V = \sqrt{(100-0)^2 + (5+5)^2 + (5+10)^2}$$

$$= \sqrt{10325}$$

Side slip angle (β)

$$\beta = \tan^{-1}\left(\frac{V_{Y_b}}{V_{X_b}}\right)$$

$$= \tan^{-1}\left(\frac{5+5}{100-0}\right)$$

$\tan\theta = 0.1$

$\theta = \tan^{-1}(0.1) + (0-1)$

$\boxed{\theta = 5.71°}$

114. Damping ratio for phugoid motion is given by

$$3 = \frac{1}{\sqrt{2}}\frac{D}{L}$$

$$= \frac{1}{\sqrt{2}} \times \frac{1}{\frac{L}{D}}$$

$\left(\frac{L}{D}\right)$ higher for glider

ξ – lesser

115. Altitude above ground is measured by static pressure

h = altitude above ground level

118. $\frac{L}{W} = n = 2$

$L = 2W$

$$L = C_L \times \frac{1}{2}\rho A v^2$$

$$C_L = \frac{2w \times 2}{\rho A v^2}$$

$$C_L \propto \frac{4w}{\rho_\infty A V^2}$$

double

119. We have

$$C_{M_{LE}} = -\frac{C_L}{4} = L_{M,\,half} - \frac{C_L}{2}$$

$$\therefore\ C_{M_{half}} = \frac{C_L}{2} - \frac{C_L}{4}$$

$$= \frac{C_L}{4} = \frac{2\pi\alpha}{4}$$

$$\therefore\ \frac{\partial C_{M,\,half}}{\partial\alpha} = \frac{\pi}{2}$$

120. 8.6 – 8.7

121. 25 sec

Given

W = 10000 N

V = 100 m/sec

$T_{R_{level}} = 1000N$

$T_A = 5000$ N

$$\frac{R}{C} = \frac{(T_A - T_R)V_\infty}{W}$$

$$= \frac{(5000-1000)\times 100}{10000} = 40$$

$$\frac{R}{C} = 40 \text{ m/sec}$$

Now

$$\frac{R}{C} = \frac{dh}{dt}$$

$$\int_0^t dt = \int_0^h \frac{dh}{\frac{R}{C}}$$

$$t = \int_0^{1000} \frac{dh}{40} = \frac{1}{40}\int_0^{1000} dh$$

$$= \frac{1}{40}|h|_0^{1000}$$

$$= \frac{1000-0}{40} = 25\text{sec}$$

122. Given [u, v, w] = [100, 10, 10]

$$V = \sqrt{u^2 + v^2 + w^2}$$

$$V = \sqrt{100^2 + 10^2 + 10^2} = 100.995$$

$$\tan\alpha = \frac{V_{zb}}{V_{xb}}$$

$\boxed{\alpha = 5.71}$

$$\tan\beta = \frac{V_{yb}}{V_{xb}} = 5.71$$

123. Given

$$\begin{bmatrix}\Delta\beta \\ \Delta r\end{bmatrix} = \begin{bmatrix}-0.26 & -1 \\ 4.49 & -0.76\end{bmatrix}\begin{bmatrix}\Delta\beta \\ \Delta r\end{bmatrix}$$

writing terms of eigen value, we get

$$\lambda^2 + 1.02\lambda + 4.6876 = 0$$

Now comparing with equation for dutch roll motion, we get

$$\lambda^2 + 2\xi W_n\lambda + W_n^2 = 0$$

$\omega_n^2 = 4.6876$

$\omega_n = 2.165$

$2\xi W_n = 1.02$

$$\xi = \frac{1.02}{2W_n} = \frac{1.02}{2\times 2.165} = 0.295$$

124. Given data

W = 3300 N

h = 1000 m

W = 14.1 cm²

ρ_∞ = 1.23 149/m²

for gliding flight

$$\tan\theta = \frac{1}{\frac{L}{D}}$$

$$\frac{h}{R} = \frac{1}{\left(\frac{L}{D}\right)}$$

$$\boxed{R = \left(\frac{L}{D}\right) \times h}$$

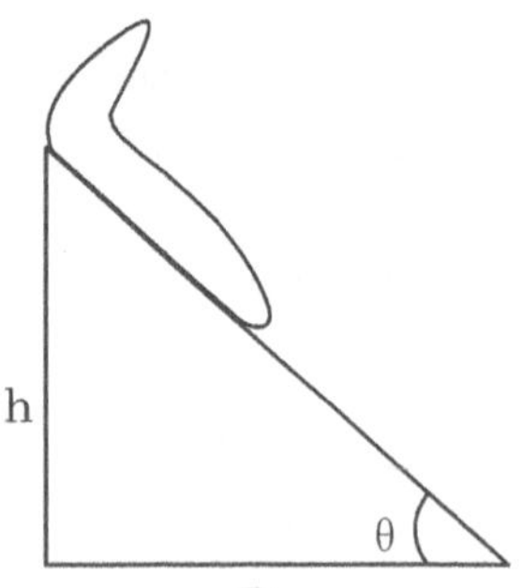

for maximum range of glide $\frac{C_L}{C_D}$ or $\left(\frac{L}{D}\right)$ must be maximum, from given data C_L/C_D C_L value is 0.9

$$L = \frac{1}{2}\rho_\infty v^2 S\, C_2$$

$$\Rightarrow 3300 = 0.5 \times 1.25 \times V^2 \times 14.1 \times 0.9$$

$$V^2 = \frac{3300}{0.5 \times 1.25} \times 14.1 \times 0.9$$

$$\boxed{V = 20.56}$$

125. By defiation the pitching moment about the aerodynamic centre of the aerofoil is independent of angle of attack.

126. $C_{m\delta} < 0$

$N_m - N_0 > 0$

$N_m > N_0$

Hence stick fixed maneuvering point N_m is aft, of the stick fixed Neutral point N.

127. In low speed aircraft at high angles of side slip rudder floats rapidly.

128. 0.05

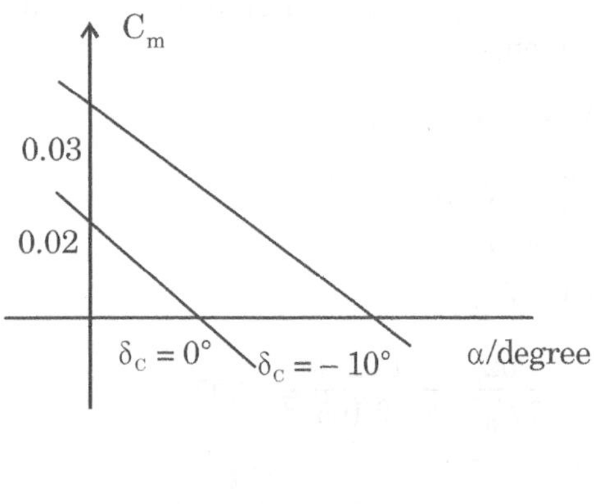

$$\frac{\partial C_m}{\partial \delta_e} = \frac{C_{m_2} - C_{m_1}}{\delta_e - \delta_e}$$

$$= \frac{0.03 - 0.02}{-10° - 0°}$$

= –0.001 per degree

= – 0.001 × 57.5

= 0.057 per radian

$$\boxed{\frac{\partial C_m}{\partial \delta_e} = -0.057 \text{ per radian}}$$

129. When aircraft landing gear is lowered which produce additional drag which acts the point below the central line or say line of CG, which developes nose down pitching moment.

In order to balance this moment elevater must be deflect upward. To over come the extra drag due to landing gear and maintain same C_L. Throttle must be increased.

130. 10.2

Given data

C_{DO} = 0.02, ρ = 1

For thrust required minimum

$C_{DO} = C_{Di}$

$$C_{DO} = \frac{C_L^2}{\pi e\, AR} \Rightarrow AR = \frac{C_L^2}{C_{DO}\pi e}$$

$$= \frac{0.8^2}{0.02 \times 3.14 \times 1}$$

AR = 10.2

131. Given complex roots of longitudinal dynamics

$\lambda_{1,2}$ = –0.02 + 0.3i

$\lambda_{3,4}$ = –2I 2.5i

Since the short period mode is highly ????? have the corresponding roots, of

$\lambda_{3,4}$ = –2± 2.5i

$$= -3\xi_n \pm i\omega_n\sqrt{1-\xi^2}$$

Now equating real and imaginary put we get

$$\xi\omega_n = 2 \quad ...(1)$$

$$\omega_n\sqrt{1-\xi^2} = 2.5 \quad ...(2)$$

dividing equation (2) by equation (1), we get

$$\sqrt{\frac{1-\xi^2}{\xi}} = \frac{2.5}{2}$$

$$\Rightarrow (1-\xi^2) \times 4 = 2.5^2\xi^2$$

$$\Rightarrow 1-\xi^2 = \frac{2.5^2}{4}3^2$$

$\xi = 0.6246$

Hence damping ratio = 0.6246

damped natural property

$$\omega_d = \omega_n\sqrt{1-\xi^2}$$

= 2.5 rad/h

undamped natural property

$$\omega_n = \frac{2\xi}{\sqrt{1-\xi^2}}$$

$$= \frac{2.5}{\sqrt{1-0.6296^2}}$$

= 3.2 rad/n.

132. Given data,

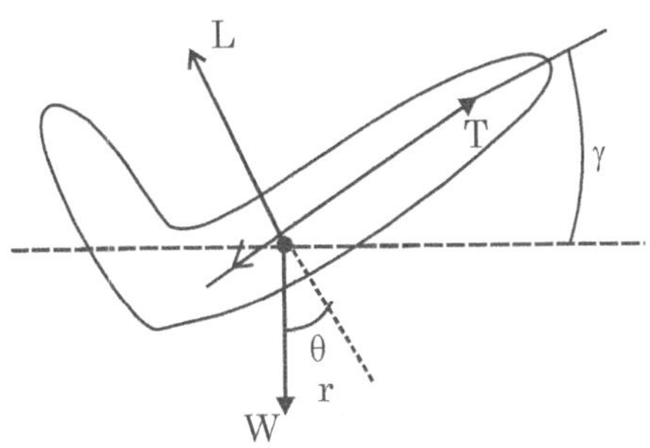

$\gamma = 10^8$

$$\frac{C_L}{C_D} = 10$$

Resolving components along the fling data.

$$(T = D + W\sin\gamma)$$

$$\frac{T}{W} = \frac{D}{W} + \sin\gamma \quad ...(1)$$

Perpendicular to flight path

$L = W\cos\gamma$

$$W = \frac{L}{\cos\gamma} \quad ...(2)$$

from equation (1) and (2) we will get

$$\frac{T}{N} = \frac{D}{L}\cos\gamma + \sin\gamma$$

$$\frac{T}{W} = \frac{\cos 10°}{10} + \sin\gamma$$

$$\boxed{\frac{T}{W} = 0.272}$$

133. Given data

$C_{mal_{ac}} = -0.02$

$C_{L_o} = 0.2$

$h_{ac,\,w} = 0.3$

$h_{cg,\,aircraft} = 0.4$

pitching moment coefficient of the wing at zero lift condition

$$C_{M_0} = C_{mac} + C_L(h - h_{ac}) - V_H C_{L\alpha}(St + i_t + \varepsilon_0)$$

pitching moment coefficient for wing body at zero angle lift condition

$$C_{M_0} = C_{Mac} + C_L(h - h_{ac})$$

$= -0.02 + 0.2\,(0.4 - 0.3)$

$$\boxed{C_{M_o} = 0}$$

134. In case of jet propelled Aircraft maximum endurance occurs at P_{min} condition. In order to move a speed $V > V_{minimum\ drag}$ & maintain steady level flight the pilot must use the eleuator to reduce the angle of attack while simultaneously increasing thrust.

137. $$S_{L_o} = \frac{1.44w^2}{\rho_\infty g\, S\, C_{L\,max} T},\ T \propto e_\infty$$

$$S_{L_o} \propto \frac{1}{e_\infty^{\ 2}}$$

on the hotter days, temperature increases density decreases.

$P = eRT$

$$ROH = \frac{P}{RT}$$

T – increases ROH – decrease

Hence the airplane requires a longer ground roll on hotter days.

139. In tropo pause regima

$$\frac{e_2}{e_1} = \left(\frac{T_2}{T_1}\right)^{\frac{g}{aR}-1}$$

Where R = 287 J/Kgk,

a = 0.0060 K/m

$$\frac{e_2}{e_1} = \left(\frac{288.16 - 0.0060 \times 3500}{288.16}\right)\frac{9.8}{0.0060 \times 287} - 1$$

$$\frac{e_{3500}}{e_{32}} = 0.7045$$

140. W = 400000 Kg

V = 240 m/sec

$$\left(\frac{L}{D}\right) = 15$$

P_R = ?

at sea level, cruising flight

L = W = 140000 × 9.8

$$T = D = \frac{W}{\left(\frac{L}{D}\right)} = \frac{40000 \times 9.8}{15}$$

= 261600N

Power required to cruise P = D × V

= 261600 × 40

= 62.784 MW

$$\boxed{P_R = 62.78 \text{ MV}}$$

141. Given

$C_{L\alpha} = 5$

$C_{M,\,cg} = 0.05 - 4\alpha.$

$$\frac{\partial C_{M,cg}}{\partial \alpha} = -4$$

$$-\left(\frac{dC_M}{d_{C_L}}\right) = X_n - X_{cg}$$

$$X_n - X_{cg} = \frac{dC_M}{d_{C_L}} \propto \frac{d\alpha}{dC_L}$$

$$= \frac{dC_M}{d_{C_L}} \propto \frac{d\alpha}{C_L} = -0.8$$

142. Given

W = 2000 Kg

V = 230 m/sec

ρ_s = 1.23 Kg/m³.

at 9000 Km

P = 0.31 atm

T = 230 K

ρ_a = 0.47 Kg/m³

Now relation between true airspeed and equivalent airspeed.

$$V_T = \frac{V_E}{\sqrt{\frac{e}{e_{SL}}}} = \frac{130}{\sqrt{\frac{0.27}{125}}} = 210 \text{m/sec}$$

143. $P_0 = 55000 \text{ N/m}^2$

$Ps = 45280 \text{ N/m}^2$

e_s = 0.6 Kg/m³

S = 1622 m²

C_L = 2

$P_0 = P_s - P_d$

$$P_0 - P_s = \frac{1}{2}\rho V_\infty^2$$

$$V_\infty = \sqrt{\frac{2(P_0 - P_s)}{\rho}}$$

At steady level flight

L = W

$$C_L \times \frac{1}{2}\rho A V_\infty^2 = W$$

$$\frac{W}{A} = C_L \times \rho \frac{1}{2} V_\infty^2$$

$$= C_L \times \frac{\rho}{2} \times \left(\frac{2(P_0 - P_s)}{\rho}\right)$$

$$= 2 \times \frac{0.6}{2} \times 2\frac{2(55000 - 45000)}{0.6}$$

$$= 19440 \text{ N/m}^2$$

144. W = 7000 N

W = 16m²

C_L = 2

e = 1.23 Kg/m³

V_{stall} = ?

$$V_{stall} = \sqrt{\frac{2W}{\rho S\, C_{L\,max}}} = \sqrt{\frac{2 \times 7000}{1.23 \times 16 \times 2}}$$

= 18.85 m/sec

Stalling velocity of aircraft

= 19 m/sec

145. 0.195 – 0.206

146. In case of head wind ground roll distance decreases and increases in case of tail wind.

147. In option (b), (c), (d) response of system is decreases or going to be constant, so considered as dynamically stable system.

148. Given $\frac{L}{D} = 18$

Gliding angle

$$\tan\theta = \frac{1}{L/D}$$

where, $\left(L/D\right)$ is Lift to drag ratio

θ is gliding angle

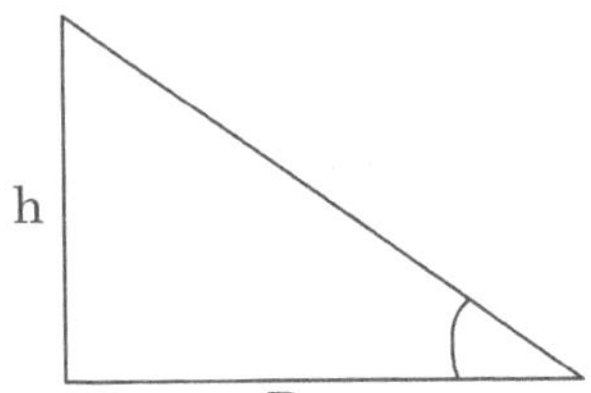

From figure

$$\tan\theta \frac{h}{R}$$

$$\frac{h}{R} = \frac{1}{L/D}$$

$$\Rightarrow R = h\left(L/D\right)$$

For R_{max} $\left(L/D\right)$ should be maximum

$$R_{max} = h\left(L/D\right)_{max}$$

$$= 4\times18 = 72 \text{ km}$$

149. Given, $\alpha = 3°$

$V_h = 300$ km/h

$V_r = 15.72$ km/h

$$\therefore \tan\beta = \frac{V_v}{V_h} = \frac{15.72}{300} = 2.99°$$

pitch attitude = $\alpha + \beta = 3 + 2.99 = 5.99° = 6°$

150. $$V_{equivalent} = V_{true}\sqrt{\frac{\rho_a}{\rho_s}}$$

$$= 50 \times \sqrt{\frac{0.91}{1.225}} = 43.09 \text{m/s}$$

152.

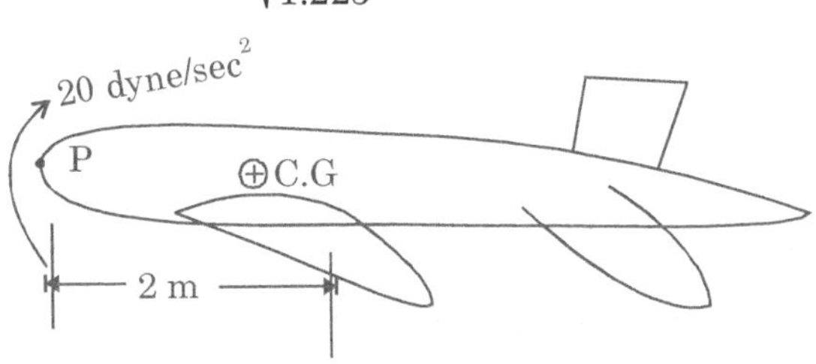

Let fr be the vertical upward force at point P due to angular acceleration α.

$$F_V = \frac{I\alpha}{r}$$

The vertical upward force at CG due to pole pace up maneuver = $F_{V.C.G} = \eta w$

Therefore

$$F_{Net} = F_V + F_{V.CG} = \frac{I\alpha}{r} + nw$$

vertical load factor at point

$$P = \eta_p = \frac{F_{net}}{W} = \frac{I\alpha}{Wr} + \eta = \frac{mr^2\alpha}{Wr} + \eta$$

$$\frac{mr^2\alpha}{mgr} + \eta$$

$$\boxed{\eta_p = \frac{r\alpha}{g} + \eta}$$

$$\eta_p = \frac{2 \times 20 \times \frac{\pi}{180}}{9.81} + 2 = 2.021$$

153. W = 8000kN

$C_{BO} = 0.02$

e = 0.8

AR = 6

$C_L = 0.375$

We have

$e_{Dj} = kC_L^2$

for same velocity flight

$$\frac{L_{climb}}{L_{steady}} = \cos\theta$$

$$\frac{C_{Lclimb}}{C_{lsteady}} = \cos\theta$$

$$\frac{C_{Dj\,steady}}{C_{Dj\,climb}} = \frac{k\,C_L^2 steady}{kC_L^2 climb}$$

$$= \left(\frac{C_{Lsteady}}{C_{Lclimb}}\right)^2$$

$$= \left(\frac{1}{\cos^2\theta}\right) = \left(\frac{1}{\cos\theta}\right)^2$$

at $\theta = 30°$

$$\frac{C_{Djsteady}}{C_{Dj\,climb}} = \left(\frac{1}{\cos\theta}\right)^2$$

$$= \left(\frac{1}{\cos 30}\right)^2$$

$$= \left(\frac{1}{\sqrt{\frac{3}{2}}}\right)^2$$

$$= \frac{4}{3} = 1.33$$

154. (1477.27 N)

$W_0 = 4905$ N

$S = 6.84\ m^2$

$$\frac{C_L^{\frac{3}{2}}}{C_D} = 13$$

$\eta_p = 0.9$

c = 0.76 kg/kw-hr

e = 1.225 kg/m³

g = 9.8 m/sec²

E = 18 hrs = 18 × 60 × 60 sec

$$E = \frac{\eta_P}{C}\frac{C_L^{\frac{3}{2}}}{C_D}\sqrt{2 l_{\propto} S}\left[\frac{1}{\sqrt{W_1}} - \frac{1}{\sqrt{W_0}}\right]$$

$$\Rightarrow 18 \times 3600 = \frac{0.9}{2.071 \times 10^{-6}} \times 13\sqrt{2 \times 1.22 \times 6.84}$$

$$\left[\frac{1}{\sqrt{W_1}} - \frac{1}{\sqrt{4905}}\right]$$

$$\Rightarrow \frac{18 \times 3600 \times 2.07 \times 10^{-6}}{0.9 \times 13 \times \sqrt{2 \times 1.225 \times 6.84}} = \frac{1}{\sqrt{W_1}} - \frac{1}{\sqrt{4905}}$$

$W_1 = 3427.72$

$W_f = W_0 - W_1 = 4905 - 3427.72 = 1477.27$ N

155. $\ell_{\eta\beta} = \frac{\partial \ell_N}{\partial \beta} = -\eta_t \overline{V_t}\left(\frac{\partial \ell_2}{\partial \propto}\right)$

$$= \frac{\frac{1}{2}eV_f^2}{\frac{1}{2}ev^2}\frac{Sv\ell v}{SwT}\left(\frac{\partial \ell_2}{\partial \propto}\right)$$

$$= \frac{\frac{1}{2}eV_f^2}{\frac{1}{2}ev^2}\frac{1.25_v \propto 0.8\ell_v}{S_w T}\left(\frac{\partial \ell_2}{\partial \propto}\right)$$

$$= 0.96 \times \frac{\frac{1}{2}ev_f^2}{\frac{1}{2}ev^2}\frac{S_v \ell_v}{S_w T}\left(\frac{\partial \ell_2}{\partial \propto}\right)$$

$$\frac{(\ell_{N\beta})_{\text{modified}}}{(\ell_{N\beta})_{\text{original}}} = \frac{0.96}{1} = 0.96$$

156. For lateral and directional stability of aircraft,

Roll control derivative, $C_{l\delta r} > 0$

Yaw control derivative, $C_{n\delta r} < 0$

157.

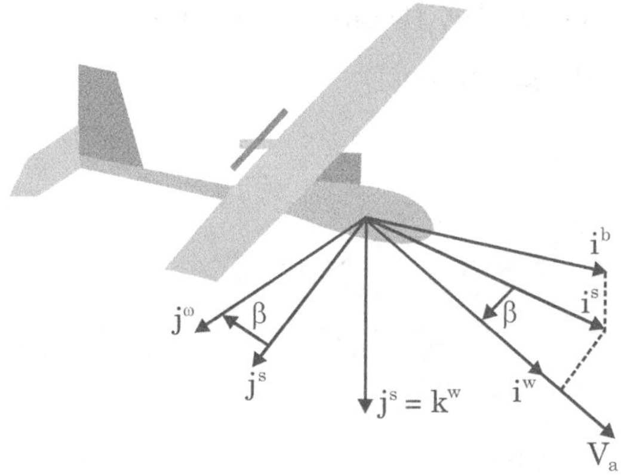

Airspeed, Angle of Attack, Sideslip Angle

$$V_a^b = \begin{pmatrix} u_r \\ v_r \\ w_r \end{pmatrix}$$

$$= R_w^b \begin{pmatrix} V_a \\ 0 \\ 0 \end{pmatrix}$$

$$= \begin{pmatrix} \cos\beta\cos\alpha & -\sin\beta\cos\alpha & -\sin\alpha \\ \sin\beta & \cos\beta & -\sin\beta\sin\alpha \\ \cos\beta\sin\alpha & 0 & \cos\alpha \end{pmatrix}\begin{pmatrix} V_a \\ 0 \\ 0 \end{pmatrix}$$

$$\begin{pmatrix} u_r \\ v_r \\ w_r \end{pmatrix} = V_a \begin{pmatrix} \cos\alpha\cos\beta \\ \sin\beta \\ \sin\alpha\cos\beta \end{pmatrix}$$

$$V_a = \sqrt{u_r^2 + v_r^2 + w_r^2}$$

$$\alpha = \tan^{-1}\left(\frac{w_r}{u_r}\right)$$

$$\beta = \sin^{-1}\left(\frac{u_r}{\sqrt{u_r^2 + v_r^2 + w_r^2}}\right)$$

$$\beta = \sin^{-1}\left(\frac{-10}{\sqrt{(100)^2 + (-10)^2 + (20)^2}}\right)$$

$\beta = 5.6°$

158. For direction stability of Aircraft $C_{\eta\beta} > 0$ plotting the given data,

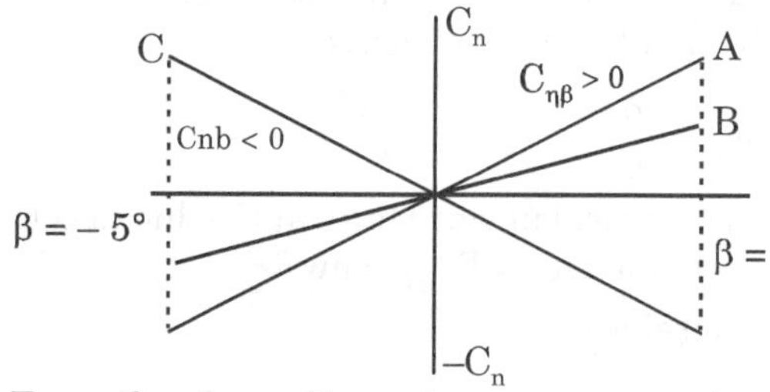

From the above discussion, we can say that C is unstable. A and B are stable with A being more stable than B.

159. Eigen value for plugoid mode $\lambda_{1,2} = -0.02 \pm i0.25$

$$g = 9.8 \text{ m/s}^2$$

$$\lambda_{1,2} = \sqrt{0.02^2 + 0.25^2} = 0.251$$

Frequency of phugoid mode is given as

$$\omega = \frac{\sqrt{2}g}{V}$$

$$V = \frac{\sqrt{2}g}{\omega}$$

$$= \frac{\sqrt{2} \times 9.8}{0.251} = 55.27 \text{ m/s}$$

160. Given

$$\eta = 0.95$$

$$V_H = 0.453$$

$$\frac{\partial \varepsilon}{\partial \alpha} = 0.35$$

$$C_{L\alpha}^{Wb} = 4.8 \text{ rad}^{-1}$$

$$C_{lx}^{t} = 4.4 \text{ rad}^{-1}$$

Expression for neutral point

$$\Rightarrow \quad x_n = x_{ac} + \eta V_H \frac{C_{L\alpha}^{Wb}}{C_{lx}^{t}}\left(1 - \frac{\partial \varepsilon}{\partial \alpha}\right)$$

$$\Rightarrow x_n - x_{ac} = \eta V_H \frac{C_{L\alpha}^{Wb}}{C_{lx}^{t}}\left(1 - \frac{\partial \varepsilon}{\partial \alpha}\right)$$

$$= 0.95 \times 0.453 \times \frac{4.8}{4.4}(1 - 0.35)$$

$$= 0.2565$$

In precentage

$$(x_n - x_{ac}) = 0.2565 \times 100$$

= 25.65% of mean Aerodynamic chord.

161. Given,

$$C_D = 0.025 + 0.049\, C_L^2$$

$$\rho_\infty = 1.225 \text{ kg/m}^3$$

$$\frac{W}{S} = 9844 \text{ N/m}^2$$

For maximum range

$$C_{Do} = C_{Di} = KC_L^2$$

$$C_L = \sqrt{\frac{C_{Do}}{K}} = \sqrt{\frac{0.025}{0.049}} = 0.7143$$

and $\quad L = W$

$$\Rightarrow \quad V_\infty = \sqrt{\frac{2W}{C_L \rho S}}$$

$$= \sqrt{\frac{2W}{S} \times \frac{1}{C_L} \times \frac{1}{\rho}}$$

$$= \sqrt{2 \times 9844 \times \frac{1}{1.225} \times \frac{1}{0.7145}}$$

$$= 150 \text{ m/s}$$

162. Given,

$$m = 4000 \text{ kg}$$

$$S = 25 \text{ m}^2$$

$$C_L = 1.65$$

$$\rho = 1.225 \text{ kg/m}^3$$

$$g = 9.8 \text{ m/s}^2$$

$$fs = 1.25$$

$$\eta = 3.2$$

$$\text{Load factor, } \eta = \frac{L}{W}$$

$$\text{Safe load factor, } \eta_{safe} = \frac{\eta}{fs}$$

$$L = \eta_{safe} W$$

$$C_L \times \frac{1}{2}\rho_\infty S V_\infty^2 = \eta_{safe} W$$

$$\Rightarrow \quad V_\infty = \sqrt{\frac{\eta_{safe} W}{C_L \rho_\infty S}}$$

$$= \sqrt{\frac{\eta W}{C_L \times \rho_\infty \times S \times fs}}$$

$$= \sqrt{\frac{3.2 \times 4000 \times 9.8}{1.65 \times 1.225 \times 25 \times 1.25}}$$

$$= 70.46 \text{ m/s}$$

163. Given,

$$h = 3 \text{ km} = 3 \times 10^3 \text{ m}$$

$$P_{Oa} = 0.72 \text{ bar}$$

$$T_O = 268.66 \text{ K}$$

$$P \text{ infinity} = 1.01 \text{ bar}$$

$$T \text{ infinity} = 288.16 \text{ K}$$

$$g = 9.8 \text{ m/s}^2$$

$$\rho_\infty = 1.225 \text{ kg/m}^3$$

$$R = 287 \text{ J/(kgK)}$$

As per formula

$$P_a = P\text{ infinity } e^{\frac{-g\Delta h}{RT\text{ infinity}}}$$

$$= 1.01 \times e^{\frac{-9.8 \times (3000-0)}{287 \times 288.16}} = 0.70 \text{ bar}$$

Equivalent velocity is given as

$$V_{eq} = \sqrt{\frac{2(P_{oa} - P_a)}{\rho_\infty}}$$

$$= \sqrt{\frac{2 \times (0.72 - 0.70) \times 10^5}{1.225}}$$

$$= 57.14 \text{ m/s}$$

164. $V_T = \sqrt{\dfrac{2(P_T - P)}{e_a}}$

$V_{eq} = \sqrt{\dfrac{2(P_T - P)}{e_5}}$

$\dfrac{V_T}{V_{eq}} = \sqrt{\dfrac{e_s}{e_a}}$

According to problem

$\dfrac{V_{TAS}}{V_{EAI}} = \sqrt{\dfrac{e_0}{e}}$

165.

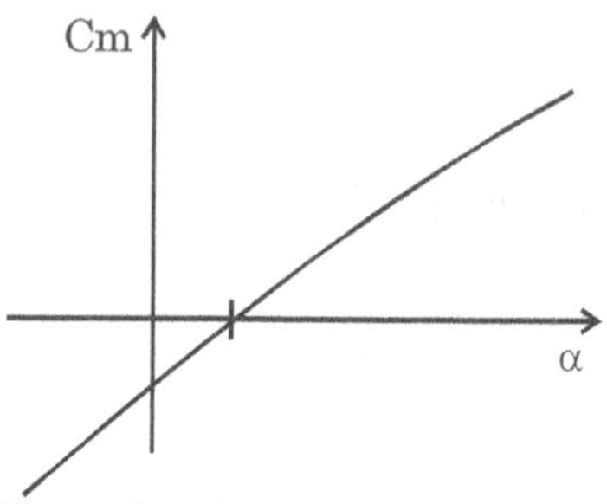

From the above figure

$\dfrac{\partial C_m}{\partial \alpha} = +ve$ or $\dfrac{\partial C_m}{\partial \alpha} > 0$

For positive α, $C_m = 0$

For stable aircraft

$C_{m\alpha} < 0$

$\dfrac{\partial C_m}{\partial \alpha} < 0$

at trimed condition $C_m = 0$

from above discussion we can say that the aircraft can trim at positive α, and Aircraft is unstable.

166. Given

$\dfrac{W}{S} = 1800\ N/m^2,\ S = 30\ m^2,\ C_D = 0.02 + 0.04\ C_L^{\ 2}$

$C_{Lmax} = 1.6,\ e_\infty = 1.225\ kg/m^3,\ C_{DO} = 0.02,\ K = 0.4$

Endurance of Jet Aircraft

$E_g = \dfrac{C_L}{C_D} \log_e \dfrac{W_0}{W_1}$

As E_g to be maximum Aircraft should fly at $\left(\dfrac{C_L}{C_D}\right)$ max

For $\left(\dfrac{C_L}{C_D}\right)$ max $\quad C_L = \sqrt{\dfrac{C_{DO}}{K}} = \sqrt{\dfrac{0.02}{0.04}} = \sqrt{\dfrac{1}{2}}$

$\therefore\ V_\infty = \sqrt{\dfrac{2W}{e_\infty S C_L}} = \dfrac{\sqrt{2 \times 1800}}{1.225 \times \sqrt{\dfrac{1}{2}}} = 64.475\ m/sec$

V∞ = 65.5 m/sec

167. Calculating

Yawing moment $= T \times \dfrac{b}{2}$

= 8000 × 5

= 40000 Nm

= 40 kNm

Coefficient of Yawing moment (C_n)

$= \dfrac{\text{Yawing moment}}{9_\infty Sb}$

$= \dfrac{40 \times 10^3}{\dfrac{1}{2} \times e_\infty \times V_\infty{}^2 Sb}$

$= \dfrac{40 \times 10^3}{\dfrac{1}{2} \times 1.225 \times 100^2 \times 50 \times 10} = 0.01308$

Now writing C_n is terms of α, δ r, β

$C_n = C_{n\alpha}\cdot\alpha'' + C_{n8_r} 8r + C_{n\beta}\beta \qquad ...(1)$

For straight and level flight

$\alpha'' = 0$

For zero side slip β = 0

From equation (1)

$\therefore\ C_n = C_{n8_r}{}^{8r}$

$\Rightarrow 0.01308 = -0.0028r$

$8r = -\dfrac{0.01308}{0.002} = -6.54°$

168.

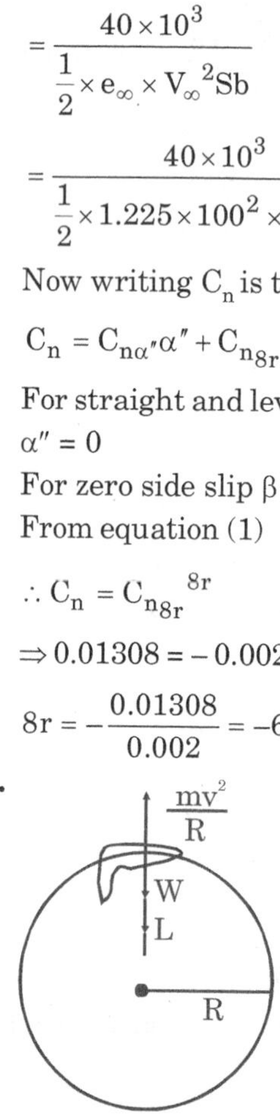

$L + W = \dfrac{mv^2}{R}$

$\left(\dfrac{L}{W} + 1\right) = \dfrac{mv^2}{RW}$

$n + 1 = \dfrac{V^2}{Rg}$

$n = \dfrac{V^2}{Rg} - 1$

$\dfrac{V^2}{Rg}$ is always positive

$\therefore\ n > -1$

$n > -1$

169. Natural frequency for short period mode is given as

$$W_n = \sqrt{\underset{(i)}{-M_\alpha} + \underset{(ii)}{Z_\alpha \frac{M_q}{u_0}}} \quad ...(1)$$

where $m_\alpha = \frac{\partial M}{\partial \alpha}$

$m_q = \frac{\partial M}{\partial q}$
↓
damping parameter

$Z_\alpha = \frac{\partial Z}{\partial \alpha}$

$C_{M\alpha} \alpha M_\alpha$

$\frac{\partial C_M}{\partial \alpha} \alpha \frac{\partial M}{\partial \alpha}$ (C_M = Pitching moment coefficient)

When $C_{M\alpha}$ will be more negative $\frac{\partial M}{\partial \alpha}$ will be more negative, then writing W_n for this case.

$$W_n = \sqrt{-(-M\alpha) + Z\alpha \frac{M_q}{u_0}}$$

$$= \sqrt{M_\alpha + Z_\alpha \frac{M_q}{u_0}}$$

So natural frquency increases.
Looking for 2nd term keeping (i) constant usually
Z_α = – ve
when $m_q \rightarrow$ –ve
then product of $Z_\alpha M_q$ = +ve
Hence W_n increase
∴ Hence natural frequency increases
Hence option (b) is incorrect.

C_{lp} = rolling moment due to rolling velocity or damping in roll parameter.
C_I = rolling moment coefficient
P = rolling velocity

$$C_{Ip} = \frac{\partial C_L}{\partial \left(\frac{Pb}{2V_\infty}\right)}$$

Usually wing contributes more in roll stability and roll damping for roll stability the primary factor is dihedral wing and swept wing. So option (c) is correct.

Increasing the size of vertical fin, it increases more yawing stability that means more Yaw damping. So option (d) is also correct.

170. Piston powered
Service ceiling → $(R/C)_{max}$ = 100 ft/min. 0.5 m/sec
Absolute ceiling → $(R/C)_{max}$ = 0

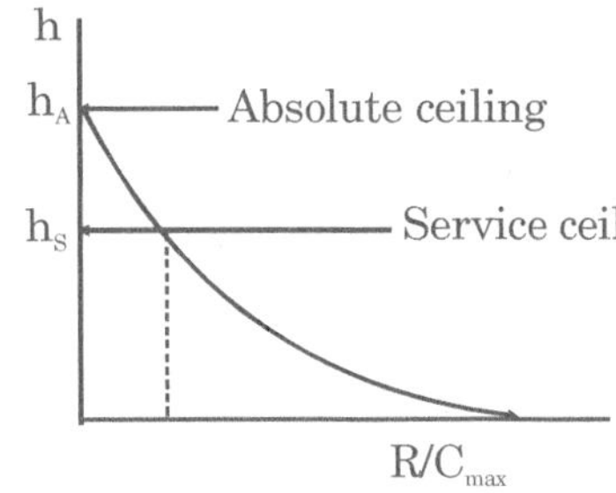

$h_A > h_S$
For
Jet powered
Service ceiling → $(R/C)_{max}$ = 500 ft/min
Absolute ceiling → $(R/C)_{max}$ = 0

Stall speed $V_S = \sqrt{\frac{2\omega}{\rho S C_{L_{max}}}}$

AS h ↑ ρ ↓ h → Increases
e → decreases
∴ AS e → decreases
V_S → Increases
In case of tail wind

$V_g = V_\infty + V_{T\omega}$
V_g = ground velocity
V_∞ + free stream velocity
$V_{T\omega}$ = tail wind velocity
General Range formula,

$$R = \frac{V_g}{C_t} = \frac{C_L}{C_D} \ln \frac{\omega_0}{\omega_1}$$

$R \propto V_g$
V_g → Increases
R → Increases
For Jet Aircraft

$$R_{Jet} = \frac{1}{C_t}\left(\frac{C_L^{1/2}}{C_D}\right) 2\sqrt{\frac{2}{e_\infty S}}\left(\omega_0^{1/2} - \omega_1^{1/2}\right)$$

R_{Jet} ↑ as e ↓
e → decrease
R_{Jet} → Increases

171. Primary function of vertical fin is to provide more directional stability but also provides lateral stability. So option (a) is correct and (b) is wrong. Presence of dihedral wing provides lateral stability and presence of anhedral wing decreases the lateral stability of Aircraft. So option (c) is incorrect. Option (d) is correct because Horizontal tail have stablizing effect on longitudinal stability.

172. Given

m = 5500 kg = 5500 × 10 = 55KN

V_∞ = 225 m/sec, R = 775 m, L_t = ?

Pull up maneuver

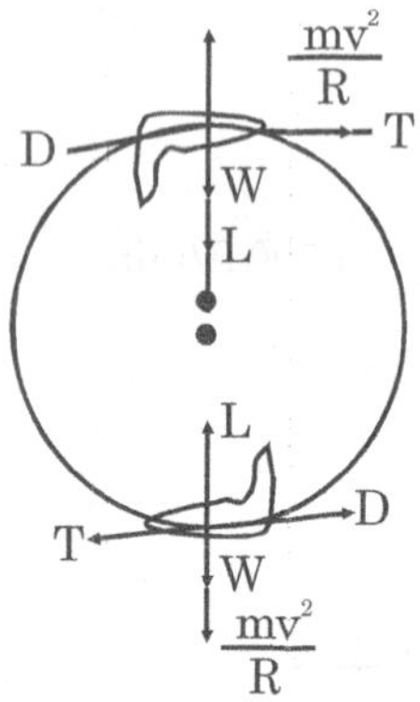

From figure given is problem moment about CP

$$-\left(W+\frac{mv^2}{R}\right)\times 0.02+L_T\times 5.5=0$$

$$-\left(55\times 10^3+\frac{5500\times 225^2}{775}\right)\times 0.02+L_t\times 5.5=0$$

L_t = 15 KN

173. Given

m = 10000 kg

W = 10000 × 10 = 10^5N

e = 1

b = 10 m

S = 30m²

$C_{DO} = 0.025$

V_∞ = 100 m/sec

e = 1.225 kg/m³

At cruise

L = W, T = D

$C_D = C_{DO} + KC_L^2$

$$= 0.025+\frac{C_L^2}{\pi eAR}$$

$$= 0.025+\frac{C_L^2}{\pi\times 1\times\frac{100}{30}} \qquad \left[\because AR=\frac{b^2}{5}\right]$$

$$= 0.025+\frac{30C_L^2}{100\pi} \qquad \ldots(1)$$

Now

$$C_L=\frac{2W}{eV_\infty^2S}=\frac{2\times 10^5}{1.225\times 100^2\times 30}=0.544$$

Putting the value of C_L in equation (1), we get

$$C_D=\frac{0.025+30\times(0.544)^2}{100\times 3.14}$$

= 0.053

$$D=C_D\times\frac{1}{2}eV_\infty^2S$$

$$= 0.053\times\frac{1}{2}\times 1.225\times 100^2\times 30 = 97.90\text{N}$$

174. $$\frac{\partial C_m}{\partial_\alpha}=a_w\left[h-h_{ae}-n_tV_H+\frac{a_t}{a_w}\left(1-\frac{\partial\varepsilon}{\partial_\alpha}\right)\right] \qquad \ldots(1)$$

Where

$$V_H=\frac{S_tl_t}{S\bar{C}}=\frac{80\times l_t}{350\times 9}$$

$l_t = (0.3 - 0.25)C$

$$l_t=\bar{l}_t-\left(\frac{x_{cg}}{\bar{c}}-\frac{x_{ac}}{\bar{c}}\right)C$$

= 28 – (0.3 – 0.25) × 9

= 28 – 0.45

= 27.55 m

$$\therefore\ V_H=\frac{80\times 27.55}{350\times 9}=0.699$$

$n_t = 1$

$\varepsilon = 0.4\alpha$

$$\frac{\partial_\alpha}{\partial_\alpha}=0.4$$

from equation (1), we get

$$\frac{\partial C_m}{\partial_\alpha}=0.1\left[0.3-0.25-1\times 0.699\times\frac{0.068}{0.1}(1-0.4)\right]$$

= – 0.024 per degree

3

CHAPTER

SPACE DYNAMICS

1. An artificial satellite remains in orbit and does not fall to earth because

(*a*) The centrifugal force acting on it balances the gravitational attraction

(*b*) The on- board rocket motor provide continuous boost to keep it in orbit.

(*c*) It's transverse velocity keeps it from hitting the earth although it falls continuously

(*d*) Due to its high speed, it derives sufficient lift from the rarefied atmosphere.

[GATE 2007, 1 M]

2. Two identical earth satellite A and B are in circular orbits at altitude h_A and h_B above the Earth's surface respectively, with $h_A > h_B$. If E denotes the total mechanical energy, T the kinetic energy and V the gravitational potential energy of a satellite, then:

(*a*) $E_A > E_B$ and $V_A < V_B$

(*b*) $E_A > E_B$ and $T_A > T_B$

(*c*) $E_A < E_B$ and $T_A > T_B$

(*d*) $E_A > E_B$ and $T_A < T_B$

[GATE 2007, 2 M]

3. The earth radius is 6.37×10^6 m and the acceleration due to gravity on its surface is 9.81 m/s^2. A satellite is in circular orbit at a height of 6.30×10^5 m above the earth's surface. The minimum additional speed it needs to escape from the earth's gravitational field is

(*a*) 3.66×10^3 m/s (*b*) 3.12×10^3 m/s

(*c*) 3.27×10^3 m/s (*d*) 3.43×10^3 m/s

[GATE 2007, 2 M]

4. The earth radius is 6.37×10^6 m and the acceleration due to gravity on its surface is 9.81 m/s^2. A satellite is in circular orbit at a height of 35.9×10^6 m above the earth's surface. This orbit is inclined at 10.5° to the equator. The velocity change needed to make the orbit equatorial is

(*a*) 561 m/s at 84.75° to the initial direction

(*b*) 561 m/s at 95.25° to the initial direction

(*c*) 281 m/s at 84.75° to the initial direction

(*d*) 281 m/s at 95.25° to the initial direction

[GATE 2007, 2 M]

5. To transfer a satellite from an elliptical orbit to a circular orbit having radius equal to the apogee distance of the elliptical orbit, the speed of satellite should be

(*a*) Increased at the apogee

(*b*) Decreased at the apogee

(*c*) Increased at the perigee

(*d*) Decreased at the perigee

[GATE 2008, 1 M]

6. Which of the following remains constant for a satellite in an elliptical orbit around the earth?

(*a*) Kinetic Energy

(*b*) Product of speed and radial distance from center of earth

(*c*) Rate of area swept by the radial vector from the center of the orbit

(*d*) Rate of area swept by the radial vector from the center of the earth

[GATE 2008, 2 M]

7. A planet is observed to be at its slowest when it is at a distance r_1 from the sun and at its fastest when it is at a distance r_2 from the sun. the eccentricity e of the planet's orbit is given by

(*a*) $e = \frac{r_1}{r_2}$ (*b*) $e = \frac{r_1 - r_2}{r_2 + r_1}$

(*c*) $e = \frac{r_2}{r_1}$ (*d*) $e = \frac{r_1 + r_2}{r_1 - r_2}$

[GATE 2008, 2 M]

8. The velocity required for a spacecraft to escape earths gravitational field depends on

(*a*) The mass of the spacecraft

(*b*) The distance between earth's Centre and the spacecraft

(*c*) The earth's rotational speed about its own axis

(*d*) The earth orbital speed

[GATE 2008, 2 M]

9. Which of the following statements is true as the altitude increases in stratosphere of international standard Atmosphere?

(*a*) Temperature increases and dynamic viscosity decreases

(*b*) Temperature remains constant and pressure increases

(*c*) Temperature decreases and sound speed decreases.

(*d*) Temperature remains constant and density decreases.

[GATE 2008, 2 M]

10. The life of a geostationary communication satellite is limited by

(*a*) The working life of the on-board electronic circuitry

(*b*) The time it takes for its orbit to decay due to atmospheric drag

(*c*) The quantity of on-board fuel available for station keeping

(*d*) The number of meteorite impacts that the satellite structure can withstand before breaking up

[GATE 2009, 1 M]

11. The acceleration due to gravity on the surface of Mars is 0.385 times that on earth, and the diameter of Mars is 0.532 times that of earth. The ratio of the escape velocity from the surface of earth is approximately

(*a*) 0.453 (*b*) 0.205

(*c*) 0.851 (*d*) 0.724

[GATE 2009, 2 M]

12. A spacecraft of mass 100 kg moving at an instantaneous speed of 1.8×10^4 m/s, picks up interstellar dust at the rate of 3.2×10^{-8} kg/s. Assuming that the dust was initially at rest, the instantaneous rate of retardation of the spacecraft is

(*a*) 7.9×10^{-3} m/s^2 (*b*) 2.3×10^{-3} m/s^2

(*c*) zero (*d*) 5.8×10^{-6} m/s^2

[GATE 2010, 2 M]

13. The angular momentum about the center of mass of the earth of an artificial satellite in a highly elliptical orbit is

(*a*) A maximum when the satellite is farthest from the earth

(*b*) A constant

(*c*) Proportional to speed of the satellite

(*d*) Proportional to the square of the speed of the satellite

[GATE 2010, 2 M]

14. The Hohmann ellipse used as Earth - Mars transfer orbit has

(*a*) Apogee at Earth and Perigee at Mars

(*b*) Apogee and perigee at Earth

(*c*) Apogee at Mars and Perigee at Earth

(*d*) Apogee and Perigee at Mars

[GATE 2012, 1 M]

15. Consider the low Earth Orbit (LEO) and Geo Synchronous Orbit (GSO). Then,

(*a*) ΔV requirement for launch to LEO is greater than that for GSO and altitude of LEO is lower than that of GSO.

(*b*) ΔV requirement for launch to LEO is lower than that for GSO and altitude of LEO is lower than that of GSO.

(*c*) ΔV requirement for launch to LEO is greater than that for GSO and altitude of LEO is greater than that of GSO.

(*d*) ΔV requirement for launch to LEO is lower than that for GSO and altitude of LEO is greater than that of GSO.

[GATE 2013, 1 M]

16. Considering R as the radius of the moon, the ratio of the velocities of two spacecraft orbiting moon in circular orbit at altitudes R and 2R above the surface of the moon is_______________.

[GATE 2014, 1 M]

17. A satellite is to be transferred from its geostationary orbit to a circular polar orbit of the same radius through a single impulse out-of-plane maneuver. The magnitude of the change in velocity required is __________ times the magnitude of the escape velocity.

[GATE 2015, 2 M]

18. A planetary probe is launched at a speed of 200 km/s and at a distance of 71,400 km from the mass center of its nearest planet of mass 1.9×10^{28} kg, The universal gravitational constant, G $= 6.67 \times 10^{-11}$ m^3/kgs^2. The ensuring path of the probe would be

(*a*) elliptic (*b*) hyperbolic

(*c*) parabolic (*d*) circular

[GATE 2015, 2 M]

19. The elliptical area swept by a satellite is 5.6×10^9 km^2 in one full orbit. Its angular speed is observed to be 0.00125 rad/s when it is at a distance of 7,200 km from the center of mass of its primary. Its orbital period (in Earth days) is________.

[GATE 2015, 2 M]

20. A satellite is injected at an altitude of 350 km above the Earth's surface, with a velocity of 8.0 km/s parallel to the local horizon. (Earth radius = 6378 km, μE (GM = Gravitational constant × Earth mass) $= 3.986 \times 10^{14}$ m^{3}s^{-2}. The satellite

(*a*) forms a circular orbit.

(*b*) forms an elliptic orbit.

(*c*) escapes from Earth's gravitational field.

(*d*) falls back to earth.

[GATE 2016, 2 M]

21. The period of revolution of earth about the sun is 365.256 days, approximately. The semi major axis of the earth's orbit is close to 1.4953×10^{11} m. The semi-major axis of the orbit of Mars is 2.2783×10^{11} m. The period of revolution of Mars, about the sun is________Earth days(in three decimal place)

[GATE 2017, 1 M]

22. The tangential velocity component 'V' of a spacecraft, which is in a circular orbit of radius 'R' around a spherical Earth (μ = GM $\rightarrow$ gravitational parameter of Earth) is given by the following expression.

(*a*) $V=\sqrt{\frac{\mu}{2R}}$ (*b*) $V=\sqrt{\frac{\mu}{R}}$

(*c*) $V=\frac{2\pi}{\sqrt{\mu}}R^{\frac{3}{2}}$ (*d*) $V=\frac{2\pi}{\sqrt{\mu}}R^{\frac{2}{3}}$

[GATE 2018, 1 M]

23. Equation of the trajectory of a typical space object around any planet, in polar coordinates (r, θ) (i.e. a general conic section geometry), is given as follows. (h is angular momentum, μ is gravitational parameter, e is eccentricity, r is radial distance from the planet center, θ, is angle between vectors $\vec{e}$ and $\vec{r}$.

(*a*) $r=\frac{\left(\frac{h^2}{\mu}\right)}{1-e\cos\theta}$ (*b*) $r=\frac{\left(\frac{h^2}{\mu}\right)}{e-e\cos\theta}$

(*c*) $r=\frac{\left(\frac{h^2}{\mu}\right)}{1+e\cos\theta}$ (*d*) $r=\frac{\left(\frac{h^2}{\mu}\right)}{e+e\cos\theta}$

[GATE 2018, 1 M]

24. In an elliptic orbit around any planet, the location at which a spacecraft has the maximum angular velocity is

(*a*) apoapsis

(*b*) periapsis.

(*c*) a point at +45° from periapsis.

(*d*) a point at –90° from apoapsis.

[GATE 2018, 1 M]

25. A spacecraft forms a circular orbit at an altitude of 150 km above the surface of a spherical Earth. Assuming the gravitational parameter, $\mu = 3.986 \times 10^{14}$ m^3/s^2 and radius of earth, R_E = 6,400 km, the velocity required for the injection of the spacecraft, parallel to the local horizon, is ____________(accurate to two decimal places).

[GATE 2018, 2 M]

26. The product of earth's mass (M) and the universal gravitational constant (G) is GM = 3.986×10^{14}m^{3}sec^2. The radius of earth is 6371 km. The minimum increment in the velocity to be imparted to a spacecraft flying in a circular orbit around the earth at an altitude of 4000 km to make it exit earth's gravitational field is ______ km/ s (round off to 2 decimal places).

[GATE 2019]

27. For hyperbolic trajectory of a satellite of mass m having velocity V at a distance r from the center of earth (G: gravitational constant, M: mass of earth), which one of the following relations is true?

(*a*) $\frac{1}{2}mV^2 > \frac{GMm}{r}$ (*b*) $\frac{1}{2}mV^2 < \frac{GMm}{r}$

(*c*) $\frac{1}{2}mV^2 = \frac{GMm}{r}$ (*d*) $\frac{1}{2}mV^2 < \frac{2GMm}{r}$

[GATE 2020]

28. Burnout velocity of a space vehicle in a circular orbit at an angle 5 degrees above the local horizon around earth is 13.5 km/s. Tangential velocity of the space vehicle in the orbit is _____ km/s (*round off to two decimal places*).

[GATE 2020]

29. The ratio of tangential velocities of a planet at the perihelion and the aphelion from the sun is 1.0339. Assuming that the planet's orbit around the sun is planar and elliptic, the value of eccentricity of the orbit is _______ (*round off to three decimal places*).

[GATE 2020]

30. The velocity required to launch a space shuttle from the surface of the earth to achieve a circular orbit of 250 km altitude is _____ (*round off to two decimal places*).

For earth, Gm_e = 398,600.4 km^3/s^2 and surface radius R_0 = 6378.14 km.

[GATE 2021]

ANSWER KEY

1. (*a*)	**2.** (*d*)	**3.** (*b*)	**4.** (*a*)	**5.** (*a*)	**6.** (*d*)	**7.** (*b*)	**8.** (*b*)	**9.** (*d*)	**10.** (*b*)
11. (*a*)	**12.** (*d*)	**13.** (*b*)	**14.** (*c*)	**15.** (*b*)	**18.** (*b*)	**20.** (*b*)	**21.** (*c*)	**22.** (*b*)	**23.** (*c*)
24. (*b*)	**26.** (2.57)	**27.** (a)	**28.** (13.42 to 13.47)		**29.** (0.016 to 0.018)	**30.** (7.75 to 7.77)			

EXPLANATIONS

1. The centrifugal force acting on it balances the gravitational attraction.

2. Potential Energy $= \frac{-GMm}{R}$

Kinetic Energy $= \frac{1}{2}mu^2 = \frac{1}{2}m\frac{g\,Re^2}{R} = \frac{GMm}{2R}$

Total energy $= \frac{-GMm}{R} + \frac{GMm}{2R}$

$$E = \frac{-GMm}{2R}$$

Potential energy increases as satellite goes away from earth and KE decreases.

$h_A > h_B$

$E_A > E_B$, $T_A < T_B$

3. $R = 6.37 \times 10^6$ m

$g = 9.81$ m/s^2

$$g = \frac{GM}{R^2}$$

Satellite altitude $h = 6.3 \times 10^5$

Escape velocity $v_e = \sqrt{\frac{2gm}{(R+h)}}$

$$v_e = \sqrt{\frac{2gR^2}{(R+h)}}$$

$$v_e = \sqrt{\frac{2 \times 9.81 \times (6.37 \times 10^6)^2}{(7 \times 10^6)}}$$

= 10.66 km/s

Orbital velocity $v = \sqrt{\frac{gR^2}{(R^2+h)}} = \frac{v_e}{\sqrt{2}} = 7.54\text{km/s}$

Increment needed $v_e - v$

$10.66 - 7.54 = 3.12$ km/s $= 3.12 \times 10^6$ m/s

4. $r = (6.378 + 35.9) \times 10^6$ m = 42270 km

$$v = \sqrt{\frac{GM}{r}} = \sqrt{\frac{398600.4418}{42270}}$$

v = 3.07 km/s

$$\Delta v = 2v\sin\frac{\theta}{2}$$

$$\Delta v = 2 \times 3.07 \sin\frac{10.5°}{2}$$

Δv= 0,561 km/s

$\Delta\theta$= 10.5°

$2\theta + 10.5° = 180°$

$\alpha = 84.75°$

5. At apogee, the satellite is farthest from earth and has it's minimum speed. To make its KE equal to that to circular orbit, speed should be increased at apogee.

Speed at apogee $= \sqrt{\left(\frac{1-e}{1+e}\right)\frac{GM}{a}}$

Required speed $= \sqrt{\frac{GM}{a}}$

6. The KE changes as the speed of satellite changes. The angular momentum remain same, but it is vector product of radial vector from earth to satellite and its velocity vector multiplied by the mass of satellite. Hence, the scaler product of speed and distance need not be constant. By kepler's second law the rate of area swept by satellite from the center of earth remains constant.

7. Planet is slowest when it is farthest and fastest when it is nearest the sun.

$$r_1 = \frac{P}{1-e} \qquad r_2 = \frac{P}{1+e}$$

$$\frac{r_1}{r_2} = \frac{1+e}{1-e}$$

$$\frac{r_1 + r_2}{r_2 - r_1} = \frac{1}{e}$$

$$e = \frac{r_1 - r_2}{r_2 + r_1}$$

8. Escape velocity of the object is given by

$$v_e = \sqrt{\frac{2GM}{r}}$$

M is mass of earth

R = distance from earth's center

9. According to ISA model temperature remain constant in stratosphere and so speed of sound and dynamic viscosity, which depends only on temperature. Pressure and density decreases with altitude.

10. R_{orbit} for geostationary satellite is high, $\rho\,(R_{orbit}) \neq 0$ and finite, though very small in magnitude. However, after a long period, the atmosphere drag can alter the trajectory of the satellite to make it useless for the purpose.

11. Escape velocity

$$v_e = \sqrt{2gr}$$

Where r is radius of the planet

$$\frac{v_{e\,Mars}}{v_{e\,Earth}} = \sqrt{\frac{g_{Mars}}{g_{Earth}} \times \frac{r_{Mars}}{r_{Earth}}}$$

= 0.453

12. We know that,

$$\frac{d}{dt}(mv) = 0$$

$$v\frac{d}{dt}m\frac{dv}{dt} = 0$$

$$-(3.2 \times 10^{-8})(1.8 \times 10^4) + 100\frac{dv}{dt} = 0$$

$$\frac{dv}{dt} = 5.76 \times 10^{-6}\,m/s^2$$

13. According to Kepler's law,

Angular Momentum H = constant

Therefore, $H = r \times \omega$

14. The hofmann elliptical transfer orbit has the perigee at the Earth and apogee at the Mars as transfer is from Earth to Mars.

15. Geo Synchronous orbit has time period of 24 hours and having an altitude of 35786 km whereas low Earth Orbit has aa altitude below 2000 km. Therefore, ΔV requirement for launch to LEO is lower than that for GSO.

16. Given,

R = radius of the moon,

Altitudes of spacecraft orbiting the moon are R and 2R

$r_1 = R + R = 2R$

and $r_2 = 2R + R = 3R$

Now,

$$V_1 = \sqrt{\frac{GM}{r_1}} \text{ and } V_2\sqrt{\frac{GM}{r_2}}$$

$$\therefore \frac{V_1}{V_2} = \sqrt{\frac{r_2}{r_1}} = \sqrt{\frac{3R}{2R}} = 1.225$$

17. As per question,

$R_{geo} = R_{polar} = R$

Escape velocity is given as,

$$V_e = \sqrt{\frac{2GM}{R}}$$

As, the radius of both the geostationary orbit and circular polar orbit is same

$$\Delta V = V_{Geo} + V_{Polar} = \sqrt{\frac{2GM}{R}}$$

$$\therefore \frac{\Delta V}{V_e} = 1$$

The magnitude of the change in velocity required 1 times the magnitude of the escape velocity.

18. (b) Given,

Velocity at orbit, $V_a = 200$ km/s $= 200 \times 10^3$ m/s

Altitude, a = 71400 km = 71400×10^3 m/s

Mass of planet, $M = 1.9 \times 10^{28}$ kg

$G = 6.67 \times 10^{-11}$ m³/kgs²

$$e = \frac{V_a{}^2 R}{GM} - 1$$

$$= \frac{(200 \times 10^3)^2 \times 71400 \times 10^3}{1.9 \times 10^{28} \times 6.67 \times 10^{-11}} - 1 = 1.25$$

e	Orbit
0	Circular
0<e<1	Elliptical
1	Parabolic
>1	Hyperbolic

Therefore, the orbit is hyperbolic.

19. Given

Area swept, $A = 5.6 \times 10^9\ km^2$.

Angular speed, $\omega = 0.00125$ rad/s

Radius, $R = 7200\ km = 7200 \times 10^3\ m$

Area swept is given as,

$$\frac{dA}{dt} = \frac{h}{2},\ \text{where h is angular momentum}$$

$$\Rightarrow \int dA = \int \frac{h}{2} dt$$

$$\Rightarrow A = \frac{h}{2} T$$

$$\Rightarrow T = \frac{A \times 2}{h}$$

Also,

$h = v \times R = \omega R \times R = \omega R^2$.

$$\therefore T = \frac{A \times 2}{\omega R^2}$$

$$= \frac{5.6 \times 10^{15} \times 2}{0.00125 \times \left(7200 \times 10^3\right)} = 155124.60s$$

Orbital period in Earth day

$$\frac{155124.60}{24 \times 60 \times 60} = 2\ \text{days}$$

20. Given,

Velocity at orbit, $Va = 8$ km/s s $= 8 \times 10^3$ m/s

Altitude, $a = 350$ km $= 350 \times 10^3$ m/s

$$V_e = \sqrt{\frac{2GM}{R}}$$

$$= \sqrt{\frac{2 \times 3.98 \times 10^{14}}{(673 + 350) \times 10^3}} = 10.88\ km/s$$

Escape velocity of the Earth is 11.2 km/s.

Thus, it does not escape Earth's gravitational field, also, $V_e > V_a$ thus it does not falls back on Earth

$$e = \frac{V_a^{\,2} r}{GM} - 1$$

$$= \frac{\left(8 \times 10^3\right)^2 \times (6378 + 350) \times 10^3}{3.986 \times 10^{14}} - 1 = 0.080$$

Here, $e \neq 0$, so orbitals is not circular,

e lies, $0 < e < 1$, thus it forms an elliptical orbit.

21. Given

$a_{earth} = 1.4953 \times 10^{11}$ m

$a_{mars} = 2.278 \times 10^{11}$ m

$$T_{planet} \propto a^{\frac{3}{2}}$$

$$\therefore \frac{T_{mars}}{T_{earth}} = \frac{a_{marss}^{\frac{3}{2}}}{a_{earth}^{\frac{3}{2}}}$$

$$\Rightarrow T_{mars} = \frac{2.278 \times 10^{11}}{1.4953 \times 10^{11}} \times 365.256$$

$= 686.80$ days

22. Given,

Tangential velocity component of a spacecraft, V

Circular orbit of radius R

$\mu = GM$, gravitational parameter of Earth

Now,

Consider a circular orbit, using Newton's law of gravitation which provides centripetal acceleration to balance centrifugal force of circular motion, we get

$$\frac{GMm}{R^2} = \frac{mV^2}{R}$$

$$\Rightarrow V = \sqrt{\frac{GM}{R}} = \sqrt{\frac{\mu}{R}}$$

23. Equation of trajectory of space objects any plant is given by equation of the conic.

Given by

$$r = \frac{\frac{h^2}{\mu}}{\left(1 + \frac{|c|}{\mu} \cos\theta\right)}$$

Also, $e = \frac{h^2}{\mu k}$ and $e = \frac{|c|}{\mu}$

$$\Rightarrow h^2 = |c|\ k$$

$$\therefore r = \frac{\frac{h^2}{\mu}}{(1 + e\cos\theta)}$$

24. In an elliptical orbit around any planet the angular moment is conserved i.e.

$I\omega$ = const

$m\,r^2\,\omega$ = const

$$\omega \propto \frac{1}{r^2}$$

$\Rightarrow$ Smaller the r larger will be ω.

Therefore, the spacecraft will have maximum angular velocity when it is nearest to periapsis.

25. Given,

Altitude, a =150 km

Radius of Earth, RE = 6400 km

$\mu = 3.986 \times 1014\ m^3/s^2$

$$V_o = \sqrt{\frac{\mu}{(R_E + a)}}$$

$$\sqrt{\frac{3.986 \times 10^{14}}{(6400+150)\times 10^3}} = 7800.95\ m/s$$

26. Given

$Gm = 3.986 \times 10^{14}\ m^3/sec^2$

R_e = 6371 km

r = 4000 km

minimum incremental velocity in partied

to the spacecraft in a circular orbit arvael the earth at the altitnde of 4000 km to make exit earth's gravitational field

$\Delta V = V_{esc} - V_{circular}$

$$= \sqrt{\frac{2le}{(R_e + h)}} - \sqrt{\frac{le}{(R_e + h)}}$$

$$= \sqrt{\frac{le}{R_e + h}}[\sqrt{2} - 1]$$

$$= \sqrt{\frac{Gm}{R_e + h}}(\sqrt{2} - 1)$$

$$= \sqrt{\frac{3.986 \times 10^{14}}{(6371+4000)\times 10^3}}(\sqrt{2} - 1)$$

$= 2.57 \times 10^3$

= 2.57 km/sec

27. If e > 1

$$\frac{1}{2}mv^2 > \frac{GMm}{r}\text{, Trajectory is a hyperbolic}$$

If e = 1

$$\frac{mv^2}{2} = \frac{GMm}{r}\text{, Trajectory is a parabola}$$

If e < 1

$$\frac{mv^2}{2} < \frac{GMm}{r}\text{, Trajectory is an ellipse}$$

28.

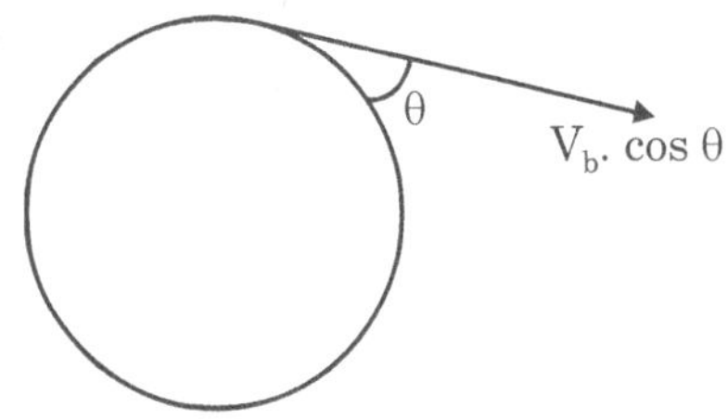

Given V_b = 13.5 km/sec

Tangential velocity of the space vehicle in orbit

$= V_b.\cos\theta$

$= 13.5 \times \cos$

= 13.448 km/sec

29. As per formula

Velocity at perihelion $(V_p) = \sqrt{\frac{\mu(1+e)}{a(1-e)}}$

Velocity at aphelion $(V_a) = \sqrt{\frac{\mu(1-e)}{a(1+e)}}$

Now, According to problem

$$\frac{V_p}{V_a} = 1.0339$$

$$\Rightarrow \quad \frac{\sqrt{\frac{\mu(1+e)}{a(1-e)}}}{\sqrt{\frac{\mu(1-e)}{a(1+e)}}} = 1.0339$$

$$\Rightarrow \quad \sqrt{\frac{(1+e)^2}{(1-e)^2}} = 1.0339$$

$$\Rightarrow \quad \frac{1+e}{1-e} = 1.0339$$

$$\Rightarrow \quad 1 + e = 1.0339(1 - e)$$

$$\Rightarrow \quad e + 1.0339\,e = 1.0339 - 1$$

$$\Rightarrow \quad e = \frac{0.0339}{2.0339} = 0.017$$

30. Given, h = 250 km

$Gm_e = 398600.4\ km^3/s^2$

R = 6378.14 km

As per formula

$$V = \sqrt{\frac{GM}{(R+h)}} = \sqrt{\frac{398600.4}{(250+6378.14)}} = 7.75\ km/s$$

4 CHAPTER AERODYNAMICS

1. Total pressure at a point is defined as the pressure when the flow is brought to rest
   (*a*) adiabatically (*b*) isentropically
   (*c*) isothermally (*d*) isobarically

   [GATE 2007, 1 M]

2. The drag divergence Mach number of an airfoil
   (*a*) is a fixed number for a given airfoil
   (*b*) is always higher than the critical Mach number
   (*c*) is equal to the critical Mach number at zero angle of attack
   (*d*) is the Mach number at which a shock wave first appears on the airfoil

   [GATE 2007, 1 M]

3. Across a normal shock
   (*a*) both total temperature and total pressure decrease
   (*b*) both total temperature and total pressure remain constant
   (*c*) total pressure remains constant but total temperature decreases
   (*d*) total temperature remains constant but total pressure decreases

   [GATE 2007, 1 M]

4. The Joukowskii airfoil is studied in aerodynamics because
   (*a*) it is used in many aircraft
   (*b*) it is easily transformed into a circle, mathematically
   (*c*) it has a simple geometry
   (*d*) it has the highest lift curve slope among all airfoils

   [GATE 2007, 1 M]

5. One of the criteria for high-speed airplanes is that the critical Mach number should be as high as possible. Therefore, high-speed subsonic airplanes are usually designed with
   (*a*) thick airfoils
   (*b*) thin airfoils
   (*c*) laminar flow airfoils
   (*d*) diamond airfoils

   [GATE 2007, 1 M]

6. Two pipes of constant sections but different diameters carry water at the same volume flow rate. The Reynolds number, based on the pipe diameter, is
   (*a*) the same in both pipes
   (*b*) is larger in the narrower pipe
   (*c*) is smaller in the narrower pipe
   (*d*) depends on the material of the pipes

   [GATE 2007, 2 M]

7. A turbulent boundary layer remains attached over a longer distance on the upper surface of an airfoil than does a laminar boundary layer, because
   (*a*) the turbulent boundary layer is more energetic and hence can overcome the adverse pressure gradient better
   (*b*) the laminar boundary layer develops more skin friction and hence slows down more rapidly
   (*c*) turbulence causes the effective coefficient of viscosity to reduce, resulting in less loss of momentum in the boundary layer
   (*d*) the turbulent boundary layer is thicker, hence the velocity gradients in it are smaller, therefore viscous losses are less

   [GATE 2007, 2 M]

8. The laminar boundary layer over a large flat plate held parallel to the freestream is 5 mm thick at a point 0.2 m downstream of the leading edge. The thickness of the boundary layer at a point 0.8 m downstream of the leading edge will be
   (*a*) 20 mm (*b*) 10 mm
   (*c*) 5 m (*d*) 2.5 mm

   [GATE 2007, 2 M]

9. For the control volume shown in the figure below, the velocities are measured both at the upstream and the downstream ends.

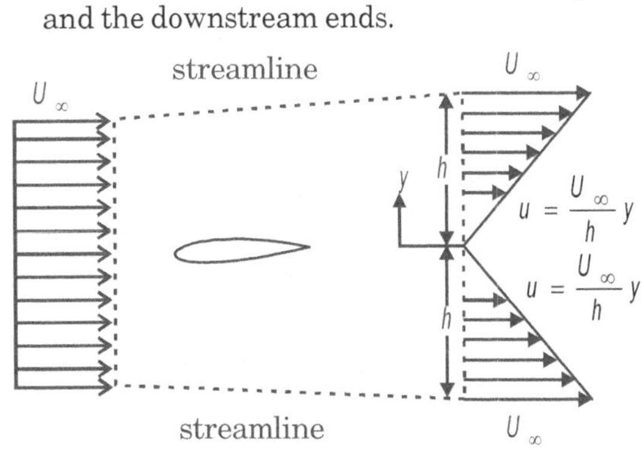

The flow of density ρ is incompressible, two dimensional and steady. The pressure is $p\infty$ over the entire surface of the control volume. The drag on the airfoil is given by,

(a) $\frac{\rho U_\infty^2 h}{3}$ (b) 0

(c) $\frac{\rho U_\infty^2 h}{6}$ (d) $2\rho U_\infty^2 h$

[GATE 2007, 2 M]

10. A circular cylinder is placed in an uniform stream of ideal fluid with its axis normal to the flow. Relative to the forward stagnation point, the angular positions along the circumference where the speed along the surface of the cylinder is equal to the free stream speed are

(a) 30, 150, 210 and 330 degrees

(b) 45, 135, 225 and 315 degrees

(c) 0, 90, 180 and 270 degrees

(d) 60, 120, 240 and 300 degrees

[GATE 2007, 2 M]

11. The maximum thickness to chord ratio for the NACA 24012 airfoil is

(a) 0.01 (b) 0.12

(c) 0.24 (d) 0.40

[GATE 2008, 1 M]

12. The maximum possible value of pressure coefficient C_p in incompressible flow is

(a) 0.5 (b) 1

(c) π (d) ∞

[GATE 2008, 1 M]

13. An irrotational and inviscid flow can become rotational on passing through a

(a) normal shock wave

(b) oblique shock wave

(c) curved shock wave

(d) March wave

[GATE 2008, 1 M]

14. Laminar flow airfoils are used to reduce

(a) trim drag (b) skin friction drag

(c) induced drag (d) wave drag

[GATE 2008, 1 M]

15. In a convergent-divergent (CD) nozzle of a rocket motor, the wall heat flux is maximum at

(a) the exit of the divergent portion of the CD nozzle

(b) the entry to the convergent portion of the CD nozzle

(c) the throat of CD nozzle

(d) the mid-length of the divergent portion of the CD nozzle

[GATE 2008, 1 M]

16. Consider 2-D flow with stream function $\psi = \frac{1}{2}\ln\left(\sqrt{x^2+y^2}\right)$. The absolute value of circulation along a unit circle centered at (x = 0, y = 0) is

(a) 0 (b) 1

(c) $\frac{\pi}{2}$ (d) π

[GATE 2008, 2 M]

17. Consider steady, inviscid flow in a convergent-divergent (CD) nozzle, with normal shock in the divergent portion. The static pressure along the nozzle downstream of the normal shock

(a) remain constant

(b) increases isentropically to the static pressure at the nozzle exit

(c) decreases isentropically to the static pressure at the nozzle exit

(d) can increase or decrease, depending on the magnitude of the static pressure at the nozzle exit

[GATE 2008, 2 M]

18. For a free stream Mach number of 0.7 the critical pressure coefficient ($C_{p,cr}$) is –0.78. I the minimum pressure coefficient for a given airfoil in incompressible flow is –0.6, then the flow over the airfoil at a free stream Mach number o 0.7 is

(a) 0 (b) 1

(c) $\frac{\pi}{2}$ (d) π

[GATE 2008, 2 M]

19. If the flow Mach number in a turbulent boundary layer over a flat plate is increased keeping the Reynolds number unchanged, the skin friction coefficient C_f.

(a) decreases

(b) increases

(c) remains constant

(d) initially decreases, followed by a rapid increase

[GATE 2008, 2 M]

20. In supersonic wind-tunnel design, an oblique diffuser is preferred over a normal shock diffuser because

(*a*) it reduces total pressure loss

(*b*) the flow is slowed down more rapidly

(*c*) the flow is accelerated more rapidly

(*d*) it increases total pressure loss

[GATE 2008, 2 M]

21. The variation of downwash along the span of an untwisted wing of elliptic planform is

(*a*) sinusoidal (*b*) parabolic

(*c*) elliptic (*d*) constant

[GATE 2008, 2 M]

22. Flow post an airfoil is to be modeled using a vortex sheet. The strength of the vortex sheet at the trailing edge will be

(*a*) 0 (*b*) 1

(*c*) 2π (*d*) ∞

[GATE 2008, 2 M]

23. Consider a 2-D body inn supersonic flow with an attached oblique shock as shown below

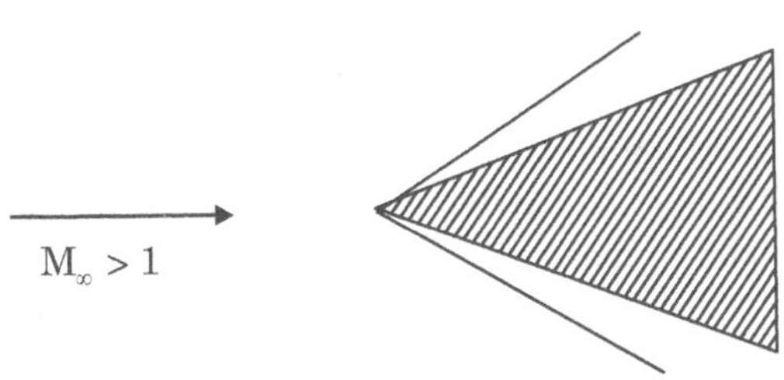

An increase in free stream Mach number M_∞ will cause the oblique shock wave to

(*a*) move closer to the body

(*b*) move away from the body

(*c*) detach from the body

(*d*) become a normal shock

[GATE 2008, 2 M]

24. The geometrical features of a supercritical airfoil are

(*a*) rounded leading edge, flat upper surface and high camber at the rear

(*b*) sharp leading edge, curved upper surface and high camber at the rear

(*c*) rounded leading edge, curved upper surface and no camber at the rear

(*d*) sharp leading edge, flat upper surface and no camber at the rear

[GATE 2008, 2 M]

25. For a flow through a Prandtl-Meyer expansion wave

(*a*) Mach number stays constant

(*b*) Entropy stays constant

(*c*) Temperature stays constant

(*d*) Density stays constant

[GATE 2009, 1 M]

26. For two-dimensional irrotational and incompressible flows

(*a*) Both potential and stream functions satisfy the Laplace equation.

(*b*) Potential function must satisfy the Laplace equation but the stream function need not.

(*c*) Stream function must satisfy the Laplace equation but the potential function need not.

(*d*) Neither the stream function nor the potential function need to satisfy the Laplace equation.

[GATE 2009, 1 M]

27. Thin airfoil theory predicts that the lift slope is -

$\frac{dc_1}{d\alpha} = 2\pi$ for

(*a*) Symmetric airfoils only.

(*b*) Cambered airfoils only.

(*c*) Any airfoil shape.

(*d*) Joukowski airfoils only.

[GATE 2009, 1 M]

28. An aircraft is flying at M = 2 where the ambient temperature around the aircraft is 250 K. If the specific heat ratio for air $\gamma = 1.4$, the stagnation temperature on the surface of the aircraft is

(*a*) 200 K (*b*) 450 K

(*c*) 350 K (*d*) 1450 K

[GATE 2009, 1 M]

29. Which of the following statements are true for flow across a stationary normal shock?

P. Stagnation temperature stays constant.

Q. Stagnation pressure decreases.

R. Entropy increases.

S. Stagnation pressure increases.

T. Stagnation temperature increases.

(*a*) P, Q, R

(*b*) Q, R, S

(*c*) R, S, T

(*d*) S, T, P

[GATE 2009, 2 M]

30. A model airfoil in a wind tunnel that is operating at 50 m/s develops a minimum pressure coefficient of - 6.29 at some point on its upper surface. The local airspeed at that point is

(*a*) 50 m/s (*b*) 125 m/s

(*c*) 135 m/s (*d*) 150 m/s

[GATE 2009, 2 M]

31. A symmetrical airfoil section produces a lift coefficient of 0.53 at an angle of attack of 5 degrees measured from its chord line. An untwisted wing of elliptical planforrn and aspect ratio 6 is made of this airfoil. At an angle of attack of 5 degrees relative to its chordal plane, this wing would produce a lift coefficient of

(*a*) 0.53 (*b*) 0.48

(*c*) 0.40 (*d*) 0.36

[GATE 2009, 2 M]

32. Consider an ideal flow of density ρ through a variable area duct as shown in the figure below:

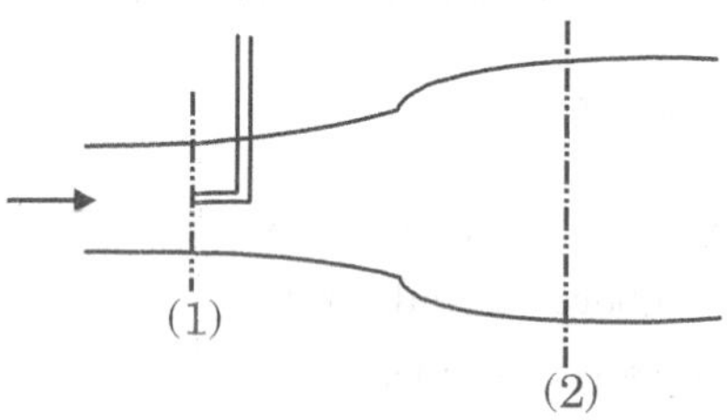

Let the cross-sectional are as at sections (1) and (2) be A_1 and A_2 respectively. The velocity measured at section (1) using a Pitot static probe is V_1. Then the static pressure drop $p_2 - p_1$ is

(*a*) $-\frac{1}{2}\rho\left(1-\frac{A_1^2}{A_2^2}\right)V_1^2$ (*b*) $\frac{1}{2}\rho\left(1-\frac{A_1^2}{A_2^2}\right)V_1^2$

(*c*) $\frac{1}{2}\rho\left(1+\frac{A_1^2}{A_2^2}\right)V_1^2$ (*d*) $-\frac{1}{2}\rho\left(1+\frac{A_1^2}{A_2^2}\right)V_1^2$

[GATE 2009, 2 M]

33. Two vortices of the same strength and sign are placed a distance d apart as shown below. Assume that the vortices are free to move and the fluid is ideal.

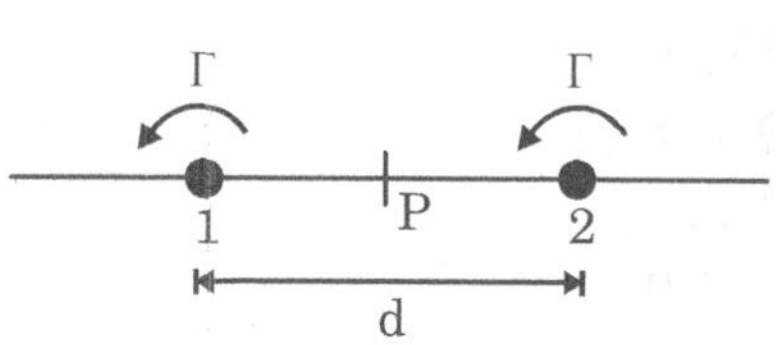

Which of the following statements is true?

(*a*) Vortices 1 and 2 spiral inwards with an initial angular speed $\frac{\Gamma}{2\pi d^2}$ to finally merge and form one vortex of twice the strength.

(*b*) Vortices 1 and 2 spiral inwards with an initial angular speed $\frac{\Gamma}{\pi d^2}$ to finally merge and form one vortex of twice the strength.

(*c*) Vortices 1 and 2 perpetually revolve about the midpoint P with radius of revolution $\frac{d}{2}$ and angular 2 speed $\frac{\Gamma}{2\pi d^2}$.

(*d*) Vortices 1 and 2 perpetually revolve about the midpoint P with radius of revolution $\frac{d}{2}$ and angular speed $\frac{\Gamma}{\pi d^2}$.

[GATE 2009, 2 M]

34. The laminar boundary layer over a large flat plate held parallel to the flow is 7.2 mm thick at a point 0.33 m downstream of the leading edge. If the free stream speed is increased by 50%, then the new boundary layer thickness at this location will be approximately

(*a*) 10.8 mm (*b*) 8.8 mm

(*c*) 5.9 mm (*d*) 4.8 mm

[GATE 2009, 2 M]

35. Consider an incompressible 2-D Coutte flow of water between two walls spaced lm apart. The lower wall is kept stationary. What is the shear mess acting on the lower wall if the upper wall is moving at a constant speed of 2 m/s? ($\mu_{water} = 7 \times 10^{-1}$ N.s/m^2)

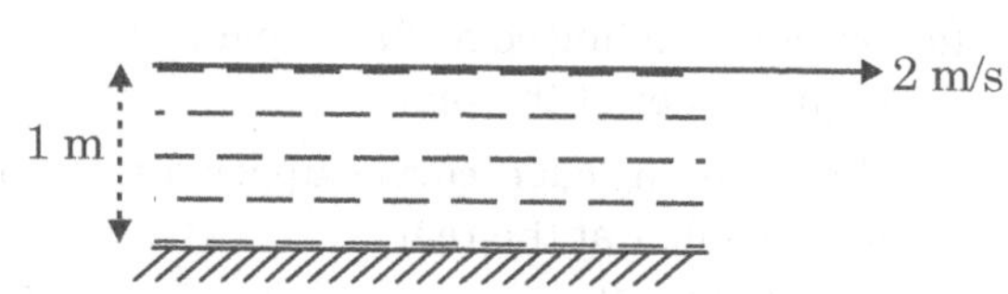

(*a*) 3.5×10^{-3} N/m^2 (*b*) 7×10^{-3} N/m^2

(*c*) 10.5×10^{-3} N/m^2 (*d*) 14×10^{-3} N/m^2

[GATE 2010, 1 M]

36. Consider an incompressible 2-D viscous flow over a curved surface. Let the pressure distribution on the surface be $p(s) = 2 + \sin\left(\frac{\pi}{2} + s\right) N/m^2$, where s is the distance along the curved surface from the leading edge. The flow separates at

(*a*) $s = (2/3)\,\pi$ m (*b*) $s = (3/2)\,\pi$ m

(*c*) $s = (\pi/2)$ m (*d*) $s = \pi$ m

[GATE 2010, 1 M]

37. In a 2-D, steady, fully developed, laminar boundary layer over a flat plate. If is the stream-wise coordinate, y is the wall normal coordinate and u is the stream-wise velocity component, which of lhe following is true:

(*a*) $\frac{\partial u}{\partial x} >> \frac{\partial u}{dy}$ (*b*) $\frac{\partial u}{\partial y} >> \frac{\partial u}{dx}$

(*c*) $\frac{\partial u}{\partial x} = \frac{\partial u}{dy}$ (*d*) $\frac{\partial u}{\partial x} = -\frac{\partial u}{dy}$

[GATE 2010, 1 M]

38. If ϕ is the potential function for an incompressible irrotational flow, and u and v are the Cartesian velocity components, then which one of the following combinations is correct.

(*a*) $u = \frac{\partial \phi}{\partial x}, v = \frac{\partial \phi}{\partial x}$ (*b*) $u = -\frac{\partial \phi}{\partial y}, v = \frac{\partial \phi}{\partial x}$

(*c*) $u = -\frac{\partial \phi}{\partial y}, v = \frac{\partial \phi}{\partial y}$ (*d*) $u = \frac{\partial \phi}{\partial x}, v = \frac{\partial \phi}{\partial y}$

[GATE 2010, 1 M]

39. For a flow across an oblique shock which of the following statements is true?

(*a*) Component of velocity normal to shock decreases while tangential component increases.

(*b*) Component of velocity normal to shock increases while tangential component decreases.

(*c*) Component of velocity normal to shock is unchanged while tangential component decreases.

(*d*) Component of velocity normal to shock decreases while tangential component is unchanged.

[GATE 2010, 1 M]

40. An aircraft is cruising at an altitude of 9 km. The free-stream static pressure and density at this altitude are 3.08×10^4 N/m² and 0.467 kg/m³ respectively. A Pitot tube mounted on the wing senses a pressure of 3.31×10^4 N/m². Ignoring compressibility effects, the cruising speed of the aircraft is approximately

(*a*) 50 m/s (*b*) 100 m/s

(*c*) 150 m/s (*d*) 200 m/s

[GATE 2010, 2 M]

41. Consider the flow of air (ρ = 1.23 kg/m³) over a wing of chord length 0.5 m and span 3m. Let the free stream velocity be U = 100 m/s and the average circulation around the wing be r = 10 m/s per unit span. The lift force acting on the wing is

(*a*) 615 N (*b*) 1845 N

(*c*) 3690 N (*d*) 4920 N

[GATE 2010, 2 M]

42. Consider a l-D adiabatic, inviscid, compressible flow of air (R = 287 J/Kg-K, C_r = 718 J/Kg-K) through a duct of constant cross-sectional area A = 1m². If the volumetric flow rate is Q = 680 m³/s and stagnation temperature is T_0 = 580.05 K, then the air temperature inside the duct is

(*a*) 300 K (*b*) 350 K

(*c*) 400 K (*d*) 450 K

[GATE 2010, 2 M]

43. Consider a potential flow over a finite wing with the following circulation distribution

$$\Gamma(y) = 100\sqrt{1 - \left(\frac{2y}{4}\right)^2}\ m^2/s$$

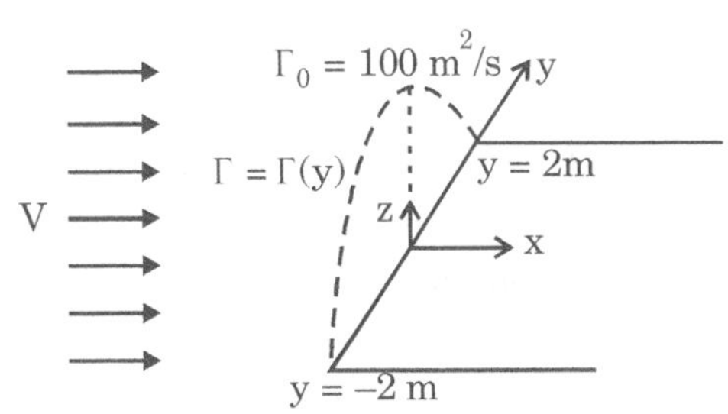

If the free stream velocity is 100 m/s, the induced angle of attack is

(*a*) 0.125 radians

(*b*) −0.125 radians

(*c*) $0.125\sqrt{1 - \left(\frac{y}{2}\right)^2}$ radians

(*d*) $-0.125\sqrt{1 - \left(\frac{y}{2}\right)^2}$ radians

[GATE 2010, 2 M]

Common Data for Questions 44 and 45:

Consider a potential flow over a spinning cylinder. The stream function is given as

$$\psi = (V_\infty r \sin\theta)\left(1 - \frac{R^2}{r^2} + \frac{\Gamma}{2\pi}\ln\frac{r}{R}\right)$$

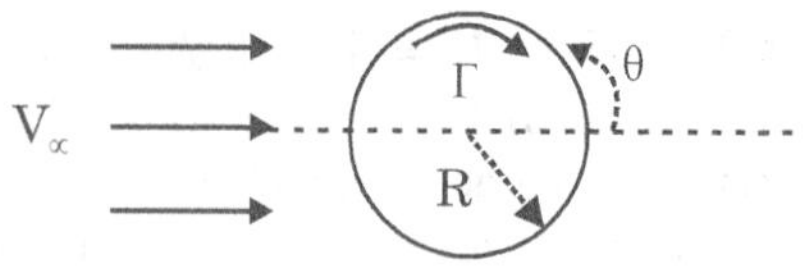

where

Free stream velocity, $V_\infty = 25$ m/s

Cylinder radius, R = 1 m

Circulation, $\Gamma = 50\pi\ m^2/s$

44. The radial and azimuthal velocities on the cylinder surface at $\theta = \pi/2$ are

(*a*) $V_1 = 0$ m/s, $V_0 = -75$ m/s

(*b*) $V_1 = 0$ m/s, $V_0 = 75$ m/s

(*c*) $V_1 = 0$ m/s, $V_0 = -25$ m/s

(*d*) $V_1 = 0$ m/s, $V_0 = 25$ m/s

[GATE 2010, 2 M]

45. The stagnation points are located at

(*a*) 210° and 330° (*b*) 240° and 300°

(*c*) 30° and 150° (*d*) 60° and 120°

[GATE 2010, 2 M]

46. It is seen that the drag polar of a certain aerofoil is symmetric about the C_d axis. This drag polar could refer to

(*a*) NACA 0012 (*b*) NACA 4415

(*c*) NACA 23012 (*d*) None of the above

[GATE 2011, 1 M]

47. The aerodynamic centre of a supersonic aerofoil, with chord c, is located at

(*a*) the leading edge (*b*) 0.25c

(*c*) 0.5c (*d*) 0.75c

[GATE 2011, 1 M]

48. A main objective of by-pass in a turbo-fan engine is to increase

(*a*) mass flow rate through engine inlet

(*b*) turbine inlet temperature

(*c*) mass flow rate through exhaust nozzle

(*d*) compressor pressure ratio

[GATE 2011, 1 M]

49. A low speed wind tunnel has a contraction ratio of 14:1 and the cross-sectional area of the test section is 1 m². The static pressure difference between the settling chamber and the test section is 40 cm of water column. Assume g = 9.81 m/s², $\rho_{air} = 1.2$ kg/m³ and $\rho_{water} = 1000$ kg/ms³. The speed of air in the test section (in m/s) is ________ .

[GATE 2011, 2 M]

50. Consider the inviscid, adiabatic flow of air at free stream conditions, $M_1 = 2$, $p_1 = 1$ atm and $T_1 = 288$ K around a sharp expansion corner ($\theta = 20°$) as shown below. The Prandtl-Meyer function v, is given as a function of Mach number, as

$$v(M) = \sqrt{\frac{\gamma+1}{\gamma-1}}\tan^{-1}\sqrt{\frac{\gamma+1}{\gamma-1}(M^2-1)} - \tan^{-1}\sqrt{M^2-1}$$

Assume air to be calorically perfect with $\gamma = 1.4$. The Mach number, M_2, downstream of the expansion corner is approximately

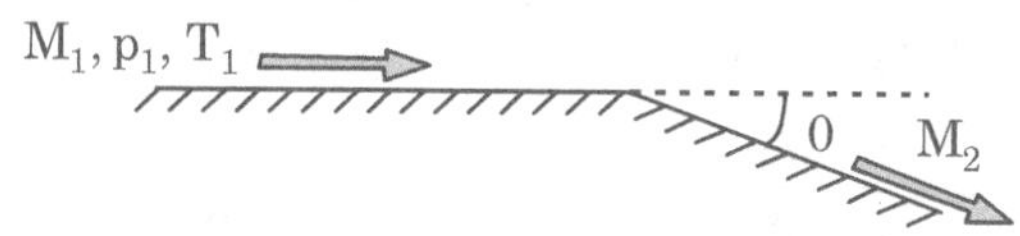

(*a*) 2.00 (c) 1.76

(*c*) 2.83 (*d*) 3.14

[GATE 2011, 2 M]

51. Consider a steady two dimensional zero-pressure gradient laminar flow of air over a flat plate as shown below. The free stream conditions are $U_\infty = 100\ ms^{-1}$, $\rho_\infty = 1.2\ kg\ m^{-3}$, $p_\infty = 1$ atm and $\mu_\infty = 1.8 \times 10^{-5}\ kg\ m^{-1}s^{-1}$. The ratio of displacement thickness to momentum thickness of the boundary layer at a distance of 2 m from the leading edge is

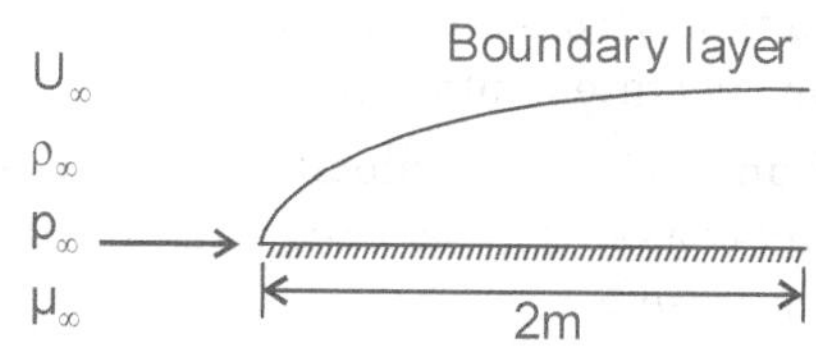

(*a*) 7.53 (*b*) 2.59

(*c*) 2.91 (*d*) 0.39

[GATE 2011, 2 M]

52. In the context of Prandtl's lifting line theory for a finite wing, which of the following combinations of statements is TRUE ?

P: The bound vortex is responsible for the lift force

Q: The trailing vortices are responsible for the induced drag

R: The bound vortex is responsible for the induced drag

S: The trailing vortices are responsible for the lift force

(a) P, Q (c) Q, R

(c) R, S (d) P, S

[GATE 2011, 2 M]

53. Consider flow over a thin aerofoil at Mach number, $M_\infty = 0.5$ at an angle of attack, α. Using the Prandtl-Glauert rule for compressibility correction, the formula for lift coefficient, c_b can be written as

(a) 5.44 α (b) 6.28 α

(c) 7.26 α (d) 14.52 α

[GATE 2011, 2 M]

Common Data for Question 54:

Consider an inviscid, adiabatic flow of air at free stream Mach Number, $M_\infty = 2$, across a compression corner (θ = 20°) as shown. The free stream total enthalpy is $h_{0\infty} = 810$ kJ kg^{-1}. Assume that air is calorically perfect with γ = 1.4, R = 287J kg^{-1} K^{-1}.

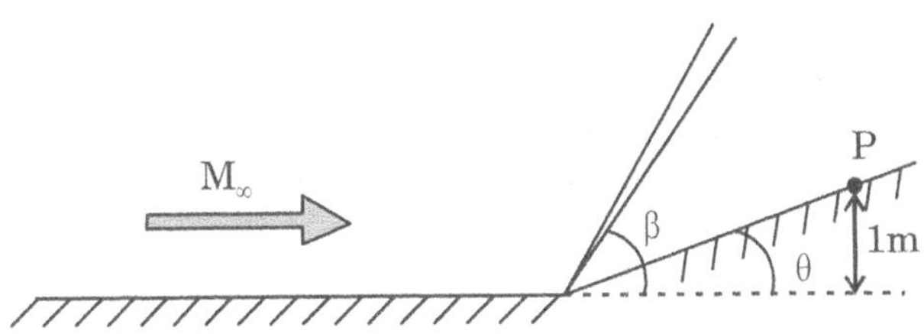

54. The shock angle β is

(a) = 20° (b) > 20° and < 30°

(c) = 30° (d) > 30°

[GATE 2011, 2 M]

Statement for Linked Answer Questions 55 and 56:

Prandtl's lifting line equation for a general wing is given by

$$\alpha(y_0) = \frac{\Gamma(y_0)}{\pi U_\infty c(y_0)} + \alpha_{L=0}(y_0) + \frac{1}{4\pi U_\infty}\int_{-\frac{b}{2}}^{\frac{b}{2}} \frac{\left(\frac{\partial\Gamma}{\partial y}\right)}{y_0 - y}dy,$$

where U_∞ is the free-stream velocity, α is the angle of attack, y_0 is the spanwise location, $\alpha_{L=0}(y_0)$ gives the spanwise variation of zero-lift angle, c is the chord, b is the span, and $\Gamma(y_0)$ gives the spanwise variation of circulation.

55. The rate of change of circulation with angle of attack $\Gamma_\alpha = \frac{\partial T}{\partial \alpha}$ is

(a) inversely proportional to α

(b) independent of α

(c) a linear function of α

(d) quadratic function of α

[GATE 2011, 2 M]

56. Given that $C_L \propto \int_{-\frac{b}{2}}^{\frac{b}{2}} \Gamma(y)\,dy$, the corresponding lift curve-slope $\frac{\partial C_L}{\partial \alpha}$ is

(a) independent of α

(b) a linear function of α

(c) a quadratic function of α

(d) a cubic function of α

[GATE 2011, 2 M]

57. For a symmetric airfoil, the lift coefficient for zero degree angle of attack is

(a) –1.0 (b) 0.0

(c) 0.5 (d) 1.0

[GATE 2012, 1 M]

58. The Mach angle for a flow at Mach 2.0 is

(a) 30° (b) 45°

(c) 60° (d) 90°

[GATE 2012, 1 M]

59. The ideal static pressure coefficient of a diffuser with an area ratio of 2.0 is

(a) 0.25 (b) 0.50

(c) 0.75 (d) 1.0

[GATE 2012, 1 M]

60. An airfoil generates a lift of 80 N when operating in a freestream flow of 60 m/s. If the ambient pressure and temperature are 100 kPa and 290 K respectively (specific gas constant is 287 J/kg-K), the circulation on the airfoil in m²/s is ____.

[GATE 2012, 2 M]

61. An oblique shock wave with a wave angle β is generated from a wedge angle of θ. The ratio of the Mach number downstream of the shock to its normal component is

(a) sin (β – θ) (b) cos (β – θ)

(c) sin (θ – β) (d) cos (θ – β)

[GATE 2012, 2 M]

62. In a closed-circuit supersonic wind tunnel, the convergent-divergent (C-D) nozzle and test section are followed by a C-D diffuser to swallow the starting shock. Here, we should have the

(*a*) diffuser throat larger than the nozzle throat and the shock located just at the diffuser throat.

(*b*) diffuser throat larger than the nozzle throat and the shock located downstream of the diffuser throat.

(*c*) diffuser throat of the same size as the nozzle throat and the shock located just at the diffuser throat.

(*d*) diffuser throat of the same size as the nozzle throat and the shock located downstream of the diffuser throat.

[GATE 2012, 2 M]

63. A vortex flowmeter works on the principle that the Strouhal number of 0.2 is a constant over a wide range of flow rates. If the bluff-body diameter in the flowmeter is 20 mm and the piezo-electric transducer registers the vortex shedding frequency to be 10 Hz, then the velocity of the flow would be measured as

(*a*) 0.1 m/s (*b*) 1 m/s

(*c*) 10 m/s (*d*) 100 m/s

[GATE 2012, 2 M]

64. Which one of the following statements is NOT TRUE for a supersonic flow?

(a) Over a gradual expansion, entropy remains constant

(b) Over a sharp expansion corner, entropy can increase

(c) Over a gradual compression, entropy can remain constant

(d) Over a sharp compression corner, entropy increases

[GATE 2013, 1 M]

65. Consider a compressible flow where an elemental volume of the fluid is $\delta\forall$, moving with velocity $\vec{V}$. Which one of the following expressions is TRUE?

(*a*) $\nabla.\vec{V} = \frac{1}{\delta\forall}\frac{D\delta\forall}{Dt}$

(*b*) $\nabla.(\nabla \times \vec{V}) = \frac{1}{\delta\forall}\frac{D\delta\forall}{Dt}$

(*c*) $\nabla.\frac{D\vec{V}}{Dt} = \frac{1}{\delta\forall}\frac{D\delta\forall}{Dt}$

(*d*) $\vec{V}.(\nabla \times \vec{V}) = \frac{1}{\delta\forall}\frac{D\delta\forall}{Dt}$

[GATE 2013, 1 M]

66. Consider a thin flat plate airfoil at a small angle α to an oncoming supersonic stream of air. Assuming the flow to be inviscid, $\frac{C_d}{C_i^2}$ is

(*a*) zero (*b*) independent of α

(*c*) proportional to α (*d*) proportional to α^2

[GATE 2013, 1 M]

67. Consider one-dimensional isentropic flow at a Mach number of 0.5. If the area of cross-section of a stream tube increases by 3% somewhere along the flow, the corresponding percentage change in density is ________

[GATE 2013, 2 M]

68. The potential flow model for a storm is represented by the superposition of a sink and a vortex. The stream function can be written in the (r, θ) system as $\Psi = -\frac{A}{2\pi}\theta + \frac{\Gamma}{2\pi}\ln r$, where ^ = Γ = 100 ms²/s. Assume a constant air density of 1.2 kg/m³. The gauge pressure at a distance of 100 m from the storm eye is

(*a*) $-\infty$ (*b*) $-\frac{1.2}{\pi^2}$

(*c*) $\frac{1.2}{2\pi^2}$ (*d*) $-\frac{1.2}{4\pi^2}$

[GATE 2013, 2 M]

69. Three identical eagles of wing span s are flying side by side in a straight line with no gap between their wing tips. Assume a single horse shoe vortex model (of equal strength Γ) for each bird. The net downwash experienced by the middle bird is

(*a*) $\frac{\Gamma}{\pi s}$ (*b*) $\frac{\Gamma}{2\pi s}$

(*c*) $\frac{\Gamma}{3\pi s}$ (*d*) $\frac{4\Gamma}{3\pi s}$

[GATE 2013, 2 M]

70. Streamline pattern of flow past a cylinder is shown in the figure below. The oncoming flow is steady, irrotational and incompressible. The flow is from left to right. Bernoulli's equation CANNOT be applied between the points

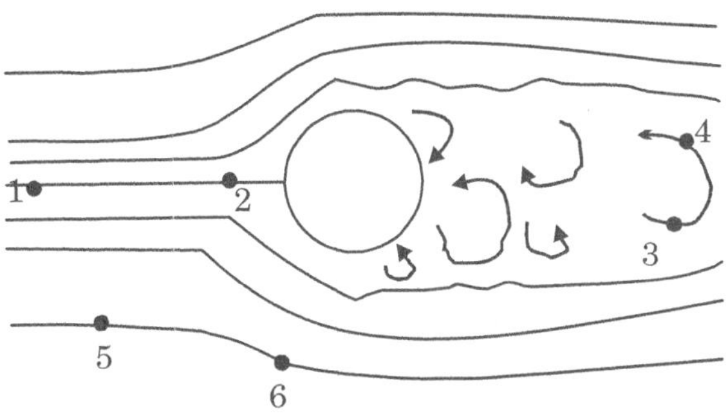

(*a*) 1 and 2 (*b*) 1 and 5
(*c*) 3 and 4 (*d*) 5 and 6

[GATE 2013, 2 M]

71. Consider a supersonic stream at a Mach number M = 2, undergoing a gradual expansion. The stream is turned by an angle of 3 degrees due to the expansion. The following data is given.

M	v(Pandtl – Meyer function)
1.8	20.73
1.9	23.59
2.0	26.38
2.1	29.10
2.2	31.73
2.3	34.28
2.4	36.75

The Mach number downstream of the expansion is

(*a*) 1.88 (*b*) 2.00
(*c*) 2.11 (*d*) 2.33

[GATE 2013, 2 M]

72. For a NACA 5-digit airfoil of chord c, the designed lift coefficient and location of maximum camber along the chord from the leading edge are denote by C_L and X_m respectively. For NACA12018 airfoil, which combination of C_L and X_m given below are correct?

(*a*) $C_L = 0.15$ and $X_m = 0.1c$
(*b*) $C_L = 0.12$ and $X_m = 0.2c$
(*c*) $C_L = 0.12$ and $X_m = 0.18c$
(*d*) $C_L = 0.15$ and $X_m = 0.2c$

[GATE 2014, 1 M]

73. For inviscid, supersonic flow over a diamond shaped airfoil, shown in the figure, which statement is correct among the following?

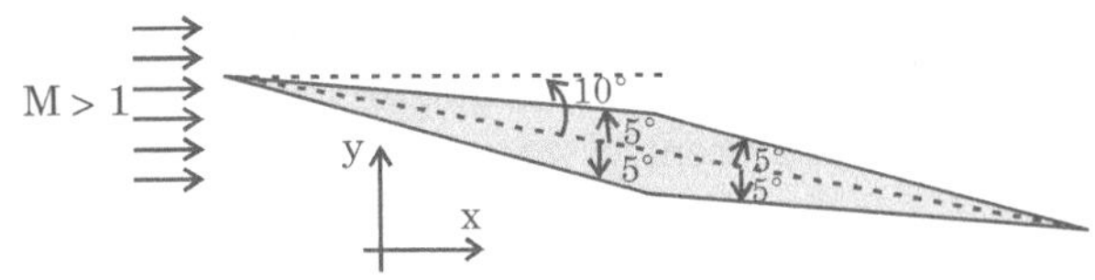

(*a*) The airfoil will experience zero lift and positive drag force
(*b*) The airfoil will experience positive lift and zero drag force
(*c*) The airfoil will experience negative lift and zero drag force
(*d*) The airfoil will experience positive lift and positive drag force

[GATE 2014, 1 M]

74. Consider supersonic flow near a corner (at an angle θ from the horizontal) with an attached oblique shock (at an angle β with horizontal) as shown in figure. If Match number M decreases gradually from a high supersonic value, which of the following statements is correct?

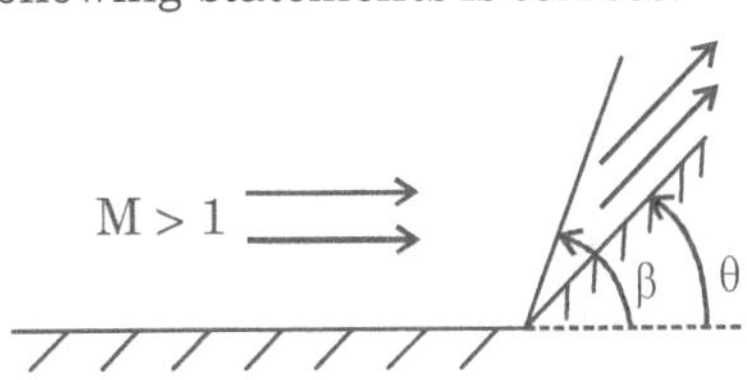

(*a*) β will decrease if the shock is a weak shock
(*b*) β will decrease if the shock is a strong shock
(*c*) β will increase for both weak and strong shocks
(*d*) β remains unchanged for both weak and strong shocks

[GATE 2014, 1 M]

75. The streamlines of a potential line vortex is concentric circles with respect to the vortex center as shown in figure. Velocity along these streamlines, outside the core of the vortex can be written as, $\upsilon_\theta = \dfrac{\Gamma}{2\pi r}$, where strength of the vortex is $\Gamma/2\pi$ and r is radial direction. The value of circulation along the curve shown in the figure is:

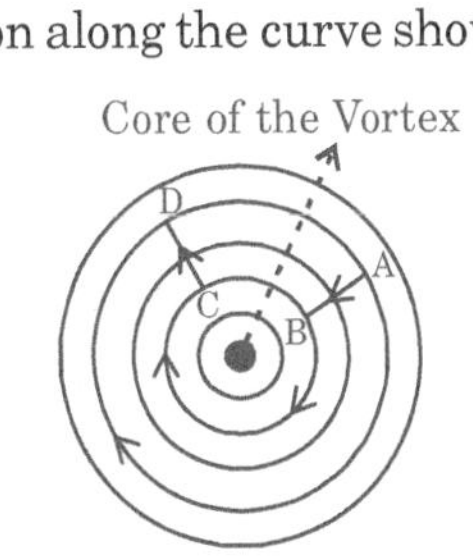

(*a*) Γ
(*b*) –2Γ
(*c*) 2Γ
(*d*) 0

[GATE 2014, 1 M]

76. To observe unsteady separated flow in a diverging channel, bubbles are injected at each 10 ms interval at point A as shown in figure. These bubbles act as tracer particles and follow the flow faithfully. The curved line AB shown at any instant represents:

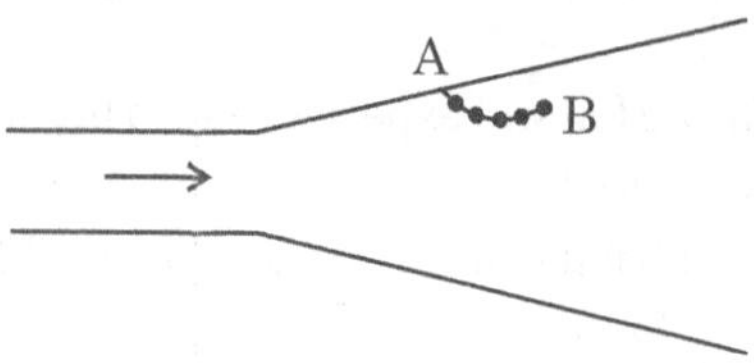

(*a*) Streamline, streakline and pathline
(*b*) Streamline and pathline
(*c*) Only a pathline
(*d*) Only a streakline

[GATE 2014, 1 M]

77. For a steady, incompressible two-dimensional flow, represented in Cartesian co-ordinates (x, y), a student correctly writes the equation of pathline of any arbitary particle as dx/dt = ax and dy/dt = by, where a and b are constants having unit of $(\text{second})^{-1}$. If value of a is 5, the value of b is ________ .

[GATE 2014, 2 M]

78. A student needs to find velocity across a stationary normal shock. He measures density and pressure across the shock as shown in the figure below. 1 bar = 10^5Pa. (No shock table is needed for the calculations). The value of u_1 in m/s is ________.

$p_1 = 1$ bar	$p_2 = 29$ bar
$\rho_1 = 1.2\text{kg/m}^3$	$\rho_2 = 6\text{kg/m}^3$
$u_1 \longrightarrow$	$\longrightarrow u_2$

[GATE 2014, 2 M]

79. For inviscid, compressible flow past a thin airfoil, shown in the figure, free-stream Mach number and pressure are denoted by M_∞ and p_∞ respectively. Ratio of pressure at point A and p_∞ is 0.8 and specific heat ratio is 1.4. If the Mach number at point A is 1.0 and rest of the flow field is subsonic, the value of M_∞ is

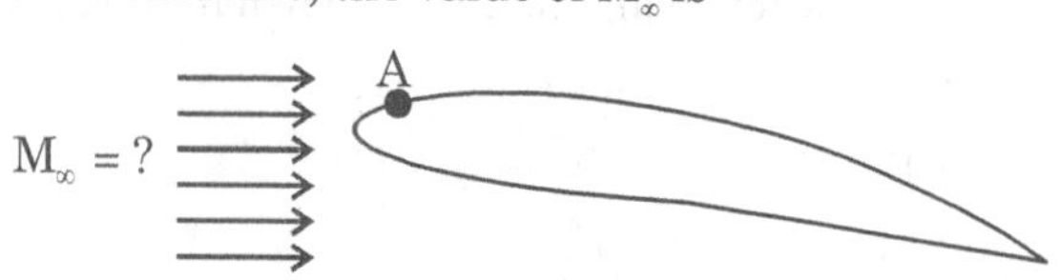

(*a*) 2.95 (*b*) 0.79
(*c*) 1.18 (*d*) 0.64

[GATE 2014, 2 M]

80. A student can measure free-stream velocity of a low-speed wind tunnel using a

i. Pitot tube alone aligned with the flow direction.
ii. Pitot tube aligned with the flow direction with static pressure measurement at an appropriate position on the tunnel wall.
iii. Pitot tube aligned with the flow direction along with barometer pressure reading of the outside ambient.
iv. Pitot static tube alone aligned with the flow direction.

Considering the above statements which of the following options is correct?

(*a*) (i) only (*b*) (i), (ii)
(*c*) (ii), (iv) (*d*) (i), (iii) & (iv)

[GATE 2014, 2 M]

81. Induced velocity w at a point $z = z_1$ along the lifting line can be calculated using the formula

$$w(z_1) = -\frac{1}{4\pi}\int_{-s}^{s}\frac{d\Gamma}{dz}\frac{1}{z-z_1}dz$$

Given $\frac{\Gamma^2}{\Gamma_0^2}+\frac{z^2}{s^2}=1$, where Γ_0 and s are given in figure below.

For the above semi-elliptic distribution of circulation, Γ, the downwash velocity at any point z_1, for symmetric flight can be obtained as $w(z_1) = \frac{\Gamma_0}{4\pi s}[\pi + z_1 I]$, where $I = z_1\int_{-s}^{s}\frac{dz}{\sqrt{s^2-z^2}(z-z_1)}$.

Which of the following options is correct if the induced drag is Di (given $\int_{-s}^{s}\sqrt{1-\frac{s^2}{z^2}}dz = \frac{\pi s}{2}$)

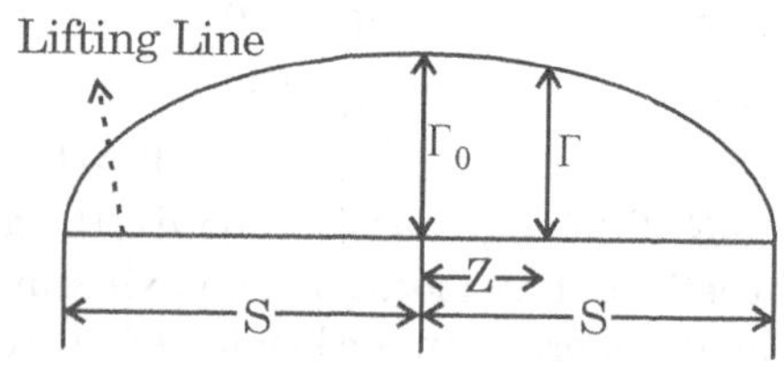

(*a*) $I = 0$ and $D_i = \frac{8\rho\Gamma_0^2}{\pi}$

(*b*) $I = 1$ and $D_i = \frac{8\rho\Gamma_0^2}{\pi}$

(*c*) $I = 0$ and $D_i = \frac{\pi\rho\Gamma_0^2}{8}$

(*d*) $I = 1$ and $D_i = \frac{\pi\rho\Gamma_0^2}{8}$

[GATE 2014, 2 M]

82. Two overflowing water reservoirs are connected with a 100m long pipe of circular cross-section (of radius, R = 0.02m), such that height difference h remains constant as shown in the figure below.

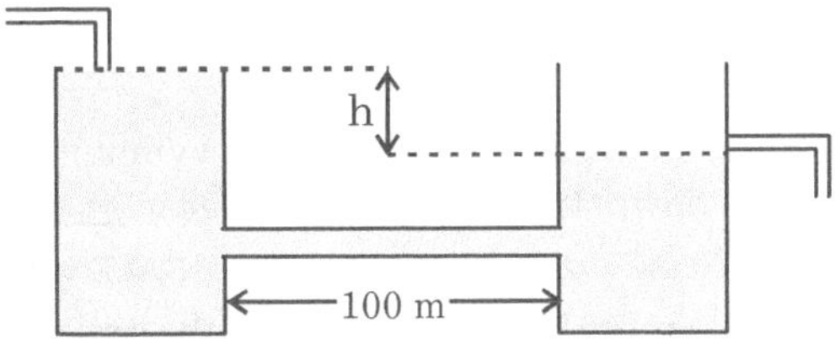

The centerline velocity in the pipe is 10m/s. The velocity profile inside the pipe over the entire length is

$$u = -\frac{R^2}{4\mu}\frac{dp}{dx}\left[1-\frac{r^2}{R^2}\right]$$

where, $\frac{dp}{dx}$ is a constant pressure gradient along the pipe length, x is measured from the left end of the pipe along its central axis and r is radial location inside the pipe with respect to its axis. (Given data : Density and kinematic viscosity of water are 1000 kg/m^3 and 1 × 10^{-6} m^2/s respectively; acceleration due to gravity is 10 m/s^2). If all other losses except the frictional losses at the pipe wall are neglected, the value of h in meter is ________.

[GATE 2014, 2 M]

83. An aircraft is flying at Mach 3.0 at the altitude where the ambient pressure and temperature are 50 kPa and 200 K respectively. If the converging-diverging diffuser of the engine (considered isentropic with ratio of specific heats, γ = 1.4 and specific gas constant R = 287 J/kgK) has a throat area of 0.05 m^2. the mass flow rate through the engine in kg/s is

(*a*) 197 (*b*) 232
(*c*) 790 (*d*) 157

[GATE 2014, 2 M]

84. As a candidate for a vertical tail, which one of the following airfoil sections is appropriate?

(*a*) NACA 0012 (*b*) NACA 2312
(*c*) NACA 23012 (*d*) Clarke Y profile

[GATE 2015, 1 M]

85. The primary purpose of a trailing edge flap is to

(*a*) avoid flow separation
(*b*) increase $C_{l,max}$
(*c*) reduce wave drag
(*d*) reduce induced drag

[GATE 2015, 1 M]

86. Consider a monoplane wing and a biplane wing with identical airfoil sections, wingspans and incidence angles in identical conditions in a wind tunnel. As compared to the monoplane, the biplane experiences

(*a*) a higher lift and a higher drag
(*b*) a higher lift and a lower drag
(*c*) a lower lift and a lower drag
(*d*) a lower lift and a higher drag

[GATE 2015, 1 M]

87. Consider a wing of elliptic planform, with its aspect ratio AR $\to \infty$. Its lift-curve slope $\frac{dC_L}{d\alpha}$ = ________.

[GATE 2015, 1 M]

88. An ideal gas in a reservoir has a specific stagnation enthalpy of h_0. The gas is is-entropically expanded to a new specific stagnation enthalpy of $\frac{h_0}{2}$ and velocity u. The flow is one-dimensional and steady. Then $\frac{u^2}{h_0}$ = ________.

[GATE 2015, 1 M]

89. The Reynolds number, Re is defined as $\frac{U_\infty L}{v}$ where L is the length scale for a flow, U_∞ its reference velocity and v is the coefficient of kinematic viscosity. In the laminar boundary layer approximation, comparison of the dimensions of the convection term $u\frac{\partial u}{\partial x}$ and the viscous term leads to the following relation between the boundary layer thickness ∂ and Re:

(*a*) $\partial \propto \sqrt{Re}$

(*b*) $\frac{\partial \propto 1}{\sqrt{Re}}$

(*c*) $\partial \propto Re$

(*d*) $\frac{\partial \propto 1}{Re}$

[GATE 2015, 1 M]

90. The velocity profile of an incompressible laminar boundary layer over a flat plate developing under constant pressure is given by $\frac{u(y)}{U_\infty} = \frac{3y}{2\delta} - \frac{1}{2}\left(\frac{y}{\delta}\right)^3$. The free stream velocity U_∞ = 10 m/ s and the dynamic viscosity of the fluid $\mu = 1.8 \times 10^{-5} \frac{kg}{ms}$. At a streamwise station where the boundary layer thickness $\delta = 5mm$, the wall shear stress is _________ $\times 10^{-3}$Pa .

[GATE 2015, 2 M]

91. The Pitot tube of an aircraft registers a pressure $p_0 = 54051\ N/m^2$. The static pressure, density and the ratio of specific heats of the freestream are $p_\infty = 45565\ N/m^2$, $\rho_\infty = 0.6417\ kg/m^3$ and γ = 1.4, respectively. The indicated airspeed (in m/s) is

(*a*) 157.6

(*b*) 162.6

(*c*) 172.0

(*d*) 182.3

[GATE 2015, 2 M]

92. Consider a NACA 0012 aerofoil of chord c in a freestream with velocity V_∞ at a non-zero positive angle of attack α. The average time-of-flight for a particle to move from the leading edge to the trailing edge on the suction and pressure sides are t_1 and t_2, respectively. Thin aerofoil theory yields the velocity perturbation to the freestream as $V_\infty \frac{(1+\cos\theta)\alpha}{\sin\theta}$ on the suction side and as $-V_\infty \frac{(1+\cos\theta)\alpha}{\sin\theta}$ on the pressure side, where θ co-rresponds to the chordwise position, $x = \frac{c}{2}(1-\cos\theta)$. then $t_2 - t_1$ is

(*a*) $-\frac{8\pi\alpha c}{V_\infty(4-\pi^2\alpha^2)}$ (*b*) 0

(*c*) $\frac{4\pi\alpha c}{V_\infty(4-\pi^2\alpha^2)}$ (*d*) $\frac{8\pi\alpha c}{V_\infty(4-\pi^2\alpha^2)}$

[GATE 2015, 2 M]

93. For a normal shock, the relation between the upstream Mach number (M_1) and the downstream Mach number (M_2) is given by

$$M_2^2 = \frac{(\gamma-1)M_1^2 + 2}{2\gamma M_1^2 + 1 - \gamma}$$

For an ideal gas with γ =1.4, the asymptotic value of the downstream Mach number is _______.

[GATE 2015, 2 M]

94. With increase in airfoil thickness, the critical Mach number for an airfoil is likely to

(*a*) decrease (*b*) increase

(*c*) remain unchanged (*d*) be undefined

[GATE 2016, 1 M]

95. Due to a body in potential flow, the velocity at a point A in the flow field is 20 m/s while the free stream velocity is only 10 m/s. The value of coefficient of pressure (Cp) at the point A is _________.

[GATE 2016, 1 M]

96. Which of the following airfoil will have location of the maximum camber at half chord length from the leading edge?

(*a*) NACA 5212 (*b*) NACA 1225

(*c*) NACA 2215 (*d*) NACA 2512

[GATE 2016, 1 M]

97. For a laminar incompressible flow past a flat plate at zero angle of attack, the variation of skin friction drag coefficient (C_f)with Reynolds number based on the chord length (Re_c) can be expressed as

(*a*) $C_f \propto \sqrt{Re_c}$ (*b*) $C_f \propto Re_c$

(*c*) $C_f \propto 1/\sqrt{Re_c}$ (*d*) $C_f \propto 1/Re_c$

[GATE 2016, 1 M]

98. Which of the following statement is NOT TRUE across an oblique shock wave?

(*a*) Static temperature increases, total temperature remains constant.

(*b*) Static pressure increases, static temperature increases.

(*c*) Static temperature increases, total pressure decreases.

(*d*) Static pressure increases, total temperature decreases.

[GATE 2016, 1 M]

99. For a completely subsonic isentropic flow through a convergent nozzle, which of the following statement is TRUE?

(*a*) Pressure at the nozzle exit > back pressure.

(*b*) Pressure at the nozzle exit < back pressure.

(*c*) Pressure at the nozzle exit = back pressure.

(*d*) Pressure at the nozzle exit = total pressure.

[GATE 2016, 1 M]

100. Air, with a Prandtl number of 0.7, flows over a flat plate at a high Reynolds number. Which of the following statement is TRUE?

(*a*) Thermal boundary layer is thicker than the velocity boundary layer.

(*b*) Thermal boundary layer is thinner than the velocity boundary layer.

(*c*) Thermal boundary layer is as thick as the velocity boundary layer.

(*d*) There is no relationship between the thicknesses of thermal and velocity boundary layers.

[GATE 2016, 1 M]

101. The maximum value of coefficient of lift (C_l) for a 2D circular cylinder, provided at least one stagnation point lies on the cylinder surface, is predicted by the potential flow theory to be

(*a*) $\frac{\pi}{2}$ (*b*) π

(*c*) 2π (*d*) 4π

[GATE 2016, 2 M]

102. The nozzle AB, as shown below, leading to the test section of a low speed subsonic wind tunnel, has a contraction ratio of 10:1. The pressure difference across the nozzle is maintained at 1000 N/m² and the density of air is 1.23 kg/m³. Assuming one-dimensional, steady, inviscid flow, the velocity in the test section as measured at point B is __________ m/s.

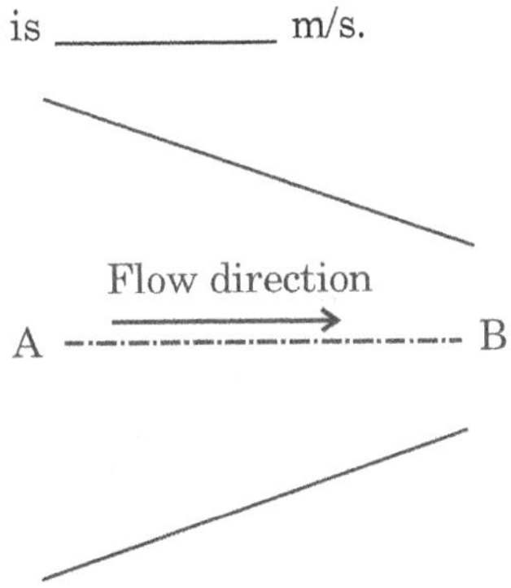

[GATE 2016, 2 M]

103. An untwisted wing of elliptic planform and aspect ratio 6 consists of thin symmetric airfoil sections. The coefficient of lift (C_L) at 10° angle of attack assuming inviscid incompressible flow is

(*a*) $\frac{\pi^2}{16}$ (b) $\frac{\pi^2}{12}$

(c) $\frac{\pi^2}{8}$ (d) $\frac{\pi}{2}$

[GATE 2016, 2 M]

104. Consider 1-D, steady, inviscid, compressible flow through a convergent nozzle. The total temperature and total pressure are T_o, P_o respectively. The flow through the nozzle is choked with a mass flow rate of $\dot{m}_0$. If the total temperature is increased to $4T_o$, with total pressure remaining unchanged, then the mass flow rate through the nozzle

(*a*) remains unchanged.

(*b*) becomes half of $\dot{m}_0$.

(*c*) becomes twice of $\dot{m}_0$.

(*d*) becomes four times of $\dot{m}_0$.

[GATE 2016, 2 M]

105. A substance experiences an entropy change of $\Delta s > 0$ in a quasi-steady process. The rise in temperature (corresponding to the entropy change Δs) is highest for the following process:

(*a*) isenthalpic (*b*) isobaric

(*c*) isochoric (*d*) isothermal

[GATE 2016, 2 M]

106. Which one of the following statements is NOT trye

(*a*) Compared to a laminar boundary layer, a turbulent boundary layer is more desirable on a wing operating at large angle of attack

(*b*) The skin friction drag for a turbulent boundary layer is larger than that for a laminar boundary layer

(*c*) The location of transition from laminar to turbulent boundary layer depends only on the operating Reyonds number

(*d*) A separated flow does not necessarily lead to a turbulent boundary layer

[GATE 2017, 2 M]

107. A De Laval nozzle is to be designed for an exit Mach number of 1.5. The reservoir conditions are given as P_0 = 1 atm (gage), T_0 = 20°C, γ = 1.4. Assuming shock free flow in the nozzle, the exit absolute pressure (in atm) is _______ (in three decimal places)

[GATE 2017, 2 M]

108. Consider a steady one dimensional flow of a perfect gas with heat transfer in a duct. The T-s diagram (shown below) shows both the static and the stagnation conditions at two locations. A and B, in the duct. A_t and B_t denote stagnation conditions for states A and B, respectively. It is known that $(\Delta T)_A = (\Delta T)_B$. M_A and M_B are the Mach numbers of the flow at locations A and B.

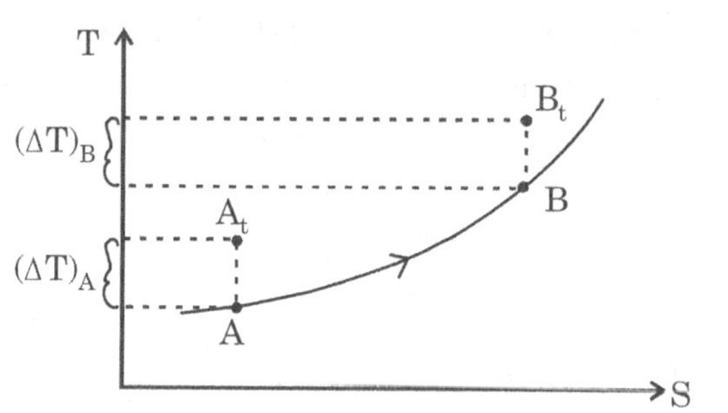

Which of the following statements is true about the flow.

(*a*) Flow is subsonic and $M_A < M_B$

(*b*) Flow is supersonic and $M_A > M_B$

(*c*) Flow is subsonic and $M_A > M_B$

(*d*) Flow is supersonic and $M_A < M_B$

[GATE 2017, 2 M]

109. Consider a system consisting of certain amount of perfect gas enclosed in a cylinder fitted with a frictionless piston. This system can undergo following processes

(i) Expansion with finite pressure difference with the surrounding.

(ii) Compression with infinitesimal pressure difference with the surrounding.

(iii) Heat transfer with finite temperature difference with the reservoir.

(iv) Heat transfer with infinitesimal temperature difference with the reservoir.

Out of these which process are reversible?

(*a*) (i) and (iii) (*b*) (i) and (iv)

(*c*) (ii) and (iii) (*d*) (ii) and (iv)

[GATE 2017, 2 M]

110. Consider a straight wing with rectangular planform of aspect ratio 10 and with a NACA 0012 airfoil. The span effectiveness factor for this wing is 0.95. Assume the flow to be incompressible and governed by thin airfoil theory. The lift coefficient of this wing, at an angle of attack of 6 deg, is _______ (in three decimal places).

[GATE 2017, 2 M]

111. Consider an incompressible flow over a flat plate with the following approximation on the velocity profile:

$$\frac{u(y)}{U} = \begin{cases} \frac{y}{\delta} \text{ for } y \le \delta \\ 1 \text{ for } y < \delta \end{cases}$$

where δ is the boundary layer thickness and U the free-steam speed. The normalized momentum thickness $\left(\frac{\theta}{\delta}\right)$ for this profile is _______ (in three decimal places)

[GATE 2017, 2 M]

112. An idealized velocity field is given by $\overline{V} = 4tx\hat{i} - 2t^2yj + 4xz\hat{k}$. At point (–1, 1, 0) and at t = 1, the magnitude of the material acceleration vector of the fluid element is _________

[GATE 2017, 2 M]

113. A trace from the schlieren photograph of the flow around corner reveals the edges of the expansion fan as shown below. The leading and trailing edges of the expansion fan make the angles as shown. Assuming γ = 1.4, the angle of the expansion fan (in degrees) is _______ (in two decimal places)

Prandtl Meyer function is given by

$$v(M) = \sqrt{\frac{y+1}{y-1}} \tan^{-1} \sqrt{\frac{y-1}{y+1}(M^2-1)} -$$

$$\tan^{-1}\sqrt{(M^2-1)}$$

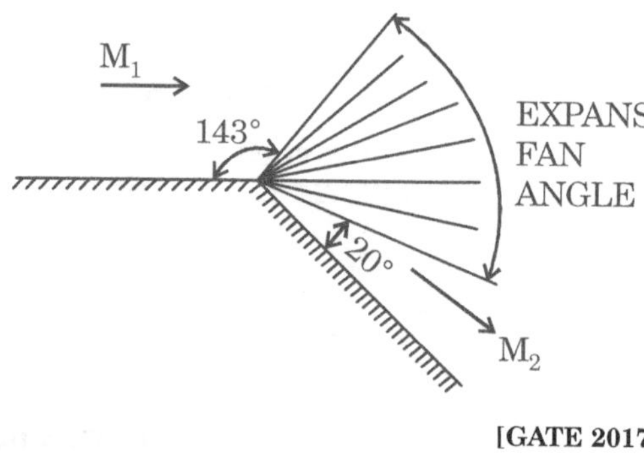

[GATE 2017, 2 M]

114. A strong normal shock wave, with a pressure ratio of 29 across it, is travelling into stationary air ($\gamma = 1.4$) at T = 280 K in a straight duct (see figure). The magnitude of the velocity of the air induced behind the shock wave is ______ m/s. (round to nearest integer)

(Gas constant = 287 J/kg K; Shock wave relations:

Pressure ratio: $\frac{p_2}{p_1} = 1 + \frac{2\gamma}{\gamma+1}(M^2 - 1)$; Density

ration: $\frac{p_2}{p_1} = \frac{(\gamma+1)M^2}{(\gamma-1)M^2 + 2}$

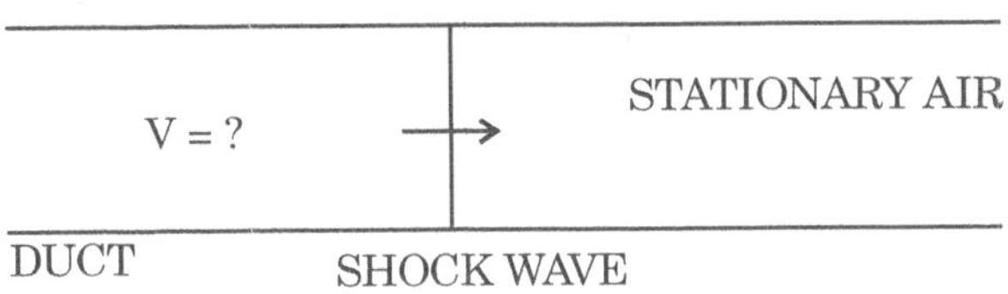

[GATE 2017, 2 M]

115. In the figure below, water exists from a nozzle into atmospheric pressure of 101 kPa. If the exit velocity is V_2 = 8m/s and friction is neglected, the magnitude of the axial force on the flange at location 1 required to keep the nozzle attached to the pipe is ______ N (round to nearest integer).

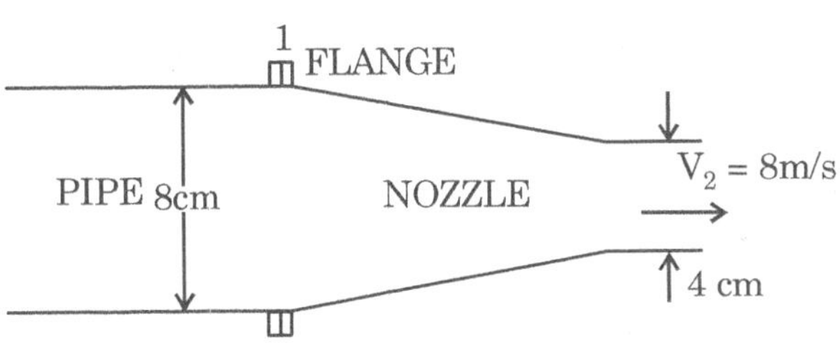

[GATE 2017, 2 M]

116. A football, meant to be thrown at 100 km/h in sea level air (ρ = 1.22 kg/m^3, μ = 1.78 × 10^{-5} N-s/m^2), is to be tested using a one-quarter scale model in a water tunnel (ρ = 1000 kg/m^3, μ = 1.78 × 10^{-3} N-s/m^2). For dynamic similarity, the ratio of the model force to the prototypal force is ______ (round to nearest integer).

[GATE 2017, 2 M]

117. The velocity profile in an incompressible, laminar boundary layer is shown in the figure below. U is the free-stream velocity, u(y) is the stream-wise velocity component. The area of the black shaded region in the figure below represents the

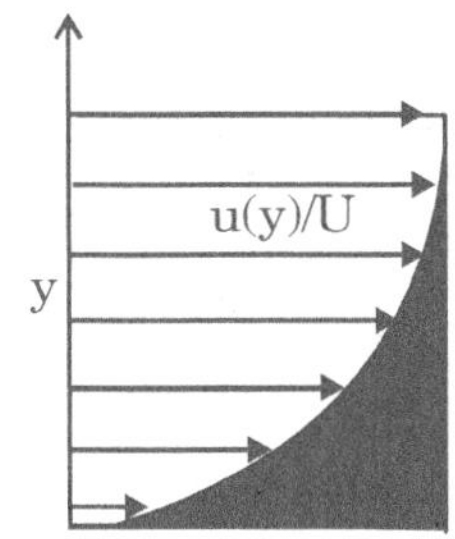

(*a*) boundary layer thickness.
(*b*) momentum thickness.
(*c*) displacement thickness.
(*d*) shape factor.

[GATE 2018, 1 M]

118. In a low-speed wind tunnel, theangular location(s) from the front stagnation point on a circular cylinder where the static pressure equals the free-stream static pressure, is

(*a*) ±38° (*b*) ±30°
(*c*) ±60° (*d*) 0°

[GATE 2018, 1 M]

119. A thermocouple, mounted flush in an insulated flat surface in a supersonic laminar flow of air measures the

(*a*) static temperature.
(*b*) temperature greater than static but less than total temperature.
(*c*) total temperature.
(*d*) temperature greater than total temperature.

[GATE 2018, 1 M]

120. A shock wave is moving into still air in a shock tube. Which one of the following happens to the air?

(*a*) static temperature increases, total temperature remains constant.
(*b*) static temperature increases, total temperature increases.
(*c*) static temperature increases, total temperature decreases.
(*d*) static pressure increases, total temperature remains constant.

[GATE 2018, 1 M]

121. The first law of thermodynamics is also known as conservation of

(*a*) mass. (*b*) momentum.
(*c*) energy. (*d*) species.

[GATE 2018, 1 M]

122. The relation between pressure (p) and velocity (V) for a steady, isentropic flow at two points along a streamline is, (c is a constant)

(a) $c\left(p_2^{\gamma} - p_1^{\gamma}\right) = \frac{v_1^2}{2} - \frac{v_2^2}{2}$

(b) $c\left(p_2^{\frac{\gamma}{\gamma-1}} - p_1^{\frac{\gamma}{\gamma-1}}\right) = \frac{v_1^2}{2} - \frac{v_2^2}{2}$

(c) $c\left(p_2^{\frac{\gamma-1}{\gamma}} - p_1^{\frac{\gamma-1}{\gamma}}\right) = \frac{v_1^2}{2} - \frac{v_2^2}{2}$

(d) $c\left(p_2^{\gamma-1} - p_1^{\gamma-1}\right) = \frac{v_1^2}{2} - \frac{v_2^2}{2}$

[GATE 2018, 2 M]

123. A thin airfoil is mounted in a low-speed, subsonic wind tunnel, in which the Mach number is 0.1. At a point on the airfoil, the pressure coefficient is measured to be –1.2. If the flow velocity is increased such that the free-stream Mach number is 0.6, the pressure coefficient at the same point on the airfoil will approximately be:

(a) –3.5

(b) –2.9

(c) –1.5

(d) –0.75

[GATE 2018, 2 M]

124. The magnitude of the x-component of a unit vector at the point (1, 1) that is normal to equi-potential lines of the potential function $\phi(r) = \frac{1}{r^2+4}$, where $r = \sqrt{x^2+y^2}$, is ______

[GATE 2018, 2 M]

125. Consider a cubical tank of side 2 m with its top open. It is filled with water up to a height of 1 m. Assuming the density of water to be 1000 kg/m^3, g as 9.81 m/s^2 and the atmospheric pressure to be 100 kPa, the net hydrostatic force (in kN) on the side face of the tank due to the air and water is ______ (accurate to two decimal places).

[GATE 2018, 2 M]

126. Consider a 20° half-angle wedge in a supersonic flow at Mach 3.0 at standard sea-level conditions. If the shock-wave angle on the wedge is 36°, the Mach number of the tangential component of the flow post-shock is _______ (accurate to two decimal places).

[GATE 2018, 2 M]

127. The boundary layer thickness at the location of a sensor on a flat plate in an incompressible, laminar flow of air is required to be restricted to 1 mm for an effective measurement. If theflow velocity is 20 m/s with 1 bar pressure, 300 K temperature, and 1.789×10^{-5} kg/(m-s) viscosity, the maximum distance (in mm) of the sensor location from the leading edge is ________ (accurate to one decimal place).

[GATE 2018, 2 M]

128. The dimensions of kinematic viscosity of a fluid (where L is length, T is time) are

(a) LT^{-1} (b) L^2T^{-1}

(c) LT^{-2} (d) $L^{-2}T$

[GATE 2019]

129. $\phi(x, y)$ represents the velocity potential of a two-dimensional flow with velocity field $\vec{V} = u(x,y)\hat{i} + v(x,y)\hat{j}$ $\hat{i}$ and $\hat{j}$ vectors along the x and y axes, respectively. Which of the following is necessarily true?

(a) $\nabla^2\phi = 0$ (b) $\nabla \times \vec{V} = 0$

(c) $\nabla - \vec{V} = 0$ (d) $u = \frac{\partial\phi}{\partial y}, v = \frac{\partial\phi}{\partial x}$

[GATE 2019]

130. For a quasi-one-dimensional isentropic supersonic flow through a diverging duct, which of the following is true in the direction of the flow?

(a) Both the Mach number and the static temperature increase.

(b) The Mach number increases and the static temperature decreases.

(c) The Mach number decreases and the static temperature increases.

(d) Both the Mach number and the static temperature decrease.

[GATE 2019]

131. For a NACA2415 airfoil of chord length c, which of the following is true?

(a) Maximum camber is located at 0.2c from the leading edge.

(b) Maximum thickness is located at 0.15c from the leading edge.

(c) Maximum camber is 0.02c.

(d) Maximum thickness is 0.05c.

[GATE 2019]

132. To simulate the aerodynamic forces on a cylinder of 1 m diameter due to a uniform air flow of 1 m/s at standard temperature and pressure (STP), low-speed wind tunnel experiments at STP are conducted on a 0.1 m diameter cylinder. The free stream air speed in the wind tunnel experiments should be _______ m/s (round off to the nearest integer).

[GATE 2019]

133. A supersonic flow in a constant area duct at Mach number M_1 encounters a ramp of angle θ_1 (see Figure 1). The resulting oblique shock with shock angle β_1 is then reflected from the top wall. For the reflected shock, the turn angle is θ_2 and the shock angle is β_2.

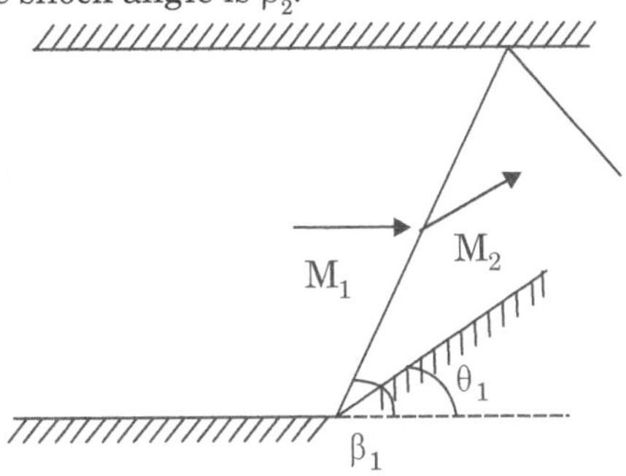

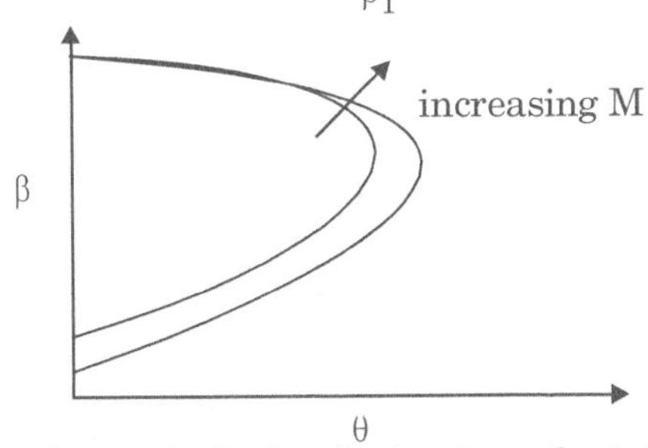

Use the weak shock solution from the θ-β-M plot shown in Figure 2 to choose the correct option from the following.

(a) $\beta_1 > \beta_2$ (b) $\beta_1 < \beta_2$

(c) $\theta_1 > \theta_2$ (d) $\theta_1 < \theta_2$

[GATE 2019]

134. The inviscid, incompressible flow field resulting from a uniform flow past a circular cylinder of radius R centered at the origin is given by:

$$u_r = U\left(1 - \frac{R^2}{r^2}\right)\cos\theta$$

$$u_\theta = -U\left(1 + \frac{R^2}{r^2}\right)\sin\theta$$

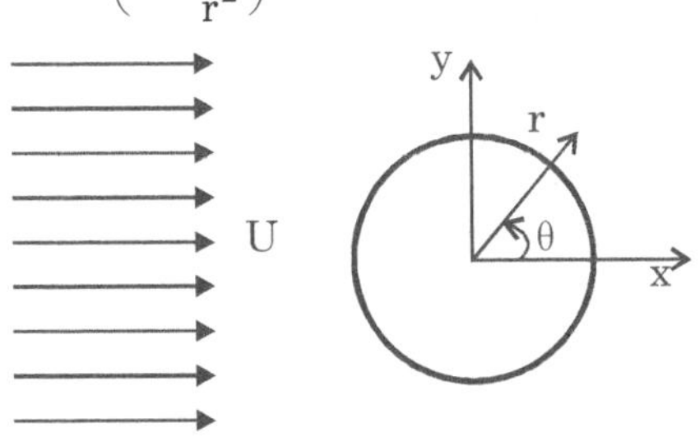

where u_r and u_θ are the radial and azimuthal velocity components in polar coordinates, (r, θ), as shown in the figure. U is the free stream speed. Ignore the effects of gravity. The azimuthal location (in the first quadrant) on the cylinder at which the pressure coefficient is zero is _____ degrees (round off to the nearest integer).

[GATE 2019]

135. A cylindrical container of radius R = 50 cm is filled with water up to a height h_0. Upon rotating the cylinder about its central axis at a constant angular speed, the free surface takes a parabolic shape (see figure), and is displaced upwards by h_1 = 10 cm at r = R. The magnitude of the downward displacement h_2 of the free surface at r = 0 is _____ cm (round off to the nearest integer).

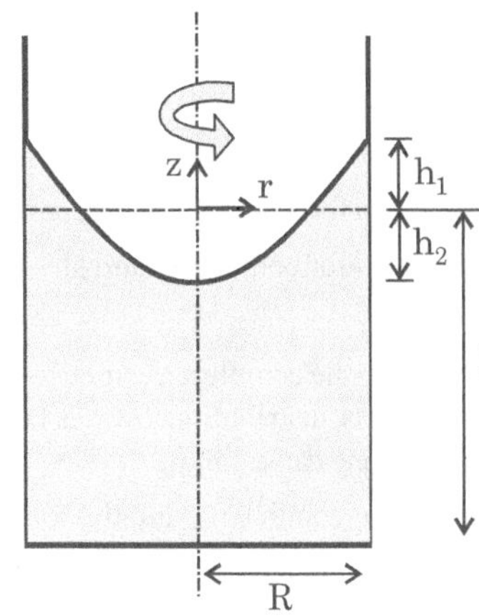

[GATE 2019]

136. A two-dimensional, incompressible fluid flow is described by the stream function $\psi = xy^3 m^2/s$ on the Cartesian x-y plane. If the density and dynamic viscosity of the fluid are 1 kg/m³ and 0.1 kg/m-s, respectively, the magnitude of the pressure gradient in the x direction at x = 1 m and y = 1 m is _______ N/m³ (round off to 1 decimal place).

[GATE 2019]

137. The static pressure ratio across a stationary normal shock is given by

$$\frac{p_1}{p_1} = 1 + \frac{2\gamma}{\gamma + 1}\left(M_1^2 - 1\right),$$

where M_1 is the upstream Mach number. For a stationary normal shock in air (γ =1.4, R = 287 J/kg-K) with upstream flow conditions given by: speed 800 m/s, static temperature 300 K and static pressure 1 atm., the static pressure downstream of the shock is ______ atm. (round off to 2 decimal places)

[GATE 2019]

138. For a symmetric airfoil at an angle of attack of 10°, assuming thin airfoil theory, the magnitude of the pitching moment coefficient about the leading edge is_______ (round off to 2 decimal places).

[GATE 2019]

139. The span-wise distribution of circulation over a finite wing of span b = 10 m is

$$\Gamma(y) = \Gamma_0\sqrt{1-\left(\frac{2y}{b}\right)^2}$$

If $\Gamma_0 = 20 m^2/s$ and the free steam density and speed are 1.2 kg/m³ and 100 m/s respectively, the total lift is _______ kN (round off to 2 decimal places)

[GATE 2019]

140. Which of the following statements is true about the effect of increase in temperature on dynamic viscosity of water and air, at room temperature?

(*a*) It increases for both water and air.

(*b*) It increases for water and decreases for air.

(*c*) It decreases for water and increases for air.

(*d*) It decreases for both water and air.

[GATE 2020]

141. Given access to the complete geometry, surface pressure and shear stress distribution over a body placed in a uniform flow, one can estimate

(*a*) the moment coefficient, and the force on the body.

(*b*) the force coefficient, and the force on the body.

(*c*) the moment coefficient, and the moment of the body.

(*d*) the force and the moment on the body.

[GATE 2020]

142. A pair of infinitely long, counter-rotating line vortices of the same circulation strength Γ are situated a distance h apart in a fluid, as shown in the figure. The vortices will

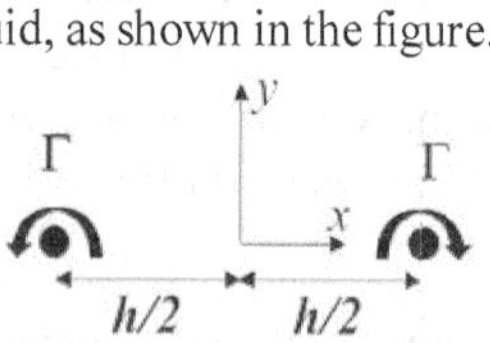

(*a*) rotate counter-clockwise about the midpoint with the tangential velocity at the line vortex equal to $\frac{\Gamma}{2\pi h}$

(*b*) rotate counter-clockwise about the midpoint with the tangential velocity at the line vortex equal to $\frac{\Gamma}{4\pi h}$

(*c*) translate along + y direction with velocity at the line vortex equal to $\frac{\Gamma}{2\pi h}$

(*d*) translate along + y direction with velocity at the line vortex equal to $\frac{\Gamma}{4\pi h}$

[GATE 2020]

143. The streamlines of a steady two dimensional flow through a channel of height 0.2 m are plotted in the figure, where ψ is the stream function in m²/s. The volumetric flow rate per unit depth is

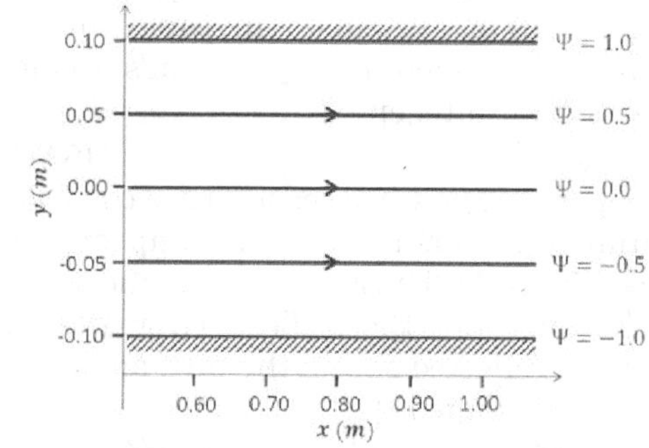

(*a*) 1.0 m²/s (*b*) 2.0 m²/s

(*c*) 0.5 m²/s (*d*) 0.1 m²/s

[GATE 2020]

144. Which of the following options can result in an increase in the Mach number of a supersonic flow in a duct?

(*a*) Increasing the length of the duct

(*b*) Adding heat to the flow

(*c*) Removing heat from the flow

(*d*) Inserting a convergent-divergent section with the same cross-sectional area at its inlet and exit planes

[GATE 2020]

145. In aircraft engine thermodynamic cycle analysis, *perfectly expanded flow* in the nozzle means that the static pressure in the flow at the nozzle exit is equal to

(*a*) the stagnation pressure at the engine inlet.

(*b*) the stagnation pressure at the nozzle exit.

(*c*) the ambient pressure at the nozzle exit.

(*d*) the static pressure at the nozzle inlet.

[GATE 2020]

146. A wedge shaped airfoil is placed in a supersonic flow as shown in the figure (not to scale). The corners of the wedge are at $x = x_A$, $x = x_B$, $x = x_C$, respectively.

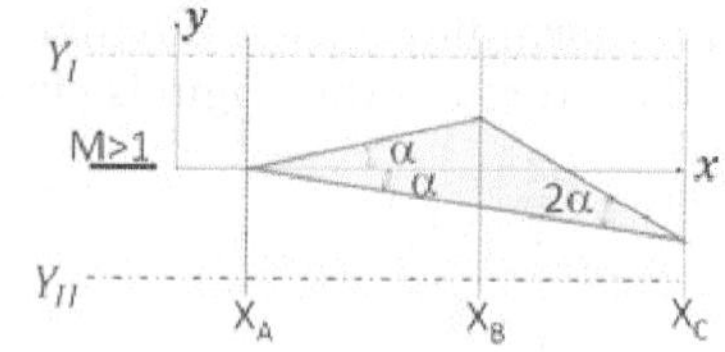

Which one of the following represents the correct static pressure profiles along $y = Y_I$ and $y = Y_{II}$?

(*a*)

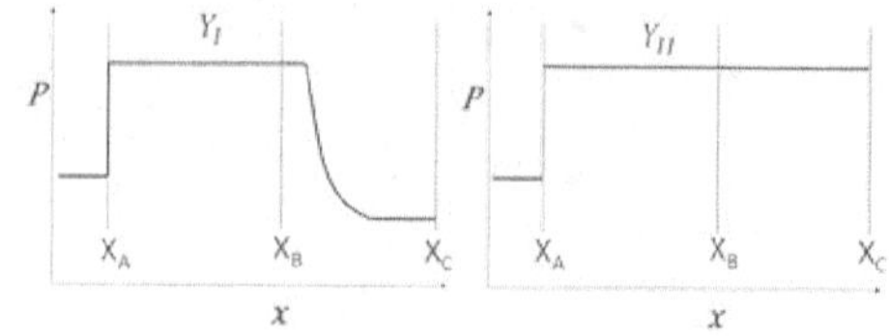

(b) 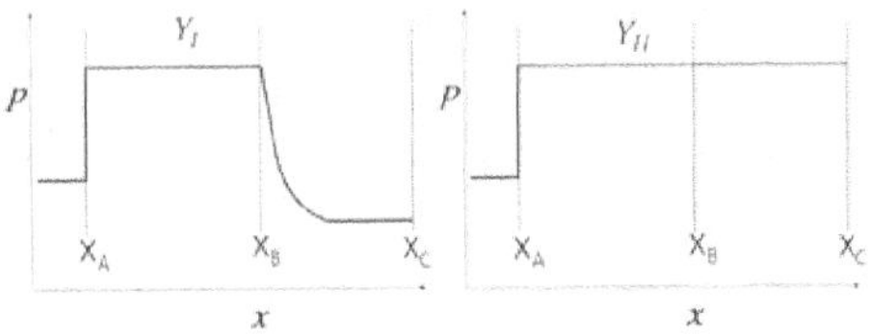

(c) 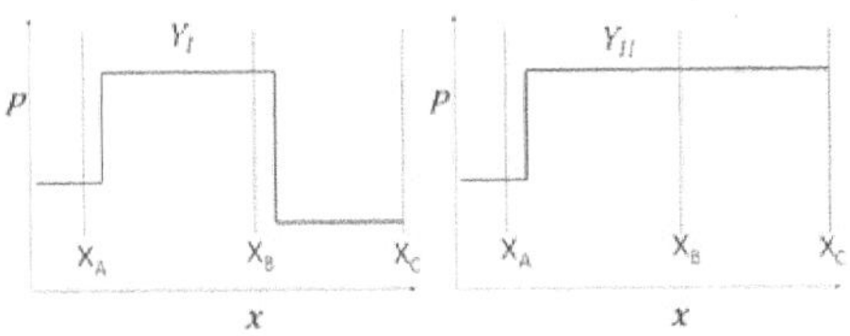

(d) 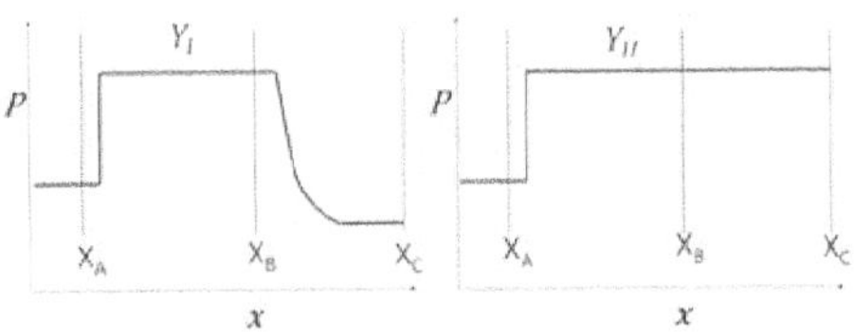

[GATE 2020]

147. The table shows the lift characteristics of an airfoil at low speeds. The maximum lift coefficient occurs at 16 degrees.

Angle of attack α (in degrees)	Lift coefficient C_l
0	0.10
4	0.53

Using Prandtl-Glauert rule, the lift coefficient for the airfoil at the angle of attack of 6 degrees and free stream Mach number of 0.6 is _______ *(round off to two decimal places)*.

[GATE 2020]

148. A low speed uniform flow U_0 is incident on an airfoil of chord c. In the figure, the velocity profile some distance downstream of the airfoil is idealized as shown for section B. The staticpressure at sections A and B is the same. The drag coefficient of the airfoil is _______ *(round off to three decimal places)*.

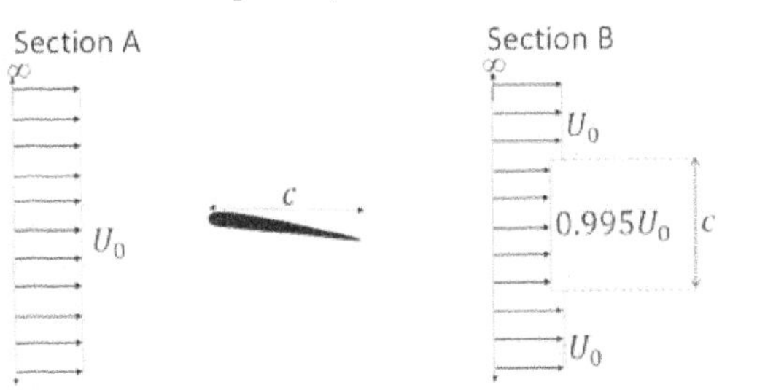

[GATE 2020]

149. An oblique shock is inclined at an angle of 35 degrees to the upstream flow of velocity 517.56 m/s. The deflection of the flow due to this shock is 5.75 degrees and the temperature downstream is 182.46 K. Assume the gas constant R = 287 J/(kg K), specific heat ratio $\gamma = 1.4$, and specific heat at constant pressure C_p = 1005 J/(kg K). Using conservation relations, the Mach number of the upstream flow can be obtained as ______ *(round off to one decimal place)*.

[GATE 2020]

150. The thickness of a laminar boundary layer (δ) over a flat plate is, $\dfrac{\delta}{x} = \dfrac{5.2}{\sqrt{Re_x}}$, where x is measured from the leading edge along the length of the plate. The velocity profile within the boundary layer is idealized as varying linearly with y. For freestream velocity of 3 m/s and kinematic viscosity of 1.5×10^{-5} m^2/s, the displacement thickness at 0.5 m from the leading edge is _________ mm *(round off to two decimal places)*.

[GATE 2020]

151. A wing of 15 m span with elliptic lift distribution is generating a lift of 80 kN at a speed of 90 m/s. The density of surrounding air is 1.2 kg/m^3. The induced angle of attack at this condition is _______ degrees *(round off to two decimal places)*.

[GATE 2020]

152. Consider the velocity field

$\vec{V} = (2x + 3y)\hat{i} + (3x + 2y)\hat{j}$. The field $\vec{V}$ is

(a) divergence-free and curl-free

(b) curl-free but not divergence-free

(c) divergence-free but not curl-free

(d) neither divergence-free nor curl-free

[GATE 2021]

153. The figure shows schematics of wave patterns at the exit of nozzles A and B operating at different pressure ratios.

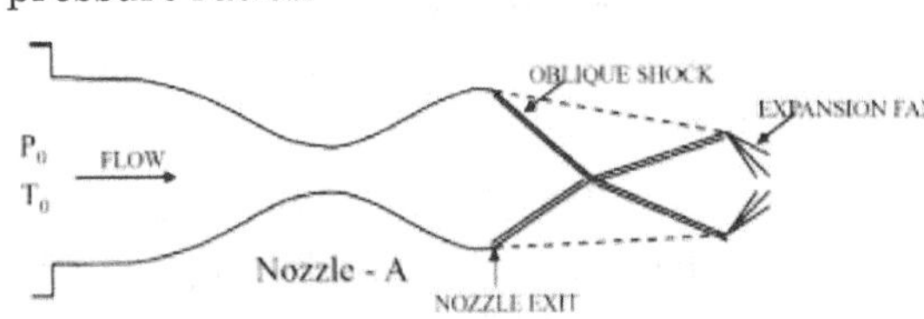

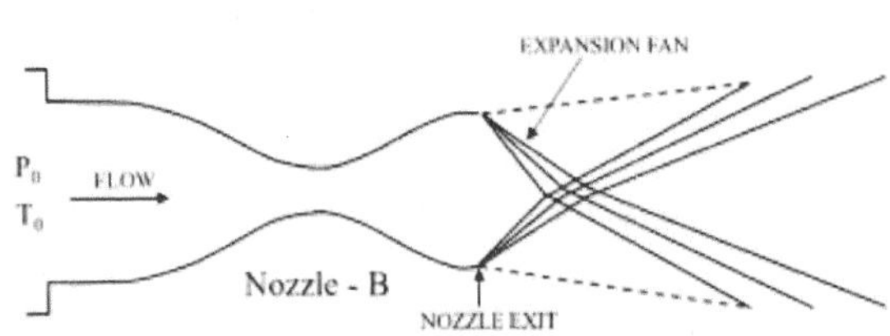

Nozzles A and B, respectively, are said to be operating in:

(a) over-expanded mode and under-expanded mode

(b) under-expanded mode and perfectly expanded mode

(c) perfectly expanded mode and under-expanded mode

(d) under-expanded mode and over-expanded mode

[GATE 2021]

154. The shape of a supersonic diffuser that slows down a supersonic flow to subsonic flow is

(a) converging

(b) diverging

(c) diverging-converging

(d) converging-diverging

[GATE 2021]

155. Which of the following statement(s) is/are true across an oblique shock (in adiabatic conditions) over a wedge shown below?

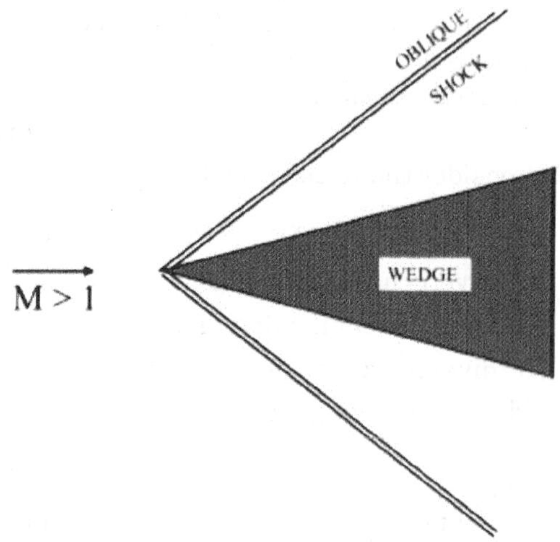

(a) Total pressure decreases

(b) Mach number based on velocity tangential to the shock decreases

(c) Total temperature remains constant

(d) Mach number based on velocity tangential to the shock remains the same and that based on velocity normal to the shock decreases

[GATE 2021]

156. Which of the following statement(s) is/are true with regards to Kutta condition for flow past airfoils?

(a) It is utilized to determine the circulation on an airfoil.

(b) It is applicable only to airfoils with sharp trailing edge.

(c) The trailing edge of an airfoil is a stagnation point.

(d) The flow leaves the trailing edge smoothly.

[GATE 2021]

157. According to the thin airfoil theory, which of the following statement(s) is/are true for a cambered airfoil?

(a) The lift coefficient for an airfoil is directly proportional to the absolute angle of attack.

(b) The aerodynamic center lies at quarter chord point.

(c) The center of pressure lies at quarter chord point.

(d) Drag coefficient is proportional to the square of lift coefficient.

[GATE 2021]

158. A free vortex filament (oriented along Z-axis) of strength K = 5 m^2/s is placed at the origin as shown in the figure. The circulation around the closed loop ABCDEFA for this flow is ______.

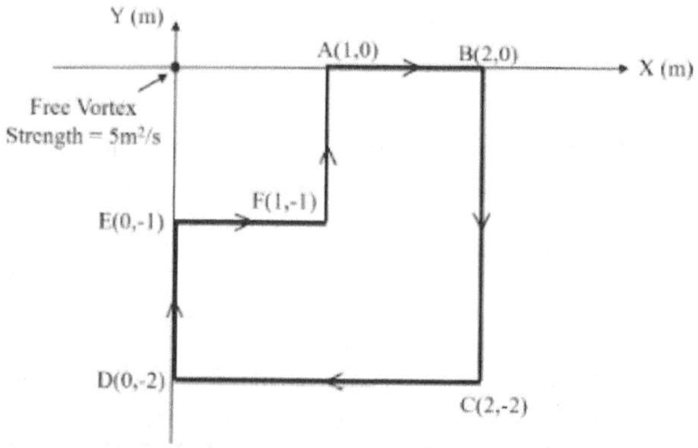

[GATE 2021]

159. The figure shows three glasses P, Q and R with water and floating ice cube. Glass P has a solid ice cube, glass Q has an ice cube with a small solid steel ball embedded in it and glass R has an ice cube with an air bubble. After the ice cube melts, the level of water in glasses P, Q and R, respectively:

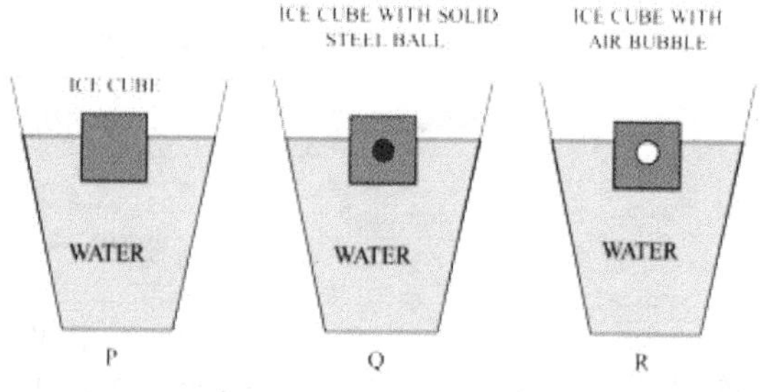

(a) remains same, increases, and decreases

(b) increases, decreases, and increases

(c) remains same, decreases, and decreases

(d) remains same, decreases, and increases

[GATE 2021]

160. To estimate aerodynamic loads on an aircraft flying at 100 km/h at standard sea-level conditions, a one-fifth scale model is tested in a variable-density wind tunnel ensuring similarity of inertial and viscous forces. The pressure used in the wind tunnel is 10 times the atmospheric pressure. Assuming ideal gas law to hold and the same temperature conditions in model and prototype, the velocity needed in the wind tunnel test-section is ______.

(a) 25 km/h (b) 50 km/h (c) 100 km/h (d) 20 km/h

[GATE 2021]

161. The figure shows schematic of a set-up for visualization of non-uniform density field in the test section of a supersonic wind tunnel. This technique of visualization of high speed flows is known as:

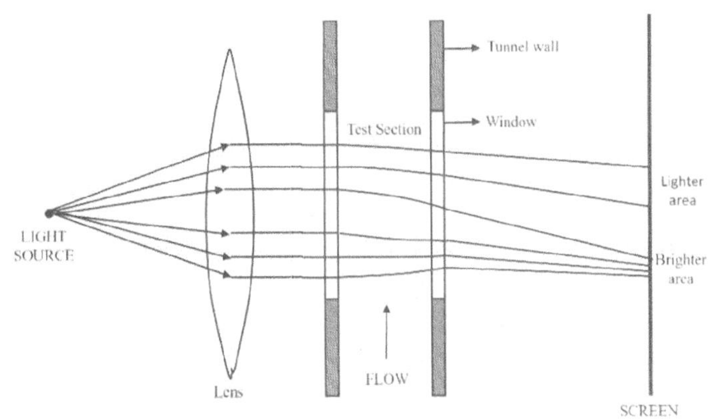

(a) schlieren (b) interferometry (c) shadowgraph (d) holography

[GATE 2021]

162. A shock moving into a stationary gas can be transformed to a stationary shock by a change in reference frame, as shown in the figure. Which of the following is/are true relating the flow properties in the two reference frames?

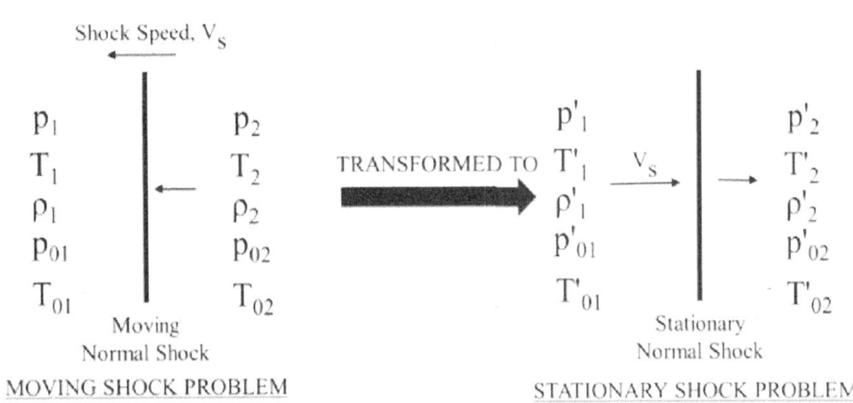

(a) $T_1' > T_1, T_{01}' > T_{01}, p_{01}' > p_{01}, \rho_2' > \rho_1'$ (b) $T_1' = T_1, T_2' < T_{01}, p_{01}' > p_{01}, \rho_2' = \rho_2$

(c) $T_1' < T_1, p_1' > p_1, p_{01}' > p_{01}, \rho_2' > \rho_1$ (d) $T_1' = T_1, p_2 > p_{01}, T_{01}' > T_{01}, p_{01}' > p_{01}$

[GATE 2021]

163. A large water tank is fixed on a cart with wheels and a vane (see figure). The wheels of the cart provide negligible resistance for rolling on the fixed support. The cart is tied to the fixed support with a thin horizontal rope. There is a hole of diameter 5 cm on the side of the tank through which a jet of constant velocity of 10 m/s emerges. The jet of water is deflected by the attached vane by 60° (see figure). Assume that the jet velocity remains constant at 10 m/s after emerging from the vane. Take density of water to be 1000 kg/m³. The tension in the connecting rope is ______ N (*round off to one decimal place*).

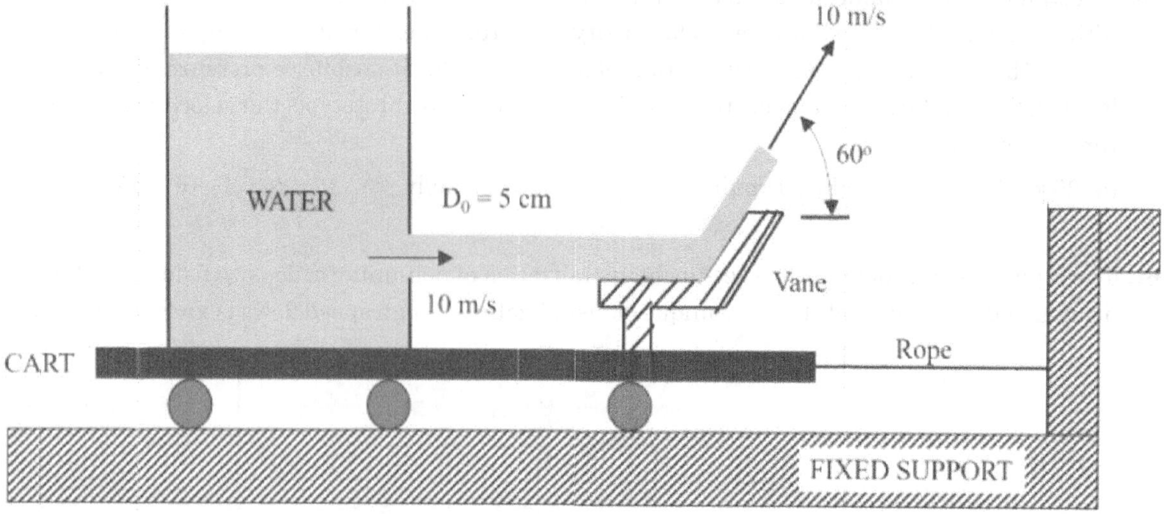

[GATE 2021]

164. A finite wing of elliptic planform with aspect ratio 10, whose section is a symmetric airfoil, is placed in a uniform flow at 5 degrees angle of attack. The induced drag coefficient for the wing is _____ (*round off to three decimal places*).

[GATE 2021]

165. Consider a model of a boundary layer with the following velocity profile:

$$\frac{u}{U} = \begin{cases} \left(\frac{y}{\delta}\right)^2 & y \leq \delta \\ 1 & y > \delta \end{cases}$$

The shape factor, defined as the ratio of the displacement thickness to momentum thickness, for this profile is ______(*round off to 2 decimal places*).

[GATE 2021]

ANSWER KEY

1. (*b*) **2.** (*b*) **3.** (*d*) **4.** (*b*) **5.** (*b*) **6.** (*b*) **7.** (*a*) **8.** (*a*) **9.** (*a*) **10.** (*a*)
11. (*b*) **12.** (*b*) **13.** (*c*) **14.** (*c*) **15.** (*b*) **16.** (*b*) **17.** (*b*) **18.** (*d*) **19.** (*a*) **20.** (*a*)
21. (*d*) **22.** (*a*) **23.** (*a*) **24.** (*a*) **25.** (*b*) **26.** (*a*) **27.** (*c*) **28.** (*b*) **29.** (*a*) **30.** (*c*)
31. (*a*) **32.** (*b*) **33.** (*d*) **34.** (*c*) **35.** (*d*) **36.** (*d*) **37.** (*b*) **38.** (*d*) **39.** (*d*) **40.** (*b*)
41. (*a*) **42.** (*b*) **43.** (*a*) **44.** (*a*) **45.** (*a*) **46.** (*a*) **47.** (*c*) **48.** (*c*) **50.** (*c*) **51.** (*b*)
52. (*a*) **53.** (*c*) **54.** (*d*) **55.** (*b*) **56.** (*a*) **57.** (*b*) **58.** (*a*) **59.** (*c*) **61.** (*a*) **62.** (*a*)
63. (*b*) **64.** (*b*) **65.** (*a*) **66.** (*b*) **68.** (*d*) **69.** (*c*) **70.** (*c*) **71.** (*c*) **72.** (*a*) **73.** (*d*)
74. (*b*) **75.** (*d*) **76.** (*a*) **79.** (*c*) **80.** (*c*) **81.** (*c*) **83.** (*d*) **84.** (*a*) **85.** (*b*) **86.** (*a*)
89. (*b*) **91.** (*b*) **92.** (*b*) **94.** (*a*) **96.** (*d*) **97.** (*c*) **98.** (*d*) **99.** (*c*) **100.** (*a*) **101.** (*d*)
103. (*b*) **104.** (*b*) **105.** (*c*) **106.** (*c*) **108.** (*b*) **109.** (*d*) **116.** (*d*) **117.** (*c*) **118.** (*b*) **119.** (*b*)
120. (*b*) **121.** (*c*) **122.** (*c*) **123.** (*c*) **128.** (b) **129.** (b) **130.** (b) **131.** (c) **132.** (10) **133.** (b)
134. (30°) **135.** (10) **136.** (2.4) **137.** (6.03 atm) **138.** (0.27) **139.** (18.85) **140.** (*c*) **141.** (*d*)
142. (*c*) **143.** (*b*) **144.** (*c*) **145.** (*c*) **146.** (d) **147.** (0.92 to 0.94) **148.** (0.009 to 0.011)
149. (1.9 to 2.1) **150.** (4.00 to 4.20) **151.** (1.29 to 1.38) **152.** (b) **153.** (a) **154.** (d)
155. (a; b; c) **156.** (a; b; d) **157.** (a; b) **158.** (0 to 0) **159.** (c)
160. (b) **161.** (c) **162.** (d) **163.** (98.0 to 98.5)
164. (0.005 to 0.009) **165.** (4.8 to 5.2)

EXPLANATIONS

1. Total pressure is defied as the pressure where flow is brought to rest at that point isentropically. Stagnation pressure needs Isotropic process.

2.

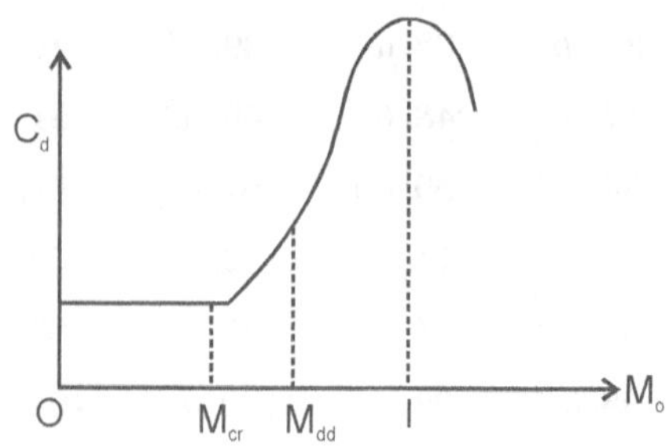

$M_{cr} < M_{dd} < 1$

The free stream mach number at which C_d begins to increase rapidly in defined as dray-divergence mach number.

Critical mach number is defined as free stream mach number at which sonic flow is first uncounter some which or the airfoil sample.

3. $M_1 < 1$ → P_{01}, T_{01}, S_1 | $M_2 < 1$ → $P_{02} P_{01} > P_{02}$, $T_{02} T_{01} > T_{02}$, $S_2 S_2 > S_1$

Normal Shockwave

Flow across shock plane is adiabatic. Total enthalpy is constant, Thus To remains constant.

4. Jowkowski transformation is used to solve two dimensional flow around a aerofoil. In Joukowski transformation the point of a circle in complex plane (Z-plane) are mapped on another complex w-plane by using conformal mapping.

$$w = Z + \frac{1}{Z}$$

where $z = \cos\theta + i \sin\theta$

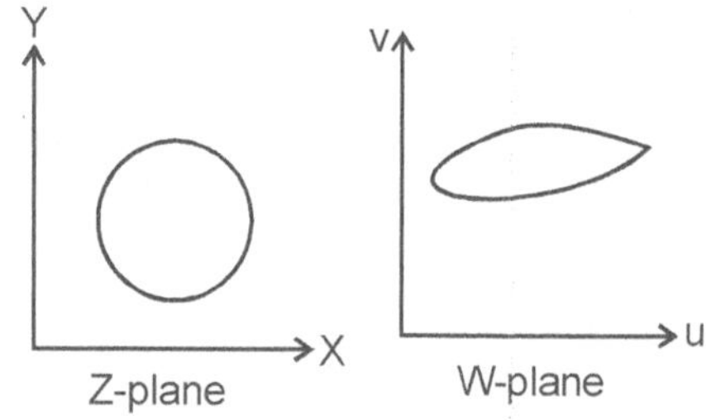

By Changing centre and radius of circle, are get different airfoils. So by inverse mapping we can easily transform joukowski airfoils to circle.

5. To keep the critical mach number high in subsonic aero planes we should deign the airplane with thin Aerofoil, because in case of thin aerofoil when flow passes over it the flow perturbed slightly from freestream values, the expansion own the top surface is mild, the velocity increase only slightly, the pressure decrease only a relatively small amount. Here the magnitnole of C_P at minimum pressure point is small. Variation of C_P with M_∞ is shown in figure.

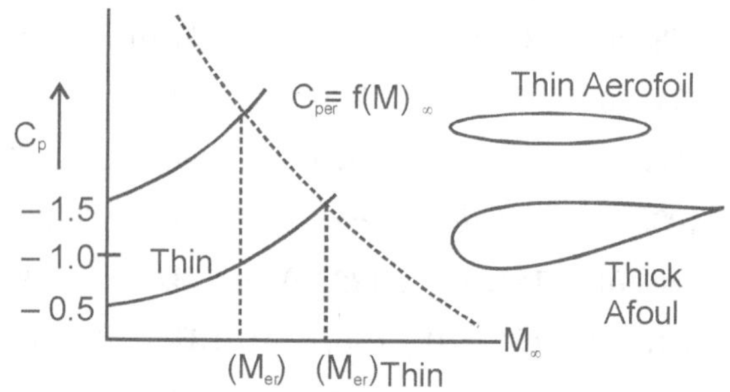

(M_{er}) Thin > (M_{er}) thick (C_P) thick > (C_S) Thin

In case of thick aerofoil the expansive oven the leading edge is stronger, the velocity in crease to larger value, the pressure will decrease to lower values and the magaithole of C_P is larger than the pressure coefficient curve for thick aerofoil lies above Thin above arofoil.

Diamond shape aerofoil is used for supersonic vehicle to reduce the wave drag.

Laminar flow Aerofoil is used to stay in larger laminar flow region on the aerofoil to reduce the skin friction drag.

6. $$Re_{(d)} = \frac{evd}{u}$$

e = density of fluid

v = velocity

d = diameter

Since the where flow rate is same is both pipes so the same fluid in flowing into them, mass flow rate will also be same.

$$m = eAV = e \times \frac{A}{4} d^2 \times V = \text{cost}$$

$$\Rightarrow e \times \frac{\pi}{4} D.D.V = C$$

$$eDV \times \frac{\pi}{4} D = C$$

$$\Rightarrow \frac{eDV}{\mu} \times \mu \times \frac{\pi}{4} D = C$$

$$\Rightarrow \text{Re}\cdot \mu \times \frac{\pi}{4} \times D = C$$

$$\Rightarrow \text{Re}\times \frac{\mu\pi}{4} \times D = C$$

$$\Rightarrow \text{Re} = \frac{4C}{\mu\pi} \times \frac{1}{D} = \frac{k}{D}$$

$$\text{Re}_1 = \frac{k_1}{D_1} \qquad \text{Re}_2 = \frac{k_2}{D_2}$$

$$\text{Re} \propto \frac{1}{D_1}$$

$D_1 \rightarrow$ larger
Re $\rightarrow$ smaller
$D_1 \rightarrow$ smaller
Re $\rightarrow$ larger
Hence the Reynolds no is longer is narrow pipe.

7. Turbulent boundary layer is mere energetic than laminar boundary layer, it has more K.E than laminar boundary layer, hence, flow separation is delayed.
   Thickness, velocity gradient and shear stresses are larger in case of turbulent boundary layer.

8. Boundary longer thickness given by Belasis solution

$$\delta = \frac{5x}{\sqrt{R_{ex}}}$$

x = distanle of point from leading edge
R_{ex} = Reynolds numbeer at x. (Local Reynolds number)

$$\delta = \frac{5x}{\sqrt{\frac{evx}{\mu}}} = 5\sqrt{\frac{x\mu}{ev}}$$

$$= 5\sqrt{\frac{x\mu}{ev}}$$

$$\delta \propto \sqrt{x}$$

$$\frac{\delta_1}{\delta_2} = \sqrt{\frac{x_1}{x_2}}$$

$$\delta_2 = \delta_1\sqrt{\frac{x_2}{x_1}}$$

$$= 5\sqrt{\frac{0.8}{0.2}}$$

$$= 5\sqrt{4}$$

$$= 10 \text{ mm}$$

$$\boxed{\delta_2 = 10\text{mm}}.$$

9.

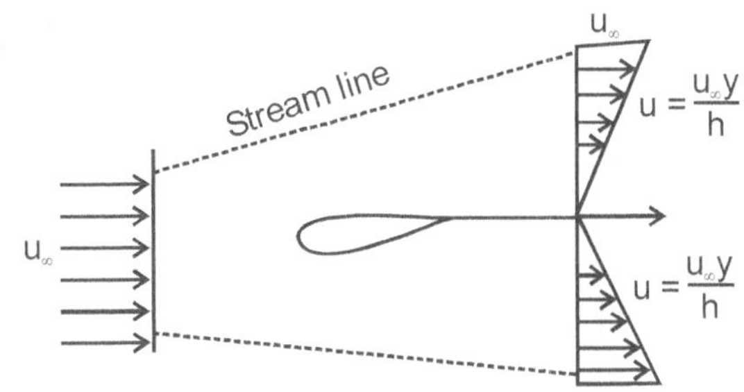

Applying continuity equations

$m_{out} : m_{in}$

$$\therefore m = \int eudA$$

taking inside width is unity

$$m_{in} = \int_{-\delta}^{\delta} eu_{\infty}dy$$

$$m_{out} = \int_{-h}^{0} e\left(-\frac{u_{\infty}}{h}y\right)dy + \int_{-h}^{h} e\left(\frac{u_{\infty}}{h}\right)y\,dy$$

$$\Rightarrow 2eu_{\infty}\delta = \frac{1}{2}\frac{eu_{\infty}}{h}\left(\frac{h}{h}\right) + \frac{1}{2}\frac{eu_{\infty}}{h}\left(\frac{h}{h}\right)$$

$$2\delta = \frac{h}{2} + \frac{h}{2}$$

$$\delta = \frac{h}{2}$$

Now momentum equation, the summation of pressure force is zero because same pressure force is acting at all faces.
Total fare on the Auofoil.

$$\sum \text{Fairfoil} = \int_{cv} ve(v \cdot n)dA$$

$\therefore$ the lower and bottom surface are streamlines so they donot contribute. Only left and right face contribute.
So

$$\sum \text{Fairfoil} = \int_{cv} ve(v \cdot n)dA - \int_{in} ve(v \cdot n)dA$$

the flow is I_r to both surface, so v . n = V

$$\sum \text{Fairfoil} = \int_{out} eu^2dA - \int_{in} ve(v \cdot n)dA$$

the flow is perpendicular in both flows. So

$V{\cdot}n = V$

$V \cdot \eta = v$

$$\sum \text{fairfoil} = \int_{out} eu^2 dA - \int_{is} ed^2 dA$$

$$= \int_{-h}^{0} e\left(-\frac{u_\infty}{h}y\right)^2 dy + \int_{0}^{h} e\left(\frac{u_\infty}{h}y\right)^2_y - \int_{-\delta}^{\delta} eu_\infty^2\, dy$$

$$= \frac{1}{3}e\left(\frac{u_\infty}{h}\right)^2 \left[-|y^3|_h^0 + (y^3)_0^h\right] - eu_\infty^2 |y|_{-\delta}^{\delta}$$

$$= \frac{1}{3}\delta\left(\frac{u_\infty}{h}\right)^2 \left[2h^3\right] - 2eu_\infty^2\delta = -\frac{eu_\infty^2 h}{3}$$

10. Coefficient of pressure C_P

$$C_P = \frac{p - p_\infty}{q_\infty} = 1 - \left(\frac{v}{v_\infty}\right)^2$$

as given $v = v_\infty$

$C_P = 1 - 1 = 0$

C_P for circular cylinder C_P is given by

$C_P = 1 - 4\sin^2\theta$

$0 = 1 - 4\sin^2\theta$

$$\sin\theta = \left(\pm\frac{1}{2}\right)$$

$$\theta = \sin^{-1}\left(\pm\frac{1}{2}\right)$$

$\theta = 30°, 150°, 210°, 330°,$

11. NACA 24012 five digit series

1st digit – When multiplied by 0.15 gives the design lift coefficient (C_L).

2nd & 3rd digits – When divided by 2 gives the distance of maximum camber from leading edge (as percent of chord).

4th & 5th digits– It gives the maximum thickness of aerofoil (as percent of chord)

Here maximum thickness of NACA 24012 series is 12% of chord.

12. C_P for incompressible, irrotational flow is given by

$$C_P = \frac{P - P_\infty}{q_\infty} = 1 - \left(\frac{V}{V_\infty}\right)^2$$

U = velocity at the point

V_∞ = free stream velocity

P_∞ = free stream pressure

P = pressure at point

q_∞ = Dynamic pressure

$$= \frac{1}{2}\rho V_\infty^2$$

For $(C_P)_{max}$, $V \to O$

$$\boxed{(C_P)_{max} = 1}$$

For viscous rotational flow, there would be further loss is total pressure decreasing the value of $(C_P)_{max}$.

13. Irrotational and inviscid flow can become rotating by passing through curved shock wave because for curved shock wave, shock angle varies due to which Δh_0 & Δs are different at different points and flow be comes rotational by (Crocco's theorem)

$$\overline{V} \times \left(\overline{\overline{V}} \times V\right) = \overline{\overline{V}}h_0 - T\overline{\overline{V}}_s$$

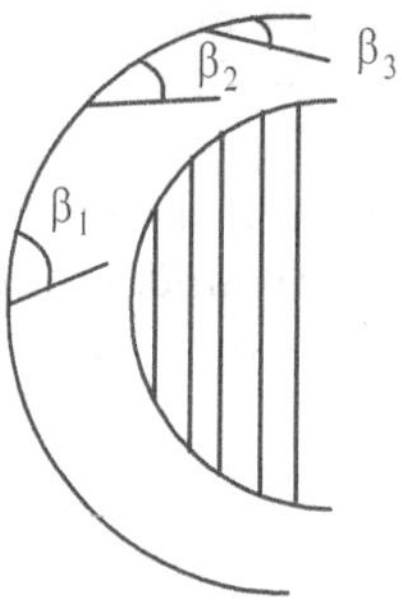

Oblique shock wave

In case of normal & oblique shock wave, wave angle (β) is fixed quantity, here Δh_0 & Δs are same at all points after shock. For a mach wave, flow is isentropic hence, no change in Δh_0 & Δs.

14. In case of Laminar flow Aerofoil is used to reduce the skin friction drag.

15. The wall heat flux due to conduction heat flow is given by

$$q = \frac{kdT}{dy} = \frac{k(T - T_a)}{t}, \ t = \text{thickness of wall}$$

T = Temperature of chamber

T_a = temperature at ambient

In case of CD nozzle, mach number increases continuously along the length towards the exit. So the temperature decreases, hence the maximum flux is at beginning.

16. $\psi = \frac{1}{2}\ln\sqrt{x^2+y^2}$

$$= \frac{1}{2}\ln r$$

$$r^2 = x^2 + y^2$$

$$r = \sqrt{x^2+y^2}$$

tangential velocity $V_\theta = \frac{\partial\psi}{\partial r} = \frac{1}{2r}$

Circulation

$\tau = \int_0^{2\pi} V_\theta . r d\theta$ [line integral of tangential velocity along closed path (contour) $\Gamma = \oint V_e \cdot dx$]

$$= \int_0^{2\pi} \frac{1}{2r} r d\theta = \frac{1}{2}\int_0^{2\pi} d\theta$$

$$= \frac{2\pi}{2} = \pi$$

17. In case of CD nozzle, the flow is supersonic till the normal shock wave after that it becomes subsonic

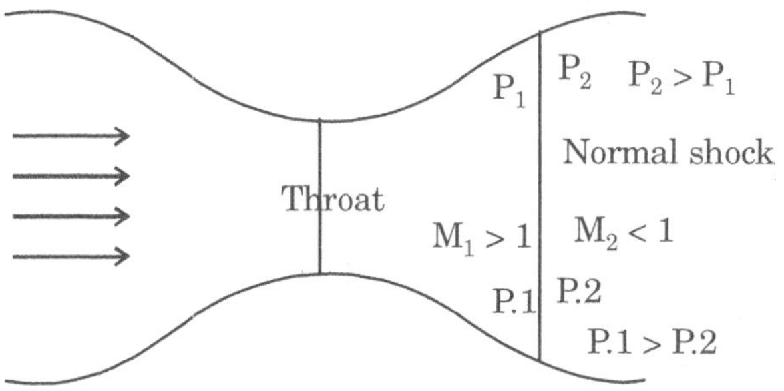

In the diverging portion, the velocity decreases to subsonic value, so the pressure should increases. As the subsonic flow is correctly expanded, flow pressure at nozzle exit should be equal to static pressure at out let.

18. From prandtl Crleurt relation, relation between pressure coefficient for incompressible and compressible flow is given by.

$$(C_P)_{Comp.} = \frac{(C_P) \text{ is comp}}{\sqrt{1-M_\infty^2}}$$

$$(C_P)_{Comp.min} = \frac{(C_P) \text{ is com, min}}{\sqrt{1-M_\infty^2}} = \frac{-0.6}{\sqrt{1-0.7^2}}$$

$$= -0.84$$

Since, the minimum pressure coefficients is below the critical pressure coefficient flow mach number will be greater than one at a point valve of minimum pressure coefficient would not be exactly –0.84, as the above formula is valid for subsonic flow over airfoil would be partly subsonic and partly supersonic.

19. Skin friction Coefficient

$$(C_f) = \frac{0.072}{(Re_L)^{1/5}} \text{ (turbulent boundary layer)}$$

Skin friction coefficient decreases with increase is mach number due to charge is value of μ. With change in temperature due to aerodynamic heating of fluid.

20. Oblique shock diffuser is preferred over normal shock diffuser because total pressure loss across oblique shock is less than that of normal shock wave. In case of normal shock wave pressure loss is larger, which cause pressure recovery to be low and objective to diffuser is lossed.

21. Downwash for the elliptical platform is given by

$$w = \frac{-\sqrt{_0}}{2b}(\text{Cons tan t})$$

22. At trailing edge, the velocity at upper and lower surface should be equal by Kutta condition, hence there is no velocity gradient and hence no vorticity.

Strength of vortex sheet should be zero.

23. For oblique shock wave, if θ is fixed

$M \rightarrow$ Increase

$\beta \rightarrow$ decrease

Hence, shock move closer to body.

24. The objective of supercritical Aerofoil is to reduce the wave drag at transonic range speeds by preventing the portion of shocks at upper surface, hence flatter upper surface, greater leading edge curvature and highly Cambered aft surface to produce lift.

25. Prandtl meyer expansion wave is an isotropic process, hence $\Delta S = 0$, $T_{01} = T_{02}$

$P_{01} = P_{02}$

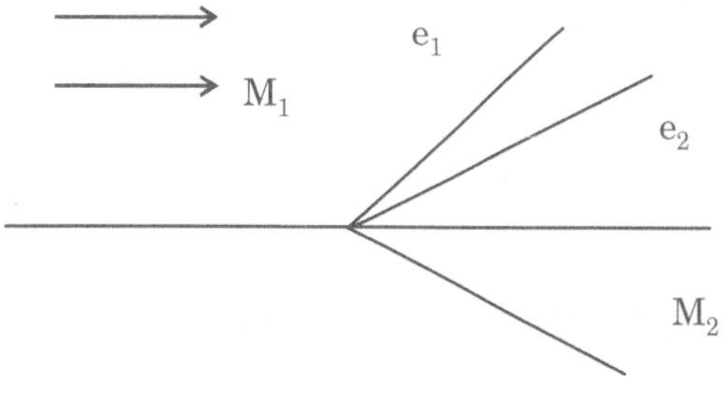

$\Delta S = 0$

$T_{01} = T_{02}$ $M_2 > M_1$

$P_{01} = P_{02}$ $e_1 > e_2$

26. For 2-D irrotational and in compressible flows, stream function, potential function both satisfy laplace equation for irrotational flow

$$\frac{\partial v}{\partial x} - \frac{\partial u}{\partial y} = 0$$

$$\frac{\partial\left(\frac{\partial \phi}{\partial x}\right)}{\partial x} - \frac{\partial\left(\frac{\partial \phi}{\partial x}\right)}{\partial y} = 0$$

$$\frac{\partial^2 \phi}{\partial x \partial y} - \frac{\partial^2 \phi}{\partial y \partial x} = 0$$

stream function

$$\frac{\partial}{\partial x}\left(-\frac{\partial \psi}{\partial x}\right) - \frac{\partial}{\partial y}\left(\frac{\partial \psi}{\partial y}\right) = 0$$

$$\frac{\partial^2 \psi}{\partial x^2} = \frac{\partial^2 \psi}{\partial y^2} = 0 = \overline{V}^2 \psi = 0$$

For In compressible flow

$$\frac{du}{\partial x} + \frac{\partial v}{\partial y} = 0$$

Potential function (ϕ)

$$\frac{\partial^2 \phi}{\partial x^2} + \frac{\partial^2 \phi}{\partial y^2} = 0 \quad \overline{V}^2 \phi = 0$$

stream function (ψ)

$$\frac{\partial}{\partial x}\left(\frac{\partial \psi}{\partial y}\right) + \frac{\partial}{\partial y}\left(\frac{-\partial \psi}{\partial x}\right) = 0$$

$$\frac{\partial^2 \psi}{\partial x\, \partial y} - \frac{\partial^2 \psi}{\partial y\, \partial x} = 0$$

27. According to Thin aerofoil theory lift coefficient at a wing with finite span as

$C_L = C_{L0} + 2\pi\alpha$

C_{L0} = lift coefficient at zero angle of attack differentiating with respect to α we get

$$\frac{dC_L}{d\alpha} = 0 + 2\pi$$

$\boxed{\frac{dC_L}{d\alpha} = 2\pi}$ valid for any aerofoil shape.

28. M = 2

T = 250 K

Stagnation temperature

$$T_0 = T\left(1 + \frac{\gamma - 1}{2} M^2\right)$$

$= 250\,(1 + 0.2 \times 2^2)$

$= 450$ K

29. Across the stationary normal shock wave.

$M_1 > 1$	$M_2 < M_1$	
V_1	V_2	$e_2 > e_1$
P_1	P_2	$T_2 > T_1$
e_1	e_2	$S_2 > S_1$
T_1	T_2	$P_{01} > P_{02}$
S_1	S_2	
P_{01}	P_{02}	$h_{01} = h_{02}$
T_{01}	T_{02}	$T_{01} = T_{02}$
		$P_2 > P_1$

Normal shock → Non – isentropic irreversible process.

$T_{01} = T_{02}$ adiabatic process

$\Delta s > 0$ Non – isentropic, irreversible

30. Given

$C_p = -6.29$

$V_\infty = 50$ m/sec

expression for C_p

$$C_p = \frac{P - P_\infty}{q_\infty}$$

$C_p\, q_\infty = P - P_\infty$

Now,

$$P + \frac{1}{2} e_\infty V^2 = P_\infty + \frac{1}{2} e_\infty V_\infty^2$$

$$\Rightarrow \quad P - P_\infty = \frac{1}{2} e_\infty \left(V_\infty^2 - V^2\right)$$

$$C_p q_\infty = \frac{1}{2} e_\infty \left(V_\infty^2 - V^2\right)$$

$$C_P \times \frac{1}{2} e_\infty V_\infty^2 = \frac{1}{2} e_\infty \left(V_\infty^2 - V^2\right)$$

$$\Rightarrow \quad C_P V_\infty^2 = V_\infty^2 - V^2$$

$$V^2 = V_\infty^2 (1 - C_P)$$

$$V = V_\infty \sqrt{(1 - C_P)}$$

$$= 50\sqrt{1 - (-6.29)}$$

$= 50\sqrt{7.29} = 135 \text{ m/sec}$

2nd method, expression of C_p in terms of V, V_∞.

$$C_P = 1 - \left(\frac{V}{V_\infty}\right)^2$$

$$-6.29 = 1 - \left(\frac{V}{50}\right)^2$$

$$\Rightarrow 7.29 = \frac{V^2}{50^2}$$

$$\Rightarrow V = 50\sqrt{7.29} = 135 \text{ m/sec}$$

31. Given, at 6°

$C_L = 0.53$ (for symmetrical airfoil)

AR = 6

$$a_0 = \frac{dC_L}{d\alpha} = \frac{0.53}{5°} = 0.106/\text{degree}$$

Lift coefficient for a finite wing

$$a = \frac{a_0}{1 + \frac{57.3a_0}{\pi eAR}}$$

$$= \frac{0.106}{1 + \frac{57.3 \times 0.106}{3.14 \times 1 \times 6}} = 0.105/\text{degree}$$

at $\alpha = 5°$

$C_L = a\alpha$

$= 0.105 \times 5 = 0.527$

$= 0.53$

32. Applying Bernoulli equation between section 1 and 2, we get

$$P_1 + \frac{1}{2}eV_1^2 = P_2 + \frac{1}{2}eV_2^2$$

$$P_2 - P_1 = -\frac{1}{2}e\left(V_2^2 - V_1^2\right)$$

$$= \frac{1}{2}e\left(V_1^2 - V_2^2\right) \quad \text{..... (1)}$$

Continuity equation between section (1) and (2), we get

$A_1V_1 = A_2V_2$

$$V_2 = \frac{A_1V_1}{A_2}$$

Putting value of V_2 is equation (1),

We get

$$P_2 - P_1 = \frac{1}{2}e\left(V_1^2 - \left(\frac{A_1V_1}{A_2}\right)^2\right)$$

$$= \frac{1}{2}eV_1^2\left(1 - \left(\frac{A_1}{A_2}\right)^2\right)$$

$$= \frac{1}{2}eV_1^2\left(1 - \frac{A_1^2}{A_2^2}\right)$$

$$\boxed{P_2 - P_1 = \frac{1}{2}eV_1^2\left(1 - \frac{A_1^2}{A_2^2}\right)}$$

33. In this problem, the system remains identical at every time step and hence, momentum & energy are conserved

Relative angular velocity

$$= \frac{\Gamma}{2\pi d} + \frac{\Gamma}{2\pi d} = \frac{\Gamma}{\pi d^2}$$

radius of revolution $= \frac{d}{2}$

34. Boundary layer thickness in case of laminar flow is given by

$$\delta = \frac{5x}{\sqrt{Re_x}}$$

$$= \frac{5x}{\sqrt{\frac{evx}{x}}} = 5\sqrt{\frac{x\upsilon}{eV}}$$

$$\delta \propto \frac{K}{\sqrt{v}}, \delta\sqrt{v} \propto K$$

$$\delta\sqrt{\upsilon} = \text{constant}$$

$$\delta_1\sqrt{v_1} = \delta_2\sqrt{v_2}$$

$$\delta_2 = \delta_1\sqrt{\frac{v_1}{v_2}}$$

$$= 7.2\sqrt{\frac{v_1}{1.5\,v_1}}$$

$$= 7.2\sqrt{\frac{1}{1.5}}$$

$$= \frac{7.2}{\sqrt{1.5}} = 5.9 \text{ mm}$$

35. Shear stress on lower wall

$$\Gamma = \mu \frac{du}{dy}$$

$$= 7 \times 10^{-3} \times \left(\frac{2-0}{1-0}\right)$$

$$= 14 \times 10^{-3} N/m^2$$

36. Pressure distribution

$$P(s) = 2 + \sin\left(\frac{\pi}{2} + 5\right)$$

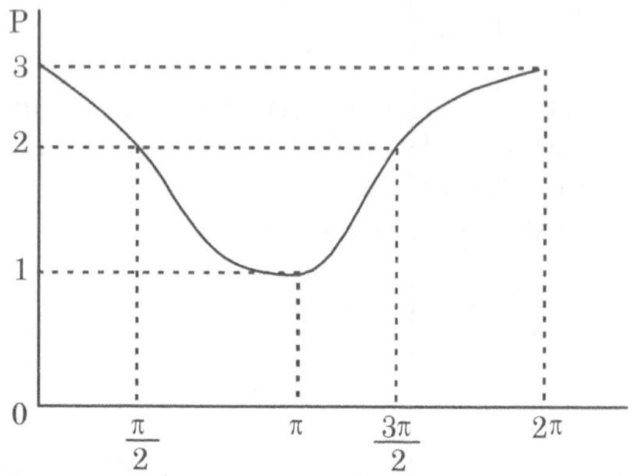

Adverse pressure gradient at S = π hence the separation will occurs at this point.

37.

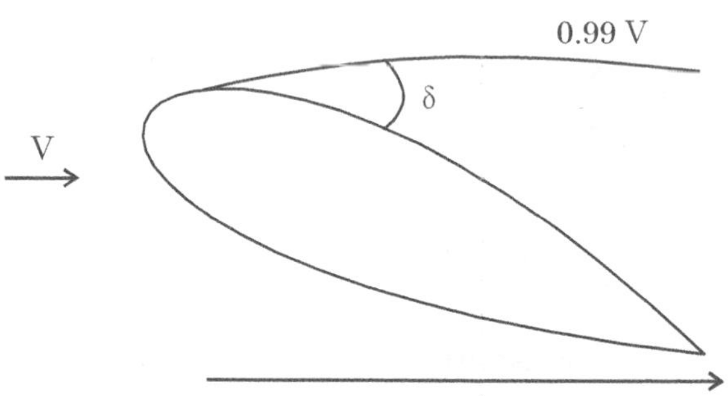

Velocity along x – direction and y - direction.

$$\frac{\partial^n v}{\partial x^n} << \frac{\partial^n v}{\partial y^n}$$

n = 1

$$\frac{\partial v}{\partial x} << \frac{\partial v}{\partial y}$$

38. u, v in terms of ϕ is given by

$$u = \frac{\partial \phi}{\partial x}$$

$$v = \frac{\partial \phi}{\partial y}$$

39.

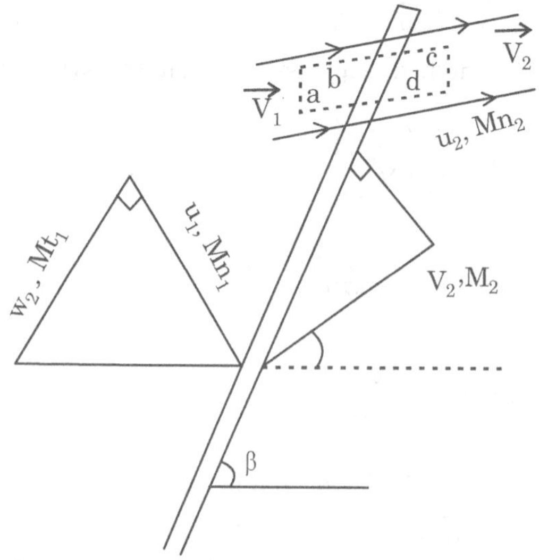

Consider the control volume shown by dashed lines in the upper part of figure sides a and d are parallel to shock wave :

Now, applying Integral form of conservation of equation for the control volume assuming flow is steady, inviscid adiabatic with no body forces. For this assumptions continuity equation.

$$\oiint ev.ds = 0$$

$$\Rightarrow - e_1 A_1 u_1 + e_2 A_2 u_2 = 0 \; [A_1 = A_2]$$

$$\boxed{e_1 u_1 = e_2 u_2}$$

where u_1 and u_2 are velocity normal to shock wave.

Now, applying integral form of momentum equation, along tangential direction

$$\oiint (ev.ds)\, w = \oiint (Pds) \text{ tangential}$$

Since ds is perpendicular to control volume then (Pds) tangential over the face a and d is zero.

$$-(e_1 u_1 A_1) w_1 + (e_2 u_2 A_2) w_2 = 0$$

$$e_1 u_1 A_1 w_1 = e_2 u_2 A_2 w_2$$

$$\boxed{w_1 = w_2}$$

The tangential component of flow velocity is constant across the oblique shock wave.

40. Given data

$P_s = 3.08 \times 10^4 \text{ N/m}^2$

$e_s = 0.467 \text{ Kg/m}^3$

$P_0 = 3.31 \times 10^4 \text{ N/m}^2$

Applying Bernoullie's equation,

We get, $P_0 = P_s + P_d$

Total pressure = static pressure + dynamic pressure

$P_0 - P_s = P_d$

$= \frac{1}{2}ev^2$

$$V = \sqrt{\frac{2(P_0 - P_s)}{e}}$$

$$= \sqrt{\frac{2(3.31 - 3.08) \times 10^4}{0.467}}$$

= 100 m/sec

41. Given

$e = 1.25\ kg/m^2$

$C = 0.5\ m$

$b = 3\ m$

$V = 100\ m/sec$

$\Gamma = 10\ m^2/sec$ per unit spans

$L = ?$

According to Kutta Joukovski theorem

$L = eV\Gamma b$

$= 1.23 \times 100 \times 10 \times 3 = 3690\ N$

42. Given

$R = 287\ J/kg–K$

$C_r = 7.8\ J/kg–K$

$A = 1\ m^2$

$Q = 680\ m3/sec$

$T_0 = 580.05\ K$

$Q = AV$

$$V = \frac{680}{1} = 680\ m/sec$$

Relation between stagnation and static temperature

$$\frac{T_0}{T} = 1 + \frac{\gamma - 1}{2}M^2$$

$$T = \frac{T_0}{1 + 0.2 \times \left(\frac{680}{\sqrt{1.4 \times 287 \times T}}\right)^2}$$

$$T = \frac{580.5}{1 + 0.2\left(\frac{(680)^2}{1.4 \times 287 \times T}\right)}$$

$$T = \frac{580.5}{1 + \frac{0.2 \times 680^2}{402T}}$$

$$T = \frac{580.5}{1 + \frac{250}{T}}$$

$$T\left(1 + \frac{250}{T}\right) = 580.5$$

$\Rightarrow T + 230 = 580.5$

$T = 349.49 = 350\ K$

$T = 300\ K$

43. $\Gamma(y) = 100\sqrt{1 - \left(\frac{2y}{4}\right)^2}$

$V_\infty = 100\ m/sec$

$\Gamma_0 = 100$

$b = 4$

Induce angle of attack $\alpha_i = \frac{100}{2 \times 4 \times 100}$

= 0.125 rad

44. $\Psi = V_\infty r \sin\theta\left(1 - \frac{R^2}{r^2}\right) + \frac{\Gamma}{2\pi}\ln\frac{r}{R}$

$V_\infty = 25\ m/sec$

$R = 1\ m$

$\Gamma = 50\ \pi m^2/sec$

$$u_0 = -\frac{\partial\psi}{\partial r} = -\left[V_\infty \sin\theta\left(1 - \frac{R^2}{r^2}\right) + V_\infty r\sin\theta.\left(\frac{+2R^2}{r^3}\right) + \frac{\Gamma}{2\pi \cancel{R}} \times \frac{1}{\frac{r}{\cancel{R}}}\right]$$

$$= -\left[V_\infty \sin\theta\left(1 - \frac{R^2}{r^2}\right) + V_\infty \sin\theta\frac{2R^2}{r^2} + \frac{\Gamma}{2\pi r}\right]$$

$$= -\left[V_\infty \sin\theta\left(1 - \frac{R^2}{r^2}\right) + V_\infty \sin\theta\frac{2R^2}{r^2} + \frac{\Gamma}{2\pi r}\right]$$

$$= -\left[V_\infty \sin\theta + V_\infty \sin\theta\frac{R^2}{r^2} + \frac{\Gamma}{2\pi r}\right]$$

$$= -\left[V_\infty \sin\theta\left(1 + \frac{R^2}{r^2}\right) + \frac{\Gamma}{2\pi r}\right]$$

On the surface of cylinder r = R

$$V_\theta - \left[V_\infty \sin\theta\left(1+\frac{R^2}{R^2}\right)+\frac{\Gamma}{2\pi R}\right]$$

$$\text{at } \theta = \frac{\pi}{r},\ r = R$$

$$V_\theta = -\left[25\sin\frac{\pi}{2}(1+1)+\frac{\Gamma}{2\pi\times 1}\right]$$

$$= -\left[25\times 1\times 2+\frac{50\pi}{2\pi}\right]$$

V_θ = – 75 m/sec

Similarly,

$$u_r = \frac{1}{r}\frac{\partial\psi}{\partial\theta}$$

$$= V_\infty\left(1-\frac{R^2}{r^2}\right)\cos\theta$$

$$\text{at } r = R,\ \theta = \frac{\pi}{2}$$

$$u_r = 25(1-1)\cos\frac{\pi}{2}$$

$u_r = 0$

$u_\theta = -75$ m/sec, $u_r = 0$

45. At stagnation point $V_\theta = 0$, $V_r = 0$ putting $V_\theta = 0$, [r = R on the surface of cylinder]

$$-\left(2V_\infty \sin\theta+\frac{\Gamma}{2\pi R}\right) = 0$$

$$2V_\infty \sin\theta = \frac{-\Gamma}{2\pi R}$$

$$\sin\theta = \left(-\frac{\Gamma}{4\pi R\, V_\infty}\right)$$

$$\theta = \sin^{-1}\left(-\frac{\Gamma}{4\pi R V_\infty}\right)$$

$$= \sin^{-1}\left(\frac{-50\pi}{4\pi\times 1\times 25}\right)$$

$$\theta = \sin^{-1}\left(-\frac{1}{2}\right)$$

θ = 210°, 330°.

46. NACA 0012, 00 indicates no camber (Symmetry)

NACA 23012, $\frac{30}{2} = 15$, maximum camber at 15% chord

NACA 4415, maximum camber 4% location 0.4 C.

C_l will be zero for symmetrical drag polar case, i.e. C_l should be zero for zero angle of attack it happen only for symmetrical Aerofoil.

For an aerofoil, drag polar expansion

$$C_D = C_{DO} + KC_L^2$$

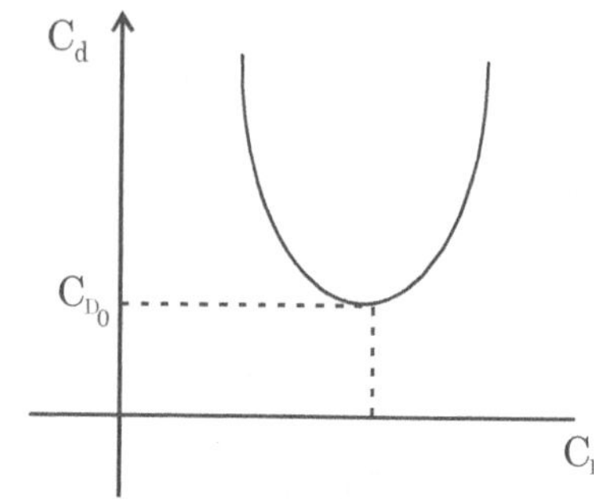

C_D dependent an angle of attack (α)

For drag polar to be symmetrical about C_d axis the minimum of drag polar lies on C_d axis.

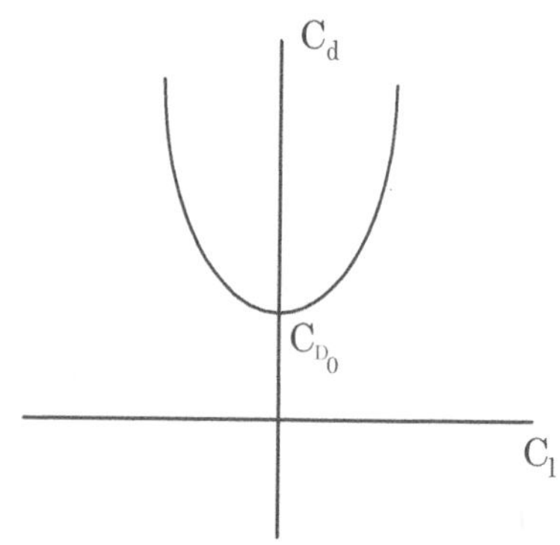

From the given option $C_l = 0$ for zero angle of attack is NACA0012 which is symmetrical.

48. Ψ for spinning circular cylinder of radius R, is given by

$$\Psi = V_\infty \sin\theta\left(1-\frac{R^2}{r^2}\right)+\frac{\Gamma}{2\pi}\ln\frac{r}{R}$$

49. (*) None of these

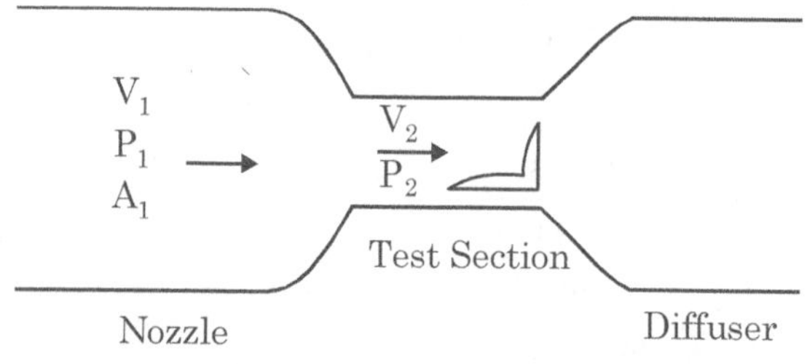

$$\frac{A_1}{A_2} = \text{Contration ratio} = 14:1$$

$A_2 = 1m^2$

Speed of air in test section is given by

$$V_2 = \sqrt{\frac{2(P_1 - P_2)}{\rho_a\left[1-\left(\frac{A_2}{A_1}\right)^2\right]}}$$

where $P_1 - P_2 = \rho_w\ g\Delta h$

ρ_w = density of water

g = 9.8 m/sec

Δh = height difference between this column.

$$\therefore\ V_2 = \sqrt{\frac{2\ P_w g\Delta h}{\rho_a\left(1-\left(\frac{A_2}{A_1}\right)^2\right)}}$$

$$= \sqrt{\frac{2\times1000\times9.8\times0.4}{1.23\left(1-\left(\frac{1}{14}\right)^2\right)}}$$

$$= \sqrt{\frac{2000\times3.92}{1.23\left(1-0.07142^2\right)}}$$

$$= \sqrt{\frac{6374}{\left(1-0.07142^2\right)}}$$

= 80.04 m/sec.

50. $\theta = 20°$

$$\nu(M) = \sqrt{\frac{\gamma+1}{\gamma-1}}\tan^{-1}\sqrt{\frac{\gamma-1}{\gamma+1}\left(M^2-1\right)} - \tan^{-1}\sqrt{M^2-1}$$

$M_1 = 2$

P_1 = 1 atm

T_1 = 288 K

$\theta = 20°$

$\gamma = 1.4$

$$\nu(M_1) = \sqrt{6}\tan^{-1}\sqrt{\frac{1}{6}\left(2^2-1\right)} - \tan^{-1}\sqrt{\left(2^2-1\right)}$$

$$= \sqrt{6}\tan^{-1}\sqrt{0.5} - \tan^{-1}\sqrt{3}$$

= 86.37 – 60

$\nu(M_1) = 26.37$

From prandtl meyer expansion theory

$\nu(M_2) - \nu(M_1) = \theta$

$\nu(M_2) = \theta + \nu(M_1)$

= 20 + 26.37

= 46.37

$\nu(M_2) = 46.37$

Around a sharp expansion corner flow expansion takes place, hence the mach number behind the expansion wave will be greater than upstream mach number. So the option (a) & (b) is incorrect. We need to check for option (c) & (d).

Option (*c*) M = 2.83

$$\therefore\ \nu(M_2) = \sqrt{6}\tan^{-1}\sqrt{\frac{2.83^2-1}{6}} - \tan^{-1}\sqrt{2.83^2-1}$$

= 46.37°

$\nu(M_2) = 46.37°$

This matches to the value of $\nu(M_2)$. Hence downstream Mach number is 2.83.

51. u_∞ = 100 m/sec

$e_\infty = 1.2\ kg/m^3$

P_∞ = 1 atm

$u_\infty = 1.8\times10^{-5}$ kg/m/sec

Boundary layer & momentum thickness given by blasius solution for steady 2-D zero pressure gradient laminar flow over the flat plate.

$$\delta^* = \frac{1.72x}{\sqrt{R_{ex}}}$$

$$\theta = \frac{0.664x}{\sqrt{R_{ex}}}$$

$$\frac{\delta^*}{\theta} = \frac{\frac{1.72x}{\sqrt{R_{ex}}}}{\frac{0.664x}{\sqrt{R_{ex}}}} = 2.59$$

$$\boxed{\frac{\delta^*}{\theta} = 2.59}$$

$\frac{\delta^*}{\theta}$ is known as shape factor (H). For laminar flow shape factor = 2.59. For turbulent flow shape factor (H) = 1.3.

52.

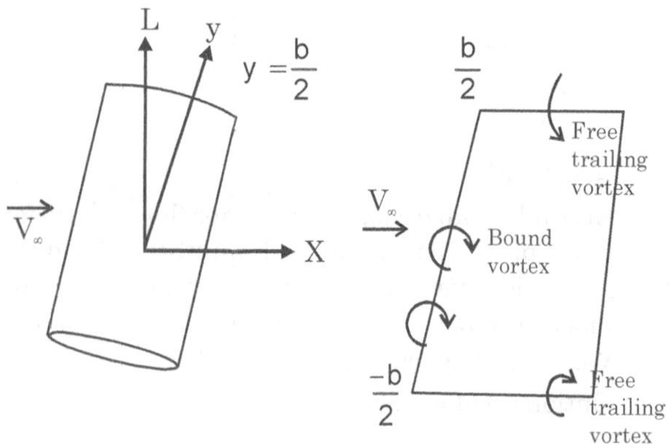

According to Prandtl's Lifting line theory for finite wing, the flow over finite wing consists of combination of bound vortex and free - trailing vortices wing extends from $\frac{b}{2}$ to $\frac{-b}{2}$ when bound vortex is responsible for lift force and free - trailing edge vortices responsible for induced drag.

53. Given

$M_\infty = 0.5$

Using prandtl - Glauert rule for compressibility correction, the variation of C_ℓ with M_∞ for high subsonic mach number.

$$C_\ell = \frac{C_{\ell_0}}{\sqrt{1-M_\infty^2}} = \frac{2\pi\alpha}{\sqrt{1-0.5^2}} = 7.28\,\alpha$$

$C_l = \frac{Cl_0}{\sqrt{1-M_\infty^2}}$ valid for $0.3 < M_\infty < 0.7$ for higher Mach number accuracy diminishes.

54. $M_\infty = 2$

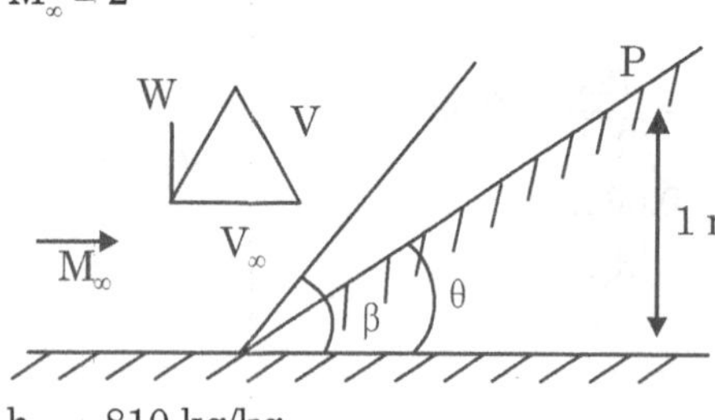

$h_{0\infty} = 810$ kg/kg

$\gamma = 1.4$

$R = 287$

In case of oblique shock wave there are two component of velocity one normal to shock wave, other in tangetial direction, the tangential velocity direction, the tangential velocity does not changes, only normal component of flow velocity changes.

The component along normal = $u_\infty \sin\beta$

The component of normal velocity should be greater than speed at sound to be exist of oblique shock wave.

$$u_\infty \sin\beta > a_\infty$$

$$\frac{u_\infty}{a_\infty}\sin\beta > 1$$

$$M_\infty \sin\beta > 1$$

$$\sin\beta > \frac{1}{M_\infty}$$

$$\beta > \sin^{-1}\left(\frac{1}{M_\infty}\right)$$

$$> \sin^{-1}\left(\frac{1}{2}\right)$$

$$\beta > 30°$$

55. Prandtl's lifting line equation for general wing is given by

$$\alpha(y_0) = \frac{\Gamma(y_0)}{\pi u_\infty C(y_0)} + \alpha_{L=0}(y_0) + \frac{1}{4\pi u_\infty}\int_{-\frac{b}{2}}^{\frac{b}{2}} \frac{\frac{\partial\Gamma}{\partial y}}{y_0 - y}dy \quad ...(A)$$

Now,

Leibniz integral rule for constant integral power.

$$\frac{d}{dx}\int_{y_0}^{y_1} f(x,y)dy = \int_{y_0}^{y_1} \frac{\partial}{\partial x} f(x,y)dy$$

By using Leibniz rule differentiating equation (A), we get

$$1 = \frac{\frac{\partial\Gamma}{\partial y}}{\pi u_\infty C(y_0)} + 0 + \frac{1}{4\pi u_\infty}\frac{d}{dx}\int_{-\frac{b}{2}}^{\frac{b}{2}} \frac{\frac{\partial\Gamma}{\partial y}}{y_0 - y}dy$$

$$1 = \frac{\Gamma_0}{\pi u_\infty C(y_0)} + \frac{1}{4\pi u_\infty}\int_{-\frac{b}{2}}^{\frac{b}{2}} \frac{\frac{\partial\Gamma}{\partial y}}{y_0 - y}dy$$

$$\Gamma_\alpha = \frac{\partial\Gamma}{\partial\alpha}$$

As the α does not come in the above expansion, Γ_α does not depend on α.

56. $C_l \propto \int_{-\frac{b}{2}}^{\frac{b}{2}} \Gamma(y)dy$

$$C_l = K\int_{-\frac{b}{2}}^{\frac{b}{2}} \Gamma(y)dy$$

differentiating with respect to α

We get,

$$\frac{\partial C_L}{\partial \alpha} = K\frac{\partial}{\partial \alpha}\int_{\frac{-b}{2}}^{\frac{b}{2}} \Gamma(y)dy$$

Applying Leibniz rule, we get

$$\frac{\partial C_L}{\partial \alpha} = K\int_{\frac{-b}{2}}^{\frac{b}{2}} \frac{\partial \Gamma}{\partial \alpha}dy$$

$\because \frac{\partial \Gamma}{\partial \alpha}$ is independent from α. Hence, $\frac{\partial C_l}{\partial \alpha}$ is also independent of α.

57. When symmetrical aerofoil placed θa zero degree angle of attack, due to symmetry of aerofoil, the $\alpha = 0$

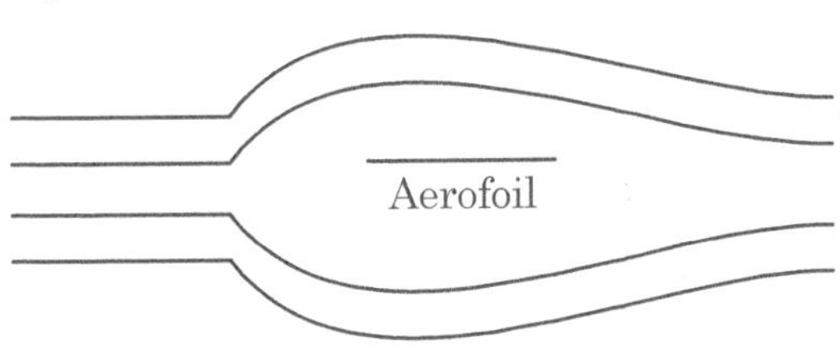

Pressure distribution on the upper and lower is symmetrical in nature hence, resulting lift is zero.

Critical mach number on a aerofoil is attained when, the mach number somewhere on the aerofoil is unity, this happen on the section surface of the aerofoil and the free stream flow many still be subsonic

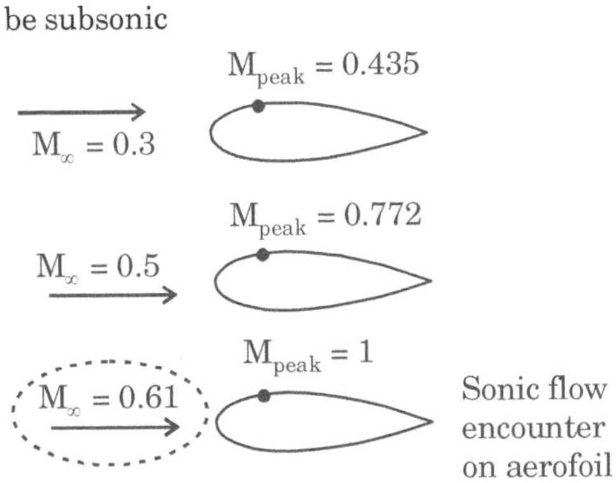

Critical mach number is denoted by M_{cr}.

M_{cr} (thin) > M_{cr} (thick)

M_{cr} is very important quantity because at some free stream mach number above M_{cr} the aerofoil will experience a dramatic increase is drag.

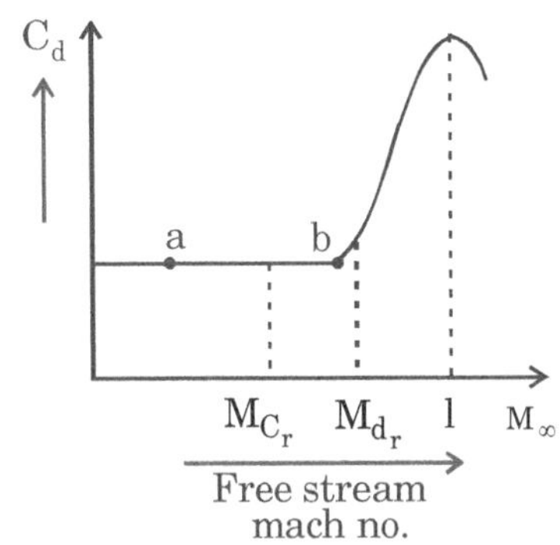

M_{dr} = drag divergence mach number b is point after that drag start dramatically increasing.

58. (a) Mach angle is given by

$$\mu = \sin^{-1}\frac{1}{M}$$

$$= \sin^{-1}\left(\frac{1}{2}\right)$$

$$= 30°$$

59. (c) Static pressure coefficient

$$C_P = 1 - \left(\frac{V}{V_\infty}\right)^2$$

$$1 - \frac{1}{AR^2}$$

$$= 1 - \frac{1}{2^2} = 1 - 0.25 = 0.75$$

60. 1.1 m^2/sec

L = 80 N

V_∞ = 60 m/sec

P_∞ = 100 KPa

T = 290 K

$$\text{Density } \rho = \frac{P}{RT} = \frac{10^5}{287 \times 290}$$

= 1.2 Kg/m^3

Circulation on the aerofoil given by Kutta Joukowski theorem

$$L = \rho V\Gamma$$

$$\Gamma = \frac{L}{\rho V} = \frac{80}{1.2 \times 60}$$

= 1.1 m^2/sec

61.

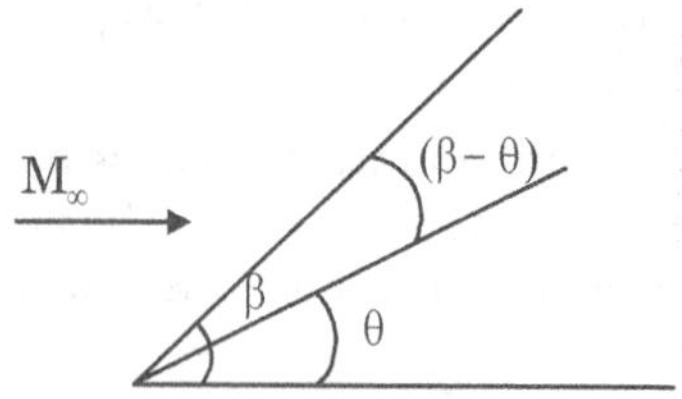

$$Mn_2 = M_2 \sin(\beta - \theta)$$

$$\frac{Mn_2}{M_2} = \sin(\beta - \theta)$$

β = wave angle.

62. The starting shock is formed at the divergent section of the nozzle A* increases across the shock due to decrease in stagnation pressure. To achieve the same mass flow rate, the throat, the throat area of the diffuses needs to be larger than that of the nozzle. The shock is swallowed when it passes through the test section and consequently the diffuser throat, then it disappears downstream of the diffuser throat

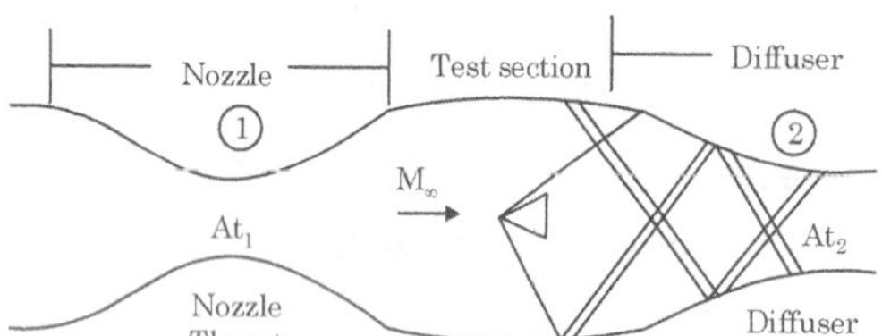

Mathematical relation

$$\frac{At_2}{At_1} = \frac{P_{01}}{P_{02}}$$

$P_{01} > P_{02}$ across shock wave

$At_2 > At_1$

63. Strouhal number = 0.2

d = 20 mm

f = 10 Hz

Strouhal number

$$St = \frac{fD}{v}$$

f = frequency

D = Diameter

v = flow velocity

$$v = \frac{fD}{St}$$

$$= \frac{10 \times 20 \times 10^{-3}}{0.2}$$

= 1 m/sec

64. For supersonic flows, over a gradual compression entropy remains constant as no shock wave is produced. Over a sharp compression corner oblique is formed and entropy increases for expansion, sharp or gradual entropy remains constant.

65. For compressible flow

$$V.\overline{V} = \frac{1}{\delta\forall}\frac{D\delta\forall}{Dt}$$

Where $\delta\forall$ = elemental volume

$\overline{V}$ = velocity of the moving object

66.

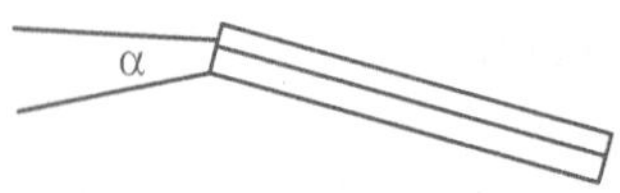

For supersonic flow,

$$C_l = \frac{4\alpha}{\sqrt{M_\infty^2 - 1}}$$

$$C_d = \frac{4\alpha^2}{\sqrt{M_\infty^2 - 1}}$$

$$C_l^2 = \frac{16\alpha^2}{\sqrt{M_\infty^2 - 1}}$$

$$\frac{C_d}{C_l^2} = \frac{\dfrac{4\alpha^2}{\sqrt{M_\infty^2 - 1}}}{\dfrac{16\alpha^2}{\left(\sqrt{M_\infty^2 - 1}\right)^2}}$$

$$= \frac{\sqrt{M_\infty^2 - 1}}{4}$$

$\frac{C_d}{C_l^2}$ independent from α.

67. Given $M_\infty = 0.5$

$$\frac{\Delta A}{A} = 3\%$$

Now continuity equation

ρAV = constant

differentiating we get

$d(\rho AV) = 0$

$$\frac{d\rho}{\rho} + \frac{dA}{A} + \frac{dv}{v} = 0 \quad ....(1)$$

$$\frac{dA}{A} = -\frac{dv}{v}\left(1 - M_\infty^2\right)$$

$$= -\frac{dv}{v}\left(1 - 0.5^2\right)$$

$$= -\frac{dv}{v}0.75$$

$$= -0.75\frac{dv}{v}$$

$$\frac{dv}{v} = -\frac{0.03}{0.075} = -0.04$$

Now,

$$\frac{d\rho}{\rho} = -\left[\frac{dA}{A} + \frac{dV}{V}\right]$$

$$= -[0.04 + 0.03] = -0.01$$

$$\frac{d\rho}{\rho} = 0.01 \times 100 = 1\%$$

68. $\Psi = -\frac{\lambda}{2\pi}\theta + \frac{\Gamma}{2\pi}\ln r$

$\lambda = \Gamma = 100\ m^2/sec$

$e = 1.2\ kg/m^3$

Radial and tangential velocity component given by.

$$V_r = \frac{1}{r}\frac{\partial\left(-\frac{\lambda}{2\pi}\theta + \frac{\Gamma}{2\pi}\ln r\right)}{\partial\theta}$$

$$= \frac{1}{r} - \frac{\lambda}{2\pi} = \frac{-\lambda}{2\pi r}$$

$$= \frac{-100}{2\pi \times 100}$$

$$= -\frac{1}{2\pi}$$

$$V_\theta = -\frac{\partial\psi}{\partial r} = \frac{-\Gamma}{2\pi r} = \frac{-1 \times 100}{2\pi \times 100}$$

$$V_\theta = \frac{-1}{2\pi}$$

Resultant velocity

$$V_R = \sqrt{V_r^2 + V_\theta^2}$$

$$= \sqrt{\left(\frac{1}{2\pi}\right)^2 + \left(\frac{-1}{2\pi}\right)^2}$$

$$= \sqrt{\frac{1}{4\pi^2} + \frac{1}{4\pi^2}}$$

$$= \sqrt{\frac{1}{2\pi^2}} = \frac{1}{\sqrt{2}\pi}$$

$$\text{Gauge pressure} = -\frac{1}{2}eV_R^2$$

$$= -\frac{1}{2}e \times \left(\frac{1}{\sqrt{2}\pi}\right)^2$$

$$= -\frac{1.23}{2} \times \frac{1}{\left(\sqrt{2} \times \pi\right)^2} = \frac{-1.2}{4\pi^2}$$

69.

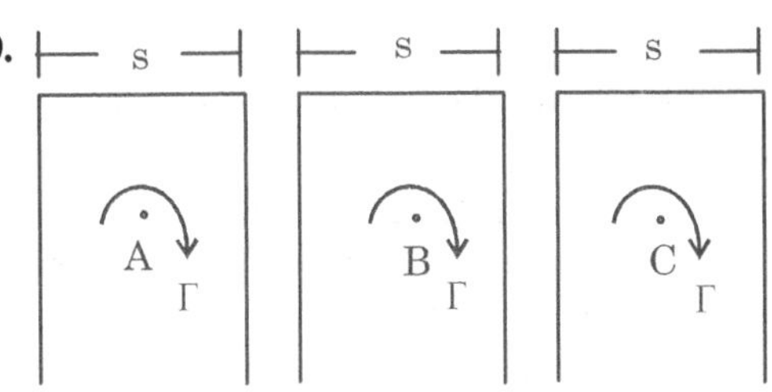

Let us consider arm (1) of the middle Vortex.

$$\text{Velocity due to Vortex A} = \frac{\Gamma}{\frac{2\pi s}{2}}$$

$$= \frac{\Gamma}{\pi s}(\downarrow)$$

$$\text{Velocity due to vortex B} = \frac{\Gamma}{\frac{2\pi s}{2}} = \frac{\Gamma}{\pi s}(\uparrow)$$

$$\text{Velocity due to vortex C} = \frac{\Gamma}{\frac{2\pi 3s}{2}} = \frac{\Gamma}{3\pi s}(\uparrow)$$

Similarly at arm (2) we will get a net velocity of $\frac{\Gamma}{3\pi s}(\downarrow)$

70. Bernoulli's equation is applicable for steady, irrotational incompressible flow here wave of the region is turbulent.

71. Given

M = 2

$\theta = 3°$

From prandlt mayer expansion theory

$\theta = \nu(M_2) - \nu(M_1)$

$\nu(M_2) = \theta + \nu(M_1)$

$= 3 + \nu(2)$

$= 3 + 26.38$

$= 29.36$

$\nu(M_2) = 29.36$

Around the expansion corner flow expansion takes place hence the Mach number behind the expansion wave will be greater upstream mach number hence the option (a) & (b) is in correct, we need to check the option (c) & (d).

Option (c) $M_2 = 2.11$

$\nu(M_2) = 29.10$ this matches to mach number value of $\nu(M_2)$ hence downstream mach number is 2.11.

72. NACA 5 – digit series

NACA 12018

first digit multiplied by 0.15 gives design lift coefficient C_L.

Second and third digit, when divided by 2, gives distance of maximum camber from the leading edge (as percent of chord).

4th & 5th digit give the maximum thickness of aerofoil (as percentage of chord).

NACA 12018

$C_L = 1 \times 0.15 = 0.15$

$$X_m = \frac{20}{2} = 10 = \frac{100}{100} = 0.1\ \%\ \text{carbon}$$

73.

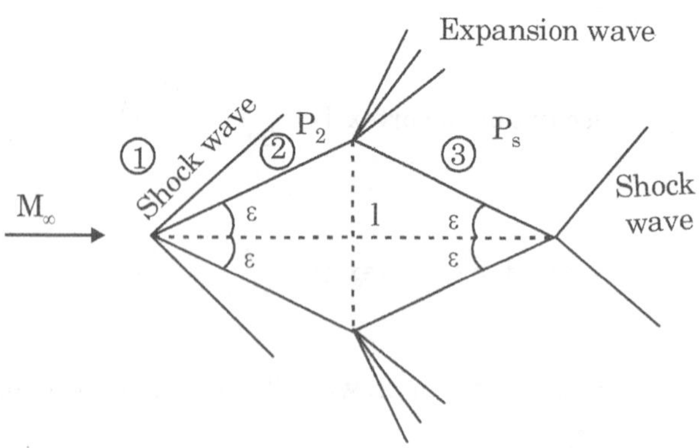

Lift is perpendicular to free stream, the pressure distribution on the top and bottom faces exactly cancel each other

$\therefore$ L = 0

$D \neq 0$

But when diamond wedge object is placed at positive angle of attack then

$L \neq 0$ (because pressure distribution on the top and bottom faces are not exactly equal it experience positive lift)

$D = (P_2 - P_3)\, t$

Drag force produced due to difference in pressure between region (2) and region (3)

$P_2 > P_3$

D = Positive

It experience positive lift and positive drag

74.

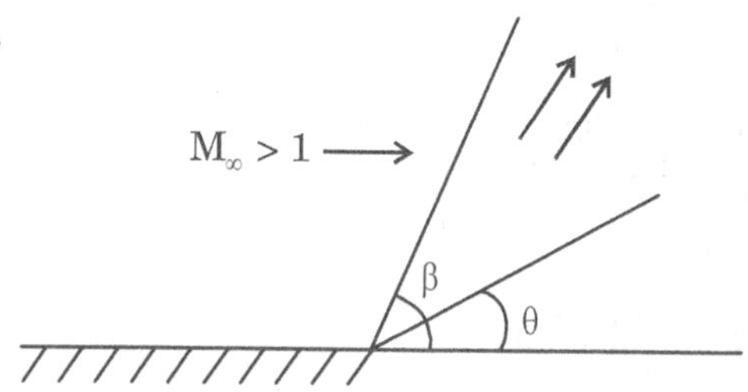

θ = deflection angle

β_2 = wave angle

the strength of shock wave is judge by the component normal to shock wave

$M_1 = M_\infty \sin\beta$

As the mach number decreased from high supersonic value. β wave angle will decrease for strong shock wave.

75. Circulation is line integral of tangential velocity along closed path (contour).

In the above problem contour is open hence $\Gamma = 0$

76. By definition of streak line.

Streak line : – The locus of particle that have earlier passed through the fixed point.

77. Equation of path line

$$\frac{dx}{dt} = ax, \frac{dy}{dt} = by$$

For steady incompressible flow continuity equation is given by

$$\frac{du}{dx} + \frac{dv}{dy} = 0$$

$$= \frac{\left(\frac{dx}{dt}\right)}{dx} + \frac{\frac{dy}{dt}}{dy} = 0$$

$$\Rightarrow \frac{d(ax)}{dx} + \frac{d(by)}{dy} = 0$$

$a + b = 0$

$5 + b = 0$

$\boxed{b = -5}$

78. $P_1 = 1$ bar | $P_2 = 29$ bar

$e_1 = 1.2\ kg/m^3$ | $e_2 = 6\ kg/m^3$

$u_1 \rightarrow$ | $\rightarrow u_2$

$$\frac{P_2}{P_1} = 1 + \frac{2\gamma}{\gamma+1}\left(M_1^2 - 1\right)$$

$$\frac{29}{1} = 1 + \frac{2.8}{2.4}\left(M_1^2 - 1\right)$$

$$\Rightarrow 28 = \frac{2.8}{2.4}\left(M_1^2 - 1\right)$$

$$\Rightarrow 28 = \frac{2.8}{2.4}\left(M_1^2 - 1\right)$$

$$\frac{28 \times 2.4}{2.8} = M_1^2 - 1$$

$$M_1^2 = \frac{28 \times 2.4}{2.8} + 1 = 25$$

$$M_1 = 5$$

$$M_1 = \frac{u_1}{\sqrt{\gamma R T_1}}$$

$$u_1 = M_1\sqrt{\gamma R T_1}$$

$$= 5\sqrt{1.4 \times \cancel{287} \times \frac{1 \times 10^5}{1.2 \times \cancel{287}}}$$

$$= 5\sqrt{\frac{1.4 \times 10^5}{1.2}} = 1707\ m/sec$$

79. $M_\infty = ?$

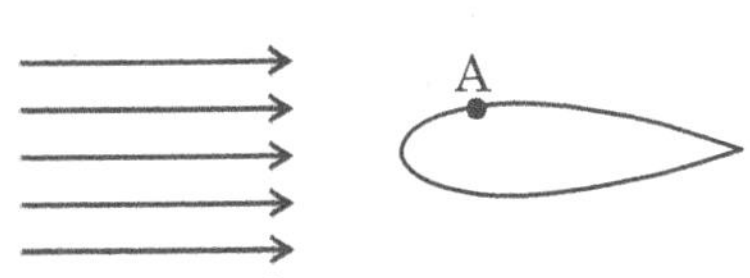

$$\frac{P_A}{P_\infty} = 0.8$$

$$M_A = 1$$

$$\frac{P_A}{P_\infty} = \left(\frac{1 + \frac{\gamma-1}{2}M_A^2}{1 + \frac{\gamma-1}{2}M_\infty^2}\right)^{\frac{\gamma}{\gamma-1}}$$

$$0.8 = \left(\frac{1 + 0.2 \times 1^2}{1 + 0.2 \times M_\infty^2}\right)^{3.5}$$

$$\frac{1 + 0.2}{1 + 0.2\,M_\infty^2} = (0.8)^{0.285}$$

$$\frac{1.2}{1 + 0.2\,M_\infty^2} = 0.9383$$

$$1 + 0.2\,M_\infty^2 = 1.278$$

$$0.2\,M_\infty^2 = .278$$

$$M_\infty^2 = \frac{.278}{0.2}$$

$$M_\infty = 1.1789$$

$$M_\infty = 1.18$$

80. In low speed subsonic tunnel free stream velocity can be measured by pitot tube aligned with flow direction with static pressure measurement at an appropriate position or the turnel wall and pitot static tube aligned with flow direction.

81. Downwash at any point $w(z_1) = \frac{\Gamma_0}{1\pi 5}[\pi + z_1 I]$

This identity is satisfied only if $I = 0$, so that any point $z - z_1$ along the span.

$$w = \frac{\Gamma_0}{4\pi S} \times A = \frac{\Gamma_0}{4\pi S} \times \pi = \frac{\Gamma_0}{4S}$$

Induce drag for elliptical distribution

$$D_j = \int_{-S}^{S} ew\Gamma\, dz \quad \left[\because \frac{\Gamma_0^2}{\Gamma_0^2} + \frac{Z^2}{S^2} = 1\right]$$

$$= \int_{-S}^{S} e\frac{\Gamma_0}{4S}\Gamma_0\sqrt{1 - \left(\frac{z}{S}\right)^2}\, dz$$

Solving this, we get

$$D_j = \frac{\pi}{8}e\Gamma_0^2$$

82. 1 m

$$u = \frac{-R^2}{4\mu}\frac{dP}{dx}\left(1 - \frac{r^2}{R^2}\right)$$

at centre line $u = u_{max}$, $r = 0$

$$u_{max} = \frac{-R^2}{4\mu}\frac{dP}{dx}$$

$$\frac{dP}{dx} = \frac{-u_{max}\,4\mu}{R^2}$$

$$dp = \frac{-u_{max} 4\mu dx}{R^2}$$

$$= -\frac{u_{max} 4ve \times L}{R^2}$$

$$= -\frac{10 \times 4 \times 1 \times 10^{-6} \times 1000 \times 100}{(0.02)^2}$$

$$= -10000 \text{ N/m}^2$$

$$\therefore h = \frac{dp}{eg} = \frac{10000}{1000 \times 10} = 1 \text{ m}$$

$$\boxed{h = 1m}$$

83. Given

$M = 3$

$P_\infty = 50$ kPa

$T = 200$ K

$R = 287$ J/kg–k

$A^* = 0.045$ m²

$$P_{0\infty} = P_\infty \left(1 + \frac{\gamma - 1}{2} M_\infty^2\right)^{\frac{\gamma}{\gamma - 1}}$$

$= 50 (1 + 0.2 \times 3^2)^{3.5}$

$= 50 \times 10^3 (36.73)$

$= 1836.6^3 \times 10^3$

$= 1.836 \times 10^6$ Pa

$T_{0\infty} = T_\infty (1 + 0.2 \times 3^2)$

$= 200 (2.8) = 560$ K

$$\text{Mass flow rate } \dot{m} = \frac{P_0 A^*}{\sqrt{T_0}} \sqrt{\frac{\gamma}{R} \left(\frac{\gamma}{\gamma + 1}\right)^{\frac{\gamma + 1}{\gamma - 1}}}$$

$$= \frac{1.836 \times 10^6 \times 0.05}{\sqrt{560}} \times \sqrt{\frac{1.4}{287} \times (0.83)^6}$$

$= 154.91$ kg/sec.

84. Symmetrical cross-section airfoil is used in the tail surface. So among the given options NACA0012 represents the symmetrical airfoil.

85. Deflection of trailing edge flaps increase the effective lamber of wing. Which results in increase in lift of the wing.

86. Lift (L) $= C_\ell \times \frac{1}{2} eV^2 S$

Drag (D) $= C_D \times \frac{1}{2} eV^2 S$

For constant dynamic pressure

$L \propto S$ and $D \propto S$

Where S is the wing span area

The weffed wing surface area of biplane wing is more than the manoplane wing. So biplan wing produce more lift and drag compare tc monoplane wing.

87. 6.28

The lift-curve slope, $\frac{dC_\ell}{d\alpha} = 2\pi$

$= 6.28$

88. 1

From Energy equation

$$ho = h_1 + \frac{v_1^2}{2}$$

$$h_1 = \frac{h0}{2} \text{ and } v_1 = u$$

$$h0 = \frac{h0}{2} + \frac{u^2}{2}$$

$$\frac{h_0}{2} = \frac{u^2}{2}$$

$$\Rightarrow \frac{u^2}{h_0} = 1$$

89. The relation between boundary layer thickness and Refnolds number in laminar boundary layer is given by

$$\delta = \frac{5.0x}{\sqrt{R_{ex}}}$$

where, x is the distance from the L.E.

$$\text{So, } \delta \propto \frac{1}{\sqrt{R_{ex}}}$$

90. 54

$$u(\tau) = U_\infty \left[\frac{3y}{2\delta} - \frac{1}{2}\left(\frac{y}{\delta}\right)^3\right]$$

$$\frac{d[u(y)]}{dy} = V_\infty \left[\frac{3}{2\delta} - \frac{3}{2}\frac{y^2}{\delta^3}\right]$$

$$\text{shear stress, } \tau = \mu \frac{du}{dy} \text{ at } \tau = 0$$

$$\left(\frac{du}{dy}\right)_{y=0} = U_\infty \times \frac{3}{2\delta}$$

$$\Gamma = \mu u_\infty \frac{3}{2\delta}$$

$$= 1.8 \times 10^{-5} \times 10 \times \frac{3 \times 1000}{2 \times 5}$$

$= 0.054$ N/m²

$$= \frac{54}{1000} \text{ N/m}^2 = 54 \times 10^{-3} \text{ Pa}$$

91. $P_0 = P_\infty + \frac{1}{2}\rho_\infty V_\infty^2$

$$V_\infty = \sqrt{\frac{2(P_0 - P_\infty)}{\rho_\infty}} = \sqrt{\frac{2\times(54051-45565)}{0.6417}}$$

= 162.63 m/s

Indicated air speed is the airspeed read by the pitot-static tube.

92. Thin airfoil theory satisfy the Kutta's condition and from Kutta's theorem we know that there should be no time delay for flow leaving at trailing edge from upward and downward surface.

93. 0.38

$$m_2^2 = \frac{(r-1)m_1^2+2}{2rm_1^2+1-r}$$

when $m_1 \to \infty$

$$\lim_{m_1\to\infty} m_2^2 = \lim_{m_1\to\infty}\frac{(r-1)+\frac{2}{m_1^2}}{2r+\frac{(1-y)}{m_1^2}} = \frac{r-1}{2r}$$

$$\therefore \lim_{m_1\to\infty} m_2 = \sqrt{\frac{2r}{r-1}} = 0.38$$

94. For thin aerofoil stream lined profile is observed, the flow over the thin aerofoil is only slightly perturbed from its free stream values the expansion over the top surface is mild, the velocity increases only slightly, and hence the magnitude at C_P at minimum pressure point is small and M_{cr} decreaseswith increase in thickness, the flow expansion over the leading edge will be stronger, the velocity will increase to larger values. The pressure will decrease to lower value and absolute magnitude of C_P is larger and M_{cr} will be decrease.

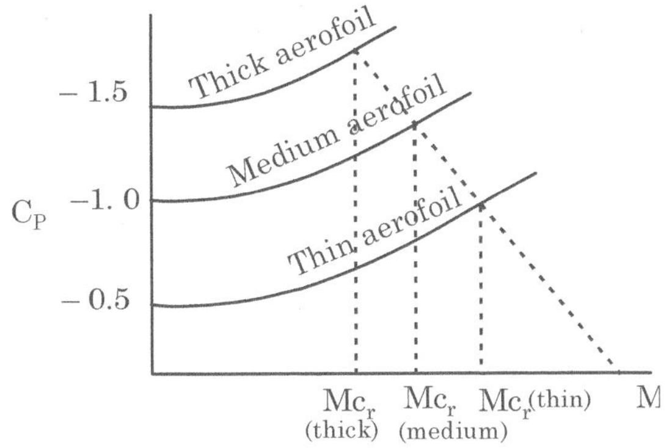

$Mc_{r\,thin} > Mc_{r\,medium} > Mc_{r\,thick}$

95. V_∞ = 10 m/sec

V = 20 m/sec

Pressure coefficient (C_P)

$$C_P = 1 - \left(\frac{V}{V_\infty}\right)^2$$

$$= 1 - \left(\frac{20}{10}\right)^2 = -3$$

96. NACA four digit series

1st digit maximum camber as percentage of the chord

2nd digit distance of maximum camber from aerofoil leading edge in tens of percents of the chord.

Last two digit maximum thickness of the aerofoil as percentage of chord.

97. $C_f \propto \frac{1}{\sqrt{Re_L}}$

98. Across the oblique shock wave, total temperature remains constant, static temperature, static pressure increases. Total pressure decreases

$$T_{01} = T_{02}, P_{01} > P_{02}, T_2 > T_1, P_2 > P_1$$

99. $P_e = P_b$

100. $P_\gamma = 0.7$

$$\frac{\delta_T}{\delta} = \frac{1}{(P_r)^{1/3}}$$

δ_T = Thermal boundary layer thickness

δ = Velocity boundary layer thickness

$\delta > \delta_T$ because $\rho_r < 1$

101. According to Kutta-Joukowski theorem

$$L = eVT \quad ....(1)$$

and

$$L = \frac{1}{2}\, eAV^2 \times C_L \quad ....(3)$$

from equation (1) and (2)

$$C_L \times \frac{1}{2} eAV^2 = eVT$$

$$C_L \times \frac{1}{2}\not{e}V^2 \times \not{2}R = \not{e}V\Gamma$$

$$C_L = \frac{\Gamma}{VR} \quad ....(3)$$

For maximum lift and atleast one stagnation point on the cylinder the total velocity generated by the circulation has to be equal to the maximum velocity by free stream hence

$\Gamma = 4V\pi R$

Putting the value of Γ we will get

$$C_L = \frac{4\pi \cancel{VR}}{\cancel{VR}}$$

$$\boxed{C_L = 4\pi}$$

102. 40.51 m/sec

Form continuity equation,

$A_A V_A = A_B V_B$

$$V_B = \frac{A_A}{A_B} V_A$$

$= 10\ V_A$

Velocity in test section

$$V_2 = \sqrt{\frac{2(P_1 - P_2)}{e\left(1 - \frac{A_2}{A_1}\right)^2}}$$

$$= \sqrt{\frac{2 \times 1000}{1.23\left(1 - \left(\frac{1}{10}\right)^2\right)}}$$

$= 40.52$ m/sec

103. AR = 6

$\alpha = 10°$

For elliptical winge = 1

We have,

$$a = \frac{a_\infty}{1 + \frac{57.3\, a_\infty}{\pi e AR}}$$

$$= \frac{0.1095}{1 + \frac{57.3 \times 0.109}{3.14 \times 1 \times 6}}$$

$= 0.0821$ /degree

$$a = \frac{dC_L}{d\alpha}$$

$C_L = a\alpha$

$= 0.0821 \times 10°$

$$= 0.82 \text{ or } \frac{\pi^2}{12}$$

104. Expression for mass flow rate in chocked condition is given by

$$\dot{m} = \frac{P_0 A^*}{\sqrt{T_0}} \sqrt{\frac{\gamma}{R}\left(\frac{2}{\gamma+1}\right)^{\frac{\gamma+1}{\gamma-1}}}$$

$$= \frac{P_0 A^*}{\sqrt{T_0}} \times \text{const.}$$

$$\dot{m}\, \alpha \frac{1}{\sqrt{T_0}} \text{ for } P_0 A^* \text{ const.}$$

when $T = 4\ T_0$

$$\dot{m}\, \alpha \frac{1}{\sqrt{4T_0}}$$

$$\alpha \frac{1}{2\sqrt{T_0}}$$

New mass flow rate becomes half.

105. Change in entropy given by

$$\Delta s = C_V \ln \frac{T_2}{T_1} + R \ln \frac{V_2}{V_1}$$

$$= C_P \ln \frac{T_2}{T_1} - R \ln \frac{P_2}{P_1}$$

For Isobaric process, $P \rightarrow$ constant

$$\Delta s = C_P \ln \frac{T_2}{T_1}$$

For Isochoric process $V \rightarrow$ const.

$$\Delta s = C_V \ln \frac{T_2}{T_1} + R \ln\left(\frac{V_2}{V_1}\right)$$

$$= C_V \ln \frac{T_2}{T_1}$$

For isentropic process

$$\Delta s = C_P \ln \frac{T_{02}}{T_{01}} - R \ln \frac{P_{02}}{P_{01}}$$

$$= -R \ln \frac{P_{02}}{P_{01}} \quad (T_{01} = T_{02})$$

$$\Delta s = -R \ln \frac{P_{02}}{P_{01}}$$

For Isothermal process

T = constant

$$\Delta s = -R \ln \frac{P_2}{P_1}$$

Out of this four maximum ratio is for Isochoric process.

106.

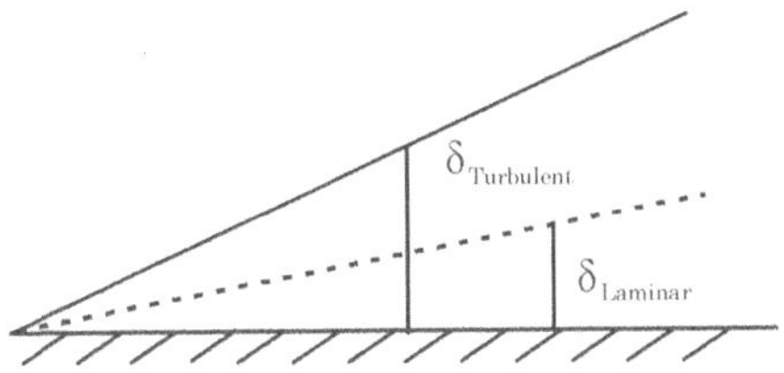

Skin friction coefficient

$$C_{f\ laminar} = \frac{1.328}{(Re_L)^{1/2}}$$

$$C_{f\ turbulent} = \frac{0.074}{(Re_L)^{1/5}}$$

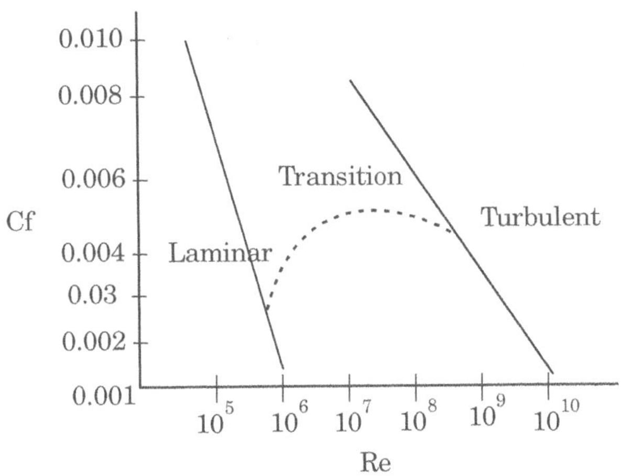

C_f varies as $1-\frac{1}{5}$ for turbulent boundary layer,

where as for laminar boundary layer $1-\frac{1}{2}$, hence

C_f is large for turbulent boundary large

$T_{w(turbulent)} > T_{w(laminer)}$

⇒ Separated flow doesnot necessarily lead to turbulent boundary layer because, turbulent flow is more energetic than laminar boundary layer for turbulent boundary layer there is more flow kinetic energy nearer to the surface hence less severve to separation seperation occurs due to adverse pressure gradient, called as the region

when $\frac{dP}{dx}$ is positive.

⇒ Location of transition point depends upon many quantities, such as Re, M, heat transfer to or from the stream, surface rough ness, pressure gradient.

107. 0.5448 atm

M = 1.5

P_0 = 1 atm (gauge) = (1 + 1) absolute

$T_0 = 20°$

$\gamma = 1.4$

Now,

Isentropic flow relation

$$\frac{P_0}{P} = \left(1 + \frac{\gamma-1}{2}\cdot M_\infty^2\right)^{\frac{\gamma}{\gamma-1}}$$

$\gamma = 1.4$

$$P = \frac{P_0}{\left(1+0.2\,M_\infty^2\right)^{3.5}}$$

$$= \frac{2}{\left(1\times 0.2\times 1.5^2\right)^{3.5}}$$

$$= \frac{2}{3.6} = 0.544.8 \text{ atm}$$

P_{static} = 0.5448 atm

108.

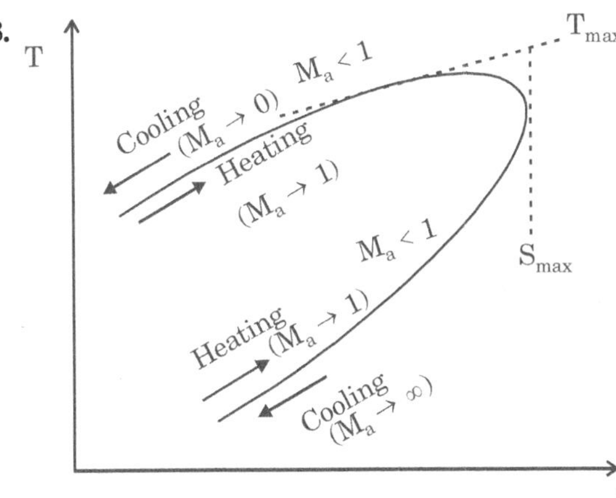

Heating increases the mach number for the subsonic flow, but decreases it for supersonic flow. a to b temperature is increased, it is being a supersonic flow upon heating will decelerate.

110. Given AR = 10

NACA 0012

$\rho = 0.95$

$\alpha = 6°$ C_L = ?

$a_0 = 2\pi$/rad = 0.1095/rad

We have

$$a = \frac{a_0}{1+\frac{57.3\,a_0}{\pi eAR}}$$

$$= \frac{0.1095}{1+\frac{57.3\times 0.1095}{3.14\times 10\times 0.95}}$$

= 0.09047/degree

$$a = \frac{dC_\ell}{d\alpha}$$

$C_L = a\alpha$

$= 0.09 \times 6°$

$C_L = 0.543$

111. $\frac{u(y)}{V} \left| \begin{array}{l} \frac{y}{\delta} \text{ for } y \le \delta \\ 1 \text{ for } y > \delta \end{array} \right.$

momentum thickness $\theta = \int_0^\delta \frac{u}{v}\left(1 - \frac{u}{v}\right) dy$

$$\theta = \int_0^\delta \frac{y}{\delta}\left(1 - \frac{y}{\delta}\right) dy$$

$$= \frac{1}{\delta}\int_0^\delta \left(\frac{y}{\delta} - \frac{y^2}{\delta}\right) dy$$

$$= \frac{1}{\delta}\int_0^\delta y\, dy - \frac{1}{\delta^2}\int_0^\delta y^2\, dy$$

$$= \frac{1}{\delta}\left|\frac{y^2}{2}\right|_0^\delta - \frac{1}{\delta^2}\left|\frac{y^3}{3}\right|_0^\delta$$

$$= \frac{1}{\delta} \times \frac{\delta^2}{2} - \frac{1}{\delta^2} \times \frac{\delta^3}{3}$$

$$= \frac{\delta}{2} - \frac{\delta}{3} = \frac{\delta}{3 \times 2}$$

$$\frac{\theta}{\delta} = \frac{1}{6} = 0.1666$$

112. $\vec{V} = 4t\,x\,\hat{i} - 2\,t^2 y\,\hat{j} + 4xz\,\hat{k}$

acceleration

$$\vec{a} = \frac{d\vec{v}}{dt} = \frac{u\partial\vec{v}}{\partial x} + \frac{v\partial\vec{v}}{\partial y} + \frac{w\partial\vec{v}}{\partial z} + \frac{\partial\vec{v}}{\partial t}$$

$$= 4\,tx\left(4t\,\hat{i} + 4z\,\hat{k}\right) + -2\,t^2 y\left(-2t^2\hat{j}\right)$$
$$+ 4x2\left(4xi - 4t\,y\,\hat{j}\right)$$

$$= 4tx\left(4\,t\hat{i} + 4\,z\,\hat{k}\right) + -2\,t^2 y\left(-2t^2\hat{j}\right)$$
$$+ 4xz\left(4x\hat{k}\right) + 4x\hat{i} - 4t\,y\hat{j}$$

$$= 4tx\left(4t\hat{i} + 4z\hat{k}\right) + -4\,t^4 y\,\hat{j}$$
$$+ 16x^2 z\hat{k} + 4x\hat{i} - 4t\,y\hat{j}$$

$\vec{a}$ at (–1, 1, 0) and t = 1 sec

$$\vec{a} = \left(4\hat{i}\right)(4 \times 1 \times -1) + \left(4\hat{j} - 4\hat{j} - 4\hat{j}\right)$$

$$= -16\hat{j} + 4\hat{j} - 4\hat{i} - 4\hat{j}$$

$$= -20\hat{i}$$

$$\vec{a} = -20\hat{i}$$

$$|\vec{a}| = 20$$

113. $\nu(M) = \sqrt{\frac{\gamma+1}{\gamma-1}} \tan^{-1}\sqrt{\frac{\gamma-1}{\gamma+1}\left(M^2 - 1\right)}$

$$- \tan^{-1}\sqrt{M^2 - 1}$$

for $\nu = 1.4$

$$\nu(M) = \sqrt{6}\tan^{-1}\sqrt{\frac{M^2-1}{6}} - \tan^{-1}\sqrt{M^2 - 1}$$

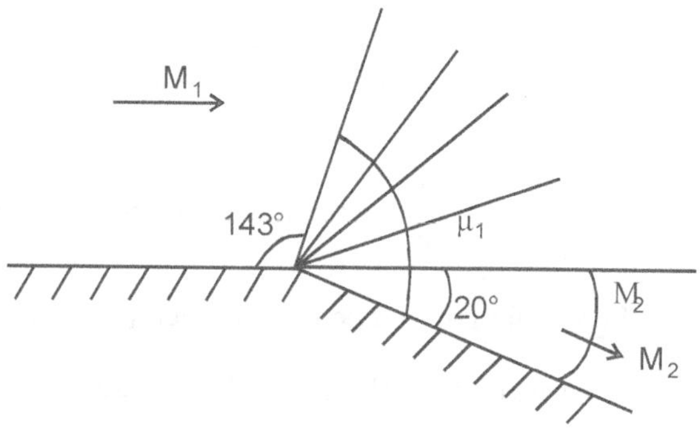

Mach angle behind and ahead of shock wave

$\mu_2 = 20°$

$\mu_1 = 180° - 143°$

$\nu(M_2) = 20°$

$\nu(M_1) = 180° - 143° = 37°$

$\nu(M_2)$

$\mu_1 = 180° - 143° = 37°$

$$\sin\mu_1 = \frac{1}{M_1}$$

$$M_1 = \frac{1}{\sin\mu_1} = 1.66$$

$\mu_2 = 20°$

$$\sin\mu_2 = \frac{1}{M_2}$$

$$M_2 = \frac{1}{\sin \mu_2} = \frac{1}{\sin 20°} = 2.923$$

deflection angle

$d\theta = \nu(M_2) - \nu(M_1)$

$$= \sqrt{6}\tan^{-1}\sqrt{\frac{2.92^2 - 1}{6}} - \tan^{-1}\sqrt{2.92^2 - 1}$$

$$-\sqrt{6}\tan^{-1}\sqrt{\frac{1.66^2 - 1}{6}} - \tan^{-1}\sqrt{1.66^2 - 1}$$

$= 48.26 - 16.68 = 31.58$

$\therefore$ For angle ϕ

$= \mu_1 + d\theta - \mu_2$

$= 37 + 31.58 - 20°$

$= 48.584°$

114. 1341.6 m/sec

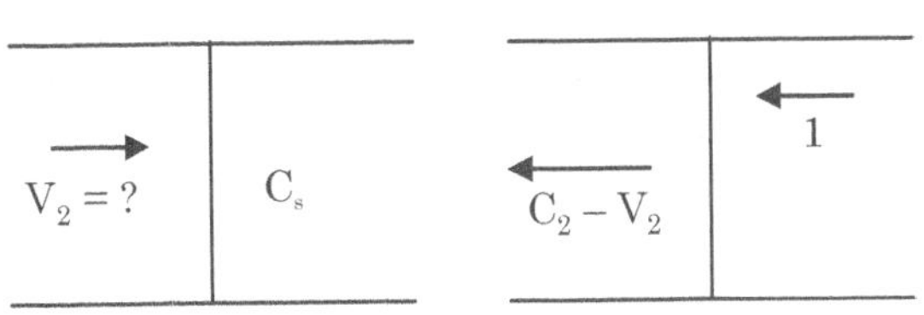

Pressure ratio

$$\frac{P_2}{P_1} = 1 + \frac{2\gamma}{\gamma + 1}\left(M_1^2 - 1\right)$$

$$29 = 1 + \frac{2.8}{2.4}\left(M_1^2 - 1\right)$$

$$\Rightarrow \quad 28 = \frac{2.8}{2.4}\left(M_1^2 - 1\right)$$

$$\Rightarrow M_1^2 = \frac{28 \times 2.4}{2.8} + 1$$

$$M_1^2 = 25$$

$$M_1 = \sqrt{25} = 5$$

$$M_2 = \sqrt{\frac{1 + \frac{\gamma}{2} + M_1^2}{\gamma M_1^2 - \frac{\gamma - 1}{2}}} = \sqrt{\frac{1 + 0.2 \times 5^2}{1.4 \times 5^2 - 0.2}}$$

$$\frac{e_2}{e_1} = \frac{(\gamma + 1)M_1^2}{2 + (\gamma - 1)M_1^2} = \frac{(1.4 + 1) \times 5^2}{2 + (1.4 - 1) \times 5^2} = 5$$

$$\frac{T_2}{T_1} = \frac{P_2}{P_1} \times \frac{e_1}{e_2} = \frac{29}{5} = 5.8$$

$$\frac{T_2}{T_1} = 5.8$$

$T_2 = 5.8 \times 280 = 1624$ K

$$M_1 = \frac{C_s}{\sqrt{\gamma R T_1}}$$

$$C_s = M_1\sqrt{\gamma R T_1}$$

$= 5\sqrt{1.4 \times 287 \times 280}$

$= 1.677$ m/sec

Now mach number, behind the shock wave

$$M_2 = \frac{C_s - V_2}{\sqrt{\gamma R T_2}}$$

$$C_s - V_2 = M_2\sqrt{\gamma R T_2}$$

$$V_2 = C_s - M_2\sqrt{\gamma R T_2}$$

$= 1677 - 0.4152\sqrt{1.4 \times 287 \times 1624}$

$= 1341.6$ m/sec

115. 90 N

Reaction force on the duct is due to change in impulse forces.

$F_R = F_2 - F_1$

$= \left(P_2A_2 + \dot{m}V_2\right) - \left(P_1A_1 + \dot{m}V_1\right)$

Continuity equation,

$A_1V_1 = A_2V_2$

$$V_1 = V_2 \frac{\frac{\pi}{4}D_2^2.V_2}{\frac{\pi}{4}D_1^2.V_1}$$

$$= V_2.\frac{D_2^2}{D_1^2}$$

$$V_1 = \frac{4^2}{8^2} \times 8 = 2 \text{ m/sec}$$

Now applying bernoullie's equation we get

$$P_1 + \frac{1}{2}\rho V_1^2 = P_2 + \frac{1}{2}\rho V_2^2$$

$$P_1 = P_2 + \frac{1}{2}\rho\left(V_2^2 - V_1^2\right)$$

$$= 101 \times 10^3 + \frac{1}{2} \times 1000\left(8^2 - 2^2\right)$$

$\boxed{P_1 = 131\text{ KPa}}$

Mass flow rate

$$\dot{m} = \rho_1 A_1 V_1 = 1000 \times \frac{\pi}{4} \times \left(8 \times 10^{-2}\right)^2 \times 2$$

$$\dot{m} = 10.057 \text{ kg/sec}$$

Now, Reaction force

$$F_R = \left(P_2A_2 - P_1A_1\right) + \dot{m}\left(V_2 - V_1\right)$$

$$= \left[10 \times \frac{\pi}{4}(0.04)^2 - 131 \times \frac{\pi}{4}(0.08)^2\right] \times 10^3 + 10.057(8-2)$$

$$= -90 \text{ N}$$

116. $\dfrac{Lm}{L_P} = 4$

For dynamic similarly between model and prototype their reynolds no. and force coefficients must be equal.

$$Re_m = Re_P$$

$$\left(\frac{\rho VL}{\mu}\right)_m = \left(\frac{eVL}{\mu}\right)_P$$

$$\frac{V_m}{V_P} = \frac{L_P}{L_m} - \frac{e_P}{e_m}\frac{\mu_m}{\mu_P}$$

$$\frac{V_m}{V_P} = 4\left(\frac{1.22}{1000}\right)\left(\frac{10^{-3}}{1.78 \times 10^{-5}}\right)$$

$$= 0.27415$$

$$F \propto \rho V^2(S)$$

$$\frac{F_m}{F_P} = \frac{\rho_m}{\rho_P}\left(\frac{V_m}{V_P}\right)^2\left(\frac{L_m}{L_P}\right)^2$$

$$= \frac{1000}{1.22} \times (0.27415)^2\left(\frac{1}{4}\right)^2$$

$$= 3.805$$

$$\frac{F_m}{F_P} = 4(\text{closest integer})$$

118. $C_P = \dfrac{P - P_\infty}{9_\infty} = 1 - 4\sin^2\theta(\text{cylinder})$

$$P = P_\infty$$

$$1 - 4\sin^2\theta = 0$$

$$\sin^2\theta = \frac{1}{4}$$

$$\sin\theta = \pm\frac{1}{2}$$

$\boxed{\theta = \pm 30^\circ}$

119. Measurement of temperature of high speed streams is made with help of thermo couples. Thermocouple is a device which is operates on the seeback principle, which states that the flow of heat in a metal is always accompained by a flow of electromotive force (emf) it measures temperature, greater than static but less than total.

120. Across moving shock static as well as total temperature both increases.

121. 1st law of thermodynamics known as conservation of energy.

122. Eulers equation of motion for steady Isentropic flow is given by

$$\frac{dP}{\rho} + vdv = 0$$

For Isentropic flow $\dfrac{P}{e^r} = C$

$$\rho^\gamma = \frac{P}{C}$$

$$\Rightarrow \rho = \frac{P^{\frac{1}{\gamma}}}{C^{\frac{1}{\gamma}}}$$

Now Integrating the above relation we get

$$C^{\frac{1}{\gamma}} \int_{P_1}^{P_2} \frac{dP}{P^{\frac{1}{\gamma}}} + \int_{V_1}^{V_2} vdv = 0$$

$$C^{\frac{1}{\gamma}}\left[\frac{P^{-\frac{1}{\gamma}+1}}{-\frac{1}{\gamma}+1} - \frac{P^{-\frac{1}{\gamma}+1}}{-\frac{1}{\gamma}+1}\right] + \frac{V_2^2}{2} - \frac{V_1^2}{2} = 0$$

$$C^{\frac{1}{\gamma}}\left[P_2^{\frac{\gamma-1}{\gamma}} - P_1^{\frac{\gamma-1}{\gamma}}\right] = \frac{V_1^2}{2} - \frac{V_2^2}{2}$$

where

$$C = \frac{C^{\frac{1}{\gamma}}}{-\frac{1}{\gamma}+1}$$

123. $C_P = \dfrac{C_{P_0}}{\sqrt{1-M_\infty^2}} = \dfrac{-1.2}{\sqrt{1-0.6^2}}$

$= -1.5$

124. 0.707

$d\phi$ can be written as

$$d\phi = \frac{d\phi}{\partial x}dx + \frac{\partial \phi}{\partial y}dy$$

Along an equipotential line $\phi = C$

$d\phi = 0$

$$0 = \frac{\partial \phi}{\partial x}dx + \frac{\partial \phi}{\partial y}dy$$

$$0 = \frac{\partial}{\partial x}\left[\frac{1}{x^2+y^2+4}\right]dx + \frac{\partial}{\partial y}\left[\frac{1}{x^2+y^2+4}\right]dx$$

$$0 = \left[\frac{-2x}{\left(x^2+y^2+4\right)^2}\right]dx + \left[\frac{2y}{x^2+y^2+4}\right]dy$$

$0 = -2xdx + xydy$

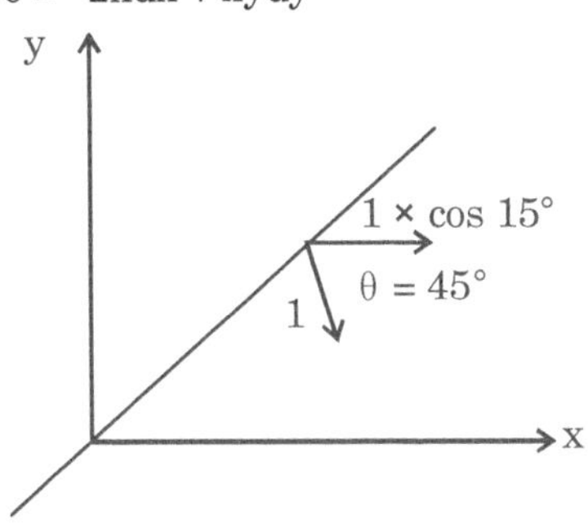

$\Rightarrow \quad xdx = ydy$

$$\Rightarrow \quad \frac{x^2}{2} = \frac{y^2}{2} + C \qquad ...(1)$$

when equation (1) passes (1, 1) then C = 0

Therefore the equipotential line

$y = \pm x$

125. 409.81 kN

$e = 1000\ kg/m^3$

$g = 9.8\ m/sec^2$

Pa = 100 KPa

Net hydrostatic force on the side face of tank

$$F_{Net} = F_{P_{air}} + F_{b_{water}} = P_{atm} \times (1 \times 2 \times m^2) +$$

$$\left(P_{atm} + 1000 \times 9.81 \times 0.5\right)\left(1 \times 2\right)$$

$= 4\ Patm + 2 \times 1000 \times 9.8 \times 0.5 \times 2$

= 409.81 kN

$\boxed{F_{Net} = 410\ kN}$

126. Tangential component ahead of shock wave = $M_1 \cos \beta$

= 3 cos 36

= 2.42

for shock wave we know tangential component of mach number remains some only normal component changes

$\boxed{wt_1 = wt_2 = 2.42}$

127. Given :

$V_\infty = 20$ m/sec

P = 1 bar

T = 300 K

$\mu = 1.789 \times 10^{-5}$ kg/m-sec

$\delta = 1$ mm

density

$$\rho = \frac{\rho}{RT} = \frac{1 \times 10^5}{287 \times 300}$$

$= 1.16\ kg/m^3$

Boundary layer thickness (δ)

$$\delta = \frac{5x}{\sqrt{R_e x}}$$

$$= \frac{5x}{\sqrt{\dfrac{\rho v x}{\mu}}} = 5\sqrt{\frac{x}{\dfrac{\rho v}{\mu}}}$$

$$\delta = 5\sqrt{\frac{x}{\dfrac{\rho v}{\mu}}}$$

$$\Rightarrow \quad \frac{\delta}{5} = \sqrt{\frac{x\mu}{\rho v}}$$

Squaring we get

$$\frac{\delta^2}{5} = \frac{x\mu}{\rho v}$$

$$x = \frac{\delta^2 \rho v}{5\mu} = \frac{\left(1 \times 10^{-3}\right)^2 \times 1.16 \times 20}{5 \times 1.785 \times 10^{-5}}$$

= 51.93 mm

$\boxed{x = 51.93\ mm}$

128. According to Newton's law of viscosity

$$\tau = \mu \frac{du}{dy}$$

where μ is Co–efficient of viscosity

$\frac{du}{dy}$ is rate of shear stress or rate of shear deformation

τ is shear stress

$$\therefore \mu = \frac{\tau}{\frac{du}{dy}} = \frac{\frac{MLT^{-2}}{L^2}}{\frac{L}{T} \times \frac{1}{L}} = ML^{-1}T^{-2}$$

Kinematic viscosity

$$\nu = \frac{\text{vis cos ity}}{\text{density}} = \frac{\mu}{\rho} = \frac{ML^{-1}T^{-2}}{ML^{-3}} = L^2T^{-1}$$

129. ϕ exists only for irrational flow and condition for irrotation flow is given by $\nabla \times V = 0$ other options are necessarily true because flow is not given to be incompressible.

130. For quasi-one dimensional isentropic supersonic flow through a diverging duet, area velocity relation

$$\frac{dA}{A} = \left(M^2 - 1\right)\frac{dV}{V}$$

$M > 1$ for supersonic flow

$\therefore M^2 - 1 > 0$ for supersonic flow

$$\therefore \frac{dA}{A} = (+ve)\frac{dV}{V}$$

dA will be positive for divegent section

So, Mach mnumber increases

Expression for static pressure is given by (for quasi 1D, isentropic flow)

$$T = \frac{T_0}{\left(1 + \frac{\gamma - 1}{2} M^2\right)}$$

where, T_0 is total temperature

M is free stream Mach number

T is static temperature

Here, When, M increases T decreases

131. NACA 2415 Aero foil

1st digit- maximum camber in percentage of chord

2nd digit-location of maximum camber in percentage of chord last two digit- maximum thickness in percentage of chord

$$\therefore \frac{2}{100} \times C0.02C$$

132. Given $(Re)_p = (Re)_m$

$$\therefore \frac{\rho_p V_p D_p}{\mu_p} = \frac{\rho_m V_m D_m}{\mu_m} \begin{Bmatrix} \text{As } \mu_p = \mu_m = \mu \\ \rho_p = \rho_m = \rho \end{Bmatrix}$$

$$V_p D_p = V_m D_m$$

$$\Rightarrow V_m = \frac{V_p D_p}{D_m} = 1 \times \frac{1}{0.1} = 10 m/s$$

133. For a given Q

Weak shoe solutions are consided as an practical solution.

As We know across oblique shock wave mach number decreases $M_2 < M_1$ hence $B_2 > B_1$.

134. $U_r = V\left(1 - \frac{R^2}{r^2}\right)\cos\theta$

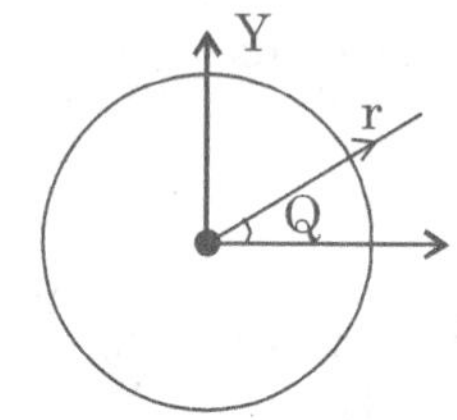

$$U_Q = -V\left(1 + \frac{R^2}{r^2}\right)\sin\theta$$

at r = R. $V_r = 0$

at r = R $U_Q = -V(1 + 1) \sin\theta$

$= -2V \sin\theta$

$C_p = 1 - 4 \sin^2\theta$

when $C_p = 0$

$1 - 4 \sin^2\theta = 0$

$$\sin\theta = +\frac{1}{-2}$$

$$\sin\theta = \frac{1}{2}$$

$\theta = 30°$

135.

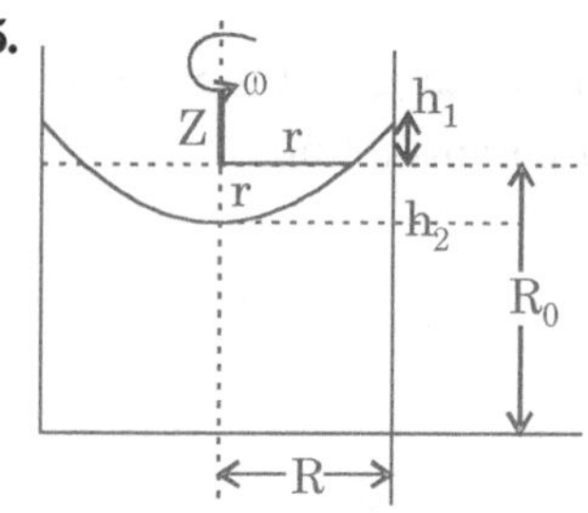

Boundrany Coaditon

r = 0, Z = h, p = P_a

$$p - P_a = \delta g(h - z) + \frac{1}{2}\delta\omega^2 r^2$$

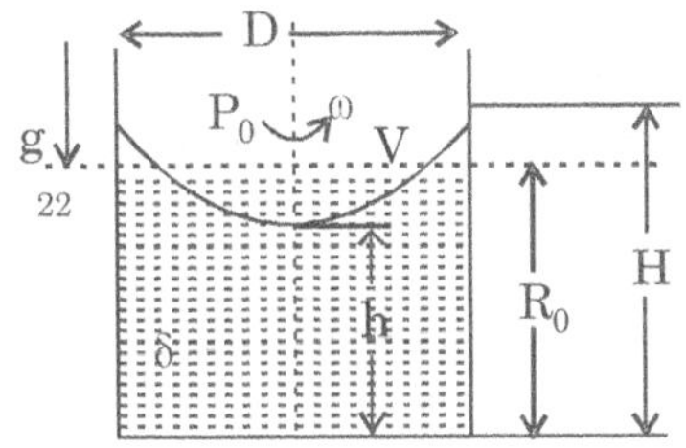

at surface

$p = P_a$

$z_0(r) = h + \frac{w^2 r^2}{2g}$ ...(1)

at $r = R$, $Z_0 = H$

$\omega^2 = 2g\frac{(H-R)}{R^2}$

Substituting in equation we get

$z_0(r) = h + (H-h)\frac{r^2}{R^2}$

Volume of the water

$\pi R^2 h_0 = \int_0^R z_0(r) 2\pi r dr$

$= \int_0^R \left(h + (H-h)\frac{r^2}{R^2} \right) 2\pi r dr$

$= \pi R^2 h^0 + \frac{1}{2}\pi R^2 (H-h)$

$h = 2h_0 - H$

$h_0 - h_2 = 2h_1 - (h_0 + h_1)$

$h_1 - h_2 = 2h_0 - h_0 - h_1$

$\boxed{h_1 = h_2 = 10 \text{ cm}}$

136. $\psi = xy^3$

$\rho = 1 \text{ k/m}^3$

$\mu = 0.1$ kg/m-sec

$u = \frac{\partial \psi}{\partial y} = 3xy^2$

$v = -\frac{\partial \psi}{\partial x} = -y^2$

Applying (N.V) momentum equation for x direction

$$u\frac{\partial u}{\partial x} + v\frac{\partial u}{\partial y} = -\frac{1}{\rho}\frac{\partial p}{\partial x} + \frac{\mu}{\rho}\left(\frac{\partial^2 u}{\partial x^2} + \frac{\partial^2 u}{\partial y^2}\right)$$

$$\Rightarrow 3xy^2\frac{\partial(3xy^2)}{\partial x} + (-y^3)\frac{\partial(3xy^2)}{\partial y} = -\frac{1}{1}\frac{\partial p}{\partial x} + \frac{0.1}{1}\left(\frac{\partial^2(3xy^2)}{\partial x^2} + \frac{\partial^2(3xy)}{\partial y^2}\right)$$

$$\Rightarrow 3xy^2(3y^2) + (-y^3)(6xy) = \frac{-\partial p}{\partial x} + 0.1(0 + 6x)$$

$$\Rightarrow \frac{-\partial p}{\partial x} = 9xy^4 - 6xy^4 - 0.6x$$

at x = 1m. y = 1 m

$$\frac{\partial p}{\partial x} = -(9 \times 1 \times 1^4 - 6 \times 1 \times^4 - 0.6 \times 1) = -2.4$$

$$\left|\frac{\partial p}{\partial x}\right| = 2.4 \text{ N/m}^3$$

137. Given $\gamma = 1.4$, R = 287 J/kg K,

$M_1 = \frac{v}{a}$, $P_1 = 1$ atm, $T_1 = 300$ k

$= \frac{800}{\sqrt{1.4 \times 287 \times 300}} = 2.309$

$\frac{p_2}{p_1} = 1 + \frac{2\gamma}{\gamma+1}(m_1^2 - 1)$

$= 1 + \frac{1.4 \times 2}{2.4}(2.304^2 - 1) = 1 + \frac{2.8}{2.4}(2.304^2 - 1)$

$\frac{p_2}{p_1} = 6.027$

$p_2 = 6.027\, p_1 = 6.03\, p_1 = 6.03 \times 1 = 6.03$ atm

138. Given $\alpha = 10°$

$C_{M_{LE}} = 2$

for symmetrical aerofoil

$A_0 = \alpha = 10° = 10 \times \frac{\pi}{180}$ rad

$A_2 = 0$

$\therefore C_{M_{LE}} = -\frac{\pi}{2}\left[A_0 + A_1 - \frac{A_2}{2}\right]$

$= -\frac{\pi}{2} \times [A_0 + 0 - 0]$

$= -\frac{\pi}{2} A_0$

$= -\frac{\pi}{2} \times \alpha$

$= -\frac{\pi}{2} \times \frac{10 \times \pi}{180}$

$= -\frac{\pi^2}{36} = 0.27$

$|C_{M_{LE}}| = 0.27$

139. Given, $\Gamma_0 = 20 \text{ m}^2/\text{sec}$, $\rho = 1.2 \text{ kg/m}^3$, v = 100 m/sec, b = 10 m

$\Gamma(y) = \Gamma_0\sqrt{1 - \left(\frac{2y}{b}\right)^2}$

b = wing span

$L = \rho v \Gamma$

$= \rho_\infty v_\infty \Gamma_0 \frac{(6\pi)}{4}$

$= \frac{\rho_\infty v_\infty \Gamma_0 . b\pi}{4}$

$= \frac{1.2 \times 100 \times 20 \times 10 \times 3.14}{4} = 18.85 \text{ kN}$

140. Variation of viscosity with temperature :

In case of liquids, the inter mdecular distance is small and hence cohesive forces are large. With increase in temperature cohesive forces decrease and hence desistance to flow also decreases.

So, viscosity of a liquid decreases with increase in temperature.

In case of Gases, as the inter molecular distance is large and hence cohesive forces are negligible. With increase in temperature the molecular agitation (or) molecular disturbance increase and the resistance to flow also increase.

Viscosity of a gas increases with increase in temperature.

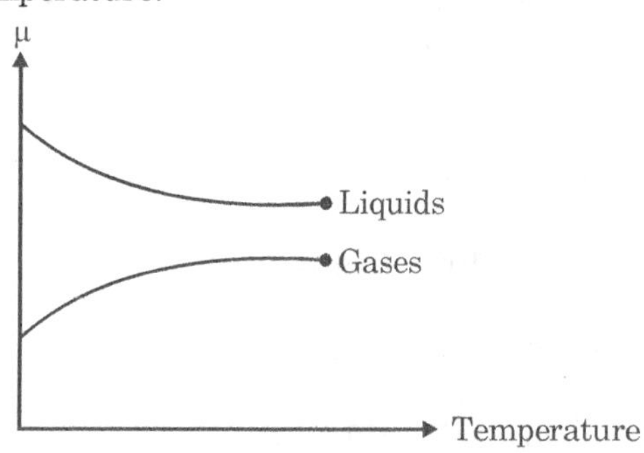

141. Given, access to complete geometry, surface pressure and shear stress distribution over a body in uniform flow, we can estimate only force and moment on body. If we want to get coefficients, we need parameter of flow like density, velocity for normalizing.

142. $\text{Velocity} = \frac{\Gamma}{4\pi h} + \frac{\Gamma}{4\pi h} = \frac{\Gamma}{2\pi h}$

143.

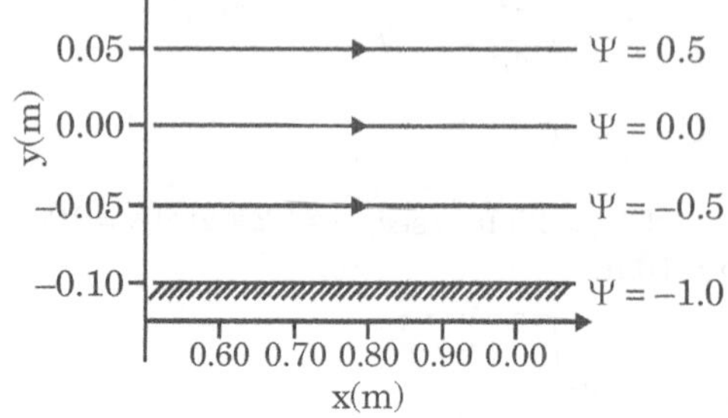

Change in stream function between two streamlines, $d\Psi$

= Volumetric flow rate between two streamlines

$= 1 - (-1) = 2 \text{ m}^3/\text{s}$

144. From Rayleigh flow the mach number in supersonic flow can be increased by removing heat from the flow.

145. When perfectly expanded condition exit in the nozzle, means that static pressure in the flow at the exit of nozzle is equal to ambient pressure at the nozzle exit.

$P_a = P_e$ (optimum expended)

P_a = Ambient pressure

P_e = static pressure in the flow at nozzle exit.

If $P_a > P_e$ (over – expanded condition)

$P_a < P_e$ (under – expanded condition)

146.

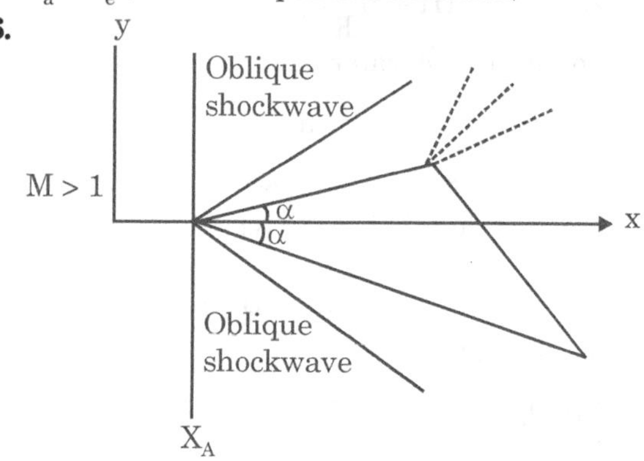

When the formation of shock waves takes place, the static pressure increases discontinuously across the shock wave and across the expansion wave the static pressure decreases continuously. Thus, option D gives the clear picture of the variation of the static pressure.

147. Given,

$C_L = 0.10$ at $\alpha = 0$

$\therefore\ C_L = C_{Lo} = 0.10$

$M = 0.6$

Now, $C_{Lo(\alpha = 6°)} = C_{Lo} + 2\pi\alpha$

$= 0.10 + 2\pi \times \frac{6\pi}{180} = 0.75$

As per Prandtl – Glauert rule

$C_L = \frac{C_{Lo}}{\sqrt{1 - M^2}}$

$= \frac{0.75}{\sqrt{1 - 0.6^2}}$

$= 0.937$

148. Given,

Static pressure are same

$P_A = P_B = P$

Total pressure $\quad P_o = P + \frac{\rho U_o^2}{2}$

As per formula

pressure force = drag force

$$(P_{oA} - P_{oB})A = \frac{1}{2}\rho U_o^2 C_d A$$

$$P_A + \frac{\rho U_o^2}{2} - \left(P_B + \frac{\rho(0.99U_o)^2}{2}\right) = \frac{1}{2}\rho U_o^2 C_d A$$

$$U_0^2(1-0.9801) = C_d U_0^2$$

$$\Rightarrow \qquad C_d = 0.01$$

149. Given,

$\beta = 35°$

$V = 517.56$ m/s

$\theta = 5.75°$

$T_2 = 182.46$ K

$R = 287$ J/kg K

$\gamma = 1.4$

Using θ-β relation for oblique shock wave, we get

$$\tan\theta = 2\cot\beta\left[\frac{M_1^2\sin^2\beta - 1}{M_1^2(\gamma + \cos 2\beta) + 2}\right]$$

$$\Rightarrow \tan 5.75° = 2\cot(35°)\times\left[\frac{M_1^2\sin^2 35° - 1}{M_1^2(1.4 + \cos 70°) + 2}\right]$$

$$\Rightarrow \quad 0.035 = \left(\frac{0.35M_1^2 - 1}{1.74M_1^2 + 1}\right)$$

$$\Rightarrow \quad M_1 = 1.99 \approx 2$$

150. Given,

$v = 1.5 \times 10^{-5}$ m²/s

$x = 0.5$ m

$V = 3$ m/s

$$\frac{\delta}{x} = \frac{5.2}{\sqrt{Re}}$$

$$Re = \frac{V \times x}{v}$$

$$= \frac{3 \times 0.5}{1.5 \times 10^{-5}} = 1 \times 10^{-5}$$

$$\delta = \frac{5.2 \times x}{\sqrt{Re}}$$

$$= \frac{5.2 \times 0.5}{\sqrt{1 \times 10^{-5}}} = 8.22 \times 10^{-3} \text{ m}$$

As per formula

Displacement thickness

$$\delta^* = \frac{\delta}{2}$$

$$= \frac{8.22 \times 10^{-3}}{2}$$

$$= 4.11 \times 10^{-3} \text{ m} = 4.11 \text{ mm}$$

151. Given,

$b = 15$ m

$L = 80$ kN

$V_\infty = 90$ m/s

$\rho_\infty = 1.2$ kg/m³

For elliptical lift distribution, induce angle of attack is

$$\alpha_i = \frac{\Gamma}{2bV_\infty}$$

$$= \frac{L}{\frac{\rho_\infty V}{2bV_\infty}}$$

$$= \frac{L}{\rho_\infty V_\infty^2 b}$$

$$= \frac{80}{1.2 \times 90^2 \times 15} = 1.334$$

152. Given,

$$\vec{V} = (2x+3y)\hat{i} + (3x+2y)\hat{j}$$

Case-I

Divergence

$$\nabla.\vec{V} = \frac{\partial}{\partial x}(2x+3y) + \frac{\partial}{\partial y}(3x+2y) = 2+2 = 4 \neq 0$$

Case-II

$$\nabla \times \vec{V} = \begin{vmatrix} \hat{i} & \hat{j} & \hat{k} \\ \frac{\partial}{\partial x} & \frac{\partial}{\partial y} & \frac{\partial}{\partial z} \\ (2x+3y) & (3x+2y) & 0 \end{vmatrix} = 0\hat{i} + 0\hat{j} + \hat{k}(3-3) = 0$$

Thus, the equation is curl free but not divergence free.

153. If $\rho_e = \rho_a$ = Optimum expanded

$\rho_e < \rho_a$ = Under expanded

$\rho_e > \rho_a$ = Over expanded

154. We have area mach number relation

$$\frac{dA}{A} = -\frac{dv}{v}(1 - M^2)$$

$$\Rightarrow \frac{dv}{v} = -\frac{dA}{A} \times \frac{1}{(1 - M^2)}$$

$$dv = -\frac{dA}{A} \times \frac{1}{(1-M^2)} \times \frac{1}{v}$$

If $M > 1$

dv = positive (Area should decrease)

If $M < 1$

dv = Negative (Area should increase)

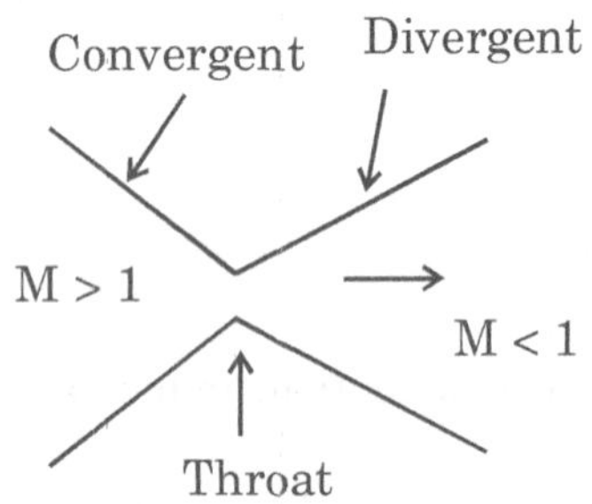

The shape of supersonic diffuser that slows down a supersonic flow to subsonic floe is converging – diverging diffuser.

155. In adiabatic process, total temperature will remains constant.

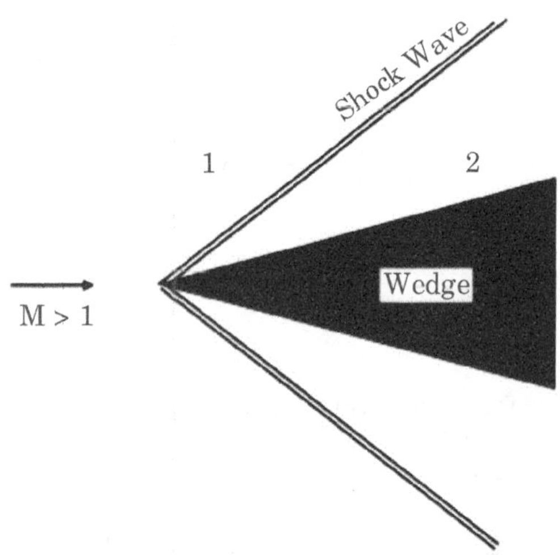

flow parameters before and after shock wave

$T_{O_1} = T_{O_2}$

$\rho_{O_1} > \rho_{O_2}$

M> M2 $\mu_1 < \mu_2$

$\omega_1 = \omega_2$ = tangential velocity remains constant.

Tangential mach number

$$M_{T_1} = \frac{\omega_1}{\sqrt{rRT_1}}$$

$$M_{T_2} = \frac{\omega_2}{\sqrt{rRT_2}} \quad [T_2 > T_1]$$

$\omega_1 = \omega_2$, $M_{T_2} < M_{T_1}$

Mach number based on tangential velocity of the shock decreases.

156. Kutta condition says the flow will leaves smoothly trailing edge of aerofoil. Kutta condition is applicable for sharp as well as cusp aerofoil, Where as trailing edge is stagnation point in case of sharp aerofoil.

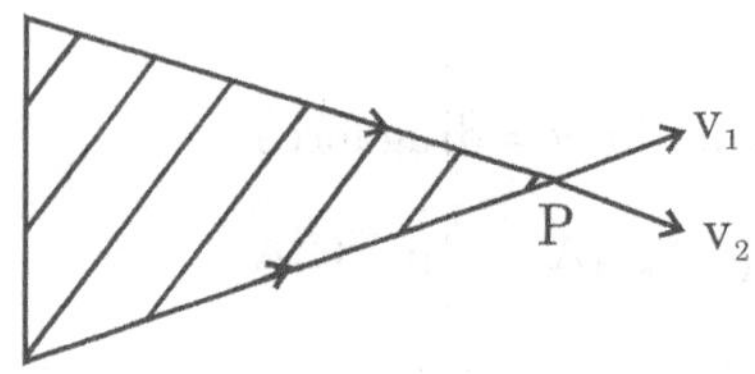

At point P

$v_1 = v_2 = 0$ [stagnation point]

In case of Cusp aerofoil

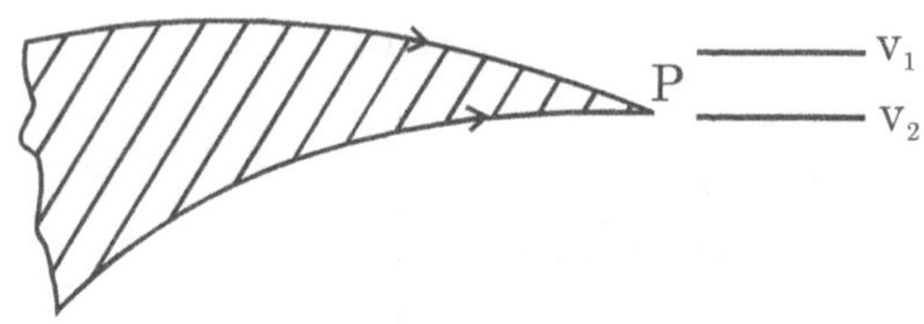

at point P

$v_1 = v_2 \neq 0$

at trailing edge it should have one single velocity $v_1 = v_2$.

Circulation $(\delta) = v_T - v_B$

$= v_2 - v_1$

Sharp aerofoil

$\delta(T.E.) = 0$ [As $v_1 = v_2 = 0$]

Cusp aerofoil

$\delta(T.E.) = 0$ [As $v_1 = v_2$]

157. Thin aerofoil theory for cambered aerofoil does not discussed about drag. Prandtt lifting line theory discussed about lift induced drag. Option (d) is incorrect.

According to thin aerofoil theory

$C_l \propto \alpha$

$C_l = 2\pi\alpha$

Aerodynamic centre is a point where moment is indepent of angle of attack.

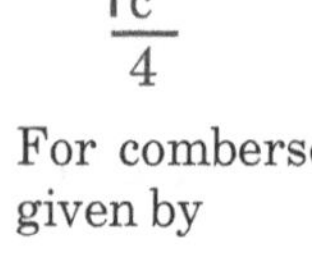

For combersed aerofoil, moment coefficient is given by

$$C_{m_{le}} = -\left[\frac{C_l}{4} + \frac{\pi}{4}(A_1 + A_2)\right]$$

$$C_{m_{c/4}} = C_{m_{le}} + \frac{C_l}{4}$$

$$= -\frac{C_l}{4} - \frac{\pi}{4}(A_1 - A_2) + \frac{C_l}{4}$$

$$= -\frac{\pi}{4}(A_1 - A_2)$$

A_1, A_2 will not change as α changes. So option (B) is correct.

Centre of pressure is a point about which moment is zero, but in the above question.

Moment about quarter chord

$$C_{m_{c/4}} = -\frac{\pi}{4}(A_1 - A_2) \text{ which is constant not a zero.}$$

So the centre of pressure will not lies on quarter chord point. Hence, option (c) is incorrect.

158. Definition of circulation

$$\tau = \int_c \vec{v}.dl$$

In the above question the position of origin is outside of closed loop ABCDEFA.

$\therefore \tau = 0$

159. In the case of P if the ice cube melts, the amount of water that gets displaced in P is amount of the volume of this ice cube. As the ice cube will melt and convert into water.

There is a increase in water level but it will never increase more than that, it will remains the same because the whatever the volume that get displaced that is going to come as the water again after it gets melt. So in P its remains the same.

In Q, volume of ice cube is volume of cube minus volume of solid ball. So the water that will go down because of melting of ice is volume of ice cube, so water will not come out it will decreased. Same stories in R, Q so it will also decreased.

160. Given, subscript a used for aircraft, m used for model.

Va = 100 km/h

$$= 1000\ 100 \times \frac{5}{18} = 27.78 = 28 \text{ m/sec}$$

$e_a = 1.225$ kg/m^3

$P_a = 1.01$ bar $\qquad P_m = 10$ Pa

$T_a = 288$ k $\qquad T_a = T_m$

According to problem

$Re_a = Re_m$

$$\frac{e_a v_a l_a}{\mu_a} = \frac{e_m v_m l_m}{\mu_m}$$

$$\because \mu_a = \mu_m,\ l_m = \frac{l_a}{5}$$

$$\Rightarrow \frac{e_a v_a l_a}{\mu_a} = \frac{e_m v_m \frac{l_a}{5}}{\mu_a}$$

$$\Rightarrow e_a v_a = \frac{e_m v_m}{5}$$

$$\Rightarrow v_m = \frac{e_a v_a \times 5}{e_m}$$

$$= \frac{e_a \times 28 \times 5}{\frac{P_m}{RT_m}}$$

$$= \frac{28 \times 5 \times 287 \times 288 \times 1.225}{10 \times 1.01 \times 10^5}$$

$$v_m = 14 \text{ m/sec} = \frac{14 \times 18}{5} = 50 \text{ km/h.}$$

161. Shadow graph – Non-uniform density

Shlieren – Uniform density

Interferometry – Optical beans.

162. When fluid is stationary, moving shock condition

$P_{O_1} = P_1$, $T_{O_1} = T_1$ but $P_2 > P_1$, $T_2 > T_1$ but, when fluid is moving. Stationary shock condition

$$P'_{O_1} = P'_1,\ T'_{O_1} = T_1' \left(\text{As } T'_{O_1} = T_1' + \frac{1}{2Cp} Vs^2\right)$$

$\therefore T_1 = T'_1 P,\ T'_{O_1} = T_{O_1}$

So option (a) and (c) are wrong.

$P'_{O_1} > P_{O_1}$ which is correct and given in option (b) and (d). In the case of stationary shock P'_{O_1} (Total pressure) which will be always greater than that of moving shock.

$P_2 > P_{O_1}$ $\left(\text{As } P_{O_1} = P_1\right)$

After the shock goes in the case of moving normal shock goes in the case of moving normal shock pressure P_2 will increase, so it will be greater than P_{O_1}.

So all the condition is satisfied by option (D).

163. In the above problem tension in connecting rope will be due to change in momentum.

Let F_x be the force in x-direction.

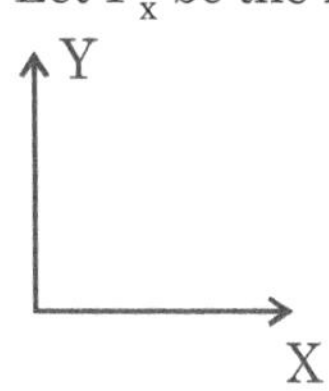

$$F_x = -\left\{\dot{m} v_{out} - \dot{m} v_{in}\right\}$$

[–ve sign because tension in negative direction]

$$= -\left\{eA_{out} v_{out}^2 - eA_{in} v_{in}^2\right\}$$

$A_{out} = A_{in} = A$

$= -\left[eAv_{out}^2 - eAv_{in}^2\right]$

$= -eA\left[(10\cos 60)^2 - 10^2\right]$

$= -1000 \times 3.14 \times \left(\frac{0.05}{2}\right)^2 \times 100\left[\cos^2 60 - 1\right]$

$= -98.4$ N

Tension in connecting rod = 98.4N

164. Given

AR = 10

$\alpha = 5°$

For ellipilical wing, e = 1

From prandtl lifting line theory, we have

$$C_{Dj} = \frac{C_I^2}{\pi eAR} = \frac{C_I^2}{\pi \times 1 \times 10}$$

$$C_{Dj} = \frac{C_I^2}{10\pi} \qquad ...1$$

From thin aerofoil theor

CI = $2\pi\alpha$ (for 2D wing) for 3D wing (finite wing)

$$a = \frac{a_0}{1+\frac{a_0}{\pi eAR}} = \frac{2\pi}{1+\frac{2\pi}{\pi \times 1 \times 10}}$$

$$= \frac{2\pi}{1+\frac{2\pi}{10\pi}} = \frac{2\pi}{1.2} = 5.23 \text{ rod}^{-1}$$

$$C_l = a\alpha = 5.23 \times \frac{5\pi}{180} = 0.4564$$

$\therefore$ From equation (1)

$$C_{Dj} = \frac{(0.4564)^2}{10\pi} = 0.0066.$$

165. Shape factor = $\frac{\text{Displacement thickness}}{\text{Momentum thickness}}$

$$= \frac{\delta *}{Q}$$

$$\delta* = \int_0^\delta \left(1-\left(\frac{u}{v}\right)\right)dy = \int_0^\delta \left(1-\frac{y^2}{\delta^2}\right)dy = \left[y - \frac{y^3}{3\delta^2}\right]_0^\delta$$

$$\delta* = \delta - \frac{\delta^3}{3\delta^2} = \frac{2\delta}{3}$$

$$\theta = \int_0^\delta \frac{u}{v}\left(1-\left(\frac{u}{v}\right)\right)dy = \int_0^\delta \frac{y^2}{\delta^2}\left(1-\frac{y^2}{\delta^2}\right)dy$$

$$= \int_0^\delta \left(y^2 - \frac{y^2}{\delta^2}\right)dy = \frac{1}{\delta^2}\left[\frac{\delta^3}{3} - \frac{\delta^5}{5}\right]$$

$$= \left(\frac{\delta}{3} - \frac{\delta}{5}\right) = \frac{2\delta}{15}$$

$$\text{Shope factor} = \frac{\frac{2\delta}{3}}{\frac{2\delta}{15}} = 5$$

5 CHAPTER AIRCRAFT STRUCTURE

1. A spring-mass-damper system with a mass of 1 kg is found to have a damping ratio of 0.2 and a natural frequency of 5 rad/s. The damping of the system is given by

   (*a*) 2 Ns/m (*b*) 2 N/s

   (*c*) 0.2 kg/s (*d*) 0.2 N.s

[GATE 2007, 1 M]

2. The sum of natural frequencies of an elastic beam with cantilever boundary conditions is

   (*a*) 1 (*b*) 3

   (*c*) 1000 (*d*) Infinite

[GATE 2007, 1 M]

3. For a plane strain problem in the $x-y$ plane, in general, the non-zero stress terms are

   (*a*) $\sigma_{zz}, \sigma_{xz}, \sigma_{yz}, \sigma_{xy}$ (*b*) $\sigma_{zz}, \sigma_{xz}, \sigma_{yz}, \sigma_{xy}$

   (*c*) $\sigma_{xx}, \sigma_{xy}, \sigma_{yy}, \sigma_{xz}$ (*d*) $\sigma_{xx}, \sigma_{yy}, \sigma_{xy}, \sigma_{zz}$

[GATE 2007, 1 M]

4. For an elastic anisotropic solid, the number of independent elastic constants in its constitutive equations is

   (*a*) 2 (*b*) 9

   (*c*) 21 (*d*) 36

[GATE 2007, 1 M]

5. A 1 kg mass attached to a spring elongates it by 16 mm. The mass is then pulled from its equilibrium position by 10 mm and released from rest. Assuming the acceleration due to gravity of 9.81 m/s^2, the response of the mass in mm is given by

   (*a*) $x = 10 \sin 24.76t$ (*b*) $x = 10\cos 24.76t$

   (*c*) $x = \sin 16t$ (*d*) $x = 10 \cos 16t$

[GATE 2007, 2 M]

6. Shown in the figure below is a model of an Euler-Bernoulli beam made up of two materials subjected to pure bending moment M. The Young's modulus of material A and B are E_A and E_B, respectively. The sectional moment of area, about the neutral axis, of the cross-sectional areas made of materials A and B, are I_A and I_B, respectively. The radius of curvature ρ of the flexural deflection of this composite beam to the bending moment M is then

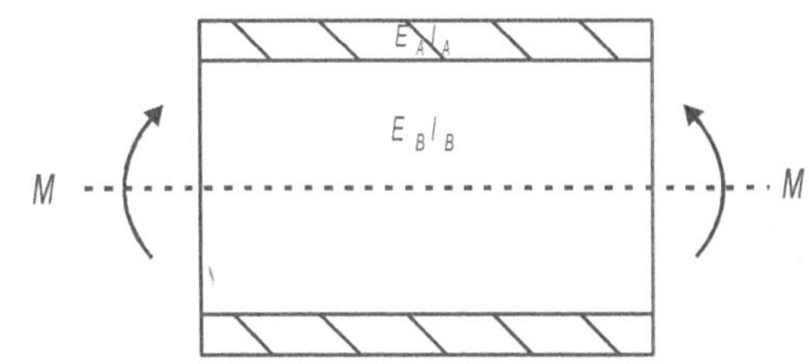

   (*a*) $\rho = \dfrac{E_A I_A + E_B I_B}{M}$

   (*b*) $\rho = \dfrac{E_A I_A + E_B I_A}{M}$

   (*c*) $\rho = \dfrac{M}{E_A I_A + E_B I_B}$

   (*d*) $\rho = \dfrac{(E_A + E_B) + (I_A + I_B)}{M}$

[GATE 2007, 2 M]

7. A spring-mass-damper system is excited by a force $F_0 \sin \omega t$. The amplitude at resonance is measured to be 1 cm. At half the resonant frequency, the amplitude is 0.5 cm. The damping ratio of the system is

   (*a*) 0.1026 (*b*) 0.3242

   (*c*) 0.7211 (*d*) 0.1936

[GATE 2007, 2 M]

8. A circular shaft is made-up of two materials A and B. The inner core is made-up of material A with diameter d_A, torsion constant J_A, and shear modulus G_A. The outer sleeve is made-up of material B with diameter d_B, torsion constant J_B, and shear modulus G_B. The composite shaft is of length L and is subjected to pure torsion moment T. The torsional stiffness, $\dfrac{T}{\phi}$, where ϕ is the angle of twist, of this composite shaft is then

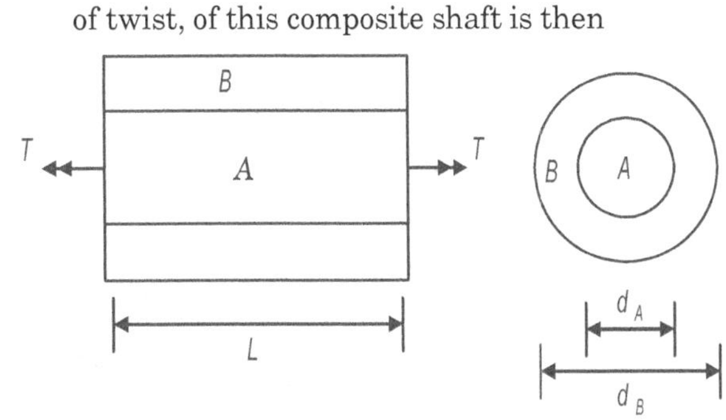

(a) $\dfrac{\left(\frac{G_A J_A}{L} \frac{G_B J_B}{L}\right)}{\left(\frac{G_A J_A}{L} \frac{G_B J_B}{L}\right)}$ (b) $\dfrac{G_A J_A}{L} + \dfrac{G_B J_B}{L}$

(c) $\dfrac{(G_A + G_B)(J_A + J_B)}{L}$ (d) $\dfrac{G_A J_B}{L} + \dfrac{G_B J_A}{L}$

[GATE 2007, 2 M]

9. The boundary conditions for an Euler-Bernoulli column are given in column X and the critical buckling loads are given in column Y. Match the boundary condition of the column to its corresponding buckling load. P_{cr} is the critical buckling load, E is the Young's modulus of the column material, I its sectional moment of area, and L is the length of the column.

X. Boundary condition	Y. Critical buckling load
$X1$. Pinned-pinned column	$Y1.\ P_{cr} = \dfrac{4\pi^2 EI}{L^2}$
$X2$. Fixed-free (cantilevered) column	$Y2.\ P_{cr} = \dfrac{2.046\pi^2 EI}{L^2}$
$X3$. Fixed-fixed column	$Y3.\ P_{cr} = \dfrac{\pi^2 EI}{4L^2}$
$X4$. Fixed-pinned column	$Y4.\ P_{cr} = \dfrac{\pi^2 EI}{L^2}$

(a) $X1$ - $Y4$, $X2$ - $Y3$, $X3$ - $Y1$, $X4$ - $Y2$
(b) $X1$ - $Y4$, $X2$ - $Y2$, $X3$ - $Y3$, $X4$ - $Y1$
(c) $X1$ - $Y4$, $X2$ -$Y1$, $X3$ - $Y2$, $X4$ - $Y3$
(d) $X1$ - $Y4$, $X2$ - $Y3$, $X3$ - $Y2$, $X4$ - $Y1$

[GATE 2007, 2 M]

10. Which one of the following materials should be selected to design an axial flow turbine operating at high temperatures?

(a) Steel alloy (b) Titanium alloy
(c) Nickel alloy (d) Aluminum alloy

[GATE 2007, 2 M]

11. The von Mises yield criterion or the maximum distortion energy criterion for a plane stress problem with σ_1 and σ_2 as the principal stresses in the plane, and s_Y as the yield stress, requires

(a) $\sigma_1^2 - \sigma_1\sigma_2 + \sigma_2^2 \le \sigma_r^2$
(b) $|\sigma_1 - \sigma_2| \le \sigma_r$
(c) $|\sigma_1| \le \sigma_r$
(d) $|\sigma_2| \le \sigma_r$

[GATE 2007, 2 M]

12. An Euler-Bernoulli beam having a rectangular cross-section, as shown in the figure, is subjected to a non-uniform bending moment along its length. $V_z = \dfrac{dM_y}{dx}$. The shear stress distribution τ_{xz} across its cross-section is given by

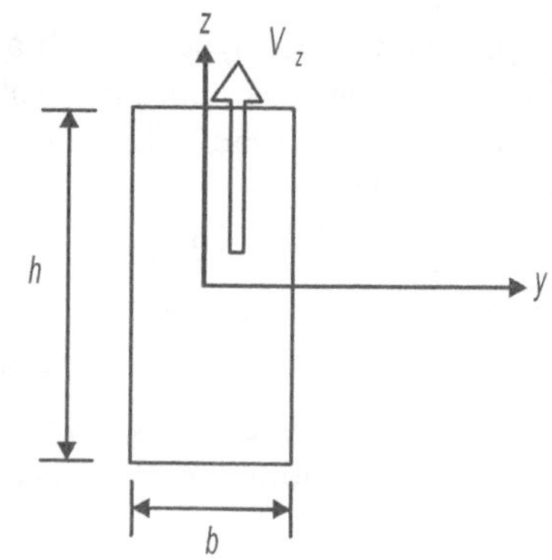

(a) $\tau_{xz} = \dfrac{V_z}{2I_y}\dfrac{z}{(h/2)}$

(b) $\tau_{xz} = \dfrac{V_z (h/1)^2}{2I_y}\left(1 - \dfrac{z^2}{(h/2)^2}\right)$

(c) $\tau_{xz} = \dfrac{V_z}{2I_y}\left(\dfrac{z}{(h/2)^2}\right)^2$

(d) $\tau_{xz} = \dfrac{V_z (h/2)^2}{2I_y}$

[GATE 2007, 2 M]

13. The torsion constant J of a thin-walled closed tube of thickness t and mean radius r is given by

(a) $J = 2\pi r t^2$ (b) $J = 2\pi r^2 t$
(c) $J = 2\pi r^2 t^2$ (d) $J = 2\pi r^4$

[GATE 2007, 2 M]

14. Let a dynamical system be described by the differential equation $2\dfrac{dx}{dt} + \cos x = 0$. Which of the following differential equations describes this system in a close approximation sense for small perturbation about $x = \pi/4$?

(a) $2\dfrac{dx}{dt} + \sin x = 0$ (b) $2\dfrac{dx}{dt} - \dfrac{1}{\sqrt{2}}x = 0$

(c) $\dfrac{dx}{dt} + \cos x = 0$ (d) $\dfrac{dx}{dt} + x = 0$

[GATE 2007, 2 M]

Common Data for Questions 15 & 16:

Consider the spring mass system shown in the figure below. This system has two degrees of freedom representing the motions of the two masses.

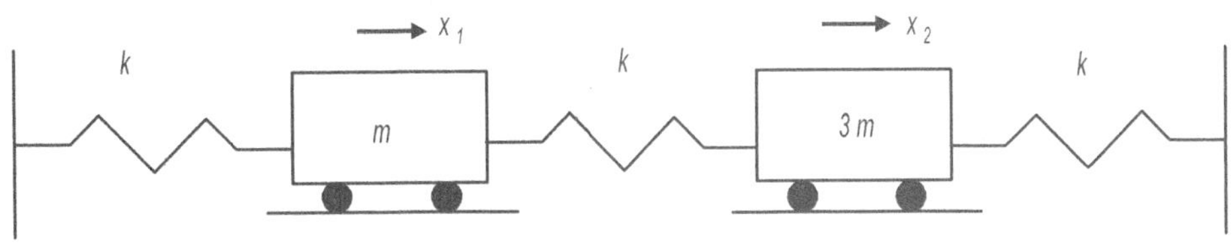

15. The system shows the following type of coordinate coupling

(*a*) static coupling

(*b*) dynamic coupling

(*c*) static and dynamic coupling

(*d*) no coupling

[GATE 2007, 2 M]

16. The two natural frequencies of the system are given as

(*a*) $\sqrt{\frac{4 \pm \sqrt{5}}{3} \frac{k}{m}}$ (*b*) $\sqrt{\frac{4 \pm \sqrt{3}}{3} \frac{k}{m}}$

(*c*) $\sqrt{\frac{4 \pm \sqrt{7}}{3} \frac{k}{m}}$ (*d*) $\sqrt{\frac{4 \pm \sqrt{11}}{3} \frac{k}{m}}$

[GATE 2007, 2 M]

Statement for Linked Answer Questions 17 & 18: The equation of motion of a vibrating rod is iven by $\frac{\partial^2 u}{dx^2} = \frac{l}{c^2} \frac{\partial^2 u}{\partial t^2}$. Here u is the displacement along the rod and is a function of both position x and time t. To find the response of the vibrating rod, we need to solve this equation using boundary conditions and initial conditions.

17. The boundary conditions needed for a rod fixed at the root ($x = 0$) and free at the tip ($x = l$) are

(*a*) $u(x = 0) = 0, \frac{\partial u}{\partial x}(x = l) = 0$

(*b*) $u(x = l) = 0, \frac{\partial u}{\partial x}(x = l) = 0$

(*c*) $u(x = 0) = 0, u(x = l) = 0$

(*d*) $\frac{\partial u}{dx}(x = 0) = 0, \frac{\partial u}{\partial x}(x = l) = 0$

[GATE 2007, 2 M]

18. The natural frequencies ω of the fixed-free rod can then be obtained using

(*a*) $\cos(\frac{\omega l}{c}) = 0$ (*b*) $\sin(\frac{\omega l}{c}) = 0$

(*c*) $\cos(\frac{\omega c}{l}) = 0$ (*d*) $\cos(\frac{\omega}{c}) = 0$

[GATE 2007, 2 M]

19. In the absence of body moments, the symmetry of the stress tensor is derived from

(*a*) force equilibrium conditions

(*b*) momentum equilibrium conditions

(*c*) linear relations between stresses and strains

(*d*) compatibility conditions

[GATE 2008, 1 M]

20. In a 3-D orthotropic material, the number of elastic constants in linear stress-strain relationship is

(*a*) 3 (*b*) 5

(*c*) 9 (*d*) 21

[GATE 2008, 1 M]

21. The compatibility conditions in theory of elasticity ensure that

(*a*) three is compatibility between various direct and shear stresses

(*b*) relationships between stresses and strains are consistent with constitutive relations

(*c*) displacements are single-valued and continuous

(*d*) stresses satisfy bi-harmonic equation

[GATE 2008, 1 M]

22. In a spring-mass-damper single degree of freedom system, the mass is 2 kg and the undamped natural frequency is 20 Hz. The critical damping constant of the system is

(*a*) 160π N.s/m (*b*) 80π N.s/m

(*c*) 1 N. s/m (*d*) 0 N. s/m

[GATE 2008, 1 M]

23. A parallelogram shaped plate of dimensions 'a' and 'b' as shown in the figure, is subjected to a uniform loading of normal stresses σ_1 and σ_2. The plate is in equilibrium for

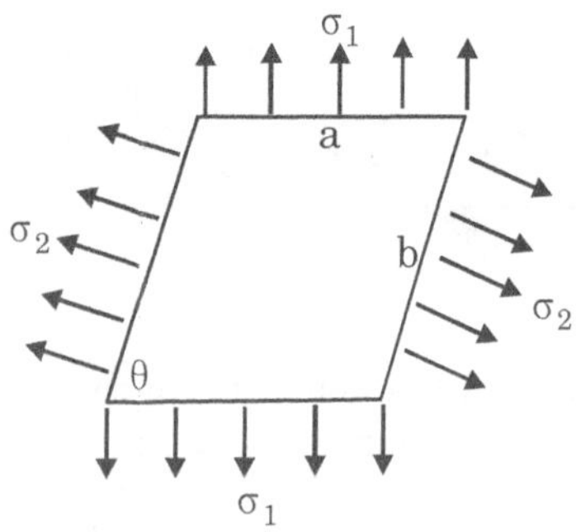

(*a*) any value of σ_1 and σ_2

(*b*) $\sigma_2 = \sigma_1 \cos\theta$

(*c*) $\sigma_1 = \sigma_2 \cos\theta$ (*d*) $\sigma_2 = \sigma_1$

24. A column of solid circular cross-section length L can have various end conditions. Choose the correct set that matches the end conditions (listed in Group I) with the corresponding effective length for bucking (listed in Group II).

Group I (end conditions)	**Group II** (effective length)
P. one end built-in and other end free	1. 1.0 L
Q. both ends pinned	2. 0.7 L
R. both ends built-in	3. 2.0 L
S. one end built-in and other end pinned	4. 0.5 L

(*a*) P-3, Q-1, R-4, S-2 (*b*) P-4, Q-1, R-2, S-3

(*c*) P-2, Q-1, R-3, S-4 (*d*) P-3, Q-1, R-2, S-4

[GATE 2008, 2 M]

25. A thin walled tube of circular cross-section with mean radius r has a central web which divides it into two symmetric cells as shown. A torque m is acting on the section. The shear flow q in the central web is

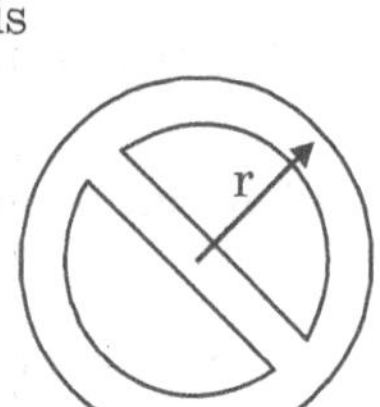

(*a*) $q = \dfrac{M}{2\pi r^2}$ (*b*) $q = 0$

(*c*) $q = \dfrac{M}{4\pi r^2}$ (*d*) $q = \dfrac{M}{\pi r^2}$

[GATE 2008, 2 M]

26. A concentrated bending moment M is acting at mid-span of a beam as shown. The shear force diagram for the beam is:

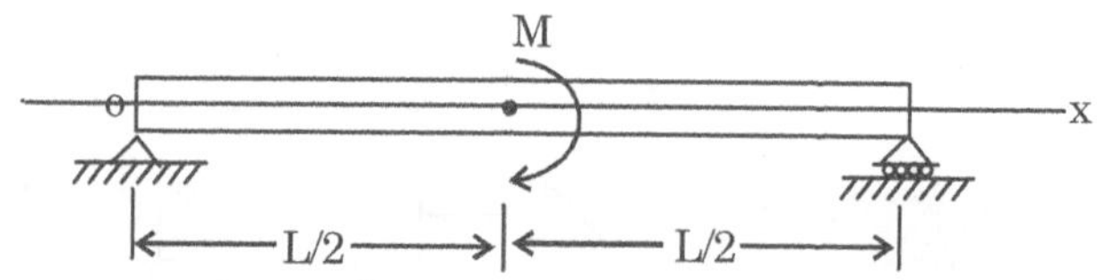

(*a*) 

(*b*) 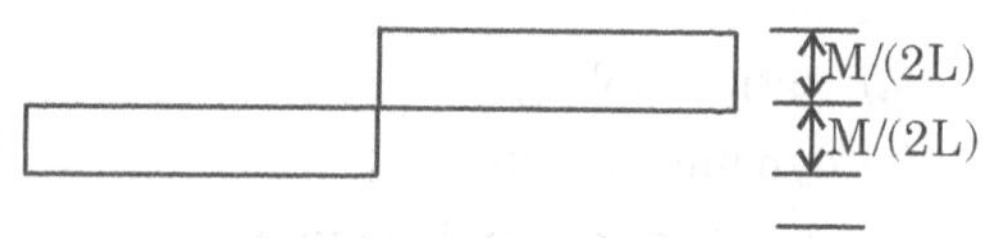

(*c*) 

(*d*) 

[GATE 2008, 2 M]

27. An idealized thin-walled cross-section of a beam and the respective areas of the booms are as shown. A bending moment M, is acting on the cross-section. The ratio of the magnitude of normal stress in the top booms to the that of the bottom boom is

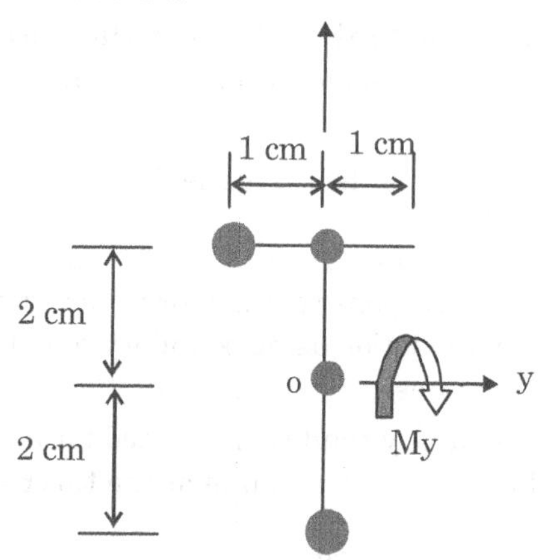

(*a*) $\dfrac{5}{11}$ (*b*) $\dfrac{2}{5}$

(*c*) 1 (*d*) $\dfrac{5}{2}$

[GATE 2008, 2 M]

28. An engineer is asked to test a system which can be idealized as SDOF (single degree of freedom) with viscous damping. A frequency response test was conducted and it is found that the quality factor Q is equal to 10. What will be the logarithmic decrement if a free vibration test is performed?

(*a*) $\frac{\pi}{40}$ (*b*) $\frac{\pi}{20}$

(*c*) $\frac{\pi}{10}$ (*d*) $\frac{\pi}{5}$

[GATE 2008, 2 M]

29. A beam occupies a region $0 \le x \le L; -c \le y\ c; -0.5 \le z \le 0.5$ as shown below. The beam can be considered to be in plane stress condition in x-y plane. Airy's stress function for the beam is given as:

$$\phi(x, y) = \frac{Pxy^3}{4c^3} + \frac{3Pxy}{4c}$$

where P is a constant

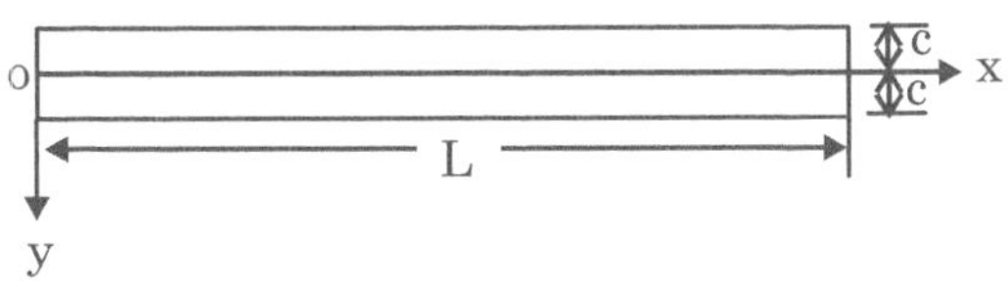

The above stress function pertains to a

(*a*) simply supported beam carrying a point load P at mid span

(*b*) simply supported beam carrying a uniform distributed load of intensity P per unit length

(*c*) cantilever beam clamped at end x = L and carrying a shear load P at x = 0

(*d*) cantilever beam clamped at end x = 0 and carrying a shear load P at x = L

[GATE 2008, 2 M]

30. The equation of motion of a uniform slender beam of length L in flexural vibration is given as $EI\frac{\partial^4 w}{\partial x^4} + \rho A\frac{\partial^2 w}{\partial t^2} = 0$, where EI is the flexural rigidity, w is the lateral displacement and ρA is the mass per unit length. The beam is simply supported at the two ends x = 0 and x = L. Assuming the mode shape in fundamental mode to be $\sin\left(\frac{\pi x}{L}\right)$, the natural frequency in fundamental mode is

(*a*) $0.5\sqrt{\frac{EI}{\rho AL^4}}\pi^2$ (*b*) $\sqrt{\frac{EI}{\rho AL^4}}\pi^2$

(*c*) $2\sqrt{\frac{EI}{\rho AL^4}}\pi^2$ (*d*) $4\sqrt{\frac{EI}{\rho AL^4}}\pi^2$

[GATE 2008, 2 M]

Common Data for Questions 31, 32 and 33: A two-dimensional state of stress in an isotropic materials is given by

$$[\sigma] = \begin{bmatrix} -8 & 5 \\ 5 & 16 \end{bmatrix} \text{MPa}$$

where c is linearly proportional to the applied loading. The failure stress is $\sigma_f = 350$ Mpa (which is 0.2% offset yield stress).

31. The principal stresses are

(*a*) $\sigma_1 = 17c$ MPa, $\sigma_2 = -9c$ MPa

(*b*) $\sigma_1 = 9c$ MPa, $\sigma_2 = -17c$ MPa

(*c*) $\sigma_1 = -17c$ MPa, $\sigma_2 = -9c$ MPa

(*d*) $\sigma_1 = 17c$ MPa, $\sigma_2 = 9c$ MPa

[GATE 2008, 2 M]

32. The maximum shear stress is

Which of the following statements is true?

(*a*) $\tau_{max} = 7c$ MPa

(*b*) $\tau_{max} = 10c$ MPa

(*c*) $\tau_{max} = 13c$ MPa

(*d*) $\tau_{max} = 15c$ MPa

[GATE 2008, 2 M]

33. The maximum value of c or safe loading of the structure, based on von-mises failure criterion is

(*a*) 10.2 (*b*) 15.3

(*c*) 25.4 (*d*) 31.8

[GATE 2008, 2 M]

Statement for Linked Answer Questions 34 and 35: An idealized thin walled two cell symmetric box beam is as shown. The shear flows in the walls are due to the applied shear forces $V_y = 480$ N, $V_z = 300$ N, and a torque M, all acting at the shear center.

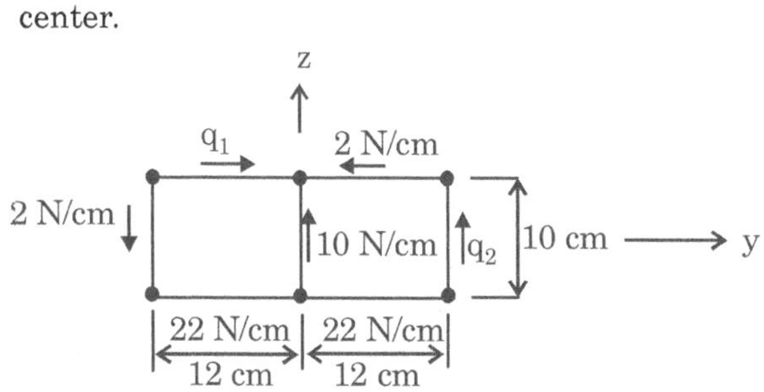

34. The shear flows q_1 and q_2 are

(*a*) $q_1 = -2$ N/cm, $q_2 = +22$ N/cm

(*b*) $q_1 = +2$ N/cm, $q_2 = +22$ N/cm

(*c*) $q_1 = +2$ N/cm, $q_2 = -22$ N/cm

(*d*) $q_1 = -2$ N/cm, $q_2 = -22$ N/cm

[GATE 2008, 2 M]

35. The torque M is

(*a*) 3360 N. cm

(*b*) 5760 N. cm

(*c*) 6960 N. cm

(*d*) 8160 N. cm

[GATE 2008, 2 M]

36. For a plane strain problem, the stresses satisfy the condition

(*a*) $\tau_{xz} = \tau_{yz} = \sigma_z = 0$

(*b*) $\tau_{xz} = \tau_{yz} = 0, \sigma_z = \upsilon(\sigma_x + \sigma_y)$

(*c*) $\tau_{xz} = \tau_{yz} = 0, \sigma_z = \upsilon(\sigma_x + \sigma_y)$

(*d*) $\tau_{xz} = \tau_{yz} = 0, \sigma_z = \upsilon(\sigma_x + \sigma_y) + (1 - v)\tau_{xy}$

[GATE 2009, 1 M]

37. For a critically damped single degree of freedom spring - mass - damper system with a damping constant c of 4 Ns/m and spring constant k of 16 N/m, the system mass m is

(*a*) 0.5 kg (*b*) 0.25 kg

(*c*) 2 kg (*d*) 4 kg

[GATE 2009, 1 M]

38. In a thin walled rectangular tube subjected to equal and oppos1te forces P as shown in the figure, the shear stress along leg AB is

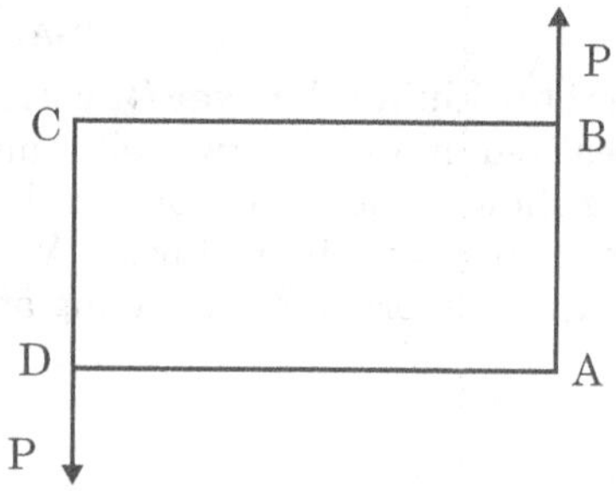

(*a*) zero

(*b*) constant non-zero

(*c*) varies linearly

(*d*) varies parabolically

[GATE 2009, 1 M]

39. For the thin walled beam cross section as shown in the figure, the shear centre lies at

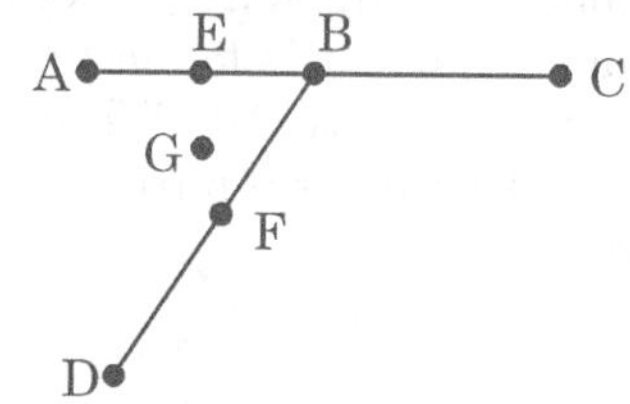

(*a*) Mid point of AB, i.e. at pointE

(*b*) Mid point of BD, i.e. at point F

(*c*) Junction point B

(*d*) at a point G lying within the area ABC

[GATE 2009, 1 M]

40. Consider a simply supported beam of length with an overhang of length loaded by an end moment M, as shown below.

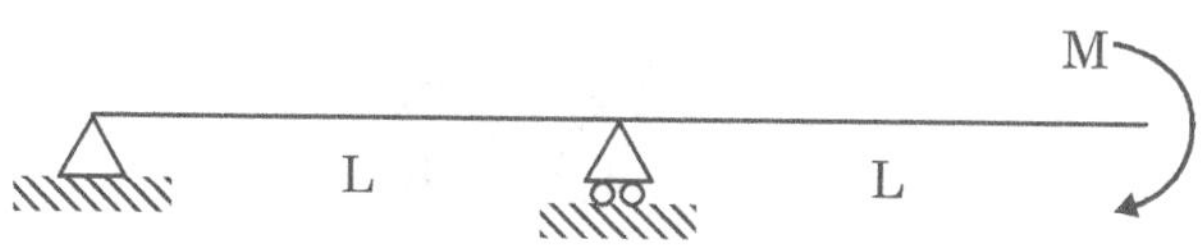

The bending moment distribution for this beam is

(*a*)

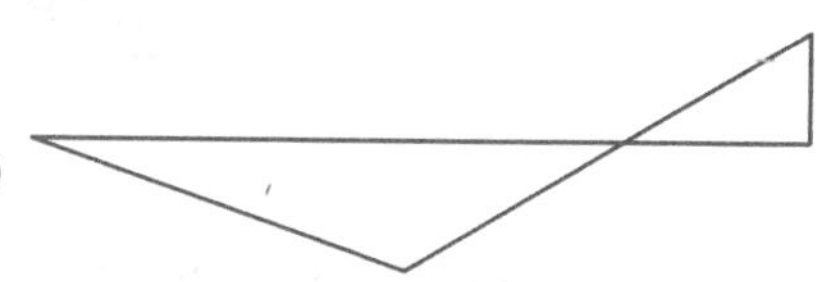

(*b*)

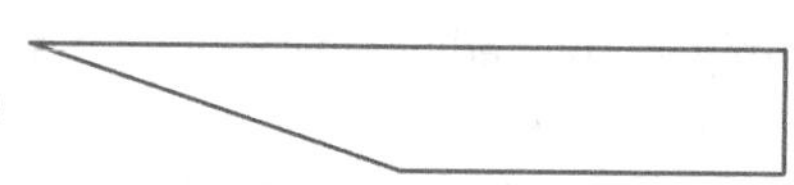

(*c*)

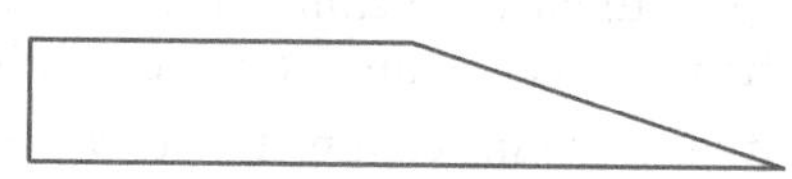

(*d*)

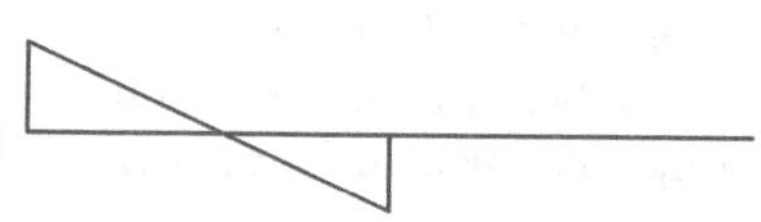

[GATE 2009, 2 M]

41. For the spring-mass system shown below, the natural frequencies are

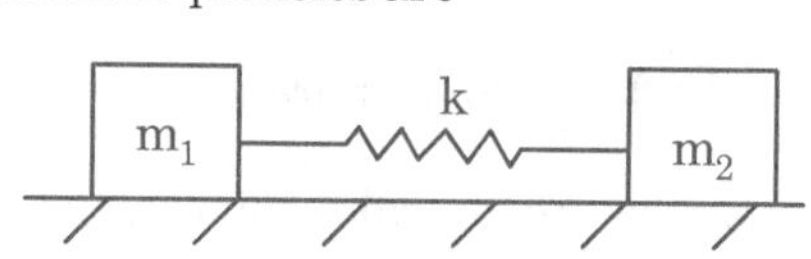

(*a*) 0 and $\sqrt{\dfrac{k(m_1 + m_2)}{m_1 m_2}}$ (*b*) 0 and $\sqrt{\dfrac{k(m_1 + m_2)}{2m_1 m_2}}$

(*c*) 0 and $\sqrt{\dfrac{k}{(m_1 m_2)}}$ (*d*) 0 and $\sqrt{\dfrac{k}{2(m_1 m_2)}}$

[GATE 2009, 2 M]

42. The buckling load for a simply supported column of rectangular cross section of dimensions 1 cm × 1.5 cm and length 0.5 m made of steel ($E = 210 \times 10^9$ N/m²) is approximately

(*a*) 10 kN (*b*) 4 kN

(*c*) 23 kN (*d*) 46 kN

[GATE 2009, 2 M]

43. A wing root cross section is idealized using lumped areas (booms) as shown below.

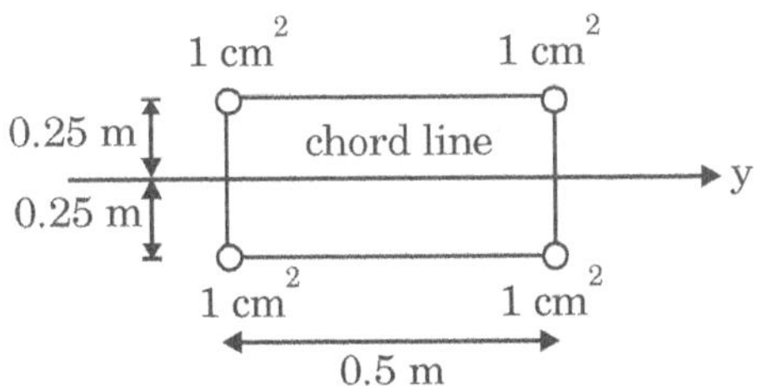

The wing root bending moment in steady level flight is M_y = 10 N-m. If the airplane flies at a load factor = 3.5, the maximum bending stress at the root is

(*a*) 1×10^6 N/m²

(*b*) 3.5×10^6 N/m²

(*c*) 7×10^6 N/m²

(*d*) 0.286×10^6 N/m²

[GATE 2009, 2 M]

44. A uniform rigid bar of mass m = 1 kg and length L = 1 m is pivoted at A. It is supported by a spring of stiffness k = 1 N/m and a viscous damper of damping constant C = 1 N-s/m, with $a = 1/\sqrt{3}$ m as shown below. The moment of inertia of the rigid bar is $I_A = \frac{mL^3}{3}$.

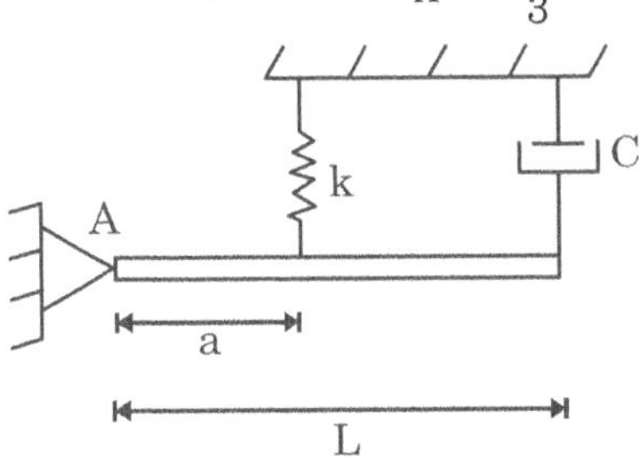

The system is

(*a*) overdamped

(*b*) underdamped with natural frequency ω = 1 rad/s

(*c*) critically damped

(*d*) underdamped with natural frequency ω_n = 2 rad/s

[GATE 2009, 2 M]

45. A 2-celled tube with wall thickness 0.5 mm is subjected to a torque of 10 N-m. The resulting shear flows in the two cells are q_1 and q_2 as shown below.

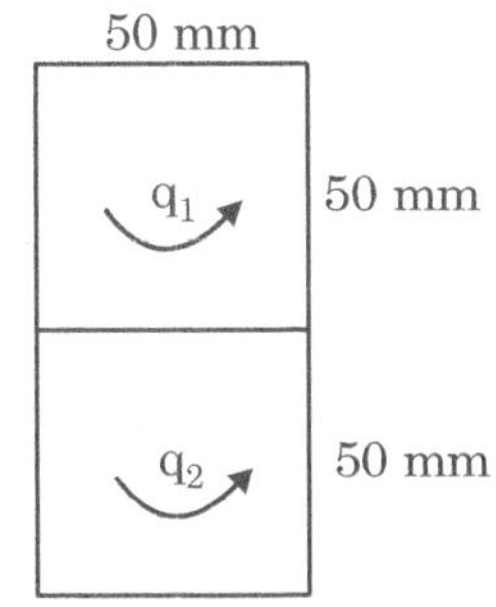

The torque balance equation (Bredt-Batho formula) for this section leads to

(*a*) $q_1 - q_2 = 2000$ N/m

(*b*) $q_1 + 2q_2 = 2000$ N/m

(*c*) $q_1 + q_2 = 2000$ N/m

(*d*) $2q_1 + q_2 = 2000$ N/m

[GATE 2009, 2 M]

Common Data for Questions 46 and 47:

The partial differential equation for the torsional vibration of a shaft of length L, torsional rigidity GJ, and mass polar moment of inertia per unit length I, is $I\frac{\partial^2\theta}{\partial t^2} = GJ\frac{\partial^2\theta}{dx^2}$, were θ is the twist.

46. If the shaft is fixed at both ends, the boundary conditions are:

(*a*) $\left.\frac{\partial\theta}{\partial x}\right|_{x=0} = 0$ and $\left.\frac{\partial\theta}{\partial x}\right|_{x=L} = 0$

(*b*) $\theta(0) = 0$ and $\theta(L) = 0$

(*c*) $\left.\frac{\partial\theta}{\partial x}\right|_{x=0} = 0$ and $\theta(L) = 0$

(*d*) $\theta(0) = 0$ and $\left.\frac{\partial\theta}{\partial x}\right|_{x=L} = 0$

[GATE 2009, 2 M]

47. If the nth mode shape of torsional vibration of the above shaft is $\sin\left(\frac{n\pi x}{L}\right)$ then the nth natural frequency of vibration, i.e., ω_n, is given by

(*a*) $\omega_n = \frac{n\pi}{L}\sqrt{\frac{GJ}{I}}$ (*b*) $\omega_n = \frac{(2n+1)\pi}{2L}\sqrt{\frac{GJ}{I}}$

(*c*) $\omega_n = \frac{n\pi}{2L}\sqrt{\frac{GJ}{I}}$ (*d*) $\omega_n = \frac{(2n+1)\pi}{L}\sqrt{\frac{GJ}{I}}$

[GATE 2009, 2 M]

48. In a general case of a homogeneous material under thermo-mechanical loading the number of distinct components of the state of stress is

(*a*) 3 (*b*) 4

(*c*) 5 (*d*) 6

[GATE 2010, 1 M]

49. A vertical slender rod is suspended by a hinge at the top and hangs freely. It is healed until it attains a uniform temperature, T Neglecting the effect of gravity, the rod has

(*a*) Stress but no strain

(*b*) Strain but no mess

(*c*) Both stress and strain

(*d*) Neither stress nor strain **[GATE 2010, 1 M]**

50. A thin rectangular plate made of isotropic material which satisfies the octahedral (i.e., Von Mises/ Distortion energy) failure criterion has yield strength of 200 MPa under uniaxial tension. As shown in the figure. If it is loaded with uniform tension of 150 MPa along the x-direction, the maximum uniform tensile stress that can be applied along the y-direction before the plate starts yielding is about

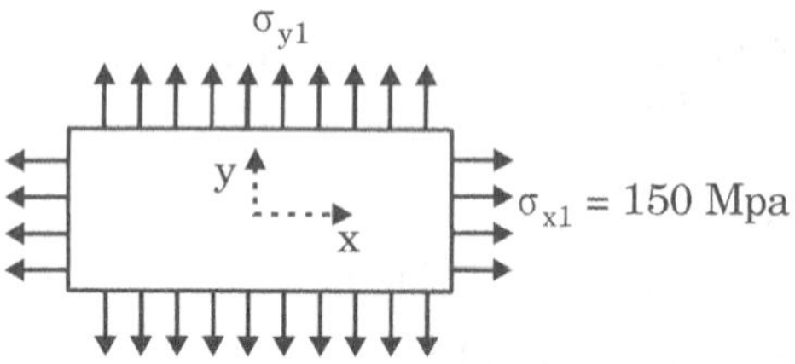

(*a*) 227 MPa (*b*) 77 MPa

(*c*) 87 MPa (*d*) 114 MPa

[GATE 2010, 1 M]

51. A column of length *l* and flexural rigidity E*l*. has one end fixed and the other end hinged. The critical buckling load for the column is

(*a*) $\dfrac{\pi^2 El}{(0.51)^2}$

(*b*) $\dfrac{\pi^2 El}{(0.71)^2}$

(*c*) $\dfrac{\pi^2 El}{1^2}$

(*d*) $\dfrac{\pi^2 El}{(21)^2}$

[GATE 2010, 1 M]

52. A horizontal cantilevered steel beam of rectangular cross-section having width b and depth d is vibrating in the vertical plane. The natural frequency of bending vibration is highest when

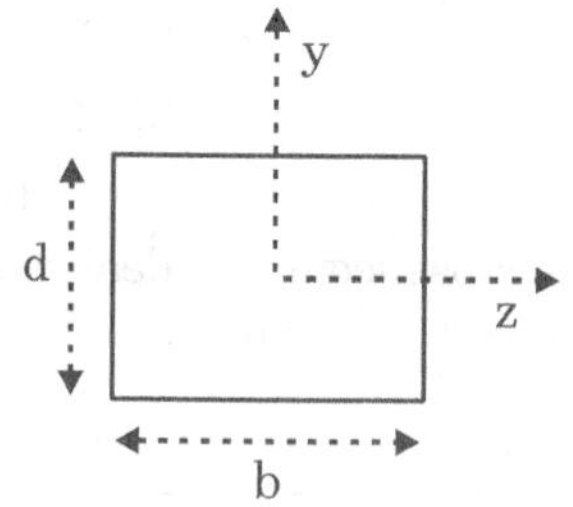

(*a*) b = 10, d = 10

(*b*) b = 20, d = 5

(*c*) b = 5, d = 20

(*d*) b = 25, d = 4

[GATE 2010, 1 M]

53. Following stress state is proposed for a 2-D problem with no body forces:

$\sigma_{xx} = 3x^2y + 4y^2$, $\sigma_{yy} = y^3 + 14xy$, $\tau_{xy} = -3xy^2 - 7x^2$.
It satisfies

(*a*) Equilibrium equations but not compatibility equation

(*b*) Compatibility equation but not equilibrium equations

(*c*) Neither equilibrium equations nor compatibility equation

(*d*) Both equilibrium equations and compatibility equation

[GATE 2010, 2 M]

54. A unifonn cross-section rigid rod of mass m and length I, is hinged at its upper end and suspended like a pendulum. Its natural frequency for small oscillations is

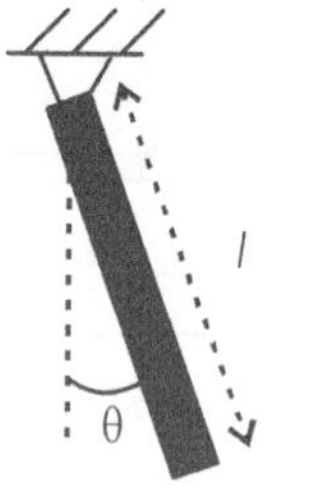

(*a*) $\sqrt{\dfrac{g}{21}}$ (*b*) $\sqrt{\dfrac{g}{1}}$

(*c*) $\sqrt{\dfrac{2g}{1}}$ (*d*) $\sqrt{\dfrac{3g}{21}}$

[GATE 2010, 2 M]

55. The thin rectangular plate shown in the figure is loaded with uniform shear, τ_0, along all edges and uniform uniaxial tension in the y-direction. The appropriate Airy's stress function to solve for stresses is given by

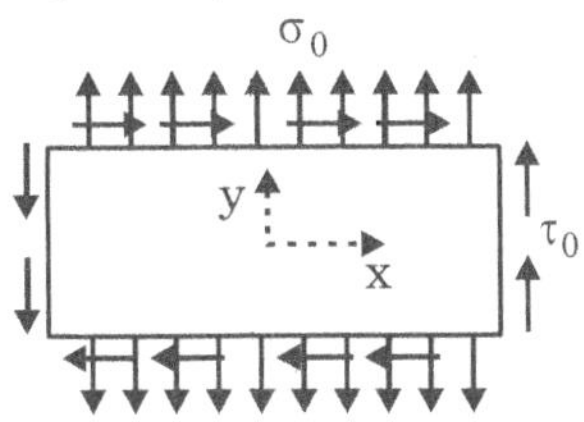

(*a*) $-\tau_0 xy - \sigma_a \frac{x^2}{2} + \sigma_a \left(x^4 - y^4\right)$

(*b*) $\tau_0 xy - \sigma_a \frac{x^2}{2}$

(*c*) $\tau_0 xy + \sigma_a \frac{x^2}{2}$

(*d*) $\tau_0 xy + \sigma_a \frac{x^2}{2} + \sigma_a \left(x^4 - y^4\right)$

[GATE 2010, 2 M]

56. The given thin wall section of uniform thickness, r. is symmetric about x-axis. Moment of inertia is given to be $I_{xy} = \frac{35}{12} th^3$. Shear center for this section is located at

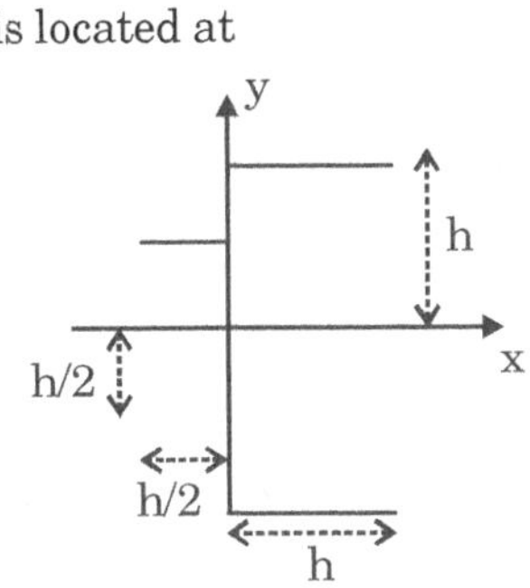

(*a*) $x = -\frac{3}{8}h$ (*b*) $x = -\frac{9}{28}h$

(*c*) $x = -\frac{35}{36}h$ (*d*) $x = -\frac{17}{35}h$

[GATE 2010, 2 M]

57. During an under-damped oscillation of a single degree of freedom system, in the time displacement plot the third peak is of magnitude 100 and the tenth peak is of magnitude 10. 1he damping ratio ζ is approximately.

(*a*) 0.052 (*b*) 0.023

(*c*) 0.366 (*d*) 0.159

[GATE 2010, 2 M]

Common Data for Questions 58 and 59:

Consider a simply supported beam of length L, carrying a bracket welded at its center. The bracket carries a vertical load P, as shown in the figure. Dimensions of bracket are a = 0.1L. The beam has a square cross section of dimension h × h.

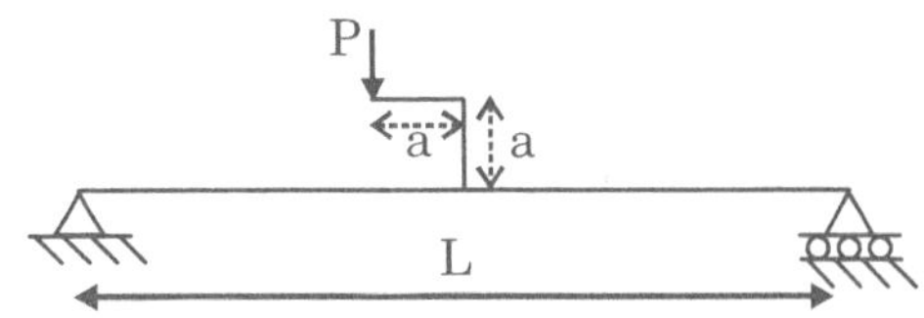

58. Bending moment diagram is given by

(*a*)

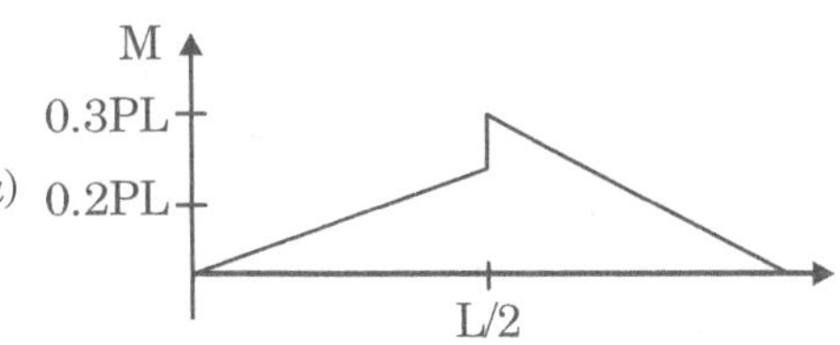

(*b*)

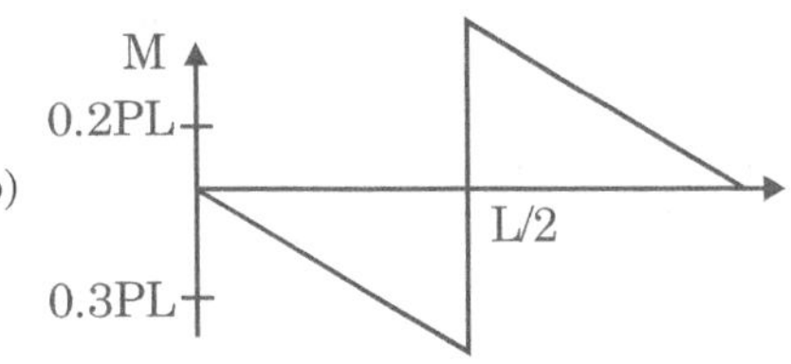

(*c*)

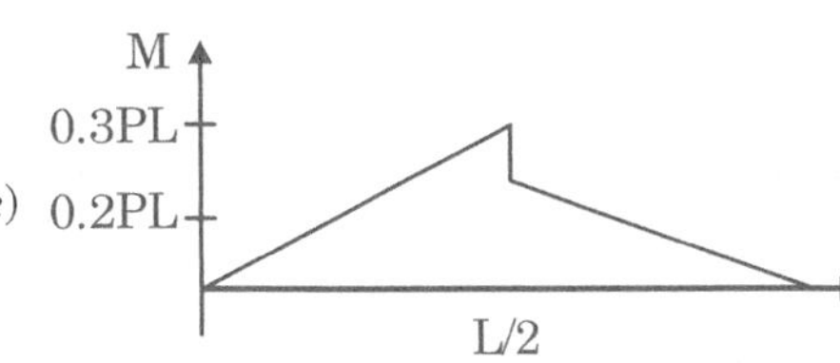

(*d*)

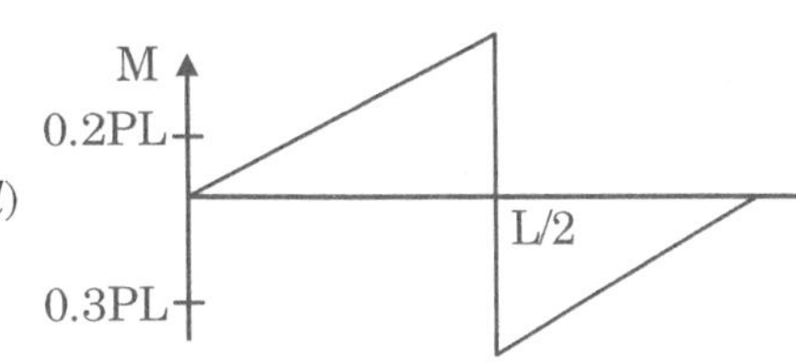

[GATE 2010, 2 M]

59. Maximum value of shear stress is

(*a*) 0.67 P/h^2 (*b*) 1.33 P/h^2

(*c*) 1.5 P/h^2 (*d*) 0.9 P/h^2

[GATE 2010, 2 M]

60. In three-dimensional linear elastic solids, the number of non-trivial stress-strain relations, strain-displacement equations and equations of equilibrium are, respectively,

(*a*) 3, 3 and 3 (*b*) 6, 3 and 3

(*c*) 6, 6 and 3 (*d*) 6, 3 and 6

[GATE 2011, 1 M]

61. An Euler-Bernoulli beam in bending is assumed to satisfy

(*a*) both plane stress as well as plane strain conditions

(*b*) plane strain condition but not plane stress condition

(*c*) plane stress condition but not plane strain condition

(*d*) neither plane strain condition nor plane stress condition

[GATE 2011, 1 M]

62. A statically indeterminate frame structure has

(*a*) same number of joint degrees of freedom as the number of equilibrium equations

(*b*) number of joint degrees of freedom greater than the number of equilibrium equations

(*c*) number of joint degrees of freedom less than the number of equilibrium equations

(*d*) unknown number of joint degrees of freedom, which cannot be solved using laws of mechanics

[GATE 2011, 1 M]

63. Consider a single degree of freedom spring-mass-damper system with mass, damping and stiffness of m, c and k, respectively. The logarithmic decrement of this system can be calculated using

(*a*) $\dfrac{2\pi c}{\sqrt{4mk - c^2}}$

(*b*) $\dfrac{\pi c}{\sqrt{4mk - c^2}}$

(*c*) $\dfrac{2\pi c}{\sqrt{mk - c^2}}$

(*d*) $\dfrac{2\pi c}{\sqrt{mk - 4c^2}}$

[GATE 2011, 1 M]

64. Consider a single degree of freedom spring-mass system of spring stiffness k_1 and m mass which has a natural frequency of 10 rad/s. Consider another single degree of freedom spring-mass system of spring stiffness k_2 and m mass which has a natural frequency of 20 rad/s. The spring stiffness is equal to

(*a*) k_1 (*b*) $2k_1$

(*c*) $\dfrac{k_1}{4}$ (*d*) $4k_1$

[GATE 2011, 1 M]

65. Consider a simply supported two-dimensional beam

If the beam is converted into a fixed-fixed beam as

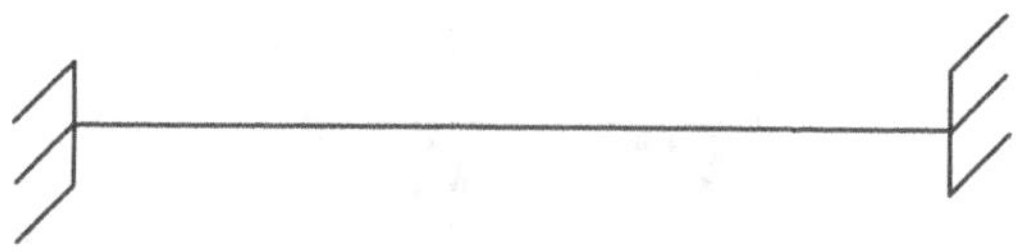

then the degree of static indeterminacy will

(a) increase by 3 (b) increase by 2

(c) decrease by 1 (d) decrease by 3

[GATE 2011, 1 M]

Questions Q.66 to Q. 68 are numericall answer type. The answer to each of these questions is either a positive whole number, or a positive real number with maximum of 2 decimal places.

66. Consider a cantilever beam having length L =1 m, square cross-section (width = depth = 0.01 m) and Young's modulus 50 GPa. The beam is subjected to a transverse load P = 1 N at the mid-span $\left(\dfrac{L}{2}\right)$ at the center of the cross-section. Under the small deformation theory, the transverse deflection of the beam (in mm) at its free-end is

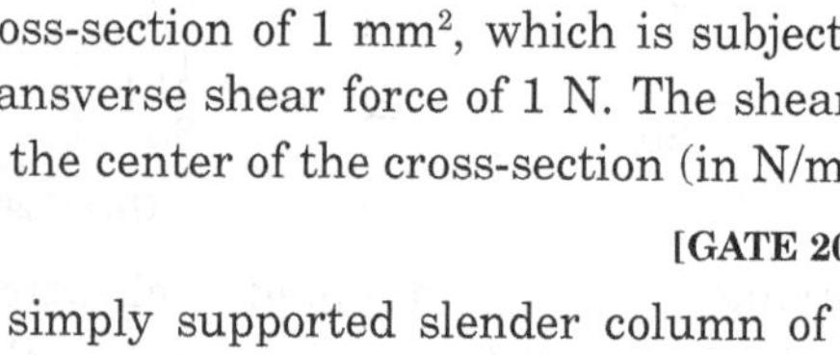

[GATE 2011, 2 M]

67. Consider a beam in bending with a solid circular cross-section of 1 mm², which is subjected to a transverse shear force of 1 N. The shear stress at the center of the cross-section (in N/mm²) is

[GATE 2011, 2 M]

68. A simply supported slender column of square cross section (width = depth = d) has to be designed such that it buckles at the same instant as it yields. Length of the column is given to be 1.57 m and it is made of a material whose Young's modulus is 200 GPa and yield stress is 240 MPa. The width, d, of the column (in cm) should be

[GATE 2011, 2 M]

69. A body undergoes deformation under plane strain conditions when subjected to the following stresses (in MPa): $\sigma_{xx} = 450$, $\sigma_{yy} = 450$, $\tau_{xy} = 75$, $\tau_{xz} = 0$, $\tau_{yz} = 0$. What are the remaining components of stresses (in MPa) and strains? Assume the material to be isotropic and linear-elastic with Young's modulus E = 200 GPa and Poisson's ratio $v = \frac{1}{3}$.

(*a*) $\sigma_{zz} = 0, \varepsilon_{xx} = 0.00225, \varepsilon_{yy} = 0.00225,$
$\gamma_{xy} = 0.002, \gamma_{xz} = 0, \gamma_{yz} = 0$

(*b*) $\sigma_{zz} = 300, \varepsilon_{xx} = 0.001, \varepsilon_{yy} = 0.001, \gamma_{xy} = 0.001,$
$\gamma_{xz} = 0, \gamma_{yz} = 0$

(*c*) $\sigma_{zz} = 300, \varepsilon_{xx} = 0.00225, \varepsilon_{yy} = 0.00225, \gamma_{xy} = 0.001,$
$\gamma_{xz} = 0, \gamma_{yz} = 0$

(*d*) $\sigma_{zz} = 0, \varepsilon_{xx} = 0.001, \varepsilon_{yy} = 0.001, \gamma_{xy} = 0.002,$
$\gamma_{xz} = 0, \gamma_{yz} = 0$

[GATE 2011, 2 M]

70. Which of the following Airy's stress functions could satisfy the given boundary conditions, assuming constant values of $\sigma_{xx} = P$, $\sigma_{yy} = Q$ and $\tau_{xy} = R$, along the boundary?

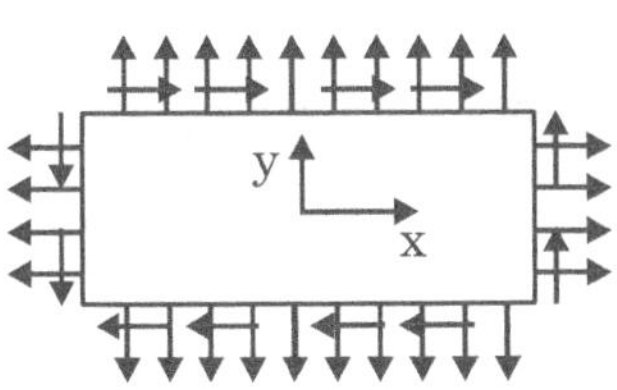

(*a*) $\phi = P\frac{x^2}{2} + Q\frac{y^2}{2} - Rxy$

(*b*) $\phi = P\frac{y^2}{2} + Q\frac{x^2}{2} + Rxy$

(*c*) $\phi = P\frac{y^2}{2} + Q\frac{x^2}{2} - Rxy$

(*d*) $\phi = P\frac{x^2}{2} + Q\frac{y^2}{2} + Rxy$

[GATE 2011, 2 M]

Common Data for Questions 71 and 72:

The partial differential equation (PDE) governing free vibrations of a uniform Euler-Bernoulli beam is given by

$EI\frac{\partial^4 w}{\partial x^4} + m\frac{\partial^2 w}{\partial t^2} = 0$, where EI is the flexural stiffness, m is the mass per unit length, is the bending displacement, m w(x, t) is the bending displacement x is the coordinate along the beam length, t is time and L is the beam length.

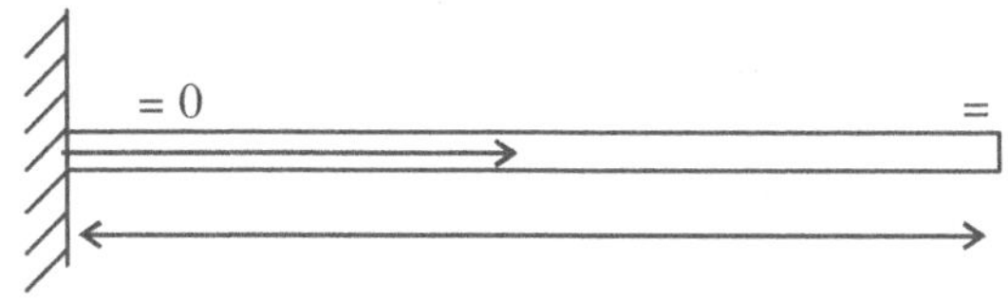

71. To solve the PDE, the number of boundary conditions (BC) and initial conditions (IC) needed are

(*a*) 4 BC, 3 IC (*b*) 2 BC, 2 IC
(*c*) 2 BC, 4 IC (*d*) 4 BC, 2 IC

[GATE 2011, 2 M]

72. For the cantilever beam shown in the figure, which of the following CANNOT be a possible boundary condition?

(*a*) $w(0, t) = 0$ (*b*) $\frac{\partial^2 w}{\partial x^2}(L, t) = 0$

(*c*) $\frac{\partial^2 w}{\partial x^2}(0, t) = 0$ (*d*) $\frac{\partial^3 w}{\partial x^3}(L, t) = 0$

[GATE 2011, 2 M]

Statement for Linked Answer Questions 73 and 74: A thin-walled (thickness << radius), hollow shaft of length 1m and mean radius, R = 5 cm has to be designed such that it can transmit a torque, T = 7 kN-m. A survey of different commercially available materials was made and following data was obtained from the suppliers (E : Young's modulus, τ_y : yield stress in shear, ρ : density):

Material	E(GPa)	τ_y (MPa)	$\rho(kg/m^3)$
X	200	550	7700
Y	70	22.5	2700
Z	110	375	4875

73. Which of the above materials would you choose such that weight of the shaft is minimum?

(*a*) X only (*b*) Y only
(*c*) Z only (*d*) X or Y

[GATE 2011, 2 M]

74. If you assume a factor of safety of 2, what should be the approximate thickness of such a shaft?

(*a*) 0.5 mm (*b*) 1 mmy
(*c*) 2 mm (*d*) 4 mm

[GATE 2011, 2 M]

75. The governing equation for the static transverse deflection of a beam under an uniformly distributed load, according to Euler-Bernoulli (engineering) beam theory, is a

(*a*) 2nd order linear homogenous partial differential equation.

(*b*) 4th order linear non-homogenous ordinary differential equation.

(*c*) 2nd order linear non-homogenous ordinary differential equation.

(*d*) 4th order nonlinear homogenous ordinary differential equation.

[GATE 2012, 1 M]

76. The Poisson's ratio, ν of most aircraft grade metallic alloys has values in the range:

(*a*) $-1 \le \nu \le 0$ (*b*) $0 \le \nu \le 0.2$

(*c*) $0.2 \le \nu \le 0.4$ (*d*) $0.4 \le \nu \le 0.5$

[GATE 2012, 1 M]

77. In a semi-monocoque construction of an aircraft wing, the skin and spar webs are the primary carriers of

(*a*) shear stresses due to an aerodynamic moment component alone.

(*b*) normal (bending) stresses due to aerodynamic forces.

(*c*) shear stresses due to aerodynamic forces alone.

(*d*) shear stresses due to aerodynamic forces and a moment component.

[GATE 2012, 2 M]

78. The logarithmic decrement measured for a viscously damped single degree of freedom system is 0.125. The value of the damping factor in % is closest to

(*a*) 0.5 (*b*) 1.0

(*c*) 1.5 (*d*) 2.0

[GATE 2012, 2 M]

A question 79 is numerical answer type. The answer to each of these questions is either a positive whole number, or a positive real number with maximum of 2 decimal places.

79. The mode shapes of an un-damped two degrees of freedom system are $\{1\ 0.5\}^T$ and $\{1 - 0.675\}^T$. The corresponding natural frequencies are 0.45 Hz and 1.2471 Hz. The maximum amplitude (in mm) of vibration of the first degree of freedom due to an initial displacement of $\{2\ 1\}^T$ (in mm) and zero initial velocities is ____.

[GATE 2012, 2 M]

80. The Airy stress function, $\phi = \alpha x^2 + \beta xy + \gamma y^2$ for a thin square panel of size $l \times l$ automatically satisfies compatibility. If the panel is subjected to uniform tensile stress, σ_o on all four edges, the traction boundary conditions are satisfied by

(*a*) $\sigma = \frac{\sigma_o}{2}; \beta = 0; \gamma = \frac{\sigma_o}{2}$ (c) $\alpha = \sigma_o; \beta = 0; \gamma = \sigma_o$

(*c*) $\alpha = 0; \beta = \frac{\sigma_o}{4}; \gamma = 0$ (*d*) $\alpha = 0; \beta = \frac{\sigma_o}{2}; \gamma = 0$

[GATE 2012, 2 M]

81. The boundary condition of a rod under longitudinal vibration is changed from fixed-fixed to fixed-free. The fundamental natural frequency of the rod is now k times the original frequency, where k is

(*a*) $\frac{1}{2}$ (*b*) 2

(*c*) $\frac{1}{\sqrt{2}}$ (*d*) $\sqrt{2}$

[GATE 2012, 2 M]

82. A spring-mass system is viscously damped with a viscous damping constant c. The energy dissipated per cycle when the system is undergoing a harmonic vibration $X \cos \omega_d t$ is given by

(*a*) $\pi c \omega_d X^2$ (*b*) $\pi \omega_d X^2$

(*c*) $\pi c \omega_d X$ (*d*) $\pi c \omega_d^2 X$

[GATE 2012, 2 M]

83. Buckling of the fuselage skin can be delayed by

(*a*) increasing internal pressure.

(*b*) placing stiffeners farther apart.

(*c*) reducing skin thickness

(*d*) placing stiffeners farther and decreasing internal pressure.

[GATE 2012, 2 M]

Statement for Linked Answer Questions 84 and 85: A thin-walled spherical vessel (1 m inner diameter and 10 mm wall thickness) is made of a material with $|\sigma_Y| = 500\text{MPa}$ in both tension and compression.

84. The internal pressure p_Y at yield, based on the von Mises yield criterion, if the vessel is floating in space, is approximately

(*a*) 500MPa (*b*) 250MPa

(*c*) 100MPa (*d*) 20MPa

[GATE 2012, 2 M]

85. If the vessel is evacuated (internal pressure = 0) and subjected to external pressure, yielding according to the von Mises yield criterion (assuming elastic stability until yield)

(*a*) occurs at about half the pressure p_Y.

(*b*) occurs at about double the pressure p_Y.

(*c*) occurs at about the same pressure p_Y.

(*d*) never occurs.

[GATE 2012, 2 M]

86. A damped single degree-of-freedom system is vibrating under a harmonic excitation with an amplitude ratio of 2.5 at resonance. The damping ratio of the system is ____________

[GATE 2013, 1 M]

87. The cross-section of a long thin-walled member is as shown in the figure. When subjected to pure twist, point A

(*a*) does not move horizontally or axially, but moves vertically

(*b*) does not move axially, but moves both vertically and horizontally

(*c*) does not move horizontally, vertically or axially

(*d*) does not move vertically or axially, but moves horizontally

[GATE 2013, 1 M]

88. The channel section of uniform thickness 2mm shown in the figure is subjected to a torque of 10 Nm. If it is made of a material with shear modulus of 25 GPa, the twist per unit length in radians/m is ________

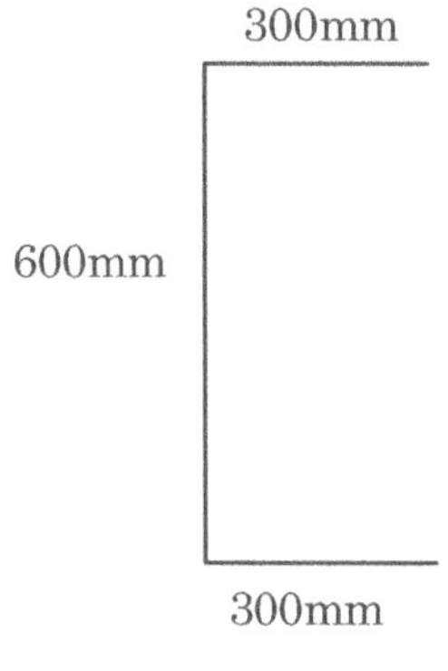

[GATE 2013, 1 M]

89. The stiffened cross-section of a long slender uniform structural member is idealized as shown in the figure below. The lumped areas at A, B, C and D have equal cross-sectional area of 3 cm^2. The webs AB, BC, CD and DA are each 5 mm thick. The structural member is subjected to a twisting moment of 10 kNm. The magnitudes of the shear flow in the webs, q_{AB}, q_{BC}, q_{CD}, and q_{DA} in kN/m are, respectively

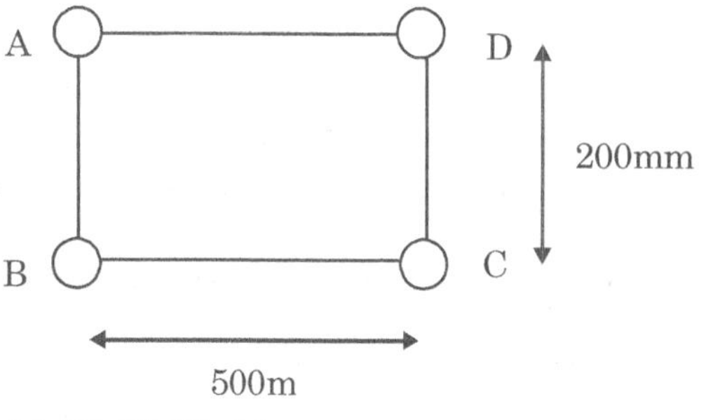

(*a*) 20, 20, 20, 20

(*b*) 0, 0, 50, 50

(*c*) 40, 40, 0, 0

(*d*) 50, 50, 50, 50

[GATE 2013, 1 M]

90. The idealized cross-section of a beam is comprised of four identical booms connected by shear webs. The beam is subjected to a bending moment M as shown in the figure. The inclination of the neutral axis to the x-axis in degrees is

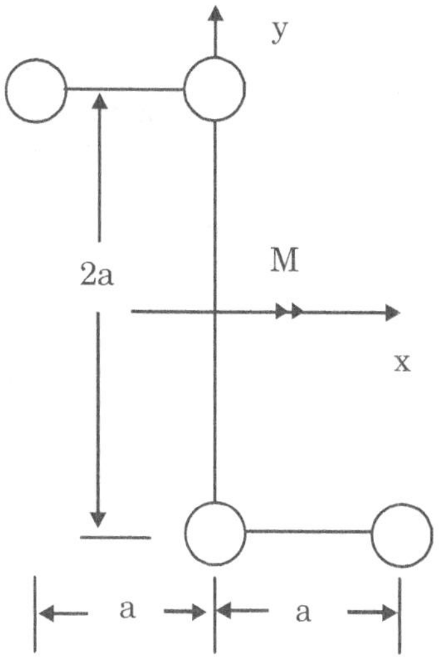

(*a*) 45 CW

(*b*) 45 CCW

(*c*) 26.6 CW

(*d*) 63.4 CCW

[GATE 2013, 2 M]

91. A composite circular shaft is comprised of a steel core surrounded by an aluminum annulus, perfectly bonded to each other as shown in the figure. If it subjected to a pure torque, which one of the following statements is TRUE?

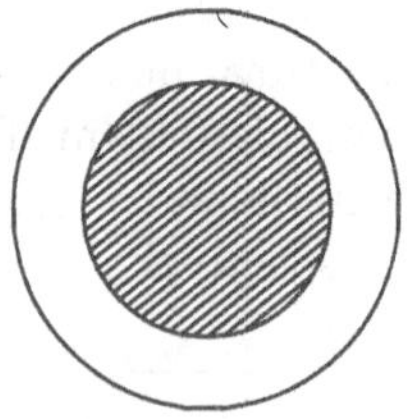

(*a*) Only shear stress is continuous across the steel–aluminum interface

(*b*) Only shear strain is continuous across the steel–aluminum interface

(*c*) Both shear stress and shear strain are continuous across the steel–aluminum interface

(*d*) Both shear stress and shear strain are discontinuous across the steel–aluminum interface

[GATE 2013, 2 M]

92. A horizontal rectangular plate ABCD is hinged at points A, B and C. AC and BD are diagonals of the plate. Downward force P is applied at D. The upward reactions R_A, R_B, and R_C at points A, B and C, respectively, are

(*a*) indeterminate (*b*) P, –P, P

(*c*) 0, P, 0 (*d*) $\frac{P}{3}, \frac{P}{3}, \frac{P}{3}$

[GATE 2013, 2 M]

93. In the steel structure (Young's modulus = 200 GPa) shown in the figure, all members have a circular cross-section of radius 10 mm. Column BD is pinned at B and D. The support at A is hinged. The minimum value of load P at which the column BD may buckle in Newtons is approximately ________

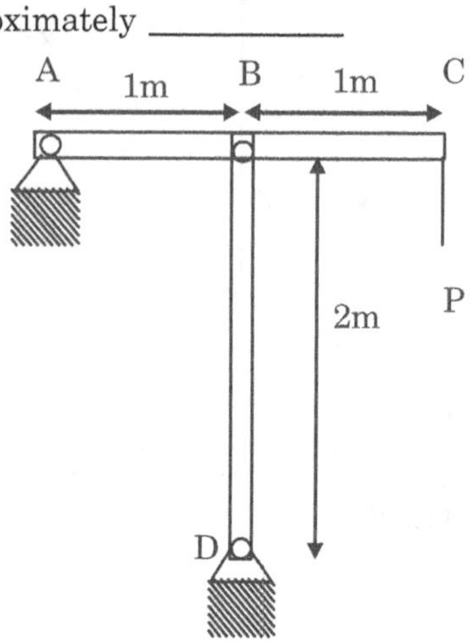

[GATE 2013, 2 M]

94. The thin rectangular plate has dimensions L × b × t. It develops a stress field corresponding to an applied bending moment M as shown in the figure. A valid Airy's stress function is

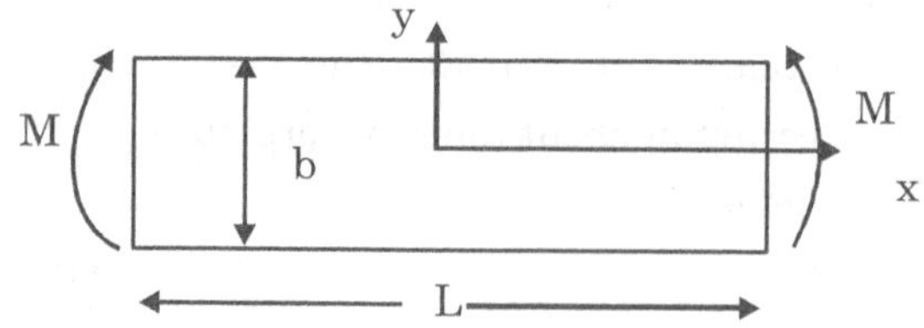

(*a*) $\frac{2M}{tb^3}x^3$ (c) $\frac{2M}{tb^3}y^3$

(*c*) $\frac{2M}{tb^3}\left(x^3 + y^3\right)$ (*d*) $\frac{2M}{tb^3}y^4$

[GATE 2013, 2 M]

95. A cantilever beam of negligible mass is 0.6 m long. It has a rectangular cross-section of width 8 mm and thickness 6 mm and carries a tip mass of 1.4 kg. If the natural frequency of this system is 10 rad/s, Young's modulus of the material of the beam in GPa is ________

[GATE 2013, 2 M]

96. A simply supported beam with overhang is loaded by uniformly distributed load of intensity q as shown in the figure. The bending moment at the mid-point of AB is

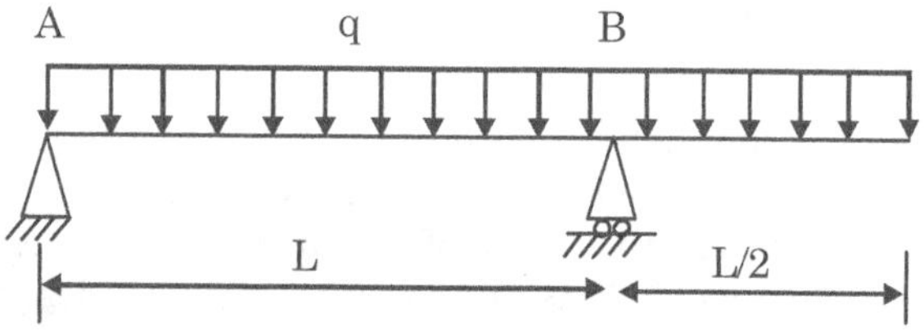

(*a*) $\frac{qL^2}{15}$ sagging

(b) $\frac{qL^2}{16}$ hogging

(*c*) $\frac{3qL^2}{16}$ hogging

(*d*) $\frac{3qL^2}{16}$ sagging

[GATE 2013, 2 M]

97. It is desired to measure the Young's modulus and the Poisson's ratio of a given homogeneous, isotropic material. A bar of length 20 cm and square cross-section (10mm × 10mm) of this material is subjected to a tensile load of 40kN. Under this load, length increases to 20.1 cm while the cross-section reduces to 9.98 mm × 9.8 mm. Young's modulus and Poisson's ratio of the material are:

(*a*) 80GPa & 0.4

(*b*) 40GPa & –0.4

(*c*) 80GPa & –0.2

(*d*) 40GPa & 0.2

[GATE 2014, 1 M]

98. In general, for any given solid subjected to arbitary loading, which of the following statements is always true:

(*a*) Volume does not vary with loading

(*b*) Mass does not vary with loading

(*c*) Density does not vary with loading

(*d*) Volume, mass and density vary with loading

[GATE 2014, 1 M]

99. Which one of the following objects with inclined face at 45°, subjected to the given stresses, are in static equilibrium

(*a*)

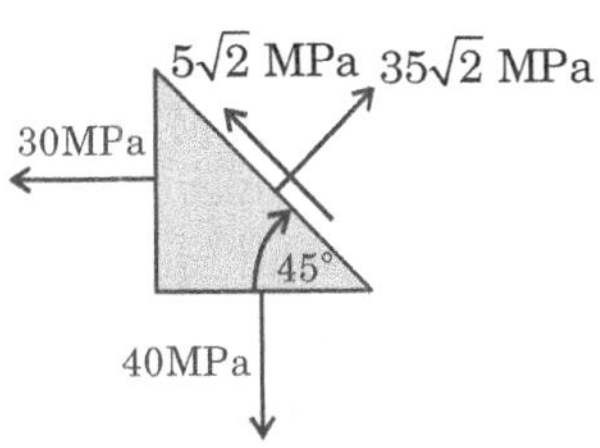

(*b*)

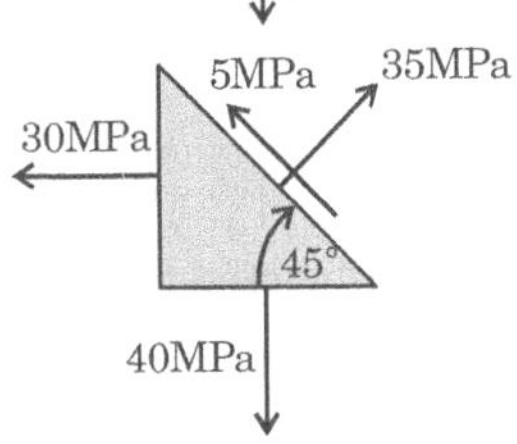

(*c*)

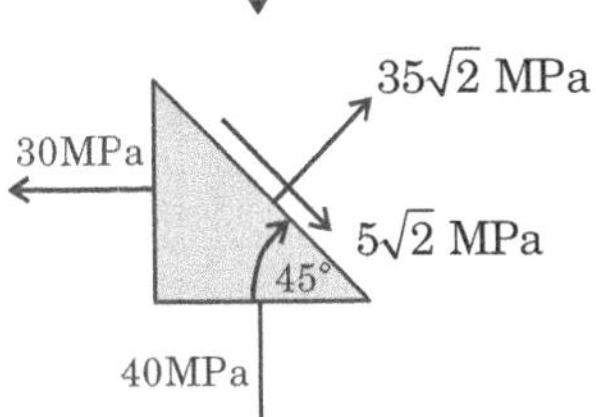

(*d*)

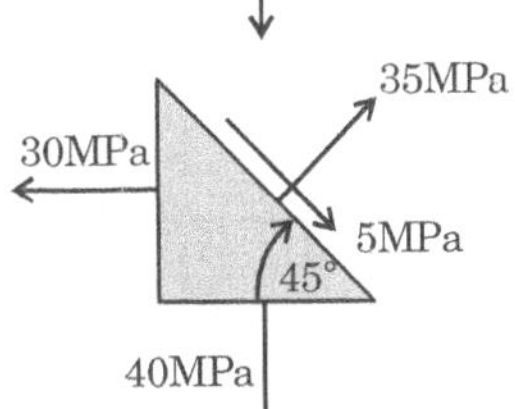

[GATE 2014, 1 M]

100. A damped single degree of freedom system whose undamped natural frequency, ω_n = 10Hz, is subjected to sinusoidal external force. Power is half of the maximum for the two frequencies of 60.9469 rad/s and 64.7168 rad/s. The damping factor associated with the vibrating system (in %) is _______.

[GATE 2014, 1 M]

101. The boundary conditions for a rod with circular cross-section, under torsional vibration, are changed from fixed-free to fixed-free. The fundamental natural frequency of the fixed-fixed rod is k times that of fixed-free rod. The value of k is

(*a*) 1.5 (*b*) π

(*c*) 2.0 (*d*) 0.5

[GATE 2014, 1 M]

102. A 1.8 m long steel beam of rectangular cross section (10mm × 6mm) is simply supported with a length of 1.2m between the supports and an overhang of 0.3m on either side. Young's modulus for the material of the beam is 200GPa. For a 50N load applied at the center of the beam, magnitude of the slope of the beam at tip S is _________.

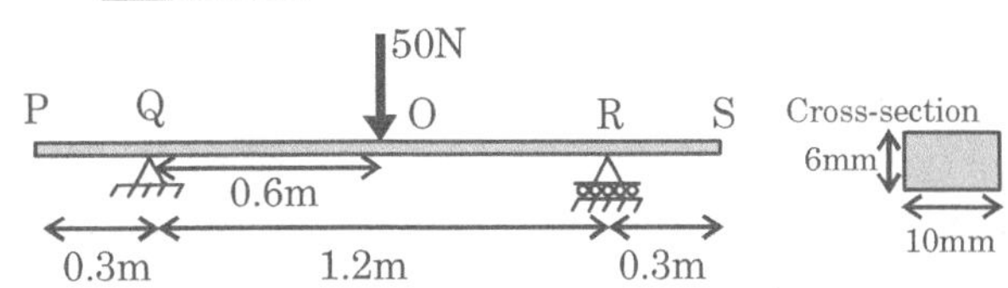

[GATE 2014, 2 M]

103. There are 2 designs proposed for a shaft of length 1, with a torque carrying capacity of T. Design I is a solid circular cross-section shaft of diameter 30mm. Design II is a thin-walled circular shaft of average diameter 40mm. Thickness of the wall in Design II has to be determined such that maximum shear stress is the same in both the designs for the given torque T (so that same material can be used for manufacturing both the shafts). Ratio of mass of shaft using Design I to the mass of shaft using Design II is

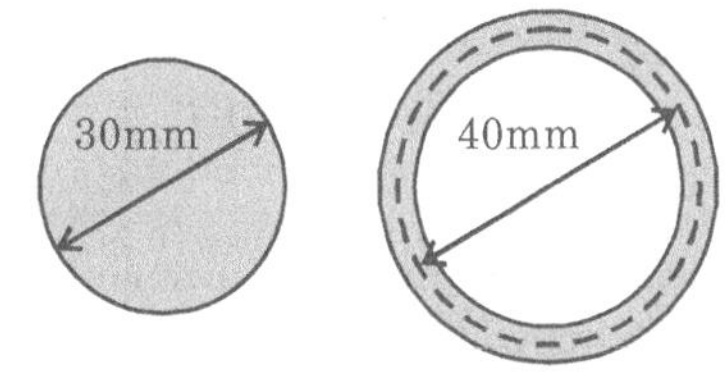

Design I Design II

(*a*) 2.68 (*b*) 5.36

(*c*) 1.79 (*d*) 3.58

[GATE 2014, 2 M]

104. A structural member of rectangular cross-section 10mm × 6mm and length 1m is made of steel (Young's modulus is 200GPa and coefficient of thermal expansion is $12 \times 10^{-6}/°C$). It is rigidly fixed at both the ends and then subjected to a gradual increase in temperature. Ignoring the three dimensional effects, the structural member will buckle if the temperature is increased by ΔT °C which is

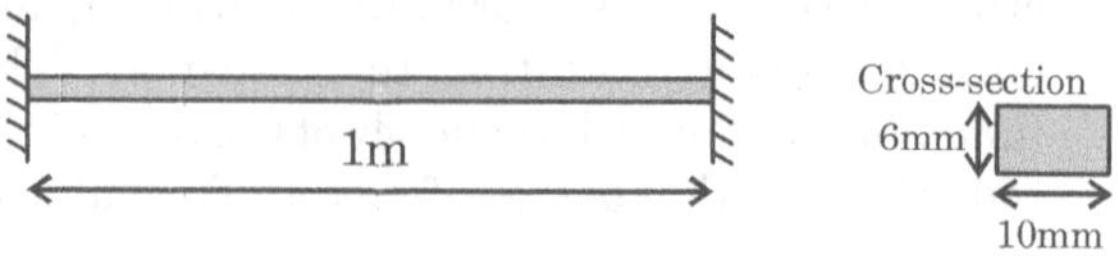

(*a*) 19.74 (*b*) 9.87

(*c*) 78.96 (*d*) 39.48

[GATE 2014, 2 M]

105. A gas cylinder (closed thin-walled cylindrical pressure vessel) of diameter 30cm and wall thickness 1mm is subjected to a design maximum internal pressure of 5 bar (0.5MPa). The material used for manufacturing this cylinder has a failure stress of 260MPa. Assuming von Mises failure criterion, the factor of safety (with respect to maximum allowable stress) for this cylinder is

(*a*) 2.8 (*b*) 2.0

(*c*) 6.9 (*d*) 4.0

[GATE 2014, 2 M]

106. A cantilevered beam is subjected to a parabolic distribution of shear traction at the right edge while the top and bottom surfaces are traction free. To solve this problem, following Airy's stress function is proposed: $\phi = C_1xy + C_2xy^3 + C_3x^2y^2 + C_4x^3y$. This is an admissible Airy's function that would satisfy the bi-harmonic equation as well as the boundary conditions if and only if

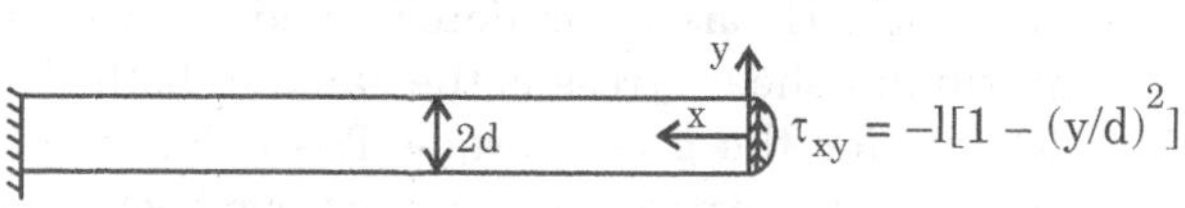

(*a*) $C_1 = 0, C_2 = \lambda, C_3 = 0, C_4 = \lambda/3d^2$

(*b*) $C_1 = \lambda, C_2 = \lambda/3d^2, C_3 = 0, C_4 = 0$

(*c*) $C_1 = 0, C_2 = 0, C_3 = \lambda, C_4 = -\lambda/3d^2$

(*d*) $C_1 = \lambda, C_2 = -\lambda/3d^2, C_3 = 0, C_4 = 0$

[GATE 2014, 2 M]

107. 1kg mass is hanging from a spring of stiffness 500N/m attached to a massless, symmetric beam of length 0.6m, moment of inertia about the bending axis $I = 8.33 \times 10^{-4}$ mm^4 and Young's modulus E = 210GPa as shown in the figure. The fundamental natural frequency (in rad/s) of the system is

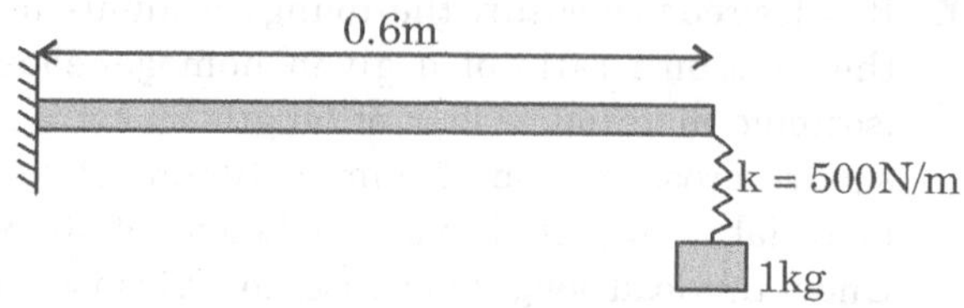

(*a*) 3.24 (*b*) 20.36

(*c*) 22.36 (*d*) 3.56

[GATE 2014, 2 M]

108. A single degree of freedom system is vibrating with initial (first cycle) amplitude of 5cm. The viscous damping factor associated with the vibrating system is 2%. Vibration amplitude of the fifth cycle (in cm) is

(*a*) 1.65 (*b*) 4.41

(*c*) 2.67 (*d*) 3.02

[GATE 2014, 2 M]

109. A cantilever with thin-walled channel cross section is subjected to a lateral force at its shear center. The cantilever undergoes

(*a*) bending without twisting

(*b*) bending and twisting

(*c*) neither bending nor twisting

(*d*) twisting without bending

[GATE 2015, 1 M]

110. The two non-zero principal stresses at a point in a thin plate are $\sigma_1 = 25$ MPa and $\sigma_2 = -25$ MPa. The maximum shear stress (in MPa) at this point is _____.

[GATE 2015, 1 M]

111. A 0.5 kg mass is suspended vertically from a point fixed on the Earth by a spring having a stiffness of 5 N/mm. The static displacement (in mm) of the mass is _____.

[GATE 2015, 1 M]

112. A slender structure is subjected to four different loading cases (I, II, III and IV) as shown below (Figures not to scale). Which pair of cases results in identical stress distribution at section S – S located far away from both ends?

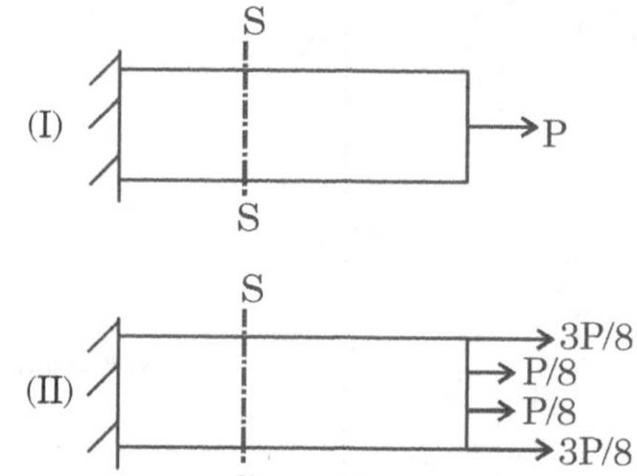

(III) 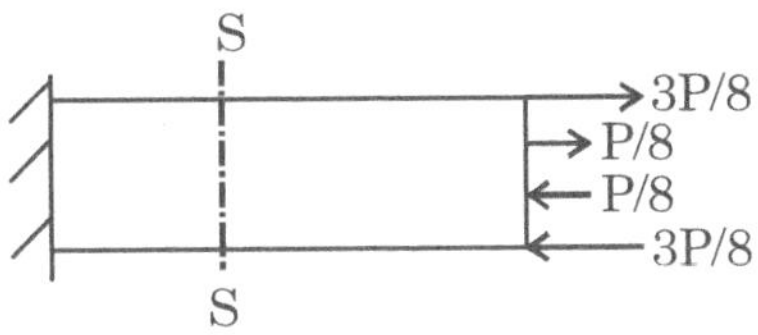

(IV) 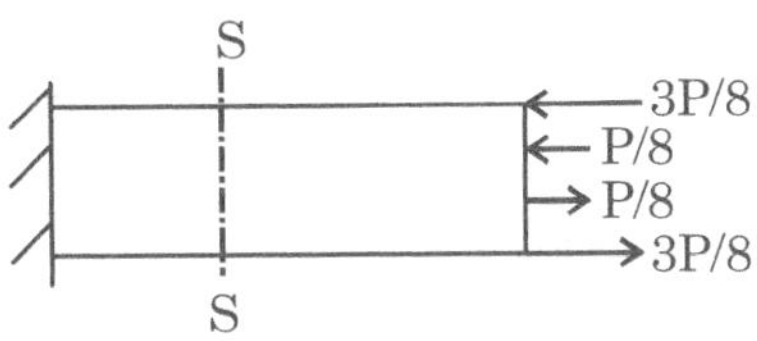

(*a*) I and II (*b*) II and III
(*c*) III and IV (*d*) IV and I

[GATE 2015, 1 M]

113. The 2-D stress state at a point P in the x-y coordinate system is $\begin{bmatrix} 60 & 50 \\ 50 & -40 \end{bmatrix}$ MPa. The magnitude of the tangential stress (in MPa) on a surface normal to the x-axis at P is _____.

[GATE 2015, 2 M]

114. A cube made of a linear elastic isotropic material is subjected to a uniform hydrostatic pressure of 100 N/mm^2. Under this load, the volume of the cube shrinks by 0.05%. The Young's modulus of the material, E = 300 GPa The Poisson's ratio of the material is _____.

[GATE 2015, 2 M]

115. A massless cantilever beam PQ has a solid square cross section (10 mm × 10 mm). This beam is subjected to a load W through a rigid massless link at the point Q, as shown below (figure not to scale). If the Young's modulus of the material E = 200 GPa, the deflection (in mm) at point Q is _____.

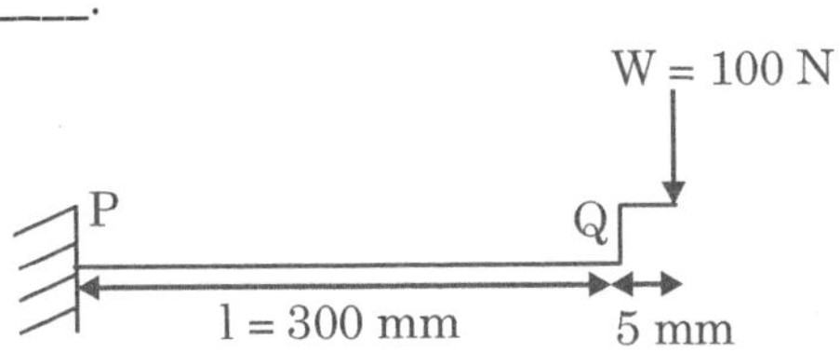

[GATE 2015, 2 M]

116. The following data is for a single degree of freedom system with viscous damping:

mass, m = 10 kg; spring stiffness, k = 2.25 N/mm; damping coefficient, c = 0.0125 N s/mm.

The ratio of any two successive amplitudes is _____.

[GATE 2015, 2 M]

117. The thin rectangular tube shown below is made of a material with shear modulus, G = 80 GPa. The shear flow is calculated based on the mid-thickness dimensions. If the free end is allowed to twist no more than 0.0727 rad, then the maximum torque (in Nm) which the tube can be subjected to at its free end is _______

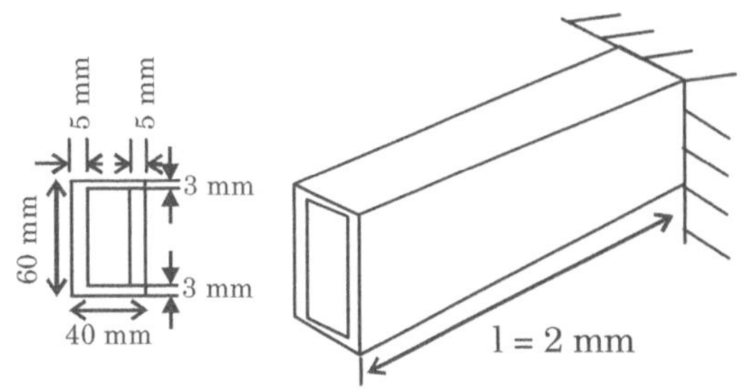

[GATE 2015, 2 M]

118. A 200 mm long simply-supported column has a 5mm × 10mm rectangular cross section. The Young's modulus of the material, E = 200 GPa. Assuming a factor of safety of 2.5 corresponding to the buckling load, the maximum load (in N) the column can support in compression is _______.

[GATE 2015, 2 M]

119. The total number of material constants that are necessary and sufficient to describe the three dimensional Hooke's law for an isotropic material is ____.

[GATE 2016, 1 M]

120. Determine the correctness or otherwise of the following statements, [a] and [r]:

[a]: In a plane stress problem, the shear strains along the thickness direction of a body are zero but the normal strain along the thickness is not zero.

[r]: In a plane stress problem, Poisson effect induces the normal strain along the thickness direction of the body.

(*a*) Both [a] and [r] are true and [r] is the correct reason for [a].

(*b*) Both [a] and [r] are true but [r] is not the correct reason for [a].

(*c*) Both [a] and [r] are false.

(*d*) [a] is true but [r] is false.

[GATE 2016, 1 M]

121. Consider four thin-walled beams of different open cross-sections, as shown in the cases (i-iv). A shear force of magnitude 'F' acts vertically downward at the location 'P' in all the beams. In which of the following case, does the shear force induce bending and twisting?

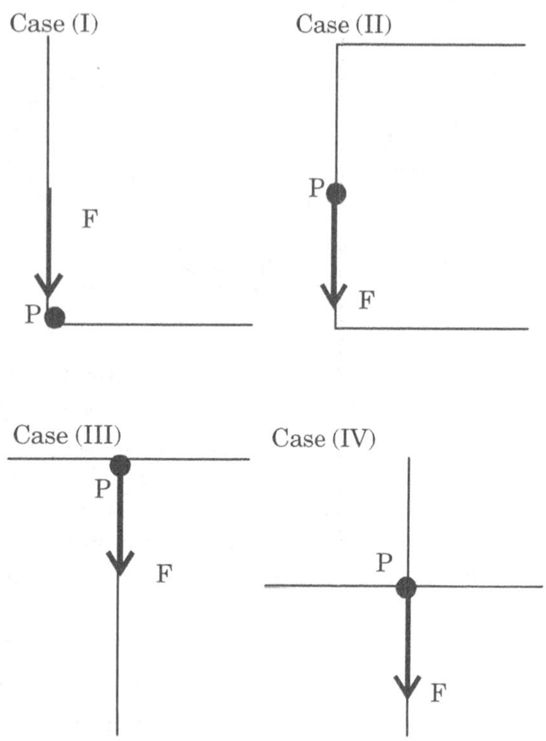

(*a*) (i) (*b*) (ii)
(*c*) (iii) (*d*) (iv)

[GATE 2016, 1 M]

122. The effective stiffness of the spring-mass system as shown in the figure below is ____ N/mm.

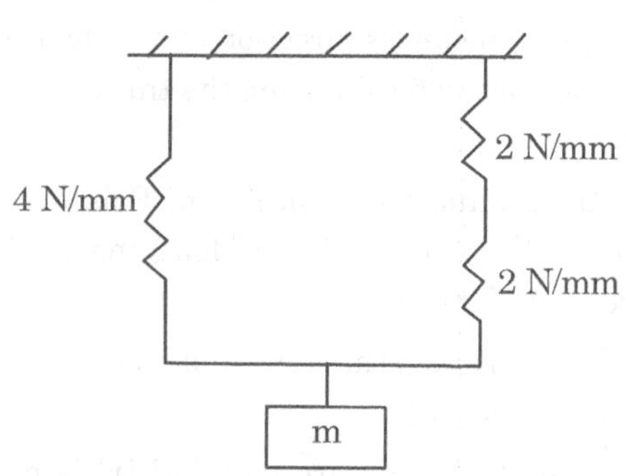

[GATE 2016, 1 M]

123. A structural member supports loads, which produce at a particular point, a state of pure shear stress of 50 N/mm^2. At what angles are the principal planes oriented with respect to the plane of pure shear?

(*a*) $\frac{\pi}{6}$ and $\frac{2\pi}{3}$ (*b*) $\frac{\pi}{4}$ and $\frac{3\pi}{4}$

(*c*) $\frac{\pi}{4}$ and $\frac{\pi}{2}$ (*d*) $\frac{\pi}{2}$ and π

[GATE 2016, 1 M]

124. A bar made of linear elastic isotropic material is fixed at one end and subjected to an axial force of 1 kN at the other end. The cross-sectional area of the bar is 100 mm^2, length is 100 mm and the Young's Modulus is 1×10^5 N/mm^2 . The strain energy stored in the bar is ____ Nmm.

[GATE 2016, 2 M]

125. A cantilever beam-spring system is shown in the figure. The beam is made with a material of Young's modulus 1×10^5 N/mm^2 and geometry such that its moment of inertia is 100 mm 4 and length l = 100 mm. It is supported by a spring of stiffness K = 30 N/mm and subjected to a load of P = 100 N at the point 'B'. The deflection at the point 'B' due to the load P is ____mm.

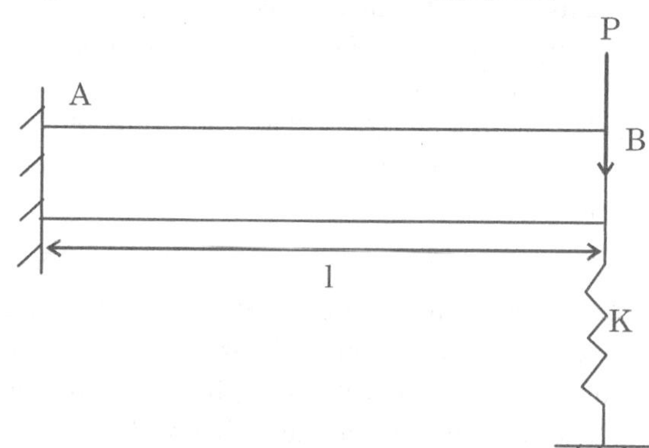

[GATE 2016, 2 M]

126. Determine the correctness or otherwise of the following statements, [a] and [r],

[a]: Ribs, used in airplane wings, increase the column buckling strength of the longitudinal stiffeners.

[r]: Ribs distribute concentrated loads into the structure and redistribute stresses around discontinuities.

(*a*) Both [a] and [r] are true and [r] is the correct reason for [a]

(*b*) Both [a] and [r] are true but [r] is not the correct reason for [a]

(*c*) Both [a] and [r] are false

(*d*) [a] is true but [r] is false

[GATE 2016, 2 M]

127. A channel section shown in the figure has uniform thickness. It is subjected to an anticlockwise torque of 62.5×10^3 Nmm. The maximum possible thickness of the channel section, such that the shear stress induced in it does not exceed 100 N/mm^2, is ____ mm.

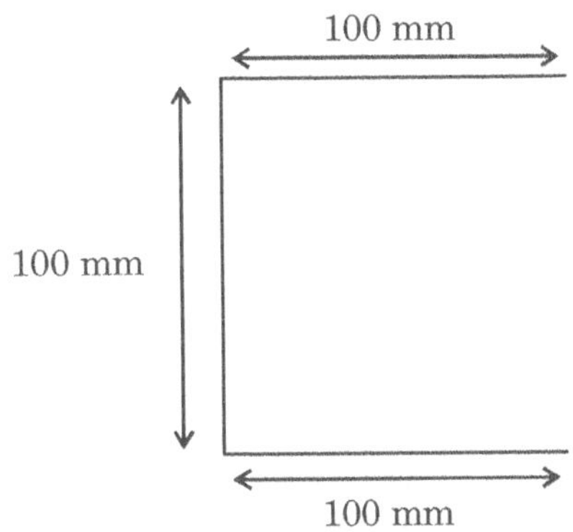

[GATE 2016, 2 M]

128. The governing differential equation of motion of a damped system is given by $m\frac{d^2x}{dt^2} + c\frac{dx}{dt} + kx = 0$. If m = 1 kg, c = 2 Ns/m and k = 2 N/m then the frequency of the damped oscillation of this system is ____ rad/s.

[GATE 2016, 2 M]

129. The two dimensional state of stress in a body is described by the Airy's stress function:

$$\phi = 5\frac{x^4}{10} + \frac{x^3y}{6} + 3\frac{x^2y^2}{2} + 7\frac{xy^3}{6} + E\frac{y^4}{12}.$$

The Airy's stress function will satisfy the equilibrium and the compatibility requirements if and only if the value of the coefficient E is ____.

[GATE 2016, 2 M]

130. Which of the following statement(s) is/are true about the shear centre of a cross-section:

P: It is that point in the cross-section through which shear loads produce no twisting.

Q: This point is also the centre of twist of sections subjected to pure torsion.

R: The normal stress at this point is always zero.

(*a*) P, Q and R (*b*) P only

(c) P and Q only (*d*) P and R only

[GATE 2017, 1 M]

131. Which of the following statements about the neutral axis of a beam with unsymmetrical cross section is true:

(*a*) The product of second moment of area about the neutral axis is always zero.

(*b*) The normal stress along the neutral axis is always zero.

(*c*) The shear stress along the neutral axis is always zero.

(*d*) The product of second moment of area about the neutral axis and the normal stress about the neutral axis always zero.

[GATE 2017, 1 M]

132. Assuming that the aircraft is flying straight, the top spar cap/flange of a wing is most likely to fail in:

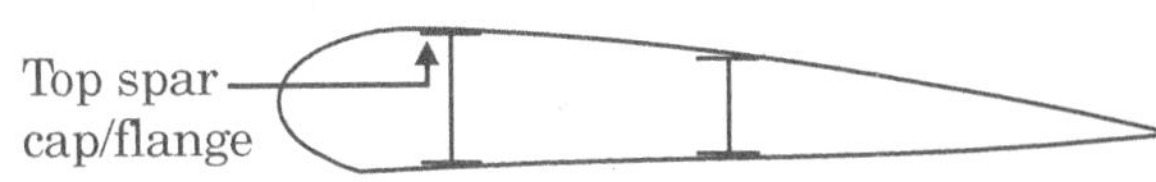

(*a*) Yielding (*b*) Buckling

(*c*) Crushing (*d*) Creep

[GATE 2017, 2 M]

133. A 2-DOF undamped spring-mass system with two masses and two springs has natural frequencies ω_1 = 0.79 rad/s and ω_2 = 1.58 rad/s. The mode shapes for the system are given by $\phi_1 = 90.732\ 1]^T$ and $\phi_2 = [-2.73\ 1]^T$. If the first mass is displaced by 1 cm, the minimum displacement in cms to the given to the second mass to make the system vibrate in first mode alone is = ________ (in three decimal place).

[GATE 2017, 2 M]

134. An aircraft landing great can be idealized as single degree of freedom spring-mass-damper system. The desirable damping characteristics of such a system is:

(*a*) Under damped (*b*) Over damped

(*c*) Critically damped (*d*) Undamped

[GATE 2017, 2 M]

135. A single degree of freedom spring-mass system of natural frequency 5 Hz is modified in the following manners:

Case 1:

Viscous damping with damping ratio ς - 0.2 is introduced in parallel to the spring.

Case 2 :

The original undamped spring-mass system is moved to a surface with coefficient of friction, μ = 0.01.

The ratio of the damped natural frequency for the cases 1 and 2 is given by ______(in three decimal places).

[GATE 2017, 2 M]

136. Which of the following statements about the compatibility equations are true:

P: Six strains are defined in terms of three displacement functions and can have arbitrary values.

R: Compatibility equations are an expression of the continuity of displacement.

(*a*) P and Q (*b*) Q and R

(*c*) P and R (*d*) P, Q and R

[GATE 2017, 2 M]

137. A batch of aluminium allow yields in uniaxial tension at the stress of 330 MN/m². If this material is subjected to the following state of stress: σ_x = 140 MN/m², σ_y = = −70 MN/m². σ_z = 0, τ_{xy} = x MN/m², σ_{yz} = 0 and τ_{zx} = 0. The value of x that would result in yielding according to the Von Mises failure criterion is _______ (in three decimal places).

[GATE 2017, 2 M]

138. An aircraft wing is idealized as a cantilever beam of constant width and length l with a tip mass of weight W (Newtons) and has a uniformly distributed loading of q_o (Newtons/m) as shown in the figure. Flexural rigidity = EI and $q_o l$ = 10 W.

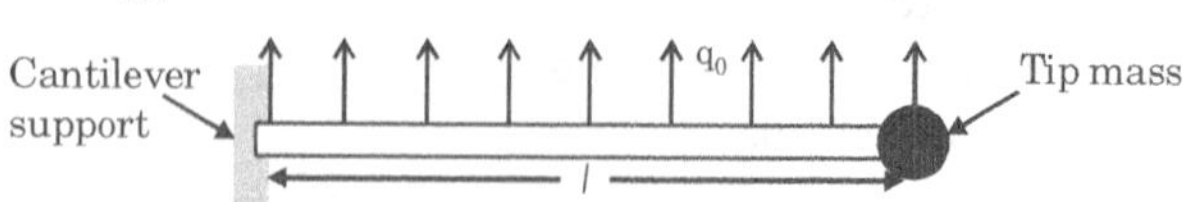

The upward deflection of the tip of the aircraft wing under the given loading can be expressed as:

$$\delta = k\frac{Wl^3}{EI}$$

The value of k is ________ (in three decimal places)

[GATE 2017, 2 M]

139. For the state of plane stress shown in the figure, the minimum principal stress is −7MN/m². The normal stress σ_x in MN/m² is equal to _______ (round to nearest integer).

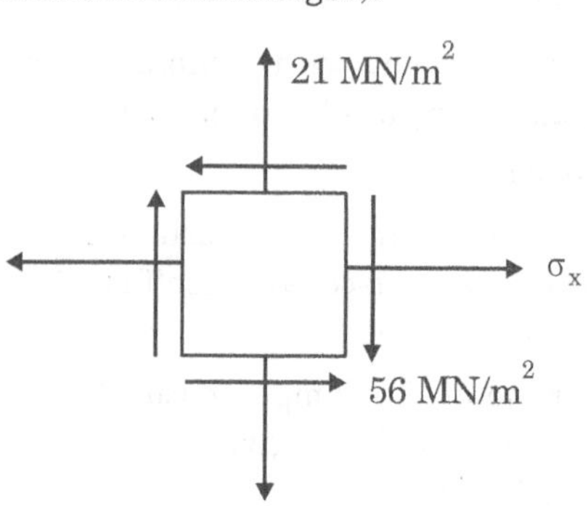

[GATE 2017, 2 M]

140. The maximum normal stress in MN/m² for the thin walled beam of square cross section of outer dimension 120 mm x 120 mm and wall thickness 1 mm under the action of moment M_o = 96 Nm as shown in the figure is _______ (in three decimal places)

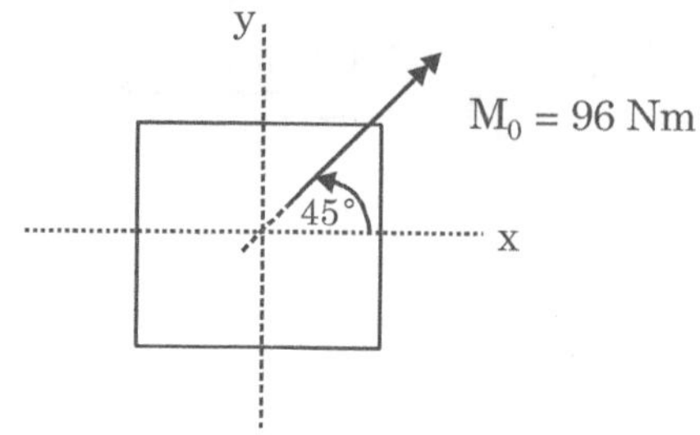

[GATE 2017, 2 M]

141. The idealized cross section of a thin-walled box structure shown in the figure is subjected to an anticlockwise torque of 10 kNm. The co-rresponding shear-flow distribution under this loading condition is shown in the figure. The area of each cell i s A_1 = 300 × 10³ mm² and A2 = 250 × 10³ mm².

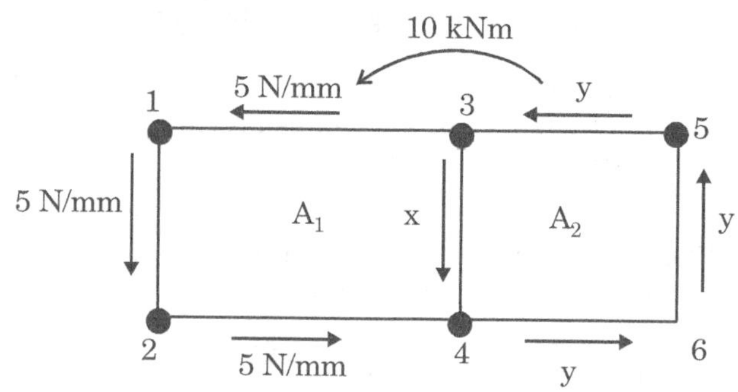

[GATE 2017, 2 M]

142. The natural frequency of the system suspended by two decimal springs of stiffness k as shown in the figure is given by $\omega_n = a\sqrt{\frac{k}{m}}$ for small displacement. Both the springs make an angle of 45° with the horizontal. The value of a is ________ (in two decimal places).

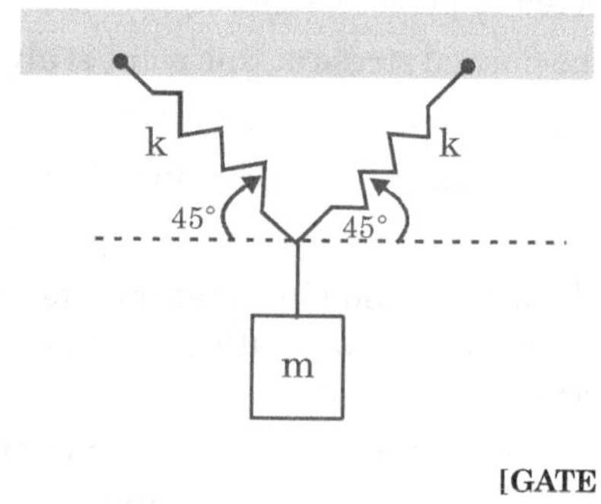

[GATE 2017, 2 M]

143. Determine the correctness or otherwise of the following statements, [a] and [r]:

[a] A closed-section box beam configuration is used in aircraft wings.

[r] Closed-section box beam configuration is capable of resisting torsional loads.

(*a*) Both [a] and [r] are true and [r] is the correct reason for [a].

(*b*) Both [a] and [r] are true but [r] is not the correct reason for [a].

(*c*) Both [a] and [r] are false.

(*d*) [a] is true but [r] is false.

[GATE 2018, 1 M]

144. For a damped single degree of freedom system with damping ratio of 0.1, ratio of twosuccessive peak amplitudes of free vibration is ________ (accurate to two decimal places).

[GATE 2018, 2 M]

145. The natural frequency (in rad/s) of the spring-mass system shown in the figure below is _____ (accurate to one decimal place).

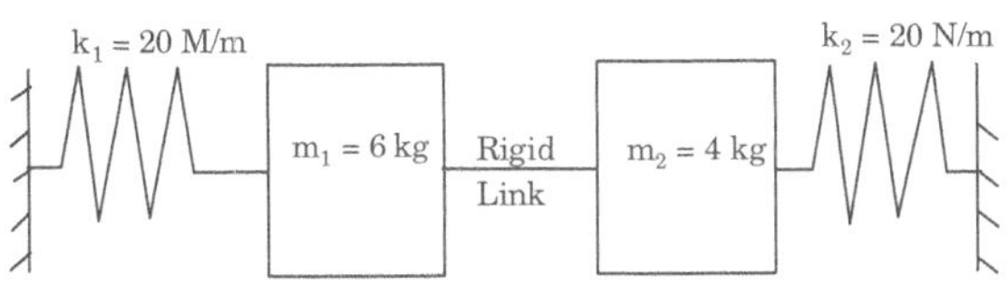

146. A solid circular shaft of diameter d is under pure torsion of magnitude T. The maximum tensile stress experienced at any point on the shaft is

(*a*) $\frac{32T}{\pi d^3}$ (*b*) $\frac{16T}{\pi d^4}$

(*c*) $\frac{32T}{\pi d^4}$ (*d*) $\frac{16T}{\pi d^3}$

[GATE 2018, 2 M]

147. A clamped-clamped beam, subjected to a point load P at the midspan, is shown in the figure below. The magnitude of the moment reaction at the two fixed ends of the beam is

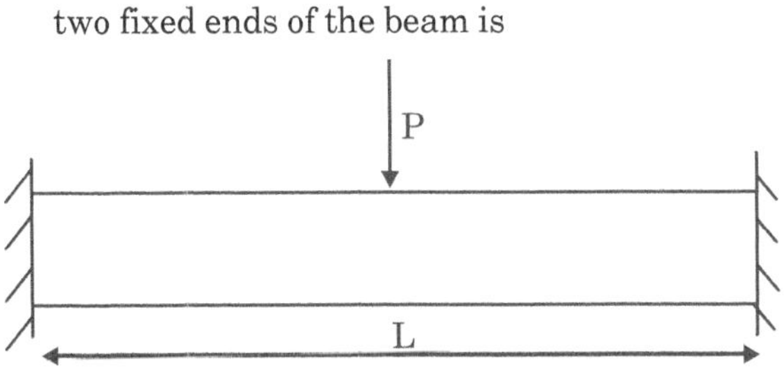

(*a*) $\frac{PL}{2}$ (*b*) $\frac{PL}{4}$ (*c*) $\frac{PL}{8}$ (*d*) $\frac{PL}{16}$

[GATE 2018, 2 M]

148. Which of the following statement(s) is/are true about the state of a body in plane strain condition?

P: All the points in the body undergo displacements in one plane only, for example the x-y plane, leading to $\varepsilon_{zz} = \gamma_{xz} = \gamma_{yz} = 0$.

Q: All the components of stress perpendicular to the plane of deformation, for example the x-y plane, of the body are equal to zero, i.e. $\sigma_{zz} = \tau_{xz} = \tau_{yz} = 0$.

R: Except the normal component, all the other components of stress perpendicular to the plane of deformation of the body, for example the x-y plane, are equal to zero, i.e.

$\sigma_{zz} \neq 0,\ \tau_{xz} = \tau_{yz} = 0.$

(*a*) P only (*b*) Q only

(*c*) P and Q (*d*) P and R

[GATE 2018, 2 M]

149. A thin-walled tube with external radius of 100 mm and wall thickness of 2 mm, is fixed at one end. It is subjected to a compressive force of 1 N acting at a point on the circumference parallel to its length. The maximum normal stress (in kPa) experienced by the structure is ____________ (accurate to two decimal places).

[GATE 2018, 2 M]

150. A 1 m long massless cantilever beam oscillates at 2Hz, while a 60 kg mass is attached at the tip of it. The flexural rigidity of the beam (in kN-m^2) is ________________ (accurate to two decimal places).

[GATE 2018, 2 M]

151. A cantilever beam having a rectangular cross-section of width 60 mm and depth 100 mm, is made of aluminum alloy. The material mechanical properties are: Young's modulus, E = 73 GPa and ultimate stress, σ_u = 480 MPa. Assuming a factor of safety of 4, the maximum bending moment (in kN-m) that can be applied on the beam is _____ (accurate to one decimal place).

[GATE 2018, 2 M]

152. The components of stress in a body under plane stress condition, in the absence of body forces, is given by:

$\sigma_{xx} = Ax^2; \sigma_{yy} = 12x^2 - 6y^2$ and $\sigma_{xy} = 12xy.$

The coefficient, A, such that the body is under equilibrium is _____ (accurate to one decimal place).

[GATE 2018, 2 M]

153. For a beam subjected to a transverse shear load through its shear center,

(a) the twist per unit length is zero.

(b) the shear stress is uniform throughout the cross-section.

(c) the bending stresses in the cross section are zero.

(d) the shear strain is zero at the shear center.

[GATE 2019]

154. The number of independent elastic constants for a homogeneous isotropic linear elastic material is ________.

[GATE 2019]

155. A thin plate with Young's modulus 210 GPa and Poisson's ratio 0.3 is loaded as shown in the figure. The change in length along the y-direction is __________ mm (round off to 1 decimal place).

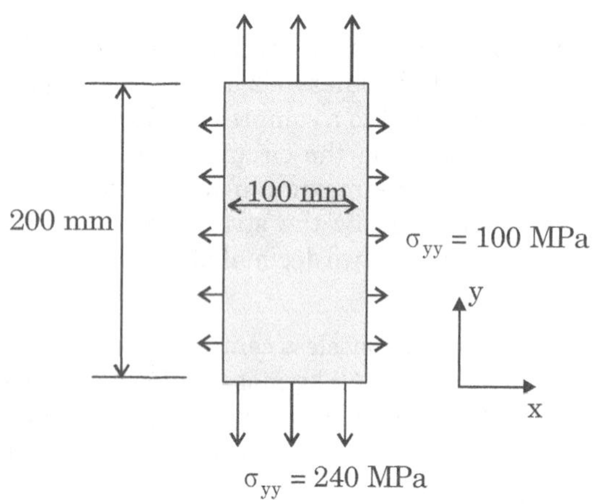

[GATE 2019]

156. For the state of stress shown in the figure, the normal stress, sn on a plane inclined at 45 degrees to the x-axis is _______MPa (round off to the nearest integer).

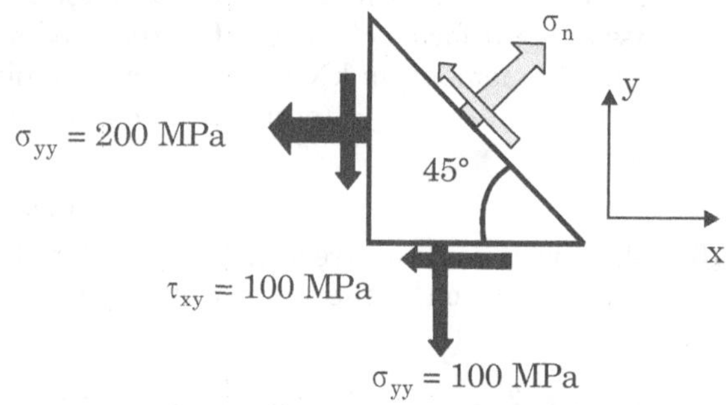

[GATE 2019]

157. In the spring-mass system, shown in the figure, mass m = 3 kg and the spring stiffness k = 20 kN/m. The natural frequency of the system is _______ Hz (round off to the nearest integer).

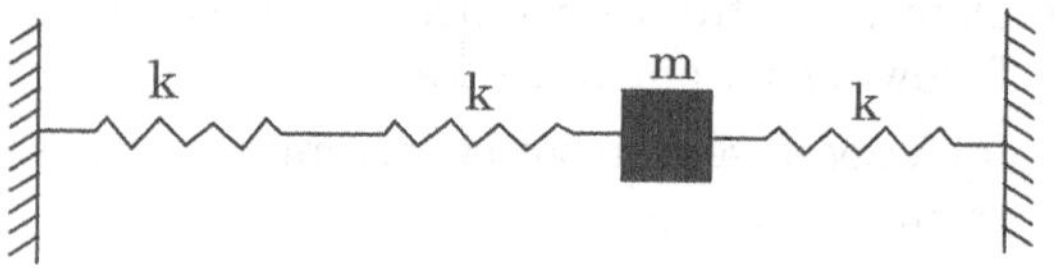

[GATE 2019]

158. For a 1 m long simply supported beam with a concentrated vertical load of 200 N and a concentrated bending moment of 100 Nm at the center as shown in the figure, the correct bending moment diagram is:

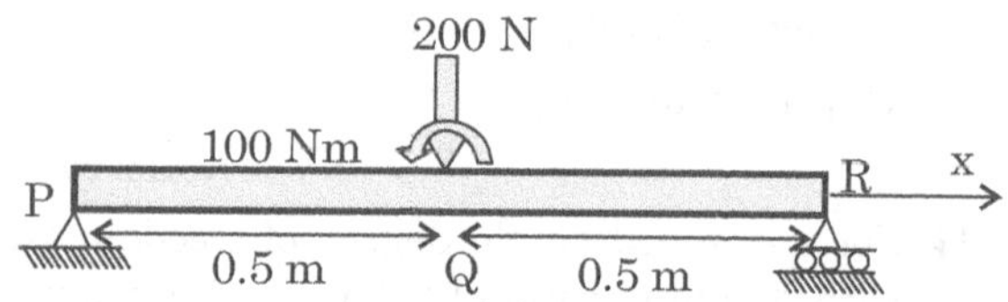

(a)

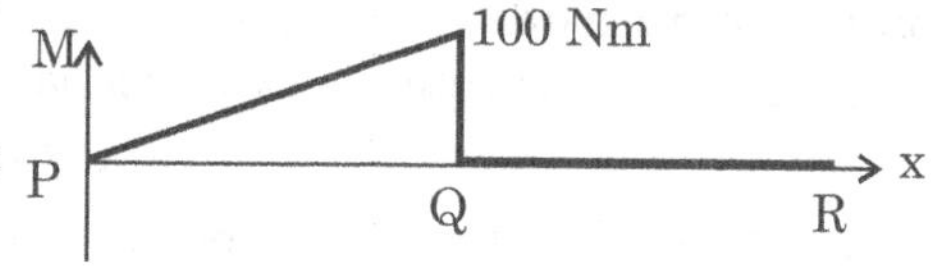

(b)

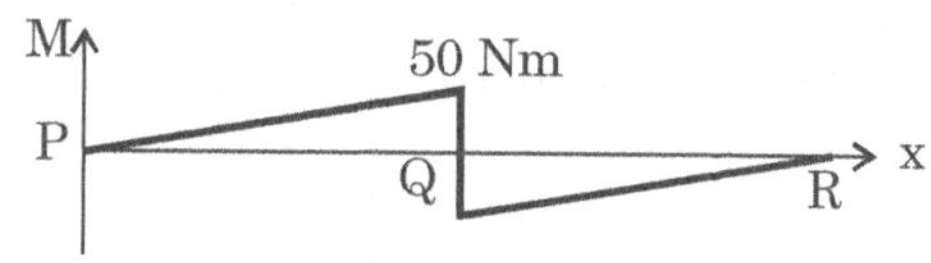

(c)

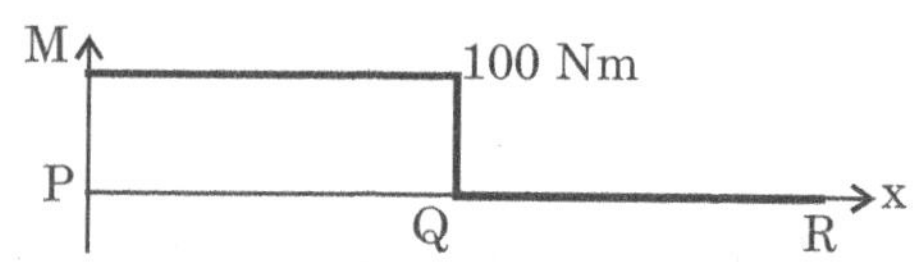

(d)

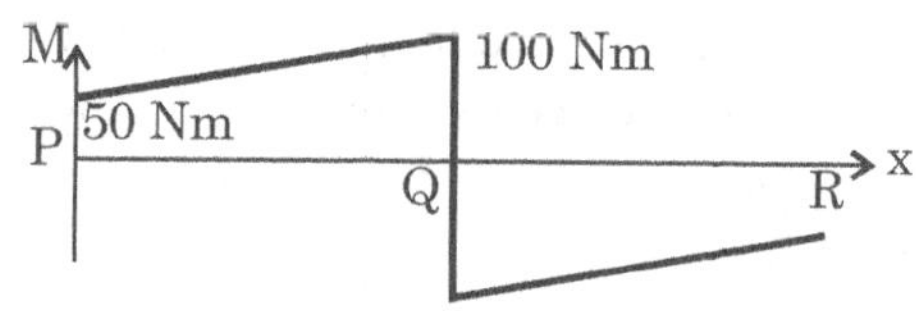

[GATE 2019]

159. As shown in the figure, a rigid slab CD of weight W (distributed uniformly along its length) is hung from a ceiling using three cables of identical length and cross-sectional area. The central cable is made of steel (Young's modulus = 3E) and the other two cables are made of aluminium (Young's modulus = E). The percentage of the total weight taken by the central cable is ________ % (round off to the nearest integer).

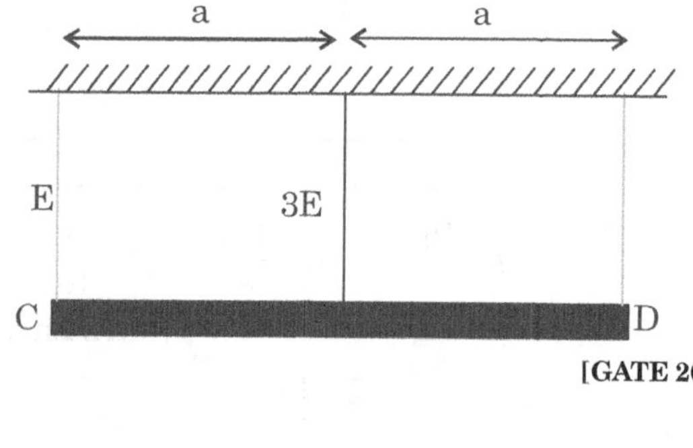

[GATE 2019]

160. All the bars in the given truss are elastic with Young's modulus 200 GPa, and have identical cross-sections with moment of inertia 0.1 cm^4. The lowest value of the load P at which the truss fails due to buckling is ______ kN (round off to the nearest integer).

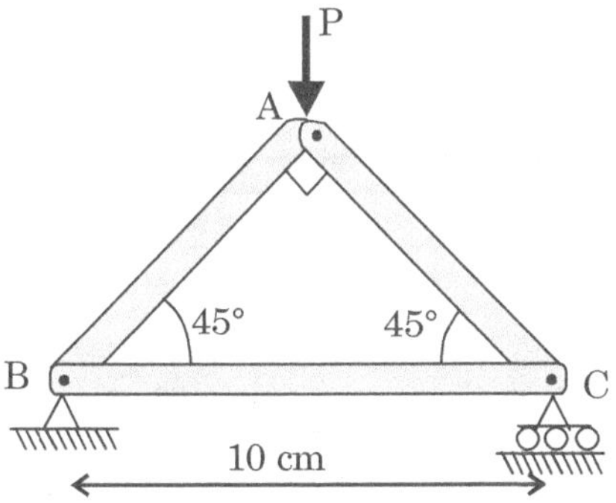

[GATE 2019]

161. A solid circular shaft is designed to transmit a torque T with a factor of safety of 2. It is proposed to replace the solid shaft by a hollow shaft of the same material and identical outer radius. If the inner radius is half the outer radius, the factor of safety for the hollow shaft is ________ (round off to 1 decimal place).

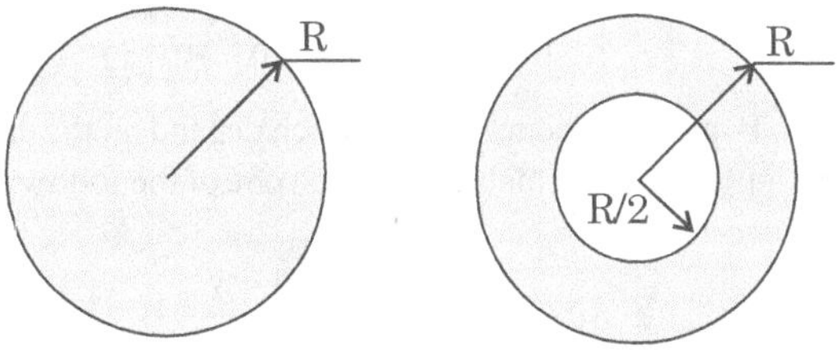

[GATE 2019]

162. In the structure shown in the figure, bars AB and BC are made of identical material and have circular cross-sections of 10 mm radii. The yield stress of the material under uniaxial tension is 280 MPa. Using the von Mises yield criterion, the maximum load along the z-direction (perpendicular to the plane of paper) that can be applied at C, such that AB does not yield is _______ N (round off to the nearest integer).

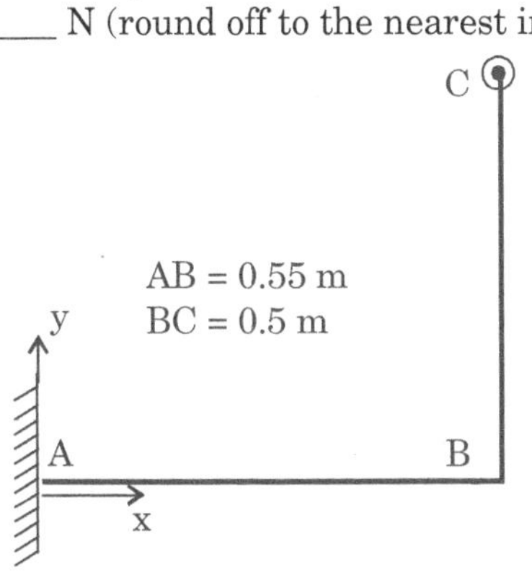

[GATE 2019]

163. A thin-walled tube, with the cross-section shown in the figure, is subjected to a torque of T = 1 kN-m. The walls have uniform thickness t = 1 mm and shear modulus G = 26 GPa.

Assume that the curved portion is semi-circular. The shear stress in the wall is _____ MPa (round off to 1 decimal place).

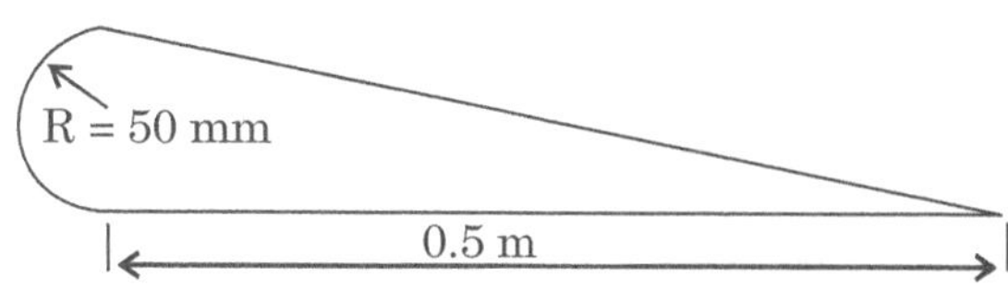

[GATE 2019]

164. For a damped spring-mass system, mass m = 10 kg, stiffness k = 10^3 N/m, and damping coefficient c = 20 kg/s. The ratio of the amplitude of oscillation of the first cycle to that of the fifth cycle is _______ (round off to 1 decimal place).

[GATE 2019]

165. For the system of springs and masses shown below, k = 1250 N/m and m = 10 kg. The highest natural frequency, ω of the system is ____ radian/s (round off to the nearest integer)

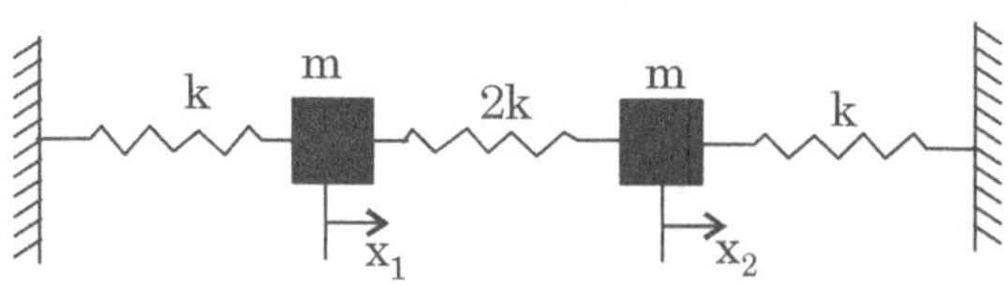

[GATE 2019]

166. Which one of the following conditions needs to be satisfied for $\phi = Ax^4 + By^4 + Cxy^3$ to be considered as an Airy's stress function?

(*a*) A − B = 0 (*b*) A + B = 0
(*c*) A − C = 0 (*d*) A + C = 0

[GATE 2020]

167. Consider the plane strain field given by $\varepsilon_{xx} = Ay^2 + x$, $\varepsilon_{yy} = Ax^2 + y$, $\gamma_{xy} = Bxy + y$. The relation between A and B needed for this strain field to satisfy the compatibility condition is

(*a*) B = A (*b*) B = 2A
(*c*) B = 3A (*d*) B = 4A

[GATE 2020]

168. Three long and slender aluminum bars of identical length are subjected to an axial tensile force. These bars have circular, triangular and rectangular cross sections, with same cross sectional area. If they yield at F_{circle}, $F_{triangle}$ and $F_{rectangle}$, respectively, which one of the following is true?

(*a*) $F_{circle} > F_{triangle} > F_{rectangle}$
(*b*) $F_{circle} < F_{triangle} < F_{rectangle}$
(*c*) $F_{triangle} > F_{circle} > F_{rectangle}$
(*d*) $F_{circle} = F_{triangle} = F_{rectangle}$

[GATE 2020]

169. The positive high angle-of-attack condition is obtained in a steady pull-out maneuver at the largest permissible angle-of-attack of the wing. Under this condition, at which of the following regions of the wing does the maximum tension occur?

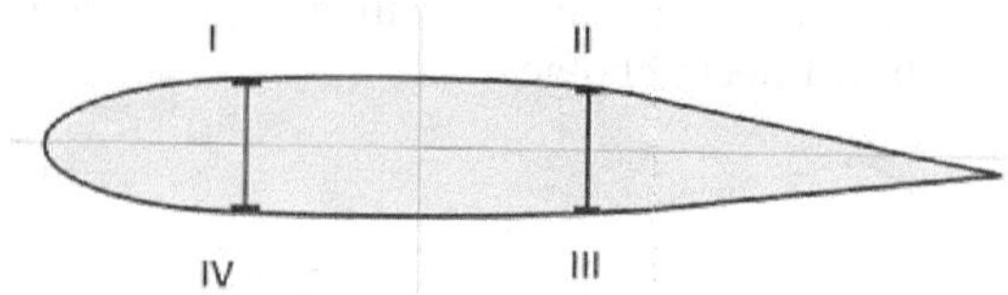

(*a*) I (*b*) II

(*c*) III (*d*) IV

[GATE 2020]

170. The natural frequency of the first mode of a rectangular cross section cantilever aluminum beam is ω rad/s. If the material and cross-section remain the same, but the length of the beam is doubled, the first mode frequency will become

(*a*) $\frac{\omega}{4}$ rad/s (*b*) 4 ω rad/s

(*c*) $\frac{\omega}{16}$ rad/s (*d*) 16 ω rad/s

[GATE 2020]

171. A thin walled beam of constant thickness shown in the figure is subjected to a torque of 3.2 kNm. If the shear modulus is 25 GPa, the angle of twist per unit length is _____ rad/m (*round off to three decimal places*).

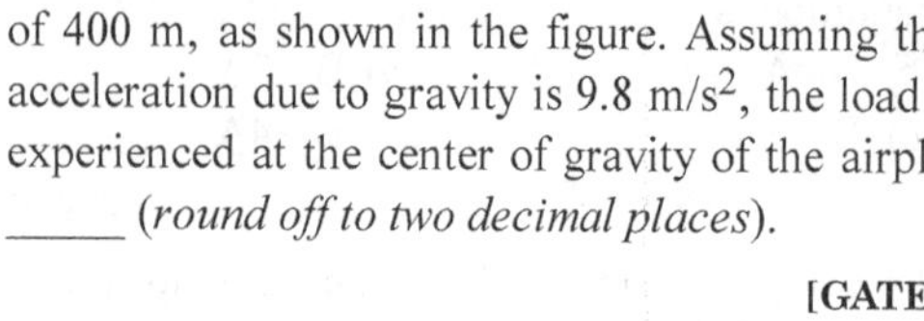

[GATE 2020]

172. An airplane of mass 5000 kg is flying at a constant speed of 360 km/h at the bottom of a vertical circle with a radius of 400 m, as shown in the figure. Assuming that the acceleration due to gravity is 9.8 m/s^2, the load factor experienced at the center of gravity of the airplane is _____ (*round off to two decimal places*).

[GATE 2020]

173. The value of Poisson's ratio at which the shear modulus of an isotropic material is equal to the bulk modulus is

(*a*) $\frac{1}{2}$ (*b*) $\frac{1}{4}$

(*c*) $\frac{1}{6}$ (*d*) $\frac{1}{8}$

[GATE 2020]

174. A load P is applied to the free end of a stepped cantilever beam as shown in the figure. The Young's modulus of the material is E, and the moments of inertia of the two sections of length 2 m and 1 m are I and 3I, respectively. Ignoring transverse shear and stress concentration effects, the deflection at the point where the load is applied at the free end of the cantilever is

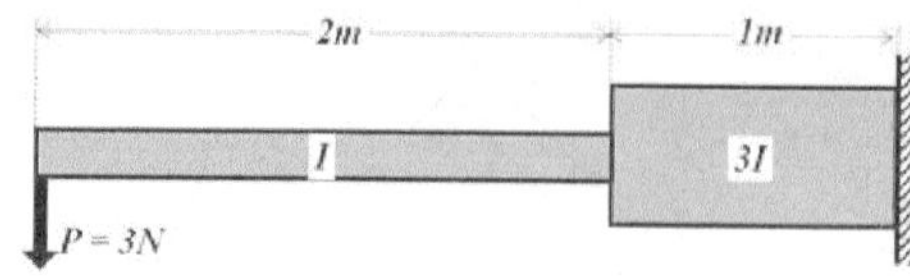

(*a*) $\frac{23}{243\text{EI}}$ (*b*) $\frac{1}{3\text{EI}}$

(*c*) $\frac{43}{3\text{EI}}$ (*d*) $\frac{23}{3\text{EI}}$

[GATE 2020]

175. The three dimensional strain-stress relation for an isotropic material, written in a general matrix form, is

$$\begin{Bmatrix} \varepsilon_{xx} \\ \varepsilon_{yy} \\ \varepsilon_{zz} \\ \gamma_{yz} \\ \gamma_{xz} \\ \gamma_{xy} \end{Bmatrix} = \begin{bmatrix} A & C & C & 0 & 0 & 0 \\ C & A & C & 0 & 0 & 0 \\ C & C & A & 0 & 0 & 0 \\ 0 & 0 & 0 & B & 0 & 0 \\ 0 & 0 & 0 & 0 & B & 0 \\ 0 & 0 & 0 & 0 & 0 & B \end{bmatrix} \begin{Bmatrix} \sigma_{xx} \\ \sigma_{yy} \\ \sigma_{zz} \\ \tau_{yz} \\ \tau_{xz} \\ \tau_{xy} \end{Bmatrix}$$

A, B and C are compliances which depend on the elastic properties of the material. Which one of the following is correct?

(*a*) $C = \frac{A}{2} - B$ (*b*) $C = \frac{A}{2} + B$

(*c*) $C = A + \frac{B}{2}$ (*d*) $C = A - \frac{B}{2}$

[GATE 2020]

176. For the state of stress shown in the figure, which one of the following represents the correct free body diagram showing the maximum shear stress and the associated normal stresses?

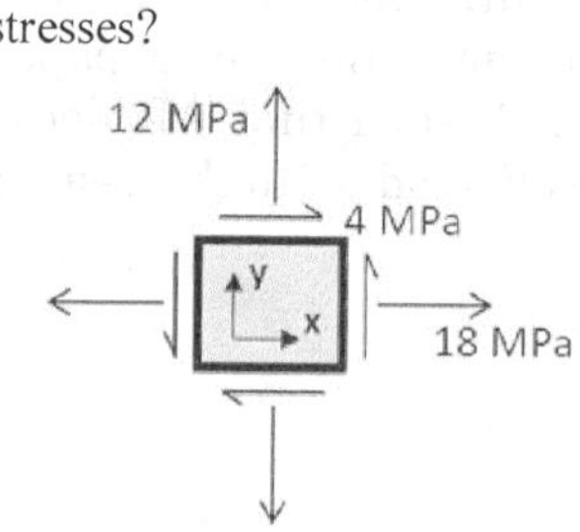

(*a*)

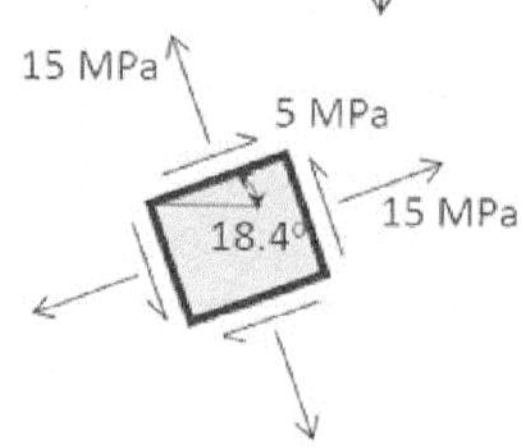

(b)

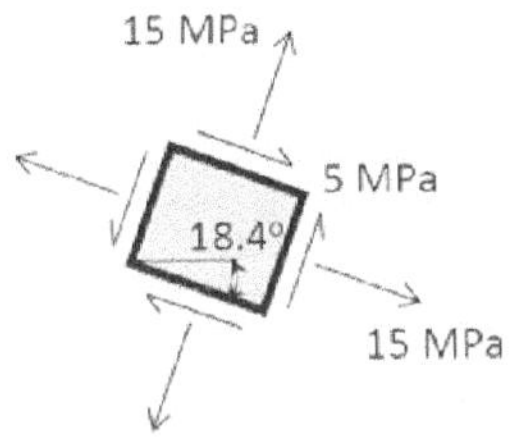

(c)

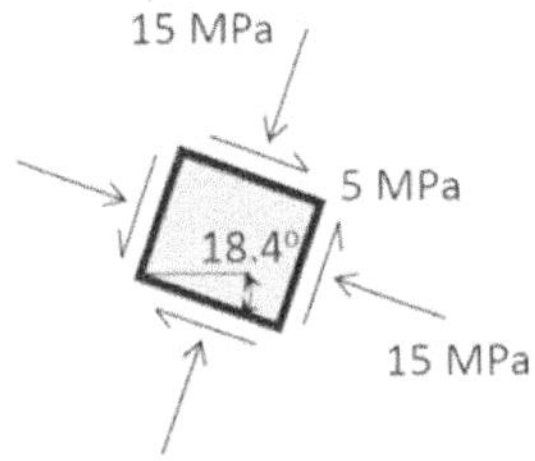

(d)

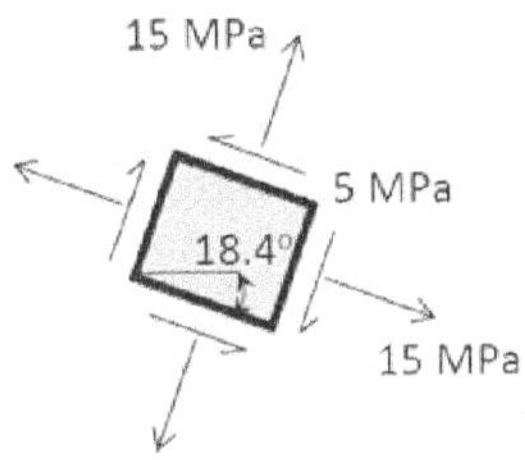

[GATE 2020]

177. A solid circular shaft, made of ductile material with yield stress $\sigma_y = 280$ MPa, is subjected to a torque of 10 kNm. Using the Tresca failure theory, the smallest radius of the shaft to avoid failure is ______ cm (*round off to two decimal places*).

[GATE 2020]

178. A critically damped single degree of freedom spring-mass-damper system used in a door closing mechanism becomes overdamped due to softening of the spring with extended use. If the new damping ratio (ξ_{new}) for overdamped condition is 1.2, the ratio of the original spring stiffness to the new spring stiffness (k_{org}/k_{new}), assuming that the other parameters remain unchanged, is _______ (*round off to two decimal places*).

[GATE 2020]

179. The two masses of the two degree of freedom system shown in the figure are given initial displacements of 2 cm (x_1) and 1.24 cm (x_2). The system starts to vibrate in the first mode. The first mode shape of this system is $\phi_1 = [1 \;\; a]^T$, where a = ____ (*round off to two decimal places*).

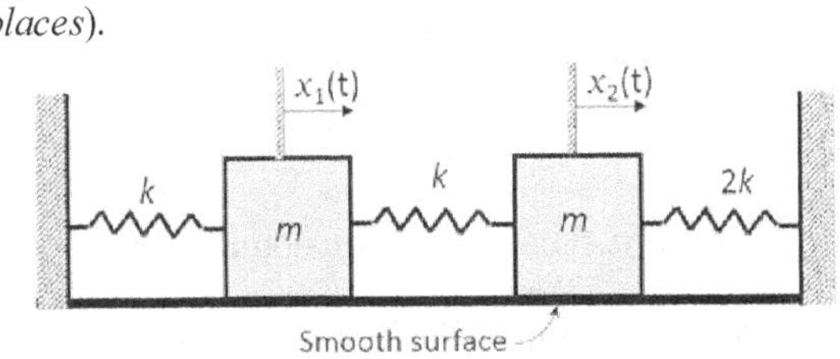

[GATE 2020]

180. As shown in the figure, a beam of length 1 m is rigidly supported at one end and simply supported at the other. Under the action of a uniformly distributed load of 10 N/m, the magnitude of the normal reaction force at the simply supported end is _______ N (*round off to two decimal places*).

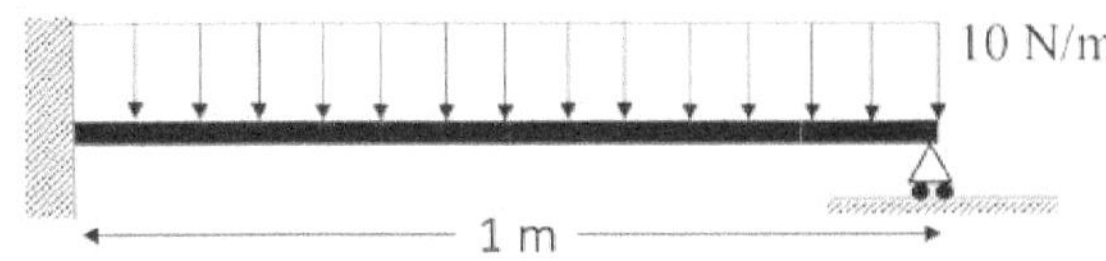

[GATE 2020]

181. Uniaxial tension test (see the figure) is conducted on two different samples prepared with homogeneous, Isotropic materials. One of the materials is brittle, whereas the other is ductile.

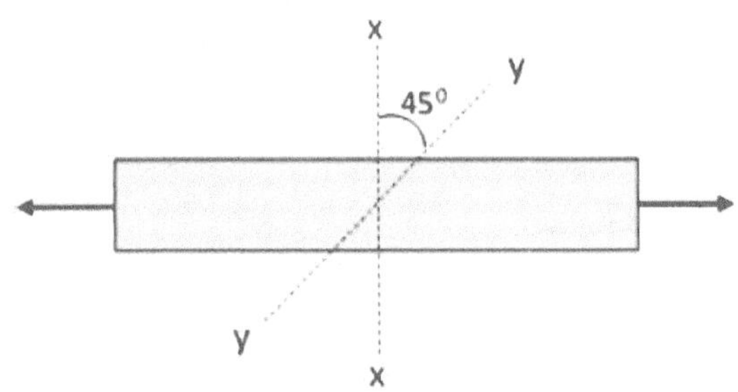

Assuming that there is no stress concentration at loading points, the failure would initiate:

(a) along x-x in both materials

(b) along x-x in brittle material and along y-y in ductile material

(c) along y-y in brittle material and along x-x in ductile material

(d) along y-y in both materials

[GATE 2021]

182. For the state of stress as shown in the figure, what is the orientation of the plane with maximum shear stress with respect to the x-axis?

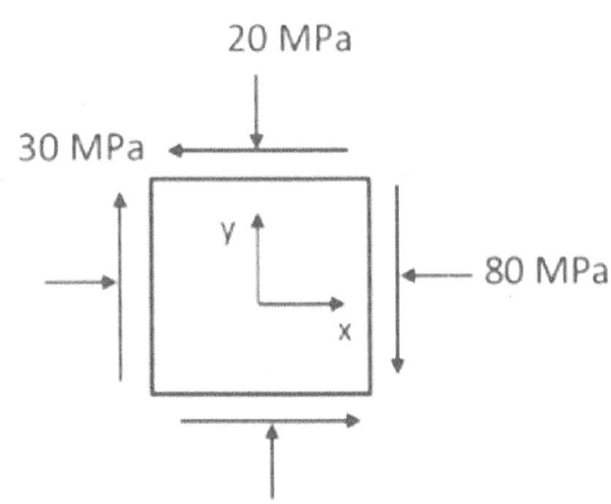

(a) 45° (b) –45°

(c) 22.5° (d) –22.5°

[GATE 2021]

183. A single degree of freedom spring-mass-damper system is designed to ensure that the system returns to its original undisturbed position in minimum possible time without overshooting. If the mass of the system is 10 kg, spring stiffness is 17400 N/m and the natural frequency is 13.2 rad/s, the coefficient of damping of the system in Ns/m is ________ *(round off to nearest integer)*.

[GATE 2021]

184. Two cantilever beams (Beam 1 and Beam 2) are made of same homogenous material and have identical cross sections. Beam 1 has length l and Beam 2 has length $2l$. Ratio of the first natural frequency of Beam 1 to that of Beam 2 is ______*(round off to nearest integer)*.

[GATE 2021]

185. A thin-walled cylindrical tank with closed ends, made of homogeneous and isotropic material, is pressurized internally. If the hoop (circumferential) strain developed in the material is thrice the value of the axial strain then the Poisson's ratio of the material is ______ *(correct up to one decimal place)*.

[GATE 2021]

186. A rigid massless rod pinned at one end has a mass m attached to its other end. The rod is supported by a linear spring of stiffness k as shown in the figure.

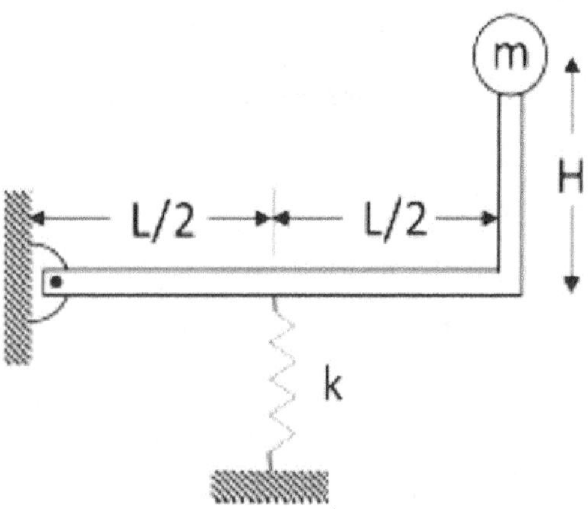

The natural frequency of this system is:

(a) $\frac{1}{2\pi}\sqrt{\frac{kL^2}{4m(L^2+H^2)}}$

(b) $\frac{1}{2\pi}\sqrt{\frac{kL^2}{m(L^2+H^2)}}$

(c) $\frac{1}{2\pi}\sqrt{\frac{4kL^2}{m(L^2+H^2)}}$

(d) $\frac{1}{2\pi}\sqrt{\frac{k(L^2+H^2)}{4mL^2}}$

[GATE 2021]

187. A two degree of freedom spring-mass system undergoing free vibration with generalized coordinates x_1 and x_2 has natural frequencies $\omega_1 = 233.9$ rad/s and $\omega_2 = 324.5$ rad/s, respectively. The corresponding mode shapes are $\phi_1 = \begin{bmatrix} 1 \\ -3.16 \end{bmatrix}$ and $\phi_2 = \begin{bmatrix} 1 \\ 3.16 \end{bmatrix}$. If the system is disturbed with certain deflections and zero initial velocities, then which of the following statement(s) is/are true?

(a) An initial deflection of $x_1(0) = 6.32$ cm and $x_2(0) = -3.16$ cm would make the system oscillate with only the second natural frequency.

(b) An initial deflection of $x_1(0) = 2$ cm and $x_2(0) = -6.32$ cm would make the system oscillate with only the first natural frequency.

(c) An initial deflection of $x_1(0) = 2$ cm and $x_2(0) = -2$ cm would make the system oscillate with a linear combination of first and second natural frequencies.

(d) An initial deflection of $x_1(0) = 1$ cm and $x_2(0) = -6.32$ cm would make the system oscillate with only the first natural frequency.

[GATE 2021]

188. The deflection y of a certain beam of length l and uniform weight per unit length w, is given as $y = \frac{\omega}{48EI}\left(2x^4 - 3lx^3 + l^3x\right)$, where x is the distance from the point of support and EI is a constant. The non-dimensional location $\frac{x}{l}$, where the deflection of the beam is maximum, is ______ *(round off to two decimal places)*.

[GATE 2021]

189. A rigid horizontal bar ABC, with roller support at A, is pinned to the columns BD and CE at points B and C, respectively as shown in figure. The other end of the column BD is fixed at D, whereas the column CE is pinned at E. A vertical load P is applied on the bar at a distance 'a' from point B. The two columns are made of steel with elastic modulus 200 GPa and have a cross section of 1.5 cm × 1.5 cm. The value of 'a' for which both columns buckle simultaneously, is ______ cm (*round off to one decimal place*).

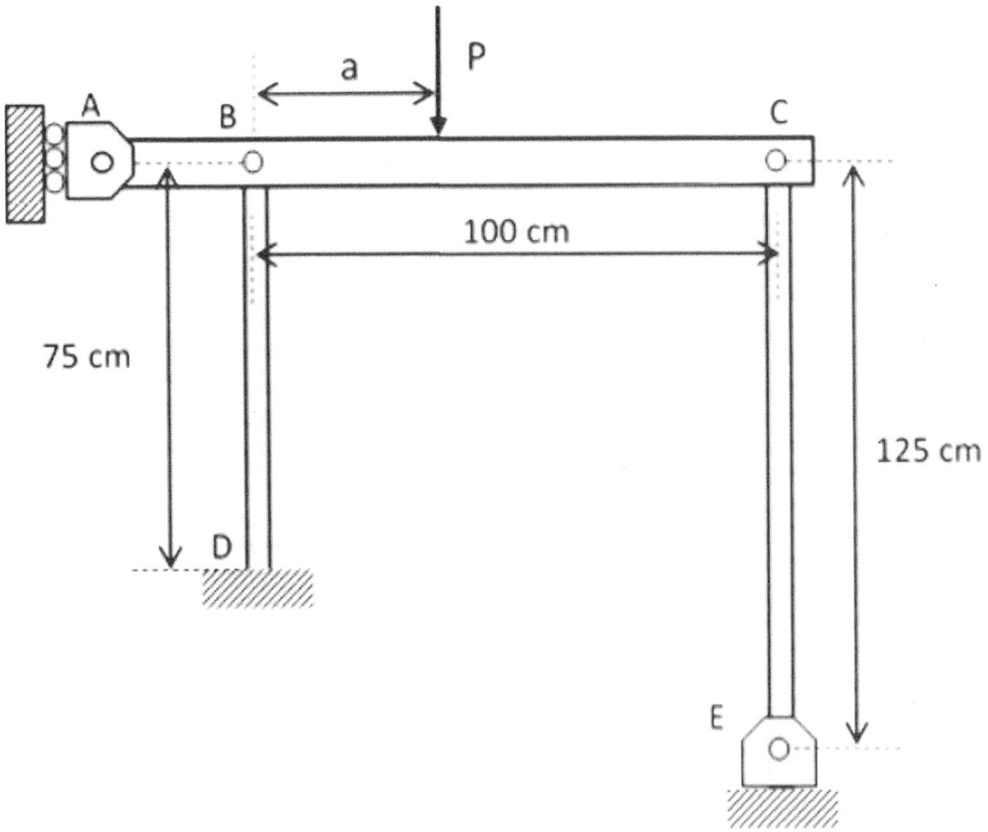

190. A two-cell wing box is shown in the figure. The cell walls are 1.5 mm thick and the shear modulus G = 27 GPa. If the structure is subjected to a torque of 12 kNm, then the wall AD will experience a shear stress of magnitude ______ MPa (*round off to one decimal place*).

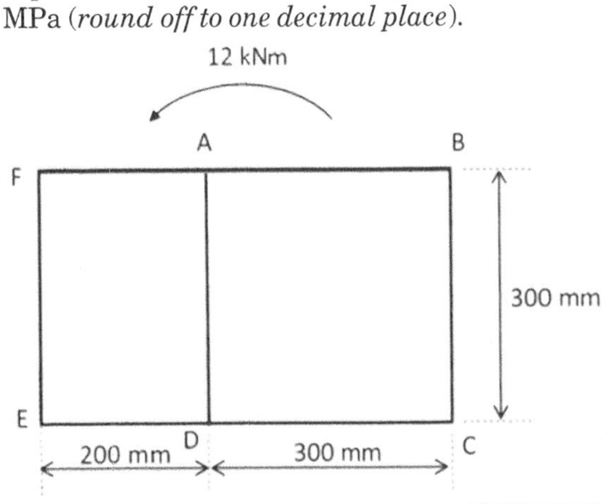

[GATE 2021]

191. Two cantilever beams AB and DC are in contact with each other at their free ends through a roller as shown in the figure. Both beams have a square cross section of 50 mm × 50 mm, and the elastic modulus E = 70 GPa. If beam AB is subjected to a uniformly distributed load of 20 kN/m, then the compressive force experienced by the roller is ______ kN (*round off to one decimal place*).

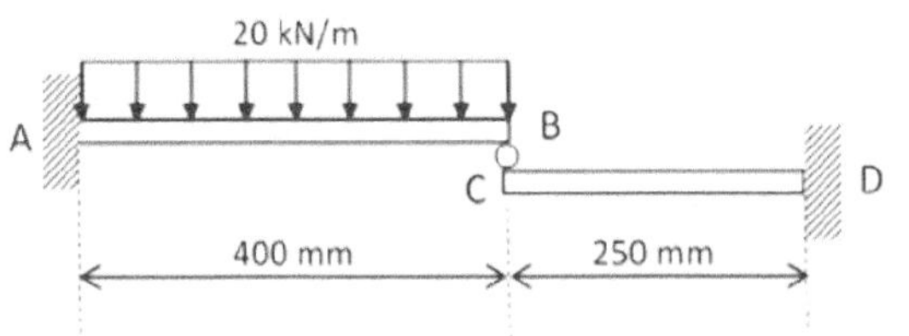

[GATE 2021]

192. A 3 m × 1 m signboard is supported by a vertical hollow pole that is fixed to the ground. The pole has a square cross section with outer dimension 250 mm. The yield strength of the pole material is 240 MPa. To sustain a wind pressure of 7.5 kPa, the dimension d of the pole is ______ mm (*round off to nearest integer*).

(Neglect the effect of transverse shear and load due to wind pressure acting on the pole)

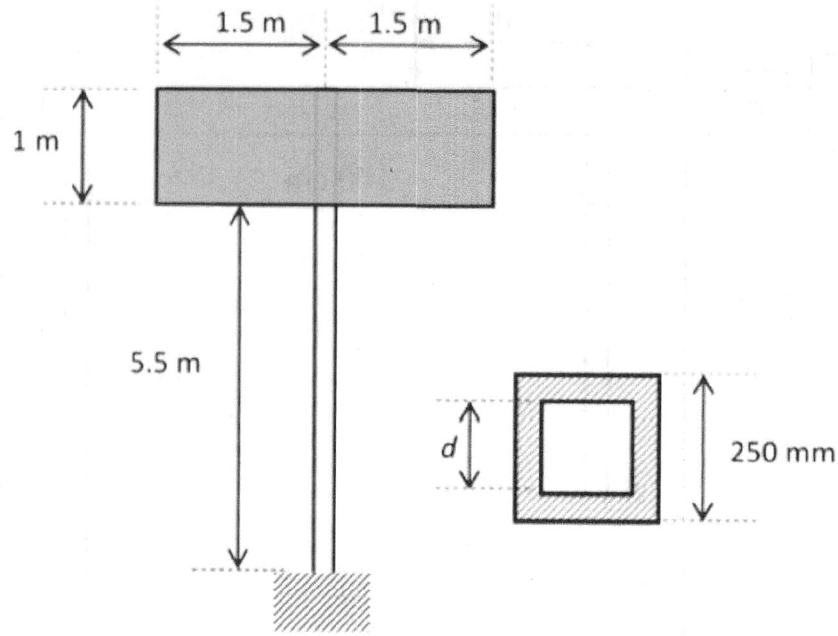

[GATE 2021]

ANSWER KEY

1. (*a*)	**2.** (*d*)	**3.** (*d*)	**4.** (*c*)	**5.** (*b*)	**6.** (*a*)	**7.** (*d*)	**8.** (*b*)	**9.** (*a*)	**10.** (*c*)
11. (*a*)	**12.** (*b*)	**13.** (*b*)	**14.** (*a*)	**15.** (*a*)	**16.** (*c*)	**17.** (*a*)	**18.** (*d*)	**19.** (*b*)	**20.** (*c*)
21. (*c*)	**22.** (*a*)	**23.** (*d*)	**24.** (*a*)	**25.** (*b*)	**26.** (*c*)	**27.** (*c*)	**28.** (*c*)	**29.** (*c*)	**30.** (*b*)
31. (*a*)	**32.** (*c*)	**33.** (*b*)	**34.** (*a*)	**35.** (*b*)	**36.** (*b*)	**37.** (*b*)	**38.** (*a*)	**39.** (*d*)	**40.** (*b*)
41. (*a*)	**42.** (*a*)	**43.** (*b*)	**44.** (*a*)	**45.** (*c*)	**46.** (*b*)	**47.** (*a*)	**48.** (*a*)	**49.** (*b*)	**50.** (*d*)
51. (*b*)	**52.** (*c*)	**53.** (*a*)	**54.** (*d*)	**55.** (*c*)	**56.** (*b*)	**57.** (*a*)	**58.** (*c*)	**59.** (*d*)	**60.** (*c*)
61. (*c*)	**62.** (*c*)	**63.** (*a*)	**64.** (*d*)	**65.** (*a*)	**69.** (*b*)	**70.** (*b*)	**71.** (*d*)	**72.** (*c*)	**73.** (*b*)
74. (*d*)	**75.** (*b*)	**76.** (*c*)	**77.** (*d*)	**78.** (*d*)	**80.** (*a*)	**82.** (*a*)	**83.** (*a*)	**84.** (*d*)	**85.** (*c*)
86. (*a*)	**87.** (*b*)	**89.** (*a*)	**90.** (*a*)	**91.** (*b*)	**92.** (*b*)	**93.** (*a*)	**94.** (*b*)	**95.** (*b*)	**97.** (*a*)
98. (*b*)	**99.** (*a*)	**101.** (*c*)	**103.** (*a*)	**104.** (*b*)	**105.** (*d*)	**106.** (*d*)	**107.** (*b*)	**108.** (*d*)	**109.** (*a*)
112. (*a*)	**120.** (*a*)	**121.** (*b*)	**123.** (*b*)	**126.** (*b*)	**130.** (*c*)	**131.** (*b*)	**132.** (*b*)	**134.** (*c*)	**136.** (*c*)

143. (*a*) **146.** (*d*) **147.** (*c*) **148.** (*c*) **153.** (a) **154.** (2) **155.** (0.2) **156.** (250 MPa.) **157.** (16)

158. (a) **159.** (–60%) **160.** (395 kN) **161.** (1.9) **162.** (314.159 N) **163.** (0.028925 m²) **164.** (12.52)

165. (25 md/mm) **166.** (*b*) **167.** (d) **168.** (d) **169.** (c) **170.** (a) **171.** (0.009 to 0.011) **172.** (3.50 to 3.60)

173. (*d*) **174.** (*c*) **175.** (*d*) **176.** (b) **177.** (3.55 to 3.58) **178.** (1.43 to 1.45) **179.** (0.61 to 0.63)

180. (3.74 to 3.76) **181.** (b) **182.** (d) **183.** (264 to 264) **184.** (4 to 4) **185.** (0.2 to 0.2) **186.** (a)

187. (b; c) **188.** (0.41 to 0.43) **189.** (14.0 to 16.0) **190.** (2.1 to 2.5) **191.** (2.2 to 2.6)

192. (232 to 238)

EXPLANATIONS

1. Given:

mass m = 1 kg

damping ratio $\xi = 0.2$

Natural frequency $\omega_n = 5$ rad/sec.

We know that damping ratio of system is given by

$$\xi = \frac{C}{C_c} = \frac{C}{2\sqrt{km}}$$

$$\Rightarrow C = \xi \times 2\sqrt{km}$$

For calculating stiffness k,

$$\omega_n = \sqrt{\frac{k}{m}}$$

$$k = \omega_n^{\ 2} \times m$$

$$= 5^2 \times 1 = 25 \text{ N/m}$$

$\therefore$ Damping coefficient, $c = 0.2 \times 2 \times \sqrt{25 \times 1}$

$= 2$ Ns/m

2. As elastic beam is a continuous system it always has ∞ degree of freedom hence infinite natural frequeny will exist. No matter what boundary condition is.

3. For plane strain problem, say in X-Y plane

$$\varepsilon_{zz} = 0,\ \varepsilon_{xz} = 0,\ \varepsilon_{yz} = 0$$

$$\varepsilon_{xx},\ \varepsilon_{yy},\ \varepsilon_{xy} \neq 0$$

thus,

$$\varepsilon_{xz} = \frac{\sigma_{xz}}{G} = 0$$

$$\Rightarrow \boxed{\sigma_{xz} = 0}$$

$$\varepsilon_{yz} = \frac{\sigma_{yz}}{G} = 0$$

$$\boxed{\sigma_{yz} = 0}$$

$$\varepsilon_{zz} = \frac{1}{E}\left[\sigma_{zz} - \upsilon\left(\sigma_{xx} + \sigma_{yy}\right)\right] = 0$$

$$\Rightarrow \boxed{\sigma_{zz} = \upsilon\left(\sigma_{xx} + \sigma_{yy}\right)} \neq 0$$

Similarly,

$$\sigma_{xx} \neq 0$$

$$\sigma_{yy} \neq 0$$

$$\sigma_{xy} \neq 0$$

Hence, Non zero stresses are σ_{xx}, σ_{yy}, σ_{xy} and σ_{zz}.

4. For an elastic anisotropic solid, no. of independent elastic constants are 21. This happens due to stress-strain symmetry.
5. Given $g = 9.8\ m/sec^2$

mass = 1 kg

(Static deflection) spring elongation $x_0 = 16$ mm = 0.016 m

Spring Pull = 10 mm = 0.01 m

Governing equation of motion is

$$m\ddot{x} + kx = 0 \qquad \dots (i)$$

Solving equation, we get

$$\left(D^2 + \frac{k}{m}\right)x = 0$$

$$D = \pm\sqrt{\frac{k}{m}}$$

$$\therefore x(t) = C_1 \cos\sqrt{\frac{k}{m}}\,t + C_2 \sin\sqrt{\frac{k}{m}}\,t$$

$$= C_1 \cos\omega_n t + C_2 \sin\omega_n t \qquad \dots (1)$$

Initial conditions on x at t = 0 is 10 mm or 0.01 m. Mass was initially at rest hence $\dot{x} = 0$ at t = 0.

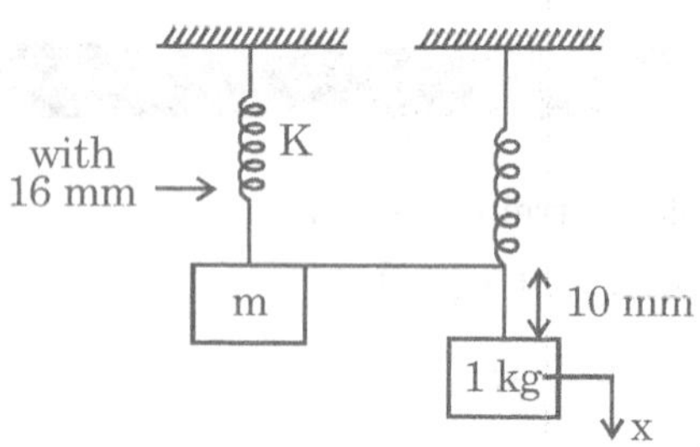

Static deflection $x_0 = 16$ mm

$$\therefore \text{Stiffness } k = \frac{mg}{x_0} = 613.13 \text{ N/m}$$

$$\text{Natural frequency, } \omega_n = \sqrt{\frac{k}{m}} = \sqrt{\frac{613.13}{1}}$$

= 24.76 rad/sec.

Apply

x = 0 at t = 0 on eq. (1)

$$0.01 = C_1 \cos(\omega_n \times 0) + C_2 \sin(\omega_n \times 0)$$

$$\boxed{C_1 = 0.01}$$

apply

$\dot{x} = 0$ at t = 0 on eq. (1)

$$\dot{x} = -C_1\omega_n \sin\omega_n t + C_2\omega_n \cos\omega_n t$$

$$0 = -C_1\omega_n \sin(\omega_n \times 0) + C_2\omega_n \cos\omega_n \times 0$$

$$\Rightarrow \boxed{C_2 = 0}$$

hence:

using (1)

$$x(t) = 0.01\cos(24.748 \cdot t)$$

or in mm units

$$\boxed{x(t) = 10\cos(24.76 \cdot t)}$$

6. For such beams with more material

Bending moment resisted is given by

$M = M_A + M_B$

Using mechanics of solids in bending

$$M = \frac{E_A I_A}{\rho} + \frac{E_B I_B}{\rho}$$

$$\text{So, } \boxed{\rho = \frac{E_A I_A + E_B I_B}{M}}$$

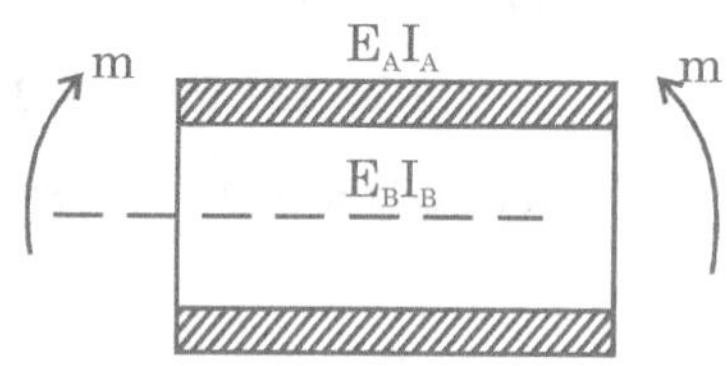

Radius of curvature conuctor is same for both material,

ρ = radius of curvature.

7. Given, spring mass damper system

Amplititude of resonance = 1 cm

[Resonance happens when applied frequency or forced frequency is equal to natural frequency of system]

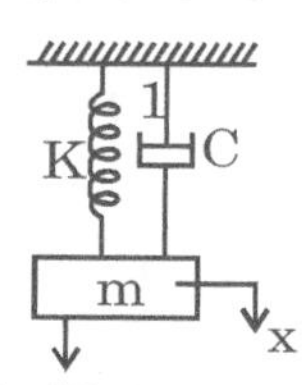

$F = F_0 \sin\omega t$

At half of resonant frequency, amplitude = 0.5 cm

Governing Differential equation of Motion of System (GDEOM)

$$m\ddot{x} + c\dot{x} + kx = F_0 \sin\omega t$$

$$\ddot{x} + 2\xi\omega_n\dot{x} + \omega_n^2 x = F_0 \sin\omega t$$

$$\text{where } \omega_n = \sqrt{\frac{k}{m}} \qquad \xi = \frac{C}{2\sqrt{km}}$$

After solving equation by assuming a response x of the form $x \sin\omega t$

We will get amplitude

$$x = \frac{\left|\frac{F_0}{k}\right|}{\sqrt{(1-\Omega^2)^2 + (2\xi\Omega)^2}} \quad \ldots (i)$$

where $\Omega = \frac{\omega}{\omega_n}$

at resonance $\omega = \omega_n$ hence $\Omega = 1$ and x = 1 cm (given) ... (A)

at half of resonant frequency $\omega = \frac{\omega_n}{2}$ hence

$\Omega = 0.5$, x = 0.5 cm (given) ... (B)

Using condition (A) & (B)

$$1 = \frac{\left|\frac{F_0}{k}\right|}{\sqrt{(1-\xi)^2 + (2\times 1\times \xi)^2}} = \frac{\left|\frac{F_0}{k}\right|}{\sqrt{(2\xi)^2}} \quad \ldots (ii)$$

and

$$0.5 = \frac{\left|\frac{F_0}{k}\right|}{\sqrt{\left(1-\left(\frac{1}{4}\right)\right)^2 + \left(2\times\xi\times\frac{1}{2}\right)^2}} = \frac{\left|\frac{F_0}{k}\right|}{\sqrt{\frac{9}{16}+\xi^2}} \quad \ldots (iii)$$

Dividing (ii) by (iii)

$$2 = \sqrt{\frac{\frac{9}{16}+\xi^2}{4\xi^2}} \Rightarrow 16\xi^2 = \frac{9}{16}+\xi^2$$

$$\Rightarrow \xi = \sqrt{\frac{9}{16\times 15}}$$

$$\Rightarrow \boxed{\xi_2 = 0.1936}$$

8. For such shafts, twist per unit length is same

i.e. $\frac{\phi}{L}$

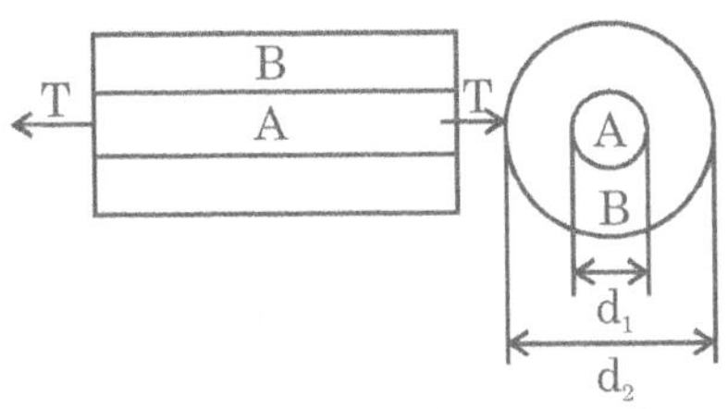

Total torque resisted is given by

$T = T_A + T_B$

Using solid mechanics approach for torsion,

$$T = \frac{G_A\phi J_A}{L} + \frac{G_B\phi J_B}{L}$$

$$\Rightarrow \boxed{\frac{T}{\phi} = \frac{G_A J_A}{L} + \frac{G_B J_B}{L}}$$

9. For Euler-Bernoulli column, general formula For critical Buckling load is given by

$$P_{cr} = \frac{\pi^2 EI}{\ell_{eq}^2}$$

ℓ_e = equivalent length

(i) For Pinned-Pinned column

$\ell_{eq} = \ell$

(ii) Fixed-Free

$\ell_{eq} = 2\ell$

(iii) Fixed Fixed

$\ell_{eq} = \frac{\ell}{2}$

(iv) Fixed pinned

$\ell_{eq} = \frac{\ell}{\sqrt{2}}$

Therefore,

Pinned-Pinned column

$$\boxed{P_{cr} = \frac{\pi^2 EI}{\ell^2}}$$

Fixed-Free

$$\boxed{P_{cr} = \frac{\pi^2 EI}{4\ell^2}}$$

Fixed-Fixed

$$\boxed{P_{cr} = \frac{4\pi^2 EI}{\ell^2}}$$

Fixed pinned

$$\boxed{P_{cr} = \frac{2\pi^2 EI}{\ell^2} \approx \frac{2.046\pi^2 EI}{\ell^2}}$$

10. Axial flow turbine is subjected to very high temperature ranges upto 2000 K. Nickle alloy is used in such high temperature application because of its high temperature resistance.

11. Von-Mises yield criterion for 3-D case, is given by

$$(\sigma_1-\sigma_2)^2 + (\sigma_2-\sigma_3)^2 + (\sigma_3-\sigma_1)^2 \le 2\sigma_y^2 \quad \ldots (A)$$

σ_1, σ_2 and σ_3 are principle stresses

σ_y is yield limit of stress-strain relationship

For plane stress $\sigma_3 = 0$

Equation reduces to

$$\boxed{\sigma_1^2 - \sigma_1\sigma_2 + \sigma_2^2 \le \sigma_y^2}$$

12. Non uniform bending moment along length given by

$$V_2 = \frac{dMy}{dx}$$

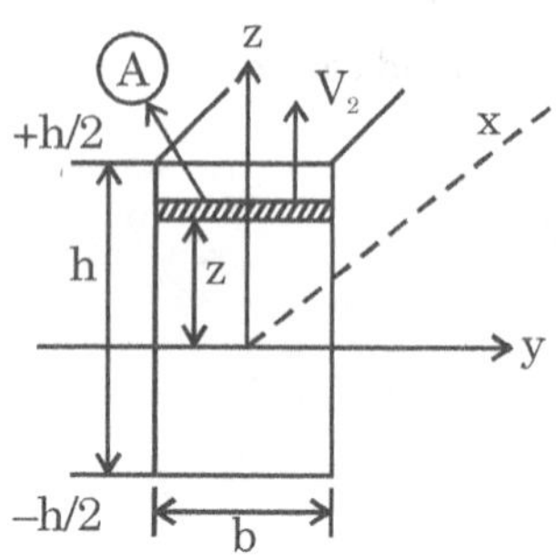

Using equilibrium equation

$$\frac{\partial \sigma_{xx}}{\partial x} + \frac{\partial \tau_{xz}}{\partial z} = 0$$

$$\sigma_{xx} = \frac{My}{I_y},\ y = \frac{h}{2}$$

$$\frac{y}{I_y}\frac{\partial M_y}{\partial x} + \frac{\partial \tau_{xz}}{\partial z} = 0$$

$$\frac{\partial \tau_{xz}}{\partial z} = -\frac{y}{I_y}\frac{\partial M_y}{\partial x}$$

$$\Rightarrow \frac{\partial \tau_{xz}}{\partial z} = \frac{-y}{I_y}V_z = \frac{-h}{2I_y}V_z \qquad \ldots(1)$$

Integrating (1) from z to $\frac{h}{2}$ for element (A)

$$\tau_{xz}(z) - \tau_{xz}\left(\frac{h}{2}\right) = -\frac{V_z}{2I_y}\left[z^2 - \frac{h^2}{4}\right]$$

As we know,

At Boundaries shear stress vanishes (Top & Bottom Faces)

$$\therefore \tau_{xz}\left(\frac{h}{2}\right) = 0$$

$$\tau_{xz}(z) = +\frac{V_z}{2I_y}\left[\frac{h^2}{4} - z^2\right]$$

Rearranging,

$$\boxed{\tau_{x2}(z) = \frac{V_z\left(\frac{h}{2}\right)^2}{2I_y}\left[1 - \frac{z^2}{\left(\frac{h}{L}\right)^2}\right]}$$

13. Torsion constant J of a thin walled close tube of thickness t and mean radius r is given by:

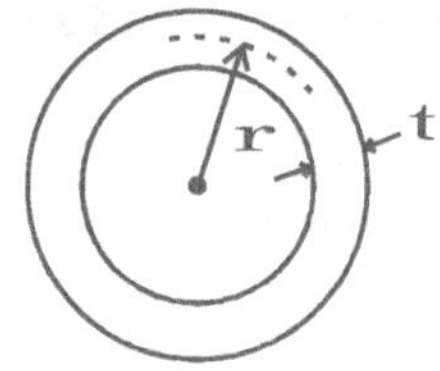

$$J = \frac{4\bar{A}^2}{\oint \frac{ds}{t}} = \frac{4 \times (\pi r^2)^2}{\frac{2\pi r}{t}} = 2\pi r^3 t$$

$$\boxed{J = 2\pi r^3 \cdot t}$$

14. Dynamical system:

$$2\frac{dx}{dt} + \cos x = 0$$

For $x = \frac{\pi}{4}$ small perturbation,

$$2\frac{dx}{dt} + \sin x = 0$$

can also describe system in a close approximation.

$\because$ Both cosx and sinx function hold same values at $\frac{\pi}{4}$.

Common direction question: (15-16) (GDEOM)

15. $K \cdot E = \frac{1}{2}m\dot{x}_1^2 + \frac{1}{2}(3m)\dot{x}_2^2$

$$P \cdot E = \frac{1}{2}kx_1^2 + \frac{1}{2}k[x_1 - x_2]^2 + \frac{1}{2}k[x_2]^2$$

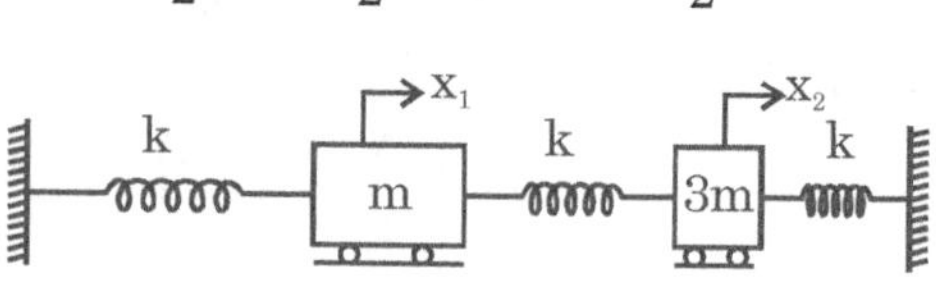

Using lagrangian equation

$$\frac{d}{dt}\left(\frac{\partial K \cdot E}{\partial \dot{x}}\right) + \frac{\partial P \cdot E}{\partial x} = 0$$

For x_1

$$m\ddot{x}_1 + kx_1 + kx_1 - kx_2 = 0 \qquad \ldots(A)$$

For x_2

$$3m\ddot{x}_2 + kx_2 + kx_2 - kx_1 = 0 \qquad \ldots(B)$$

In Matrix form

$$\begin{bmatrix} m & 0 \\ 0 & 3m \end{bmatrix}\begin{bmatrix} \ddot{x}_1 \\ \ddot{x}_2 \end{bmatrix} + \begin{bmatrix} 2k & -k \\ -k & 2k \end{bmatrix}\begin{bmatrix} x_1 \\ x_2 \end{bmatrix} = \begin{bmatrix} 0 \\ 0 \end{bmatrix}$$

As we see our stiffness matrix is non-diagonal or in other words in both the equation (A) and (B), Cordinates are coupled due to stiffness and not mass or inertia. That's why system shows static coupling.

16. Natural frequencies can be calculated using eigen values analysis of Matrix $A = M^{-1}K$

$$= \begin{bmatrix} \frac{1}{m} & 0 \\ 0 & \frac{1}{3}m \end{bmatrix}\begin{bmatrix} 2k & -k \\ -k & 2k \end{bmatrix} = \begin{bmatrix} \frac{2k}{m} & -\frac{k}{m} \\ -\frac{k}{3m} & \frac{2k}{3m} \end{bmatrix}$$

Eigen value Analyses

$$|A - \lambda I| = 0$$

$$A - \lambda I = \begin{bmatrix} \frac{2k}{m} & -\frac{k}{m} \\ -\frac{k}{3m} & \frac{2k}{3m} \end{bmatrix} - \begin{bmatrix} \lambda & 0 \\ 0 & \lambda \end{bmatrix}$$

$$= \begin{bmatrix} \left(\frac{2k}{m}\right) - \lambda & -\frac{k}{m} + 0 \\ -\frac{k}{3m} + 0 & \left(\frac{2k}{3m}\right) - \lambda \end{bmatrix}$$

$$|A - \lambda I| = \left[\left(\frac{2k}{m}\right) - \lambda\right]\left[\left(\frac{2k}{m}\right) - \lambda\right] - \frac{1}{3}\left(\frac{k}{m}\right)^2 = 0$$

$$\Rightarrow \frac{1}{3}\left(\frac{2k}{m}\right)^2 - \left(\frac{2k}{m}\right)\lambda - \lambda\left(\frac{2k}{m}\right) + \lambda^2 - \frac{1}{3}\left(\frac{k}{m}\right)^2 = 0$$

$$\Rightarrow \lambda^2 - \left[\frac{2k}{m} + \frac{2k}{3m}\right]\lambda + \left[\left(\frac{4}{3} - \frac{1}{3}\right)\left(\frac{k}{m}\right)^2\right] = 0$$

$$\Rightarrow \lambda^2 - \left[\frac{2k}{m} + \frac{2}{3}\frac{k}{m}\right]\lambda + \left[1\left(\frac{k}{m}\right)^2\right] = 0$$

$$\therefore \lambda_{1,2} = \frac{-\left(-\left(\frac{2k}{m} + \frac{2}{3}\frac{k}{m}\right)\right) \pm \sqrt{\left(\frac{2k}{m} + \frac{2}{3}\frac{k}{m}\right)^2 - 4\left(\frac{k}{m}\right)^2}}{2}$$

$$= \frac{+\left[\frac{8}{3}\frac{k}{m}\right] \pm \sqrt{\left(\frac{8}{3}\frac{k}{m}\right)^2 - 4\left(\frac{k}{m}\right)^2}}{2}$$

$$= \frac{\frac{8}{3}\left(\frac{k}{m}\right) \pm \sqrt{\left(\frac{8}{3}\right)^2\left(\frac{k}{m}\right)^2 - 4\left(\frac{k}{m}\right)^2}}{2}$$

$$= \frac{\frac{8}{3}\left(\frac{k}{m}\right) \pm \left(\frac{k}{m}\right)\sqrt{\left(\frac{8}{3}\right)^2 - 4}}{2}$$

$$\lambda_{1,2} = \frac{4}{3}\left(\frac{k}{m}\right) \pm \left(\frac{k}{m}\right)\frac{1}{2}\sqrt{\frac{28}{9}}$$

Now Natural frequencies are $\omega_{1,2} = \sqrt{\lambda_{1,2}}$

$$\Rightarrow \omega_{1,2} = \sqrt{\left(\frac{k}{m}\right)\left(\frac{4}{3} \pm \frac{1}{2}\sqrt{\frac{28}{3}}\right)}$$

$$= \sqrt{\frac{4 \pm \frac{1}{2}\sqrt{7 \times 4}}{3}}$$

$$\boxed{\omega_{1,2} = \sqrt{\frac{4 \pm \sqrt{7}}{3}}}$$

Linked Answer Question 17 and 18.

17. Equation of motion of a vibrating rod is given by

$$\frac{\partial^2 u}{\partial x^2} = \frac{1}{C^2}\frac{\partial^2 u}{\partial t^2} \qquad \ldots(1)$$

Boundary condition needed for a rod fixed at the root (x = 0) and free at tip (x = l)

As fixed end displacement is zero

$$\therefore u(x = 0) = 0$$

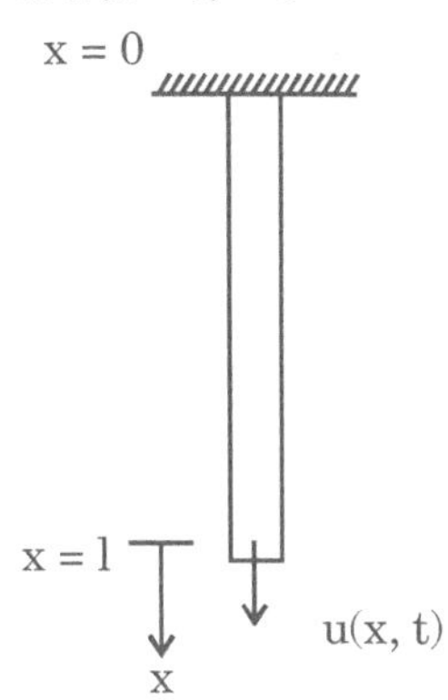

at free end, shear force is zero (No vertical forces)

$$\frac{\partial u}{\partial x}(x = \ell) = 0$$

18. d Assuming solution of (1)

$u(x, t) = U(x)G(t)$

Differentiation w.r.t. x

$$\left|\frac{d^2u(x,t)}{dx^2} = \frac{d^2U}{dx^2}G(t)\right|$$

Differentiation w.r.t. t

$$\frac{d^2u(x,t)}{dt^2} = U\frac{d^2G}{dt^2}$$

Putting in equation (1)

$$G(t)\frac{d^2U}{dx^2} = \frac{1}{C^2}\frac{d^2G}{dt^2}U(x)$$

$$\Rightarrow \boxed{\frac{C^2}{U(x)}\frac{d^2U}{dx^2} = \frac{1}{G(t)}\frac{d^2G}{dt^2} = -\omega^2} \quad \text{... (A)}$$

Now solving equation (A) in space,

$$\frac{C^2}{U(x)}\frac{d^2U}{dx^2} = -\omega^2$$

$$\Rightarrow \frac{d^2U}{dx^2} - \frac{\omega^2}{C^2}U(x) = 0$$

Solution of this equation is in the form

U(x) = C.F + P.I.

But P.I. = 0

$$C.F. \Rightarrow D^2 + \left(\frac{\omega}{C}\right)^2 = 0$$

$$D = \pm\frac{i\omega}{C}$$

$$\therefore U(x) = C_1\cos\frac{\omega}{C}x + C_2\sin\frac{\omega}{C}x \quad \text{... (i)}$$

Using fixed-free – Boundary condition

$U(0, t) = 0$

$$0 = C_1\cos\left(\frac{\omega}{C}\right)(0) + C_2\sin\left(\frac{\omega}{C}\right)(0)$$

$$\boxed{C_1 = 0}$$

$$U(x) = C_2\sin\left(\frac{\omega}{C}\right)x$$

Now,

$$\frac{dU}{dx}(x) = C_2\left(\frac{\omega}{C}\right)\cos\left(\frac{\omega}{C}\right)x \quad \text{... (ii)}$$

Using 2nd Boundary condition

$$\frac{\partial u}{\partial x}(\ell, t) = 0$$

From equation. (ii)

$$0 = C_2\left(\frac{\omega}{C}\right)\cos\left(\frac{\omega}{C}\right)\ell$$

Now

C_2 cannot be zero for solution to exist

$$\therefore \boxed{\cos\left(\frac{\omega}{C}\right) = 0} \rightarrow \text{we can find Natural frequency}$$

using this expression.

19. In absence of body moments, symmetry of stress tensor is derived using moment equilibrium conditions

$$\sum M_x = 0, \sum M_y = 0, \sum M_z = 0$$

Symmetry of stress tensor comes from

$$\tau_{xy} = \tau_{yx},\ \tau_{yz} = \tau_{zy},\ \tau_{xz} = \tau_{zx}$$

20. No. of elastic constants in orthotropic material are 9 because it possesses three mutually orthogonal planes of symmetry.

21. Compatibility conditions in theory of elasticity ensures that we are able to find unique displacement field. In other words, diplacements are single valued and continuous.

22. Given, mass m = 2 kg

Undamped Natural frequency $\omega_n = 20$ Hz

Critical damping constant $C_c = 2\sqrt{km}$

$= 2m\omega_n = 2\times2\times2\times\pi\times20$

$= 160\pi$ N-s/m

$$\omega_n = \sqrt{\frac{k}{m}}$$

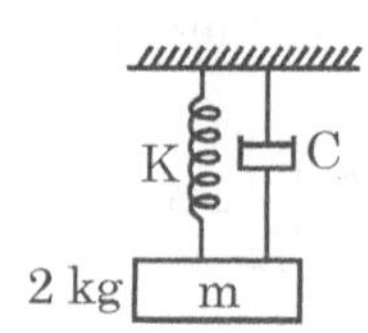

23. For equilibrium of plates, we will have to satisfy force and moment equilibrium. Forces are already in equilibrium due to uniform loading.

$$\Sigma F_x = \Sigma F_y = 0$$

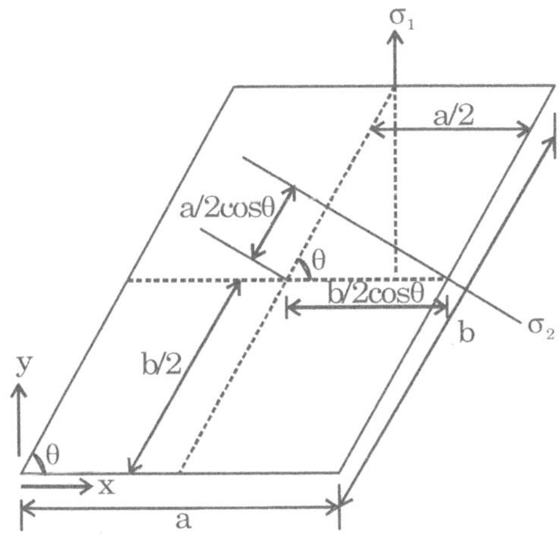

Using moment equilibrium condition, we get

$$(\sigma_1 \times a) \times \frac{b\cos\theta}{2} - (\sigma_2 \times b) \times \frac{a\cos\theta}{2} = 0$$

which gives $\boxed{\sigma_1 = \sigma_2}$ ⇒ condition for equilibrium.

24. Column with various end conditions, effective lengths are:

(i) For Pinned-Pinned (Both ends Pinned)

$l_{eq} = l$

(ii) Both ends built in or fixed

$$\ell_{eq} = \frac{\ell}{2} = 0.5\ell$$

(iii) One end built in (fixed), other free

$\ell_{eq} = 2\ell$

(iv) One end built in other pinned

$\ell_{eq} = 0.707\ell$

25. As we know that shear flow at webs, dividing two sections is given by

$q_{12} = q_1 - q_2$

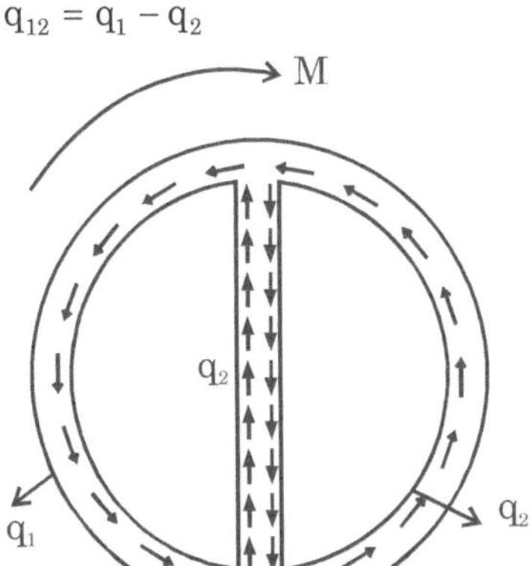

Due to symmetry of sections around the web torque M will cause equal shear force q_1 & q_2 thus

$$\boxed{q_{12} = q_1 - q_2 = 0}$$

26.

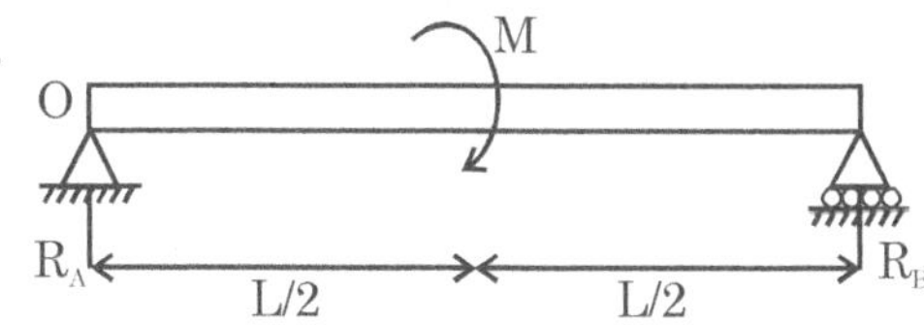

Using force equilibrium

$R_A + R_B = 0$

$$\boxed{R_A = -R_B}$$

∴ Shear force diagram

Using moment equilibrium about 0

$R_B \cdot L = M$

$$\boxed{R_B = \frac{M}{L}}$$

$$\Rightarrow \boxed{R_A = \frac{-M}{L}}$$

27. Normal stresses on beam under bending moment M_y is given by

$$\sigma_{zx} = \frac{M_y . y}{Izz}$$

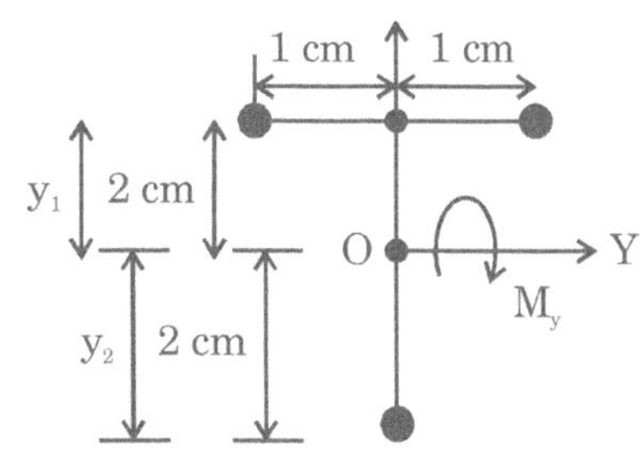

● $A = 0.2\ cm^2$

● $A = 0.1\ cm^2$

$$\frac{|\sigma_{xx}|_{top}}{|\sigma_{xx}|_{bottom}} = \frac{\left|\frac{M_y \cdot y_1}{I_{zz}}\right|}{\left|\frac{M_y \cdot y_2}{I_{zz}}\right|} = \left|\frac{y_1}{y_2}\right| = |-1|$$

$$\Rightarrow \frac{|\sigma_{xx}|_{top}}{|\sigma_{xx}|_{bottom}} = 1$$

M_y and I_{zz} are constant for across section. Only distance y is varying for different booms.

$y_1 = 2$ cm $\quad\quad y_2 = -2$ cm

28. Given, idealized SDOF system
free vibration test is performed
Quality factor Q = 10

Quality factor for SDOF system $Q = \frac{1}{2\xi}$

ξ = damping ratio

$\Rightarrow \xi = \frac{1}{20}$

Logarithmic decrement

$$\boxed{\delta = \frac{2\pi\xi}{\sqrt{1-\xi^2}} = \frac{2\pi \times \frac{1}{20}}{\sqrt{1-\left(\frac{1}{20}\right)^2}} = \frac{\pi}{10}}$$

29. Airy stress function

$$\phi(x,y) = -\frac{P_{xy}{}^3}{4c^3} + \frac{3P_{xy}}{4c}$$

$$\sigma_{xx} = \frac{\partial^2\phi}{\partial y^2} = \frac{-3P_{xy}}{2c^3} \quad \text{...(A)}$$

Moment for a beam

$$\sigma_{xx} = \frac{My}{I_{zz}} \quad \text{...(B)}$$

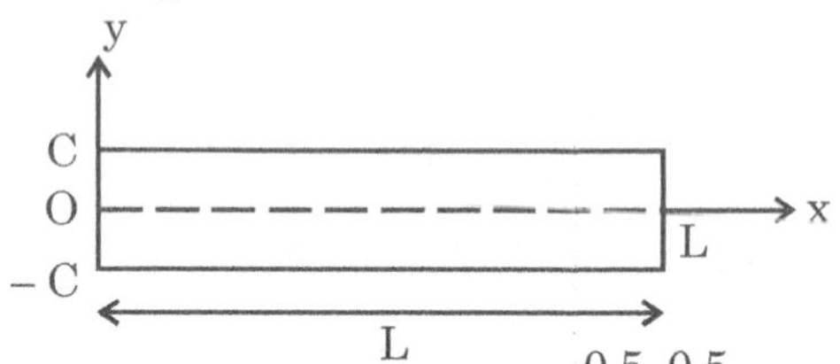

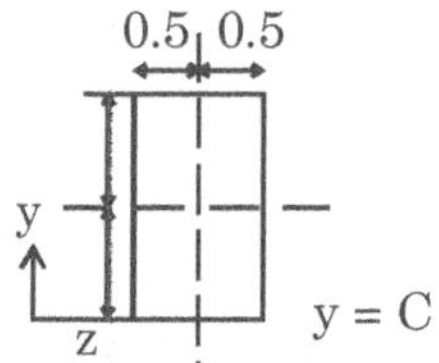

$$I_{zz} = \frac{1 \times (2C)^3}{12} = \frac{2}{3}C^3 \quad \text{...(C)}$$

Using (A), (B) & (C)

$$\frac{M_z \cdot y}{\frac{2}{3}C^3} = \frac{-3\rho xy}{2C^3}$$

$$\Rightarrow \boxed{M_z = -Px} \quad \text{...(1)}$$

Shear force $V_x = -\frac{dM}{dx}$

$$\boxed{V_x = P}$$

- Constant shear force throughout the span
- Bending moment varying as function of length
- Bending moment is maximum at x = L

These conditions tell us that it is cantilever beam clamped at end x = L and carrying shear load at x = 0

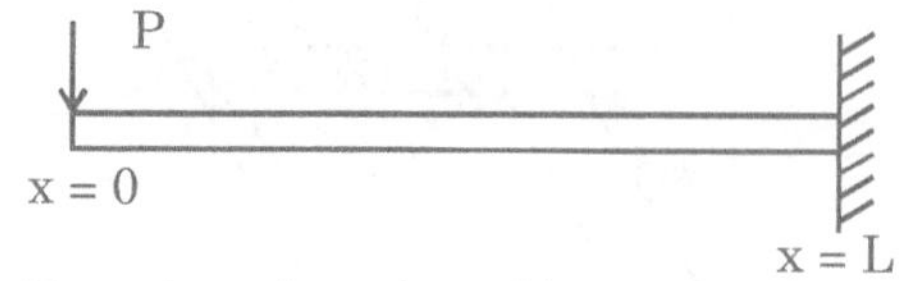

30. Equation of motion of beam (uniform slender) of length L in flexural vibration:

$$EI \cdot \frac{\partial^4\omega}{\partial x^4} + \rho A \frac{\partial^2\omega}{\partial t^2} = 0$$

Simply suported beam at x = 0 and x = L

$$\omega^{IV} + \frac{\rho A}{EI}\ddot{\omega} = 0 \quad \text{...(A)}$$

if $\frac{\rho A}{EI} = k^2$ and assuming $\omega(x,t) = \phi(x)G(t)$

Using this in (A)

$$G(t)\phi^{IV} + K^2\phi(x)\ddot{G} = 0$$

$$\Rightarrow \frac{\phi^{IV}}{\phi} = -\frac{K^2\ddot{G}}{G(t)} = \text{cons tan t} = \omega^2K^2 = a^4$$

{∵ L.H.S. is function of x alone & R.H.S. is function of t alone hence it can be possible only if both are equal to constant)

Using equation

$$\phi^{IV} - a^4\phi = 0$$

Solution of this equation

$$D^4 - a^4 = 0$$

$$D = \pm ai, \pm a$$

Solution can be written as

$$\phi(x) = C_1\sin(ax) + C_2\cos(ax) + C_3\sinh(ax) + C_4\cosh(ax)$$

Constant can be found using B.C.S.

ϕ at x = 0, L $\quad \omega = 0 \Rightarrow \phi(0) = \phi(L) = 0$

at x = 0, L $\quad \omega'' = 0 \Rightarrow \phi''(0) = \phi''(L) = 0$

Now mode shape given is

$$\phi(x) = \sin\left(\frac{\pi x}{L}\right)$$

Comparing it with sin(ax)

$$\Rightarrow a = \frac{\pi}{L}, \text{ also} \qquad a = \sqrt{k\omega}$$

$$\sqrt{k\omega} = \frac{\pi}{L} \text{ and} \qquad K = \sqrt{\frac{\rho A}{EI}}$$

$$\boxed{\omega = \frac{\pi^2}{L^2}\frac{1}{k} = \frac{\pi^2}{L^2}\sqrt{\frac{EI}{\rho A}}}$$

$$= \pi^2\sqrt{\frac{EI}{\rho A L^4}}$$

31-33. Common data questions:

2-D state of stress in isotropic material is given by

$$[\sigma] = C\begin{bmatrix} -8 & 15 \\ 5 & 16 \end{bmatrix} \text{MPa}$$

$\sigma_{xx} = -8c$ MPa

$\sigma_{yy} = 16c$ MPa

$\sigma_{xy} = 5c$ MPa

31. Principle streses are given by

$$\sigma_{1,2} = \left(\frac{\sigma_{xx} + \sigma_{yy}}{2}\right) \pm \sqrt{\left(\frac{\sigma_{xx} - \sigma_{yy}}{2}\right)^2 + \sigma_{xy}^{\ 2}}$$

$$\Rightarrow \sigma_{1,2} = \left[\left(\frac{-8+16}{2}\right) \pm \sqrt{\left(\frac{-8-16}{2}\right)^2 + 5^2}\right] c \text{ MPa}$$

$$= (4 \pm 13)c \text{ MPa}$$

$$\sigma_1 = 17c, \ \sigma_2 = -9c$$

32. Maximum shear stress is given by

$$T_{max} = \left|\frac{\sigma_1 - \sigma_2}{2}\right|$$

$$= \left|\frac{17-(-9)}{2}\right| c = 13 \text{ c MPa}$$

33. Vonn mises failure criterion:

Given, Failure stress

$$\boxed{\sigma_y = 350 \text{ MPa}}$$

$$\sigma_1^{\ 2} - \sigma_1\sigma_2 + \sigma_2^{\ 2} = \sigma_y^{\ 2}$$

$$(17c)^2 - (17c)(-7c) + (-9c)^2 = 350^2$$

$$523c^2 = 122500 \Rightarrow \boxed{c = 15.3}$$

Question 34 & 35:

34. Symmetric box beam

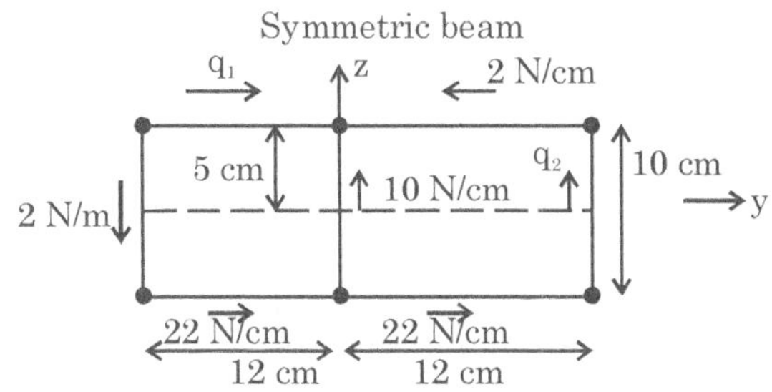

Applied shear stresses

$V_y = 480$ N

$V_z = 300$ N

Torque M acting in shear center.

Shear flow $\boxed{q = \frac{V_2}{\ell}}$

ℓ = length of member

Cross section is symmetric about y and z axis. Thus shear center and all the applied force and moments are acting through the shear center. Thus shear center lies at the origin 0. Using eq.

$$\boxed{\begin{aligned} &\Sigma F_Y = V_y \ldots(1),\ \Sigma F_z = V_z \ldots(2) \\ &\Sigma M_X = V_y \cdot e_z + V_z \cdot e_y + M \ldots(3) \end{aligned}}$$

Here e_y & e_2 are zero because v_y & v_2 acts through shear center.

using first

$$\Sigma F_y = V_y$$

$$q_1 \times 12 - 2 \times 12 + 22 \times 12 + 22 \times 12 = 480$$

$$\Rightarrow \boxed{q_1 = \frac{-504 + 480}{12} = -2 \text{ N/cm}}$$

Now,

$$\Sigma F_z = V_z$$

$$-2 \times 10 + 10 \times 10 + q_2 \times 10 = 300$$

$$\Rightarrow \boxed{q_2 = \frac{300 - 80}{10} = \frac{220}{10} = 22 \text{ N/cm}}$$

35. Using eq. (iii)

$\Sigma M_{s.c} = M$ as e_y, $e_z = 0$

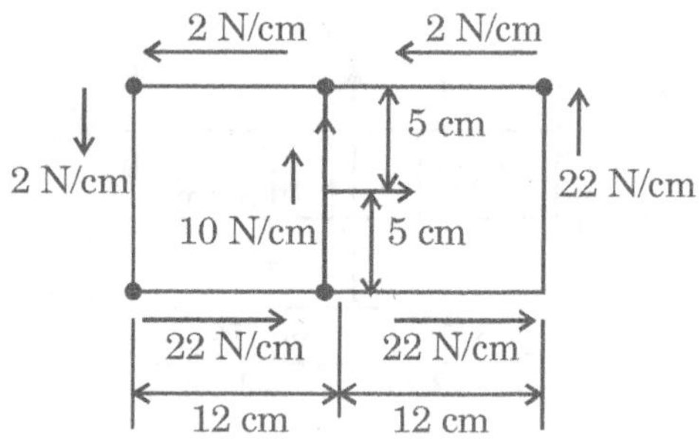

$$\Rightarrow 2\times12\times5+2\times12\times5+22\times12\times5+22\times12\times5$$
$$+22\times10\times12+2\times10\times12 = M$$

$$\Rightarrow \boxed{M = 5760 \text{ N-cm}}$$

36. Plain strain:

For plain stain

$\varepsilon_{zz} = \gamma_{yz} = \gamma_{xz} = 0$

(i) $\gamma_{yz} = \dfrac{\tau_{yz}}{G} = 0$

$$\Rightarrow \boxed{\tau_{yz} = 0}$$

(ii) $\gamma_{xz} = \dfrac{\tau_{xz}}{G} = 0$

$$\Rightarrow \boxed{\tau_{xz} = 0}$$

(iii) $\varepsilon_{zz} = 0$

$$\varepsilon_{zz} = \frac{\sigma_{zz}}{E} - \upsilon\frac{\sigma_{xx}}{E} - \upsilon\frac{\sigma_{yy}}{E} = 0$$

$$\Rightarrow \boxed{\sigma_{zz} = \upsilon(\sigma_{xx} + \sigma_{yy})}$$

37. Critically damped SDOF system

damping constant c = 4 N-s/m

K = 16 N/m

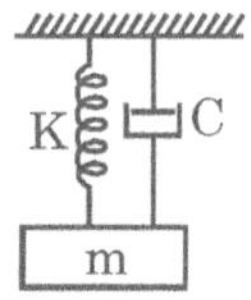

as system is critically damped

$\therefore \xi = 1$

$$\Rightarrow \frac{c}{c_c} = 1$$

$$\Rightarrow c = c_c = 2\sqrt{km}$$

$$\Rightarrow \boxed{m = \frac{1}{K}\left(\frac{c}{2}\right)^2 = 0.25 \text{ kg}}$$

38. Thin walled Rectangular tube with equal and opposite force P will not have any stress untill & unless there is any constrain to motion of section. In this case, rectangular tube will posses rigid body motion.

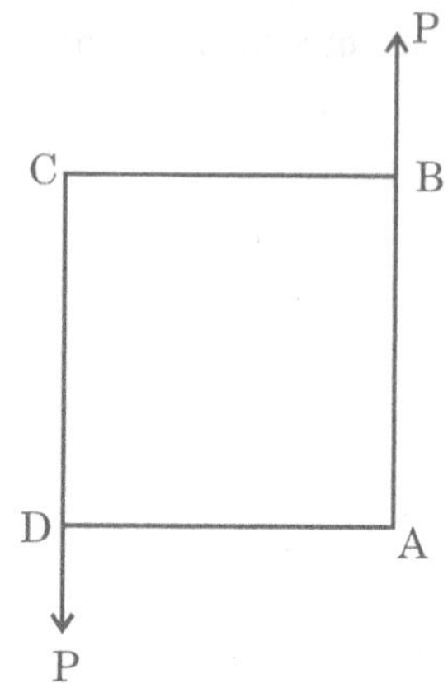

Therefore, it will have zero shear stresses along leg AB.

39. If cross section is symmetric about any axis, then shear center lies on that axis. In given cross section if section was straight T section then it was symmetric about the vertical member hence shear center would have been on BD line. But as member, BD is at angle.

$\therefore$ section is no more symmetric.

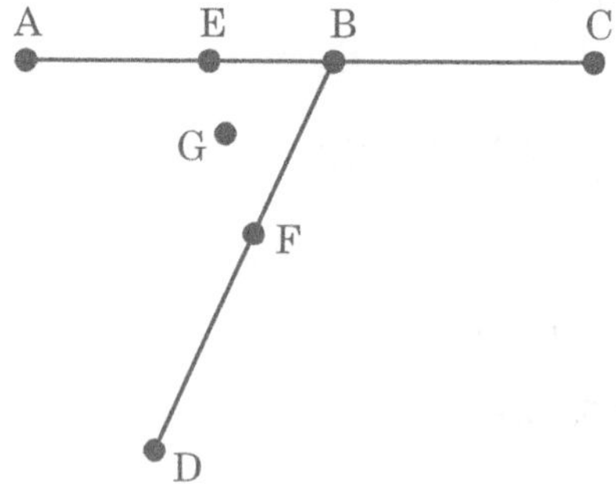

Now shear center cannot be on this member. Section is not symmetric about AB.

$\therefore$ E can't be shear center, so as point B.

Only possible shear center can be shifted point from vertical member that is G.

40. Simply supported beam of length 2L

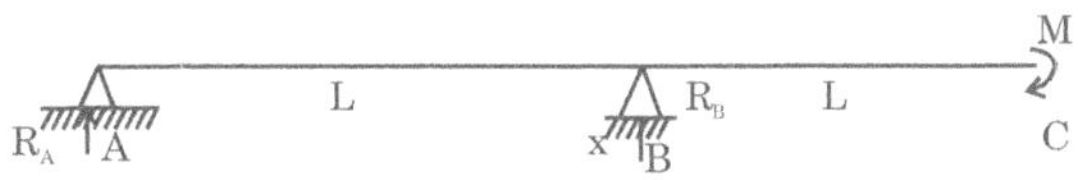

If we move from left to right between C and B there is no other moment than end moment M. Thus this part is having a constant Bending Moment. After this due to Reaction R_B. Bending moment will start reducing lineary $(M - R_B.x)$. which becomes zero at R_A.

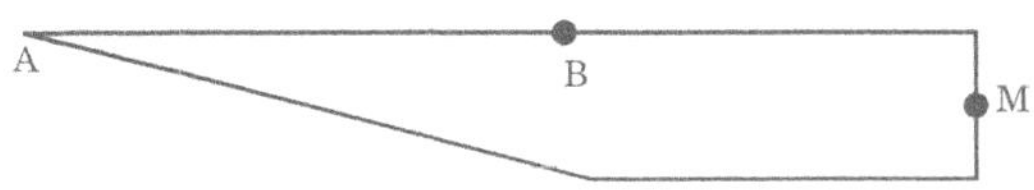

41.

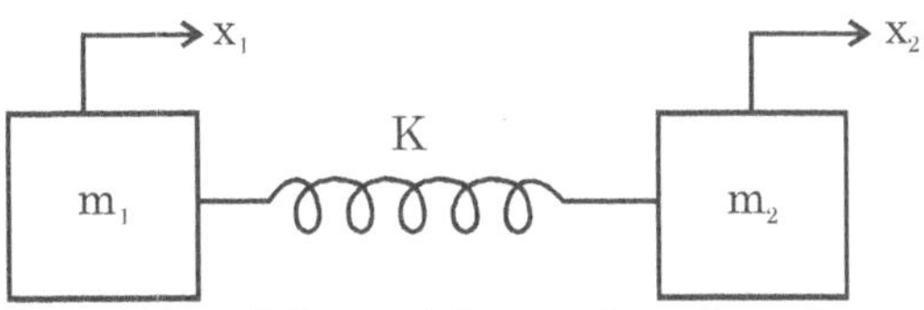

Governing differential equation of motion :

$$K.E. = \frac{1}{2}m_1\dot{x}_1^{\,2} + \frac{1}{2}m_2\dot{x}_2^{\,2}$$

$$P.E. = \frac{1}{2}k[x_1 - x_2]^2$$

Using lagrangian equation,

$$m_1\ddot{x}_1 + Kx_1 - Kx_2 = 0 \quad \ldots (i)$$

$$m_2\ddot{x}_2 + Kx_2 - Kx_1 = 0 \quad \ldots (ii)$$

$$\begin{bmatrix} m_1 & 0 \\ 0 & m_2 \end{bmatrix}\begin{bmatrix} \ddot{x}_1 \\ \ddot{x}_2 \end{bmatrix} + \begin{bmatrix} k & -k \\ -k & k \end{bmatrix}\begin{bmatrix} x_1 \\ x_2 \end{bmatrix} = 0$$

$$\text{let } A = M^{-1}K = \begin{bmatrix} \frac{1}{m_1} & 0 \\ 0 & \frac{1}{m_2} \end{bmatrix}\begin{bmatrix} k & -k \\ -k & k \end{bmatrix}$$

$$= \begin{bmatrix} \frac{k}{m_1} & -\frac{k}{m_1} \\ -\frac{k}{m_2} & \frac{k}{m_2} \end{bmatrix}$$

Solving given value problem

$$|A - \lambda I| = \begin{vmatrix} \frac{k}{m_1} - \lambda & \frac{-k}{m_1} - 0 \\ -\frac{k}{m_2} - 0 & \frac{k}{m_2} - \lambda \end{vmatrix}$$

$$\left(\frac{k}{m_1} - \lambda\right)\left(\frac{k}{m_2} - \lambda\right) - \frac{k^2}{m_1m_2} = 0$$

$$\left(\frac{k^2}{m_1m_2}\right) - \frac{k}{m_1}\lambda - \frac{k}{m_2}\lambda + \lambda^2 - \frac{k^2}{m_1m_2} = 0$$

$$\lambda^2 - \left[\frac{k}{m_1} + \frac{k}{m_2}\right]\lambda = 0$$

$$\lambda\left[\lambda - \left(\frac{k}{m_1} + \frac{k}{m_2}\right)\right] = 0$$

$$\lambda_1 = 0,\ \lambda_2 = \frac{k}{m_1} + \frac{k}{m_2}$$

Natural frequency $\omega_{1,2} = \sqrt{\lambda_{1,2}}$

$$\boxed{\omega_1 = \sqrt{\lambda_1} = 0}$$

$$\omega_2 = \sqrt{\lambda_2} = \sqrt{\frac{k}{m_1} + \frac{k}{m_2}} = \sqrt{\frac{km_2 + km_1}{m_1m_2}}$$

$$\Rightarrow \boxed{\omega_2 = \sqrt{\frac{k(m_1 + m_2)}{m_1m_2}}}$$

42. Buckling load

Given; Rectangular cross section [1 cm x 1.5 cm]

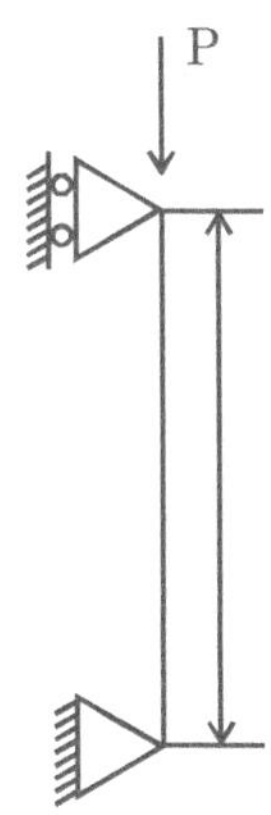

For simply supported case, critical Buckling load is given by

$$P_{cr} = \frac{c\pi^2EI}{\ell^2} \qquad \boxed{c = 1}$$

$$E = 210 \times 10^9\ N/m^2$$

$$= \frac{\pi^2EI}{\ell^2}$$

h = 1 cm

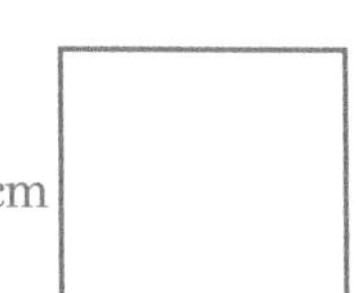

b - 1.5 cm

$$I = \frac{bh^3}{12} = \left[\frac{1.5 \times 1^3}{12}\right]\frac{1}{(100)^4} m^4$$

$$= 1.25 \times 10^{-4}\ m^4$$

$$P_{cr} = \frac{\pi^2 \times 210 \times 10^9 \times 1.25 \times 10^{-9}}{(0.5)^2} = 10363.084\ N$$

$$\approx 10\ kN$$

43. Wing root

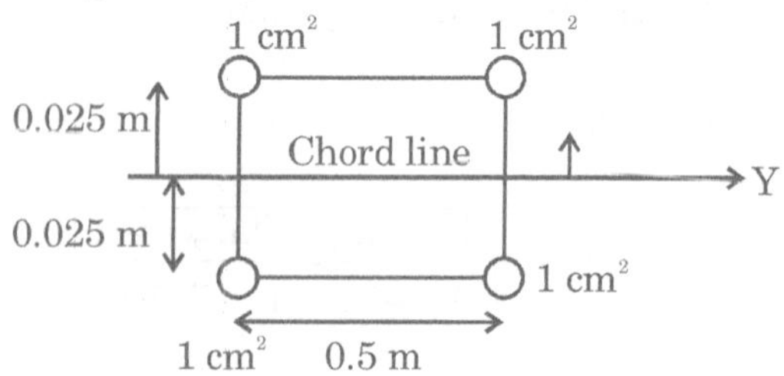

Bending moment $M_y = 10$ N-m

Forced factor n = 3 . 5

Maximum bending stress at root

$$\sigma = \frac{M_Y}{I_{YY}} x = \frac{10 \times 0.025}{4 \times 10^{-4} \times 0.025} = 10^6\ Pa$$

Load factor = 3.5

$\therefore$ Maximum stress $\sigma_{max.} = 3.5 \times 10^6$ Pa

44. Mass of bar m = 1 kg

L = 1 m

K = 1 N/m

C = 1 N-s/m

$$a = \frac{1}{\sqrt{3}}$$

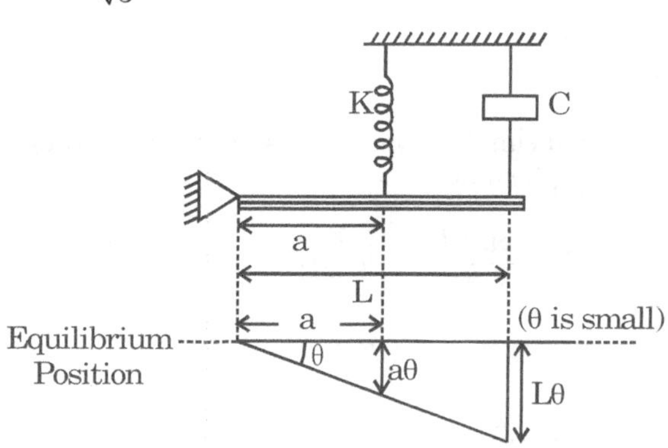

$$\text{Moment of inertia } I_A = \frac{mL^2}{3} = \frac{1}{3}\ kgm^2$$

$$K.E. = I_A\dot{\theta}^2,\ P.E. = \frac{1}{2}K[a\theta]^2 = \frac{1}{2}Ka^2\theta^2$$

$$P.E. = \frac{1}{2}C[\ell\dot{\theta}]^2 = \frac{1}{2}C\ell^2\dot{\theta}^2$$

Using lagrangian equation,

$$I_A\ddot{\theta} + Cl^2\ddot{\theta} + Ka^2\theta = 0$$

$$\ddot{\theta} + \frac{Cl^2}{I_A}\dot{\theta} + \frac{Ka^2}{I_A}\theta = 0$$

Comparing with

$$\ddot{\theta} + 2\xi\omega_n\dot{\theta} + \omega_n^2\theta = 0$$

$$\omega_n = \sqrt{\frac{Ka^2}{I_A}} = \sqrt{\frac{1 \times \left(\frac{1}{\sqrt{3}}\right)^2}{\frac{1}{3}}} = 1 \text{ rad/sec.}$$

$$\text{Now, } 2\xi\omega_n = \frac{CI^2}{I_A} = \frac{1 \times 1^2}{\frac{1}{3}} = 3$$

$$\xi = \frac{3}{2 \times \omega_n} = \frac{3}{2 \times 1} = \frac{3}{2} = 1.5$$

$$\text{Also, } \xi = \frac{Cl^2}{2\sqrt{Ka^2 I_A}} = \frac{1 \times 1^2}{2\sqrt{1 \times \frac{1}{3} \times \frac{1}{3}}} = \frac{3}{2} = 1.5$$

$$\boxed{\xi = 1.5}$$

as system have damping ratio more than 1. Therefore it is over damped system. Option (a) is correct.

45. Two cells tube thickness t = 0.5 mm

Torque = 10 N-m

Applying Bredt-Batho formula,

$$2\Sigma q\overline{A} = T$$

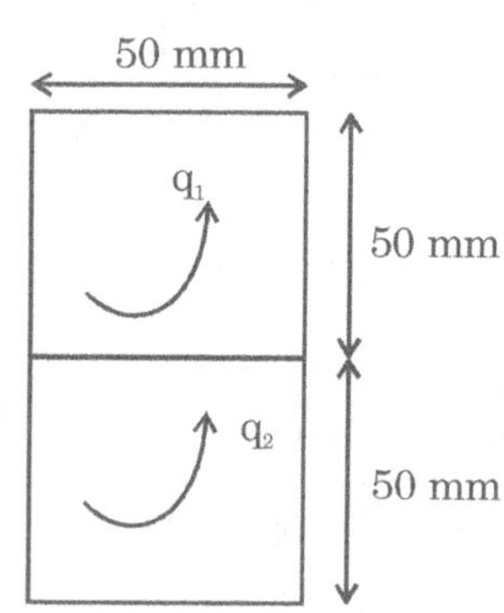

$$\frac{2q_1[50 \times 50]}{10^6} + \frac{2q_2[50 \times 50]}{10^6} = 10$$

$$q_1 + q_2 = \frac{10^2 \times 10^{16}}{5000} = 2000$$

$$\boxed{q_1 + q_2 = 2000\ N/m}$$

Common data question 46 and 47

46. Partial differential equation for torsional shaft

$$I\frac{\partial^2\theta}{\partial t^2} = GJ\frac{\partial^2\theta}{\partial x^2} \quad \ldots (1)$$

GJ = Torsional Rigidity

I = Mass polar moment of inertia

For Shaft fixed at both ends

Boundary conditions are given by

$\theta(0) = 0 \qquad \theta(L) = 0$

47. Equation (1) can be written as

$$\frac{\partial^2\theta}{\partial t^2} = \frac{GJ}{I}\frac{\partial^2\theta}{\partial x^2}$$

$$\Rightarrow \frac{\partial^2\theta}{\partial t^2} = C^2\frac{\partial^2\theta}{\partial x^2} \quad \ldots (A)$$

$$\boxed{C^2 = \frac{GJ}{I}}$$

Assume a solution

$\theta(x, t) = P(x)\theta(t)$

Put in equation (A)

$$P(x)\frac{\partial^2\theta}{\partial t^2} = C^2\theta(t)\frac{\partial^2 P}{\partial x^2}$$

[L.H.S. & R.H.S. are of different variables so they both are equal to constant]

$$\frac{\ddot{\theta}}{\theta} = \frac{c^2P''}{P} = \text{constant} = -\omega^2$$

Using equation

$$c^2p'' + \omega^2 p = 0$$

$$p'' + \left(\frac{\omega^2}{c^2}\right)P = 0 \quad \ldots (ii)$$

Solution of (B) gives mode shape

$$D^2 + \left(\frac{\omega^2}{c^2}\right) = 0$$

$$D = \pm i\left(\frac{\omega}{c}\right)$$

$$P(x) = A_1\cos\left(\frac{\omega}{c}\right)x + A_2\sin\left(\frac{\omega}{c}\right)x \quad \ldots (C)$$

A_1 & A_2 can be found out using Boundary condition.

Given Mode shape in question:

$$\sin\left(\frac{n\pi x}{L}\right)$$

Comparing argument of sin with Previous equation (C)

$$\frac{\omega}{c} = \frac{n\pi}{L} \qquad c = \sqrt{\frac{G1}{I}}$$

n^{th} natural frequency

$$\boxed{\omega = \frac{n\pi}{L}\sqrt{\frac{GJ}{I}}}$$

48. Thermomechanical loading cause three normal stresses and three shear stress. Total 6 components of state of stress exist.

49. Vertical slender rod is suspended by a hinge at the top and hangs freely. Because there is no prevention for expansion of bar in hinge, Boundary condition at one end is hinged and free at other, thus there will not be any stress developed. But rod is free to expand due to temperature; it will have

$$\text{strain} = \frac{\alpha L(T - T_0)}{L}$$

T_0 = initial Temperature

T = final temperature

50. Given: yield strength 6_y = 200 MPa

Using Vonn-mises Criterion for plane

$$\sigma_1^2 + \sigma_2^2 - \sigma_1\sigma_2 = \sigma_y^2$$

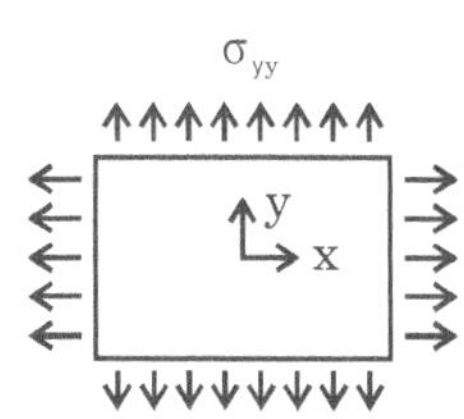

$\sigma_{xx} = 150$ MPa $\begin{pmatrix} \sigma_{xx} = \sigma_2 \\ \sigma_{yy} = \sigma_1 \end{pmatrix}$

$$\sigma_1^2 + (150)^2 - 150\,\sigma_1 = 200$$

$$\boxed{\sigma_2 = 227.06 \text{ MPa}}$$

Maximum tensile stress along y direction.

51. Column of length ℓ and flexural rigidity EI

One end fixed other hinged.

Critical Building load is given by

$$P_{cr} = \frac{\pi^2 EI}{(0.7\ell)^2}$$

52. Natural frequency of beam is given by

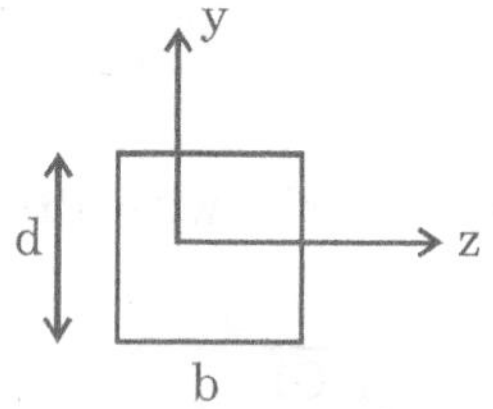

$$\omega = \frac{\pi^2}{L^2}\sqrt{\frac{EI}{\rho A}} = \frac{\pi^2}{L^2}\sqrt{\frac{E\frac{bd^3}{12}}{\rho b \cdot d}}$$

$$\omega = \frac{\pi^2}{L^2}\sqrt{\frac{Ed^2}{12\rho}}$$

$$\Rightarrow \boxed{\omega \propto d}$$

if d is highest, Natural frequency is higher.

53. 2-D problem:

No body forces

Stress state:

$\sigma_{xx} = 3x^2y + 4y^2$

$\sigma_{yy} = y^3 + 14xy$

$\tau_{xy} = -3xy^2 - 7x^2$

Equilibrium Equations

$$\frac{\partial \sigma_{xx}}{\partial x} + \frac{\partial \tau_{xy}}{\partial y} = 0$$

L.H.S. $\Rightarrow \sigma xy + (-\sigma xy) = 0$ = R.H.S.

$$\frac{\partial \tau_{xy}}{\partial x} + \frac{\partial \sigma_{yy}}{\partial y} = 0$$

L.H.S. $\Rightarrow (-3y^2 - 14x) + (3y^2 + 14x) = 0 =$ R.H.S.

Stress state satisfy equilibrium equations.

Compatibility equation in terms of stress are

$$\left(\frac{\partial^2}{\partial x^2} + \frac{\partial^2}{\partial y^2}\right)(\sigma_{xx} + \sigma_{yy}) = 0$$

$$\Rightarrow \frac{\partial^2 \sigma_{xx}}{\partial x^2} + \frac{\partial^2 \sigma_{xx}}{\partial y^2} + \frac{\partial^2 \sigma_{yy}}{\partial y^2} + \frac{\partial^2 \sigma_{yx}}{\partial x^2} = 0$$

L.H.S. = $6 + 8 + 6y + 0 \neq 0$

Hence, compatibility equations are not satisfied.

54.

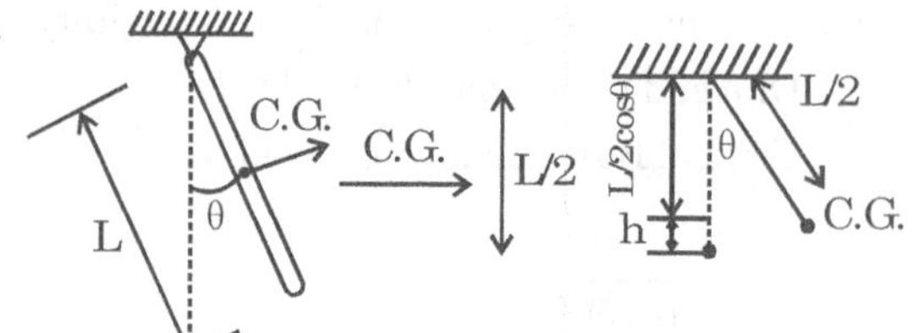

GDEOM:

$$K.E = \frac{1}{2}I\dot{\theta}^2$$

$$P \cdot E = mgh = mg\left(\frac{L}{2} - \frac{L}{2}\cos\theta\right)$$

Using Lagrangian

$$\frac{d}{dt}\left(\frac{\partial K \cdot E}{\partial \dot{\theta}}\right) - \frac{d(K \cdot E)}{\partial \theta} + \frac{\partial P \cdot E}{\partial \theta} = 0$$

For small θ

$\sin\theta \approx \theta$, $\cos\theta = 1$

I for rigid rod

$$I = \frac{mL^2}{3}$$

$$\Rightarrow I\ddot{\theta} + mg\frac{L}{2}\sin\theta = 0$$

$$\Rightarrow \frac{mL^2}{3}\ddot{Q} + mg\frac{L}{2}\theta = 0$$

$$\Rightarrow \ddot{\theta} + \frac{mg\frac{L}{2}}{\frac{mL^2}{3}}\theta = 0$$

$$\Rightarrow \ddot{\theta} + \frac{3g}{2L}\theta = 0$$

$$\Rightarrow \text{Natural Frequency } \omega_n = \sqrt{\frac{3g}{2L}}$$

55.

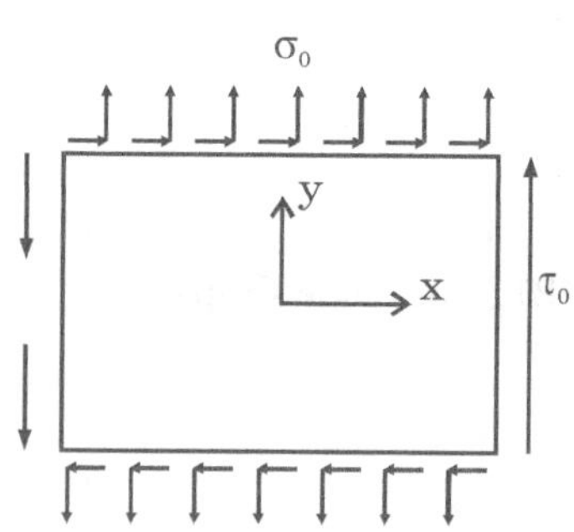

For given stress field, Appropriate airy stress function can be calculated using

$$\sigma_{xx} = \frac{\partial^2 \phi}{\partial y^2}$$

$$\sigma_{yy} = \frac{\partial^2 \phi}{\partial x^2}$$

$$6_{xy} = \frac{-\partial^2\phi}{\partial x \partial y}$$

Taking ϕ from given options

$$\phi = -\tau_{0xy} + \sigma_0 \cdot \frac{x^2}{2}$$

$$\sigma_{xx} = \frac{\partial^2\phi}{\partial y^2} = 0$$

$$\sigma_{yy} = \frac{\partial^2\phi}{\partial x^2} = \sigma_0$$

$$\sigma_{xy} = \frac{-\partial^2\phi}{\partial x \partial y} = \tau_0$$

Hence ϕ is a valid Airy stress functions.

56. Given section is symmetric about x-axis. therefore shear center lies on x-axis.

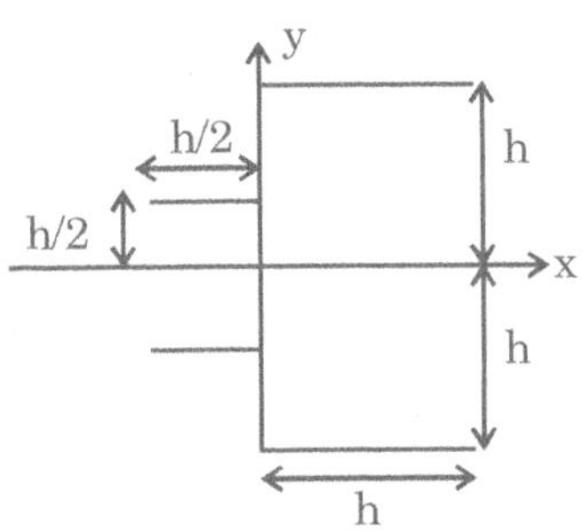

Moment of inertia is given by

$$I_{xx} = \frac{35}{12} th^3$$

For only C section

shear center lies on axes about which it is symmetric i.e. x axis and given by

$$C = \frac{.3b^2t_1}{6bt_1 + t_2H} = \frac{-3h^2t}{6ht + 2ht} = \frac{-3}{8}h$$

here $b = h$, $H = 2h$, $t_1 = t_2 = t$

$$\therefore \boxed{e = \frac{-9}{28}h}$$

But due to extra section = whole section will become more symmetric hence shear center will move right side of e.

[All other option i.e. $-\frac{35}{36}$h and $-\frac{17}{5}$h are left side of $-\frac{3}{8}$h]

57. Underdamped system

Amplitude of 3rd Peak = 100

10th Peak = 10

No. of peaks in between = 10 – 3 = 7

Using logarithms decrement

Damping ratio

$$\Rightarrow \xi = \frac{1}{2\pi n} \ell n\left(\frac{A_1}{A_n}\right)$$

$$= \frac{1}{2\pi 7} \ell n\left(\frac{100}{10}\right) = 0.052$$

Common Data Questions: (58-59)

Moment induced due to Bracket is

$$M = P \times a = P \times 0.1 \times \ell$$

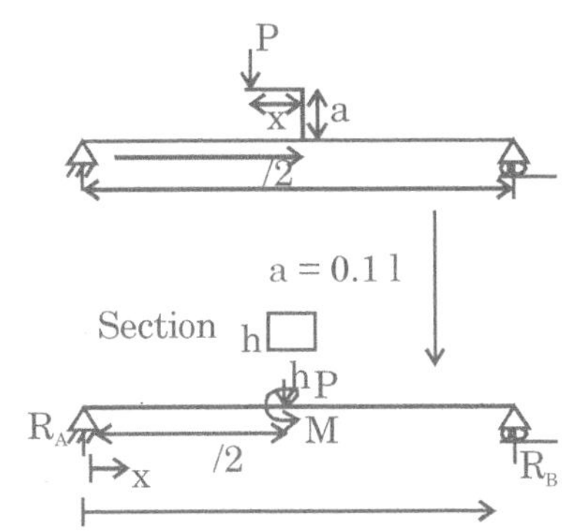

Packet can be replaced by force an a moment at piont $x = \frac{\ell}{2}$

Force balance:

$$R_A + R_B = P \qquad \dots(1)$$

Moment balance about A

$$R_B \times \ell = P \times \frac{\ell}{2} - M$$

$$\Rightarrow R_B = \frac{P}{2} - \frac{M}{\ell} = \frac{P}{2} - P \times 0.1 = \frac{P}{2} - \frac{P}{10}$$

$$\Rightarrow R_B = \frac{8P}{20} = \frac{2}{5}P = 0.4P$$

$$\Rightarrow R_A = P - \frac{2}{5}P = \frac{3}{5}P = 0.6P$$

58. Bending moment diagram
From R.H.S.
moment will be $R_B \times x$
$\therefore$ will increase linearly.

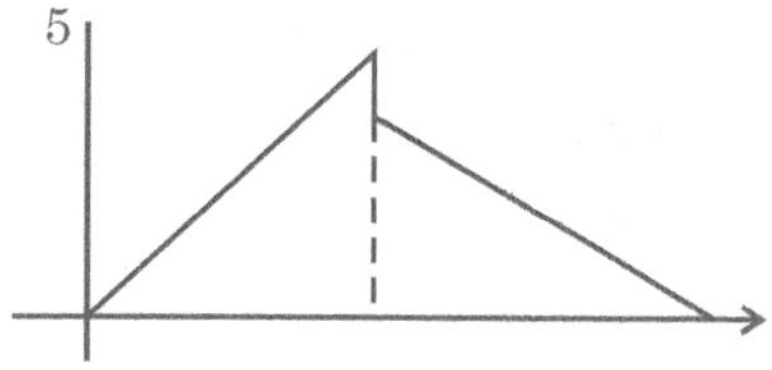

as it reaches at $\frac{\ell}{2}$ it become = $0.4P \times \frac{\ell}{2} = 0.2\ell P$

at $\frac{\ell}{2}$ we have Pure moment M = $0.1P\ell$

$\therefore$ Total moment at x = $\frac{\ell}{2}$ will be = $0.3P\ell$

59. For Maximum value of shear stress:
we need, shear force diagram:

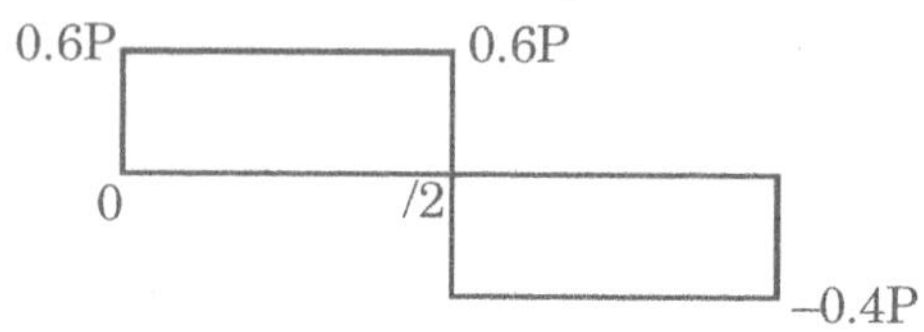

$V_{max} = 0.6P$
Cross section:

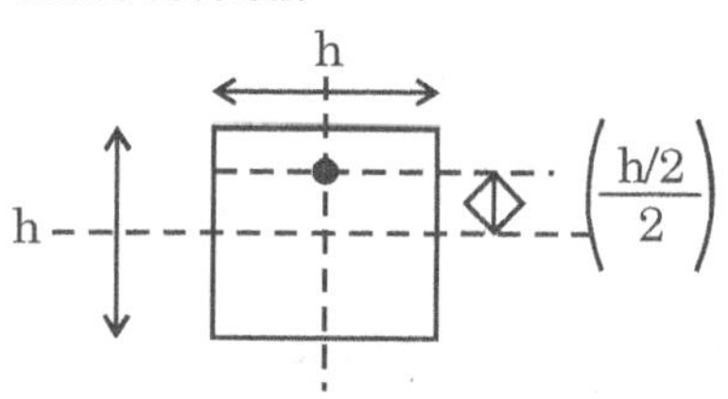

Shear stress is given by $\tau = \frac{VQ}{It}$
I = moment of inertia about neutral axes

$$I = \frac{bt^3}{12} = \frac{h^4}{12}$$

First moment of area $Q = A \cdot \bar{y} = \frac{h}{2} \times h \times \frac{h/2}{2}$

$$= \frac{h^3}{8}$$

$$\therefore \tau_{max} = \frac{0.6 \times P \times \frac{h^3}{8}}{\frac{h^4}{12} \times h} = \frac{0.9}{h^2}P$$

60. For 3-D linear elastic solids:
6 –Stress - strain relations
6 – Strain displacement equation
3 – Equation of equilibrium

61. For Euler Bernoulli beam, non zero stress and strains are σ_{xx}, τ_{xy}, ε_{xx} and all others are zero.
For Plain stress : $\sigma_{zz}, \tau_{xz}, \tau_{yz} = 0$
For Plain strain : $\epsilon_{zz}, \epsilon_{xz}, \epsilon_{yz} = 0$
Thus it satisfies both plain stress and plane strain condition.

62. For statically indeterminate frame structure no. of unknowns (i.e. reactive force and moments) are more than no. of equations. But no. of joints degree of freedoms are less than no. of equations.
For example :

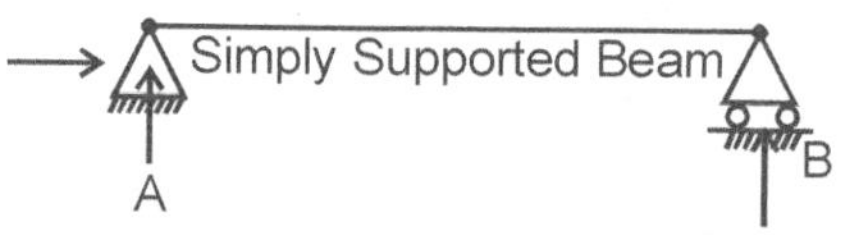

Joint degree of freedom = Rotational A + Rotational B + Translation at B = 3
No. of equilibrium equations = 3
No. of Reaction = 3 ($\therefore$ statically determinate)

Fixed-fixed

No. of degree of freedom = 0
Equilibrium equation = 3
Reaction = 6 ($\therefore$ statically indeterminate)

63. Logarithmic determent is given by :

$$\delta = \frac{2\pi\xi}{\sqrt{1-\xi^2}} \quad \text{... (A)}$$

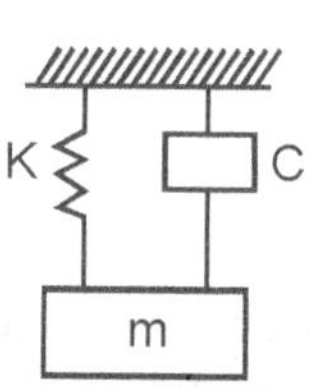

Now damping ratio, $\xi = \frac{C}{2\sqrt{km}}$ put in (A)

$$\delta = \frac{2\pi \frac{C}{2\sqrt{km}}}{\sqrt{1-\left(\frac{C}{2\sqrt{km}}\right)^2}}$$

$$\delta = \frac{2\pi C}{\sqrt{4mk - C^2}}$$

64. Natural frequency $\omega_{n1} = 10$ rad/sec

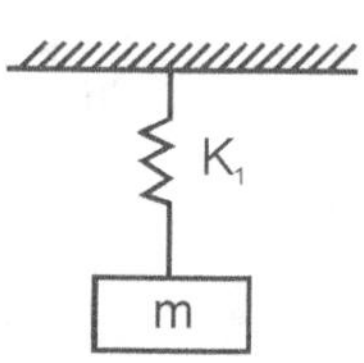

Natural frequency ω_{n2} = 12 rad/sec

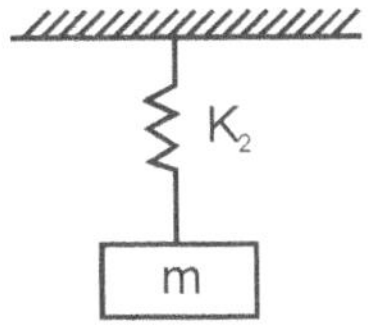

$$\omega_{n_1} = \sqrt{\frac{k_1}{m}} = 10 \text{ rad/sec} \quad \text{... (i)}$$

$$\omega_{n_2} = \sqrt{\frac{k_2}{m}} = 20 \text{ rad/sec} \quad \text{... (ii)}$$

$$\Rightarrow m = \frac{k_2}{400} \text{ put in (i)}$$

$$\sqrt{\frac{k_1}{\frac{k_2}{40}}} = 10$$

$$k_1 = \frac{100 \times k_2}{400} = \frac{k_2}{4}$$

$$\Rightarrow \boxed{k_2 = 4k_1}$$

65. Simply supported beam is converted to fixed

Fixed beam

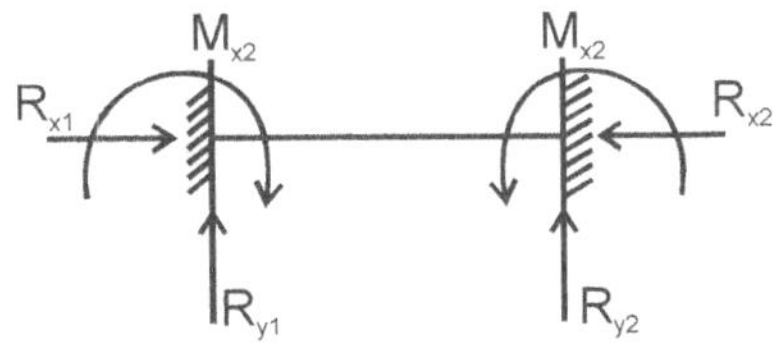

No. of unknown reactions are 6.

No. of equlibrium equation are 3.

Degree of static indeterminancy

= 6 – 3 = 3

In simply supported beam degree of static indeterminency is zero.

Hence it is increased by 3.

66. P = 1 N

E = 50 × 10^9 Pa

Transverse deflaction at free end = y

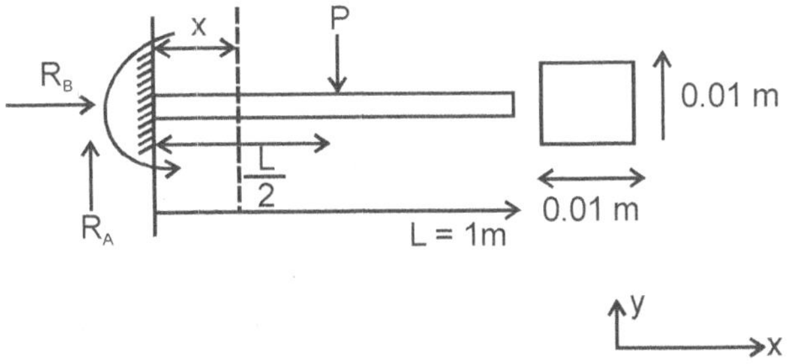

We know that

$$EI\frac{d^2y}{dx^2} = P\left[\frac{l}{2} - x\right]$$

By integrating twice, we get

$$EIy = P\left[\frac{l}{2} \times \frac{x^2}{2} - \frac{x^3}{6}\right] + c_1x + c_2$$

y = 0 at x = 0

$$\Rightarrow \boxed{c_2 = 0}$$

y′ = 0 at x = 0

$$EIy' = P\left[\frac{1}{2}x - \frac{x^2}{2}\right] + c_1$$

$c_1 = 0$

$$\therefore \; y = \frac{1}{EI} \times P\left[\frac{l}{4}x^2 - \frac{x^3}{6}\right]$$

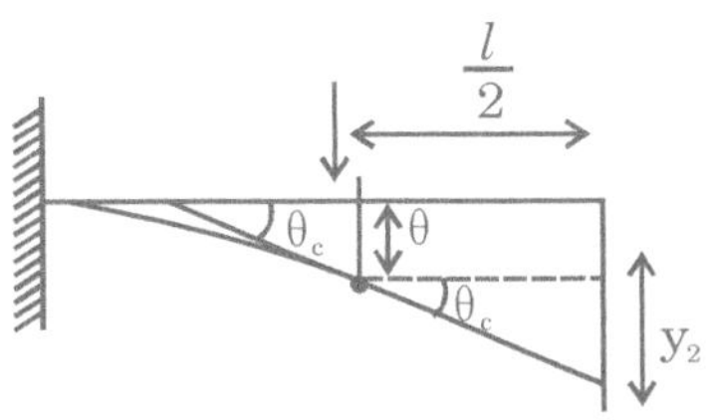

$$\theta_c = \left.\frac{dy}{dx}\right|_{x=\frac{1}{2}}$$

$$\frac{y_2}{1/2} = \tan\theta_c$$

for small angle

$\tan\theta_c = \theta_c$

$$\Rightarrow y_2 = \frac{1}{2}\theta_c$$

This equation is valid till load P.

Deflection at end :

$$Y = y_1\Big|_{x=\frac{1}{2}} + y_2$$

$$= \frac{1}{EI}P\left[\frac{l}{4} \times \frac{l^2}{4} - \frac{l^3}{8 \times 6}\right] + \frac{l}{2} \times \left[\frac{1}{EI} \times P\left[\frac{l}{2} \times \frac{l}{2} - \frac{l^2}{2 \times 4}\right]\right]$$

= 2.50 × 10^{-3} m

= 2.5 mm

67. For circular cross-section

Shear stress is given by

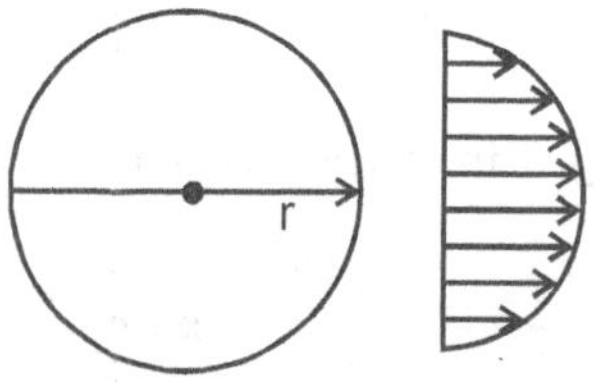

$$\tau = \frac{PA\bar{y}}{I.b}$$

Maximum shear stress at centre of cross section is given by :

$$\tau_{max} = \frac{4}{3}\frac{P}{\pi r^2}$$

Given Area $(\pi r^2) = 1\ mm^2$

Load P = 1N

$$\boxed{\tau_{max} = \frac{4}{3} = 1.33\ N/mm^2}$$

68. Buckling load is s ame as yield stress = 240 MPa

Critical bucking load for simply supposed column is given by:

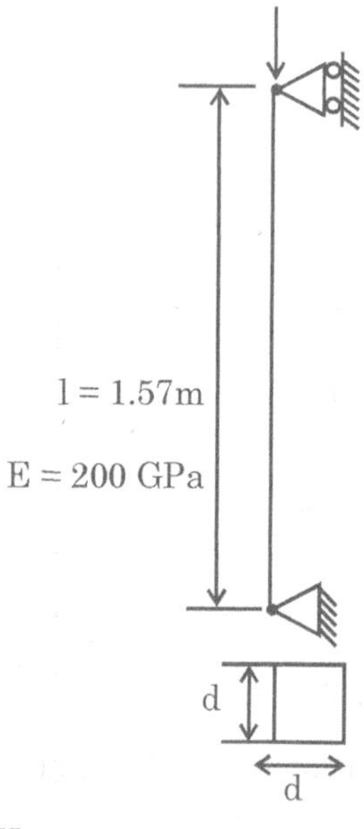

$$p_{cr} = \frac{\pi^2 EI}{l^2}$$

$$\sigma_{cr} = \frac{Pcr}{A}$$

$$\Rightarrow \frac{\sigma_{cr} A \times l^2}{\pi^2 E} = \frac{1}{12}dd^3$$

$$\Rightarrow d^4 = \frac{12 \times 240 \times 10^6 \times d^2 \times 1.57}{\pi^2 \times 200 \times 10^9}$$

$$\Rightarrow d = \sqrt{\frac{12 \times 240 \times 10^6 \times (1.57)^2}{\pi^2 \times 200 \times 10^9}}$$

= 6 cm

69. Body undergoes deformation under plane strain condition :

$\therefore \in_{zz} = \in_{zx} = \in_{zy} = 0$

$\in_{zz} = 0$

$\therefore \sigma_{zz} = \nu(\sigma_{xx} + \sigma_{yy})$ $\quad \nu = \frac{1}{3}$; E = 200 GPa

= 200 × 10^3 MPa

$\in_{zx} = 0 \quad \Rightarrow \tau_{zx} = 0$

$\in_{zy} = 0 \quad \Rightarrow \tau_{zy} = 0$

$\sigma_{xx} = \sigma_{yy} = 450$ MPa, $\tau_{xy} = 75$ MPa

$$\boxed{\sigma_{zz} = \frac{1}{3}[450 + 450] = 300\ MPa}$$

Now, $\boxed{\in_{xx} = \frac{1}{E}\left[\sigma_{xx} - \upsilon\left[\sigma_{yy} + \sigma_{zz}\right]\right] = \frac{1}{10^3} = 0.001}$

Similarly, $\boxed{\in_{zz} = 0.001}$

$\therefore \boxed{\gamma_{xy} = \frac{\tau_{xy}}{G} = \frac{75}{200 \times 10^3} \times 2\left(1 + \frac{1}{3}\right) = 0.001}$

70.

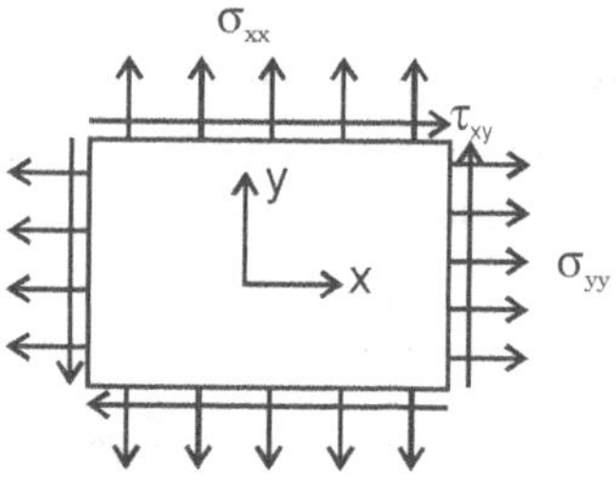

$\sigma_{xx} = P$

$\sigma_{yy} = Q$

$\tau_{xy} = R$

if $\phi = P\frac{y^2}{2} + Q\frac{x^2}{2} + Rxy$

We know that,

$$\left.\begin{aligned} \sigma_{xx} &= \frac{d^2\phi}{dy^2} = P = \sigma_{xx} \\ \sigma_{yy} &= \frac{\partial^2\phi}{\partial x^2} = Q = \sigma_{yy} \\ \tau_{xy} &= \frac{\partial^2\phi}{\partial x \partial y} = R = \tau_{xy} \end{aligned}\right| \text{ Satisfied}$$

Common data question for 71 to 72:

71. $EI\frac{\partial^4 w}{\partial x^4} + m\frac{\partial^2 w}{\partial t^2} = 0$

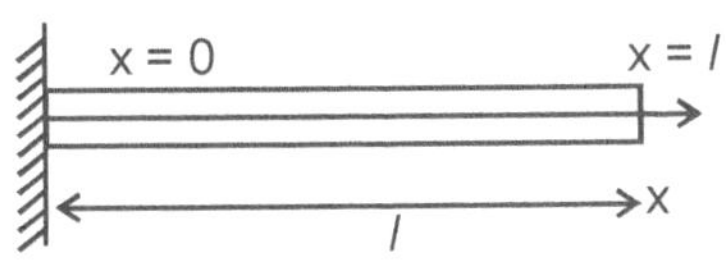

To solve equation we need 4 boundary conditions and 2 initial condition.

72. At x = 0 there is always a reactive moment which is proportional to $\frac{d^2w}{dx^2}$

thus condition (C) $\frac{d^2w}{dx^2}(0, t) = 0$ is not possible.

Linked answer 73 and 74:

73. Thin walled hollow shaft

Length l = 1m

Mean Radius R = 5 cm = 5×10^{-2}m

Torque to be transmitted = 7 kN – m

= 7×10^3 N – m

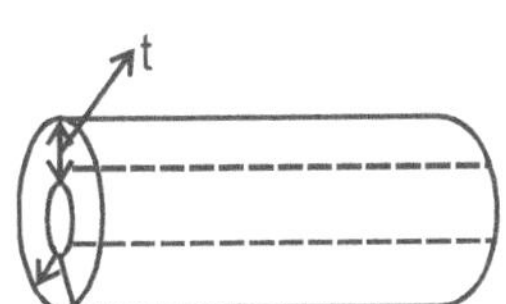

Shear stress developed in shaft is given by

$$\tau_{xy} = \frac{T}{I_p}\times r$$

Polar moment of inertia $I_p = \frac{4\bar{A}^2}{\oint \frac{ds}{t}}$

$$= \frac{4\times\pi R^2}{\frac{2\pi R}{t}} = 2\pi R^3 t$$

$$(\tau_{xy})_{max} = \frac{T}{2\pi R^3.t}R = \frac{T}{2\pi R^2 t}$$

Weight of shaft is given by w = m.g. = $\rho 2\pi R g l$

$$t = \frac{w}{\rho 2\pi R g l}$$

$$\Rightarrow (\tau_{xy})_{max} = \frac{T}{2\pi R^2\left[\frac{w}{\rho 2\pi R g \ell}\right]}$$

$$w = \left(\frac{T}{R}\times l g\right)\cdot\frac{\rho}{\tau_{max}}$$

Now given shaft ω is minimum if $\frac{\rho}{\tau_{max}}$ is minimum.

For X:

$$\frac{\rho}{(\tau_{max})} = \frac{\rho}{\tau_{yield}} = \frac{7700}{550\times10^6} = 14\times10^{-6}$$

For Y :

$$\frac{\rho}{\tau_{yield}} = \frac{7700}{225\times10^6} = 12\times10^{-6}$$

For Z :

$$\frac{\rho}{\tau_{yield}} = \frac{4875}{375\times10^6} = 13\times10^{-6}$$

Thus if we will choose Y, we will get minimum weight.

74. FOS = 2

$$\frac{\tau_{max}}{\tau_{Design}} = 2$$

$$\tau_{Design} = \frac{225}{2} = 112.5 \text{ MPa}$$

$M_t = T$

$$\frac{M_t\cdot R}{2\pi R^3 t^6} = 112.5$$

$$\Rightarrow t = \frac{7000}{2\times\pi\times112.5\times5\times10^{-2}}$$

$$= 3.961 \text{ mm} \approx 4\text{mm}$$

75. 4th order linear non-homogenous ordinary differential equation

$$\frac{d^2}{dx^2}\left(EI\frac{d^2w}{dx^2}\right) = q(x)$$

76. $0.2 \le \upsilon \le 0.4$

77. Semi monoceoque construction :

Skin and spars are primary carriers of shear stress due aerodynamic forces and moments.

78. Logarithmic decrement, δ = 0.125 (For SDOF) damping factor is given by

$$\xi = \frac{1}{\sqrt{1+\left(\frac{2\pi}{\delta}\right)^2}} = \frac{1}{\sqrt{1+\left(\frac{2\pi}{0.125}\right)^2}}$$

$$= 1.98\% \approx 2.0\%$$

79. Mode shape of an unamped 2DOFs are :

$$X_1 = \begin{Bmatrix} 1 \\ 0.5 \end{Bmatrix} \qquad X_2 = \begin{Bmatrix} 1 \\ -0.67 \end{Bmatrix}$$

$f_1 = 0.45$ Hz $\qquad f_2 = 1.2471$ Hz

$\omega_1 = 2\pi(0.45)$ rad/sec $\qquad \omega_2 = 2\pi(1.2471)$ rad/sec

Initial displacements $x(0) = \begin{Bmatrix} 2 \\ 1 \end{Bmatrix}$

Initial velocity $\dot{x}(0) = \begin{Bmatrix} 0 \\ 0 \end{Bmatrix}$

Solution is written as :

$$x = (A_1 \cos \omega_1 t + B_1 \sin \omega_1 t)\begin{bmatrix} 1 \\ 0.5 \end{bmatrix} + (A_2 \cos \omega_2 t + B_2 \sin \omega_2 t)$$

Using initial condition :

$$x(0) = \begin{bmatrix} 2 \\ 1 \end{bmatrix}$$

$$\begin{bmatrix} 2 \\ 1 \end{bmatrix} = A_1 \begin{bmatrix} 1 \\ 0.5 \end{bmatrix} + A_2 \begin{bmatrix} 1 \\ -0.67 \end{bmatrix}$$

$$\begin{bmatrix} 2 \\ 1 \end{bmatrix} = \begin{bmatrix} A_1 + A_2 \\ 0.5\,A_1 - 0.67\,A_2 \end{bmatrix}$$

Solution is :

$A_1 = 2$, $A_2 = 0$

Using other initial condition

$$\dot{x}(0) = \begin{bmatrix} 0 \\ 0 \end{bmatrix}$$

$$x(t) = (2\cos \omega_1 t + B_1 \sin \omega_1 t)\begin{bmatrix} 1 \\ 0.5 \end{bmatrix} + (B_2 \sin \omega_2 t)\begin{bmatrix} 1 \\ -0.67 \end{bmatrix}$$

$$\dot{x}(t) = [-2\omega_1 \sin \omega_1 t + B_1\omega_1 \cos \omega_1 t]\begin{bmatrix} 1 \\ 0.5 \end{bmatrix} + (B_2\omega_2 \cos \omega_2 t)\begin{bmatrix} 1 \\ -0.67 \end{bmatrix}$$

$$\begin{bmatrix} 0 \\ 0 \end{bmatrix} = [B_1\omega_1 \cos(0)]\begin{bmatrix} 1 \\ 0.5 \end{bmatrix} + B_2\omega_2 \cos(0)\begin{bmatrix} 1 \\ -0.67 \end{bmatrix}$$

$$B_1\omega_1 + B_2\omega_2 = 0$$

$$\Rightarrow \boxed{B_1 = \frac{-B_2\omega_2}{\omega_1}}$$

$$0.5\,B_1\omega_1 - 0.67\,B_2\omega_2 = 0$$

$$\Rightarrow 0.5\left[\frac{-B_2\omega_2}{\omega_1}\right]\omega_1 - 0.67\,B_2\,\omega_2 = 0$$

$\Rightarrow B_2 = 0$

$\Rightarrow B_1 = 0$

Solution

$$x(t) = 2(\cos w_1 t)\begin{bmatrix} 1 \\ 0.5 \end{bmatrix}$$

$\therefore$ Maximum amplitude in mm is 2 mm.

80. Airy stress functions

$$\phi = \alpha x^2 + \beta xy + \gamma y^2$$

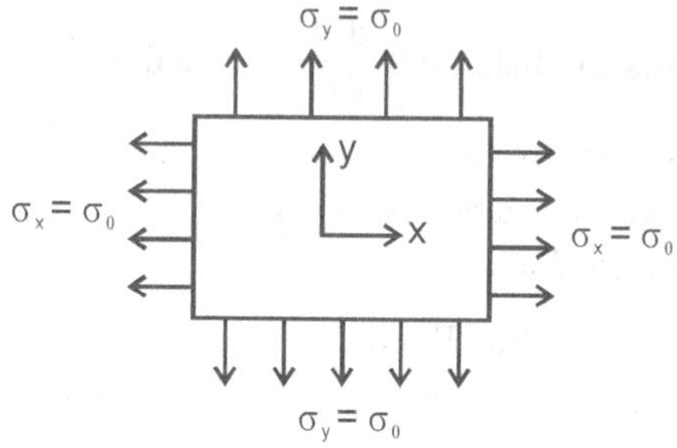

Plate subjected to uniform tensile stress σ_0 at four edge

$$\frac{\partial^2 \phi}{\partial x^2} = \sigma_y = 2\alpha$$

If $\alpha = \frac{\sigma_0}{2}$ then $\boxed{\sigma_y = \sigma_0}$

$$\frac{\partial^2 \phi}{\partial y_2} = \sigma_x = 2\gamma$$

If $\gamma = \frac{\sigma_0}{2} \Rightarrow \boxed{\sigma_y = \sigma_0}$

$$\frac{\partial^2 \phi}{\partial x 2y} = \sigma_{xy} = -\beta$$

if $\beta = 0$ $\boxed{\sigma_{xy} = 0}$

81. Boundary condition of rod under longitudinal vibration.

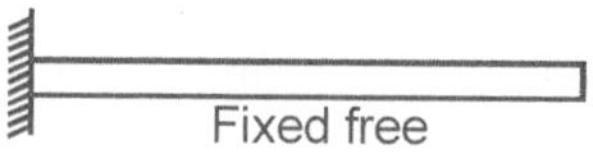

For fixed-fixed case

Natural frequency is given by

$w_n = \frac{n\pi c}{l} = \frac{\pi c}{l}$ (n = 1) n = 1, 2, 3, ... First natural frequency

For fixed-free

$w_n = \frac{(2n+1)}{2l}\pi C = \frac{\pi c}{2l}$ (n = 0)

n = 0, 1, 2 .....

Fixed free natural frequency is $\frac{1}{2}$ times fixed-fixed natural frequency

$k = \frac{1}{2}$

82. Spring mass system with viscous damper

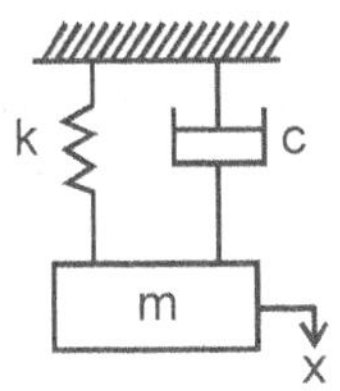

Harmonic vibration

$x = X\cos\omega_d t$

$v = \dot{x} = -X\omega_d \sin\omega_d t$

Total force resisting motion

$F = -kx - cv = -10x - c\dot{x}$

$= -kX\cos\omega_d t + c\omega_d X\sin\omega_d t$

Energy dissipated is given by

$$\Delta\omega = \int_{t=0}^{2\pi/\omega_d} Fv\,dt$$

$$= \int_0^{2\pi/\omega_d} kX^2\omega_d \sin\omega_d t.\cos\omega_d t.d(\omega_d t)$$

$$= \int_0^{2\pi/\omega_d} c\omega_d \sin^2\omega_d t.\cos\omega_d t.d(\omega_d t)$$

$$\boxed{\Delta\omega = \pi c\omega_d X^2}$$

83. Only possible way from given option is increasing internal pressure which will resist buckling loads of fuselage skin.

Other way is by placing stifferss closer to each other.

Linked questions 84 & 85 :

84. Thin walled spherical vessel

(internal diameter = 1m, wall thickness 10 mm = 0.01 m)

Stresses are given by

$$\sigma_c = \frac{pd}{4t}$$

$\sigma_{yield} = 500$ MPa

For a - thin walled spherical vessel, stress other than $\sigma_{\theta\theta}$ and $\sigma_{\phi\phi}$ are zero (spherical coordinate r, θ, ϕ)

$$\sigma_{\theta\theta} = \sigma_{\phi\phi} = \frac{pd}{4t}$$

According to von-misses criterion.

$$\sigma_{\theta\theta}^2 + \sigma_{\phi\phi}^2 - \sigma_{\theta\theta}\sigma_{\phi\phi} = \sigma_y^2$$

$$\left(\frac{pd}{4t}\right)^2 + \left(\frac{pd}{4t}\right)^2 - \left(\frac{pd}{4t}\right)^2 + \left(\frac{pd}{4t}\right)^2 = (500)^2$$

$$\left(\frac{pd}{4t}\right) = 500$$

$$P = \frac{500 \times 4t}{d} = \frac{500 \times 4 \times 0.01}{1}$$

= 20 MPa

85. Yielding of material will occur at same pressure P_y. As yield strength is material property it will not change due to loading condition.

86. Damped SDOFs

Amplitude ratio = 2.5 at resonance.

For SDOF's amplitude ratio

$$\left|\frac{x_2}{x}\right| = \frac{1}{\sqrt{(1-\Omega^2)^2 + (2\xi\Omega)^2}}$$

at resonance $\Omega = \frac{\omega}{\omega_n} = 1$

$$\Rightarrow \frac{1}{\sqrt{(2\xi)^2}} = 2.5$$

$$\Rightarrow 2\xi = \frac{1}{2.5}$$

$$\Rightarrow \xi = \frac{1}{2.5 \times 2} = 0.2$$

87. Does not move axially, but can move horizontally and vertically

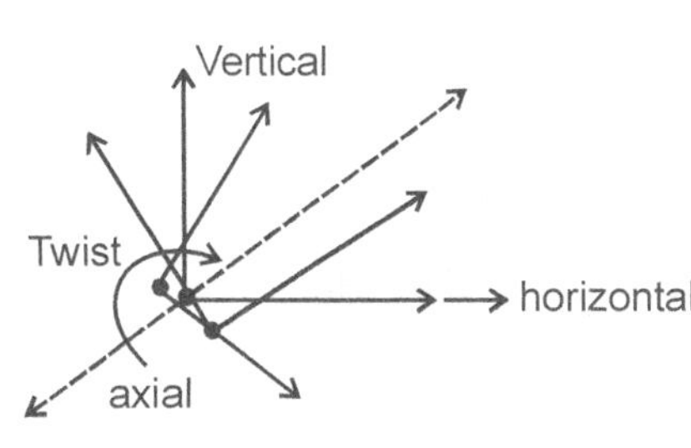

88. Uniform thickness = 2mm

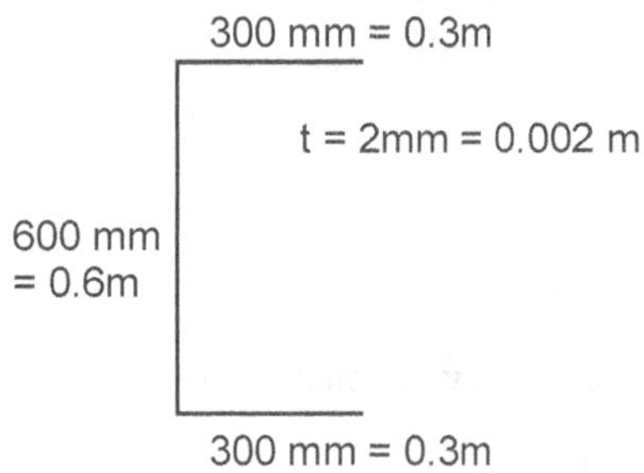

Torque = 10 Nm

Shear modulus G = 25 GPa

Twist per unit length is given by

$T = GJ\theta$

Polar moment of inertia J, For channel section

$$J = \sum \frac{1}{3} bt^3 = \frac{1}{3}[0.3 \times t^3 + 0.6 \times t^3 \times 0.3 \times t^3]$$

$$= 3.2 \times 10^{-9} \, m^4$$

$$\therefore \text{ Twist per unit length } \theta = \frac{T}{GJ}$$

$$= \frac{10}{(25 \times 10^9) \times (3.2 \times 10^{-9})}$$

= 0.125 rad which lies in between 0.12 & 0.13.

89. Cross-section A, B, C, D are of equal area = 3 cm^2

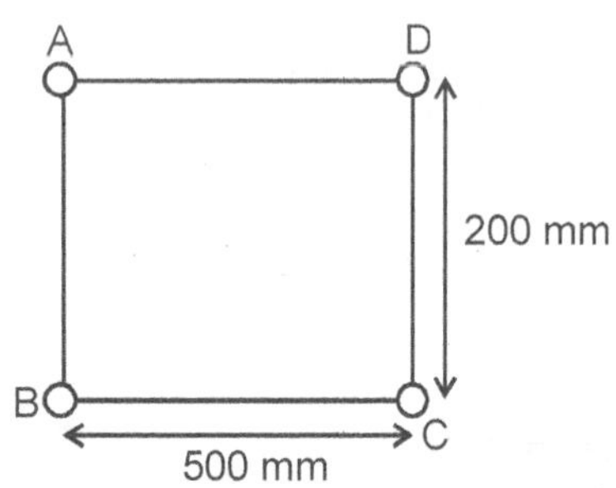

Thickness of each web AB, BC, CD, DA = 5 mm.

Twishing moment = 10 kNm

A = Area enclosed

= 200 × 500 mm^2

$$= \frac{200 \times 500}{(1000)^2} m^2$$

Shear flow q in such member is given by

$$q = \frac{T}{2A} = \frac{10 \times 10^3}{\frac{2 \times (200 \times 500)}{(1000)^2}}$$

= 50000 N/m = 50 KN/m

90. Angle of inclination between neutral Axes and z axis is given by

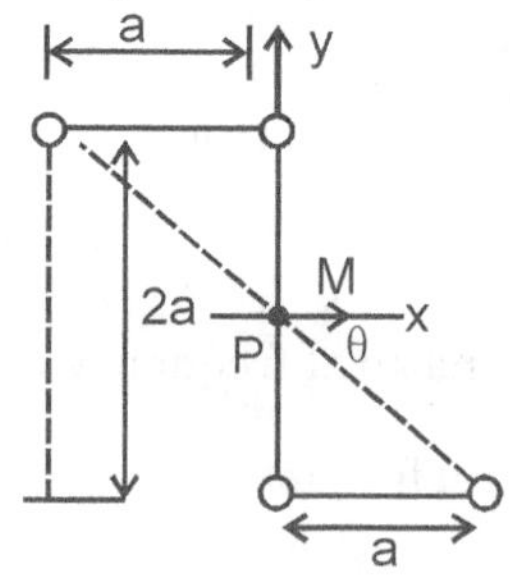

$$\theta = \tan^{-1}\left(\frac{y}{x}\right)$$

Point P is centroid of section

As each boom is at equal distance of a.

x, y are co-ordinates of centroid point P

$$\theta = \tan^{-1}\left(\frac{a}{a}\right) = 45° \text{ clockwise}$$

91. As we know twist per unit length in outer and inner material is same.

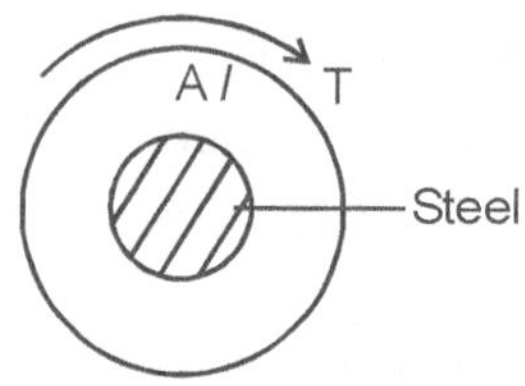

∴ Shear strains are same in both material.

But shear modulus of inner and outer material is different

$\therefore T_{outer} = G_{out} \times \gamma_{outer} = G_{Al} \times \gamma$

$\tau_{inner} = G_{in} \times \gamma_{inner} = G_{stre} \times \gamma$

Here stresses are discontinous.

Thus, only strain is continuous.

92. Equilibrium of force

$R_A + R_B + R_C = P$ ... (A)

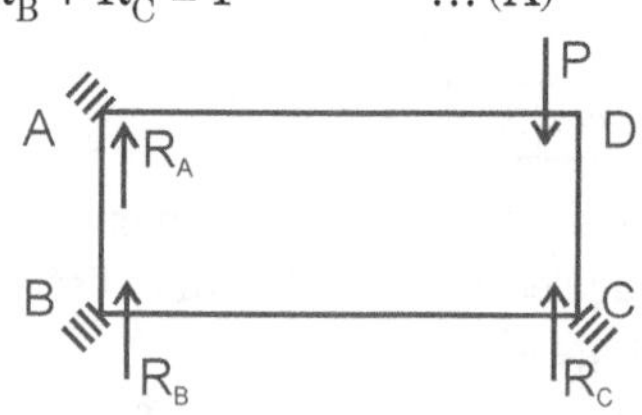

Moment about AB

$R_C \times BC = P \times AD$

$\boxed{R_C = P}$

Momentum about BC

$R_A \times AB = P \times DC$

Using (A)

$$\boxed{R_A = P} \Rightarrow \boxed{R_B = -P} \quad (P, -P, P)$$

93. E = 200 GPa

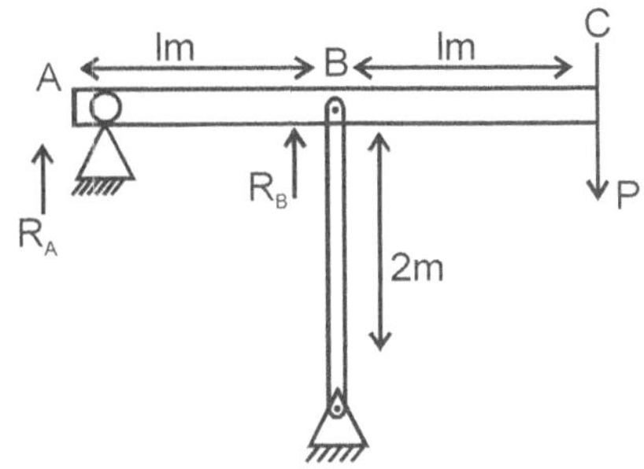

Radius r = 10 mm = 0.01 m

Solving Beam AC first :

Force equilibrium

$\Sigma F_y = 0$

$R_A + R_B = P$

Taking moment about A

$R_B \times 1 = P \times 2$

$\Rightarrow \boxed{R_B = 2P}$

$\therefore \boxed{R_A = -P}$

For pinned-pinned

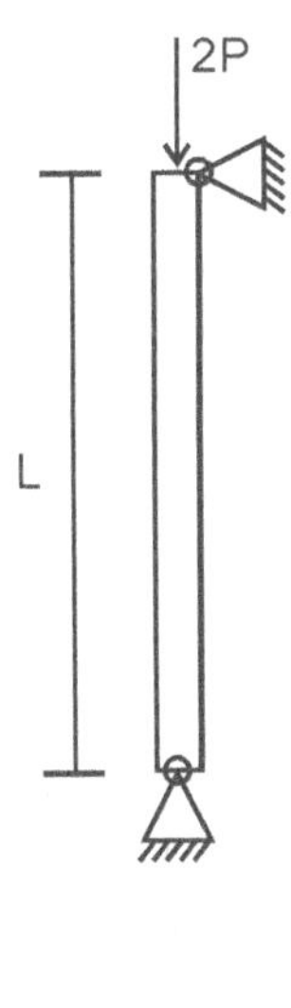

$$(2P_{cr}) = \frac{\pi^2 EI}{L^2}$$

$$P_{cr} = \frac{\pi^2 \times 200 \times 10^9 \times \frac{\pi}{64} \times (2 \times 0.01)^4}{2^2 \times 2}$$

= 1934.94 N

$\therefore P_{cr} = (1900 - 2000)$ N

94. Pure bending moment

Stresses developed is given by

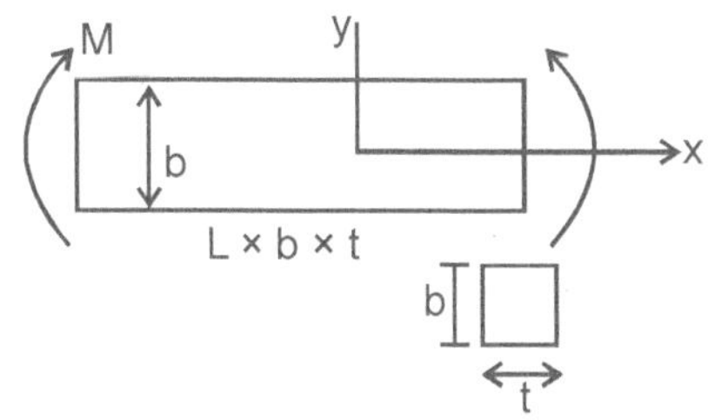

$$\sigma_{xx} = \frac{My}{I} \qquad y = \frac{b}{2} \text{ at upper surface}$$

$$= \frac{M\frac{b}{2}}{\frac{1}{12} \times tb^3}$$

$$\boxed{\sigma_{xx} = \frac{6M}{tb^2}}$$

All other stress σ_{yy} & σ_{xy} are zero

If we take Airy stress function

$$\phi = \frac{2M}{tb^3} y^3$$

$$\sigma_{xx} = \frac{\partial^2 \phi}{\partial y^2} = \frac{2M}{tb^3} \times \sigma y\Big|_{y=\frac{b}{2}} = \frac{2M \times \sigma \times \frac{b}{2}}{tb^3} = \frac{6M}{tb^2}$$

95. Cantilever beam :

Length = 0.6 m

Natural freqency = 10 rad/sec

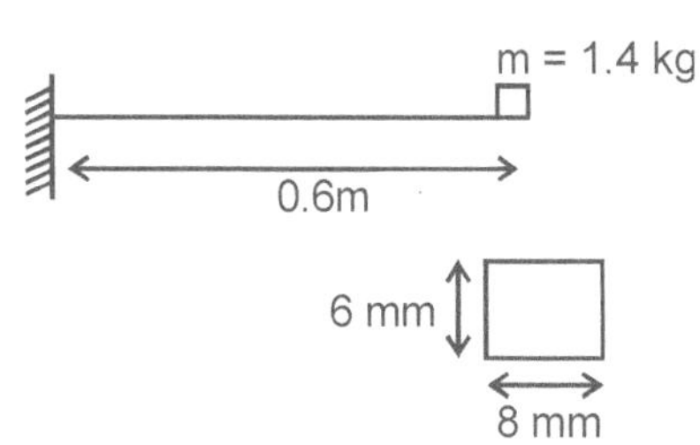

$$I = \frac{8 \times 6^3}{12}$$

Equivalent stiffness of cantilever beam

$$k = \frac{3EI}{L^3}$$

Natural frequency $\omega_n = \sqrt{\frac{k}{m}} = \sqrt{\frac{3EI}{mL^3}}$

$$\Rightarrow E = \frac{\omega_n^2 \times m \times L^3}{3 \times I} = \frac{10^2 \times 1.4 \times 0.6^3}{\dfrac{3 \times \left(\dfrac{8 \times 6^3}{12}\right)}{10^{12}}}$$

$\Rightarrow E = 70 \times 10^9$ Pa

$= 70$ GPa

96.

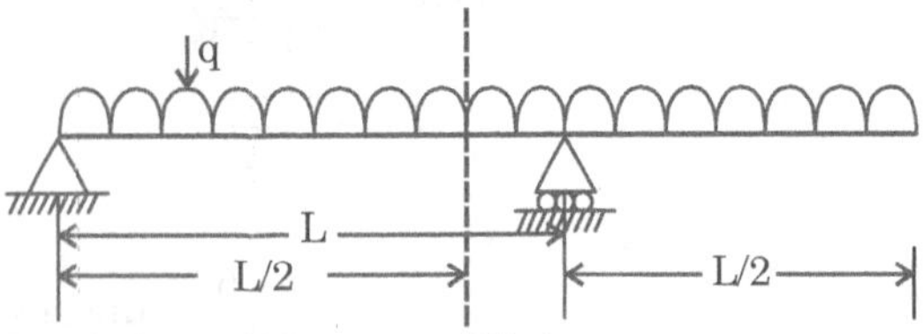

Reactions : Force equilibrium

$$R_A + R_B = \left(L + \frac{L}{2}\right)q = \frac{3L}{2}q$$

Moment equilibrium

Taking moment about A

$$R_B \times L = q \times \frac{3L}{2} \times \frac{1}{2}\left(\frac{3L}{2}\right)$$

$$= \frac{9q}{8}L^2$$

$$\Rightarrow \boxed{R_B = \frac{9q}{8}L}$$

$$\Rightarrow R_A = \frac{3L}{2}q - \frac{9qL}{8} = \frac{12Lq - 9Lq}{8} = \frac{3Lq}{8}$$

Bending moment at midpoint of A – B is given by

$$M = R_A \times \frac{L}{2} - q \times \frac{L}{2} \times \frac{1}{2}\frac{L}{2}$$

$$= \frac{3Lq}{8} \times \frac{L}{2} - \frac{q}{8}L^2$$

$$= \frac{3qL^2}{16} - \frac{q}{8}L^2 = \frac{3qL^2 - 2qL^2}{16} = \frac{qL^2}{16}$$

$$\boxed{M = \frac{qL^2}{16}}$$

Sagging

97. Tensile load

F = 49 kN = 49000 N

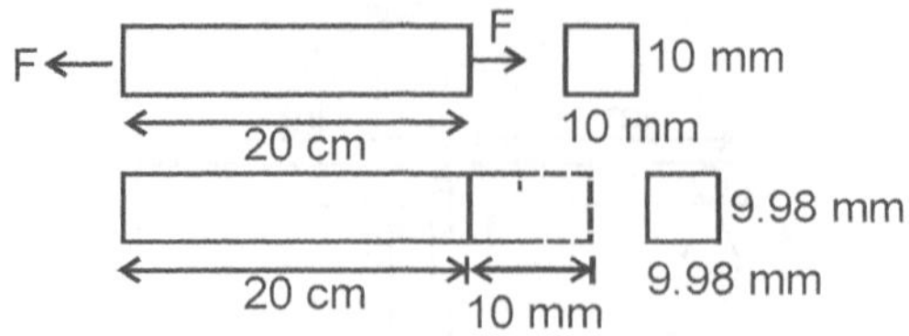

Orignal cross section = 10 × 10 mm

= 0.01 × 0.01 m

Length l = 20 cm = 0.2 m

We know poisson ratio υ

$$\upsilon = \frac{-\text{Lateral strain}}{\text{Longitudinal strain}}$$

$$= \frac{\left(\dfrac{10 - 9.98}{10}\right)}{\left(\dfrac{20.1 - 20}{20}\right)}$$

$$\boxed{\upsilon = 0.4}$$

and Young's modulus = 80 GPa

98. Mass does not vary with loading

99.

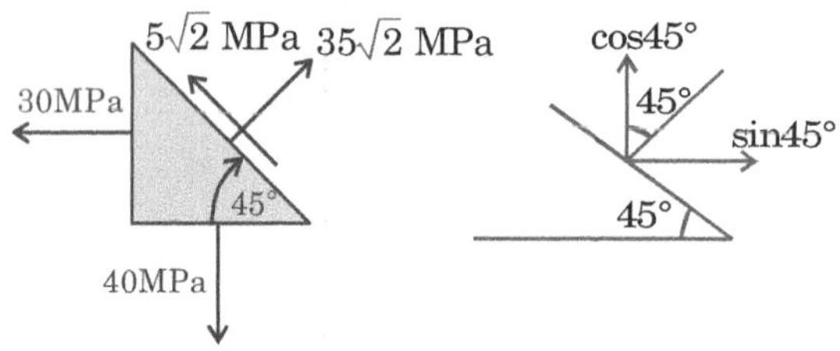

For static equlibrium $\Sigma F_{x,y} = 0$

x-direction

$$\Sigma F_x = 30 - 35\sqrt{2}\sin 45° + 5\sqrt{2}\sin 45° = 0$$

$$\Sigma F_y = 40 - 35\sqrt{2}\cos 45° - 5\sqrt{2}\sin 45° = 0$$

100. Undamped natural frequency

ω_n = 10Hz = 2π 10 rad/sec

According to half power width method,

$$2\xi = \frac{\omega_a - \omega_b}{\omega_r}$$

$$\omega_r = \omega_n\sqrt{1 - \xi^2}$$

$\omega_a = 64.7168$

$\omega_b = 60.9469$

$$\xi\sqrt{1-\xi^2} = \frac{64.7168 - 60.9469}{2 \times 2 \times \pi \times 10}$$

Solving equation

$\xi = 0.03001$

$\xi = 3\%$

101. Torsional vibration of bar

Natural frequency for fixed free case is:

$$w_n = \frac{(2n+1)}{2l}\pi C$$

n = 0, 1, 2, ...

$$\boxed{w_1 = \frac{\pi c}{2l}}$$

where

$$C = \sqrt{\frac{GJ}{I_0}}$$

$$= \sqrt{\frac{GJ}{\rho J}}$$

Fixed-fixed case :

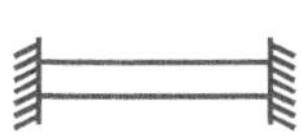

$$w_n = \frac{n\pi C}{l}$$

n = 1, 2, 3

$$\boxed{w_1 = \frac{\pi C}{l}}$$

Which is 2 times the fixed-free case

102. Slope which is given by $\frac{dy}{dx}$ can be easily calculated using bending moment equation and Macaulay's methods.

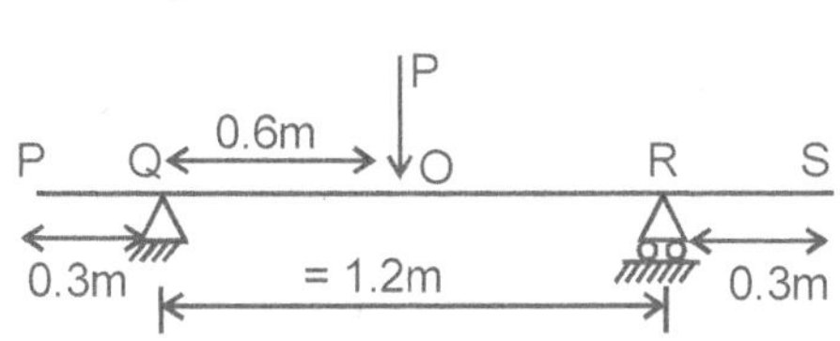

$$EI\frac{d^2y}{dx^2} = M_x$$

In our case, as there is no loading on the overhang.

∴ Slope at S will be same as that of R which is given by

$$\frac{dy}{dx} = \frac{Pl^2}{16EI}$$

$$= \frac{50 \times (1.2)^2}{16 \times 200 \times 10^9 \times \frac{1}{12} 0.01 \times 0.006^3}$$

= 0.125

103. Shear stress on first shaft

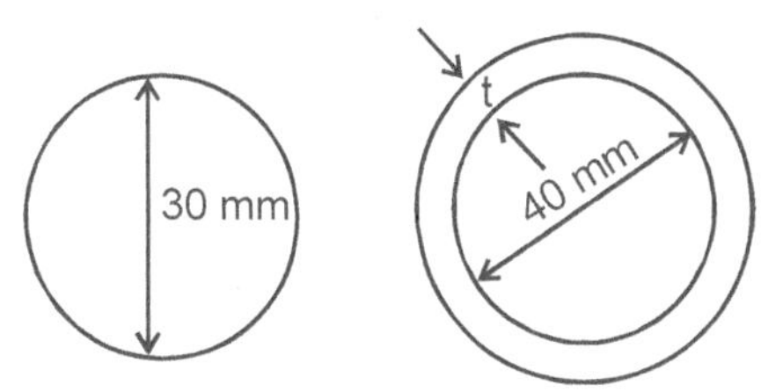

$$\text{mass } m_f = \rho\left(\frac{\pi d_1^{\,2}}{4}\right) \times \ell$$

$$\text{mass } m_s = \rho\frac{\pi\left(D^2 - d_2^{\,2}\right)}{4} \times \ell$$

$$\text{Mass Ratio} = \frac{m_f}{m_s} = \frac{d_1^{\,2}}{D^2 - d_2^{\,2}}$$

Using torque T is given by

$$\tau = \frac{T}{J} \times r = \frac{T}{\frac{\pi}{32}\left(d_1^4\right)} \times \left(\frac{d_1}{2}\right) \qquad \text{... (i)}$$

For 2nd shaft

$$\tau = \frac{T}{J_2} \times r = \frac{T}{\frac{\pi}{32}\left(D^4 - d_2^4\right)} \times \left(\frac{D}{2}\right) \qquad \text{... (ii)}$$

Dividing (i) by (ii)

$$\frac{d_1}{d_1^4} \times \frac{D^4 - d_2^4}{D} = 1$$

$D^4 - (40)^4 = (30)^3 D$

D = 43.99 mm

$$\therefore \text{Mass ratio} = \frac{(30)^2}{(43.99)^2 - (40)^2} = 2.68$$

104. Critical buckling load for fixed-fixed column is given by :

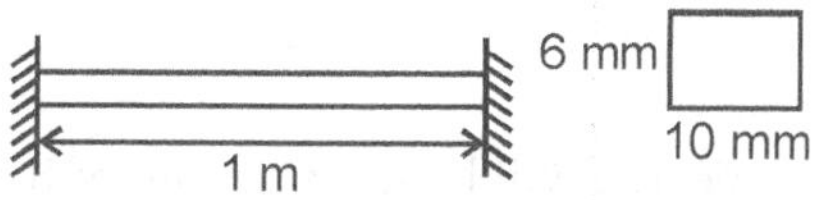

$$p_{cr} = \frac{4\pi^2 EI}{l}$$

$$= \frac{4 \times \pi^2 \times 200 \times 10^9 \times \frac{1}{12} \times 0.01 \times 0.06^3}{1}$$

(E = 200 GPa)

= 1421.223 N

$$\sigma_{cr} = \frac{P_{cr}}{A} = \frac{1421.223}{(0.01 \times 0.006)} = 23687050 \text{ Pa}$$

= 23.687 MPa

Coefficient of thermal expansion α

$= 12 \times 10^{-6}/°C$

Thermal stresses are given by

$\sigma_{th} = E \times \varepsilon_{th} = E \times \alpha \, \Delta T$

Critical $\sigma_{th} = \sigma_{cr} = 23.687$ MPa

$$\Delta T = \frac{23.68 \times 10^6}{200 \times 10^9 \times 12 \times 10^{-6}}$$

= 9.869 = 9.87 °C

105. Cylinder subjected to pressure

$P = 5 \text{ bar} = 5 \times 10^5$ Pa

Hoop stress,

$$\sigma_c = \frac{pd}{2t} = \frac{p(d_{out} - 2t)}{2t} \qquad \text{dia } d = d_{out} - 2t$$

$$= \frac{5 \times 10^5 (30 \times 10^{-2} - 2 \times 0.001)}{2 \times 0.001}$$

= 74.5 MPa

Longitudinal stresses

$$\sigma_L = \frac{pd}{4t} = \frac{1}{2} \times \sigma_c = 37.25 \text{ MPa}$$

$\sigma_y = 260$ MPa

According to Von-miss criteria :

$(\sigma_{eq} = \sigma_c^2 + \sigma_l^2 - \sigma_c\sigma_l)$

$$\text{FOS} = \frac{\sigma_y}{\sigma_{eq}}$$

$$\text{FOS} = \frac{260}{\sqrt{(74.5)^2 + (37.25)^2 - (74.5)(37.25)}}$$

= 4.0625 = 4

106. Airy stress tention

$$\phi = c_1xy + c_2xy^3 + c_3x^2y^2 + c_4x^3y$$

$$\text{if } c_1 = \lambda, c_2 = -\frac{\lambda}{3d^2}, c_3 = 0, c_4 = 0$$

$$\phi = \lambda xy + \left(\frac{-\lambda}{3d^2}\right)xy^3 \qquad \text{... (A)}$$

$$\sigma_{xx} = \frac{\partial^2\phi}{\delta y^2} = \left(\frac{-\lambda}{3d^2}\right)(\sigma_{xy}) = \frac{-2\lambda xy}{d^2}$$

$$\sigma_{yy} = \frac{\partial^2\phi}{\partial x^2} = 0$$

$$\tau_{xy} = \frac{-\partial^2\phi}{\partial x \partial y} = -\left(\lambda - \frac{\lambda}{d^2}y^2\right)$$

$$= -\lambda\left(1 - \frac{y^2}{d^2}\right)$$

Equilibrium equation

$$\frac{\partial\sigma_{xx}}{\partial x} + \frac{\partial\tau_{xy}}{\partial y} = 0$$

$$\frac{-2\lambda y}{d^2} + \frac{2\lambda y}{d^2} = 0$$

Hence sutisfied.

Biharmonic equation:

$$\frac{\partial^4\phi}{\partial y^4} + \frac{\partial^4\phi}{\partial xy} + 2\frac{\partial^2}{\partial y^2}\frac{\partial^2\phi}{\partial x^2} = 0$$

In equation (A) variable of x are of only degree 1 i.e., x^1 [all 2nd dirivatives of x are zero and 4th dirivatives of y are zero]

$\Rightarrow 0 + 0 + 2 \times 0 = 0$

It is also satisfied.

107. Length of beam $l = 0.6$ m

$I = 8.33 \times 10^{-10}$ m^4

E = 210 GPa

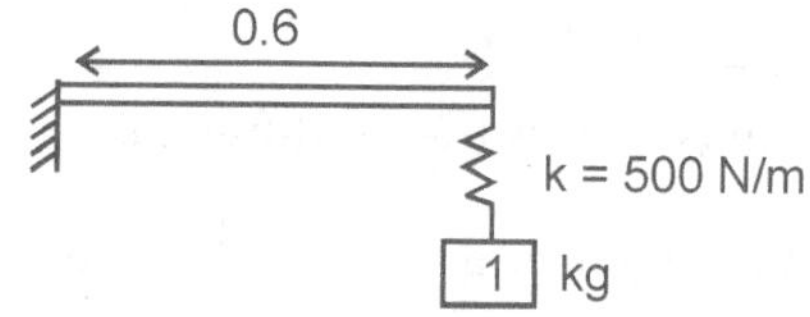

As beam is massless, whole system can be converted into two springs in series

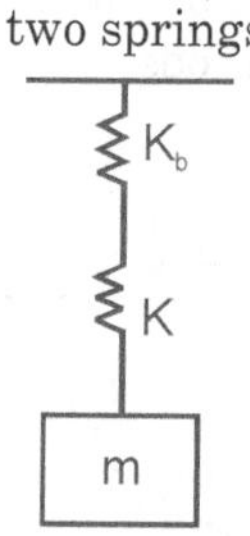

Stiffness of beam

$$k_b = \frac{3EI}{l^3}$$

$$= \frac{3 \times 210 \times 10^9 \times 8.33 \times 10^{-10}}{(0.6)^3}$$

= 2429.58 N/m

Fundamental natural frequency is given by

$$\omega_n = \sqrt{\frac{k_{eq.}}{m}} = \sqrt{\frac{\frac{1}{\left(\frac{k_b + k}{k_b \times k}\right)}}{1}}$$

$$= \sqrt{\frac{\frac{1}{\left(\frac{2429.58 + 500}{2429.58 \times 500}\right)}}{1}}$$

= 20.36 rad/sec.

108. SDOF's : Initial amplitude = 5 cm

$\xi = 2\% = 0.02$

Amplitude of 5th cycle

$$\ell n\left(\frac{A_1}{A_5}\right) = 2\pi(n-1) \times \xi$$

$$\ell n\left(\frac{A_1}{A_5}\right) = 2\pi(5-1) \times 0.02$$

$$\Rightarrow \frac{A_1}{A_5} = e^{(2 \times \pi \times 4 \times 0.02)}$$

$$\Rightarrow \boxed{A_5 = 3.02 \text{ cm}}$$

109. If a lateral force is applied at shear centre of beam it will always bend without causing any twist.

110. Principal stresses

$\sigma_1 = 25$ MPa

$\sigma_2 = -25$ MPa

Maximum shear stress is given by

$$\tau_{max} = \frac{\sigma_1 - \sigma_2}{2} = \frac{25 - (-25)}{2} = 25 \text{ MPa}$$

111. Static displacement of mass m is given by

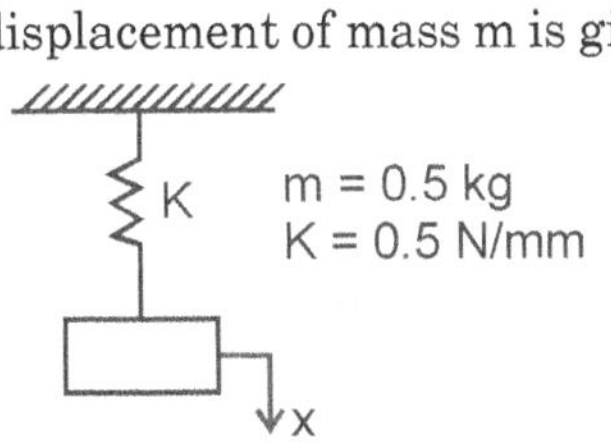

$$x_{st} = \frac{m.g.}{k} = 0.981 \text{ mm}$$

112. I and II result in identical stress distribution at section located far away

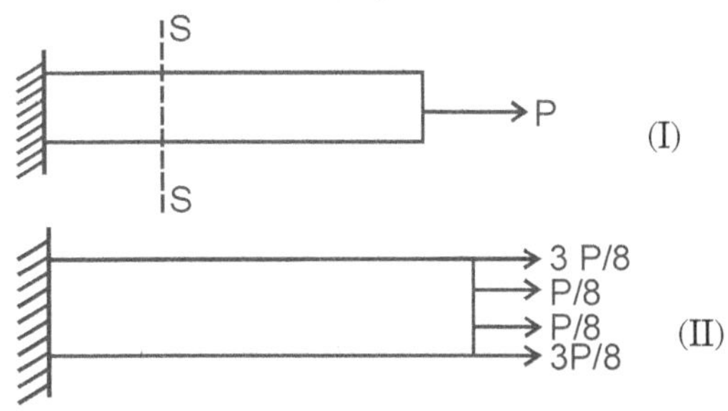

Sum of 3P/8, P/8, P/8, 3P/8 these all and direction is same as I i.e. P → and all loads are symmetric about axis of bar

113. 2–D stress state at a point P in xy co-ordinate is given by :

$$\begin{bmatrix} 60 & 50 \\ 50 & -40 \end{bmatrix} \text{MPa} = \begin{bmatrix} \sigma_{xx} & \tau_{xy} \\ \tau_{xy} & \sigma_{yy} \end{bmatrix}$$

$$\boxed{\tau_{xy} = 50 \text{ MPa}}$$

114. Cube subjected to uniform hydrostatic pressure

P = 100 N/mm^2 = 100 × 10^6 Pa

Initial volume = V

Volume after shrinking = V – 0.05% V

Young's Modulus E = 300 GPa

Bulk Modulus of elasticity

$$k = \frac{\text{Volumetric stress}}{\text{Volumetric strain}}$$

$$= \frac{P}{\frac{\text{change in volume}}{\text{orignal volume}}}$$

$$= \frac{100 \times 10^6}{\frac{\frac{0.05}{100}V}{v}}$$

= 200 × 10^9 Pa = 200 GPa

Now relation between bulk modulus, young's modulus and position ratio.

$$k = \frac{E}{3(1-2v)}$$

$$\Rightarrow 1 - 2\upsilon = \frac{E}{3k}$$

$$\Rightarrow \upsilon = \frac{\left(1 - \frac{E}{3k}\right)}{2} = \frac{\left(1 - \frac{300 \times 10^9}{3 \times 200 \times 10^9}\right)}{2}$$

$\Rightarrow \boxed{\upsilon = 0.25}$ Poisson ratio.

115.

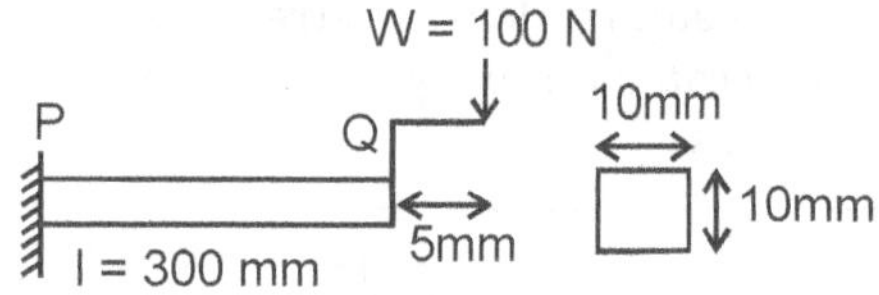

$$K_{eq} = \frac{3EI}{l^3}$$

$$= \frac{3 \times 200 \times 10^9 \times \frac{1}{12} \times (0.01)^4}{(300 \times 10^{-3})^3}$$

= 18518.51852 N/m

Deflection

$$x_{st.} = \frac{w}{k_{eq}} = \frac{100}{18518.51852} = 5.45 \times 10^{-3}\,\text{m}$$

= 5.45 mm = 5.5 mm

Another method

E = 200 GPa

Massless beam can be converted to equivalent spring.

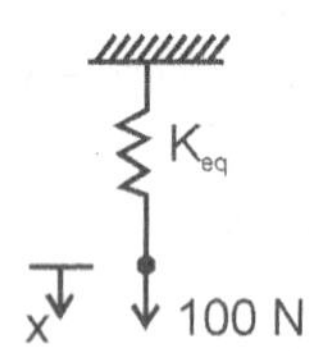

It can also be calculated as :

Used bending movement equation .

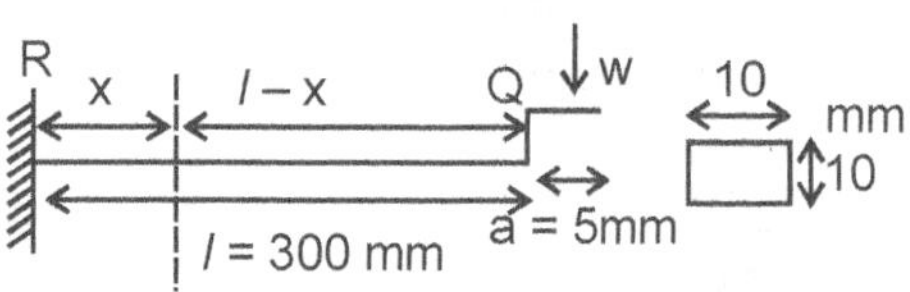

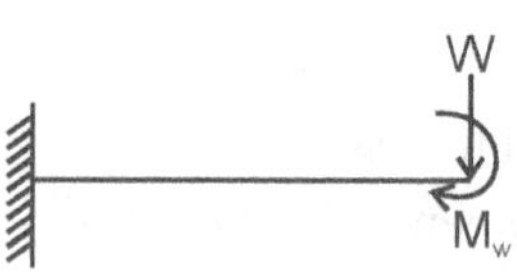

$M_x = -w(l - x) - M_w$

$= -w(l - x) - w \times a$

$$EI\frac{d^2y}{dx^2} = -w(l - x) - w \times a$$

Integrating it once,

$$EI\frac{dy}{dx} = -w\left(lx - \frac{x^2}{2}\right) - w \times a \times x + c_1 \qquad \text{... (A)}$$

B.C. at x = 0 $\quad \frac{dy}{dx} = 0$

$\Rightarrow \boxed{C_1 = 0}$

again integrating equation (A)

$$EIy = -w\left(l\frac{x^2}{2} - \frac{x^3}{6}\right) - wa\frac{x^2}{2} + c_2$$

B.C. at x = 0, $\quad$ y = 0

$\Rightarrow \boxed{C_2 = 0}$

$$y = \frac{1}{EI}\left[\left[-w\left(l\frac{x^2}{2} - \frac{x^3}{6}\right)\right] - wa\frac{x^2}{2}\right]$$

Now deflection at point x = l i.e. Q

$$y = \frac{1}{EI}\left[\left[-w\left(\frac{l^3}{2} - \frac{l^3}{6}\right)\right] - wa\frac{l^2}{2}\right]$$

$$= \frac{1}{EI}\left[-\frac{wl^3}{3} - wa\frac{l^2}{2}\right]$$

$$= \frac{1}{200 \times 10^9 \times \frac{1}{12} \times (0.01)^4}$$

$$\left[\frac{-100 \times 0.3^3}{3} - \frac{100 \times 0.005 \times 0.3^2}{2}\right]$$

= –5.535 × 10^{-3} m

= –5.5 mm

$\boxed{|y| = 5.5\text{ mm}}$

116. SDOF system:

mass m = 10 kg.

stiffens k = 2.25 N/mm

damping cofficient c = 0.0125 N-s/mm

Ratio of two successive amplitude is given by

$$\delta = \left|\frac{x_1}{x_2}\right| = e^{\frac{2\pi\xi}{\sqrt{1-\xi^2}}}$$

$$\xi = \frac{c}{2\sqrt{km}} = \frac{0.0125}{2\sqrt{2.25 \times 10}} = 0.0416$$

$\boxed{\left|\frac{x_1}{x_2}\right| = 1.29}$

117. Shear Modulus = 80 GPa
Twist angle θ = 0.0727 rad

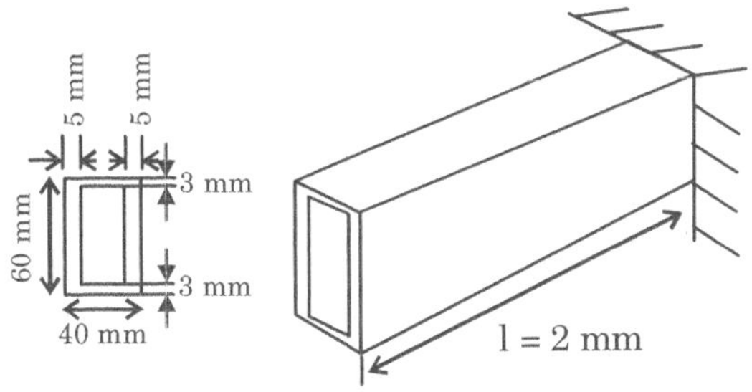

We know that

$$\frac{T}{J} = \frac{c\theta}{l}$$

$$T = \frac{c\theta}{l} \times J$$

$$= \frac{c\theta}{l} \times \frac{4\bar{A}^2}{\oint \frac{ds}{t}}$$

c = Shear modulus
l = length = 2m

$(\bar{A} = a \times b)$

$$= \frac{80 \times 10^9 \times 0.0727}{2} \times \frac{4 \times [(0.06 - 0.003) \times (0.04 - 0.005)]^2}{\left[\frac{0.04 - 0.005}{0.003} + \frac{(0.06 - 0.003)}{0.005}\right] \times 2}$$

= 1003.42 Nm

118. 200 mm long simply supported column
E = 200 GPa

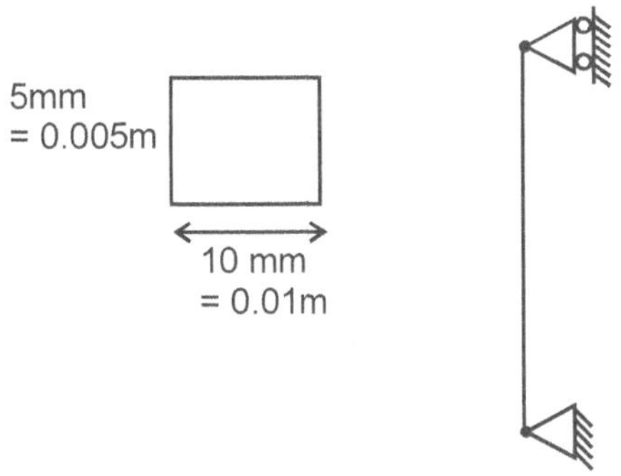

Factor of safety = 2.5
Critical load

$$p_{cr} = \frac{\pi^2 EI}{l^2}$$

$$= \frac{\pi^2 \times 200 \times 10^9 \times \frac{1}{12} \times 0.01 \times 0.005^3}{0.2^2}$$

= 5140.4189 N

$$FOS = \frac{p_{cr}}{p_{max}}$$

$$\Rightarrow P_{max} = \frac{p_{cr}}{FOS} = \frac{5140.4189}{2.5} = 2056.16 \text{ N}$$

119. 2 material constants i.e., Young's modulus E and poisson's ratio υ are needed to describe 3-D hooks law for isotropic material

120. In plane stress problem, stresses τ_{xz}, τ_{yz} are zero which emply shear strains are zero in thickness direction. But normal strains are there due to σ_{xx} and σ_{yy} because of poison's effect.

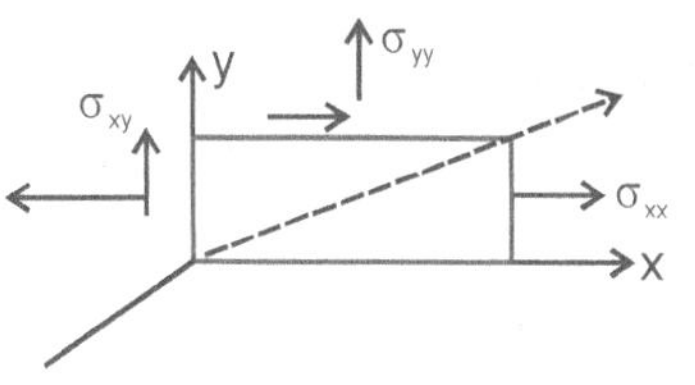

i.e., stress in one direction causes strain in other.

121.

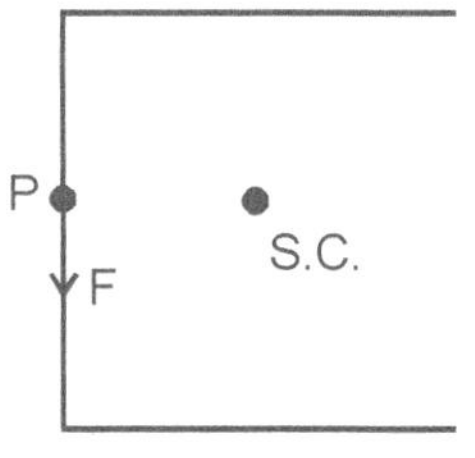

Force P is not acting at shear center as in other case. That it will produce bending and twisting.

122. Two springs are in series

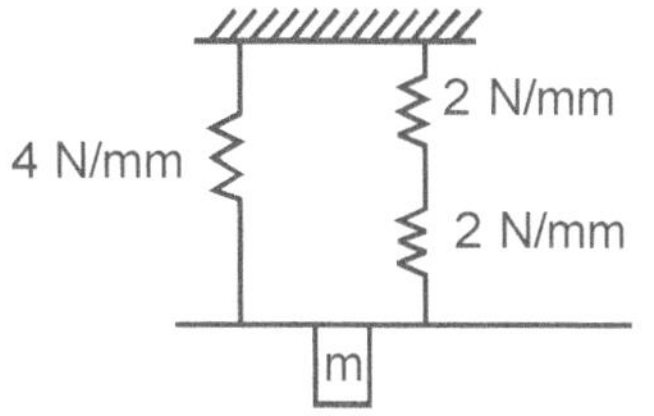

$\therefore$ equivalent stiffness

$$\frac{1}{k_{eq}} = \frac{1}{2} + \frac{1}{2}$$

$\Rightarrow k_{eq.}$ = 1 N/mm

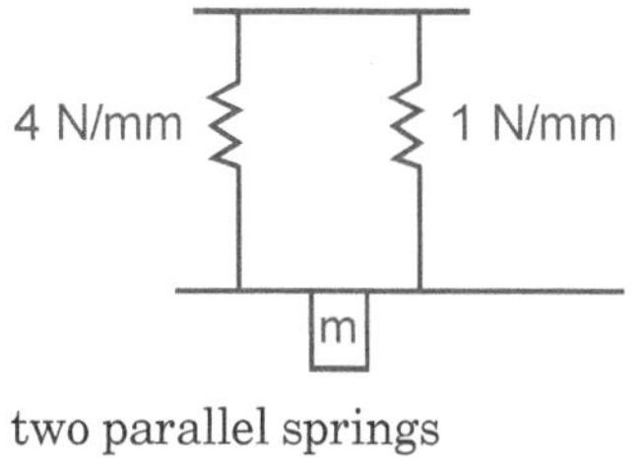

two parallel springs
$\Rightarrow k_{eq.}$ = 4 + 1 = 5 N/mm

123. State of pure shear stress of 50 N/mm^2. Principle planes are always oriented 45° or 135° to the pure stress state $\frac{\pi}{4}$ & $\frac{3\pi}{4}$

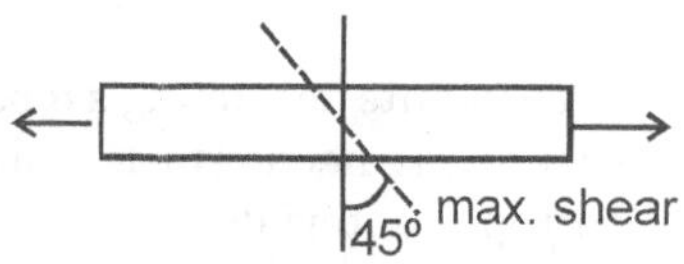

124. Strain energy density stored is given by:

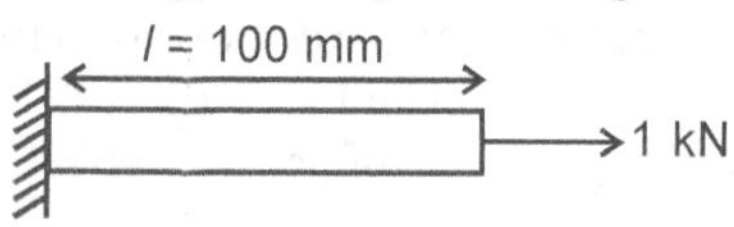

Area A = 100 mm^2

= 100 × 10^{-6} m^2

Young's modulus = 1 × 10^5 N/mm^2

$$w = \frac{1}{2}(\sigma)(E) = \frac{1}{2}(\sigma)\left(\frac{6}{E}\right)$$

$$= \frac{1}{2} \times \left(\frac{F}{A}\right)\left(\frac{F}{A}\right) \times \frac{1}{E}$$

$$= \frac{1}{2}\left(\frac{1000}{100}\right)^2 \times \frac{1}{10^5}$$

$$= 5 \times 10^{-4}\ \frac{N}{mm^2}$$

strain energy U = W × volume

= 5 × 10^{-4} × (100 × 100)

= 5 N mm

125. Beam : Young's modulus E = 10^5 N/mm^2

Moment of inertia I = 100 mm^4

Length l = 100 mm

k = 30 N/mm

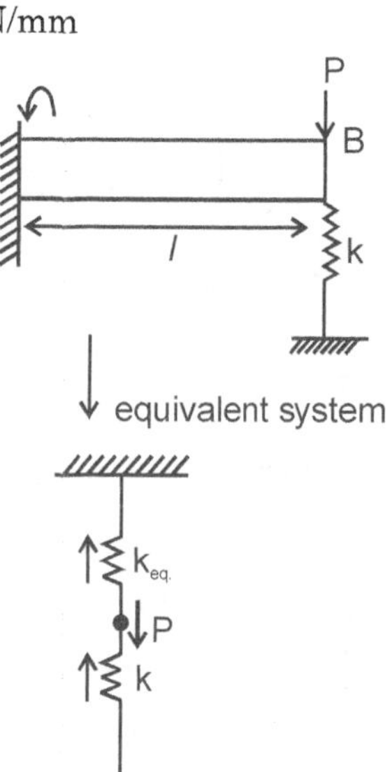

P = 100 N

$$k_{eq.} = \frac{3EI}{l^3} = \frac{3 \times 10^5 \times 100}{(100)^3}$$

= 30 N/mm

Force equilibrium

P = $k_{eq.}$ x + kx

$$\boxed{x = \frac{P}{(k_{eq.} + k)} = 1.66 \text{ mm}}$$

126. Ribs are used for increasing buckling strength of longitudnal stiffners and also distributes the concentrated loads into the structure and redistribute stress discontinuities. But IInd statement is not correct reson for (I) statement. Buckling of longitudinal stiffness increase just because ribs reduces the effective length of stiffners.

127. Anticlockwise torque

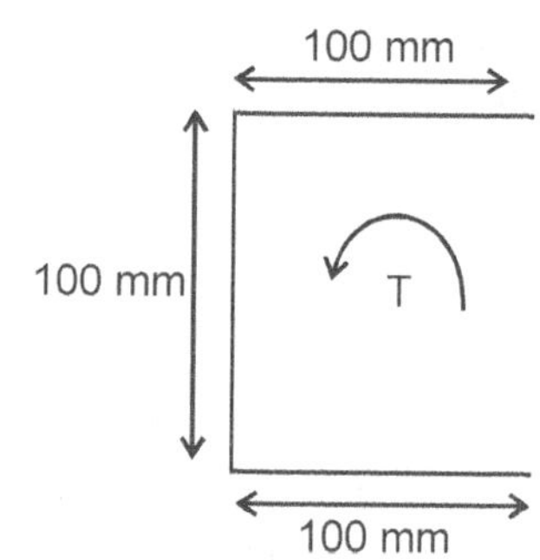

T = 62.5 × 10^3 N-mm

Max. shear stress τ = 100 N/mm^2

We know that

$$\frac{T}{J} = \frac{\tau}{t}$$

For thin reactingular section

$$J = \frac{1}{3}bt^3 = \frac{1}{3}[100t^3 + 100t^3 + 100t^3]$$

$$= \frac{1}{3}[300]t^3 = 100t^3$$

$$\frac{t \times 62.5 \times 10^3}{100 \times t^3} = 100$$

$$t^2 = \frac{62.5 \times 10^3}{100 \times 100}$$

$$t = \sqrt{6.25}$$

$$\boxed{t = 2.5 \text{ mm}}$$

128. $\boxed{m\ddot{x} + c\dot{x} + kx = 0}$

m = 1kg

c = 2 Ns/m

$$\omega_x = \sqrt{\frac{k}{m}} = \sqrt{2} \text{ rad/sec}$$

k = 2 N/m

damped natural frequency

$$\omega_d = \omega_n \sqrt{1-\xi^2}$$

$$\xi = \frac{c}{2\sqrt{km}} = \frac{2}{2\sqrt{2\times 1}} = 0.707 = \frac{1}{\sqrt{2}}$$

$$\Rightarrow \omega_2 = \sqrt{2} \times \sqrt{1-\left(\frac{1}{\sqrt{2}}\right)^2}$$

$$\boxed{\omega_d = 1 \text{ rad/sec}}$$

129. $\phi = 5\frac{x^4}{12} + \frac{x^3y}{6} + \frac{3x^2y^2}{2} + 7\frac{xy^3}{6} + E\frac{y^4}{12}$

$$\frac{\partial^4\phi}{\partial x^4} + \frac{\partial^4\phi}{\partial y^4} + 2\frac{\partial^4\phi}{\partial y^2\partial x^2} = 0$$

$$\frac{5\times4\times3\times2\times1}{12} + \frac{E\times4\times3\times2\times1}{12} + 2\times6 = 6$$

10 + 2E + 12 = 0

$$\boxed{E = \frac{-22}{2} = -11}$$

130. Shear center is a point where shear load produces no twist and also called center of twist in torsion. there can be stresses at this point.

131. Normal stress along neutral axis is always zero

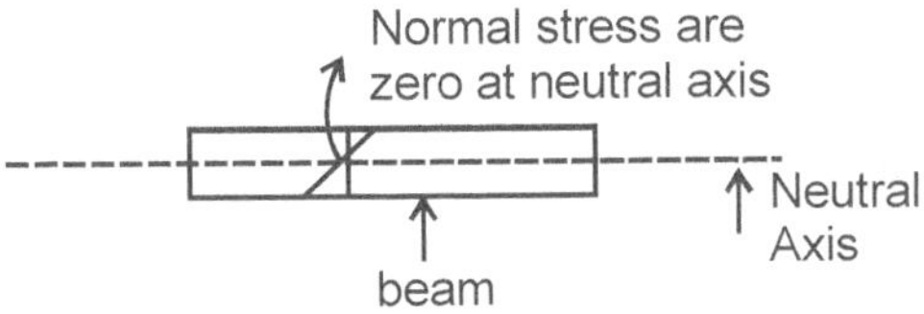

132. Spar caps are long longitudinal members along wing mainly take bending loads. But due to long length of these members those can have less buckling strength and can be prone to buckling.

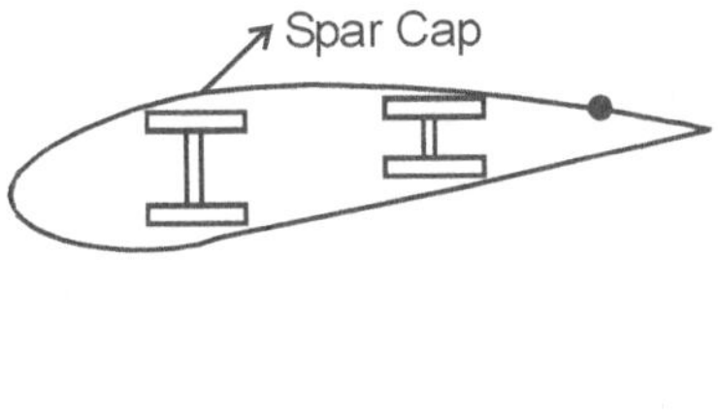

133. 2DOF's natural frequencies

$\omega_1 = 0.79$ rad/s

$\omega_2 = 1.538$ rad/s

mode stress $\phi_1 = [0.732 \quad 1]^T$

$\phi_2 = [-2.73 \quad 1]^T$

Initial displacement of first mass = 1

Initial displacement of 2nd mass = x

$$\text{mode shape } \phi_1 = \begin{bmatrix} 0.732 \\ 1 \end{bmatrix}$$

Normalizing ϕ_1 by 0.732 $\phi_1 = \begin{bmatrix} 1 \\ 1.366 \end{bmatrix}$ it shows that if mass/displaced by 1 cm mass 2 should be displace by 1.366 cm, so that system vibrates in first mode only.

134. SOOF-landing gear idealized as spring mass damper desirable damping is critical damping which give non-oscillatory-fastest decay of response to any input.

135. Natural frequency

$$\omega_n = \sqrt{\frac{k}{m}} = 5\text{Hz} = 2\pi \times 5 \text{ rad/sec}$$

$$\omega_x = \sqrt{\frac{k}{m}} = \sqrt{2} \text{ rad/sec}$$

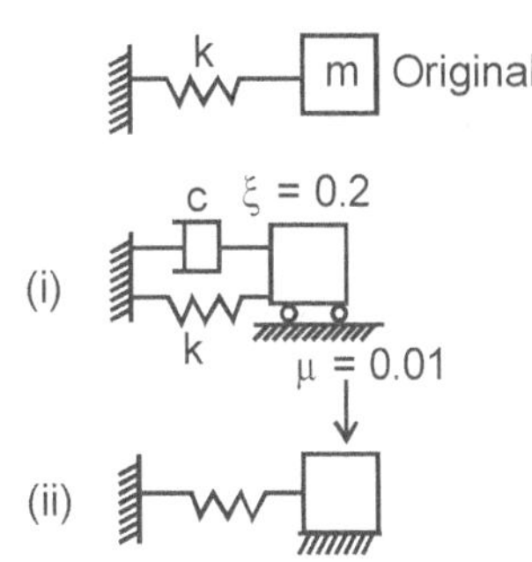

Ist case : damped natural frequency

$$\omega_{d1} = \omega_n \times \sqrt{1-\xi^2}$$

$$= 2\pi 5 \times \sqrt{1-0.2^2}$$

= 30.78119 rad/sec

IInd case :

$$\xi_{eq.} = \frac{\mu g}{2\times\omega_n} = \frac{0.01\times 9.8}{2\times2\times\pi\times5} = 1.55\times10^{-3}$$

$$\omega_{d_2} = 2\pi 5 \times \sqrt{1-(1.55\times10^{-3})}$$

= 31.4158

$$\boxed{\text{ratio} = \frac{\omega_{d1}}{\omega_{d2}} = \frac{30.7819}{31.4158} = 0.9797}$$

136. Compatability equation ensure continuity of displacements and must be satisfied for valid solution of 3-D problem of elasticity.

137. Yield stress = 330 MN/m^2 = 330 MPa

According to von-mises criteria

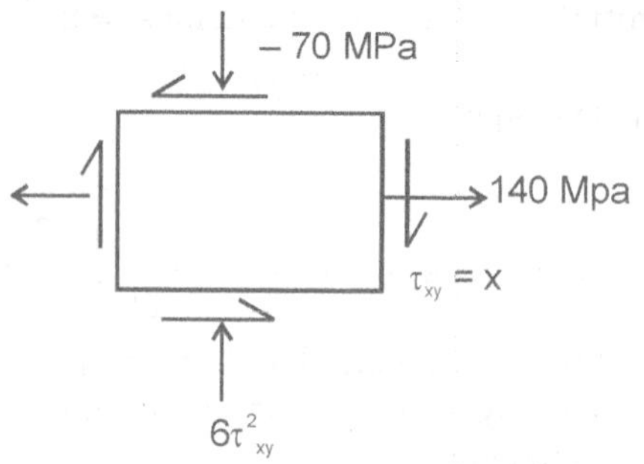

$$\frac{1}{6}\left[(\sigma_{xx}-\sigma_{yy})^2+\sigma_{xx}^2+\sigma_{yy}^2+\sigma_{xy}^2\right]$$

$$=\frac{1}{3}\sigma_y^2$$

$$\Rightarrow 6\tau_{xy}^2 = 2\sigma_y^2-\left[(\sigma_{xx}-\sigma_{yy})^2+\sigma_{xx}^2+\sigma_{yy}^2\right]$$

$$=2\times 330^3-\left[(140-(-70))^2+140^2+(-70)^2\right]$$

= 149200

$$\boxed{\tau_{xy}=157.692\text{ MPa}}$$

138. Using B.M. equation at section x

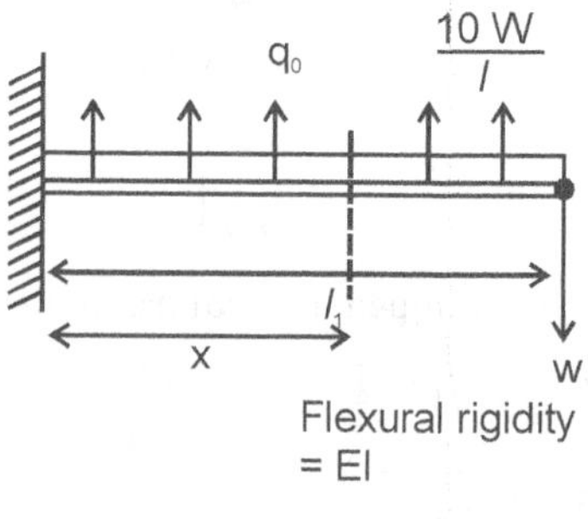

$$EI\frac{d^2y}{dx^2}=-W(l-x)+\frac{q_0(l-x)^2}{2}$$

Integrating once

$$EI\frac{dy}{dx}=-W\left(lx-\frac{x^2}{2}\right)+\frac{q_0}{2}\left(l^2x+\frac{x^3}{3}-\frac{2lx^2}{2}\right)+c_1$$

$$\frac{dy}{dx}=0 \quad \text{at} \quad x=0$$

$$\Rightarrow \boxed{c_1=0}$$

Integrating again

$$EIdy=-w\left(l\frac{x^2}{2}-\frac{x^3}{6}\right)+\frac{q_0}{2}\left(\frac{l^2x^2}{2}+\frac{x^4}{12}-\frac{2lx^3}{6}\right)+c_2$$

y = 0 at x = 0

$$\Rightarrow \boxed{c_2=0}$$

$\therefore$ at x = l

$$y=\frac{1}{EI}\left(-w\left(\frac{l^3}{2}-\frac{l^3}{6}\right)+\frac{q_0}{2}\left(\frac{l^4}{2}+\frac{l^4}{12}+\frac{l^4}{3}\right)\right)$$

$$=\frac{1}{EI}\left(\frac{-wl^3}{3}+\frac{10w}{2l}\left[\frac{6l^4+l^4-4l^4}{12}\right]\right)$$

$$=\frac{1}{EI}\left(\frac{-wl^3}{3}+\frac{5w}{l}\times\frac{3l^4}{12}\right)$$

$$=\frac{1}{EI}\left(\frac{-wl^3}{3}+\frac{5wl^3}{4}\right)$$

$$=\left(\frac{-4+15}{12}\right)\frac{wl^3}{EI}$$

$$=0.9166\left(\frac{wl^3}{EI}\right)$$

$$\therefore \boxed{K=0.9166}$$

139. Minimum principle stress = –7 MN/m^2

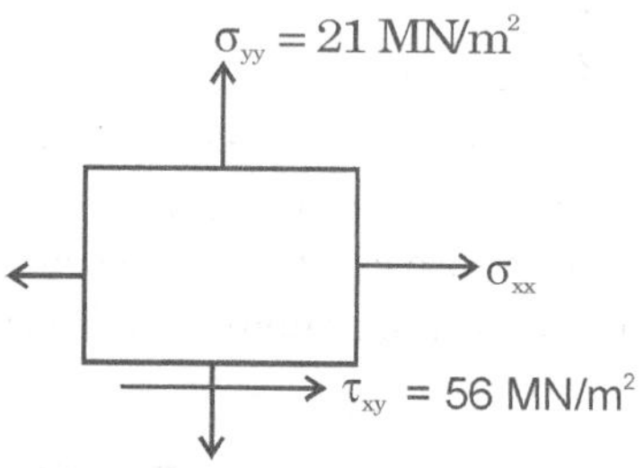

minimum principle stress

$$\sigma_{1,2}=\left(\frac{\sigma_{xx}+\sigma_{yy}}{2}\right)\pm\sqrt{\left(\frac{\sigma_{xx}-\sigma_{yy}}{2}\right)^2+\tau_{xy}^2}$$

$$-7=\left(\frac{\sigma_{xx}+21}{2}\right)-\sqrt{\left(\frac{\sigma_{xx}-21}{2}\right)^2+56^2}$$

Solving above eq.

$$\boxed{\tau_{xy}=105\text{ MN/m}^2}$$

140. Moment Mo = 96 Nm

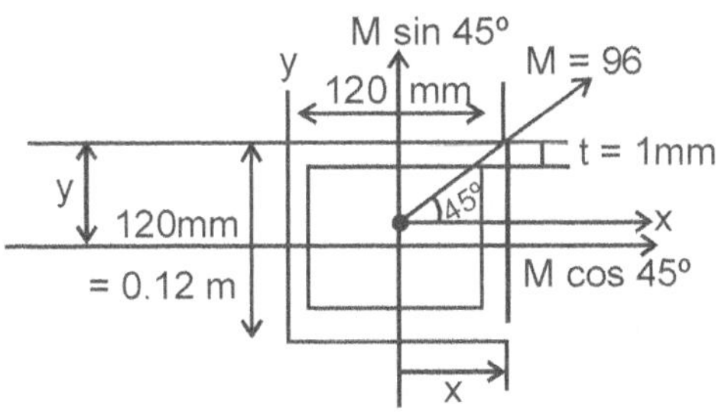

Maximum normal stress is given by

$$\frac{\sigma}{y} = \frac{M}{I}$$

as moment M when divided into x and y component stress

$$\sigma = \frac{M_{xx} \times y}{I_{xx}} + \frac{M_{yy} \times x}{I_{yy}}$$

as section is symmetric

$$I_{xx} = I_{yy} = \left(\frac{BD^3}{12}\right)_{outer} - \left(\frac{bd^3}{12}\right)_{internal}$$

$$= \frac{0.12 \times 0.12^3}{12} - \frac{0.118 \times 0.118^3}{12}$$

$= 1.12351 \times 10^{-6}\ m^4$

As x = y = 60 mm = 0.06 m

and $M_{xx} = M_{yy} = M \sin 45° = M \cos 45°$

= 96 × 0.707 = 67.88

$$\therefore\ \sigma = 2 \times \left[\frac{M_{xx} \times y}{I_{xx}}\right] = 2 \times \left[\frac{67.88 \times 0.06}{1.12351 \times 10^{-6}}\right]$$

= 7250135.735 N/m

$\boxed{6 = 7.2\ MN/m}$

141. Area $A_1 = 300 \times 10^3\ mm^2$

$= 0.3\ m^2$

$A_2 = 250 \times 10^3\ mm^2$

$= 0.25\ m^2$

T = 10 KN m = 10,000 Nm

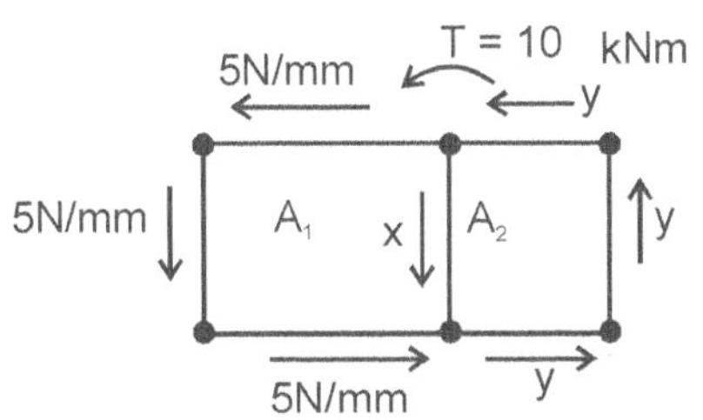

Shear flow q_1 = 5 N/mm = 500 N/m

Shear flow q_2 = y N/m

$x = q_2 - q_1$

= y – 500 N/m

We know that

$T = 2A_1q_1 + 2A_2q_2$

10,000 = 2 × 5000 × 0.3 + 2 × 0.25 × y

⇒ y = 14000 N/m

∴ x = 14000 – 5000 = 9000 N/m

$\boxed{\text{ratio}\frac{x}{y} = 0.6428}$

142. body diagram and resolving forces

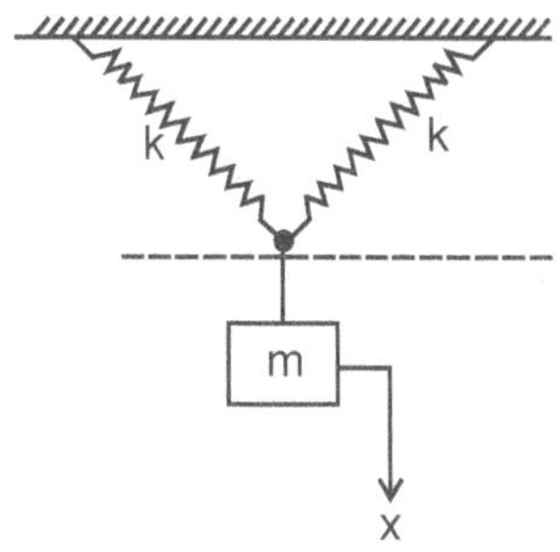

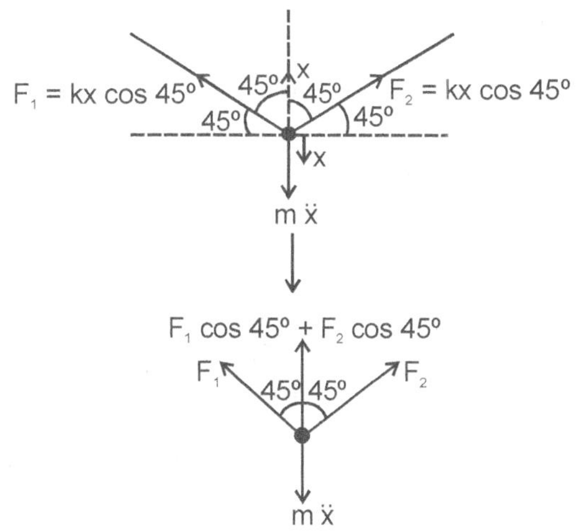

∴ equation of motion

$m\ddot{x} + F_1 \cos 45° + F_2 \cos 45° = 0$

$mx + kx(\cos 45°)^2 + kx(\cos 45°)^2 = 0$

$$\ddot{x} + \frac{2kx(\cos 45°)^2}{m} = 0$$

∴ Frequency

$$w_n = \sqrt{\frac{2\cos(45°)^2 k}{m}} = 1 \times \sqrt{\frac{k}{m}}$$

for small displacements

$\boxed{a = 1}$

143. a Aircraft wing is closed section box beam configuration which is capable of resisting torsional-loads coming due to aerodynamic forces

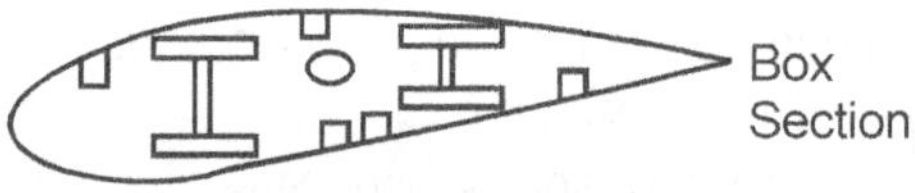

144. Damping ratio = 0.1

Ratio of two successive peaks = $\dfrac{A_1}{A_2} = e^{\frac{2\pi\xi}{\sqrt{1-\xi^2}}}$

= 1.88

145.

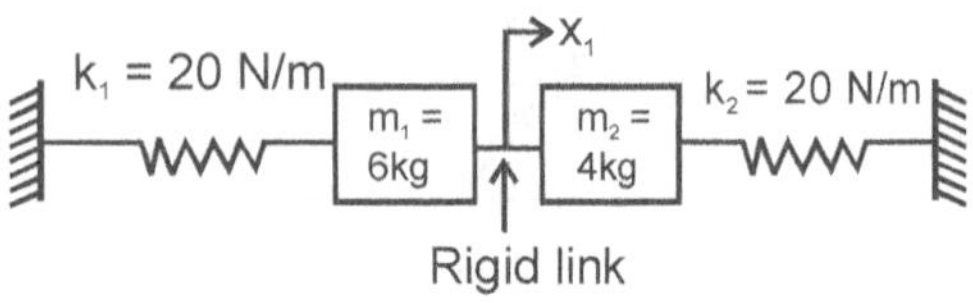

Masses are connected through a rigid link thus can be considered as

total 1 mass = $m_1 + m_2$ = 10 kg = M

Equation of motions of this system :

$$M\ddot{x}_1 + k_1x_1 + k_2x_1 = 0$$

$$\ddot{x}_1 + \left(\frac{k_1 + k_2}{M}\right)x_1 = 0$$

∴ Natural frequency

$$w = \sqrt{\frac{k_1 + k_2}{M}} = \sqrt{\frac{20+20}{10}} = 2 \text{ rad/sec}$$

146. Max stress developed at any point on shaft under torque T and dia d is

$$\tau_{max} = \frac{T}{J} \times \frac{d}{2}$$

$$\frac{T}{\frac{\pi}{32}d^4} \times \frac{d}{2} = \frac{16T}{\pi d^3}$$

147. Clamped-clamped beam

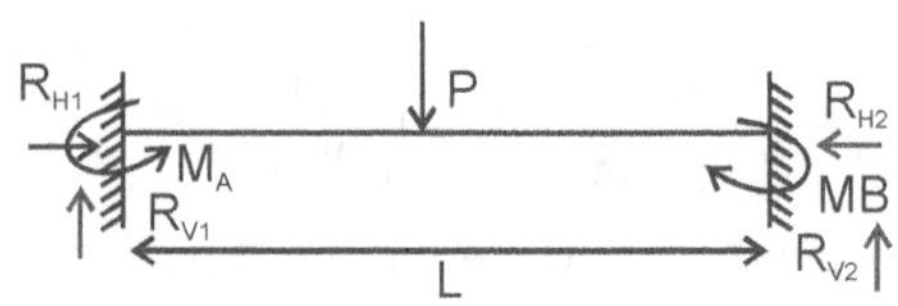

This case is equivalent to simply supported + end moment reaction

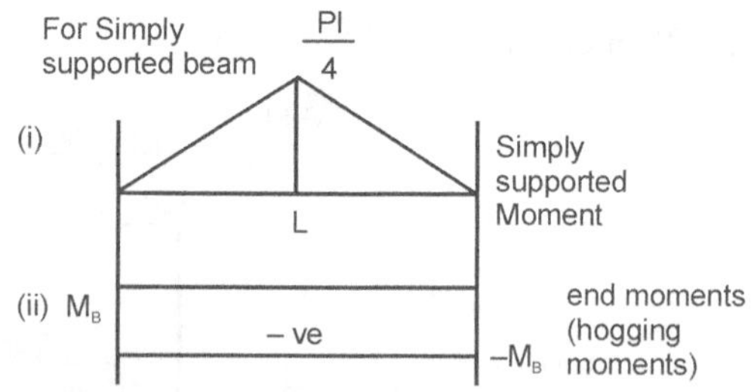

Equilibrium :

$\Sigma F_x = 0$

$Rv_1 + Rv_2 = p$

$R_{H_1} = R_{H_2} = 0$

$M_A = M_B$

Now Area of (I) + (II) =0 (Moment Equlibrium)

$$\frac{1}{2} \times \frac{PL}{4} \times L + M_B \times L = 0$$

$$LM_B = \frac{-PL^2}{8}$$

$$\boxed{M_B = \frac{PL}{8}} = M_A$$

148. In plane strain condition

$$\varepsilon_{zz} = \gamma_{xz} = \gamma_{yz} = 0$$

But $\sigma_{zz} \neq 0$ [$\sigma_{zz} = \nu(\sigma_{xx} + \sigma_{yy})$]

This can be proved

Using $\sigma_{zz} = \dfrac{1}{E}(\sigma_{zz} - \nu\sigma_{xx} + \nu\sigma_{yy}) = 0$

$\tau_{xz} = \tau_{yz} = 0$

∴ P and Q are correct

149. Thin walled tube

ext. radius R = 100 mm = 0.1 m

int. radius r = 100 – t = 100 – 2 = 98 mm = 0.098 m

Compressive force F = 1 N

Maximum normal stress $= \dfrac{-F}{\pi(R^2 - r^2)}$

$= -803.81 \text{ N/m}^2$

$= -0.803 \times 10^3 \text{ N/m}^2$

= – 0.803 KN

–ve sign indicates compression

150. Length l = 1m

m = 60 kg

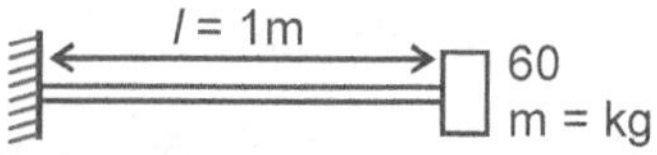

Normal frequency = 2Hz

$= 2\pi.2$ rad/sec

Natural frequency

$$\omega_n = \sqrt{\frac{k_{eq.}}{m}} = 2 \times 2\pi \text{ rad/sec}$$

$$k_{eq.} = \frac{3EI}{l^3}$$

Flexrul regidity = EI

$$\therefore \sqrt{\frac{3(EI)}{l^3.m}} = 4\pi$$

$$EI = \frac{(4\pi)^2 \times l^3 \times m}{3}$$

$$= \frac{(4\pi)^2 \times (1)^3 \times 60}{3}$$

$= 3158.27$ Nm^2

$= 3.158$ KN m^2

151. Factor of safety = 4

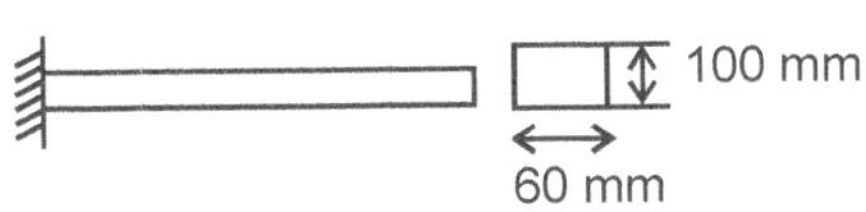

$E_{Al} = 73$ GPa

$\sigma_a = 480$ MPa

$\therefore$ Max. stress can be

$$\text{applied } \sigma_{max} = \frac{\sigma_a}{4} = \frac{480}{4}$$

= 120 MPa

= 120 MN/mm^2

Maximum moment that can be applied

$$M = \frac{\sigma_{max}}{y} \times I = \frac{120}{\left(\frac{100}{2}\right)} \times \frac{1}{12} \times (60) \times (100)^3$$

= 12000000 N mm

$$= \frac{12000000}{1000 \times 1000} = 12 \text{ kNm}$$

152. $\sigma_{xx} = Ax^2$

$\sigma_{yy} = 12x^2 - 6y^2$

$\sigma_{xy} = 12xy$

To satisfy equilibrium equation

$$\frac{\partial \sigma_{xx}}{\partial x} + \frac{\partial \sigma_{xy}}{\partial y} = 0$$

$2Ax + 12x = 0$

$$\Rightarrow \boxed{A = -6}$$

153. As the beam is subjected to transverse shear load through its shear center, the beam is subjected for bending, it may not be subjected for torsion, hence twist per unit length is zero.

154. For homogeneous isotropic linear material three are two elastic constant

Young modulus & poisson's ratio.

155.

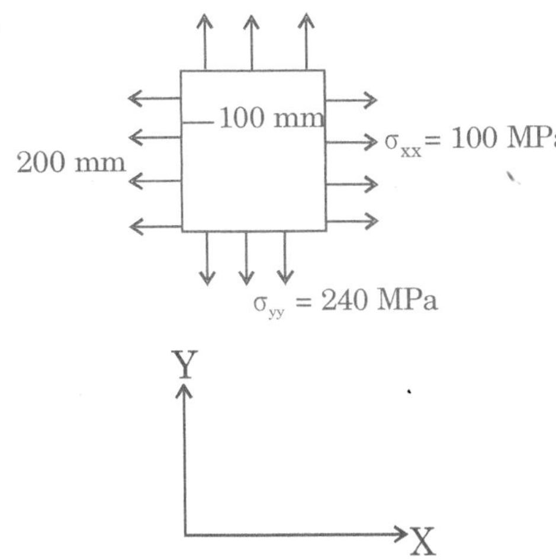

$$\epsilon_{YY} = \frac{\sigma_{yy}}{E} - \frac{\gamma\sigma_{xx}}{E}$$

$$= \frac{1}{E}(\sigma_{yy} - \gamma\sigma_{xx})$$

$$= \frac{1}{210 \times 10^9}(240 \times 10^6 - 0.3 \times 100 \times 10^6)$$

$= 1 \times 10^{-3}$

$(\delta l)_{YY} = \epsilon_{yy} \times l_{yy} = 1 \times 10^{-3} \times 210 = 0.2$ mm

156.

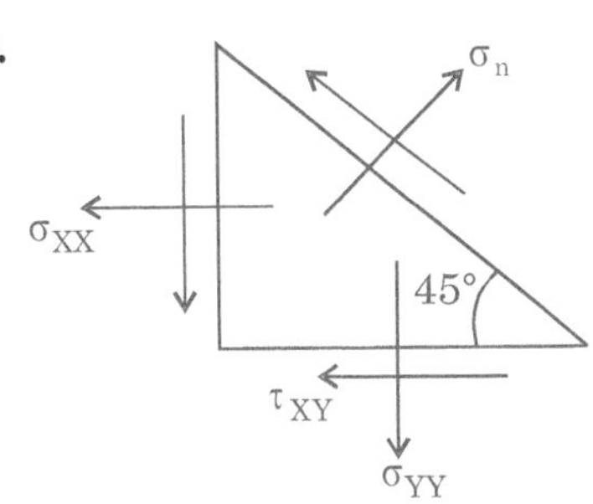

$\sigma_{XX} = 200$ MPa

$\sigma_{YY} = 100$ MPa

$\tau_{XY} = 100$ MPa

$\theta = 45°$

$$\sigma_n = \frac{\sigma_{XX} + \sigma_{YY}}{2} + \frac{\sigma_{XX} - \sigma_{YY}}{2}\cos 2\theta + \tau_{XY}\sin 2\theta$$

$$= \frac{200 + 100}{2} + \frac{200 - 100}{2}\cos 90° + 100 \sin 91°$$

$6\eta = 250$ MPa.

157.

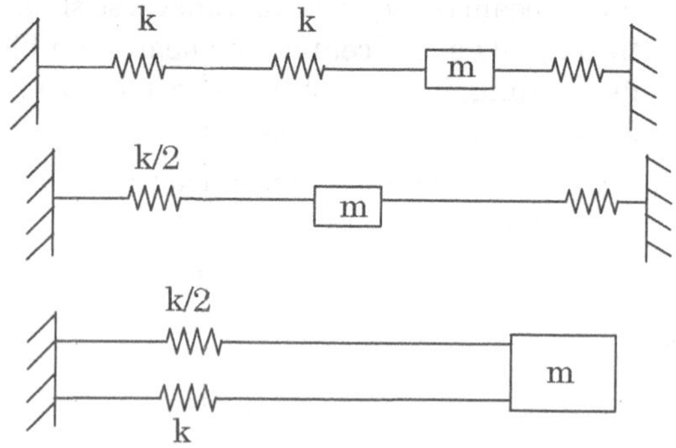

$$k_{eq} = k + \frac{k}{2} = \frac{3k}{2}$$

$$\therefore\ f = \frac{1}{2\pi}\sqrt{\frac{k_{eq}}{m}}$$

$$= \frac{1}{2\pi}\sqrt{\frac{3k}{2m}}$$

$$= \frac{1}{2\pi}\sqrt{\frac{3 \times 20 \times 10^3}{2 \times 3}}$$

$= 15.91 + 12 = 16 + 12.$

158.

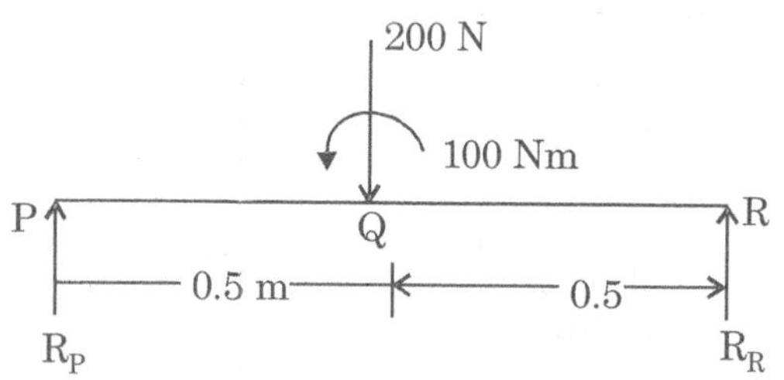

$R_p + R_R = 200$ ...(1)

$\sum M_P = 0$

$\Rightarrow R_R \times 1 + 100 - 200 \times 0.5 = 0$

$\boxed{R_R = 0}$

$\therefore$ putting value of R_p in equation (1)

We get

$R_p = 200 - 0 = 200$ N

taking moment about $Q \leq x \leq 0.5$

$M_Q = 200 \times 0.5 = 100$ Nm.

for $0.5 \leq x \leq 0.1$

$M_Q = 200 \times 0.5 - 100 = 0$

$M_Q = 0$Nm

drawing bending moment diagram for the beam

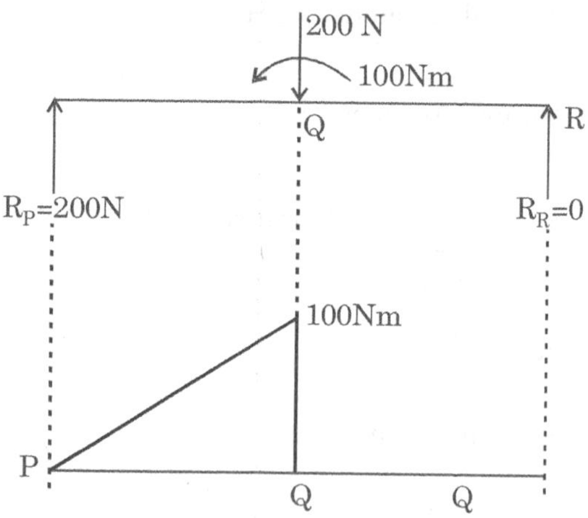

Bending moment diagram

159.

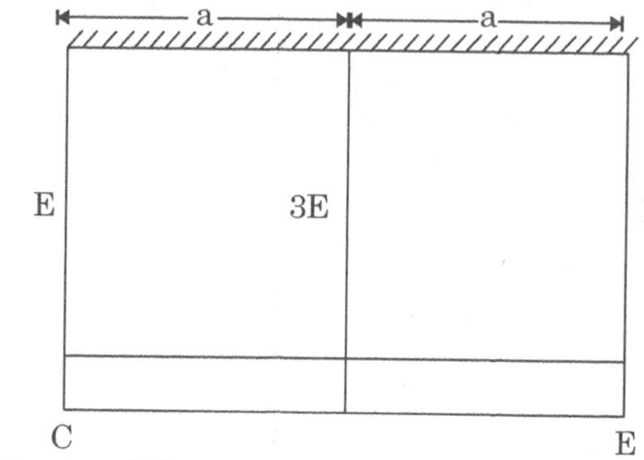

Strain will be equal

$\therefore\ \varepsilon_1 = \varepsilon_2$

$$\frac{\frac{F_2}{A}}{3E} = \frac{\frac{F_1}{A}}{E}$$

$$\Rightarrow \frac{F_2}{3} = F_1$$

$$\boxed{\Rightarrow F_1 = \frac{F_2}{3}}$$

free body diagram

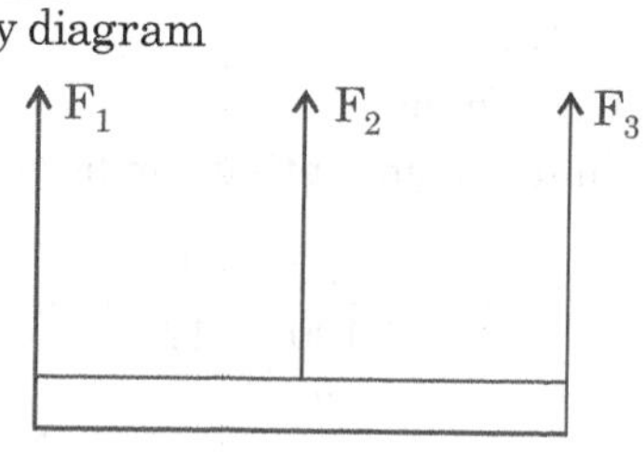

$\sum Fy = 0,\ F_2 + F_1 + F_1 - W = 0$

$\Rightarrow\ F_2 + 2F_1 = W$

$$\Rightarrow F_2 + \frac{2F_2}{3} = W$$

$$\Rightarrow F_2 + \frac{3}{5} = W$$

$= 0.6$ W

– 60% weight.

160.

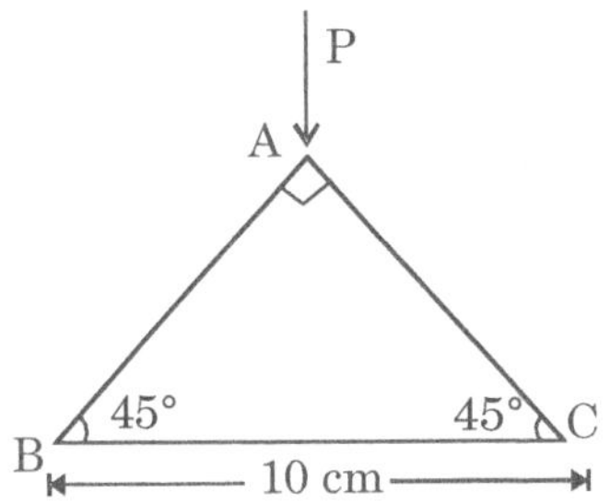

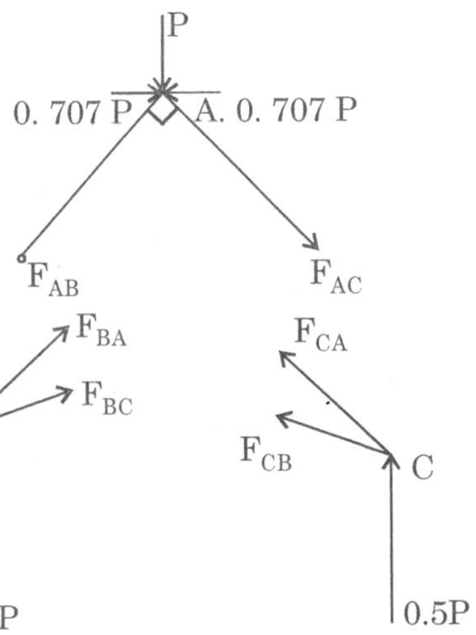

From Point A

$F_{AB} = 0.707\ P = P_{cr}$

$$P_{cr} = \frac{\pi^2 EI}{l_e^{\ 2}} = \frac{A^2 EI}{l^2}$$

$$= \frac{\pi^2 200 \times 10^9 \times 0.1 \times 10^{-4}}{(10 \times 10^{-2})2}$$

$= 395$ kN

161.

Fos = 2

We know

$$\text{Fos} = \frac{\tau_{allowable}}{\tau_{working}} = \frac{\tau_a}{\tau_w}$$

$\Rightarrow \tau_a = 2\tau_5$ (1)

$$\frac{T}{J} = \frac{\tau}{R}$$

$$T = \frac{\tau}{J}R \qquad T \propto \frac{1}{J}$$

$$T_{solid} \propto \frac{1}{J_S}$$

$$T_{hollow} \propto \frac{1}{J_H}$$

$$\frac{T_S}{T_H} = \frac{J_H}{J_S} = \frac{\left(R_0^4 - R_i^4\right)}{R^4}$$

$$\frac{T_S}{T_H} = \frac{R^4 - \left(R/2\right)^4}{R^4} = 0.9375$$

$T_H = 1.066\ T_5$ (2)

from equation (1) and (2)

$$\frac{Ta}{T_A} = \frac{2\tau_S}{1.066\tau_S} = 1.9$$

162.

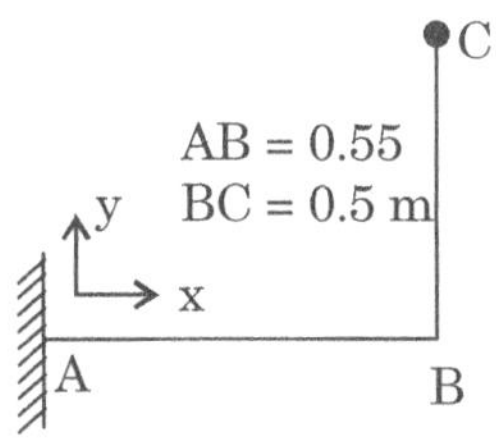

$\sigma_{xy} = \sigma_{zx} = \sigma_{xy} = 622 = 0$

$$\sigma_{zy} = \tau_{zy} = \frac{T}{J}r = \frac{pxBC}{\frac{\pi d4}{32}} \times \frac{d}{2} = \frac{16P \times BC}{\pi d^3}$$

$$\sigma_{xx} = \frac{M}{I}y = \frac{P \times AB}{\frac{\pi d^4}{64}} \times \frac{d}{2} = \frac{32PAB}{\pi d^3}$$

Now applying Von – mises yield criterion we get

$$\sigma_{yp} = \sqrt{\frac{\left(\sigma_{xp} - \sigma_{xy}\right)^2 \left(\sigma_{yy} - \sigma_{zz}\right)^2 + \left(\sigma_{zz} - \sigma_{xx}\right)^2 + \sigma\left(3_{xx}^2 + \sigma_{yz}^2 + \sigma_{zx}^2\right)}{2}}$$

substituting the above value

$$\sigma_{yp} = \sqrt{\frac{\left(\sigma_{xx} - 0\right)^2 + (0-0)^2 + \left(0 - \sigma_{xx}\right)^2 + \sigma\left(\sigma_{xx}^2 + 0.0\right)}{2}}$$

$$= \sqrt{\frac{2\sigma_{xx}^2 + 6\sigma_{xy}^2}{2}} = \sqrt{\sigma_{xx}^2 + 3\sigma_{zy}^2}$$

$$= \frac{1\sigma P}{\pi d^3}\sqrt{(2aB)^2 + 3(BC)^2}$$

$$= \frac{1\sigma P}{\pi d^3}\sqrt{(2x0.55)^2 + 3(0.5)^2}$$

$$= \frac{1\sigma P}{\pi(0.02)^3}\sqrt{(2x0.55)^2 + 3 \times 0.5^2}$$

$$\Rightarrow 280 \times 10^6 = \frac{1\sigma P}{\pi(0.02)^3}\sqrt{1.1^2 + 0.75}$$

$$\rho = \frac{250 \times 10^6 \times \pi \times 0.02^3}{16 \times \sqrt{1.96}}$$

$$\boxed{\rho = 314.159N}$$

163.

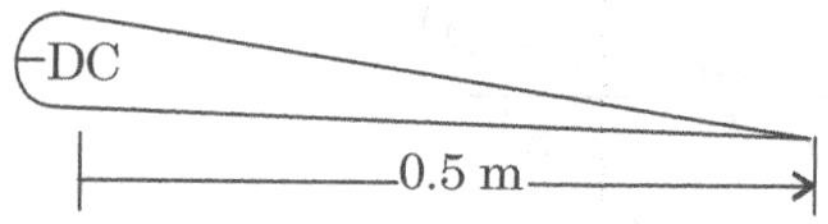

$\tau = 2\,q\,A = 2(T \times t)\,A$

$$\boxed{\tau = \frac{T}{2At}}$$

$$T = \frac{1 \times 10^3}{2 \times 0.28925 \times 1 \times 10^{-3}} = 17.5 \text{ MPa}$$

$$A = \frac{\pi r^2}{2} + \frac{1}{2} b \times R$$

$$= \left(\frac{\pi}{2} / 50 \times 10^{-3}\right)^2 \times \frac{1}{2} \times \left(100 \times 10^{-3}\right) \times 0.5$$

$= 0.028925 \text{ m}^2$

164. $m = 10$ kg

$k = 10^5$ N/m

$C = 20$ kg / sec

$$\xi_1 = \frac{C}{Cc} = \frac{C}{2m\omega} = \frac{C}{2\sqrt{km}} = \frac{20}{2\sqrt{10^3 \times 10}} = 0.1$$

Now logarithmic decrement δ

$$\gamma = \frac{2\pi\xi_1}{\sqrt{1-\xi^2}} = \frac{1}{n}\ln\left(\frac{x_1}{x_{n+1}}\right)$$

$n+1 = 5$

$\therefore n = 4$

$$\delta = \frac{1}{4}\ln\left(\frac{x_1}{x_5}\right) = \frac{2\pi\xi}{\sqrt{1-\xi^2}}$$

$$\Rightarrow \frac{2\pi \times 0.1}{\sqrt{1-0.1^2}} = \frac{1}{4}\ln\left(\frac{x_1}{x_5}\right)$$

$$\frac{x_1}{x_5} = e^{2.52} = 12.52$$

165.

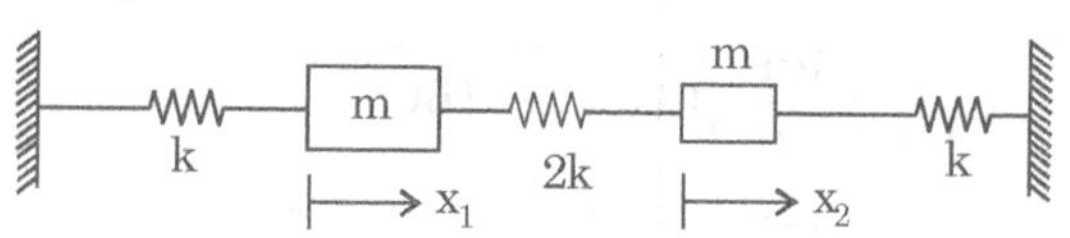

$$W_1^4 - \left(\frac{k_1+k_2}{m_1} + \frac{k_2+k_3}{m_2}\right)\omega^2 + \left(\frac{k_1k_2+k_2k_3+k_5k_4}{m_1+m_2}\right) = 0$$

$$W^4 - \left(\frac{k+2k}{m} + \frac{2k+k}{m}\right)\omega^2 + \left(\frac{k-2k+2k-k+k-k}{m-m}\right) = 0$$

$$W^4 - \frac{6k}{m}\omega^2 + \frac{5k^2}{m^2} = 0$$

Let $\omega^2 = x$

$$\lambda = \frac{-\left(\frac{6k}{m}\right) + \sqrt{\left(6k/m\right)^2 - 4 \times 1 \times \frac{5k^2}{m^2}}}{2 \times 1}$$

$$= \frac{\frac{6k}{m} + \sqrt{\frac{36k^2}{m^2} - \frac{20k^2}{m^2}}}{2}$$

$$= \frac{\frac{6k}{m} \pm \frac{4k}{m}}{2}$$

$$\lambda = \frac{5k}{m}$$

$\therefore \omega^2 = 5k/m,$

$$\omega = \sqrt{5 k/m} = \sqrt{\frac{5 \times 1250}{10}} = 25\text{md} / \text{mm}$$

166. $\phi = Ax^4 + By^4 + Cxy^3$

For Airy's stress function to be valid

$$\nabla^4\phi = 0$$

$$\therefore \quad \nabla^4\phi = \frac{\partial^4\phi}{\partial x^4} + \frac{\partial^4\phi}{\partial y^4} + \frac{2\partial^4\phi}{\partial x^2 \partial y^2} = 0 \quad ...(1)$$

$$\therefore \quad \frac{\partial^2\phi}{\partial x^4} = 24\,A$$

$$\frac{\partial^2\phi}{\partial y^4} = 24\,B$$

$$\frac{\partial^2\phi}{\partial x^2 \partial y^2} = 0$$

Put above values in equation (1), we get

$24A + 24B + 0 = 0$

$24(A + B) = 0$

$A + B = 0$

167. $\varepsilon_{xx} = Ay^2 + x;$

$\varepsilon_{yy} = Ax^2 + y;$

$\gamma_{xy} = Bxy + y$

Compatibility equation for plain strain condition is

$$\frac{\partial^2\gamma_{xy}}{\partial x \partial y} = \frac{\partial^2\varepsilon_{yy}}{\partial x^2} + \frac{\partial^2\varepsilon_{xx}}{\partial y^2}$$

$B = 2A + 2A$

$B = 4A$

168. Since tensile yield strength is a material property, therefore it remains same for all the cases.

$$F_{circle} = (\sigma_{yt})_{circle} \cdot A_{circle}$$

$$F_{triangle} = (\sigma_{yt})_{triangle} \cdot A_{triangle}$$

$$F_{rectangle} = (\sigma_{yt})_{rectangle} \cdot A_{rectangle}$$

$$A_{circle} = A_{triangle}$$

$= A_{rectangle}$ and

$$(\sigma_{yt})_{circle} = (\sigma_{yt})_{triangle}$$

$$= (\sigma_{yt})_{rectangle}$$

$$\therefore \quad F_{circle} = F_{triangle}$$

$$= F_{rectangle}$$

169. At positive high angle of attack the beading moment from the normal force produce compressive stress on the upper side of the wings flanges of the front spar and adjacent stringers and tensile stress at low after region of the wing crossections, were is case of positive low angle of attack wing bending moment produce the maximum compressive stress on the upper rear spar flanges and stringers and maximum tension stress on.

The lower front spar f largos and stagers.

170. For a cantilever beam subjected to free vibration, and the system is considered as continuous system in which the beam mass in considered as distributed along with the stiffness of the shaft, the equation of motion can be written as (Meirovitch, 1967)

$$\frac{d^2}{dx^2}\left\{EI(x)\frac{d^2Y(x)}{dx^2}\right\} = \omega^2 m(x)Y(x) \quad ...(1)$$

Where, E is the modulus of rigidity of beam material, I is the moment of inertia of the beam cross-section, Y(x) is displacement in y direction at distance x from fixed end, ω is the circular natural frequency, m is the mass per unit length, $m = \rho A(x)$, ρ is the material density, x is the distance measured from the fixed end.

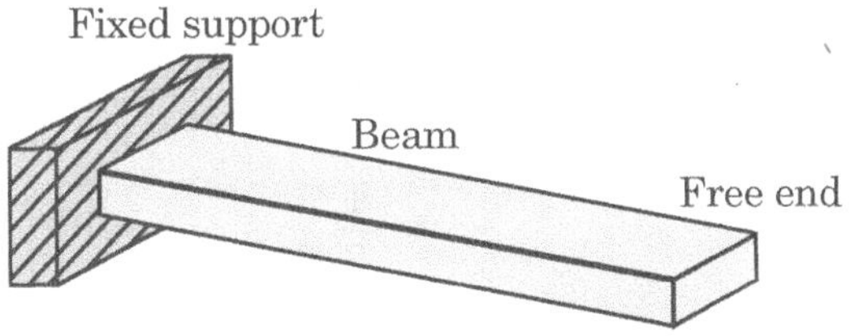

Fig. 1(a) : A cantilever beam

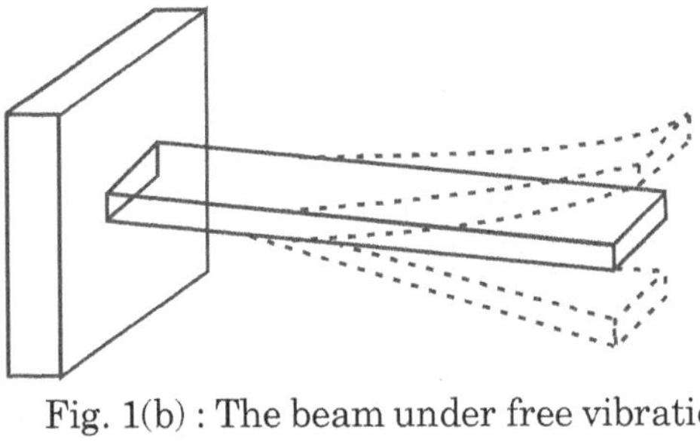

Fig. 1(b) : The beam under free vibration

Fig. 1(a) shows of a cantilever beam with rectangular cross section, which can be subjected to bending vibration by giving a small initial displacement at the free end; and Fig. 1(b) depicts of cantilever beam under the free vibration.

We have following boundary conditions for a cantilever beam (Fig. 1)

at $x = 0, Y(x) = 0$,

$$\frac{dY(x)}{dx} = 0 \quad ...(2)$$

at $x = l$,

$$\frac{d^2Y(x)}{dx^2} = 0,$$

$$\frac{d^3Y(x)}{dx^3} = 0 \quad ...(3)$$

For a uniform beam under free vibration from equation (1), we get

$$\frac{d^4Y(x)}{dx^4} - \beta^4 Y(x) = 0 \quad ...(4)$$

with $\beta^4 = \frac{\omega^4 m}{El}$

The mode shapes for a continuous cantilever beam is given as

$$f_n(x) = A_n\{(\sin\beta_nL - \sin h\,\beta_nL)(\sin\beta_nx - \sin h\,\beta_nx) + (\cos\beta_nL - \cos h\,\beta_nL)(\cos\beta_nx - \cosh\beta_nx)\} \quad ...(5)$$

Where $n = 1, 2, 3....\infty$ and $\beta_nL = n\pi$

A closed form of the circular natural frequency ω_{nf}, from above equation of motion and boundary conditions can be written as,

$$\omega_{nf} = \alpha_n^2\sqrt{\frac{El}{mL^4}} \quad ...(6)$$

Where $\alpha_n = 1.875, 4.694, 7.885$

So, First natural frequency

$$\omega_{nf} = 1.875^2\sqrt{\frac{El}{\rho AL^4}} \quad ...(7)$$

Second natural frequency

$$\omega_{nf} = 4.694^2\sqrt{\frac{El}{\rho AL^4}} \quad ...(8)$$

Third natural frequency

$$\omega_{nf} = 7.855^2\sqrt{\frac{El}{\rho AL^4}} \quad ...(9)$$

First natural frequency

$$\omega_{n1} = 1.875^2\sqrt{\frac{El}{\delta AL^4}}$$

If L becomes 2L, then

$$\omega'_{n1} = 1.875^2\sqrt{\frac{EI}{16\delta AL^4}}$$

$$\frac{\omega_{n1}}{\omega'_{n1}} = \frac{1}{\sqrt{16}}$$

$$\frac{\omega_{n1}}{\omega'_{n1}} = \frac{1}{4}$$

$$\omega_{n1} = \omega$$

$$\therefore \quad \omega'_{n1} = 4\omega \text{ rad/sec}$$

171. Given,

$G = 25$ Gpa $= 25\times 10^{-9}$ Pa

$T = 3.2$ kNm $= 3.2 \times 10^3$ Nm

$A = 0.1 \times 0.4$ m^2

$t = 2$ mm $= 2 \times 10^{-3}$ m

Angle of twist per unit length,

$$\alpha = \frac{\theta}{L}$$

$$= \frac{q}{2AG}\int_0^s \frac{ds}{t} \quad \text{...(i)}$$

From Bredth-bath theory

$$T = 2\,qA$$

$$q = \frac{T}{2A} \quad \text{...(ii)}$$

Shear flow $q = T \times t$ ...(iii)

Putting the value in equation (i), we get

$$\alpha = \frac{\theta}{L}$$

$$= \frac{T}{4A^2G}\int_0^s \frac{ds}{t}$$

$$= \frac{T}{4A^2Gt}\left[\int_0^{0.1} ds + \int_0^{0.4} ds + \int_0^{0.1} ds + \int_0^{0.4} ds\right]$$

$$= \frac{T}{4A^2Gt}[0.1 + 0.4 + 0.1 + 0.4]$$

$$= \frac{3.2\times 10^3}{4\times(0.1\times 0.4)^2\times 25\times 10^9\times 2\times 10^{-3}}\times 1$$

$= 0.010$ rad/m

172.

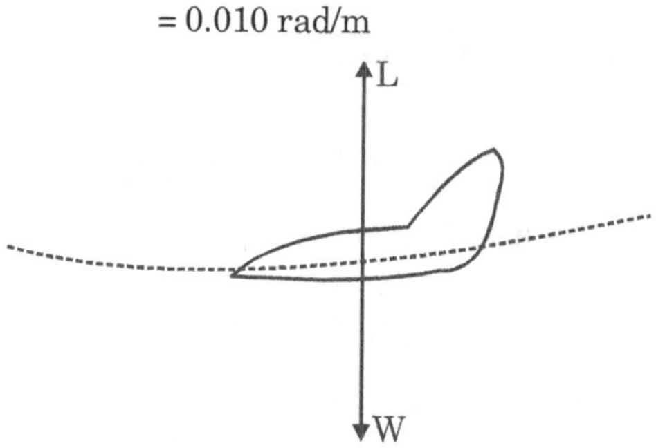

$V = 360$ km/h

$$= 360 \times \frac{5}{18}$$

$= 100$ m/sec

$g = 9.8$ m/sec^2

$R = 400$ m

Writing equation of motion of airplane

$$L - W = \frac{mv^2}{R}$$

$$\frac{L - W}{W} = \frac{v^2}{gR} \qquad [W = mg]$$

$$n - 1 = \frac{v^2}{gR}$$

$$n = \frac{v^2}{gR} + 1$$

$$n = \frac{100^2}{9.8\times 400} + 1$$

$= 3.54$

173. We have,

$$E = 2G(1 + \nu) \quad \text{...(i)}$$

$$E = 3K(1 - 2\nu) \quad \text{...(ii)}$$

From (i) and (ii), we get

$$2G(1 + \nu) = 3K(1 - 2\nu)$$

$$\Rightarrow \quad \frac{G}{K} = \frac{3}{2}\frac{(1-2\nu)}{(1+\nu)}$$

As per question

$$G = K$$

$$\therefore \quad 1 = \frac{3}{2}\frac{(1-2\nu)}{(1+\nu)}$$

$$\Rightarrow \quad 2 + 2\nu = 3 - 6\nu$$

$$\Rightarrow \quad 8\nu - 1 = 0$$

$$\Rightarrow \quad \nu = \frac{1}{8}$$

174. Given,

$W = 3$ N

Moment of inertia of CB = I

Moment of inertia of AC = 3I

Length of AC = 1 m

Length of CB = 2 m

Now as per formula

Strain Energy stored,

$$E = \int \frac{(\text{Bending moment})^2\,dx}{2(\text{Flexural rigidity})}$$

$$= \int_0^2 \frac{(Wx)^2\,dx}{2EI} + \int_2^3 \frac{(Wx)^2\,dx}{2E(3I)}$$

$$= \frac{W^2}{2EI}\int_0^2 x^2dx + \frac{W^2}{6EI}\int_2^3 x^2dx$$

$$= \frac{W^2}{2EI}\left[\frac{x^3}{3}\right]_0^2 + \frac{W^2}{6EI}\left[\frac{x^3}{3}\right]_2^3$$

$$= \frac{W^2}{2EI}\left[\frac{8}{3}\right] + \frac{W^2}{6EI}\left[\frac{27}{3} - \frac{8}{3}\right]$$

$$= \frac{8W^2}{6EI} + \frac{19W^2}{18EI} = \frac{43W^2}{18EI}$$

Also, $E = \frac{1W\delta}{2}$, where δ is the deflection

Equating both sides

$$\frac{1W\delta}{2} = \frac{43W^2}{18EI}$$

$$\Rightarrow \quad \delta = \frac{43}{3EI}$$

175.

$$\begin{bmatrix}\varepsilon_{xx}\\ \varepsilon_{yy}\\ \varepsilon_{zz}\\ \gamma_{yz}\\ \gamma_{xz}\\ \gamma_{xy}\end{bmatrix} = \begin{bmatrix}A & C & C & 0 & 0 & 0\\ C & A & C & 0 & 0 & 0\\ C & C & A & 0 & 0 & 0\\ 0 & 0 & 0 & B & 0 & 0\\ 0 & 0 & 0 & 0 & B & 0\\ 0 & 0 & 0 & 0 & 0 & b\end{bmatrix}\begin{bmatrix}\alpha_{xx}\\ \alpha_{yy}\\ \alpha_{zz}\\ \tau_{yz}\\ \tau_{xz}\\ \tau_{xy}\end{bmatrix} \quad ...(i)$$

Strain – stress relation in three dim ension

$$\begin{bmatrix}\varepsilon_{xx}\\ \varepsilon_{yy}\\ \varepsilon_{zz}\\ \gamma_{yz}\\ \gamma_{xz}\\ \gamma_{xy}\end{bmatrix} = \begin{bmatrix}\frac{1}{E} & \frac{-\gamma}{E} & \frac{-\gamma}{E} & 0 & 0 & 0\\ \frac{-\gamma}{E} & \frac{1}{E} & \frac{-\gamma}{E} & 0 & 0 & 0\\ \frac{-\gamma}{E} & \frac{-\gamma}{E} & \frac{1}{E} & 0 & 0 & 0\\ 0 & 0 & 0 & \frac{1}{G} & 0 & 0\\ 0 & 0 & 0 & 0 & \frac{1}{G} & 0\\ 0 & 0 & 0 & 0 & 0 & \frac{1}{G}\end{bmatrix}\begin{bmatrix}\sigma_{xx}\\ \sigma_{yy}\\ \sigma_{zz}\\ \tau_{yz}\\ \tau_{xz}\\ \tau_{xy}\end{bmatrix} \quad ...(ii)$$

From equation (i) and (ii), we get

$$A = \frac{1}{E}$$

$$B = \frac{1}{G}$$

$$C = \frac{-\nu}{E}$$

We have

$$E = 2G(1 + \nu)$$

$$\frac{1}{2G} = \frac{(1+\nu)}{E}$$

$$\Rightarrow \quad \frac{1}{2G} = \frac{1}{E} + \frac{\nu}{E} = A + (-C)$$

$$\Rightarrow \quad \frac{B}{2} = A - C$$

$$\Rightarrow \quad C = A - \frac{B}{2}$$

176. All normal stresses are tensile stresses, at any angle compressive normal stresses can't occur and looking direction of shear stress τ_{xy} we can directly neglect (C) and (D).

Given,

$$\sigma_{XX} = 18 \text{ MPa}$$
$$\sigma_{YY} = 12 \text{ MPa}$$
$$\sigma_{XY} = 4 \text{ MPa}$$

Plane of maximum shear stress

$$\tan 2\theta = \left(\frac{\sigma_{YY} - \sigma_{XX}}{2\tau_{XY}}\right)$$

$$= \left(\frac{18-12}{2\times 4}\right)$$

$$= \frac{3}{4}$$

$$2\theta = \tan^{-1}\left(3/4\right)$$

$$\theta = \frac{\tan^{-1}\left(3/4\right)}{2}$$

$$= 18.4°$$

Now, we know Normal stress at any angle θ for given loading is

$$\sigma_n = \tau_{XY}\sin 2\theta + \frac{\sigma_{XX} + \sigma_{YY}}{2} + \frac{\sigma_{XX} - \sigma_{YY}}{2}\cos 2\theta$$

Here, $\theta = 90° - 18.4°$

$$\Rightarrow \sigma_n = 4\times\sin\left(2(90° - 18.4°)\right) + \frac{18+12}{2} + \frac{18-12}{2}\cos\left(2(90° - 18.4°)\right)$$

$$= 14.99 \approx 15 \text{ MPa}$$

$$\tau_{max} = \left(\frac{\sigma_{XX} - \sigma_{YY}}{2}\right)\sin 2\theta - \tau_{XY}\cos 2\theta$$

$$= \left(\frac{18-12}{2}\right)\sin\left(2(90° - 18.4°)\right) - 4\times\cos\left(2(90° - 18.4°)\right)$$

$$= 4.99 \approx 5 \text{ MPa}$$

177. Given,

Yield stress, $\sigma_y = 280$ MPa

Torque, T = 10 kNm

Using tresca failure theory or maximum shear stress theory

$$\tau_{max} = \frac{\sigma_y}{2}$$

$$= \frac{280 \times 10^6}{2} = 140 \times 10^6 \text{ Pa}$$

Also, $$\tau_{max} = \frac{16T}{\pi d^3}$$

$$\Rightarrow \quad d = \sqrt[3]{\frac{16T}{\pi \times \tau_{max}}}$$

$$= \sqrt[3]{\frac{16 \times 10 \times 10^3}{\pi \times 140 \times 10^6}} = 0.071 \text{ m}$$

$$\Rightarrow \quad r = \frac{d}{2}$$

$$= \frac{0.071}{2} = 0.035 \text{ m} = 3.5 \text{ cm}$$

178. Given $\xi_{new} = 1.2$

$$\xi = \frac{C}{C_m}$$

$$= \frac{C}{2\sqrt{km}}$$

$$= \frac{C}{2\sqrt{m}} \times \frac{1}{\sqrt{k}}$$

$$= \frac{C_1}{\sqrt{k}}$$

$$\frac{\xi_{original}}{\xi_{new}} = \frac{C_1}{\sqrt{k_{org}}} \times \frac{\sqrt{k_{new}}}{C_1}$$

$$= \frac{\sqrt{k_{new}}}{\sqrt{k_{org}}}$$

$$\frac{\xi_{new}}{\xi_{original}} = \sqrt{\frac{k_{org}}{k_{new}}}$$

$$(1.2)^2 = \frac{k_{org}}{k_{new}}$$

$$\frac{k_{org}}{k_{new}} = 1.44$$

179. Equation of motion

$$K.E = \frac{1}{2}mx_1^2 + \frac{1}{2}mx_2^2$$

$$P.E = \frac{1}{2}kx_1^2 + \frac{1}{2}k(x_1 - x_2)^2 + \frac{1}{2}(2k)(x_2)^2$$

Using lagrangian equation

$$\frac{d}{dt}\left[\frac{d(K.E)}{dx}\right] + \frac{d(P.E)}{dx} = 0$$

For, x_1

$$m\ddot{x}_1 + kx_1 + k(x_1 - x_2) = 0$$

$$\Rightarrow \quad m\ddot{x}_1 + 2kx_1 - kx_2 = 0 \quad \text{...(i)}$$

For, x_2

$$m\ddot{x}_2 + k(x_1 - x_2)(-1) + 2kx_2 = 0$$

$$m\ddot{x}_2 + 3kx_2 + kx_1 = 0 \quad \text{...(ii)}$$

$$\begin{bmatrix} m & 0 \\ 0 & m \end{bmatrix}\begin{bmatrix} \ddot{x}_1 \\ \ddot{x}_2 \end{bmatrix} + \begin{bmatrix} 2k & -k \\ -k & 3k \end{bmatrix}\begin{bmatrix} x_1 \\ x_2 \end{bmatrix} = 0$$

For getting mode shape and natural frequency

$$A = M^{-1}k$$

$$A = \begin{bmatrix} \frac{1}{m} & 0 \\ 0 & \frac{1}{m} \end{bmatrix}\begin{bmatrix} 2k & -k \\ -k & 3k \end{bmatrix} = \begin{bmatrix} \frac{2k}{m} & \frac{-k}{m} \\ \frac{-k}{m} & \frac{3k}{m} \end{bmatrix}$$

$$|A - \lambda I| = 0$$

$$\begin{vmatrix} \frac{2k}{m} - \lambda & \frac{-k}{m} \\ \frac{-k}{m} & \frac{3k}{m} - \lambda \end{vmatrix} = 0$$

$$\left(\frac{2k}{m} - \lambda\right)\left(\frac{3k}{m} - \lambda\right) - \left(\frac{k}{m}\right)^2 = 0$$

$$\lambda^2 - \frac{5k}{m}\lambda + 5\left(\frac{\lambda}{m}\right)^2 = 0$$

$$\lambda_{1,2} = \frac{-\frac{5k}{m} \pm \sqrt{\left(\frac{-5k}{m}\right)^2 \pm 4 \times 1 \times 5\left(\frac{k}{m}\right)^2}}{2}$$

$$= \left(1.38\frac{k}{m}, 3.61\frac{k}{m}\right)$$

$$\Rightarrow \omega_{1,2} = \sqrt{\lambda_{1,2}} = 1.174\sqrt{\frac{k}{m}}, 1.9\sqrt{\frac{k}{m}}$$

To get 1st mode shape

Putting λ in $[A - \lambda I][\phi] = 0$

$$\begin{bmatrix} \frac{2k}{m} - \frac{1.38k}{m} & \frac{-k}{m} \\ \frac{-k}{m} & \frac{3k}{m} - \frac{1.38k}{m} \end{bmatrix}\begin{bmatrix} X_{m1} \\ X_{m2} \end{bmatrix} = \begin{bmatrix} 0 \\ 0 \end{bmatrix}$$

Where, X_{m1} and X_{m2} are amplitude of mass 1 and mass 2 respectively,

$$\begin{bmatrix} 2-1.38 & -1 \\ -1 & 3-1.38 \end{bmatrix}\begin{bmatrix} X_{m1} \\ X_{m2} \end{bmatrix} = \begin{bmatrix} 0 \\ 0 \end{bmatrix}$$

$$\begin{bmatrix} 0.62 & -1 \\ -1 & 1.62 \end{bmatrix}\begin{bmatrix} X_{m1} \\ X_{m2} \end{bmatrix} = \begin{bmatrix} 0 \\ 0 \end{bmatrix}$$

$\therefore$ $X_{m1} = 1$ and $X_{m2} = a = 0.62$

180. As we know,

$$\gamma_{udl} = \frac{wl^4}{8EI}$$

$$\gamma_{cantiliver} = \frac{Rl^3}{3EI}$$

$$\therefore \quad \gamma_{udl} = \gamma_{cantiliver}$$

$$\Rightarrow \quad \frac{wl^4}{8EI} = \frac{Rl^3}{3EI}$$

$$\Rightarrow \quad R = \frac{wl^4}{8EI} \times \frac{3EI}{Rl^3}$$

$$= \frac{3}{8}Wl$$

$$= \frac{3}{8} \times 10 \times 1 = 3.7 \text{ N}$$

181.

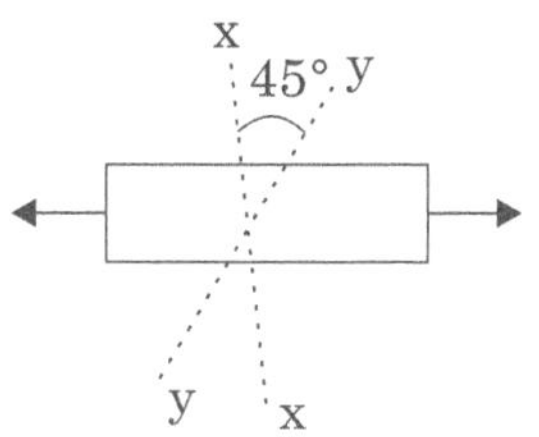

Brittle failure:

Brittle failure occures without any plastic deformation or necking

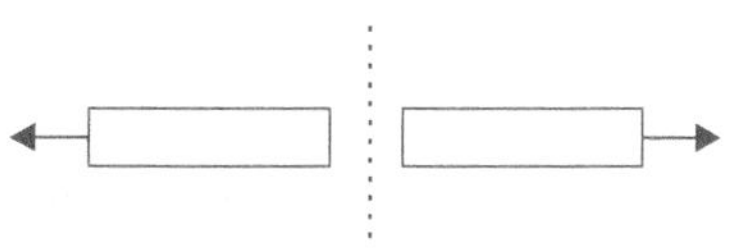

Ductile:

Ductile failure involves plastic deformation and necking mostly it will occure at 45°

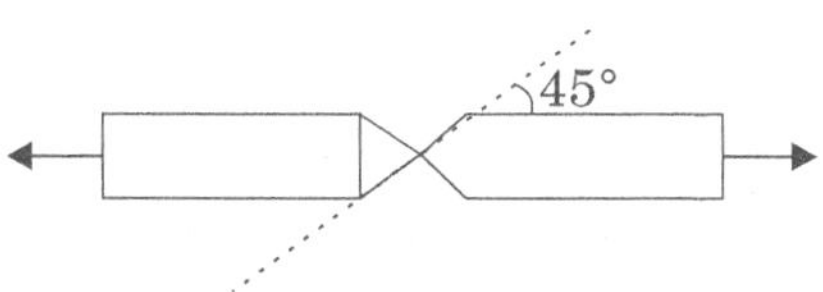

(Option B) along x-x brittle and along yy ductile material

182. State of stress

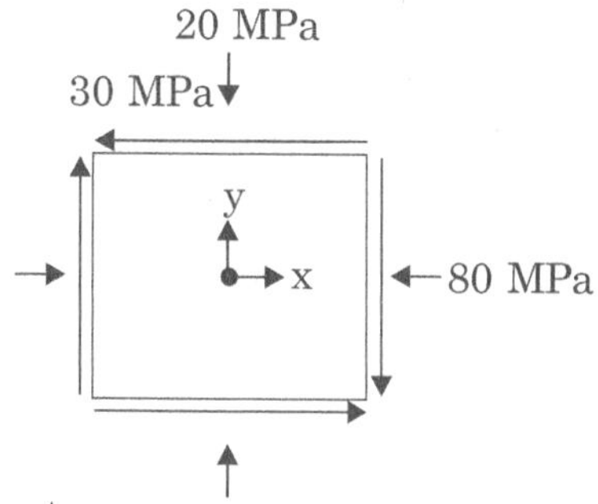

Here, stress

$6_x = -80$ MPa

$6_y = -20$ MPa

$\tau_{xy} = -30$ MPa

Angle of maximum shear stress can be calculated using

$$\tan(2\theta_s) = \frac{\frac{-(6_x - 6_y)}{2}}{\tau_{xy}}$$

$$\theta_s = \frac{1}{2}\tan^{-1}\left[\frac{\frac{-(6_x - 6_y)}{2}}{\tau_{xy}}\right]$$

$$= \frac{1}{2}\tan^{-1}\left[\frac{\frac{-[(-80)-(-20)]}{2}}{(-30)}\right]$$

$$= -\frac{1}{2}\tan^{-1}(1) = \frac{-45}{2} \qquad \left(\tan^{-1}(-x) = -\tan^{-1}(x)\right)$$

$= -22.5°$

183. We know that

System returns to its original position in minimum possible time when it is critically damped

$\therefore$ damping ratio $\zeta = 1$

$$\Rightarrow \frac{C}{C_c} = 1$$

$\Rightarrow C = 2m\,\omega_n$ $\qquad C_c = 2\,m\,\omega_n$

$= 2 \times 10 \times 13.2$ $\qquad m = 10$ kg

$= 264$ N s/m $\qquad \omega_n = 13.2$ rad/sec.

Natural frequency

184. Natural frequency for contiliver beam is given by:

$$f = \frac{k}{2\pi l^2}\sqrt{\frac{EI}{\rho A}} \qquad k = \text{constant}$$

given: both beam have same cross-section and homoganeous material

$\therefore$ E, I, & A are same for both beams

$\therefore$ Ratio of first natural frequencies

$$\frac{(f_l)}{(f_{2l})} = \frac{\left(\frac{k}{2\pi l^2}\right)\sqrt{\frac{EI}{\rho A}}}{\left(\frac{k}{2\pi(2l)^2}\right)\sqrt{\frac{EI}{\rho A}}} = \frac{4l^2}{l^2}$$

K is same for both beams for any natural frequencies

= 4

185. Cylindrical tank with closed ends:

given: $\varepsilon_c = 3\,\varepsilon_{axial}$ ....(i)

we know that

$$\text{hoop stress } 6_c = \frac{Pd}{2t}$$

$$\text{axial or longitudinal stress } 6_l = \frac{Pd}{4t}$$

writing eq. (i) in terms of stresses

$$\frac{1}{E}(6_c - \upsilon 6_l) = 3 \times \frac{1}{E}[6_l - \upsilon 6_c]$$

$$\Rightarrow \frac{1}{E}\left[\frac{Pd}{2t} - \upsilon\frac{Pd}{4t}\right] = 3 \times \frac{1}{E}\left[\frac{Pd}{4t} - \frac{Pd}{2t}\upsilon\right]$$

$$\Rightarrow \left[1 - \upsilon\frac{1}{2}\right]\frac{Pd}{2t} = 3 \times \left[\frac{1}{2} - \upsilon\right]\frac{Pd}{2t}$$

$$\Rightarrow 1 - \frac{\upsilon}{2} = \frac{3}{2} - 3\upsilon$$

$$\Rightarrow 3\upsilon - \frac{\upsilon}{2} = \frac{3}{2} - 1$$

$$\Rightarrow \frac{5\upsilon}{2} = \frac{1}{2}$$

$$\Rightarrow \boxed{\upsilon = \frac{1}{5} = 0.2}$$

186. Kinetic energy $= \frac{1}{2} I\theta^2$

Potential energy $= \frac{1}{2} kx^2$

$$= \frac{1}{2}k\left[\theta^2\right]\left[\frac{l^2}{4}\right]$$

Using logragian equation:

$$\frac{d}{dt}\left(\frac{\partial k \cdot E}{\partial \dot{\theta}}\right) + \frac{\partial (P \cdot \varepsilon)}{\partial \theta} = f(t)$$

$$I\ddot{\theta} + \frac{kl^2}{4}\theta = 0$$

Moment of inertia $I = mz^2 = m[l^2 + h^2]$

$\therefore$ Natural frequency in Hz

$$= \frac{1}{2\pi}\sqrt{\frac{\left(\frac{kl^2}{4}\right)}{I}} = \frac{1}{2\pi}\sqrt{\frac{kl^2}{4m\left[l^2 + h^2\right]}}$$

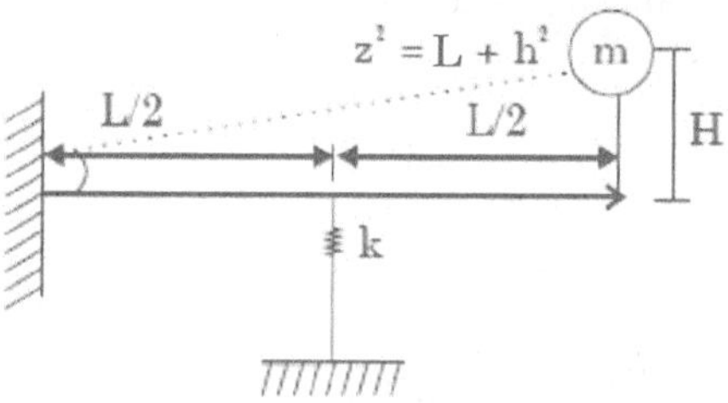

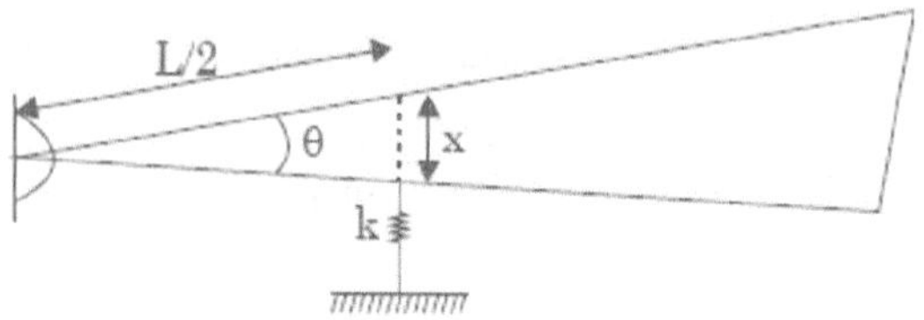

187. Natural frequency 1, $\omega_{n_1} = 233.9$ rad/s

Natural frequency 2, $\omega_{n_2} = 324.5$ rad/s

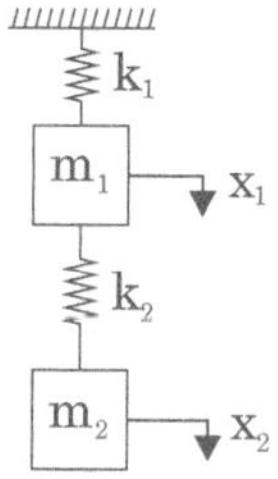

1st mode shape $\phi_1 = \begin{bmatrix} 1 \\ -3.16 \end{bmatrix}$

2nd mode shape $\phi_2 = \begin{bmatrix} 1 \\ 3.16 \end{bmatrix}$

(i) 1st mode shape tells us that if mass 1 in displaced by 1 unit deflection mass 2 will move by 3.16 times unit deflection in opposite side.

(ii) 2nd mode shape tells us if mass 1 is displaced by b unit deflection mass 2 will move with 3.16 times unit deflection in same direction.

Option (B) is correct

If mass 1 is displaced by $x_1(0) = 2$ cm then mass 2 will be displaced by $-3.16 \times 2 = -6.32$ cm

$\Rightarrow x_2(0) = -6.32$

For first natural frequency i.e. 1st mode shape.

188. Deflection function

$$y = \frac{\omega}{48EI}\left(2x^4 - 3lx^3 + l^3x\right) \quad ....(A)$$

at maximum deflection of beam slope is zero

$$\text{Slope} \Rightarrow \frac{dy}{dx}$$

$$= \frac{\omega}{48EI}\left[8x^3 - 9lx^2 + l^3\right]$$

$$\Rightarrow \left(\frac{dy}{dx}\right)_x = 0$$

$$\frac{\omega}{48EI}\left[8x^3 - 9lx^2 + l^3\right] = 0$$

$$8x^3 - 9lx^2 + l^3 = 0$$

$$8\left(\frac{x}{l}\right)^3 - 9\left(\frac{x}{l}\right)^2 + 1 = 0 \quad ....(i)$$

Solving eq. (i)

$$\frac{x}{l} = -0.296,\ 1,\ 0.4215$$

$\frac{x}{l}$ cannot be –ve

$\therefore$ –0.296 cannot be ans.

If we put $\frac{x}{l} = 1 \Rightarrow x = l$ in (A)

$$y = \frac{\omega}{48EI}\left[2l^4 - 3l^4 + l^4\right] = 0$$

$\therefore$ deflection is zero at $\frac{x}{l} = 1$

$\therefore$ $\frac{x}{l} = 1$ cannot be ans

$$\therefore \boxed{\frac{x}{l} = 0.4215}$$

189.

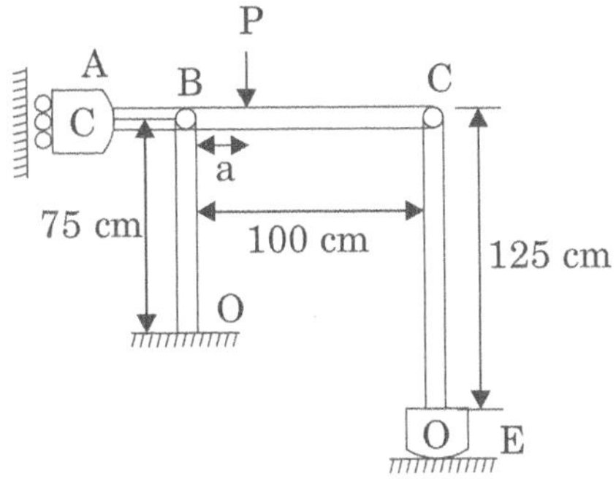

E = 200 GPa

Cross-section = 1.5 × 1.5 cm^2

Free body diagram

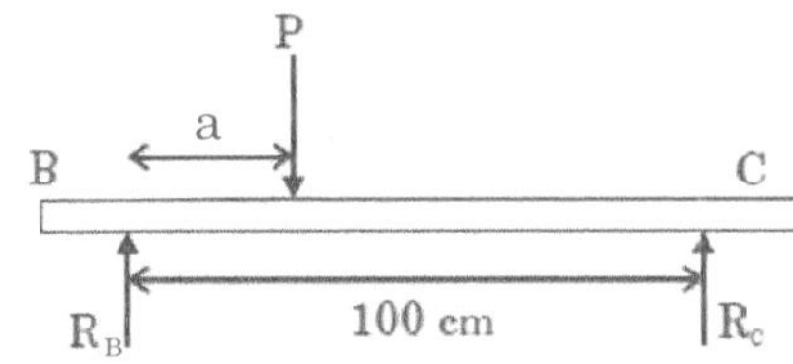

Simply supported beam

Force equilibrum $R_B + R_C = P$ ....(i)

Moment equilibrium about A

$R_C \times 100 = P \times a$

$$R_C = \frac{P \times a}{100}$$

$$\Rightarrow R_B = \frac{P(100-a)}{100}$$

$\Rightarrow$ R_B will act as a buckling load for BD column

R_C will act as a buckling load for CE column

FOr column BD which is fixed-hinged

$$\left(P_{cr}\right)_{BD} = \frac{\pi^2 EI}{\left(\frac{l}{\sqrt{2}}\right)^2}$$

For column C.E which is Hinged-Hinged

$$\left(P_{cr}\right)_{CE} = \frac{\pi^2 EI}{l^2}$$

Taking ratio

As EI are same for both materials

$$\frac{\left(P_{cr}\right)_{BD}}{\left(P_{cr}\right)_{CE}} = \frac{\dfrac{2\pi^2 EI}{(75)^2}}{\dfrac{\pi^2 EI}{(125)^2}} = 5.56$$

$\Rightarrow (P_{Cr})_{BD} = 5.56\,(P_{Cr}.)_{CE}$

Now using R_B and R_C as bukling load

$R_B = 5.56\, R_C$

$$\frac{P(100-a)}{100} = 5.56\frac{Pa}{100}$$

P(100 – a) = 5.56 Pa

$\Rightarrow \boxed{a = 15.24}$ in 1 decimal

$$\boxed{a = 15.2}$$

190. Two cell wing box

- Cell wall thickness = 1.5 mm
- Shear modulus = 127 GPa
- Applied Torque, T = 12 KN m
  = 12 × 10^6 Nm

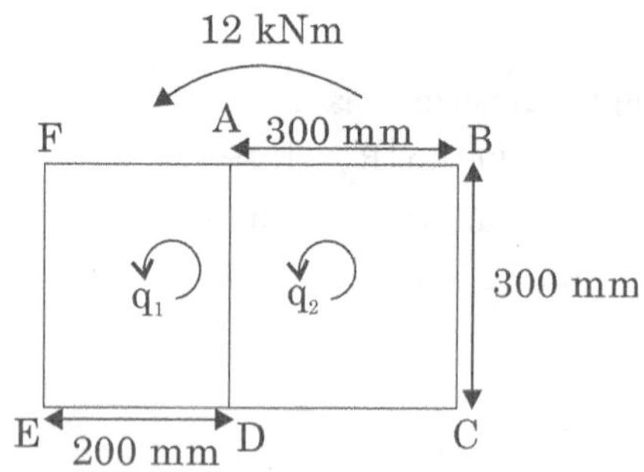

Say q_1 and q_2 are shear flows in cell 1 and cell 2

We know T = Σ2Aq

$\therefore$ T = $2A_1q_1 + 2A_2q_2$

= 2[200 × 300]q_1 + 2[300 × 300]q_2

$$\Rightarrow 12\times10^6 = 2\left[6\times10^4\right]q_1 + 2\left[9\times10^4\right]q_2$$

$$\Rightarrow 600 = 6q_1 + 9q_2$$

$$\Rightarrow 200 = 2q_1 + 3q_2 \qquad ....(i)$$

We know that

$$\beta = \frac{\theta}{l} = \frac{1}{2AG}\oint\frac{qdS}{t}$$

Cell 1 : $\beta_1 = \left(\frac{\theta}{l}\right)$

$$= \frac{1}{2A_1Gt}\left[300q_1 + 200q_1 + 300\left(q_1 - q_2\right) + 200q_1\right]$$

$$= \frac{1}{2A_1Gt}\left[1000q_1 - 300q_2\right]$$

Cell 2:

$$\beta_2 = \left(\frac{\theta}{l}\right) = \frac{1}{2A_2Gt}$$

$$\left[300q_2 + 300q_2 + 300\left(q_2 - q_1\right) + 300q_2\right]$$

$$= \frac{1}{2A_2Gt}\left[1200q_2 - 300q_1\right]$$

We know: $\beta_1 = \beta_2$

$$\Rightarrow \frac{1}{2A_1Gt}\left[1000q_1 - 300q_2\right]$$

$$= \frac{1}{2A_2Gt}\left[1200q_2 - 300q_1\right]$$

$$\Rightarrow \frac{1}{\left(6\times10^4\right)}\left[1000q_1 - 300q_2\right]$$

$$= \frac{1}{9\times10^4}\left[1200q_2 - 300q_1\right]$$

$$\Rightarrow \frac{1}{2}\left[10q_1 - 3q_2\right]\times100$$

$$= \frac{1}{3}\left[12q_2 - 3q_1\right]\times100$$

$$\Rightarrow 5q_1 - \frac{3}{2}q_2 = 4q_2 - q_1$$

$$\Rightarrow \frac{11}{2}q_2 = 6q_1$$

$$\Rightarrow q_2 = 1.09q_1 \qquad ....(ii)$$

Putting eq. (ii) in (i)

200 = $2q_1 + 3q_2$

$\Rightarrow$ 200 = $2q_1$ + (3 × 1.09q_1)

$\Rightarrow$ q_1 = 37.95 N/mm

$\therefore$ q_2 = 1.09 × 37.95 = 41.366 N/mm

Shear stress along wall AD

$$\tau_{AD} = \frac{q_2 - q_2}{t}$$

$$\frac{3.416}{1.5} = 2.277 \text{ N/mm}^2 \text{ or MPa}$$

= 2.3 MPa

191.

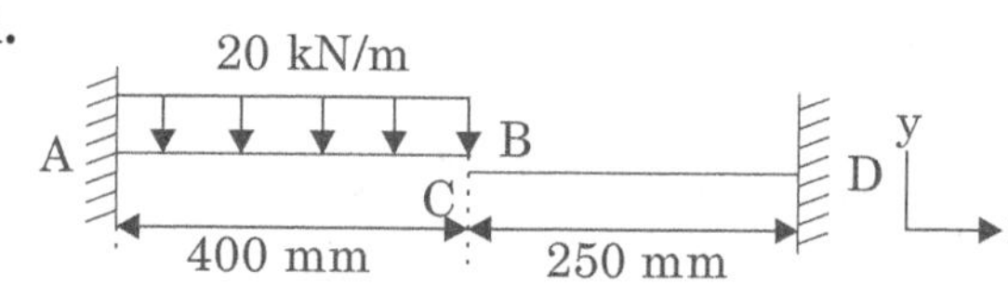

Contilever Beams AB and DC

Square cross-section 50 mm × 50 mm

Elastic modulus E = 70 GPa

= 70 × 10^3 N/mm^2

UDL on beam AB = 20 KN/m

= 20 N/mm

Maximum deflection for beam AB which is at end B is

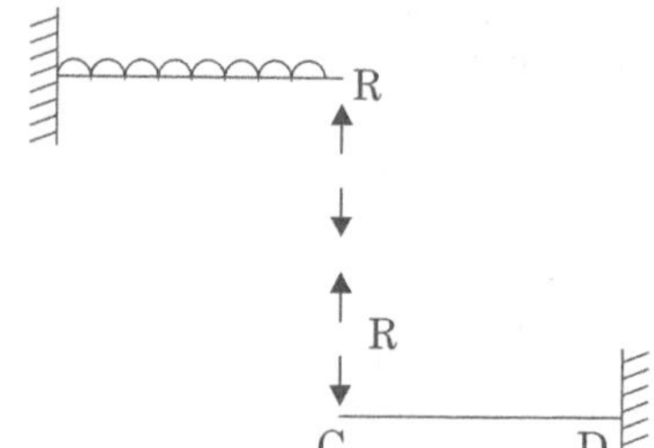

$$y_B = \frac{\omega l^4_{AB}}{8EI}$$

deflection at B should be equal to deflection at C

$[\text{deflection}]_{\text{UDL on AB}} - [\text{deflection}]_{\text{R on AB}} = [\text{deflection}]_{\text{R on DC}}$

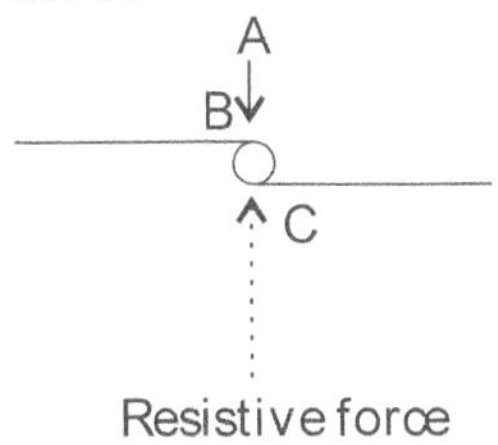

$$\frac{\omega l^4_{AB}}{8EI} - \frac{R \times l^3_{AB}}{3EI} = \frac{R \times l^3_{DC}}{3EI}$$

$$\Rightarrow \frac{20 \times (400)^4}{8} - \frac{R(400)^3}{3} = \frac{R \times [250]^3}{3}$$

$\Rightarrow$ R = 2411.3 N

= 2.4 kN

Thus compressive force roller = 2.4 kN

192. Length of Pole = 5.5 m

Yield strength = 240 MPa

= 240 N/mm2

Wind pressure, P = 7.5 kPa

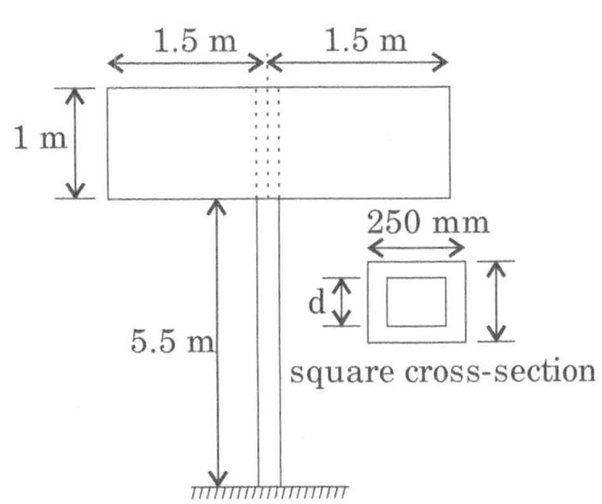

Force acting of Board

F = P × A

= $7.5 \times 10^3 \times 1 \times [1.5 + 1.5]$

= 22.5×10^3 N

This force will act at centroid of plate

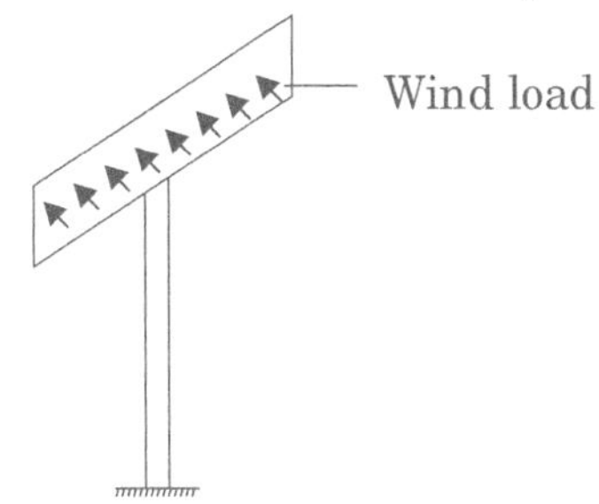

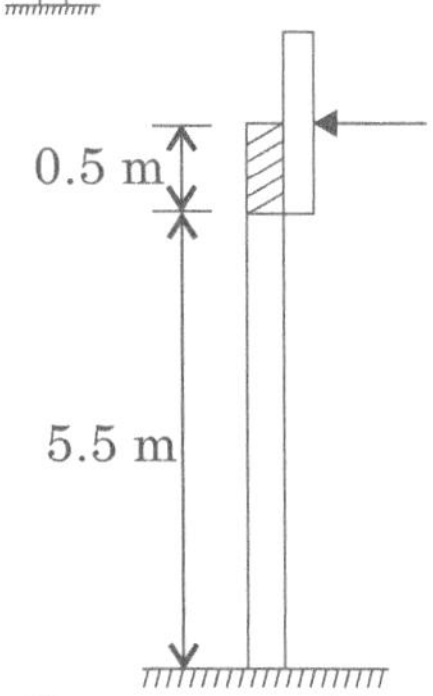

this is a contilever case with end load

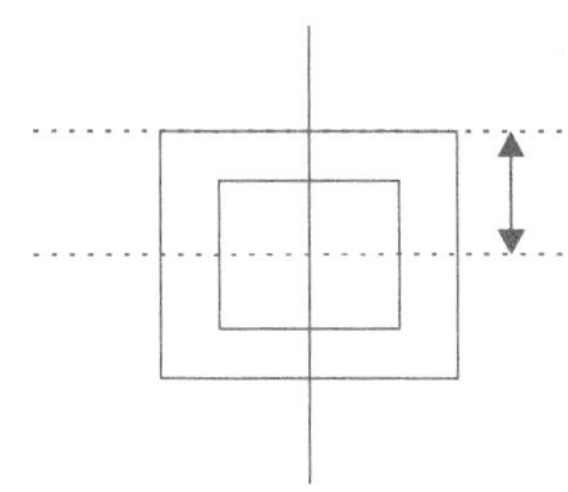

Bending stress:

$$6g = \frac{My}{I}$$

$$\Rightarrow 240 = \frac{\left[22.5 \times 10^3 \times 6 \times 10^3\right] \times \left[\frac{250}{2}\right]}{\frac{250^4}{12} - \frac{d^4}{12}}$$

$\Rightarrow$ d = 235.24 ~ 235 mn

6

CHAPTER

JET PROPULSION

1. Which one of the following engines should be used by a subsonic passenger transport airplane for minimum specific fuel consumption?
   (*a*) Turbojet engine with afterburner
   (*b*) Turbofan engine
   (*c*) Ramjet engine
   (*d*) Scramjet engine

[GATE 2007, 1 M]

2. On which one of the following thermodynamic cycles does an ideal ramjet operate?
   (*a*) The Rankine cycle (*b*) The Brayton cycle
   (*c*) The Carnot cycle (*d*) The Otto cycle

[GATE 2007, 1 M]

3. Air enters through the eye of a centrifugal compressor with a stagnation temperature 300 K and exits the compressor with a stagnation temperature 424 K. If the isentropic efficiency of the compressor is 0.81 and the ratio of specific heats of the flowing gas (assumed as constant) is 1.4, then the pressure ratio across the compressor is
   (*a*) 2.75 (*b*) 5.60
   (*c*) 65.00 (*d*) 228.00

[GATE 2007, 2 M]

4. For an impulse turbine with identical stages, the hot gas exits from the stator blades at the mean blade height at an absolute angle of 70 degrees with the axis of the turbine. If the absolute inlet blade angle with the axis of the turbine at the mean blade height for the rotor blades is 37 degree, then the absolute exit blade angle with the axis of the turbine at the mean blade height of the rotor blades is
   (*a*) 33 degrees (*b*) 37 degrees
   (*c*) 53 degrees (*d*) 53.5 degrees

[GATE 2007, 2 M]

5. Which one of the following materials should be selected to design an axial flow turbine operating at high temperatures?
   (*a*) Steel alloy (*b*) Titanium alloy
   (*c*) Nickel alloy (*d*) Aluminum alloy

[GATE 2007, 2 M]

6. Which one of the following statements is true?
   (*a*) The isentropic efficiency of a compressor is constant throughout the compressor
   (*b*) Flow separation problems are more critical for the axial compressors than for the centrifugal compressors
   (*c*) The pressure ratio of a centrifugal compressor approaches zero as the compressor mass flow rate approaches zero
   (*d*) Centrifugal compressor are always designed with multiple stages

[GATE 2007, 2 M]

7. Combustion between fuel (octane) and oxidizer (air) occurs inside a combustor with the following stoichiometric chemical reaction:

$$2C_8H_{18} + (25O_2 + 94N_2) \rightarrow 16CO_2 + 18H_2O + 94N_2$$

The atomic weight of carbon (c), Hydrogen (H), Oxygen (O) and Nitrogen (N) are 12, 1, 16 and 14 respectively. If the combustion takes place with the fuel-to-air ratio 0.028, then the equivalence ratio of the fuel-oxidizer mixture is
   (*a*) 0.094 (*b*) 0.422
   (*c*) 0.721 (*d*) 2.371

[GATE 2007, 2 M]

8. A gas turbine engine operates with a constant area duct combustor with inlet and outlet stagnation temperatures 540 K and 1104 K respectively. Assume that the flow is one dimensional, incompressible and frictionless and that the heat addition is driving the flow inside the combustor. The pressure loss factor (stagnation pressure loss non-dimensionalized by the inlet dynamic pressure) of the combustor is
   (*a*) 0 (*b*) 0.489
   (*c*) 1.044 (*d*) 2.044

[GATE 2007, 2 M]

9. The diffuser of an airplane engine decelerates the airflow from the flight Mach number 0.85 to the compressor inlet Mach number 0.38. Assume that the ratio of the specific heats is constant and equal to 1.4. If the diffuser pressure recovery ratio is 0.92, then the isentropic efficiency of the diffuser is

(*a*) 0.631 (*b*) 0.814
(*c*) 0.892 (*d*) 1.343

[GATE 2007, 2 M]

10. The purpose of a fuel injection system in the combustor is

(*a*) to accelerate the flow in the combustor
(*b*) to increase the stagnation pressure of the fuel-air mixture
(*c*) to ignite the fuel-air mixture
(*d*) to convert the bulk fuel into tiny droplets

[GATE 2007, 2 M]

Statement of linked Answer Question 11 and 12:

Air enters the compressor of a gas turbine engine with velocity 127 m/s, density 1.2 kg/m^3 and stagnation pressure 0.9 MPa. Air exits the compressor with velocity 139 m/s and stagnation pressure 3.15 Mpa. Assume that the ratio of specific heat is constant and equal to 1.4

11. The compressor pressure ratio is

(*a*) 0.22 (*b*) 0.28
(*c*) 3.50 (*d*) 3.90

[GATE 2007, 2 M]

12. If the polytropic efficiency of the compressor is 0.89, then the isentropic efficiency of the compressor is

(*a*) 0.613 (*b*) 0.869
(*c*) 0.89 (*d*) 0.98

[GATE 2007, 2 M]

13. The degree of reaction of an impulse turbine is

(*a*) 1 (*b*) 0.75
(*c*) 0.5 (*d*) 0

[GATE 2008, 1 M]

14. In a scramjet engine, the Mach number at the entry to the combustion chamber is around

(*a*) 0 (*b*) 0.3
(*c*) 2 (*d*) 6

[GATE 2008, 1 M]

15. A turbofan engine has a bypass ratio of 5 and a total mass flow rate of 120 kg/s. The mass flow rate through the bypass duct is

(*a*) 20 kg/s (*b*) 100 kg/s
(*c*) 120 kg/s (*d*) 600 kg/s

[GATE 2008, 2 M]

16. A turbojet engine is operating with afterburner off. If the afterburner is switched on, then

(*a*) both thrust and sfc decrease
(*b*) thrust increases and sfc decreases
(*c*) thrust decreases and sfc increases
(*d*) both thrust and sfc increases

[GATE 2008, 2 M]

17. A centrifugal compressor operates with a tip blade speed of 340 m/s. The air leaves the impeller with a radial velocity of 88 m/s. If the slip factor is 0.85, the relative velocity at the blade tip is

(*a*) 101.7 m/s (*b*) 120.3 m/s
(*c*) 132.6 m/s (*d*) 135.8 m/s

[GATE 2008, 2 M]

18. An ideal ramjet engine is flying at a Mach number M. The exhaust gas static temperature at the outlet of the nozzle is T_e. The ambient static temperature is T_a. Gas constant R and specific heat ratio γ do not vary through the ramjet. Assuming that nozzle exhaust static pressure is equal to the ambient pressure and fuel air ratio f <<1, the thrust per unit mass flow rate is

(*a*) $\sqrt{\gamma R T_a}\left[\sqrt{\frac{T_e}{T_a}}\right]$

(*b*) $\sqrt{\gamma R T_a}\left[\sqrt{\frac{T_e}{T_a}}-1\right]$

(*c*) $M\sqrt{\gamma R T_a}\left[\sqrt{\frac{T_e}{T_a}}-1\right]$

(*d*) $M\sqrt{\gamma R T_a}\left[\sqrt{\frac{T_e}{T_a}}\right]$

[GATE 2008, 2 M]

19. A 50 percent degree of reaction axial flow turbine operates with a mean blade speed of 180 m/s. The flow leaves the stator and enters the rotor at an angle of 60 degrees to the axial direction. The axial velocity is 150 m/s, and remains constant throughout the stage. The turbine power per unit mass flow is

(*a*) 29.76 kJ/kg (*b*) 41.12 kJ/kg
(*c*) 58.33 kJ/kg (*d*) 61.13 kJ/kg

[GATE 2008, 2 M]

20. Air enters the combustor of a gas turbine engine at total temperature of 500 K and leaves the combustor at total temperature of 1800 K. If c_p constant at 1.005 kJ/kgK and heating value of the fuel used is 44 MJ/kg, the fuel to air ratio is

(*a*) 0.003 (*b*) 0.012
(*c*) 0.031 (*d*) 0.074

[GATE 2008, 2 M]

21. An aircraft is cruising at a Mach number of 0.8 an altitude where the ambient static pressure is 95 kPa. The diffuser exit total pressure is 140 kPa. Assuming there is no change in the specific heat at constant pressure across the diffuser, and ratio of specific heats is 1.4, the adiabatic efficiency of the intake is

(*a*) 0.988 (*b*) 0.915
(*c*) 0.722 (*d*) 0.684

[GATE 2008, 2 M]

Statement for Linked Answer Questions 22 and 23: A multi-stage axial flow compressor operating at an adiabatic efficiency of 0.9 develops a total pressure ratio of 11. The total temperature at inlet to the compressor is 335 K and the stagnation enthalpy rise across each stage is 37 kJ/kg. Ratio of specific heats is 1.4 and specific heat constant pressure is 1.006 kJ/kg K.

22. The total temperature rise across the compressor is

(*a*) 310.1 K (*b*) 366.3 K
(*c*) 392.1 K (*d*) 405.4 K

[GATE 2008, 2 M]

23. The total number of stages required are

(*a*) 9 (*b*) 10
(*c*) 11 (*d*) 12

[GATE 2008, 2 M]

24. The propulsive efficiency of a turbo-jet engine moving at velocity U_∞ and having exhaust velocity U_e with respect to the engine is given by

(*a*) $\dfrac{2}{U_\infty / U_e + 1}$ (*b*) $1 - \dfrac{U_\infty}{U_e}$

(*c*) $\dfrac{2U_\infty U_e}{U_e^2 + U_\infty^2}$ (*d*) $\dfrac{2U_\infty}{U_e + U_\infty}$

[GATE 2009, 1 M]

25. An aircraft is flying at M = 2 where the ambient temperature around the aircraft is 250 K. If the specific heat ratio for air $\gamma = 1.4$, the stagnation temperature on the surface of the aircraft is

(*a*) 200 K (*b*) 450 K
(*c*) 350 K (*d*) 1450 K

[GATE 2009, 1 M]

26. The division of feed air to an aircraft gas-turbine combustor into primary and secondary streams serves which of the following purposes?

P. a flammable mixture can be formed

Q. cooling of combustor liner and flame tube can be accomplished

R. specific fuel consumption can be reduced

(*a*) P and R (*b*) Q and R
(*c*) P ans Q (*d*) P, Q and R

[GATE 2009, 1 M]

27. An ideal axial compressor is driven by an ideal turbine across which the total temperature ratio is 0.667. If the total temperature at turbine inlet is T_0 = 1500 K and specific heat of gas c_P = 1 kJ/kg/K, the power drawn by the compressor per unit mass flow rate of air is approximately

(*a*) 300 kW/kg/s (*b*) 1000 kW/kg/s
(*c*) 600 kW/kg/s (*d*) 500 kW/kg/s

[GATE 2009, 2 M]

28. An ideal ramjet is flying at an altitude of 10 km with a velocity of 1 km/s. The ambient pressure is 0.25 bar and temperature is 225 K. The exhaust gases from the engine are optimally expanded and leave the nozzle at 900 K. If the specific heat ratio (γ) remains constant, the specific thrust developed by the engine is approximately

(*a*) 1000 N-s/kg (*b*) 2000 N-s/kg
(*c*) 500 N-s/kg (*d*) 4000 N-s/kg

[GATE 2009, 2 M]

29. A combat aircraft engine is equipped with an afterburner followed by a variable area convergent nozzle (operating with the nozzle choked). The exhaust gas temperature is 750 K when afterburner is off and 3000 K when it is on. When the afterburner is turned on, (assuming the total pressure remains the same, the mass of fuel added in the afterburner is negligible i.e., the mass flow rate remains the same. and the specific heat ratio (γ) remains constant), approximately by what factor must the nozzle area be changed?

(*a*) 0.5 (*b*) 4
(*c*) 1 (*d*) 2

[GATE 2009, 2 M]

Statement for Linked Answer Questions 30 and 31: Air enters the combustor of a gas-turbine engine at a total temperature T_0 of 500 K. The air stream is split into two parts: primary and secondary streams. The primary stream reacts with fuel supplied at a fuel-air ratio of 0.05. The resulting combustion products are then mixed with the secondary air stream to obtain gas with total temperature of 1550 K at the turbine inlet. The fuel has a heating value of 42 MJ/kg. The spectfic heats of air and combustion products are taken as c_p = 1 kJ/kg/K.

30. If the sensible enthalpy of fuel is neglected, the temperature of combustion products from the reaction of primary air stream with fuel is approximately

(*a*) 2100 K (*b*) 3200 K
(*c*) 2600 K (*d*) 1800 K

[GATE 2009, 2 M]

31. The approximate ratio of mass flow rates of the primary air stream to the secondary air stream required to achieve the turbine inlet total temperature of 1550 K is

(*a*) 2 : 1 (*b*) 1 : 2
(*c*) 1 : 1.5 (*d*) 1 : 1

[GATE 2009, 2 M]

Statement for Linked Answer Questions 32 and 33: A piston compresses 1 kg of air inside a cylinder as shown

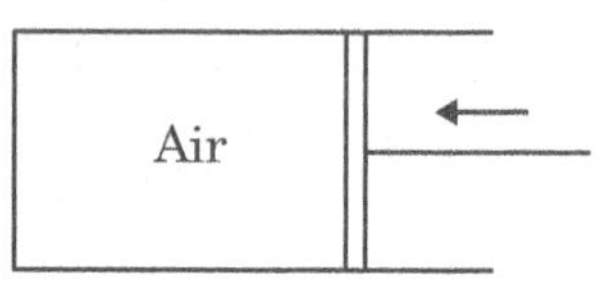

The rate at which the piston does work on the air is 3000 W. At the same time, heat is being lost through the walls of the cylinder at a rate of 847.5 W.

32. After 10 seconds, the change in specific internal energy of the air is

(*a*) 21,525 J/kg (*b*) –21,525 J/kg
(*c*) 30,000 J/kg (*d*) –8,475 J/kg

[GATE 2009, 2 M]

33. Given that the specific heats of air at constant pressure and volume are c_p = 1004.5 J/kg-K and c_v = 717.5 J/kg-K respectively, the corresponding change in the temperature of the air is

(*a*) 21.4 K (*b*) –21.4 K
(*c*) 30 K (*d*) –30 K

[GATE 2009, 2 M]

34. Isentropic efficiency n_d of a subsonic diffuser is defined as

(Note: 'a' represents the ambient, '2' represents the exit of the diffuser and 's' represents an isentropic process)

(*a*) $\frac{T_{02s} - T_a}{T_{02} - T_a}$ (*b*) $\frac{T_{02s} + T_a}{T_{02} + T_a}$

(*c*) $\frac{P_{02s} + P_a}{P_{02} + P_a}$ (*d*) $\frac{P_a - P_{02s}}{P_a - P_{02}}$

[GATE 2010, 1 M]

35. For a multi-stage axial compressor with constant diameter hub

(*a*) Blade height decreases in the flow direction
(*b*) Blade height increases in the flow direction
(*c*) Blade height remains constant
(*d*) Blade height first increases and then decreases in the flow direction

[GATE 2010, 1 M]

36. How does the specific thrust, at constant turbine inlet temperature, produced by a turbofan engine change with an increase in compressor pressure ratio?

(*a*) By increasing
(*b*) By decreasing
(*c*) First by increasing and then decreasing
(*d*) First by decreasing and then increasing

[GATE 2010, 1 M]

37. The maximum operating flow rate through a centrifugal compressor at a given RPM is limited by

(*a*) Impeller stall
(*b*) Surge
(*c*) Choking of diffuser throat
(*d*) Inlet flow distortion

[GATE 2010, 1 M]

38. A missile with a Ramjet engine is flying in air. The temperature at the inlet and the outlet of the combustor are 1200K and 2500 K respectively. The heating value of the fuel is 43 MJ/kg and the burner efficiency is 90%. Considering the working fluid to be air (C_p = 1005 J/kgK and γ = 1.4). The fuel/air ratio $\left(f = \frac{\dot{m}_f}{\dot{m}_a}\right)$ for this engine is equal to:

(*a*) 0.032 (*b*) 0.036
(*c*) 0.042 (*d*) 0.026

[GATE 2010, 2 M]

39. The inlet stagnation temperature for a single stage axial compressor is 300 K and the stage efficiency is 0.80. Following conditions exist at the mean radius of the rotor blade:

Blade speed = 200 m/s
Axial flow velocity = 160 m/s
Inlet blade angle β_1 = 44°
Outlet blade angle β_2 = 14°
C_p =1005 J/kgK and γ = 1.4

What is the stagnation pressure ratio (P_{RS}) for this compressor?

(*a*) 1.41 (*b*) 1.37
(*c*) 1.51 (*d*) 1.23

[GATE 2010, 2 M]

Statement for Linked Answer Questions 40 and 41: An aircraft with an IDEAL Turbojet engine is flying at 200 m/s at an altitude where the ambient pressure is equal to 0.8 bar. The stagnation pressure and temperature at the inlet of the turbine are 6 bar and 1400 K respectively. The change in specific enthalpy across the compressor is 335 kJ/kg. Assume the fuel flow rate to be very small in comparison to the air flow rate and consider C_p = 1117 J/kgK and γ = 1.3.

40. What is the stagnation pressure at the inlet of the nozzle?

(*a*) 2.8 bar (*b*) 5.7 bar

(*c*) 2.1 bar (*d*) 6.3 bar

[GATE 2010, 2 M]

41. What is the specific thrust produced by this engine under the given conditions?

(*a*) 586 Ns/kg (*b*) 745 Ns/kg

(*c*) 686 Ns/kg (*d*) 500 Ns/kg

[GATE 2010, 2 M]

42. A main objective of by-pass in a turbo-fan engine is to increase

(*a*) mass flow rate through engine inlet

(*b*) turbine inlet temperature

(*c*) mass flow rate through exhaust nozzle

(*d*) compressor pressure ratio

[GATE 2011, 1 M]

43. The pressure ratio in any one stage of a jet engine compressor is limited by

(*a*) entry stagnation temperature in that stage

(*b*) entry Mach number in that stage

(*c*) pressure gradient induced separation in that stage

(*d*) mass flow rate in that stage

[GATE 2011, 1 M]

44. Thermodynamic cycle on which the jet engine operates can be

(*a*) open Rankine cycle only

(*b*) either open or closed Rankine cycle

(*c*) open Brayton cycle only

(*d*) either open or closed Brayton cycle

[GATE 2011, 1 M]

45. Propulsion efficiency of a jet engine is

(*a*) directly proportional to both the thrust power and the air mass flow rate

(*b*) inversely proportional to both the thrust power and the air mass flow rate

(*c*) directly proportional to the thrust power and inversely proportional to the air mass flow rate

(*d*) inversely proportional to the thrust power and directly proportional to the air mass flow rate

[GATE 2011, 1 M]

46. A turbojet powered aircraft is flying at Mach number 0.8 at an altitude of 10 km. The inlet and exit areas of the engine are 0.7 m^2 and 0.4 m^2 respectively. The exhaust gases have velocity of 500 m/s and pressure of 60 kPa. The free stream pressure, density and speed of sound are 26.5 kPa, 0.413 kg/m^3 and 299.5 m/s respectively. The thrust of the engine (in kN)

(a) 29.50 (b) 31.44

(c) 30.19 (d) 28.00

[GATE 2011, 2 M]

47. A jet engine is operating at a Mach number of 0.8 at an altitude of 10 km. The efficiency of the air intake is 0.8 and that of the compressor is 0.87. The total temperatures (in K) at the exits of the air intake and the compressor respectively are (Ambient pressure = 26.5 kPa; Ambient temperature = 223.3 K; Gas constant, γ = 1.4; P_{rc} = 8)

(*a*) 251.9 and 458.2

(*b*) 234.9 and 486.8

(*c*) 251.9 and 486.8

(*d*) 234.9 and 458.2

[GATE 2011, 2 M]

48. The ratio of flight speed to the exhaust velocity for maximum propulsion efficiency is

(*a*) 0.0 (*b*) 0.5

(*c*) 1.0 (*d*) 2.0

[GATE 2012, 1 M]

49. The stagnation temperatures at the inlet and exit of a combustion chamber are 600 K and 1200 K, respectively. If the heating value of the fuel is 44 MJ/kg and specific heat at constant pressure for air and hot gases are 1.005 kJ/kg.K and 1.147 kJ/kg.K respectively, the fuel-to-air ratio is

(*a*) 0.0018 (*b*) 0.018

(*c*) 0.18 (*d*) 1.18

[GATE 2012, 2 M]

Statement for Linked Answer Questions 50 and 51: Air at a stagnation temperature of 15° C and stagnation pressure 100 kPa enters an axial compressor with an absolute velocity of 120 m/s. Inlet guide vanes direct this absolute velocity to the rotor inlet at an angle of 18° to the axial direction. The rotor turning angle is 27° and the mean blade speed is 200 m/s. The axial velocity is assumed constant through the stage.

50. The blade angle at the inlet of the rotor is

(*a*) 25.5° (*b*) 38.5°
(*c*) 48.5° (*d*) 59.5°

[GATE 2012, 2 M]

51. If the mass flow rate is 1 kg/s, the power required to drive the compressor is

(*a*) 50.5 kW (*b*) 40.5 kW
(*c*) 30.5 kW (*d*) 20.5 kW

[GATE 2012, 2 M]

52. The mass flow rate of air through an aircraft engine is 10 kg/s. The compressor outlet temperature is 400 K and the turbine inlet temperature is 1800 K. The heating value of the fuel is 42 MJ/kg and the specific heat at constant pressure is 1 kJ/kg-K. The mass flow rate of the fuel in kg/s is approximately

(a) 0.2 to 0.25 (b) 0.3 to 0.35
(c) 0.6 to 0.65 (d) 0.2 to 0.35

[GATE 2013, 1 M]

53. A gas turbine engine is mounted on an aircraft which can attain a maximum altitude of 11 km from sea level. The combustor volume of this engine is decided based on conditions at

(*a*) sea level (*b*) 8 km altitude
(*c*) 5.5 km altitude (*d*) 11 km altitude

[GATE 2013, 1 M]

54. Which one of the following shows the CORRECT variation of stagnation temperature along the axis of an ideal ram jet engine?

(*a*)
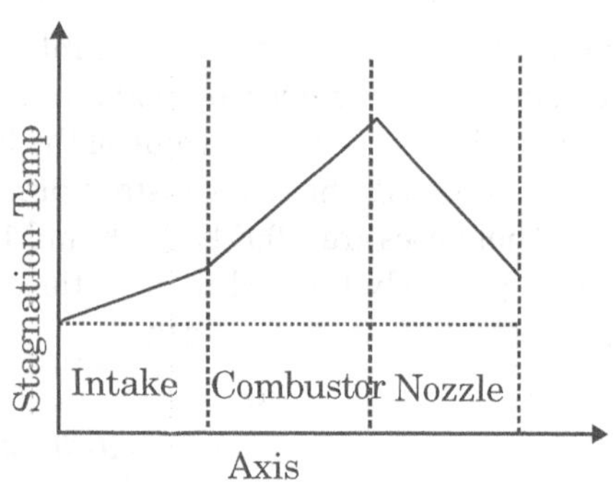

(*b*)
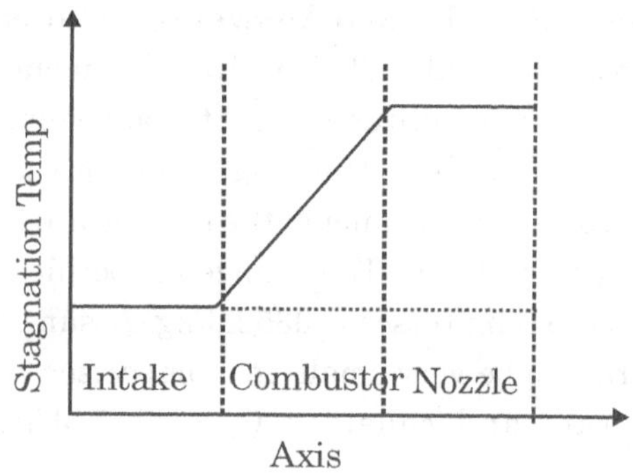

(*c*)
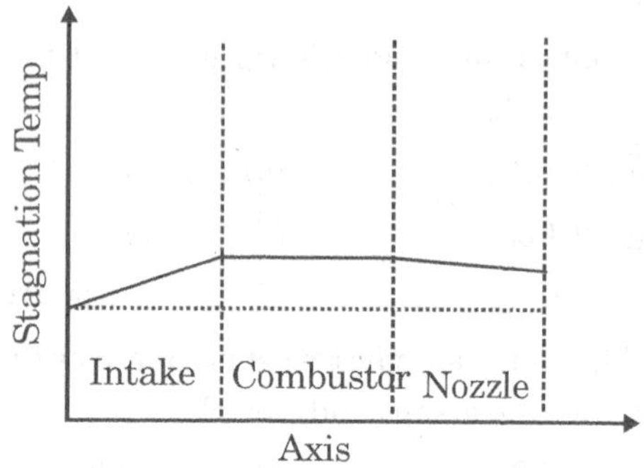

(*d*)
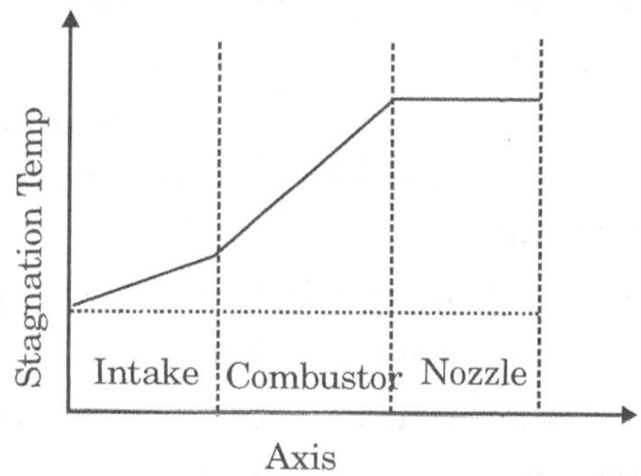

[GATE 2013, 1 M]

55. Consider two engines P and Q. In P, the high pressure turbine blades are cooled with a bleed of 5% from the compressor after the compression process and in Q the turbine blades are not cooled. Comparing engine P with engine Q, which one of the following is NOT TRUE?

(*a*) Turbine inlet temperature is higher for engine P
(*b*) Specific thrust is higher for engine P
(*c*) Compressor work is the same for both P and Q
(*d*) Fuel flow rate is lower for engine P

[GATE 2013, 1 M]

56. In a 50% reaction axial compressor stage, the local blade velocity is 300 m/s and the axial component of velocity is 100 m/s. If the absolute inlet flow angle $\alpha_1 = 45°$, the work per unit mass done on the fluid by the stage in kJ/kg is

(*a*) 30 (*b*) 40
(*c*) 50 (*d*) 60

[GATE 2013, 2 M]

Common Data for Questions 57 and 58: An aircraft is flying at Mach number M = 1.5, where the ambient temperature is 250 K. The stagnation temperature of gases at the entry to the nozzle is 800 K. The nozzle is choked and always under expanded. Assume the molecular weight of the exhaust gases to be 29, the ratio of specific heats to be 1.4 and the universal gas constant is 8314 J/kmol-K.

57. For which one of the nozzle exit Mach numbers given below is the propulsive efficiency highest?

(*a*) 1 (*b*) 1.5

(*c*) 2 (*d*) 2.5

[GATE 2013, 2 M]

58. For which one of the nozzle exit Mach numbers given below is the thrust highest?

(*a*) 1 (*b*) 1.5

(*c*) 2 (*d*) 2.5

[GATE 2013, 2 M]

59. For a given inlet condition, if the turbine inlet temperature is fixed, what value of compressor efficiency given below leads to the lowest amount of fuel added in the combustor of a gas turbine engine?

(*a*) 1 (*b*) 0.95

(*c*) 0.85 (*d*) 0.8

[GATE 2013, 1 M]

60. Match the appropriate engine (in right column) with the corresponding aircraft (in left column) for most efficient performance of the engine.

a. Low speed transport aircraft	i. Ramjet
b. High subsonic civilian aircraft	ii. Turboprop
c. Supersonic fighter aircraft	iii. Turbojet
d. Hypersonic aircraft	iv. Turbofan

(*a*) a – iv, b – iii, c – i, d – ii

(*b*) a – ii, b – i, c – iii, d – iv

(*c*) a – i, b – ii, c – iv, d – iii

(*d*) a – ii, b – iv, c – iii, d – i

[GATE 2014, 1 M]

61. For a given fuel flow rate and thermal efficiency, the take-off thrust for a gas turbine engine burning aviation turbine fuel (considering fuel-air ratio f<<1) is

(*a*) Directly proportional to exhaust velocity

(*b*) Inversely proportional to exhaust velocity

(*c*) Independent of exhaust velocity

(*d*) Directly proportional to the square of the exhaust velocity

[GATE 2014, 1 M]

62. For a fifty percent reaction axial compressor stage, following statements are given:

I. Velocity triangles at the entry and exit of the rotor are symmetrical

II. The whirl or swirl component of absolute velocity at the entry of rotor and entry of stator are same.

Which of the following options are correct?

(*a*) Both I and II are correct statements

(*b*) I is correct but II is incorrect

(*c*) I is incorrect but II is correct

(*d*) Both I and II are incorrect

[GATE 2014, 1 M]

63. The thrust produced by a turbojet engine

(*a*) Increases with increasing compressor pressure ratio

(*b*) Decreases with increasing compressor pressure ratio

(*c*) Remains constant with increasing compressor pressure ratio

(*d*) First increases and then decreases with increasing compressor pressure ratio

[GATE 2014, 1 M]

64. A cruise missile with an ideal ramjet engine is flying at Mach 4.0 at an altitude where the ambient temperature is 100K. Consider ratio of specific heats γ = 1.4 and specific gas constant R = 287J/kgK. If the stagnation temperature in the combustion chamber is equal to 2310K, the speed of the exhaust gases (in m/s) is ________.

[GATE 2014, 2 M]

65. A gas turbine engine is operating under the following conditions:

Stagnation temperature at turbine inlet = 1350K

Stagnation pressure at the turbine inlet = 10 bar

Static temperature at turbine exit = 800 K

Velocity at turbine exit = 200 m/s

Total-to-total efficiency of turbine = 0.96

γ (ratio of specific heats) = 1.33

C_P (specific heat at constant pressure) = 1.147 kJ/kgK

The stagnation pressure (in bar) in the nozzle (considering isentropic nozzle) is equal to ________.

[GATE 2014, 2 M]

66. Air at a stagnation temperature of 300K (ratio of specific heats, γ = 1.4 and specific gas constant R = 287 J/kgK) enters the impeller of a centrifugal compressor in axial direction. The stagnation pressure ratio between the diffuser outlet and impeller inlet is 4.0. The impeller blade radius is 0.3 m and it is rotating at 15000 rev/min. If the slip factor σ_s (Ratio of tangential component of air velocity at the blade tip to the blade tip speed) is 0.88, the overall efficiency (total-to-total) of the compressor (in %) is ________.

[GATE 2014, 2 M]

67. An aircraft is flying at Mach 3.0 at the altitude where the ambient pressure and temperature are 50 kPa and 200 K respectively. If the converging-diverging diffuser of the engine (considered isentropic with ratio of specific heats, γ = 1.4 and specific gas constant R = 287 J/kgK) has a throat area of 0.05 m^2. the mass flow rate through the engine in kg/s is

(*a*) 197 (*b*) 232
(*c*) 790 (*d*) 157

[GATE 2014, 2 M]

68. Which one of the following aero engine has the highest propulsive efficiency

(a) Turbojet with afterburner
(b) Turbojet without after burner
(c) Turbofan engine
(d) Ramjet engine

[GATE 2015, 1 M]

69. The stoichiometric fuel-to-air ratio in an aircraft engine combustor varies with the compressor pressure ratio as follows:

(*a*) increases linearly (*b*) decreases linearly
(*c*) is independent (*d*) increases nonlinearly

[GATE 2015, 1 M]

70. Isentropic efficiencies of an aircraft engine operating at typical subsonic cruise conditions with the following components - intake, compressor, turbine and nozzle - are denoted by n_i, n_c, n_t and n_n, respectively. Which one of the following is correct?

(*a*) $\eta_i < \eta_c < \eta_t < \eta_n$ (*b*) $\eta_t < \eta_i < \eta_c < \eta_n$
(*c*) $\eta_c < \eta_t < \eta_i < \eta_n$ (*d*) $\eta_c < \eta_i < \eta_t < \eta_n$

[GATE 2015, 1 M]

71. Air enters an aircraft engine at a velocity of 180 m/s with a flow rate of 94 kg/s. The engine combustor requires 9.2 kg/s of air to burn 1 kg/s of fuel. The velocity of gas exiting from the engine is 640 m/s. The momentum thrust (in N) developed by the engine is

(*a*) 43241 (*b*) 45594
(*c*) 47940 (*d*) 49779

[GATE 2015, 1 M]

72. Following are the operational parameters of an axial compressor stage:

Air mass flow rate	= 24 kg / s
Static temperature of air at the rotor inlet	= 278 K
Velocity of air at the rotor inlet (zero whirl velocity)	= 140 m / s
Work done on the compressor rotor	= 734 kJ
Isentropic efficiency of the compressor stage	= 0.86
Ratio of specific heats	= 1.4
Specific heat at constant pressure	= 1.005kJ / kgK

The stagnation pressure ratio across the axial compressor stage is ________

[GATE 2015, 2 M]

73. A centrifugal air compressor is operating at the following conditions:

Inlet stagnation temperature = 288 K
Inlet stagnation pressure =1.15 bar
Exit stagnation temperature = 454 K
Exit stagnation pressure = 4.8 bar

The energy loss due to non-isentropic compression per unit mass of flowing air (ratio of specific heats, γ =1.4 and specific heat at constant pressure, C_p = 1.005 KJ/Kg–K) is ______ kJ / kg.

[GATE 2015, 2 M]

74. Hot gas (ratio of specific heats, γ =1.33) at a temperature of 1450 K enters into an axial turbine and expands isentropically. Assume that the kinetic energy of the gas across the turbine is negligible. If the ratio of inlet to outlet pressures of the turbine is 9.5, then the temperature (in K) of gas exiting the turbine is _______.

[GATE 2015, 2 M]

75. For a completely subsonic isentropic flow through a convergent nozzle, which of the following statement is TRUE?

(*a*) Pressure at the nozzle exit > back pressure.
(*b*) Pressure at the nozzle exit < back pressure.
(*c*) Pressure at the nozzle exit = back pressure.
(*d*) Pressure at the nozzle exit = total pressure.

[GATE 2016, 1 M]

76. Which of the following aircraft engines has the highest propulsive efficiency at a cruising Mach number of less than 0.5?

(*a*) Turbofan engine (*b*) Turbojet engine

(*c*) Turboprop engine (*d*) Ramjet engine

[GATE 2016, 1 M]

77. Combustion in gas turbine engines is ideally represented as the following process:

(*a*) Adiabatic (*b*) Isentropic

(*c*) Isobaric (*d*) Isochoric

[GATE 2016, 1 M]

78. An un-mixed turbofan engine with a bypass ratio of 6.0, flies with a velocity of 200 m/s. The core and the bypass nozzles of the engine, that are both convergent nozzles, operate under choked condition and have exhaust static temperatures of 580 K and 295 K, respectively. The specific gas constant and the ratio of specific heats for both the streams are 287 J/kgK and 1.4, respectively. If the fuel-air ratio is negligible, the thrust per unit mass flow rate generated by the engine is _______ Ns/kg.

[GATE 2016, 2 M]

79. A single-stage gas turbine operates with an axial absolute flow at the entry and exit from the stage. The absolute flow angle at the nozzle exit is 70°. The turbine stage generates a specific work of 228 kJ/kg when operating with a mean blade speed of 440 m/s. The absolute velocity at the rotor entry is

(*a*) 275.7 m/s (*b*) 551.5 m/s

(c) 1103.0 m/s (*d*) 1654.5 m/s

[GATE 2016, 2 M]

80. An axial compressor operates such that it has an inlet and an exit total temperature of 300 K and 430 K, respectively. The isentropic efficiency of the compressor is 85 %. If the ratio of specific heats is 1.4, then the total pressure ratio across the compressor is ___________.

[GATE 2016, 2 M]

81. A gaseous mixture of air and fuel enters a constant area combustion chamber at a velocity of 100 m/s and at a static temperature of 300 K. The heat release due to combustion is 1000 kJ/kg. The specific heat at constant pressure of the calorically perfect gas is 1000 J/kgK. The total temperature of air-fuel mixture after combustion is ______ K.

[GATE 2016, 2 M]

82. A substance experiences an entropy change of $\Delta s > 0$ in a quasi-steady process. The rise in temperature (corresponding to the entropy change Δs) is highest for the following process:

(*a*) isenthalpic

(*b*) isobaric

(*c*) isochoric

(*d*) isothermal

[GATE 2016, 2 M]

83. The maximum gas flow rate that can be handled by a multistage axial compressor at a given rotational speed is dictated by

(*a*) Compressor Surge

(*b*) Rotating Stall

(*c*) Choking

(*d*) Optimum Design Pressure Ratio

[GATE 2017, 1 M]

84. For a turbine state, which one of the following losses occurs due to the turning of the wall boundary layer through an angle due to curved surface?

(*a*) Profit loss

(*b*) Annulus loss

(*c*) Tip clearance loss

(*d*) Secondary flow loss

[GATE 2017, 1 M]

85. In the vane-less space between the impeller and the diffuser vanes in a Centrifugal Compressor, the angular momentum varies in the following manner in the radial direction

(*a*) Increases

(*b*) Remains constant

(*c*) Decreases

(*d*) First increases and then decreases

[GATE 2017, 1 M]

86. In a combustor, gaseous Octane (C_8H_{18}) and air are to be burned in stoichiometric proportions. If the required flow rate of air is 1 kg/s, what should be the corresponding flow rate of Octane?

(*a*) 0.066 kg/s

(*b*) 15.15 kg/s

(*c*) 0.16 kg/s

(*d*) 6.25 kg/s

[GATE 2017, 2 M]

87.

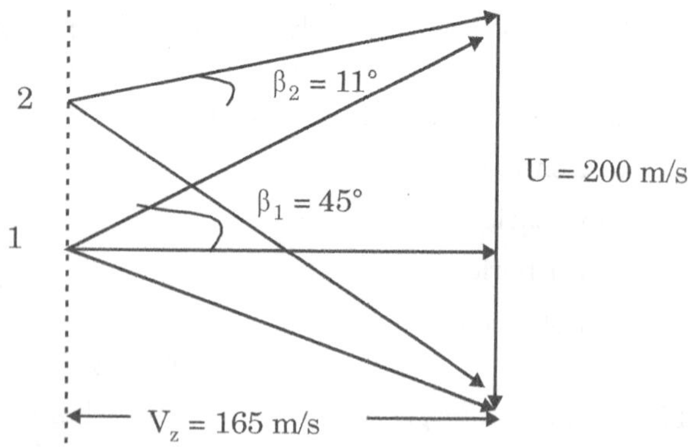

The given diagram represents the velocity triangles at the leading edge (1) and trailing edge (2) at the mean radius of a single stage axial compressor rotor. The stage efficiency of the compressor is 0.8. The stagnation temperature of air entering the stage is 298 K and the specific heat at constant pressure for air is 1.005 kJ/kg K. The ratio of specific heats for air is 1.4. Considering actual work in the rotor is equal to the ideal work, the pressure ratio for the stage is equal to ________ (in two decimal points).

[GATE 2017, 2 M]

88. An aircraft with a turbojet engine, having an inlet area of 1 m^2, is flying at 270 m/s at an altitude where the atmospheric pressure is equal to 0.9 bar and the ambient temperature is equal to 290 K. The stagnation pressure and temperature at the exit of the turbine are equal to 1.6 bar and 774 K respectively. The specific heat at constant pressure of the burned gases is equal to 1.147 kJ/kgK and the ratio of specific heats is equal to 1.33. Considering ideal expansion in the nozzle with no losses, the specific thrust produced by the engine is ________ Ns/kg (in one decimal place)

[GATE 2017, 2 M]

89. Air, at 450 K stagnation temperature and at a rate of 50 kg/s, enters the combustor of a turbofan engine and is burned with 1 kg/s of Aviation Kerosene (Heating value 44 MJ/kg). The specific heat at constant pressure for the incoming air and the burned products are 1.005 kJ/kg K and 1.147 kJ/kgK respectively. Considering 100% burner efficiency, the stagnation temperature at the exit of the combustor is equal to ________ K. (in one decimal place).

[GATE 2017, 2 M]

90. A centrifugal compressor requires 1800 kW power to compress 10 kg/s of air. Consider the whirl velocity component is equal to the impeller speed (ie., no slip) and no losses in the impeller. If the impeller has to rotate at 1900 rad/s, the diameter of the impeller is to be ________ m (in two decimal place).

[GATE 2017, 2 M]

91. The first law of thermodynamics is also known as conservation of

(*a*) mass. (*b*) momentum.

(*c*) energy. (*d*) species.

[GATE 2018, 1 M]

92. In an ideal gas turbine cycle, the expansion in a turbine is represented by

(*a*) an isenthalpic process.

(*b*) an isentropic process.

(*c*) an isobaric process

(*d*) an isochoric process.

[GATE 2018, 1 M]

93. The theoretical maximum velocity (in m/s) of air expanding from a reservoir at 700 K is ________ (accurate to two decimal places). Specific heat of air at constant pressure is 1005 J/(kg-K).

[GATE 2018, 1 M]

94. The stagnation pressures at the inlet and exit of a subsonic intake are 100 kPa and 98 kPa, respectively. The pressure recovery of this intake will be ______ (accurate to two decimal places).

[GATE 2018, 1 M]

95. A combustor is operating with a fuel-air ratio of 0.03. If the stoichiometric fuel-air ratio of the fuel used is 0.06, the equivalence ratio of the combustor will be _______ (accurate to two decimal places).

[GATE 2018, 1 M]

96. An aircraft with a turbojet engine flies at a velocity of 100 m/s. If the jet exhaust velocity is 300 m/s, the propulsive efficiency of the engine, assuming a negligible fuel-air ratio, is

(*a*) 0.33 (b) 0.50

(c) 0.67 (d) 0.80

[GATE 2018, 2 M]

97. An aircraft with a turboprop engine produces a thrust of 500 N and flies at 100 m/s. If the propeller efficiency is 0.5, the shaft power produced by the engine is

(*a*) 50 kW (b) 100 kW
(c) 125 kW (d) 500 kW

[GATE 2018, 2 M]

98. An axial compressor that generates a stagnation pressure ratio of 4.0, operates with inlet and exit stagnation temperatures of 300 K and 480 K, respectively. If the ratio of specific heats (γ) is 1.4, the isentropic efficiency of the compressor is

(*a*) 0.94 (*b*) 0.81
(*c*) 0.72 (*d*) 0.63

[GATE 2018, 2 M]

99. An axial compressor rotor with 50 % degree of reaction, operates with an axial velocity of 200 m/s. The absolute flow angle at the inlet of the rotor is 22° with reference to the axial direction. If the axial velocity is assumed to remain constant through the rotor, the magnitude of the relative velocity (in m/s) at the rotor exit is ________ (accurate to one decimal place).

[GATE 2018, 2 M]

100. The relative velocity of air leaving a straight radial impeller of a centrifugal compressor is 100 m/s. If the impeller tip speed is 200 m/s, for a slip free operation, the absolute velocity (in m/s) at the impeller exit is ________ (accurate to one decimal place).

[GATE 2018, 2 M]

101. The propulsive efficiency of a ramjet engine is lower than that of a low bypass turbofan engine operating under the same conditions and producing the same thrust, primarily because the ramjet engine

(a) has larger kinetic energy lost in the exhaust jet.
(b) has lower thrust power.
(c) is not self-starting.
(d) has higher thrust to weight ratio.

[GATE 2019]

102. While flying at Mach 2.0, 11 km altitude and producing the same thrust, what is the correct order from the lowest thrust specific fuel consumption (tsfc) to the highest tsfc?

(a) Turbofan, Ramjet, Turbojet
(b) Turbofan, Turbojet, Ramjet
(c) Ramjet, Turbojet, Turbofan
(d) Turbojet, Turbofan, Ramjet

[GATE 2019]

103. For a single stage subsonic compressor, which of the following statements about the highest possible compressor pressure ratio (CPR) is correct?

(a) CPR of an axial compressor > CPR of centrifugal compressor.
(b) CPR of an axial compressor < CPR of centrifugal compressor.
(c) CPR of an axial compressor = CPR of centrifugal compressor.
(d) CPR of any value can be attained with either an axial or a centrifugal compressor.

[GATE 2019]

104. For the complete combustion of 1 mole of ethanol (C_2H_5OH), the required number of moles of oxygen is ________.

[GATE 2019]

105. One kg of diatomic gas is heated and its temperature increases from 100 K to 600 K. The energy added at constant pressure during this process is 500 kJ. The specific heat at constant volume for the gas is ____________ kJ/kgK. (round off to 2 decimal places).

[GATE 2019]

106. In a turbojet engine, the compressor outlet temperature increases with decreasing efficiency of the compressor. If the turbine inlet temperature remains constant, with decreasing efficiency of the compressor, the thrust specific fuel consumption of the engine

(a) decreases, as the heat input is lower.
(b) remains unchanged.
(c) increases, as the compressor needs more work input from the turbine.
(d) decreases, as the thrust produced is higher.

[GATE 2019]

107. The combustion products of a gas turbine engine can be assumed to be a calorically perfect gas with γ = 1.2. The pressure ratio across the turbine stage is 0.14. The measured turbine inlet and exit stagnation temperatures are 1200 K and 900 K, respectively. The total-to-total turbine efficiency is ________ % (round off to the nearest integer).

[GATE 2019]

108. The figure shows the velocity triangles for an axial compressor stage. The specific work input to the compressor stage is _______ kJ/kg (round off to 2 decimal places).

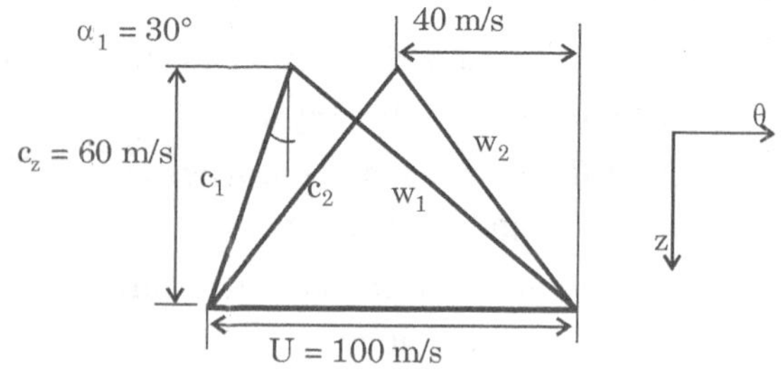

[GATE 2019]

109. The ratio of exit stagnation pressure to inlet stagnation pressure across the rotating impeller of a centrifugal compressor, operating with a closed exit, is

(*a*) 0 (*b*) 1

(*c*) >1 (*d*) 0.5

[GATE 2020]

110. Air enters the rotor of an axial compressor stage with no pre-whirl ($C_\theta = 0$) and exits the rotor with whirl velocity, $C_\theta = 150$ m/s. The velocity of rotor vanes, U is 200 m/s. Assuming $C_p = 1005$ J/(kg K), the stagnation temperature rise across the rotor is _______ K (*round off to one decimal place*).

[GATE 2020]

111. An axial compressor is designed to operate at a rotor speed of 15000 rpm and an inlet stagnation temperature of 300 K. During compressor testing, the inlet stagnation temperature of the compressor measured was 280 K. What should be the rotor speed for the compressor to develop the same performance characteristics during this test as in the design condition?

(*a*) 14000 rpm (*b*) 14491 rpm

(*c*) 15526 rpm (*d*) 16071 rpm

[GATE 2020]

112. The design flight Mach number of an ideal ramjet engine is 2.8. The stagnation temperature of air at the exit of the combustor is 2400 K. Assuming the specific heat ratio of 1.4 and gas constant of 287 J/(kg K), the velocity of air at the exit of the engine is ______ m/s (*round off to one decimal place*).

[GATE 2020]

113. The oprating conditions of an aircraft engine combustor are as follows.

The rate of total enthalpy of air entering the combustor = 28.94 MJ/s.

The rate of total enthalpy of air leaving the combustor = 115.42 MJ/s.

Mass flow rate of air = 32 kg/s.

Air to fuel mass ratio = 15.6

Lower heating value of the fuel = 46 MJ/kg.

The efficiency of the combustor is ________ % (*round off to two decimal places*).

[GATE 2020]

114. The figure shows the T-S diagram for an axial turbine stage.

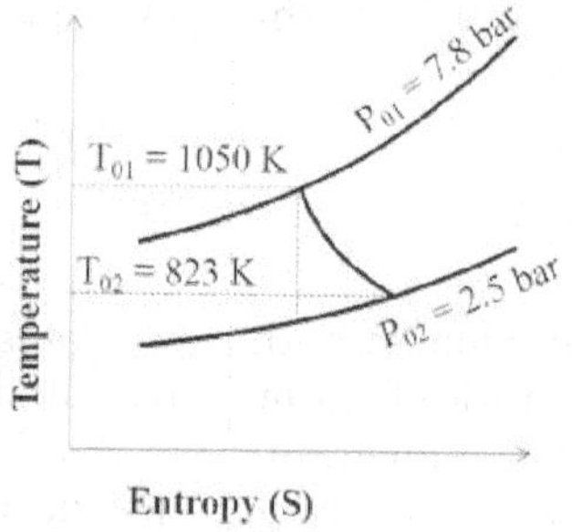

Assuming specific heat ratio of 1.33 for the hot gas, the isentropic efficiency of the turbine stage is _______ % (*round off to two decimal places*).

[GATE 2020]

115. The combustion process in a turbo-shaft engine during ideal operation is:

(a) isentropic (b) isobaric

(c) isochoric (d) isothermal

[GATE 2021]

116. How does the specific thrust of a turbojet engine change for a given flight speed with increase in flight altitude?

(a) Increases monotonically

(b) Decreases monotonically

(c) Remains constant

(d) First increases and then decreases

[GATE 2021]

117. How does the propulsion efficiency of a turbofan engine, operating at a given Mach number and a given altitude, change with increase in compressor pressure ratio?

(a) Remains constant

(b) Increases monotonically

(c) Decreases monotonically

(d) First decreases and then increases

[GATE 2021]

118. An aircraft with a turbojet engine is flying at 270 m/s. The enthalpy of the Incoming air at the intake is 260 kJ/kg and the enthalpy of the exhaust gases at the nozzle exit is 912 kJ/kg. The ratio of mass flow rates of fuel and air is equal to 0.019. The chemical energy (heating value) of fuel is 44.5 MJ/kg and the combustion process is ideal. The total loss of heat from the engine to the ambient is 25 kJ per kg of air. The velocity of the exhaust jet is ______ m/s (*round off to two decimal places*).

[GATE 2021]

119. A twin-spool turbofan engine is operated at sea level (P_a = 1 bar, T_a = 288 K). The engine has separate cold and hot nozzles. During static thrust test at sea level, the overall mass flow rate of air through the engine and the cold exhaust temperature are measured to be 100 kg/s and 288 K, respectively. The parameters for the engine are:

Fan pressure ratio = 1.6

Overall pressure ratio = 20

Bypass ratio = 3.0

Turbine entry temperature=1800 K.

The specific heat at constant pressure (C_p) is 1.005 kJ/kg-K and the ratio of specific heats (γ) is 1.4 for air.

Assuming ideal fan and ideal expansion in the nozzle, the sea-level static thrust from the cold nozzle is______ kN (*round off to two decimal places*).

[GATE 2021]

120. At the design conditions, the velocity triangle at the mean radius of a single stage axial compressor is such that the blade angle at the rotor exit is equal to 30°. The absolute velocities at the rotor inlet and exit are equal to 140 m/s and 240 m/s, respectively. The flow velocities relative to the rotor at inlet and exit of the rotor are equal to 240 m/s and 140 m/s, respectively.

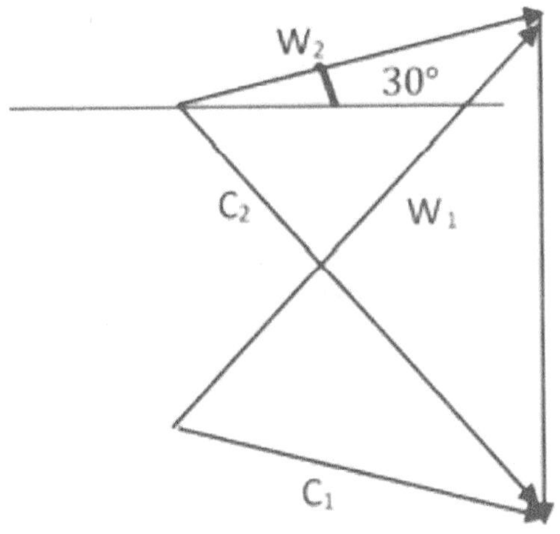

The blade speed (U) at the mean radius of the rotor is _____ m/s (*round off to two decimal places*).

[GATE 2021]

121. A single stage axial turbine has a mean blade speed of 340 m/s at design condition with blade angles at inlet and exit of the rotor being 21° and 55° respectively. The degree of reaction at the mean radius of the rotor is equal to 0.4. The annulus area at the rotor inlet is 0.08 m^2 and the density of gas at the rotor inlet is 0.9 kg/m^3. The flow rate through the turbine at these conditions is _______ kg/s (*round off to two decimal places*).

[GATE 2021]

122. The air flow rate through the gas generator of a turboprop engine is 100 kg/s. The stagnation temperatures at inlet and exit of the combustor are 600 K and 1200 K, respectively. The burner efficiency is 90% and the heating value of the fuel is 40 MJ/kg. The specific heats at constant pressure (Cp) for air and burned gases are 1000 J/kg-K and 1200 J/kg-K, respectively. The flow rate of the fuel being used is ______ kg/s (*round off to two decimal places*).

(Note: Do not neglect the fuel flow rate with respect to the air flow rate)

[GATE 2021]

ANSWER KEY

1. (*b*)	**2.** (*b*)	**3.** (*a*)	**4.** (*b*)	**5.** (*c*)	**6.** (*a*)	**7.** (*b*)	**8.** (*a*)	**9.** (*b*)	**10.** (*d*)
11. (*c*)	**12.** (*b*)	**13.** (*d*)	**14.** (*c*)	**15.** (*b*)	**16.** (*d*)	**17.** (*a*)	**18.** (*c*)	**19.** (*d*)	**20.** (*c*)
21. (*b*)	**22.** (*b*)	**23.** (*b*)	**24.** (*d*)	**25.** (*b*)	**26.** (*c*)	**27.** (*d*)	**28.** (*a*)	**29.** (*d*)	**30.** (*c*)
31. (*d*)	**32.** (*a*)	**33.** (*c*)	**34.** (*a*)	**35.** (*a*)	**36.** (*c*)	**37.** (*c*)	**38.** (*b*)	**39.** (*d*)	**40.** (*c*)
41. (*a*)	**42.** (*c*)	**43.** (*c*)	**44.** (*c*)	**45.** (*c*)	**47.** (*c*)	**48.** (*c*)	**49.** (*b*)	**50.** (*c*)	**51.** (*d*)
52. (*b*)	**53.** (*d*)	**54.** (*b*)	**55.** (*d*)	**56.** (*a*)	**57.** (*a*)	**58.** (*d*)	**59.** (*d*)	**60.** (*d*)	**61.** (*a*)
62. (*b*)	**63.** (*d*)	**67.** (*d*)	**68.** (*c*)	**69.** (*c*)	**70.** (*c*)	**71.** (*d*)	**73.** (*a*)	**74.** (*c*)	**75.** (*c*)
76. (*c*)	**77.** (*c*)	**79.** (*b*)	**82.** (*c*)	**83.** (*c*)	**84.** (*d*)	**85.** (*b*)	**86.** (*a*)	**91.** (*c*)	**92.** (*b*)
96. (*b*)	**97.** (*b*)	**98.** (*b*)	**101.** (a)	**102.** (b)	**103.** (b)	**104.** (c)	**105.** (0.714)	**106.** (c)	**107.** (89%)

108. (2.54 kJ/kg) **109.** (c) **110.** (29.8 to 30.0) **111.** (b) **112.** (1712.0 to 1719.0) **113.** (91 to 93)

114. (87 to 89) **115.** (b) **116.** (a) **117.** (d) **118.** (605 to 607) **119.** (20 to 22) **120.** (274 to 282)

121. (18.0 to 19.5) **122.** (2.35 to 2.50)

EXPLANATIONS

1. The Turbojet with afterburner is used to produce high thrust required for short duration. Basically when short take-off run required, use of afterburner increase the fuel consumption of the engine.

In turbofan, thrust is produced partly by fan and by nozzle. The contribution of fan thrust is greater than nozzle thrust hence it will reduce the fuel consumption of engine.

Ramjet and scramjet engine is used for high supersonic and hypersonic operation.

2. The Brayton cycle

3. $$\eta_{isc} = \frac{W_{ideal}}{W_{actual}} = \frac{h_{02}' - h_{01}}{h_{02} - h_{01}}$$

$$= \frac{T_{02}' - T_{01}}{T_{02} - T_{01}}$$

$$\eta_{isc} = \frac{T_{01}\left(\frac{T_{02}'}{T_{01}} - 1\right)}{T_{02} - T_{01}}$$

$$\frac{\eta_{isc} \times (T_{02} - T_{01})}{T_{01}} = r_p^{\frac{\gamma-1}{\gamma}} - 1$$

$$r_p = \left[1 + \frac{\eta_{isc} \times (T_{02} - T_{01})}{T_{01}}\right]^{\frac{\gamma}{(\gamma-1)}}$$

$$r_p = \left[1 + \frac{0.81 \times (424 - 300)}{300}\right]^{\frac{1.4}{0.4}}$$

$= 2.748 \simeq 2.75$

4. Degree of reaction (R):

$$R = \frac{C_a}{2u}(\tan\beta_3 - \tan\beta_2)$$

Ca = Axial velocity

u = Blade speed

β_3 = Rotor outlet angle

β_2 = Rotor inlet angle

Since every term is given at mean blade-height, so we can consider it as an impulse turbine.

For impulse turbine, R = O

$$\frac{C_a}{2u}(\tan\beta_3 - \tan\beta_2) = O$$

$\tan\beta_3 = \tan\beta_2$.

$\beta_3 = \beta_2 = 37°$

6. The isentropic efficiency of compressor depends on inlet and exit condition only, hence it will remain constant throughout

$$\eta_{isc} = \frac{h_{02}' - h_{01}}{h_{02} - h_{01}} = \frac{T_{02}' - T_{01}}{T_{02} - T_{01}}$$

Flow separation is more critical in cfc because flow of air takes sharp turns in cfc. Pressure ratio depends on the inlet and outlet temperatures not on the mass flow rate. Multistaging is not possible in cfc due to its complex design.

7. $2C_8H_{18} + (25O_2 + 94N_2) \longrightarrow 16CO_2 + 18H_2O + 94N_2$

$$\left(\frac{F}{A}\right)_{actual} = 0.028$$

fuel $\rightarrow C_8H_{18}$, Air $\rightarrow O_2 + N_2$.

$$\text{Equivalence ratio } (\phi); \frac{(F/A)_{actual}}{(F/A)_{stoich}}$$

$$\left(\frac{F}{A}\right)_{stoich} = \frac{2 \times \text{At.wt.of } C_8H_{18}}{25 \times \text{At. wt. of } O_2 + 94 \times \text{At. wt. of } N_2}$$

$$= \frac{2 \times (8 \times 12 + 18 \times 1)}{25 \times 2 \times 16 + 94 \times 2 \times 14} = 0.0664$$

$$\phi = \frac{0.028}{0.0664} = 0.422$$

8. $PLF = \dfrac{P_{02} - P_{01}}{P_{01}}$

Since flow is 1 – D, incompressible and frictionless, so the stagnation pressure will remain constant through the inlet

i.e., $P_{01} = P_{02}$

so, PLF = 0

9. $\eta_d = \dfrac{T_{02}' - T_a}{T_{oa} - T_a}$

$$= \frac{\left(\frac{T_{02}'}{T_a} - 1\right)}{\frac{T_{oa}}{T_a} - 1} = \frac{\frac{T_{02}'}{T_{oa}} \times \frac{T_{oa}}{T_a} - 1}{1 + \frac{\gamma - 1}{2}M^2{}_a - 1}$$

$$= \frac{r_p^{\frac{\gamma-1}{\gamma}} \times \left(1 + \frac{\gamma - 1}{2}M_a^2\right) - 1}{\left(\frac{\gamma - 1}{2}\right)M_a^2}$$

$$= \frac{0.92^{\frac{0.4}{1.4}} \times (1 + 0.2 \times 0.85^2) - 1}{0.2 \times 0.85^2} = 0.8135.$$

10. Complete combustion of fuel and air in combustion chamber can be achieved by injecting the fuel in droplet form into the combustion chamber; so injector is used to inject the tiny droplets of fuel in combustion chamber.

11. $r_p = \dfrac{P_{02}}{P_{01}} = \dfrac{3.15}{0.9} = 3.5$

12. $\eta_{isc} = \dfrac{r_p^{\frac{\gamma-1}{\gamma}} - 1}{r_p^{\frac{\gamma-1}{\gamma \times \eta_{poly}}} - 1}$

$$= \frac{3.5^{\frac{0.4}{1.4}} - 1}{3.5^{\frac{0.4}{1.4 \times 0.89}} - 1} = 0.869$$

13. $R = \dfrac{\text{static enthalpy drop in rotor}}{\text{total enthalpy drop across stage}}$

In impulse stage, the enthalpy drop across the rotor is zero which means no change in pressure across the rotor.

15. $BPR = \dfrac{\dot{m}_c}{\dot{m}_h} = \dfrac{\dot{m}_c}{\dot{m}_a - \dot{m}_c}$

$\dot{m}_c$ = mass flow rate through cold nozzle

$\dot{m}_h$ = mass flow rate through engine core

$\dot{m}_a$ = mass low rate of total air

$= \dot{m}_c + \dot{m}_h$

$BPR(\dot{m}_a - \dot{m}_c) = \dot{m}_c$

$5 \times (120 - \dot{m}_c) = \dot{m}_c$

$\dot{m}_c = 100 \text{ kg/s}$

16. In after burner additional fuel is burned with the gases coming out from the turbine, to increase the nozzle inlet temperature and hence to produce high thrust. So when atterburner is switched on the thrust produce will increase with increase of the fuel consumption.

17. Slip factor $= \dfrac{C_{t2}}{u}$

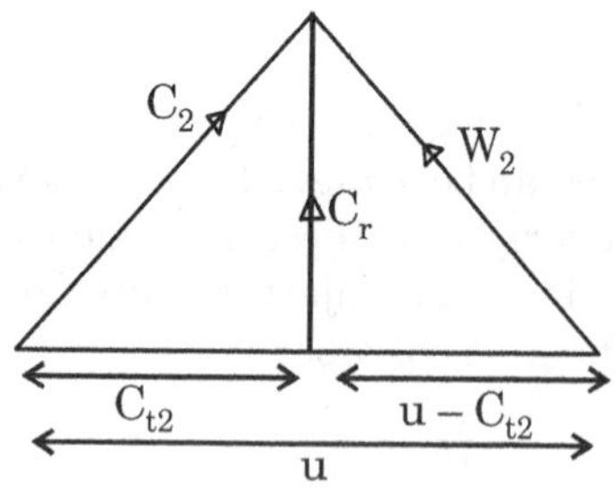

$$0.85 = \frac{C_{t2}}{u}$$

$$C_{t2} = 0.85\ u$$

$$W_2 = \sqrt{(C_r)^2 + (u - C_{t2})^2}$$

$$= \sqrt{(88)^2 + (u - 0.854)^2}$$

$$= \sqrt{(88)^2 + (51)^2}$$

$$= 101.71 \text{ m/s}$$

18. $T = (\dot{m}_a + \dot{m}_f)\, v_e - \dot{m}_a v_i$

$$= \dot{m}_a (1 + f) v_e - \dot{m}_a v_i$$

for $f <<< 1$

$$T = \dot{m}_a (v_e - v_i)$$

$$\frac{T}{\dot{m}_a} = v_e - v_i$$

for ideal ramjet engine.

$M_i = M_e$

$$\frac{v_i}{\sqrt{\gamma R T_a}} = \frac{v_e}{\sqrt{\gamma R T_e}}$$

$$\frac{T}{\dot{m}_a} = v_i \left(\frac{v_e}{v_i} - 1 \right)$$

$$= v_i \left(\sqrt{\frac{T_e}{T_i}} - 1 \right)$$

$$\frac{T}{\dot{m}_a} = M \times \sqrt{\gamma R T_a} \left(\sqrt{\frac{T_e}{T_a}} - 1 \right)$$

19. $\alpha_2 = \beta_3 = 60°$.

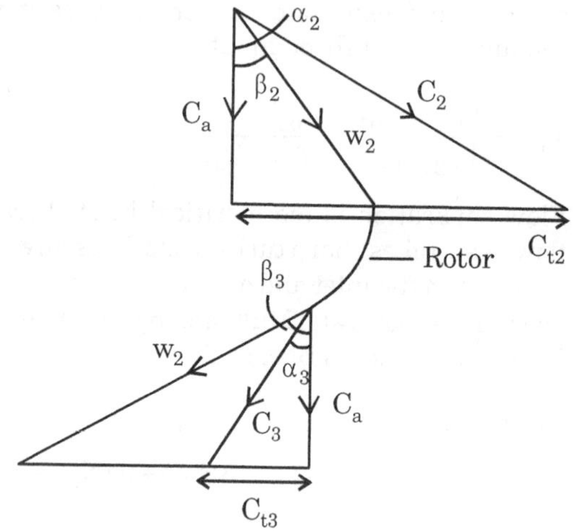

$$R = \frac{C_a}{2u} (\tan \beta_3 - \tan \beta_2)$$

$$\frac{1}{2} = \frac{150}{2 \times 180} (\tan 60° - \tan \beta_2)$$

$\beta_2 = 28.015° = \alpha_3$.

$w = u\ (C_{t2} + C_{t3})$

$= u\ C_a\ (\tan \alpha_2 + \tan \alpha_3)$

$= 150 \times 180 \times (\tan 60° + \tan 28.015°)$

$= 61.13$ kJ/kg

20.

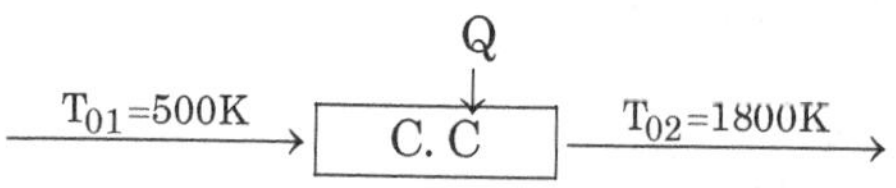

$H_{01} + Q = H_{02}$.

$$\dot{m}_a C_p T_{01} + \dot{m}_f \times CV = (\dot{m}_a + \dot{m}_f)\, C_p T_{02}$$

$$C_p T_{01} + f \times cv = (1 + f)\, C_p T_{02}$$

$1.005 \times 500 + f \times 44 \times 10^3$

$= (1 + f) \times 1.005 \times 1800$

$f = 0.031$

21. $\eta_d = \dfrac{h_{02}' - h_a}{h_{oa} - h_a}$

$$= \frac{T_{02}' - T_a}{T_{oa} - T_a} = \frac{\dfrac{T_{02}'}{T_a} - 1}{\dfrac{T_{oa}}{T_a} - 1}$$

$$= \frac{\dfrac{T_{02}'}{T_{oa}} \times \dfrac{T_{oa}}{T_a} - 1}{1 + \dfrac{\gamma - 1}{2} M_a^2 - 1}$$

$$\eta_d = \frac{(r_p)^{\frac{\gamma-1}{\gamma}} \times \left(1 + \frac{\gamma-1}{2} M_a^2\right) - 1}{\frac{\gamma-1}{2} M_a^2}$$

$$P_{oa} = P_a \left(1 + \frac{\gamma-1}{2} M_a^2\right)^{\frac{\gamma}{(\gamma-1)}}$$

$$= 95 \times (1 + 0.2 \times 0.8^2)^{\frac{1.4}{0.4}}$$

P_{oa} = 144.812 kPa

$$r_p = \frac{P_{02}}{P_{oa}} = \frac{140}{144.812} = 0.967$$

$$\eta_d = \frac{(0.967)^{\frac{0.4}{1.4}} \times (1 + 0.2 \times 0.8^2) - 1}{0.2 \times 0.8^2}$$

$$\eta_d = 0.9159$$

22. $\eta = \frac{h_{02}' - h_{01}}{h_{02} - h_{01}}$

$$= \frac{C_p (T_{02}' - T_{01})}{C_p (T_{02} - T_{01})}$$

$$\eta = \frac{T_{01}\left(r_p^{\frac{\gamma-1}{\gamma}} - 1\right)}{(\Delta T_o)_{overall}}$$

$$(\Delta T_o)_{overall} = \frac{335 \times \left(11^{\frac{1.4-1}{1.4}} - 1\right)}{0.9}$$

= 366.27 K

23. No. of stage (N):-

$$N = \frac{(\Delta T_o)_{overall}}{(\Delta T_o)_{stage}}$$

$$(\Delta T_o)_{stage} = \frac{37}{1.005} = 36.816$$

$$\therefore\ N = \frac{366.27}{36.816} = 9.9 \simeq 10 \text{ stage}$$

24. Propulsive efficiency

$$= \frac{\text{Thrust power}(F \times v_i)}{\text{propulsive power}}$$

$$= \frac{[(\dot{m}_a + \dot{m}_f)v_e - \dot{m}_a v_i]v_i}{\frac{1}{2}(\dot{m}_a + \dot{m}_f)v_e^2 - \frac{1}{2}\dot{m}_a v_i^2}$$

$$= \frac{[(1+f)v_e - v_i]v_i}{\frac{1}{2}(1+f)v_e^2 - \frac{1}{2}v_i^2}$$

$f << 1$, so

$$= \frac{2v_i(v_e - v_i)}{(v_e^2 - v_i^2)}$$

$$= \frac{2v_i}{v_e + v_i} = \frac{2}{\frac{v_e}{v_i} + 1} \text{ or } \frac{2u_\infty}{u_e + u_\infty}$$

25. $\frac{T_o}{T} = 1 + \frac{\gamma-1}{2} M^2$

$$T_o = 250(1 + 0.2 \times 2^2) = 450 \text{ K}$$

27.

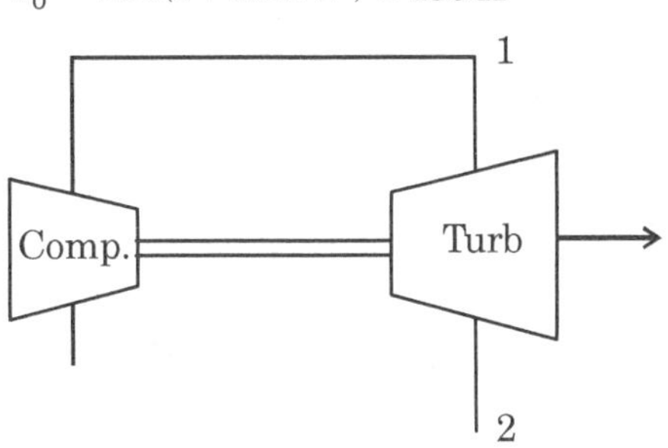

$W_C = W_T$.

$= C_p (T_{01} - T_{02})$

$$\frac{T_{o2}}{T_{o1}} = 0.667$$

$$T_{o2} = 0.667 \times 1500$$

= 1000.5 K

$$W = 1 \times (1500 - 1000.5)$$

= 499.5 kJ/kg

$$= 499.5 \frac{kJ \times S}{kg \times S}$$

= 499.5 kW/kg/s.

28. For ideal Ramjet

$M_i = M_e$

$$\frac{v_i}{\sqrt{\gamma RT_\infty}} = \frac{v_e}{\sqrt{\gamma RT_e}}$$

$$v_e = v_i\sqrt{\frac{T_e}{T_\infty}} = 1000 \times \sqrt{\frac{900}{225}}$$

$v_e = 2000$ m/s

specific thrust

$$= \frac{T}{\dot{m}_a} = \frac{\dot{m}_a(v_e - v_i)}{\dot{m}_a}$$

$= (2000 - 1000) = 1000$ N-s/kg

29. $\dfrac{\dot{m}_a}{A^*} \propto \dfrac{P_o}{\sqrt{T_{oe}}}$

where P_o and $\dot{m}$ is given constant for both case.

$$\frac{\left(\dfrac{\dot{m}}{A_1^*}\right)}{\left(\dfrac{\dot{m}}{A_2^*}\right)} \propto \sqrt{\frac{T_{oe2}}{T_{oe1}}}$$

$$\frac{A_2^*}{A_1^*} = \sqrt{\frac{T_{oe2}}{T_{oe1}}}$$

$$= \sqrt{\frac{3000}{750}} = 2$$

30.

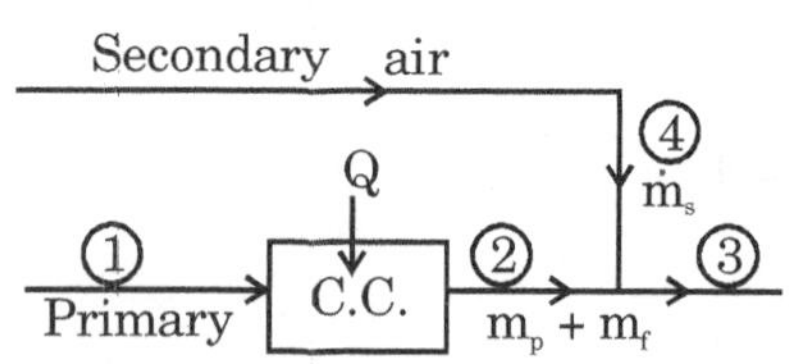

$T_{01} = 500$ K

$T_{02} = ?$

$h_{01} + Q = h_{02}$

$$\dot{m}_a C_p T_{01} + \dot{m}_f \times HV = (\dot{m}_a + \dot{m}_f)C_p T_{02}$$

$$C_p T_{01} + f \times HV = (1+f)C_p T_{02}$$

$$T_{01} = \frac{C_p T_{01} + f \times HV}{(1+f)C_p}$$

$$= \frac{1 \times 500 + 0.05 \times 42 \times 10^3}{(1+0.05) \times 1}$$

$T_{02} = 2476\text{ K} \simeq 2600\text{ K}$

31. $T_{03} = 1550$ K

$H_{02} + H_{04} = H_{03}$.

$$(\dot{m}_p + \dot{m}_f)C_p T_{02} + \dot{m}_s C_p T_{04} =$$

$$(\dot{m}_p + \dot{m}_f + \dot{m}_s)C_p T_{03}$$

$$\frac{\dot{m}_f}{\dot{m}_p} = 0.05 \Rightarrow \dot{m}_f = 0.05\,\dot{m}_p$$

$$1.05\dot{m}_p C_p T_{02} + \dot{m}_s C_p T_{04} = (1.05\dot{m}_p + \dot{m}_s)T_{03}C_p$$

$$1.05\frac{\dot{m}_p}{\dot{m}_s}T_{02} + T_{04} = \left(1.05\frac{\dot{m}_p}{\dot{m}_s} + 1\right)T_{03}$$

$$\frac{\dot{m}_p}{\dot{m}_s}(1.05 \times 2600 - 1.05 \times 1550) = 1550 - 500$$

$$\frac{\dot{m}_p}{\dot{m}_s} = 0.9524 \simeq 1:1$$

32. In 10 seconds:-

Work done $= -3000 \times 10 = -30000$ J

Heat transfer $= -847.5 \times 10 = -8475$ J

$dQ = du + dw$

$du = dQ - dw$

$= -8475 + 30000$

$u = +21525$ J

$$du = \frac{21525}{1}\text{ J/kg of air}$$

$= 21525$ J/kg

33. $du = 21525$ J

$mC_v(T_2 - T_1) = 21525$

$$T_2 - T_1 = \frac{21525}{717.5}$$

$$\boxed{T_2 - T_1 = 30K}$$

34. $\eta_{isen} = \dfrac{w_{ideal}}{w_{actual}} = \dfrac{h_{02s} - h_a}{h_{02} - h_a}$

$$= \frac{C_p(T_{02s} - T_a)}{T_{02} - T_a} = \frac{T_{o2s} - T_a}{T_{o2} - T_a}$$

35. For multistage axial flow compressor, the length of blade decrease to compensate for the reduction in volume which results in increase of pressure from stage to stage.

36. Specific thrust first increases with increase in compressor pressure ratio then decreases at fixed turbine inlet temp.

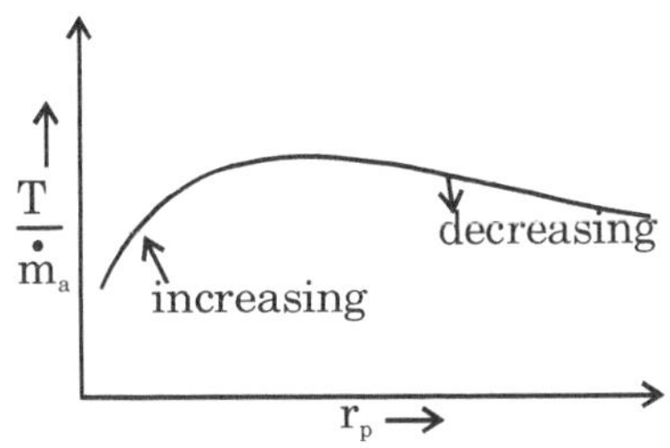

37. Surging occurs at low mass flow rate, when compressor operates at maximum mass flow rate the pressure ratio decreases and hence decrease in density. Which results in increase of absolute velocity and it moves towards the chocking condition.

38.

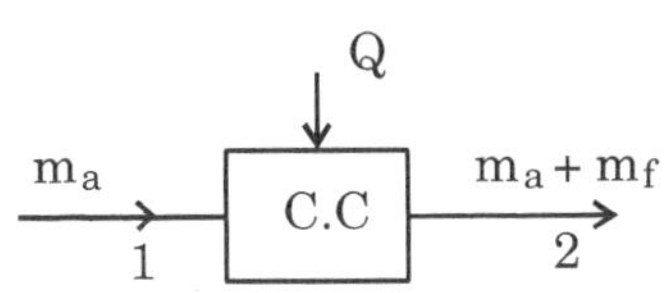

$$\eta_{cc} = \frac{h_2 - h_1}{\dot{m}_f \times HV}$$

$$\eta_{cc} = \frac{(\dot{m}_a + \dot{m}_f)C_{pg}T_2 - \dot{m}_a C_{Pa} T_1}{\dot{m}_f \times HV}$$

$$\dot{m}_f \times HV \times \eta_{cc} = (\dot{m}_a + \dot{m}_f)C_p T_2 - \dot{m}_a C_p T_1$$

$$f \times HV \times \eta_{cc} = (1+f)C_p T_2 - C_p T_1$$

$$f \times (HV \times \eta_{cc} - C_p T_2) = C_p (T_2 - T_1)$$

$$f \times (43 \times 10^3 \times 0.9 - 1.005 \times 2500)$$

$= 1.005\,(2500 - 1200)$

$f = 0.036$

39. for axial flow compressor

$w = uC_a(\tan\beta_1 - \tan\beta_2)$

$C_p \Delta T_S = uC_a(\tan\beta_1 - \tan\beta_2)$

$\Delta T_S = 22.809$ K

$$\eta_{iso} = \frac{W_{ideal}}{W_{actual}}$$

$$\frac{T_{02}' - T_{01}}{\Delta T_s} = \frac{T_{01}\left(r_p^{\frac{\gamma-1}{\gamma}} - 1\right)}{\Delta T_s}$$

$$r_p = \left[1 + \frac{\eta_{isc} \times \Delta T_s}{T_{01}}\right]^{\frac{\gamma}{(\gamma-1)}}$$

$$= \left[1 + \frac{0.8 \times 22.809}{300}\right]^{\frac{1.4}{0.4}}$$

$r_p = 1.23$

40. $P_{03} = 6$ bar

$T_{03} = 1400$ K

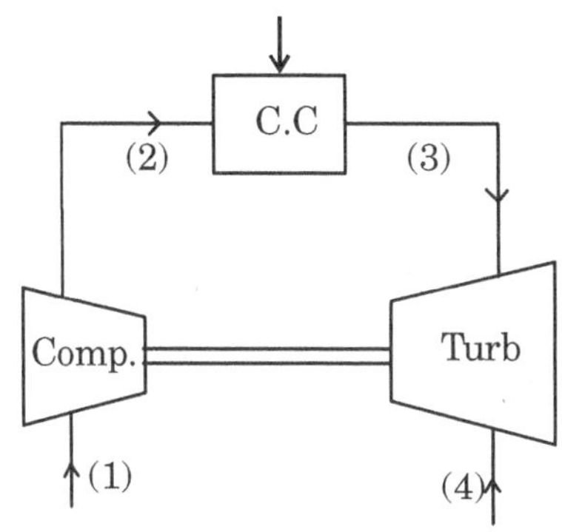

$T_{03} = 1400$ K

$h_{02} - h_{01} = 335$ KJ/kg

$\dot{m}_f << \dot{m}_a$

$w_c = w_t$

$h_{02} - h_{01} = h_{03} - h_{04}$

$$\frac{335 \times 1000}{1117} - T_{04}$$

$T_{04} = 1100.089$ K

$$P_{04} = P_{03} \times \left(\frac{T_{04}}{T_{03}}\right)^{\frac{\gamma}{(\gamma-1)}}$$

$$= 6 \times \left(\frac{1100.089}{1400}\right)^{\frac{1.3}{0.3}}$$

$P_{04} = 2.11$ bar

43. At high pressure ratio, flow will separate inside the compressor which results in increase of loss in compressor.

45. Propulsive efficiency

(η_p)

$$= \frac{\text{Thrust power}}{\text{propulsive power}}$$

So, $\eta_p \propto$ Thrust power

and propulsive power $= \dfrac{1}{2}\dot{m}_a\left[(1+f)v_e^2 - v_i^2\right]$

so, $\eta_p \propto \dfrac{1}{\dot{m}_a}$

46. $M = \dfrac{v_i}{a_\infty}$

$\Rightarrow v_i = a_\infty M = 299.5 \times 0.8$

$v_i = 239.6$ m/s

$\dot{m}_a = \rho_\infty v_i A_i$

$= 0.413 \times 239.6 \times 0.7$

$= 69.268$ Kg/s

$T = \dot{m}_a(v_e - v_i) + (p_e - p_\infty)A_e$ for $\dot{m}_f <<< \dot{m}_a$

$T = 69.286(500 - 239.6) + (60 - 26.5) \times 10^3 \times 0.4$

$= 31442.07$ N.

47. $T_{oa} = T_{01} = \left[1 + \dfrac{\gamma - 1}{2} m_\infty^2\right] T_a$

$= [1 + 0.2 \times 0.8^2] \times 223.3$

$T_{01} = 251.88$ K

$$\eta_{isen} = \frac{T_{02s} - T_{01}}{T_{02} - T_{01}}$$

$$T_{02} = T_{01} + \frac{T_{01}}{\eta_{isen}}\left[r_p^{\frac{\gamma-1}{\gamma}} - 1\right]$$

$$T_{02} = 251.9 + \frac{251.9}{0.87}\left(8^{\left(\frac{0.4}{1.4}\right)} - 1\right)$$

$T_{02} = 486.8$ K

48. $\eta_p = \dfrac{2v_i}{v_i + v_e}$

if $\dfrac{v_i}{v_e} = 1 \Rightarrow v_i = v_e$

then $\eta_p = 1$ i.e. 100%

49. $h_{01} + Q = h_{02}$

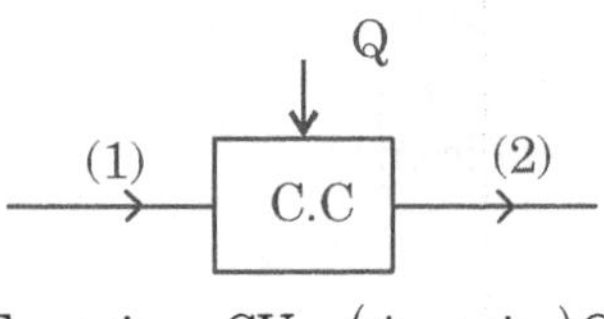

$\dot{m}_o C_{pa} T_{01} + \dot{m}_f \times CV = (\dot{m}_a + \dot{m}_f) C_{pg}\ T_{02}$

$C_{pa} T_{01} + f \times CV = (1 + f) C_{pg}\ T_{02}$

$1.005 \times 600 + f \times 44 \times 10^3$

$= (1 + f) \times 1.147 \times 1200$

$= 0.0181$

50. $\alpha_1 = 18°$

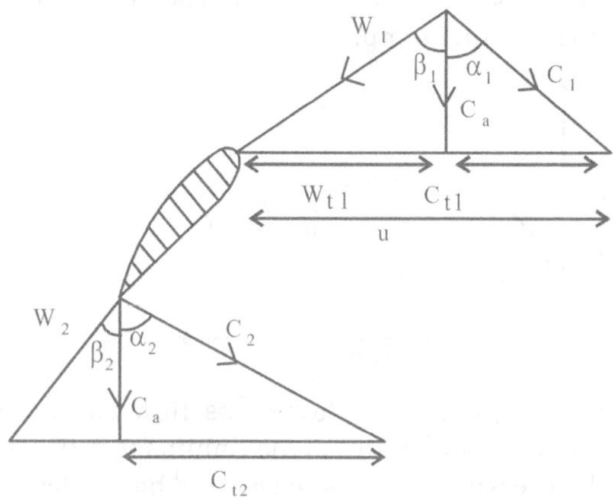

$C_a = C_1 \cos \alpha_1$

$C_{t1} = C_1 \sin \alpha_1$

$W_{t1} = u - C_{t1}$

$\tan_{\beta_1} = \dfrac{W_{t1}}{C_a} = \dfrac{u - C_{t1}}{C_a}$

$$\beta_1 = \tan^{-1}\left(\frac{u - C_{t1}}{C_a}\right)$$

$$= \tan^{-1}\left(\frac{200 - 120 \sin 18°}{120 \cos 18°}\right)$$

$\beta_1 = 54.988° \simeq 55°$

51. $\dot{m} = 1$ kg / s

$P = \dot{m}u(C_{t2} - C_{t1})$

$\beta_2 = 55° - 27° = 28°$

$P = \dot{m}u(u - C_a \tan\beta_2 - C_1 \sin\alpha_1)$

$= 1 \times 200\ (200 - 120 \times \cos 18° \times \tan 28° - 120 \times \sin 18°)$

$= 20.453$ kW.

52.

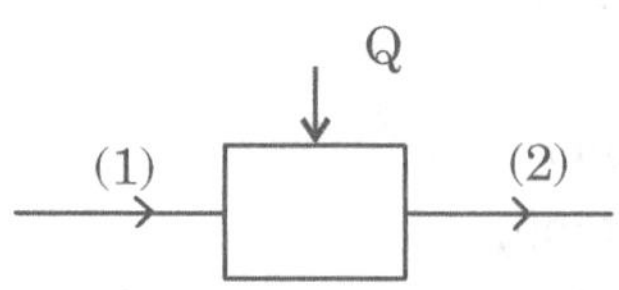

$h_1 + Q = h_2.$

$\dot{m}_a C_p T_1 + \dot{m}_f \times Hv = (\dot{m}_a + \dot{m}_f) C_p T_2$

$\Rightarrow 10 \times 1 \times 400 + \dot{m}_f \times 42 \times 10^3 = (10 + \dot{m}_f) \times 1 \times 1800$

$\dot{m}_f = 0.3483$ kg/s

53. Air density is reduces with increase in altitude. So its value is higher at sea level and lower at 11 km altitude. Design of component should be done for optimum condition. Hence combustion chamber volume will decide based on conditions at 11 km altitude.

54. For ideal condition, the losses within the component due to various flow phenomenon is neglected. So stagnation temperature will remain constant throughout the intake and nozzle. But it will increase in the combustion chamber due to energy of fuel addition.

55. Turbine blades of engine 'P' are cooled by bleeding the compressor air, so turbine inlet temperature will be greater of engine 'p'. Due to high turbine inlet temperature, the exhaust velocity from nozzle of engine 'P' will be higher and hence produce higher specific thrust.

For the same compressor inlet condition of engine 'P' and 'Q' the compressor work will be same for both engine; because of same outlet condition. Bleeding of air from compressor of engine 'P' occurs after the compression process.

Fuel flow rate is directly proportional to turbine inlet temperature. So fuel flow rate in engine 'P' will be higher than engine 'Q'.

56. $\alpha_1 = \beta_2$ and $\beta_1 = \alpha_2$

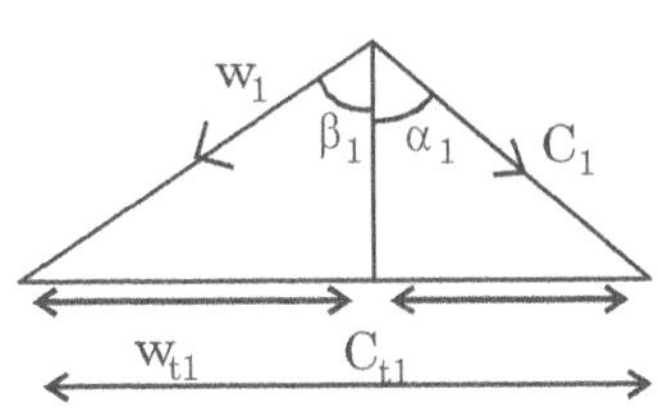

$$w = u(C_{t2} - C_{t1})$$

$$= u(u - w_{t2} - C_{t1})$$

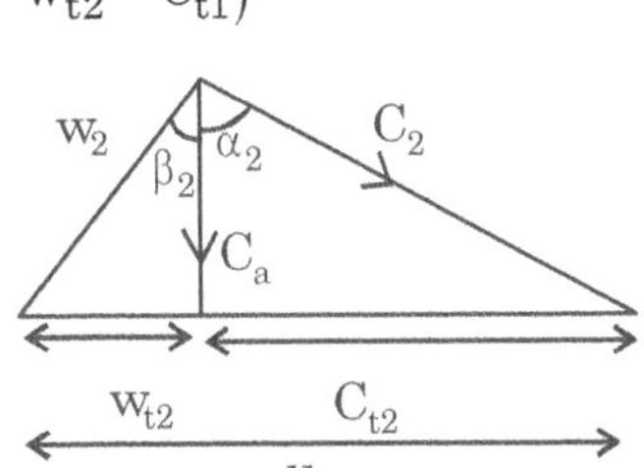

$$= u(u - C_a \tan\beta_2 - C_a \tan\alpha_1)$$

$$= 300\,(300 - 100 \times \tan 45° - 100 \times \tan 45°)$$

$$= 30 \text{ kJ/kg}$$

57. for small fuel-air ratio, the propulsive efficiency,

$$\eta_p = \frac{2v_i}{v_e + v_i}$$

for constant v_i, $\eta_p \propto \frac{1}{v_e}$

and $M_e \propto v_e$

M_e is exit mach number

M_e is exhaust velocity

Hence for smaller exhaust velocity the propulsive efficiency will be maximum.

Smaller value of mach number is '1' and also the nozzle is choked.

58. Thrust,

$$T = \dot{m}_a (v_e - v_i)$$

$\dot{m}_a$ is mass flow rate of air

so, $T \propto v_e$.

means if velocity is higher, thrust produce will also be higher. In other words for higher mach number thrust will be higher.

59. By applying Steady flow energy equation,

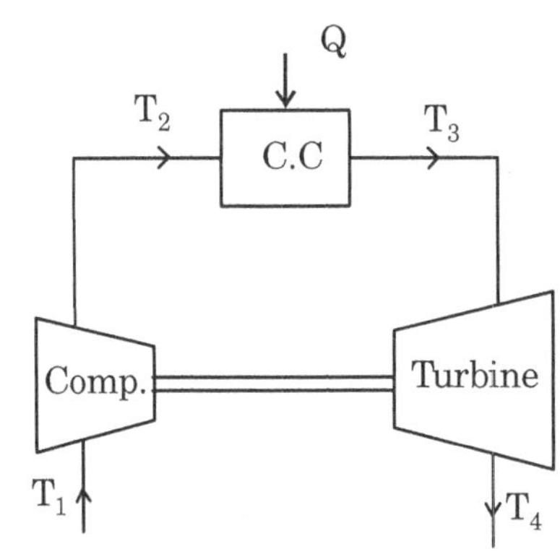

Assuming ideal condition.

$$\dot{m}_a C_p T_2 + \dot{m}_f \times cv = (\dot{m}_a + \dot{m}_f) C_p T_3$$

$$\dot{m}_f (CV - C_p T_3) = \dot{m}_a C_p (T_3 - T_2)$$

$$\dot{m}_f = \frac{\dot{m}_a C_p (T_3 - T_2)}{(CV - C_p T_3)} \quad \text{... (A)}$$

Since turbine inlet temperature (T_3) is fixed, lowest amount of fuel addition is possible at maximum 'T_2'

$$\eta_{ic} = \frac{h_2' - h_1}{h_2 - h_1} = \frac{T_2' - T_1}{T_2 - T_1}$$

So for maximum 'T_2' efficiency of compressor is minimum, since 'T_1' is also fixed.

61. $T = (\dot{m}_a + \dot{m}_f) v_e - \dot{m}_a v_i$

$= \dot{m}_a (1 + f) v_e - \dot{m}_a v_i$

$f << 1$

$T = \dot{m}_a (v_e - v_i)$

$T \propto v_e$.

62. For 50% reaction axial compressor stage

$\alpha_1 = \beta_2 \Rightarrow C_{t1} = w_{t2}$

$\alpha_2 = \beta_1 \Rightarrow C_{t2} = w_{t1}$

So, $C_{t1} \neq w_{t1}$

63.

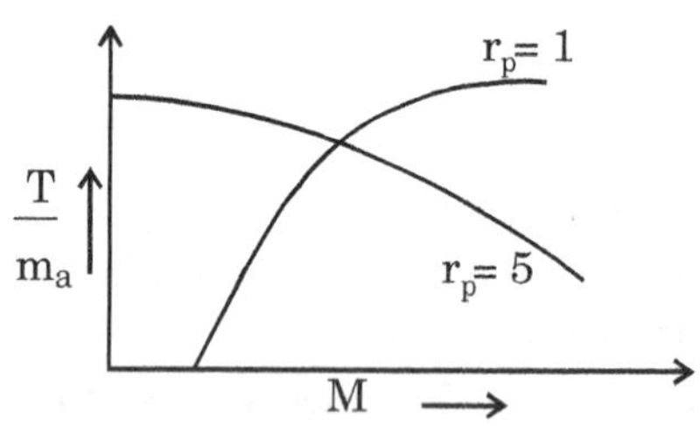

64. 1880.38 m/s

For ideal Ramjet Engine, $M_i = M_j$.

$$\frac{T_{oe}}{T_e} = 1 + \frac{\gamma - 1}{2} M_j^2 = 1 + 0.2 \times 16 = 4.2$$

$$T_e = \frac{2310}{4.2} = 550K$$

$$T_{oe} = T_e + \frac{C_e^2}{2C_p}$$

$C_e = 1880.38$ m/s

65. 1.185 bar

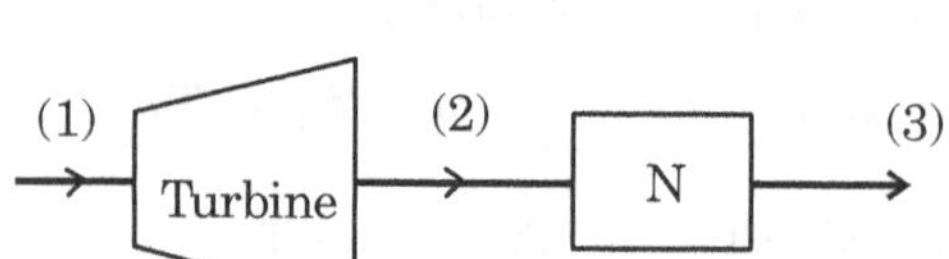

$$T_{o2} = T_2 + \frac{v_2^2}{2C_p}$$

$$= 800 + \frac{200^2}{2 \times 1147} = 817.437K$$

$$\eta = \frac{T_{01} - T_{02}}{T_{01} - T'_{02}}$$

$$0.96 = \frac{1350 - 817.437}{1350 - T'_{02}}$$

$T_{02}' = 795.247$ K

$$P_{02} = P_{01} \times \left(\frac{T_{02s}}{T_{01}}\right)^{\frac{\gamma}{(\gamma-1)}}$$

$$= 10 \times \left(\frac{795.217}{1350}\right)^{\frac{1.33}{0.33}}$$

$P_{02} = 1.185$ bar

66. 75%

$$u = \frac{\pi DN}{60}$$

$$= \frac{\pi \times 0.6 \times 15000}{60} = 150\pi$$

$$W_c = \sigma_s u^2 = C_p (T_{02} - T_{01})$$

$$0.88 \times (150\pi)^2 = 1005 (T_{02} - 300)$$

$T_{02} = 494.446$ K

$$\eta_c = \frac{T'_{02s} - T_{01}}{T_{02} - T_{01}} = \frac{T_{o1}\left(r_{po}^{\frac{\gamma-1}{\gamma}} - 1\right)}{T_{02} - T_{01}}$$

$$= \frac{300\left(4^{\frac{0.4}{1.4}} - 1\right)}{494.446 - 300} \times 100\%$$

$= 74.98\% \simeq 75\%$

67. $\dfrac{T_{oa}}{T_a} = 1 + \dfrac{\gamma - 1}{2} M^2$

$= 1 + 0.2 \times 9 = 2.8$

$T_{oa} = 2.8 \times 200 = 560$ K

$$\frac{P_{oa}}{P_a} = \left(\frac{T_{oa}}{T_a}\right)^{\frac{\gamma}{(\gamma-1)}} = (2.8)^{\frac{1.4}{0.4}}$$

$P_{oa} = 1836.636$ kPa

$\dot{m} = \rho^* A^* V^*$

$$= \frac{P^*}{RT^*}\sqrt{\gamma RT^*} \times A^*$$

$$= \sqrt{\frac{\gamma}{R}} \times \frac{P^*}{\sqrt{T^*}} \times A^*$$

$$\frac{P_o}{P^*} = \left(\frac{\gamma+1}{2}\right)^{\frac{\gamma}{\gamma-1}}$$

$$\frac{T_o}{T^*} = \frac{\gamma+1}{2}$$

$$\dot{m} = \sqrt{\frac{\gamma}{R}} \times \frac{P_o}{\sqrt{T_o}} \times \left(\frac{\gamma+1}{2}\right)^{\frac{-(\gamma+1)}{2(\gamma-1)}} \times A^*$$

$$= \sqrt{\frac{1.4}{287}} \times \frac{1836.636 \times 10^3}{\sqrt{560}} \times \left(\frac{2.4}{2}\right)^{\frac{-2.4}{2\times 0.4}} \times 0.05$$

= 156. 848 Kg/s

70. Large losses occurs in nozzle and intake due to flow separation.

71. for 9.2 kg/s of air, $\dot{m}_f = 1$ kg/s

for 1 kg/s of air, $\dot{m}_f = \frac{1}{9.2}$ kg/s

So, for 94 Kg/s of air,

$$\dot{m}_f = 94 \times \frac{1}{9.2} \text{ kg/s}$$

$$\dot{m}_f = 10.217 \text{ kg/s}$$

$$T = (\dot{m}_a + \dot{m}_f)v_e - \dot{m}_a v_i$$

= (94 + 10.217) 640 – 94 × 180

= 49778. 88 N

72. 1.132

$$T_{01} = T_1 + \frac{C_1^2}{2\,C_p}$$

$$= 278 + \frac{140^2}{2 \times 1005}$$

T_{01} = 306.66 K

$$\eta_{isc} = \frac{\dot{m}_a C_p (T_{02}' - T_{01})}{w}$$

$$= \frac{\dot{m}_a C_p\, T_{01}\left(r_p^{\frac{\gamma-1}{\gamma}} - 1\right)}{w}$$

$$r_p = \left[\frac{\eta_{isc} \times w}{\dot{m}_a C_p T_{01}} + 1\right]^{\frac{\gamma}{\gamma-1}}$$

$$= \left[\frac{0.86 \times 734}{24 \times 1.005 \times 306.66} + 1\right]^{\frac{1.4}{0.7}}$$

r_p = 1.332

73. 20.5-21.3 $\frac{kJ}{kg}$

74. 829.42 K

$$\frac{T_{01}}{P_{01}^{\frac{\gamma-1}{\gamma}}} = \frac{T_{02}}{P_{02}^{\frac{\gamma-1}{\gamma}}}$$

$$T_{02} = T_{01} \times \left(\frac{P_{02}}{P_{01}}\right)^{\frac{r-1}{r}}$$

$$= 1450 \times \left(\frac{1}{9.5}\right)^{\frac{0.33}{1.33}}$$

T_{02} = 829.42 K

76. Thrust produced by engine can be increased either by increasing the mass flow rate of air or by increasing the exhaust velocity. At low speed the turboprop engine produce highest thrust and hence thrust power. Since propulsive efficiency is directly proportional to thrust power, so propulsive efficiency of turboprop engine is larger at low speed.

77. Heating and cooling process in ideal case is always occurs at constant pressure.

78. 1148.4458 Ns/kg

$$BPR = \frac{(\dot{m}_a)_{bypass}}{(\dot{m}_a)_{engine}} = 6$$

$$(\dot{m}_a)_{bypass} = 6\,(\dot{m}_a)_{engine}$$

$$C_p = \frac{\gamma R}{\gamma - 1} = \frac{1.4 \times 287}{0.4}$$

= 1004.5 J/kg-K

For chocked condition

$$\frac{T_o}{T_c} = \frac{\gamma+1}{2} = \frac{2.4}{2} = 1.2$$

core nozzle:

$T_{oi} = T_c \times 1.2 = 580 \times 1.2 = 696$ K

For bypass nozzle : $T_{oi} = 1.2 \times 295 = 354$ K

velocity exhaust from nozzle,

$$v_e = \sqrt{2C_p(T_{oi} - T_c)}$$

$$(v_e)_{core\ nozzle} = \sqrt{2C_p(T_o - T_c)}$$

$$= \sqrt{2 \times 1004.5 \times (696 - 580)}$$

$= 482.746$ m/s

$$(v_e)_{bypass\ nozzle} = \sqrt{2 \times 1004.5 \times (354 - 295)}$$

$= 344.2883$ m/s

$$T = (\dot{m}_a)_{engine}(v_{eCN} - v_i) + (\dot{m}_a)_{bypass}(v_{e\,bN} - v_i)$$

$$= (\dot{m}_a)_{engine}(v_{eCN} - v_i) + 6(\dot{m}_a)_{engine}(v_{e\,bN} - v_i)$$

$$\frac{T}{(\dot{m}_a)_{engine}} = (482.746 - 200) + 6 \times (344.2833 - 200)$$

$$= 1148.4458 \frac{Ns}{kg}$$

79. For axial exhaust

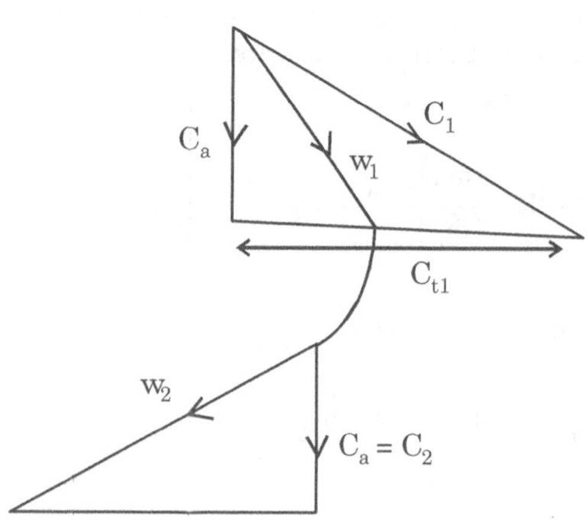

$C_2 = C_a$ and $C_{t2} = 0$

$w = u\,(C_{t1} + C_{t2})$

$w = u\,C_{t1} = u\,C_a \tan\alpha_1$

$$C_a = \frac{w}{u \tan\alpha_1}$$

$$\cos\alpha_1 = \frac{C_a}{C_1}$$

$$\Rightarrow C_1 = \frac{C_a}{\cos\alpha_1} = \frac{w}{u\tan\alpha_1 \cdot \cos\alpha_1}$$

$$C_1 = \frac{w}{u \sin\alpha_1} = \frac{228 \times 1000}{440 \times \sin 70°}$$

$C_1 = 551.437$ m/s

80. 3

$$r_p = \left[1 + \frac{\eta_{isc} \times (T_{02} - T_{01})}{T_{01}}\right]^{\frac{\gamma}{(\gamma-1)}}$$

$$= \left[1 + \frac{0.85 \times (430 - 300)}{300}\right]^{\frac{1.4}{0.4}}$$

$$r_p = 2.996 \simeq 3$$

81. 1305 K

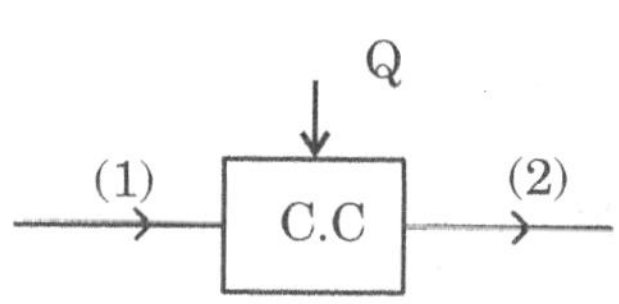

Applying S.f.E.E.

$$h_1 + \frac{v_1^2}{2} + Q = h_{02}$$

$$C_pT_1 + \frac{v_1^2}{2} + Q = C_pT_{02}$$

$$1000 \times 300 + \frac{100^2}{2} + 1000 \times 1000 = 1000 \times T_{02}$$

$T_{02} = 1305$ K

82. Give $\Delta s > 0$

$Tds = dh - vdp$

$\Rightarrow$ for Isobaric process (p = c)

$Tds = dh = mC_p\, dT$

$$\frac{dT}{T} = \frac{ds}{mc_p} > 0$$

$\Rightarrow$ For Isenthalpic process (h = c)

$Tds = -vdp$

$$T = -\frac{vdp}{ds} < 0$$

Also $Tds = du + pdv$ (2)

$\Rightarrow$ For Isochoric process (v = c)

$Tds = du = mc_v\, dT$

$$\frac{dT}{T} = \frac{ds}{mc_v} > 0$$

$\Rightarrow$ and or isothermal process temperature change is zero

since $c_p > c_v$

$$\text{so } \left(\frac{dT}{T}\right)_{\text{Isochoric}} > \left(\frac{dT}{T}\right)_{\text{Isobaric}}$$

86. Chemical reaction takes places between octane and air are written as

$$^2C_8H_{18} + 25O_2 + 94N_2 \longrightarrow 16CO_2 + 18H_2 + 94N_2$$

so,

$$\left(\frac{f}{a}\right)_{\text{stichio}} = \frac{2 \times \text{At, wt. of } C_8H_{18}}{25 \times \text{At.wt.of } O_2 + 94 \times \text{wt. of } N_2}$$

$$= \frac{2 \times (8 \times 12 + 18 \times 1)}{25 \times 2 \times 16 + 94 \times 2 \times 14} = 0.0664$$

Given $\dot{m}_a = 1$ kg/s

so, $\dot{m}_f = 0.0664 \times 1 = 0.0664$ kg/s

87. Axial velocity, $C_a = v_2$.

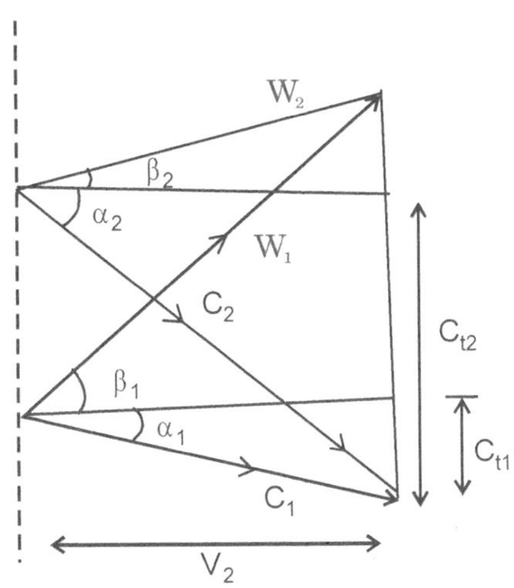

$w = u\,(C_{t2} - C_{t1})$

$= uC_a\,(\tan \alpha_2 + \tan \alpha_1)$

Since 'u' is constant

$\tan \alpha_1 + \tan \beta_1 = \tan \alpha_2 + \tan \beta_2$

$\Rightarrow \tan \alpha_2 - \tan \alpha_1 = \tan \beta_1 - \tan \beta_2$

$w = uCa\,(\tan \beta_1 - \tan \beta_2)$

$= 200 \times 165\,(\tan 45° - \tan 11°)$

$= 26585.4498$ J/kg

$$\eta = \frac{h_{02}' - h_{01}}{w} = \frac{C_pT_{01}\left(\frac{T_{02}'}{T_{01}} - 1\right)}{w' - w}$$

$$\eta = \frac{C_pT_{01}\left(r_p^{\frac{r-1}{r}} - 1\right)}{w}$$

$$r_p = \left[1 + \frac{\eta \times w}{C_pT_{01}}\right]^{\frac{r}{(r-1)}}$$

$$= \left[1 + \frac{0.8 \times 26585.4498}{1005 \times 298}\right]^{\frac{1.4}{0.4}}$$

$= 1.27$

88. 216.02 N-S/kg

$V_i = 270$ m/s

$P_\infty = 0.9$ bar

$T_\infty = 290$ K

At Nozzle inlet,

$P_0 = 1.6$ bar

$T_0 = 774$ K

Checking "choking" condition of nozzle

$$\left(\frac{P_0}{P_C}\right) = \left(\frac{r+1}{2}\right)^{\frac{\gamma}{(\gamma-1)}} = 1.85$$

$P_C = 0.86$ bar

Since, $P_\infty > P_C$, so nozzle is not choked.

$$\frac{T_0}{P_0^{\frac{\gamma-1}{\gamma}}} = \frac{T_e}{P_e^{\frac{\gamma-1}{\gamma}}}$$

$$T_e = 774 \times \left(\frac{0.9}{1.6}\right)^{\frac{0.33}{1.33}}$$

$= 671.03 \text{ K}$

$$V_e = \sqrt{2C_p(T_0 - T_e)}$$

$$= \sqrt{2 \times 1147(774 - 671.03)}$$

$= 486.02$ m/s

$$\frac{T}{\dot{m}_a} = V_e - V_i$$

$= 486.02 - 270$

$= 216.02$ N-S/kg

89. $h_{01} + Q = h_{02}$.

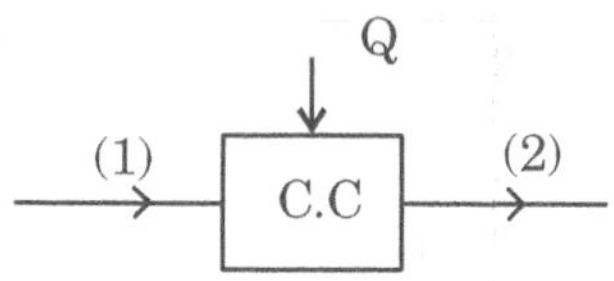

$$h_{01} + Q = h_{02}$$

$$\dot{m}_a C_{pa} T_{01} + \dot{m}_f \times cv = (\dot{m}_a + \dot{m}_f) C_{pg} T_{02}$$

$50 \times 1.005 \times 450 + 1 \times 44 \times 10^3$

$= (50 + 1) \times 1.147 \times T_{02}$

$T_{02} = 1138.734$ K

90. $P = \dot{m}\mu u^2$

μ = slip factor = 1

P = Power

$1800 \times 10^3 = 10 \times \mu^2$

$$u = \sqrt{1800 \times 100} = 424.264 \text{ m/s}$$

$$u = \frac{\pi DN}{60} \Rightarrow D = \frac{60u}{\pi N}$$

where 'N' in RPM

In 1900 $\frac{\text{rad}}{\text{s}}$ No. of revolution per minute is

$$N = \frac{1900 \times 60}{2\pi}$$

$$D = \frac{60 \times 424.264 \times 2\pi}{\pi \times 1900 \times 60}$$

$= 0.446$ m

93. From energy equation;

$$h_o = h + \frac{v^2}{2}$$

Expansion of air for maximum velocity is achieve when the Total enthalpy is converted into kinetic energy of air.

$$\text{so, } h_o = \frac{v^2}{2}$$

$$C_p T_o = \frac{v^2}{2}$$

$$v = \sqrt{2C_p T_o} = \sqrt{2 \times 1005 \times 700}$$

$= 1186.17$ m/s

94. 0.98

$$PR = \frac{P_{o2}}{P_{o1}} = \frac{98}{100}$$

PR = 0.98

95. 0.50

$$\text{Equivalence ratio } (\phi) = \frac{\left(\frac{f}{a}\right)_{\text{actual}}}{\left(\frac{f}{a}\right)_{\text{stichiometric}}}$$

$$\phi = \frac{0.03}{0.06} = 0.5$$

96. For negligible fuel - air ratio.

$$\eta_p = \frac{2v_i}{v_e + v_i}$$

$$= \frac{2 \times 100}{300 + 100} = \frac{200}{400} = 0.5$$

$\eta_p = 0.5$

97. Propeller efficiency

$$(\eta_p) = \frac{\text{Thrust power}}{\text{Shat power}}$$

$$(\eta_p) = \frac{f \times v_i}{\text{Shaft power}}$$

$$\text{Shaft power} = \frac{500 \times 100}{0.5}$$

$= 100000$ watt

$= 100$ kW

98. $\eta_{is} = \dfrac{h_{02}' - h_{01}}{h_{02} - h_{01}}$

$= \dfrac{T_{02}' - T_{01}}{T_{02} - T_{01}}$

$= \dfrac{T_{01}\left(r_{po}^{\frac{r-1}{r}} - 1\right)}{T_{02} - T_{01}}$

$\eta_{isc} = \dfrac{300 \times \left(4^{\frac{0.4}{1.4}} - 1\right)}{480 - 300}$

$= 0.809 \simeq 0.81$

99. For 50% reaction stage

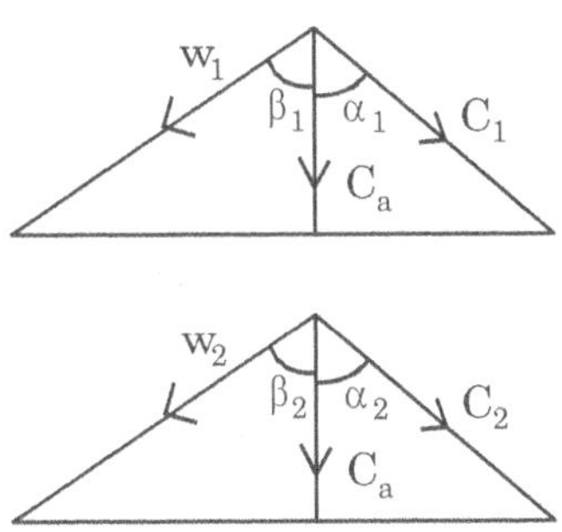

$\alpha_1 - \beta_2 = 22°.$

$\cos\beta_2 = \dfrac{C_a}{w_2}$

$w_2 = \dfrac{C_a}{\cos\beta_2}$

$= \dfrac{200}{\cos 22°} = 215.707 \text{ m/s}$

100. 223.6 m/s

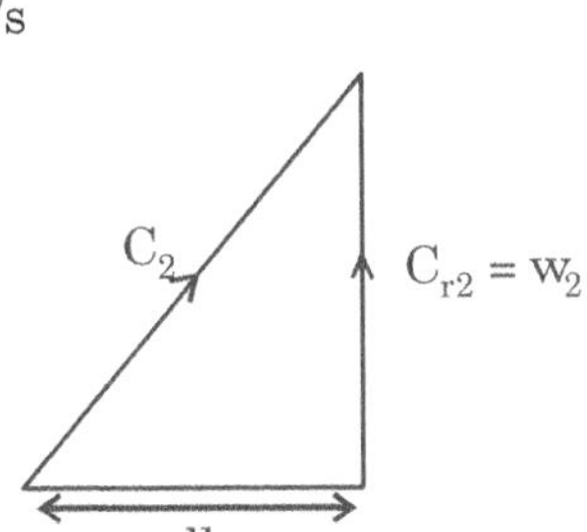

$C_2 = \sqrt{(w_2)^2 + (u)^2}$

$= \sqrt{(100)^2 + (200)^2}$

$= 223.607$ m/s

101. Propulsive efficiency (η_p) is given by

$$\eta_p = \frac{\text{thrust power}}{\text{propulsive power}} = \frac{P_T}{P_p}$$

Propulsive power is due to change is Kinetic Energy at the exit and at the inlet and thrust power is given as useful work done by the system in the product of thrust and flight speed.

η_p is measure of the effectiveness with which the Kinetic Energy imparted to the fluid is transported into useful work. In case of Ramjet propulsive efficiency is lower than low bypass transfers because has a layer Kinetic Energy loss in exhaust jets.

102. Thrust specific fuel consumption (TSFC) is the measure of thermal efficiency of an engine for a given air speed and jet speed.

$$TSFC = \frac{3600 \times m_f}{F} = \frac{3600 \times f}{I_{sp}} \text{ where } f = \frac{m_f}{m_a}$$

$$\therefore \frac{(TSFC)_{Lowest}}{(TSFC)_{highest}} = \alpha$$

Therefore, $\alpha_{\text{transfer}} < \alpha_{\text{Turbojet}} < \alpha_{\text{Ramjet}}$

103. CPR for axial compressor < CPR for Centrifugal compressor

104. $C_2H_5OH + O_2 \rightarrow CO_2 + H_2O + \text{Heat}$

when we balance this chemical equation

$C_2H_5OH + 3O_2 \rightarrow 2CO_2 + 3H_2O + \text{Heat}$

Required no. of mole of oxygen = 3.

105. $dQ = mCpdt$

$\Rightarrow 500 = 1 \times Cp \times (600 - 100)$

$\Rightarrow Cp = 1 \text{ kJ/kg} - \text{k}$

for diatomic gas, $\gamma = 1.4$

$\Rightarrow \dfrac{Cp}{Cv} = 1.4$

$\Rightarrow Cv = \dfrac{Cp}{1.4} = \dfrac{1}{1.4}$

$\Rightarrow Cv = 0.714$

106.

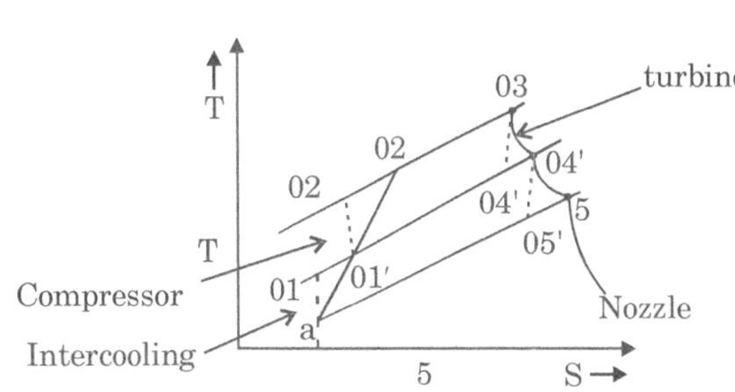

T – S diagram for turbojet
for compressor

$$T_{02}-T_{01} = \frac{T_{01}}{\eta_c}\left(\gamma_p \frac{\gamma-1}{\gamma-1}\right)$$

$$T_{02}-T_{01} = \left[1+\frac{1}{\eta_c}\left(\gamma \frac{\gamma-1}{\gamma-1}\right)\right]$$

i.e η_p – increases

T_{02} – decrease

107. $r_t = 0.14$

$T_{0_3} = 1200$ K

$T_{0_4} = 900$ K

$\eta_t = ?$

for turbine stage

$$T_{0_3} - T_{0_4} = \eta_T T_{0_3}\left(1-\frac{1}{(r_t)^{\frac{\gamma-1}{\gamma}}}\right)$$

$$\Rightarrow 1200 - 900 = n_t \times 1200\left(1-\frac{1}{(0.14)^{\frac{1.2-1}{1.2}}}\right)$$

$\eta_t = 0.08947 = 89\%$

108. Given $\alpha_i = 30°$

$$\alpha_2 = \tan^{-1}\left(\frac{60}{60}\right)$$

$\boxed{\alpha_2 = 45°}$

Work done by compressor stage (W_e)

$W_C = V\,C_a(\tan\alpha_2 - \tan\alpha_1)$

$= 100 \times 60\,(\tan 45° - \tan 30°)$

$= 2054 \times 10^3 = 2.54$ kJ/kg

109. For centrifugal compressor,

$$\frac{P_{oe}}{P_{oi}} > 1$$

110.

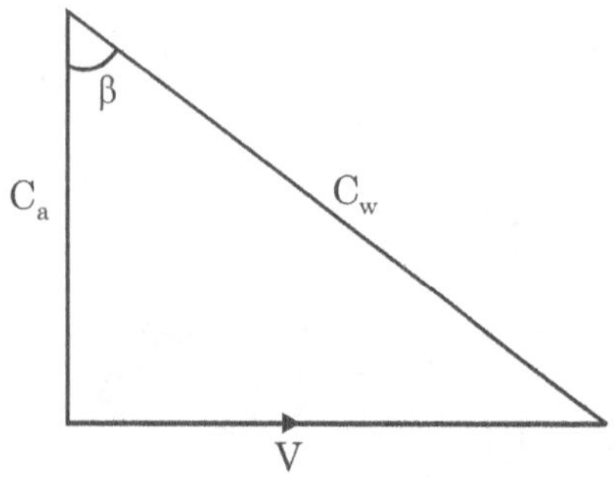

$C_\theta = 0$

$C_{w1} = 0$

$C_{w2} = 150$ m/sec

$V = 200$ m/sec

$$C_p = 1005\frac{J}{kg\,K}$$

Work done by the compressor (W)

$$= \dot{m}V(C_{w2} - C_{w1})$$

$$\Rightarrow \quad \dot{m}C_p.\Delta T = \dot{m}V(C_{w2} - C_{w1})$$

$$\Delta T = \frac{V(C_{w2} - C_{w1})}{C_p}$$

$$= \frac{200 \times (150 - 0)}{1005}$$

$\Delta T = 30°C$ or 30 K

111. Given,

$N_1 = 15000$ rpm

$T_1 = 300$ K

$T_2 = 280$ K

As per formula

$$\frac{N_1}{\sqrt{\gamma R T_1}} = \frac{N_2}{\sqrt{\gamma R T_2}}$$

$$\Rightarrow \quad N_2 = N_1\sqrt{\frac{T_2}{T_1}}$$

$$= 15000 \times \sqrt{\frac{280}{300}}$$

$= 14491$ rpm

112. For ramjet engine

$M_i = M_e$[Mach number at inlet = Mach number at exit]

$\therefore \quad T_{oe} = T_{oi}$

$$\Rightarrow \quad M_i = \frac{V_e}{\sqrt{\gamma R T_e}}$$

$$\Rightarrow \quad V_e = M_i\sqrt{\gamma R T}$$

$$= M_i\sqrt{\gamma R\left(\frac{T_{oe}}{1+\frac{\gamma-1}{2}M_e^2}\right)}$$

$$= 2.8 \times \sqrt{1.4 \times 287 \times \left(\frac{2400}{1+0.2\times 2.8^2}\right)}$$

$= 1715.82$ m/s

113. Given,

$h_{03} = 115.42$ MJ/sec

$h_{02} = 28.94$ MJ/sec

$m_a = 32$ kg/s

$$\frac{\dot{m}_f}{\dot{m}_a} = 15.6 \text{ kg/s}$$

$$m_f = \frac{32}{15.6} = 2.05 \text{ kg/s}$$

CV = 46 MJ/kg

Efficiency of combustor is given by

$$\eta_c = \frac{(\dot{m}_a + \dot{m}_f)h_{03} - \dot{m}_a h_{02}}{\dot{m}_f \, CV}$$

$$= \frac{(32+2.05)\times 115.42 - 32\times 28.94}{2.05\times 46\times 10^6}$$

$= 91.65\% \approx 92\%$

114. Efficiency, $\eta_T = \dfrac{T_{01} - T_{02}}{T_{01} - T_{02}'}$

$$= \frac{1050 - 825}{1050 - T_{02}'}$$

Now, isentropic relation

$$\left(\frac{P_{01}}{P_{02}}\right)^{\frac{\gamma}{\gamma-1}} = \frac{T_{01}}{T_{02}'}$$

$$\Rightarrow \quad T_{02}' = \frac{T_{01}}{\left(\frac{P_{01}}{P_{02}}\right)^{\frac{\gamma-1}{\gamma}}}$$

$$= \frac{1050}{\left(\frac{7.8}{2.5}\right)^{\frac{1.33-1}{1.33}}} = 789.5 \text{ K}$$

$$\therefore \quad \eta_T = \frac{1050 - 823}{1050 - 789.5}$$

$= 0.8789 = 87.89 \approx 88\%$

116. Specific thrust $= \dfrac{T}{\dot{m}_a} = (V_e - V_i)$

For constant V_i, with increase in altitude ambient pressure decreases. In other words magnitude of exhaust velocity from the nozzle increases due to increase in expansion rate with altitude and hence specific thrust increases.

118. 606.83

Given, $f = \dfrac{\dot{m}_f}{\dot{m}_a} = 0.019$, $h_1 = 260$ KJ/kg,

$h_2 = 912$ KJ/kg, HV = 44.5 MJ/kg

Q_L = heat loss to the surrounding = 25 KJ/kg

From law of conservation of energy,

$(\text{Energy})_{in} = (\text{Energy})_{out}$

$$\dot{m}_a h_1 + \frac{\dot{m}_a v_1^2}{2} + \dot{m}_f \times HV$$

$$= \left(\dot{m}_a + \dot{m}_f\right)h_2 + \left(\dot{m}_a + \dot{m}_f\right)\frac{v_2^2}{2} + \dot{m}_a Q_L$$

Dividing by $\dot{m}_a$ on both sides

$$h_1 + \frac{v_1^2}{2} + f\times HV(1+f)h_2 + (1+f)\frac{v_2^2}{2} + Q_L$$

$$260\times 1000 + \frac{270^2}{2} + 0.019\times 44.5\times 10^6$$

$$= (1+0.019)\,(912\times 1000 + \frac{v_2^2}{2}) + 25\times 1000$$

$V_2 = 606.83$ m/s

119. 21.64

$$\dot{m}_a = \dot{m}_c + \dot{m}_h$$

$$B = \text{by-pass ratio} = \frac{\dot{m}_c}{\dot{m}_h}$$

$$3 = \frac{\dot{m}_c}{\dot{m}_a - \dot{m}_c} = \frac{\dot{m}_c}{100 - \dot{m}_c}$$

$\dot{m}_c = 75$ kg/s

P_{oa} = Total ambient pressure = 1 bar

$P_{o,aftfan}$ = Total pressure after fan = Total pressure at exit of cold nozzle

r_F = Fan pressure ratio $= \dfrac{P_{o,aftfan}}{P_{oa}} = 1.6 = 1.6$

$P_{o,aftfan} = 1.6$ bar

$$\frac{P_{oe}}{P_e} = \left[1 + \frac{(\gamma-1)M_e^2}{2}\right]^{\frac{\gamma}{\gamma-1}}$$

$$\frac{1.6}{1} = \left[1 + \frac{(1.4-1)M_e^2}{2}\right]^{\frac{1.4}{1.4-1}}$$

$M_e = 0.848$

$$C_P = \frac{\gamma R}{\gamma - 1}$$

$$1005 = \frac{1.4R}{1.4-1}$$

R = 287.143 J/kg-K

$$V_{e,cold\ nozzle} = M_e\sqrt{\gamma R T_e} = 288.539 \text{ m/s}$$

Thrust from cold nozzle,

$$T_{cold} = \dot{m}_c (V_{e,cold\ nozzle} - V_{intake})$$

Since test is static so, $V_{intake} = 0$

$$T_{cold} = \dot{m}_c V_{e,cold\ nozzle} = 21640.539 \text{ N}$$

120. 277.125

$C_1 = W_2 = 140$ m/s

$C_2 = W_1 = 240$ m/s

Hence, Degree of reaction, R = ½

And, $\beta_2 = \alpha_1 = 30°$

$C_a = W_2 \times \cos\beta_2 = 121.243$ m/s

$$\beta_1 = \alpha_2 = \cos^{-1}\left(\frac{C_a}{C_2}\right) = 59.657°$$

$$R = \frac{1}{2} = \frac{C_a}{2U}(\tan\beta_1 + \tan\beta_2)$$

U = 277.125 m/s

121. 19.12

$\beta_2 = 22°$, $\beta_3 = 55°$

U = 340 m/s, R = 0.4

$$R = \frac{C_a}{2U}(\tan\beta_3 - \tan\beta_2)$$

$C_a = 265.593$ m/s

$$\dot{m} = \rho_i A_i C_a = 0.9 \times 0.08 \times 265.593$$

$$\dot{m} = 19.123 \text{ kg/s}$$

122. 2.43

Applying Law of conservation of energy across combustion chamber,

$(\text{Energy})_{in} = (\text{Energy})_{out}$

$H_{03} + \eta_{cc} \times Q = H_{04}$

$$\dot{m}_a C_{Pa} T_{03} + \eta_{cc}\, \dot{m}_f HV = (\dot{m}_a + \dot{m}_f) C_{Pg} T_{04}$$

$$\dot{m}_f = \frac{\dot{m}_a C_{pg} T_{04} - \dot{m}_a C_{Pg} T_{03}}{HV \times \eta_{cc} - C_{Pg} T_{04}}$$

$$\dot{m}_f = \frac{100 \times 1200 \times 1200 - 100 \times 1000 \times 600}{40 \times 10^6 \times 0.9 - 1200 \times 1200}$$

$$\dot{m}_f = 2.43 \text{ kg/s}$$

7 CHAPTER ROCKET PROPULSION

1. The on-board rocket motor of a satellite of initial mass 2000 kg provides a specific impulse of 280 seconds. If this motor is fired to give a speed increment of 500 m/s along the direction of motion, the mass of propellant consumed is:

(*a*) 685 kg (*b*) 333 kg

(*c*) 1666 kg (*d*) 167 kg

[GATE 2007, 2 M]

2. In a rocket engine, the hot gas generated in the combustion chamber exits the nozzle with a mass flow rate 719 kg/sec and velocity 1794 m/s. The area of the nozzle exit section is 0.635 m^2. If the nozzle expansion is optimum, then the thrust procuced by the engine is

(*a*) 811 kN (*b*) 1290 kN

(*c*) 1354 kN (*d*) 2172 kN

[GATE 2007, 2 M]

3. Which one of the following values is nearer to the vacuum specific impulse of a rocket engine using liquid hydrogen and liquid oxygen as propellants?

(*a*) 49 sec (*b*) 450 sec

(*c*) 6000 sec (*d*) 40000 sec

[GATE 2007, 2 M]

4. In a convergent-divergent (CD) nozzle of a rocket motor, the wall heat flux is maximum at

(*a*) the exit of the divergent portion of the CD nozzle

(*b*) the entry to the convergent portion of the CD nozzle

(*c*) the throat of CD nozzle

(*d*) the mid-length of the divergent portion of the CD nozzle

[GATE 2008, 1 M]

5. DB denotes double base solid propellant.

LOX-RP1 denotes liquid oxygen - kerosene combination.

LOX-LH_2 denotes liquid oxygen - hydrogen combination.

The correct order of increasing specific impulse is

(*a*) DB < LOX-RP1 < LOX-LH_2

(*b*) LOX-RP1 < DB < LOX-LH_2

(*c*) LOX-LH_2 < DB < LOX-RP_1

(*d*) DB < LOX-LH_2 < LOX-RP_1

[GATE 2008, 1 M]

6. The chamber stagnation temperature inside a rocket motor is T_e. Only a convergent nozzle is used, and the flow at the exit of this nozzle is choked. Assume that the nozzle exhaust static pressure is equal to ambient static pressure. Gas constant for exhaust gases is R and ratio of specific heats is γ. The specific impulse of the rocket motor is

(*a*) $\sqrt{\dfrac{2\gamma RT_e}{\gamma - 1}}$ (*b*) $\sqrt{\dfrac{\gamma RT_e}{\gamma - 1}}$

(*c*) $\sqrt{\dfrac{\gamma RT_e}{\gamma + 1}}$ (*d*) $\sqrt{\dfrac{2\gamma RT_e}{\gamma + 1}}$

[GATE 2008, 2 M]

7. The initial temperature sensitivity of burn rate of a solid rocket motor propellant is positive. If the initial temperature increases then

(*a*) thrust increases but burn time decreases

(*b*) thrust decreases and burn time decreases too

(*c*) thrust remains same but burn time increases

(*d*) thrust increases but burn time remains same

[GATE 2008, 2 M]

Common Data for Questions 8 and 9: A liquid rocket engine with oxidizer to fuel ratio of 5 : 1 produces a thrust 1 MN. The initial mass of the rocket engine is 100,000 kg and its mass at burn out is 10,000 kg. The characteristic velocity C and thrust C_F for the engine are 2386 m/s and 1.4 respectively.

8. The mass flow rate of fuel is

(*a*) 300.3 kg/s (*b*) 269.5 kg/s

(*c*) 87.4 kg/s (*d*) 49.9 kg/s

[GATE 2008, 2 M]

9. Neglecting gravity and drag effects, if the initial velocity of the liquid rocket engine is 2.5 km/s, the velocity of the rocket at burnout is

(*a*) 1.2 km/s (*b*) 2.5 km/s

(*c*) 10.2 km/s (*d*) 11.8 km/s

[GATE 2008, 2 M]

10. Classify the following propellants as: cryogenic (C), semi-cryogenic (SC), compressed gas (CG) and earth storable (ES)

N_2O_4-UDMH (nitrogen tetra oxide and unsymmetrical di-methyl hydrazine)

LOX-RPI (liquid oxygen and kerosene)

LOX-LH_2 (liquid oxygen and liquid hydrogen)

N_2 (nitrogen gas)

(*a*) N_2O_4-UDMH (ES), LOX-RP1 (C), LOX-LH_2 (C), N_2(C)

(*b*) N_2O_4-UDMH (ES), LOX-RP1 (SC), LOX-LH_2 (C), N_2(C)

(*c*) N_2O_4-UDMH (ES), LOX-RP1 (SC), LOX-LH_2 (C), N_2(CG)

(*d*) N_2O_4-UDMH (ES), LOX-RP1 (C), LOX-LH_2 (C), N_2(CG)

[GATE 2009, 1 M]

11. Let M_0 be the total mass of a single stage rocket, M_P be the total mass of propellant, M_L be the mass of payload carried by the rocket and M_S be the mass of inert structural components. If I_{sp} is the specific impulse of the propulsion system (in seconds) and g is the acceleration due to gravity, then the maximum velocity that can be attained by the rocket vehicle in the absence of gravity and atmospheric drag is given by

(*a*) $gI_{sp} \ln\left(\frac{M_0}{M_P}\right)$ (*b*) $gI_{sp} \ln\left(\frac{M_0}{M_L + M_S} - 1\right)$

(*c*) $gI_{sp} \ln\left(\frac{M_0}{M_S}\right)$ (*d*) $gI_{sp} \ln\left(\frac{M_0}{M_0 - M_P}\right)$

[GATE 2009, 2 M]

12. The performance of a solid rocket motor is improved by replacing the old propellant with a new one. The new propellant gives a combustion temperature 40% higher than the previous propellant without appreciable change in molecular weight of combustion products and other operating parameters. By approximately what percentage is the specific impulse of the new motor higher than the old one?

(*a*) 18% (*b*) 96%

(*c*) 42% (*d*) 112%

[GATE 2009, 2 M]

13. A solid rocket motor has an end burning grain of cross-sectional area Acs = 0.4 m^2. The density of propellant is ρ_p = 1500 kg/m^3 and has linear regression rate $\dot{r}$ = 5 mm/s. If the specific impulse of the propulsion system is I_{sp} = 200 seconds, the thrust produced by the motor is approximately

(*a*) 3 kN (*b*) 6 kN

(*c*) 1.5 kN (*d*) 12 kN

[GATE 2009, 2 M]

14. Among the choices given below, the Specific Impulse is maximum for a

(*a*) Cryogenic (*b*) Solid Rocket

(*c*) Liquid Rocket (*d*) Ramjet

[GATE 2010, 1 M]

15. The stagnation pressure and stagnation temperature inside the combustion chamber of a liquid rocket engine are 15 MPa and 2500 K respectively. The burned gases have γ = 1.2 and R = 692.83 J/kgK. The rocket has a Convcrging-diverging nozzle with a throat area of 0.025 m^2 and the flow at the exit of the nozzle is supersonic. If the flow through the nozzle is isentropic. what is the mass flow rate of the gases out of the nozzle?

(*a*) 18.5 kg/s (*b*) 31.2 kg/s

(*c*) 29.7 kg/s (*d*) 19.4 kg/s

[GATE 2010, 2 M]

16. A two stage chemical rocket, having the same specific impulse (1_{sp}) of 300s for both the stages is designed in such a way that the payload ratio and the structural ratio are same for both the stages. The second stage of the rocket has following mass distribution :

Propellast Mass = 10208 kg

Structural Mass = 1134 kg

Payload Mass = 1700 kg

g_e =9.8 m/s^2

If the rocket is fired from rest and it flies in a zero gravity field and a drag free environment, the final velocity attained by the payload is

(*a*) 9729.3 m/s (*b*) 897.3 m/s

(*c*) 9360.2 m/s (*d*) 8973.2 m/s

[GATE 2010, 2 M]

17. An impulsive launch of a rocket minimizes the loss of burn-out velocity due to

(*a*) aerodynamic drag force only

(*b*) gravitational force only

(*c*) both aerodynamic drag and gravitational forces

(*d*) reaction jet control force

[GATE 2011, 1 M]

18. Multi-staging in rockets improves the burn-out performance by increasing mainly stage-wise

(*a*) payload mass ratios

(*b*) structural mass efficiencies

(*c*) propellant masses

(*d*) control system masses

[GATE 2011, 1 M]

19. A rocket engine is tested on a test bed under the ideal condition of fully expanded jet. The exhaust velocity is 2 km/s through a nozzle of area 2.5 m^2. The mass flow rate is 200 kg/s. The specific impulse of the propellant and the thrust developed respectively are (assume g = 9.81 m/s^2)

(*a*) 175.87 s and 200 kN

(*b*) 203.87 s and 400 kN

(*c*) 231.87 s and 200 kN

(*d*) 280.87 s and 400 kN

[GATE 2011, 2 M]

20. A rocket motor has combustion chamber temperature of 2600 K and the products have molecular weight of 25 g/mol and ratio of specific heats 1.2. The universal gas constant is 8314 J/kg-mole-K. The value of theoretical c* (in m/s) is ____.

[GATE 2012, 2 M]

21. A solid propellant of density 1800 kg/m^3 has a burning rate law $r = 6.65 \times 10^{-3}p^{0.45}$ mm/s, where p is pressure in Pascals. It is used in a rocket motor with a tubular grain with an initial burning area of 0.314 m^2. The characteristic velocity is 1450 m/s. What should be the nozzle throat diameter to achieve an equilibrium chamber pressure of 50 bar at the end of the ignition transient?

(*a*) 35 mm (*b*) 38 mm

(*c*) 41 mm (*d*) 45 mm

[GATE 2012, 2 M]

22. A bipropellant liquid rocket motor operates at a chamber pressure of 40 bar with a nozzle throat diameter of 50 mm. The characteristic velocity is 1540 m/s. If the fuel-oxidizer ratio of the propellant is 1.8, and the fuel density is 900 kg/m^3, what should be the minimum fuel tank volume for a burn time of 8 minutes

(*a*) 1.65 m^3 (c) 1.75 m^3

(*c*) 1.85 m^3 (*d*) 1.95 m^3

[GATE 2012, 2 M]

23. The propellant in a single stage sounding rocket occupies 60% of its initial mass. If all of it is expended instantaneously at an equivalent exhaust velocity of 3000 m/s, what would be the altitude attained by the payload when launched vertically?

[Neglect drag and assume acceleration due to gravity to be constant at 9.81 m/s^2.]

(*a*) 315 km (c) 335 km

(*c*) 365 km (*d*) 385 km

[GATE 2012, 2 M]

24. A rocket motor has a chamber pressure of 100 bar and chamber temperature of 3000 K. The ambient pressure is 1 bar. Assume that the specific heat at constant pressure is 1 kJ/kg-K. Also assume that the flow through the nozzle is isentropic and optimally expanded. The exit static temperature in K is

(*a*) 805 (*b*) 845

(*c*) 905 (*d*) 945

[GATE 2013, 2 M]

25. Thrust of liquid oxygen - liquid hydrogen rocket engine is 300 kN. The O/F ratio used is 5. If the fuel mass flow rate is 12.5 kg/s, the specific impulse of the rocket motor in Ns/kg is

(*a*) 400 (*b*) 4000

(*c*) 4200 (*d*) 4400

[GATE 2013, 2 M]

26. Consider two rockets P and Q fired vertically up with identical specific impulse and a payload of 2 kg. Rocket P has 2 identical stages, and each stage has 200 kg of propellant and 20 kg of structural weight. Rocket Q has a single stage with 400 kg of propellant and 40 kg of structural weight. Neglecting drag and gravity effects, the ratio of the change in velocity of P to that attained by Q is

(*a*) 1.13 (*b*) 1.23

(*c*) 1.33 (*d*) 1.43

[GATE 2013, 2 M]

27. A small rocket having a specific impulse of 200s produces a total thrust of 98 kN, out of which 10kN is the pressure thrust. Considering the acceleration due to gravity to be 9.8m/s^2, the propellant mass flow rate in kg/s is

(*a*) 55.1 (*b*) 44.9

(*c*) 50 (*d*) 60.2

[GATE 2014, 1 M]

28. A stationary two stage rocket with initial mass of 16000 kg, carrying a payload of 1000kg, is fired in a vertical trajectory from the surface of the earth. Both the stages of the rocket have same specific impulse, I_{sp}, of 300s and same structural coefficient of 0.14. The acceleration due to gravity is 9.8m/s^2. Neglecting drag and gravity effects and considering both the stages with same payload ratio, the terminal velocity attained by the payload in m/s is __________.

[GATE 2014, 2 M]

29. A cryogenic rocket has a specific impulse of 455s and characteristic velocity of 2386 m/s. The value of thrust coefficient for this rocket is

(*a*) 1.78 (*b*) 1.73

(*c*) 1.87 (*d*) 1.95

[GATE 2014, 2 M]

30. A rocket engine produces a total impulse of 112 kN.s in a burn time period of 3.5 minutes with a propellant mass flow rate of 0.25 kg/s. The effective exhaust velocity (in m/s) of gas ejecting from the engine is _____.

[GATE 2015, 1 M]

31. A solid rocket motor is designed with a cylindrical end-burning propellant grain of length 1 m and diameter 32 cm. The density of the propellant grain is 1750 kg/m^3. The specific impulse of the motor is 190 s and the acceleration due to gravity is 9.8 m/s^2. If the propellant burns for a period of 150 s, then the thrust (in N) produced by the rocket motor is ________.

[GATE 2015, 2 M]

32. A liquid propellant rocket has the following component masses:

Mass of payload	= 180
Mass of fuel	= 470
Mass of oxidizer	= 1170
Mass of structures	= 150
Mass of guidance systems	= 20

The effective exhaust velocity is 3136 m /s. The velocity increment (in km/ s) of the rocket at burnout, while operating in outer space, is ________

[GATE 2015, 2 M]

33. A rocket nozzle is designed to produce maximum thrust at an altitude, H = 8 km from the sea level. The nozzle operates in

(*a*) under-expanded condition for H > 8 km

(*b*) under-expanded condition for H < 8 km

(*c*) sonic exit condition for H > 8 km

(*d*) unchoked condition for H < 8 km

[GATE 2015, 2 M]

34. For a given chamber pressure, the thrust of a rocket engine is highest when

(*a*) the rocket is operating at its design altitude.

(*b*) the rocket is operating in vacuum.

(*c*) the rocket is operating at sea-level.

(*d*) there is a normal shock in the rocket nozzle.

[GATE 2016, 1 M]

35. A launch vehicle has a main rocket engine with two identical strap-on motors, all of which fire simultaneously during the operation. The main engine delivers a thrust of 6300 kN with a specific impulse of 428 s. Each strap-on motor delivers a thrust of 12000 kN with specific impulse of 292 s. The acceleration due to gravity is 9.81 m/s^2. The effective (combined) specific impulse of the vehicle is _______ s.

[GATE 2016, 2 M]

36. In a particular rocket engine, helium propellant is heated to 6000 K and 95% of its total enthalpy is recovered as kinetic energy of the nozzle exhaust. Consider helium to be a calorically perfect gas with specific heat at constant pressure of 5200 J/kgK. The exhaust velocity for such a rocket for an optimum expansion is _____ m/s.

[GATE 2016, 2 M]

37. A rocket, with a total lift-off mass of 10000 kg, moves vertically upward from rest under a constant gravitational acceleration of 9.81 m/s^2. The propellant mass of 8400 kg burns at a constant rate of 1200 kg/s. If the specific impulse of the rocket engine is 240 s, neglecting drag, the burnout velocity in m/s is

(*a*) 3933.7 (*b*) 4314.6

(*c*) 4245.9 (*d*) 4383.3

[GATE 2016, 2 M]

38. A single stage chemical rocket, having an initial mass of 10,000 kg and specific impulse of 450 s, is launched from the surface of the earth and has to reach the escape velocity (11 km/s) at burn out. Consider g_e = 9.8 m/s^2. If the atmospheric drag and the effect of gravity are to be neglected, the mass of propellant to be carried by the rocket is equal to ________ kg (in one decimal places).

[GATE 2017, 2 M]

39. Among the following engines, which one is expected to have the maximum Specific Impulse?

(*a*) Cryogenic Rocket

(*b*) Solid Propellant Rocket

(*c*) Liquid Propellant Rocket

(*d*) SCRAM Jet

[GATE 2017, 1 M]

40. A rocket has an initial mass of 150 kg. After operating for a duration of 10 s, its final mass is 50 kg. If the acceleration due to gravity is 9.81 m/s^2 and the thrust produced by the rocket is 19.62 kN, the specific impulse of the rocket is

(*a*) 400 s (*b*) 300 s

(*c*) 200 s (*d*) 100 s

[GATE 2017, 2 M]

41. For a rocket engine, the velocity ratio r is V_a/V_e, where V_a is the vehicle velocity and V_e is the exit velocity of the exhaust gases. Assume the flow to be optimally expanded through the nozzle. For r = 2, if F is the thrust produced and $\dot{m}$ is the mass flow rate of exhaust gases, then, $\frac{F}{\left(\dot{m}\, V_e\right)}$ is _____.

[GATE 2019]

42. The specific impulse of a rocket engine is 3000 Ns/kg. The mass of the rocket at burnout is 1000 kg. The propellant consumed in the process is 720 kg. Assume all factors contributing to velocity loss to be negligible. The change in vehicle velocity ?u is _____ km/s (round off to 2 decimal places).

[GATE 2019]

43. Which one of the following is a hypergolic propellant combination used in rocket engines?

(*a*) Liquid hydrogen – liquid oxygen

(*b*) Unsymmetrical dimethyl hydrazine – nitrogen tetroxide

(*c*) Rocket fuel RP-1 – liquid oxygen

(*d*) Liquid hydrogen – liquid fluorine

[GATE 2020]

44. A rocket engine has a sea level specific impulse of 210 s and a nozzle throat area of 0.005 m^2. While testing at sea level conditions, the characteristic velocity and pressure for the thrust chamber are 1900 m/s and 50 bar, respectively. Assume the acceleration due to gravity to be 9.8 m/s^2. The thrust produced by the rocket engine is _____ kN (*round off to one decimal place*).

[GATE 2020]

45. A solid propellant rocket producing 25 MN thrust is fired for 150 seconds. The specific impulse of the rocket is 2980 Ns/kg. How much propellant is burned during the rocket operation?

(a) 8390 kg (b) 82300 kg

(c) 1.26×10^6 kg (d) 11.2×10^6 kg

[GATE 2021]

46. Hot gases are generated at a temperature of 2100 K and a pressure of 14 MPa in a rocket chamber. The hot gases are expanded ideally to the ambient pressure of 0.1 MPa in a convergent-divergent nozzle having a throat area of 0.1 m^2. The molecular mass of the gas is 22 kg/kmol. The ratio of specific heats (γ) of the gas is 1.32. The value of the universal gas constant (R_0) is 8314 J/kmol-K. The acceleration due to gravity, g, is 9.8 m/s^2. The specific impulse of the rocket is _____ seconds (*round off to two decimal places*).

[GATE 2021]

ANSWER KEY

1. (*b*)	**2.** (*b*)	**3.** (*b*)	**4.** (*b*)	**5.** (*a*)	**6.** (*a*)	**7.** (*a*)	**8.** (*d*)	**9.** (*c*)	**10.** (*c*)
11. (*d*)	**12.** (*a*)	**13.** (*b*)	**14.** (*d*)	**15.** (*a*)	**16.** (*d*)	**17.** (*b*)	**18.** (*a*)	**19.** (*b*)	**20.** (*c*)
21. (*b*)	**22.** (*b*)	**23.** (*d*)	**24.** (*a*)	**25.** (*b*)	**26.** (*b*)	**27.** (*b*)	**29.** (*c*)	**33.** (*a*)	**34.** (*b*)
39. (*d*)	**40.** (*c*)	**41.** (1)	**42.** (1.63 km/sec)		**43.** (b)	**44.** (27.0 to 27.2)			**45.** (*c*)
46. (216 to 220)									

EXPLANATIONS

1. $\Delta v_b = gI_{sp} \ln\left(\frac{m_i}{m_f}\right)$

$$\frac{\Delta v_b}{gI_{sp}} = \ln\left(\frac{m_i}{m_f}\right)$$

$$e^{\left(\frac{\Delta v_b}{gI_{sp}}\right)} = \left(\frac{m_i}{m_f}\right)$$

$$m_f = \frac{m_i}{e^{\frac{\Delta v_b}{gI_{sp}}}}$$

$$m_p = m_i - m_f$$

$$= 2000 - \frac{2000}{e^{\left(\frac{500}{9.81\times 280}\right)}}$$

$$m_p = 332.847 \text{ kg}$$

2. $T = \dot{m}_e \times v_e + (P_e - P_a) A_e$

for optimum expansion, $P_e = P_a$

$T = \dot{m}_e \times v_e$

$= 719 \times 1794 = 1289.886$ kN

3. The value of I_{sp} of liquid propellant rocket engine is moderate. The given value of 49S is vary low while 6000S and 40,000 S is very high. Therefore the correct answer will be 450S.

4. Heat flux, $(q) = -K\left(\frac{dT}{dx}\right)$

$$q = K\frac{(T - T_a)}{t}$$

t is wall thickness

T = Temperature inside the wall

T_a = Ambience temperature

Expansion of hot gas from combustion chamber takes place in the nozzle, so along the length of nozzle the temperature will decrease, it will be maximum at nozzle inlet and minimum at outlet. Therefore heat flux is maximum at nozzle inlet.

5. Solid propellant has lower specific impulse than liquid propellant, so option (b) and (c) is incorrect. Also the specific impulse of liquid hydrogen is greater than the hydrocarbon fuels.

6. $T = \dot{m}_e V_e + (P_e - P_a) A_e$

Given, $P_e = P_a$

$T = \dot{m}_e v_e$

$$\Rightarrow \frac{T}{\dot{m}_e} = v_e$$

$I_{sp} = v_e$ (Nozzle is choked Hence, $v_e = a_e$)

$$I_{sp} = a_e = \sqrt{\gamma RT_e}$$

$$\frac{T_c}{T_e} = \frac{\gamma+1}{2}$$

$$I_{sp} = \sqrt{\frac{2\gamma RT_c}{\gamma+1}}$$

7. For solid rocket, burn rate, $r = ap_o^n$

where a is a function of initial temperature.

If initial temperature increases then burn rate also increases and burn time decreases because it is inversely proportional to burn rate kg/sec.

Also with increase in burns rate the kinetic energy of exhaust gases from nozzle increases and hence thrust will increase.

8. Thrust coefficient

$$C_f = \frac{T}{P_c \times A^*}$$

A^* = Throat area

$$P_C = \frac{T}{C_f \times A^*}$$

$$\text{Also, } C^* = \frac{P_c \times A^*}{\dot{m}_p}$$

$$\Rightarrow \dot{m}_p = \frac{P_C \times A^*}{C^*}$$

$$\dot{m}_p = \frac{T \times A^*}{C_f \times C^* \times A^*}$$

$$\dot{m}_p = \frac{1 \times 10^6}{1.4 \times 2386}$$

= 299.365 Kg/s

$$\dot{m}_p = \dot{m}_f + \dot{m}_o$$

$$= \dot{m}_f + 5\dot{m}_f$$

$$\dot{m}_f = \frac{\dot{m}_p}{6} = \frac{299.365}{6} = 49.894 \text{ kg/s}$$

9. $\Delta v_b = gI_{sp} \ln \frac{m_i}{m_f}$

$$= g \times \frac{T}{g \times \dot{m}_p} \ln \frac{m_i}{m_f}$$

$$v_f - v_i = \frac{T}{\dot{m}_p} \ln \frac{m_i}{m_f}$$

$$= \frac{1 \times 10^6}{299.365} \ln \frac{100000}{10000}$$

= 7691.564 m/s = 7.691 km/s

v_f = 7.691 + v_i.

= 7.691 + 2.5 = 10.191 km/s

11. $\Delta v_b = gI_{sp} \ln \frac{m_i}{m_f}$

$m_f = m_i - m_p$

$$\Delta v_b = gI_{sp} \ln\left(\frac{m_i}{m_i - m_p}\right)$$

$m_i = m_o$.

12. $T_{CN} = \left(1 + \frac{40}{100}\right) T_{CO}$

$T_{CN} = 1.4\, T_{CO}$

For solid rocket,

$$I_{sp} \propto \sqrt{T_C}$$

$$I_{sp(New)} \propto \sqrt{1.4 T_{CO}}$$

$$I_{sp(Old)} \propto \sqrt{T_{CO}}$$

Change in percentage of I_{sp}

$$= \frac{I_{spN} - I_{spo}}{I_{spo}} \times 100$$

$$= \frac{\sqrt{1.4 T_{CO}} - \sqrt{T_{CO}}}{\sqrt{T_{CO}}} \times 100$$

= 18.32%

13. $\dot{m}_p = \rho_p A_{cs} \dot{r}$

$$= 1500 \times 0.4 \times \frac{5}{1000}$$

= 3 kg/s

$$T = \dot{m}_p \times v_e$$

$$= \dot{m}_p \times I_{sp} \times g$$

= 3 × 200 × 9.81

= 5886 N

≃ 5.9 kN

15. $\dot{m} = \rho^* A^* v^*$

$$= \frac{P^*}{RT^*} \times A^* \sqrt{\gamma R T^*}$$

$$= \sqrt{\frac{\gamma}{R}} \times \frac{P^*}{\sqrt{T^*}} \times A^*$$

$$= \sqrt{\frac{\gamma}{R}} \times \frac{P_{OC}}{\sqrt{T_{OC}}} \times \left(\frac{\gamma+1}{2}\right)^{-\frac{(\gamma+1)}{2(\gamma-1)}} \times A^*$$

$$\dot{m} = \sqrt{\frac{1.2}{692.83}} \times \frac{1.5 \times 10^6}{\sqrt{2500}} \times \left(\frac{2.2}{2}\right)^{-5.5} \times 0.025$$

$\dot{m}$ = 18.479 Kg/s

16. $\Delta v_b = \Delta v_1 + \Delta v_2$.

$$\Delta v_b = gI_{sp} \ln\left(\frac{m_i}{m_f}\right)_1 + gI_{sp} \ln\left(\frac{m_i}{m_f}\right)_2$$

Since both stages are identical

so, $\Delta v_b = 2gI_{sp} \ln\left(\frac{m_i}{m_f}\right)$

$$\Delta v_b = 2gI_{sp} \ln\left(\frac{m_p + m_s + m_{pay}}{m_s + m_{pay}}\right)$$

$$= 2 \times 9.81 \times 300 \times \ln\left(\frac{10208 + 1134 + 1700}{1134 + 1700}\right)$$

= 8984 . 895 m/s

17. for rocket operation

$$dv_b = -v_{eg} \times \frac{dm}{m} - \frac{D}{m}dt - g\cos\theta dt$$

An impulsive launch of rocket-means very short burning time and very high acceleration.

Aerodynamic drag impose penalty if the vehicle is accelerated quickly because 'D' is a function of velocity square.

For very short burning time the term g cos θt will reduce and hence minimising the loss of burn-out velocity.

18. The burn-out velocity of rocket increases with increasing the number of stages due to increasing payload mass ratios of each stage.

19. $T = \dot{m}_p v_e + (P_e - P_a) A_e$

$= 200 \times 2 \times 1000 = 400$ kN

$$I_{sp} = \frac{f}{\dot{m}_p \times g}$$

$$I_{sp} = \frac{400 \times 10^3}{200 \times 9.81}$$

$I_{sp} = 203.874$ s

21. $\dot{m} = \rho \times A \times \dot{r}$

$$\dot{m} = 1800 \times 0.314 \times 6.65 \times 10^{-3} \times \frac{(50 \times 10^5)^{0.45}}{1000}$$

= 3.886 kg/s

$$C^* = \frac{P_{OC} \times A^*}{\dot{m}_p} = \frac{P_{OC} \times \pi d^2}{4\dot{m}_p}$$

$$d = \sqrt{\frac{4\dot{m}_p C^*}{P_{OC}\pi}}$$

$$d = \sqrt{\frac{4 \times 3.886 \times 1450}{50 \times 10^5 \times \pi}}$$

$d = 37.88$ mm $\simeq$ 38 mm

22. $C^* = \frac{P_{OC} \times A^*}{\dot{m}_p}$

$$\dot{m}_p = \frac{P_{OC} \times \pi d_t^2}{4C^*}$$

$$\dot{m}_p = \frac{40 \times 10^5 \times \pi \times (0.05)^2}{4 \times 1540}$$

$\dot{m}_p = 5.0999$ kg/s

$\dot{m}_p = \dot{m}_f + \dot{m}_o$

$$5.0999 = \dot{m}_f + \frac{\dot{m}_f}{1.8}$$

$\dot{m}_f = 3.278$ Kg/s

Volumes, $V = \frac{\dot{m}_f \times t}{\rho_f}$

$$= \frac{3.278 \times 8 \times 60}{900}$$

$V = 1.748$ m³.

23. $\dot{m}_p = \frac{60}{100} m_i = 0.6\, m_i$

$$\Delta v_b = gI_{sp} \ln\frac{m_i}{m_f}$$

$$v_e = \ln\frac{m_i}{m_f}$$

$$= v_e \ln\frac{m_i}{m_i - m_p}$$

$$\Delta v_b = 3000 \times \ln\frac{m_i}{m_i - 0.6m_i}$$

$$= 3000 \times \ln\frac{m_i}{m_i - (1 - 0.6)} = 2748.872 \text{ m/s}$$

max. height attained (4.0)

$$h = \frac{v^2}{2g} = \frac{(2778.872)^2}{2 \times 9.91} = 381.246 \text{ km}$$

24. $\dfrac{T_{OC}}{P_{OC}^{\frac{\gamma-1}{\gamma}}} = \dfrac{T_e}{P_e^{\frac{\gamma-1}{\gamma}}}$

assuming $\gamma = 1.4$

$P_e = P_a$

$$T_e = T_{OC} \times \left(\frac{P_e}{P_{OC}}\right)^{\frac{\gamma-1}{\gamma}}$$

$$= 3000 \times \left(\frac{1}{100}\right)^{\frac{0.4}{1.4}}$$

$= 804.808$ K

25. $\dot{m}_f = 12.5$ kg/s

$$\frac{\dot{m}_o}{\dot{m}_f} = 5$$

$$\dot{m}_o = 5 \times 12.5 = 62.5 \text{ kg/s}$$

$$\dot{m}_p = \dot{m}_f + \dot{m}_o = 12.5 + 62.5$$

$= 75$ kg/s

$$I_{sp} = \frac{T}{\dot{m}_p} = \frac{300 \times 10}{75} = 4000 \frac{N-s}{Kg}$$

26. For Engine 'P' $\rightarrow$

$(m_i)_1 = m_{pay} + (m_s + m_p) \times 2$

$= 2 + (20 + 200) \times 2$

$= 442$ Kg

$(m_f)_1 = (m_i)_1 - (m_p)_1$

$= 442 - 200 = 242$ Kg

$(m_i)_2 = (m_f)_1 - (m_s)_1$

$= 242 - 20 = 222$ Kg

$(m_f)_2 = (m_i)_2 - (m_p)_2$

$= 222 - 200 = 22$ Kg

$\Delta v_p = \Delta v_1 + \Delta v_2$

$$= gI_{sp}\left[\ln\frac{(m_i)_1}{(m_f)_1} + \ln\left(\frac{m_i}{m_f}\right)_2\right]$$

$$\Delta v_p = gI_{sp} \ln\frac{(m_i)_1}{(m_f)_1} \times \frac{(m_i)_2}{(m_f)_2}$$

$$\Delta v_p = gI_{sp} \ln\left(\frac{442}{242} \times \frac{222}{22}\right)$$

For Engine Q.

$$\Delta v_Q = gI_{sp} \ln\left(\frac{400 + 40 + 2}{40 + 2}\right)$$

$$= gI_{sp} \ln\left(\frac{442}{42}\right)$$

$$\frac{\Delta v_p}{\Delta v_Q} = \frac{gI_{sp} \ln\left(\frac{442}{242} \times \frac{222}{22}\right)}{gI_{sp} \ln\left(\frac{442}{42}\right)} = 1.23$$

27. Total thrust = momentum thrust + pressure thrust

98 = Momentum thrust + 10

Momentum thrust = 88 KN $= \dot{m}_p \times v_e$

$v_e = g \times I_{sp}$

$$\dot{m}_p = \frac{88 \times 10^3}{9.81 \times 200} = 44.852 \text{ kg/s}$$

28. $m_L = 1000$ kg

$m_i = 16000$ kg

$m_i = m_p + m_s + m_L$

$m_i - m_L = m_p + m_s$

$16000 - 1000 = m_p + m_s$

$15000 = m_p + m_s$

$$\text{Payload ratio } (\lambda) = \frac{m_L}{m_p + m_s}$$

$$\lambda = \frac{1000}{15000} = 0.0666$$

Terminal velocity attained by the payload is

$$= nJ_{sp} g \ell_n \left(\frac{1 + \lambda}{\varepsilon + \lambda}\right)$$

$$= 2 \times 300 \times 9.8 \times \ell n\left(\frac{1 + 0.066}{0.14 + 0.066}\right)$$

$= 9666$ m/s

29. $$C_f = \frac{T}{P_{OC} \times A^*}$$

$$C^* = \frac{P_{OC} \times A^*}{\dot{m}_p}$$

$$C_f = \frac{T}{\frac{C^* \dot{m}_p}{A^*} \times A^*}$$

$$C_f = \frac{\frac{T}{\dot{m}_p}}{C^*} = \frac{I_{sp} \times g}{C^*}$$

$$= \frac{455}{2386} \times 9.81$$

$C_f = 1.87.$

30. 2133.33 m/s

Total Impulse = Thrust × Time

$$T = \frac{112 \times 1000}{3.5 \times 60}$$

$= 533.333$ N

$$T = \dot{m}_p \times v_e$$

$$v_e = \frac{533.333}{0.25} = 2133.33 \text{ m/s}$$

31. 1748.897N

Volume of propellant $= \frac{\pi}{y} d^2 \ell$

Mass flow rate of propellant, $(\dot{m}_p)$

$$= \frac{\text{volume} \times \text{density}}{\text{time}}$$

$$\dot{m}_p = \frac{\frac{\pi}{4} \times (0.32)^2 \times 1 \times 1750}{150}$$

$= 0.9383$ kg/s

$$I_{sp} = \frac{T}{\dot{m}_p \times g}$$

$$T = I_{sp} \times \dot{m}_p \times g$$

$= 190 \times 0.9383 \times 9.81$

$= 1748.897$ N

32. $m_i = 180 + 470 + 1170 + 150 + 20$

$= 1990$ kg

$m_f = 1990 - (m_f + m_o)$

$= 1990 - (470 + 1170)$

$= 350$ kg

$$\Delta v_b = v_e \ln \frac{m_i}{m_f}$$

$$= 3136 \times \ln\left(\frac{1990}{350}\right)$$

$= 5450.23$ m/s

$\simeq 5.45$ Km/s

33. For under expanded nozzle, the exit pressure of nozzle is greater than the ambient pressure while in overexpanded nozzle, the exit pressure is smaller than the ambient pressure.

Nozzle of rocket engine is designed for some intermediate back pressure. With increase in altitude the ambient pressure reduces, it means $P_e < P_a$ at lower altitude and $P_e > P_a$ at higher altitude. So rocket nozzle operates as under expanded at high altitude and overexpanded at low altitude.

34. Thrust equation of rocket is

$$T = \dot{m} v_e + (P_e - P_a) A_e$$

If rocket is operating at design altitude, then $P_e = P_a$ and their is only momentum thrust produced by the rocket

In vacuum, $P_a = 0$, so pressure thrust $= P_e A_e$ will be add up with the momentum thrust.

35. 312.657 sec

Mass flow rate of propellant from main rocket engine,

$$\dot{m}_{p1} = \frac{f}{I_{sp} \times g}$$

$$= \frac{6300 \times 10^3}{428 \times 9.81} = 1500.4716 \text{ kg/s}$$

mass flow rate of propellant from one strap-on motor,

$$\dot{m}_{p2} = \frac{1200 \times 10^3}{292 \times 9.81} = 4189.1835 \text{ kg/s}$$

Total mass flow rate of propellant

$$= \dot{m}_{p1} + 2 \times \dot{m}_{p2}$$

$= 1500.4716 + 2 \times 4189.1835$

$= 9878.8386$ kg/s

Total thrust $= 6300 + 2 \times 12000$

$= 30300$ kN

$$\therefore \text{ Combined } I_{sp} = \frac{f_{net}}{\dot{m}_{pnet} \times g}$$

$$= \frac{30300 \times 10^3}{9878.8386 \times 9.81} = 312.657 \text{ s}$$

36. 7699.35 m/s

$h_o = C_p T_o = 5200 \times 6000$

$$K.E. = \frac{95}{100} \times 5200 \times 6000$$

$= 29640$ kJ/kg

for m = 1 kg

$$\frac{1}{2}v^2 = 29640$$

$$v = \sqrt{2 \times 29640 \times 1000}$$

$= 7699.35$ m/s

37. $$\Delta v_b = gI_{sp} \ln\left(\frac{m_i}{m_f}\right)$$

$m_f = m_i - m_p$.

$= 10000 - 8400 = 1600$ kg

$$\Delta v_b = 9.81 \times 240 \times \ln\left(\frac{10,000}{1600}\right)$$

$= 4314.63$ m/s

38. 9174.5 kg/s

$$\Delta v_b = gI_{sp} \ln\left(\frac{m_i}{m_i - m_p}\right)$$

$$v_f - v_i = gI_{sp} \ln\left(\frac{m_i}{m_i - m_p}\right)$$

$$\frac{m_i}{m_i - m_p} = e^{\frac{V_f}{gI_{sp}}}$$

$$m_p = m_i - \frac{m_i}{e^{\frac{V_f}{gI_{sp}}}}$$

$$= 10000 - \frac{10000}{e^{\frac{11 \times 1000}{9.8 \times 450}}}$$

$m_p = 9174.5$ kg/s

39. $I_{sp} \propto$ Thrust

and $I_{sp} \times \frac{1}{\dot{m}_p}$

The thrust produced by the SCRAMJET Engine is very high and mass low rate of burned gases from the nozzle of SCRAM jet engine is lower due to lower density of air compared to other given engine oxidizer.

40. mass flow rate of propellant

$$(\dot{m}_p) = \frac{m_i - m_f}{t}$$

$$\dot{m}_p = \frac{150 - 50}{10} = 10 \text{ kg/s}$$

$$I_{sp} = \frac{f}{\dot{m}_p \times g}$$

$$= \frac{19.62 \times 1000}{10 \times 9.81}$$

$I_{sp} = 200$ sec

41. $F = mv_e + (P_e - P_a)A_e$

$$\frac{F}{mVe} = 1$$

42. $I_{sp} = 3000$ Ns/kg

$M_f = 1000$ kg

$M_o = 1000 + 720 = 1720$ kg

Change in vehicle velocity (Δu)

$$= I_{sp} \ln\left(\frac{M_i}{M_f}\right)$$

$$= 3000 \ln\left(\frac{1720}{1000}\right)$$

$= 1.63 \times 10^3$m/sec

$= 1.63$km/sec

$$\boxed{\Delta u = 1.63 \text{ km/sec}}$$

43. A hypergolic propellant combination used in a rocket engine is one whose components spontaneously ignite when they come into contact with each other.

The two propellant components usually consist of a fuel and an oxidizer. Although commonly used hypergolic propellants are difficult to handle because of their extreme toxicity and/or corrosiveness, they can be stored as liquids at room temperature and hypergolic engines are easy to ignite reliably and repeatedly.

In contemporary usage, the terms "hypergol" or "hypergolic propellant" usually mean the most common such propellant combination, dinitrogen tetroxide plus hydrazine and/or its relatives monomethylhydrazine and unsymmetrical demethylhydrazine.

44. Given,

$I_{Sp} = 210$ sec

$A_t = 0.005$ m^2

$C^* = 1900$ m/s

$P_C = 50$ bar $= 50 \times 10^5$ bar

$g = 9.8$ m/s^1

We have

$$I_{Sp} = \frac{T}{\dot{m}g}$$

$$T = m\,gI_{sp} \quad ...(i)$$

and expression for m is given by

$$m = \frac{P_o A_t}{C^*} \quad ...(ii)$$

Putting value of m in equation (i) we get,

$$T = \frac{P_o A_t I_{sp}\, g}{C^*}$$

$$= \frac{50 \times 10^5 \times 0.005 \times 210 \times 9.8}{1900}$$

$= 27.1$ kN

45. Given, T = 25 MN, t = 150 Sec, $I_{sp} = 2980$ Ns/kg

I_{sp} = Thrust/mass flow rate of propellant

$$\frac{T}{\dot{m}_p}$$

$$\dot{m}_p = \frac{25 \times 10^6}{2980} = 8389.2617 \text{ kg/s}$$

Total mass of propellant burned in 150 seconds

$= \dot{m}_p \times t = 8389.2617 \times 150 = 1258389.255$ kg

46. 218.16

$$R = \frac{\bar{R}}{M} = \frac{8314}{22} = 377.909 \text{ KJ/kg}$$

$$\frac{T_{OC}}{T^*} = \frac{\gamma+1}{2} = \frac{1.32+1}{2}$$

$T^* = 1810.345$ K

$$\frac{P_{OC}}{P^*} = \frac{\gamma+1^{\frac{\gamma}{\gamma-1}}}{2}$$

$P^* = 1.367$ MPa

$$\rho^* = \frac{P^*}{RT^*} = 1.9981 \text{ kg/m}^3$$

$$a^* = \sqrt{\gamma RT^*} = 950.301 \text{ m/s}$$

$$\dot{m}_p = \rho^* A^* a^* = 1.9981 \times 0.1 \times 950.301$$

$= 189.879$ kg/s

$$\frac{P_{OC}}{P_e}\left[1 + \frac{(\gamma-1)M_e^2}{2}\right]^{\frac{\gamma}{\gamma-1}}$$

$$\frac{14}{0.1}\left[1 + \frac{(1.32-1)M_e^2}{2}\right]^{\frac{1.32}{1.32-1}}$$

$M_e = 3.802$

$$\frac{T_{OC}}{T_e}\left[1 + \frac{(\gamma-1)M_e^2}{2}\right]$$

$T_e = 633.899$ K

$$V_e = M_e\sqrt{\gamma RT_e} = 2137.975 \text{ m/s}$$

$$T = \dot{m}_p V_e = 405956.555 \text{ N}$$

$$I_{sp} = \frac{T}{\dot{m}_p g} = 218.16 \text{ sec}$$

8 CHAPTER GENERAL APTITUDE

1. Which of the following options is the closest in meaning to the word below:

   Circuitous

   (*a*) cyclic (*b*) indirect
   (*c*) confusing (*d*) crooked

   [GATE 2010, 1 M]

2. The question below consists of a pair of related words followed by four pairs of words. Select the pair that best expresses the relation in the original pair.

   (*a*) fallow : land (*b*) unaware : sleeper
   (*c*) wit : jester (*d*) renovated : house

   [GATE 2010, 1 M]

3. Choose the most appropriate word from the options given below to complete the following sentence:

   If we manage to ______ our natural resources, we would leave a better planet for our children.

   (*a*) uphold (*b*) restrain
   (*c*) cherish (*d*) conserve

   [GATE 2010, 1 M]

4. Choose the most appropriate word from the options given below to complete the following sentence:

   His rather causal remarks on politics ________ his lack of seriousness about the subject.

   (*a*) masked (*b*) belied
   (*c*) betrayed (*d*) suppressed

   [GATE 2010, 1 M]

5. 25 persons are in a room. 15 of them play hockey. 17 of them play football and 10 of them play both hockey and football. Then the number of persons playing neither hockey nor football is:

   (*a*) 2 (*b*) 17
   (*c*) 13 (*d*) 3

   [GATE 2010, 1 M]

6. Modern warfare has changed from large scale clashes of armies to suppression of civilian populations. Chemical agents that do their work silently appear to be suited to such warfare; and regretfully, there exist people in military establisbmenls who think that chemical agents are useful tools for their cause.

   Which of the following statements best sums up the meaning of the above passage:

   (*a*) Modern warfare has resulted in civil strife.
   (*b*) Chemical agents are useful in modem warfare.
   (*c*) Use of chemical agents in warfare would be undesirable.
   (*d*) People in military establishments like to use chemical agents in war.

   [GATE 2010, 2 M]

7. If 137 + 276 = 435 how much is 731 + 672?

   (*a*) 534 (*b*) 1403
   (*c*) 1623 (*d*) 1513

   [GATE 2010, 2 M]

8. 5 skilled workers can build a wall in 20 days; 8 semi-skilled workers can build a wall in 25 days; 10 unskilled workers can build a wall in 30 days. If a team has 2 skilled. 6 semi-skilled and unskilled workers. how long will it take to build the wall?

   (*a*) 20 days (*b*) 18 days
   (*c*) 16 days (*d*) 15 days

   [GATE 2010, 2 M]

9. Given digits 2, 2, 3, 3, 3, 4, 4, 4, 4 how many distinct 4 digit numbers greater than 3000 can be fonned?

   (*a*) 50 (*b*) 51
   (*c*) 52 (*d*) 54

   [GATE 2010, 2 M]

10. Hari (H), Gita (G), Irfan (I) and Saira (S) are siblings (i.e brolhers and sisters). All were born on I[st] January. The age difference between any two successive siblings (that is born one after another) is less than 3 years. Given the following facts:

    i. Hari's age + Gita's age > lrfan's age + Saira's age.
    ii. The age difference between Gita and Saira is 1 year. However, Gita is not the oldest and Saira is not the youngest.
    iii. There are no twins.

    In what order were they born (oldest first)?

    (*a*) HSIG (*b*) SGHI
    (*c*) IGSH (*d*) IHSG

    [GATE 2010, 2 M]

11. Choose the word from the options given below that is most nearly opposite in meaning to the given word:

Deference

(*a*) aversion (*b*) resignation
(*c*) suspicion (*d*) contempt

[GATE 2011, 1 M]

12. Choose the most appropriate word(s) from the options given below to complete the following sentence.

We lost confidence in him because he never ________ the grandiose promises he had made.

Deference

(*a*) delivered (*b*) delivered on
(*c*) forgot (*d*) reneged on

[GATE 2011, 1 M]

13. Choose the word or phrase that best completes the sentence below.

__________ in the frozen wastes of Arctic takes special equipment.

(*a*) To survive (*b*) Surviving
(*c*) Survival (*d*) That survival

[GATE 2011, 1 M]

14. In how many ways 3 scholarships can be awarded to 4 applicants, when each applicant can receive any number of scholarships?

(*a*) 4 (*b*) 12
(*c*) 64 (*d*) 81

[GATE 2011, 1 M]

15. Choose the most appropriate word from the options given below to complete the following sentence.

The ________ of eviden ce was on the side of the plaintiff since all but one witness testified that his story was correct.

(*a*) paucity (*b*) propensity
(*c*) preponderance (*d*) accuracy

[GATE 2011, 1 M]

16. If $\frac{(2y+1)}{(y+2)} < 1$, then which of the following alternatives gives the CORRECT range of y?

(*a*) $-2 < y < 2$ (*b*) $-2 < y < 1$
(*c*) $-3 < y < 1$ (*d*) $-4 < y < 1$

[GATE 2011, 2 M]

17. A student attempted to solve a quadratic equation in x twice. However, in the first attempt, he incorrectly wrote the constant term and ended up with the roots as (4, 3). In the second attempt, he incorrectly wrote down the coefficient of x and got the roots as (3, 2). Based on the above information, the roots of the correct quadratic equation are

(*a*) (-3, 4) (*b*) (3, -4)
(*c*) (6, 1) (*d*) (4, 2)

[GATE 2011, 2 M]

18. L, M and N are waiting in a queue meant for children to enter the zoo. There are 5 children between L and M, and 8 children between M and N. If there are 3 children ahead of N and 21 children behind L, then what is the minimum number of children in the queue?

(*a*) 28 (*b*) 27
(*c*) 41 (*d*) 40

[GATE 2011, 2 M]

19. Four archers P, Q, R and S try to hit a bull's eye during a tournament consisting of seven rounds. As illustrated in the figure below, a player receives 10 points for hitting the bulls' eye, 5 points for hitting within the inner circle and 1 point for hitting within the outer circle.

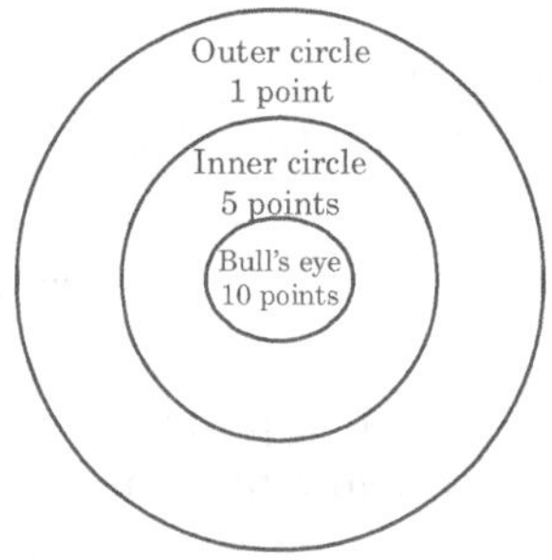

The final scores received by the players during the tournament are listed in the table below.

Round	P	Q	R	S
1	1	5	1	10
2	5	10	10	1
3	1	1	1	5
4	10	10	1	1
5	1	5	5	10
6	10	5	1	1
7	5	10	1	1

The most accurate and the most consistent players during the tournament are respectively

(*a*) P and S (*b*) Q and R
(*c*) Q and Q (*d*) R and Q

[GATE 2011, 2 M]

20. Nimbus clouds are dark and ragged, stratus clouds appear dull in colour and cover the entire sky. Cirrus clouds are thin and delicate, whereas cumulus clouds look like cotton balls.

It can be inferred from the passage that

(*a*) A cumulus cloud on the ground is called fog

(*b*) It is easy to predict the weather by studying clouds

(*c*) Clouds are generally of very different shapes, sizes and mass

(*d*) There are four basic cloud types: stratus, nimbus, cumulus and cirrus

[GATE 2011, 2 M]

21. Choose the most appropriate alternative from the options given below to complete the following sentence:

I ___ to have bought a diamond ring.

(*a*) have a liking (*b*) should have liked

(*c*) would like (*d*) may like

[GATE 2012, 1 M]

22. Choose the most appropriate alternative from the options given below to complete the following sentence:

Food prices ___ again this month.

(*a*) have raised (*b*) have been raising

(*c*) have been rising (*d*) have arose

[GATE 2012, 1 M]

23. Choose the most appropriate alternative from the options given below to complete the following sentence:

The administrators went on to implement yet another unreasonable measure, arguing that the measures were already ___ and one more would hardly make a difference.

(*a*) reflective (*b*) utopian

(*c*) luxuriant (*d*) unpopular

[GATE 2012, 1 M]

24. Choose the most appropriate alternative from the options given below to complete the following sentence:

To those of us who had always thought him timid, his ___ came as a surprise.

(*a*) intrepidity (*b*) inevitability

(*c*) inability (*d*) inertness

[GATE 2012, 1 M]

25. The arithmetic mean of five different natural numbers is 12. The largest possible value among the numbers is

(*a*) 12 (*b*) 40

(*c*) 50 (*d*) 60

[GATE 2012, 1 M]

26. Two policemen, A and B, fire once each at the same time at an escaping convict. The probability that A hits the convict is three times the probability that B hits the convict. If the probability of the convict not getting injured is 0.5, the probability that B hits the convict is

(*a*) 0.14 (*b*) 0.22

(*c*) 0.33 (*d*) 0.40

[GATE 2012, 2 M]

27. The total runs scored by four cricketers P, Q, R, and S in years 2009 and 2010 are given in the following table:

Player	2009	2010
P	802	1008
Q	765	912
R	429	619
S	501	701

The player with the lowest percentage increase in total runs is

(*a*) P (*b*) Q

(*c*) R (*d*) S

[GATE 2012, 2 M]

28. If a prime number on division by 4 gives a remainder of 1, then that number can be expressed as

(*a*) sum of squares of two natural numbers

(*b*) sum of cubes of two natural numbers

(*c*) sum of square roots of two natural numbers

(*d*) sum of cube roots of two natural numbers

[GATE 2012, 2 M]

29. Two points (4, p) and (0, q) lie on a straight line having a slope of $\frac{3}{4}$. The value of (p – q) is

(*a*) –3 (*b*) 0

(*c*) 3 (*d*) 4

[GATE 2012, 2 M]

30. In the early nineteenth century, theories of social evolution were inspired less by Biology than by the conviction of social scientists that there was a growing improvement in social institutions. Progress was taken for granted and social scientists attempted to discover its laws and phases.

Which one of the following inferences may be drawn with the greatest accuracy from the above passage?

Social scientists

(*a*) did not question that progress was a fact.

(*b*) did not approve of Biology.

(*c*) framed the laws of progress.

(*d*) emphasized Biology over Social Sciences.

[GATE 2012, 2 M]

31. If $3 \le X \le 5$ and $8 \le Y \le 11$ then which of the following options is TRUE?

(*a*) $\frac{3}{5} \le \frac{X}{Y} \le \frac{8}{5}$

(*b*) $\frac{3}{11} \le \frac{X}{Y} \le \frac{5}{8}$

(*c*) $\frac{3}{11} \le \frac{X}{Y} \le \frac{8}{5}$

(*d*) $\frac{3}{11} \le \frac{X}{Y} \le \frac{8}{11}$

[GATE 2013, 1 M]

32. The Headmaster __________ to speak to you.

Which of the following options is incorrect to complete the above sentence?

(*a*) is wanting (*b*) wants

(*c*) want (*d*) was wanting

[GATE 2013, 1 M]

33. Mahatama Gandhi was known for his humility as

(*a*) he played an important role in humiliating exit of British from India.

(*b*) he worked for humanitarian causes.

(*c*) he displayed modesty in his interactions.

(*d*) he was a fine human being.

[GATE 2013, 1 M]

34. All engineering students should learn mechanics,
I II
mathematics and how to do computation.
III IV

Which of the above underlined parts of the sentence is not appropriate?

(*a*) I (*b*) II

(*c*) III (*d*) IV

[GATE 2013, 1 M]

35. Select the pair that best expresses a relationship similar to that expressed in the pair:

water: pipe::

(*a*) cart: road

(*b*) electricity: wire

(*c*) sea: beach

(*d*) music: instrument

[GATE 2013, 1 M]

36. Velocity of an object fired directly in upward direction is given by $V = 80 - 32\,t$, where t (time) is in seconds. When will the velocity be between 32 m/sec and 64 m/sec

(*a*) $\left(1, \frac{3}{2}\right)$ (*b*) $\left(\frac{1}{2}, 1\right)$

(*c*) $\left(\frac{1}{2}, \frac{3}{2}\right)$ (*d*) $(1, 2)$

[GATE 2013, 2 M]

37. In a factory, two machines M1 and M2 manufacture 60% and 40% of the auto-components respectively. Out of the total production, 2% of M1 and 3% of M2 are found to be defective. If a randomly drawn autocomponent from the combined lot is found defective, what is the probability that it was manufactured by M2?

(*a*) 0.35 (*b*) 0.45

(*c*) 0.5 (*d*) 0.4

[GATE 2013, 2 M]

38. Following table gives data on tourists from different countries visiting India in the year 2011.

Country	Number of Tourists
USA	2000
England	3500
Germany	1200
Italy	1100
Japan	2400
Australia	2300
France	1000

Which two countries contributed to the one third of the total number of tourists who visited India in 2011?

(*a*) USA and Japan (*b*) USA and Australia

(*c*) England and France (*d*) Japan and Australia

[GATE 2013, 2 M]

39. If $|-2X + 9| = 3$ then the possible value of $|-X| - X^2$ would be:

(*a*) 30 (*b*) –30

(*c*) –42 (*d*) 42

[GATE 2013, 2 M]

40. All professors are researchers

Some scientists are professors

Which of the given conclusions is logically valid and is inferred from the above arguments:

(*a*) All scientists are researchers

(*b*) All professors are scientists

(*c*) Some researchers are scientists

(*d*) No conclusion follows **[GATE 2013, 2 M]**

41. A student is required to demonstrate a high level of comprehension of the subject, especially in the social sciences. The word closest in meaning to comprehension is

(*a*) understanding (*b*) meaning
(*c*) concentration (*d*) stability

[GATE 2014, 1 M]

42. Choose the most appropriate word from the options given below to complete the following sentence. One of his biggest ________ was his ability to forgive.

(*a*) vice (*b*) virtues
(*c*) choices (*d*) strength

[GATE 2014, 1 M]

43. Rajan was not happy that Sajan decided to do the project on his own. On observing his unhappiness, Sajan explained to Rajan that he preferred to work independently. Which one of the statements below is logically valid and can be inferred from the above sentences?

(*a*) Rajan has decided to work only in a group.
(*b*) Rajan and Sajan were formed into a group against their wishes.
(*c*) Sajan had decided to give in to Rajan's request to work with him.
(*d*) Rajan had believed that Sajan and he would be working together.

[GATE 2014, 1 M]

44. If $y = 5x^2 + 3$, then the tangent at $x = 0$, $y = 3$

(*a*) passes through $x = 0$, $y = 0$
(*b*) has a slope of +1
(*c*) is parallel to the x-axis
(*d*) has a slope of −1

[GATE 2014, 1 M]

45. A foundry has a fixed daily cost of Rs. 50,000 whenever it operates and a variable cost of Rs. 8000 Q, where Q is the daily production in tonnes. What is the cost of production in Rs per tonne for a daily production of 100 tonnes?

[GATE 2014, 1 M]

46. Find the odd one in the following group: ALRVX, EPVZB, ITZDF, OYEIK

(*a*) ALRVX (*b*) EPVZB
(*c*) ITZDF (*d*) OYEIK

[GATE 2014, 2 M]

47. Anuj, Bhola, Chandan, Dilip, Eswar and Faisal live on different floors in a six-storeyed building (the ground floor is numbered 1, the floor above it 2, and so on). Anuj lives on an even-numbered floor. Bhola does not live on an odd numbered floor. Chandan does not live on any of the floors below Faisal's floor. Dilip does not live on floor number 2. Eswar does not live on a floor immediately above or immediately below Bhola. Faisal lives three floors above Dilip. Which of the following floor-person combinations is correct?

	Anuj	Bhola	Chandan	Dilip	Eswar	Faisal
(a)	6	2	5	1	3	4
(b)	2	6	5	1	3	4
(c)	4	2	6	3	1	5
(d)	2	4	6	1	3	3

[GATE 2014, 2 M]

48. The smallest angle of a triangle is equal to two thirds of the smallest angle of a quadrilateral. The ratio between the angles of the quadrilateral is 3 : 4 : 5 : 6. The largest angle of the triangle is twice its smallest angle. What is the sum, in degrees, of the second largest angle of the triangle and the largest angle of the quadrilateral?

[GATE 2014, 2 M]

49. One percent of the people of country X are taller than 6 ft. Two percent of the people of country Y are taller than 6 ft. There are thrice as many people in country X as in country Y. Taking both countries together, what is the percentage of people taller than 6 ft.

(*a*) 3.0 (*b*) 2.5
(*c*) 1.5 (*d*) 1.25

[GATE 2014, 2 M]

50. The monthly rainfall chart based on 50 years of rainfall in Agra is shown in the following figure. Which of the following are true? (k percentile is the value such that k percent of the data fall below that value)

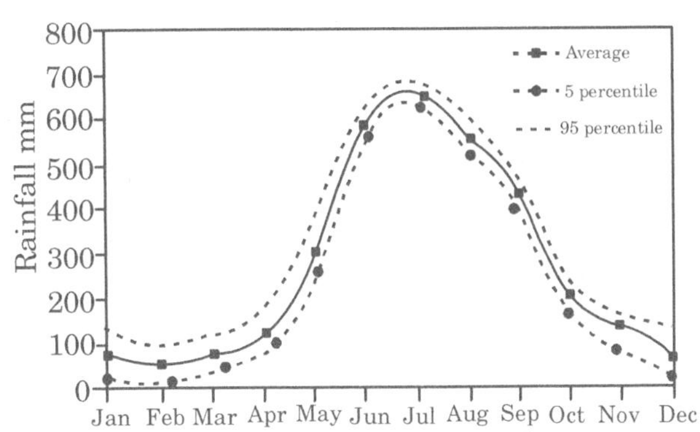

(i) On average, it rains more in July than in December
(ii) Every year, the amount of rainfall in August is more than that in January
(iii) July rainfall can be estimated with better confidence than February rainfall

(iv) In August, there is at least 500 mm of rainfall

(*a*) (i) and (ii) (*b*) (i) and (iii)

(*c*) (ii) and (iii) (*d*) (iii) and (iv)

[GATE 2014, 2 M]

51. Choose the most appropriate word'phrase, out of the four options given below, to complete the following sentence:

Apparent lifelessness _____ dormant life.

(*a*) harbours (*b*) leads to

(*c*) supports (*d*) affects

[GATE 2015, 1 M]

52. Fill in the blanks with correct idiom/phrase

That boy from the town was a _____ in the sleepy village.

(*a*) dog out of herd (*b*) sheep from the heap

(*c*) fish out of water (*d*) bird from the flock

[GATE 2015, 1 M]

53. Choose the statement where underlined word is used correctly:

(*a*) When the teacher eludes to different authors, he is being <u>elusive</u>.

(*b*) When the thief keeps eluding from the police, he is being <u>elusive</u>.

(*c*) Matters that are difficult to understand, identify or remember are <u>allusive</u>.

(*d*) Mirages can be <u>allusive</u>, but a better way to express them is illusory.

[GATE 2015, 1 M]

54. Tanya is older than Eric.

Cliff is older than Tanya.

Eric is older than Cliff.

If the first two statements are true, then the third statement is

(*a*) true (*b*) false

(*c*) uncertain (*d*) data insufficient

[GATE 2015, 1 M]

55. Five team have to compete in a league, with every team playing every other team exactly once, before going to the next round. How many matches will have to be held to complete the league round of matches.

(*a*) 20 (*b*) 10

(*c*) 8 (*d*) 5

[GATE 2015, 1 M]

56. Select the appropriate option in place of underlined part of sentence:

<u>Increased productivity necessary</u> reflects greater efforts made by the employees.

(*a*) Increase in productivity necessary

(*b*) Increase productivity is necessary

(*c*) Increase in productivity necessarily

(*d*) No improvement required.

[GATE 2015, 2 M]

57. Given below are two statements followed by two conclusions. Assuming these statements to be true, decide which one logically follows.

Statements:

I. No manager is a leader

II. All leaders are executives

Conclusions:

I. No manager is an executive

II. No executive is a manager

(*a*) Only conclusion I follows

(*b*) Only conclusion II follows

(*c*) Neither conclusion I nor II follows

(*d*) Both conclusions I and II follow

[GATE 2015, 2 M]

58. In the given figure angle Q is a right angle, PS:QS = 3 : 1, RT : QT = 5 : 2 and PU : UR = 1 : 1. If the area of triangle QTS is 20 cm^2, then the area of triangle PQR in cm^2 is _____.

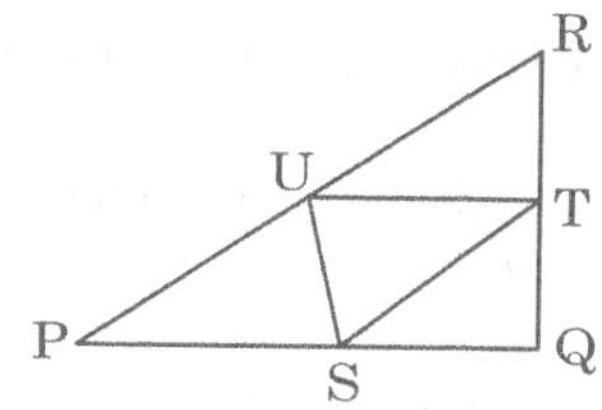

[GATE 2015, 2 M]

59. Right triangle PQR is to be constructed in the xy plane so that the right angle is at P and line PR is parallel to the x-axis. Then x and y coordinates of P, Q and R are to beintegers that satisfy the inequalities: $-4 \le x \le 5$ and $6 \le y \le 16$. How many triangles could be constructed with these properties?

(*a*) 110 (*b*) 1,100

(*c*) 9,900 (*d*) 10,000

[GATE 2015, 2 M]

60. A coin is tossed thrice. Let X be the event that head occurs in each of the first two tosses. Let Y be the event that a tail occurs on the third toss. Let Z be the event that two tails occur in three tosses. Based on the above information, which one of the following statements is true?

(*a*) X and Y are not independent

(*b*) Y and Z are independent

(*c*) Y and Z are not independent

(*d*) X and Z are independent

[GATE 2015, 2 M]

61. The chairman requested the aggrieved shareholders to ______________ him.

(*a*) bare with

(*b*) bore with

(*c*) bear with

(*d*) bare

[GATE 2016, 1 M]

62. Identify the correct spelling out of the given options:

(*a*) Managable

(*b*) Manageable

(*c*) Mangaeble

(*d*) Managible

[GATE 2016, 1 M]

63. Pick the odd one out in the following:
13, 23, 33, 43, 53

(*a*) 23 (*b*) 33

(*c*) 43 (*d*) 53

[GATE 2016, 1 M]

64. R2D2 is a robot. R2D2 can repair aeroplanes. No other robot can repair aeroplanes.

Which of the following can be logically inferred from the above statements?

(*a*) R2D2 is a robot which can only repair aeroplanes.

(*b*) R2D2 is the only robot which can repair aeroplanes.

(*c*) R2D2 is a robot which can repair only aeroplanes.

(*d*) Only R2D2 is a robot.

[GATE 2016, 1 M]

65. If $|9y - 6| = 3$, then $y^2 - \frac{4y}{3}$ is ________.

(*a*) 0 (*b*) $+\frac{1}{3}$

(*c*) $-\frac{1}{3}$ (*d*) undefined

[GATE 2016, 1 M]

66. The following graph represents the installed capacity for cement production (in tonnes) and the actual production (in tonnes) of nine cement plants of a cement company. Capacity utilization of a plant is defined as ratio of actual production of cement to installed capacity. A plant with installed capacity of at least 200 tonnes is called a large plant and a plant with lesser capacity is called a small plant. The difference between total production of large plants and small plants, in tonnes is ____.

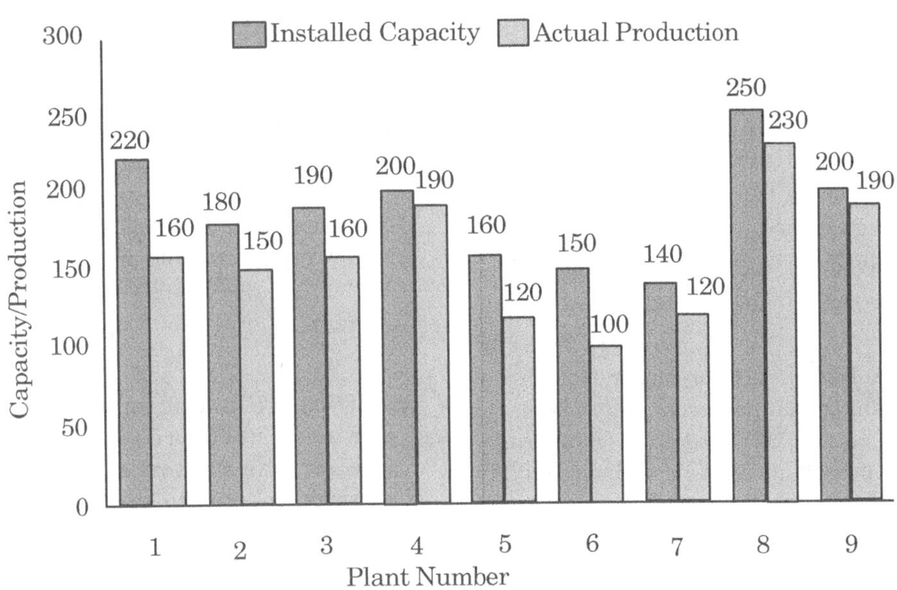

[GATE 2016, 2 M]

67. A poll of students appearing for masters in engineering indicated that 60 % of the students believed that mechanical engineering is a profession unsuitable for women. A research study on women with masters or higher degrees in mechanical engineering found that 99% of such women were successful in their professions.

Which of the following can be logically inferred from the above paragraph?

(*a*) Many students have misconceptions regarding various engineering disciplines.

(*b*) Men with advanced degrees in mechanical engineering believe women are well suited to be mechanical engineers.

(*c*) Mechanical engineering is a profession well suited for women with masters or higher degrees in mechanical engineering.

(*d*) The number of women pursuing higher degrees in mechanical engineering is small.

[GATE 2016, 2 M]

68. Sourya committee had proposed the establishment of Sourya Institutes of Technology (SITs) in line with Indian Institutes of Technology (IITs) to cater to the technological and industrial needs of a developing country.

Which of the following can be logically inferred from the above sentence?

Based on the proposal,

(i) In the initial years, SIT students will get degrees from IIT.

(ii) SITs will have a distinct national objective.

(iii) SIT like institutions can only be established in consultation with IIT.

(iv) SITs will serve technological needs of a developing country.

(*a*) (iii) and (iv) only (*b*) (i) and (iv) only

(*c*) (ii) and (iv) only (*d*) (ii) and (iii) only

[GATE 2016, 2 M]

69. Shaquille O' Neal is a 60% career free throw shooter, meaning that he successfully makes 60 free throws out of 100 attempts on average. What is the probability that he will successfully make exactly 6 free throws in 10 attempts?

(*a*) 0.2508 (*b*) 0.2816

(*c*) 0.2934 (*d*) 0.6000

[GATE 2016, 2 M]

70. The numeral in the units position of $211^{870} + 146^{127} \times 3^{424}$ is _____.

[GATE 2016, 2 M]

71. The ninth and the tenth of this month are Monday and Tuesday _____________.

(*a*) figuratively (*b*) restrospectively

(*c*) respectively (*d*) rightfully

[GATE 2017, 1 M]

72. It is _________ to read this years textbook ______ the last year's

(*a*) easier, than (*b*) most easy, than

(*c*) easier, from (*d*) easiest, from

[GATE 2017, 1 M]

73. A rule states that in order to drink beer, one must be over 18 years old. In a bar, there are 4 people. P is 16 years old, Q is 25 years old, R is drinking milkshake and S is drinking a beer. What must be checked to ensure that the rule is being followed?

(*a*) Only P's drink

(*b*) Only P's drink and S's age

(*c*) Only S's age

(*d*) Only P's drink. Q's drink and S's age

[GATE 2017, 1 M]

74. Fatima starts from point P, goes North for 3 km, and then East for 4 km to reach point Q. She then turns to face point P and goes 15 km in that direction. She then goes North for 6 km. How far is the from point P, and in which direction should she go to reach point P?

(*a*) 8 km. East (*b*) 12 km, North

(*c*) 6 km, East (*d*) 10 km, North

[GATE 2017, 1 M]

75. 500 students are taking one more course out of chemistry, physics, and mathematics. Registration records indicates course enrolment as follows: chemistry (329), physics (186), mathematics (63). How many students are taking all 3 subjects?

(*a*) 37 (*b*) 43

(*c*) 47 (*d*) 53

[GATE 2017, 1 M]

76. "If you are looking for a history of India, or for an account of the rise and fall of the British Raj, or for the reason of the cleaving of the subcontinent into two mutually antagonistic parts and the effects this mutilation will have in the respective sections, and ultimately on Asia, you will not find it in these pages: for though I have spent a lifetime in the country, I lived too near the seat of events, and was too intimately associated with the actors, to get the perspective needed for the impartial recording of these matters."

Which of the following statements best reflects the author's opinion?

(*a*) An intimate association does not allow for the necessary perspective.

(*b*) Matters are recorded with an impartial perspective.

(*c*) An intimate association offers an impartial perspective.

(*d*) Actors are typically associated with the impartial recording of matters.

[GATE 2017, 2 M]

77. Each of P, Q, R, S, W, X, Y and Z has been married at most once. X and Y are married and have two children P and Q. Z is the grandfather of the daughter S of P. Further, Z and W are married and are parents of R. Which one of the following must necessarily be FALSE?

(*a*) X is the mother-in-low of R

(*b*) P and R are not married to each other

(*c*) P is a son of X and Y

(*d*) Q cannot be married to R

[GATE 2017, 2 M]

78. 1200 men and 500 women can build a bridge in 2 weeks. 900 men and 250 women will take 3 weeks to build the same bridge. How many men will be needed to build the bridge in one week?

(*a*) 3000 (*b*) 3300

(*c*) 3600 (*d*) 3900

[GATE 2017, 2 M]

79. The number of 3-digit numbers such that the digit 1 is never to the immediate right of 2 is

(*a*) 781 (*b*) 791

(*c*) 881 (*d*) 891

[GATE 2017, 2 M]

80. A contonour line joins locations having the same height above the mean sea level. The following is a contour plot of a geographical region. Contour lines are shown at 25 m intervals in this plot.

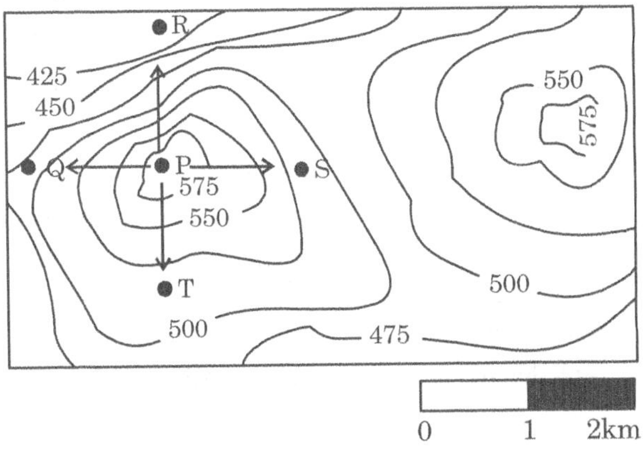

Which of the following is the steepest path leaving from P?

(*a*) P to Q (*b*) P to R

(*c*) P to S (*d*) P to T

[GATE 2017, 2 M]

81. "The dress ________ her so well that they all immediately ________ her on her appearance."

The words that best fill the blanks in the above sentence are

(*a*) complemented, complemented

(*b*) complimented, complemented

(*c*) complimented, complimented

(*d*) complemented, complimented

[GATE 2018, 1 M]

82. "The judge's standing in the legal community, though shaken by false allegations of wrongdoing, remained ________."

The word that best fills the blank in the above sentence is

(*a*) undiminished

(*b*) damaged

(*c*) illegal

(*d*) uncertain

[GATE 2018, 1 M]

83. Find the missing group of letters in the following series:

BC, FGH, LMNO, _____

(*a*) UVWXY (*b*) TUVWX

(*c*) STUVW (*d*) RSTUV

[GATE 2018, 1 M]

84. The perimeters of a circle, a square and an equilateral triangle are equal. Which one of the following statements is true?

(*a*) The circle has the largest area.

(*b*) The square has the largest area.

(*c*) The equilateral triangle has the largest area.

(*d*) All the three shapes have the same area.

[GATE 2018, 1 M]

85. The value of the expression $\frac{1}{1+\log_u vw}+\frac{1}{1+\log_v wu}+\frac{1}{1+\log_w uv}$ is ______.

(*a*) –1 (*b*) 0

(*c*) 1 (*d*) 3

[GATE 2018, 1 M]

86. Forty students watched films A, B and C over a week. Each student watched either only one film or all three. Thirteen students watched film A, sixteen students watched film B and nineteen students watched film C. How many students watched all three films?

(*a*) 0 (*b*) 2

(*c*) 4 (*d*) 8

[GATE 2018, 2 M]

87. A wire would enclose an area of 1936 m^2, if it is bent into a square. The wire is cut into two pieces. The longer piece is thrice as long as the shorter piece. The long and the short pieces are bent into a square and a circle, respectively. Which of the following choices is closest to the sum of the areas enclosed by the two pieces in square meters?

(*a*) 1096 (*b*) 1111

(*c*) 1243 (*d*) 2486

[GATE 2018, 2 M]

88. A contract is to be completed in 52 days and 125 identical robots were employed, each operational for 7 hours a day. After 39 days, five-seventh of the work was completed. How many additional robots would be required to complete the work on time, if each robot is now operational for 8 hours a day?

(*a*) 50 (*b*) 89

(*c*) 146 (*d*) 175

[GATE 2018, 2 M]

89. A house has a number which needs to be identified. The following three statements are given that can help in identifying the house number.

i. If the house number is a multiple of 3 , then it is a number from 50 to 59.

ii. If the house number is NOT a multiple of 4, then it is a number from 60 to 69.

iii. If the house number is NOT a multiple of 6, then it is a number from 70 to 79.

What is the house number?

(*a*) 54 (*b*) 65

(*c*) 66 (*d*) 76

[GATE 2018, 2 M]

90. An unbiased coin is tossed six times in a row and four different such trials are conducted. One trial implies six tosses of the coin. If H stands for head and T stands for tail, the following are the observations from the four trials:

(1) HTHTHT (2) TTHHHT

(3) HTTHHT (4) HHHT__ __.

Which statement describing the last two coin tosses of the fourth trial has the highest probability of being correct?

(*a*) Two T will occur.

(*b*) One H and one T will occur.

(*c*) Two H will occur.

(*d*) One H will be followed by one T.

[GATE 2018, 2 M]

91. The fishermen, _______ the flood victims owed their lives, were rewarded by the government.

(a) whom (b) to which

(c) to whom (d) that

[GATE 2019]

92. Some students were not involved in the strike.

If the above statement is true, which of the following conclusions is/are logically necessary?

1. Some who were involved in the strike were students.
2. No student was involved in the strike.
3. At least one student was involved in the strike.
4. Some who were not involved in the strike were students.

(a) 1 and 2 (b) 3

(c) 4 (d) 2 and 3

[GATE 2019]

93. The radius as well as the height of a circular cone increases by 10%. The percentage increase in its volume is ______.

(a) 17.1 (b) 21.0

(c) 33.1 (d) 72.8

[GATE 2019]

94. Five numbers 10, 7, 5, 4 and 2 are to be arranged in a sequence from left to right following the directions given below:

1. No two odd or even numbers are next to each other.
2. The second number from the left is exactly half of the left-most number.
3. The middle number is exactly twice the right-most number.

Which is the second number from the right?

(a) 2 (b) 4

(c) 7 (d) 10

[GATE 2019]

95. Until Iran came along, India had never been ______________ in kabaddi.

(a) defeated (b) defeating

(c) defeat (d) defeatist

[GATE 2019]

96. Since the last one year, after a 125 basis point reduction in repo rate by the Reserve Bank of India, banking institutions have been making a demand to reduce interest rates on small saving schemes. Finally, the government announced yesterday a reduction in interest rates on small saving schemes to bring them on par with fixed deposit interest rates.

Which one of the following statements can be inferred from the given passage?

(a) Whenever the Reserve Bank of India reduces the repo rate, the interest rates on small saving schemes are also reduced

(b) Interest rates on small saving schemes are always maintained on par with fixed deposit interest rates

(c) The government sometimes takes into consideration the demands of banking institutions before reducing the interest rates on small saving schemes

(d) A reduction in interest rates on small saving schemes follow only after a reduction in repo rate by the Reserve Bank of India

[GATE 2019]

97. In a country of 1400 million population, 70% own mobile phones. Among the mobile phone owners, only 294 million access the Internet. Among these Internet users, only half buy goods from e-commerce portals. What is the percentage of these buyers in the country?

(a) 10.50 (b) 14.70

(c) 15.00 (d) 50.00

[GATE 2019]

98. The nomenclature of Hindustani music has changed over the centuries. Since the medieval period dhrupad styles were identified as baanis. Terms like gayaki and baaj were used to refer to vocal and instrumental styles, respectively. With the institutionalization of music education the term gharana became acceptable. Gharana originally referred to hereditary musicians from a particular lineage, including disciples and grand disciples.

Which one of the following pairings is NOT correct?

(a) dhrupad, baani (b) gayaki, vocal

(c) baaj, institution (d) gharana, lineage

[GATE 2019]

99. Two trains started at 7AM from the same point. The first train travelled north at a speed of 80km/h and the second train travelled south at a speed of 100 km/h. The time at which they were 540 km apart is ______ AM.

(a) 9 (b) 10

(c) 11 (d) 11.30

[GATE 2019]

100. "I read somewhere that in ancient times the prestige of a kingdom depended upon the number of taxes that it was able to levy on its people. It was very much like the prestige of a head- hunter in his own community."

Based on the paragraph above, the prestige of a head-hunter depended upon __________

(a) the prestige of the kingdom

(b) the prestige of the heads

(c) the number of taxes he could levy

(d) the number of heads he could gather

[GATE 2019]

101. The untimely loss of life is a cause of serious global concern as thousands of people get killed _____ accidents every year while many other die ______ diseases like cardio vascular disease, cancer, etc.

(*a*) in,of (*b*) from, of

(*c*) during, from (*d*) from, from

[GATE 2020]

102. He was not only accused of theft______of conspiracy.

(*a*) rather (*b*) but also

(*c*) but even (*d*) rather than

[GATE 2020]

103. Select the word that fits the analogy :

Explicit : Implicit : : Express :_____

(*a*) Impress (*b*) Repress

(*c*) Compress (*d*) Suppress

[GATE 2020]

104. The Canadian constitution requires that equal importance be given to English and French. Last year, Air Canada lost a lawsuit, and had to pay a six-figure fine to a French-speaking couple after they filed complaints about formal in-flight announcements in English lasting 15 seconds, as opposed to informal 5 second messages in French.

The French-speaking couple were upset at _____.

(*a*) the in-flight announcements being made in English.

(*b*) the English announcements being clearer than the French ones.

(*c*) the English announcements being longer than the French ones.

(*d*) equal importance being given to English and French.

[GATE 2020]

105. A supperadditive function $f(\cdot)$ satisfies the following property

$$f(x_1 + x_2) \geq f(x_1) + f(x_2)$$

Which of the following functions is a supperadditive function for > 1?

(*a*) e^x (*b*) $\sqrt{x}$

(*c*) $1/x$ (*d*) e^{-x}

[GATE 2020]

106. The global financial crisis in 2008 is considered to be the most serious world -wide financial crisis, which started with the sub-prime lending crisis in USA in 2007. The sub-prime lending crisis led to banking crisis in 2008 with the collapse of Lehman Brothers in 2008. The sub-prime lending refers to the provision of loans to those borrowers who may have difficulties in repaying loans, and it arises because of excess liquidity following the east Asian crisis.

Which one of the following sequences show the correct precedence as per the given passage?

(*a*) East Asian crisis → subprime lending crisis → banking crisis → global financial crisis.

(*b*) Subprime lending crisis → global financial crisis → banking crisis → East Asian crisis.

(*c*) Banking crisis → subprime lending crisis → global financial crisis → East Asian crisis..

(*d*) Global financial crisis → East Asian crisis → banking crisis → subprime lending crisis.

[GATE 2020]

107. It is a quarter past three in your watch. The angle between the hour hand and the minute hand is

(*a*) 0° (*b*) 7.5°

(*c*) 15° (*d*) 22.5°

[GATE 2020]

108. A circle with centre O is shown in figure. A rectangle PQRS of maximum possible area in inscribed in the circle. If the radius of the circle is *a*, then the area of the shaded portion is______.

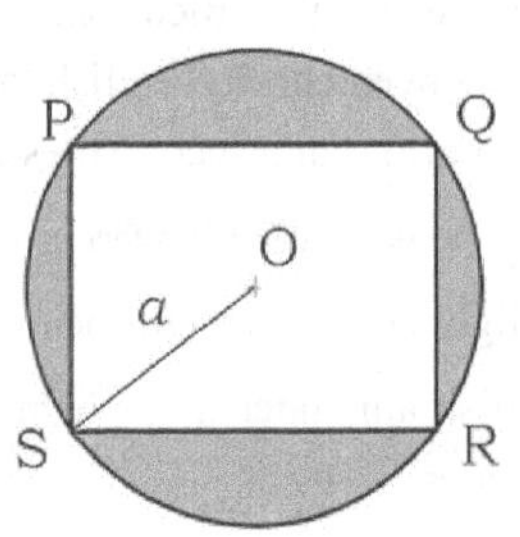

(*a*) $\pi a^2 - a^2$ (*b*) $\pi a^2 - \sqrt{2}a^2$

(*c*) $\pi a^2 - 2a^2$ (*d*) $\pi a^2 - 3a^2$

[GATE 2020]

109. a, b, c are the real numbers. The quadratic equation $ax^2 - bx + c = 0$ has equal roots, which is β, then

(*a*) $\beta = b/a$ (*b*) $\beta^2 = ac$

(*c*) $\beta^3 = bc/(2a^2)$ (*d*) $b^2 \neq 4ac$

[GATE 2020]

110. The following figure shows the data of students enrolled in 5 years (2014 to 2018) for two schools P and Q. During this period, the ratio of the average number of the students enrolled in school P to the average of the difference of the number of students enrolled in schools P and Q is _____.

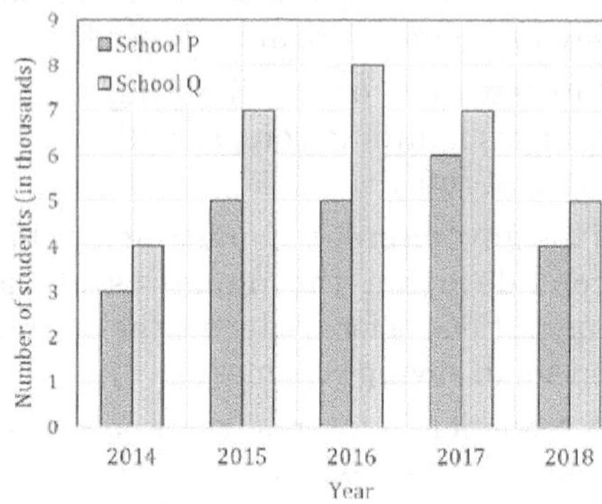

(*a*) 8 : 23 (*b*) 23 : 8

(*c*) 23 : 31 (*d*) 31 : 23

[GATE 2020]

111. (i) Arun and Aparna are here.
(ii) Arun and Aparna is here.
(iii) Arun's families is here.
(iv) Arun's family is here.
Which of the above sentences are grammatically CORRECT?

(a) (i) and (ii) (b) (i) and (iv)

(c) (ii) and (iv) (d) (iii) and (iv)

[GATE 2021]

112.

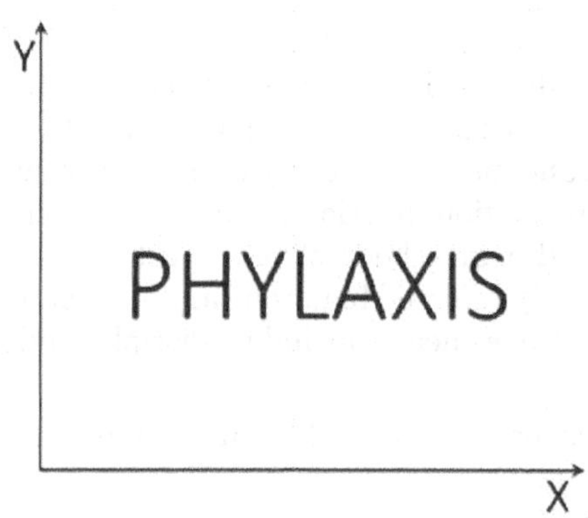

The mirror image of the above text about the x-axis is

(a)  (b) 

(c) 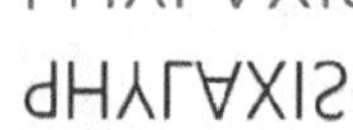 (d) 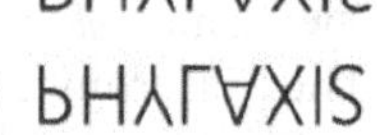

[GATE 2021]

113. Two identical cube shaped dice each with faces numbered 1 to 6 are rolled simultaneously. The probability that an even number is rolled out on each dice is:

(a) $\frac{1}{36}$ (b) $\frac{1}{12}$

(c) $\frac{1}{8}$ (d) $\frac{1}{4}$

[GATE 2021]

114. $\oplus$ and $\odot$ are two operators on numbers p and q such that $p \odot q = p - q$, and $p \oplus q = p \times q$

Then, $(9 \odot (6 \oplus 7)) \odot (7 \oplus (6 \odot 5)) =$

(a) 40 (b) – 26

(c) – 33 (d) – 40

[GATE 2021]

115. Four persons P, Q, R and S are to be seated in a row. R should not be seated at the second position from the left end of the row. The number of distinct seating arrangements possible is:

(a) 6 (b) 9

(c) 18 (d) 24

[GATE 2021]

116. On a planar field, you travelled 3 units East from a point O. Next you travelled 4 units South to arrive at point P. Then you travelled from P in the North-East direction such that you arrive at a point that is 6 units East of point O. Next, you travelled in the North-West direction, so that you arrive at point Q that is 8 units North of point P.

The distance of point Q to point O, in the same units, should be ______

(a) 3 (b) 4

(c) 5 (d) 6

[GATE 2021]

117. The author said, "Musicians rehearse before their concerts. Actors rehearse their roles before the opening of a new play. On the other hand, I find it strange that many public speakers think they can just walk on to the stage and start speaking. In my opinion, it is no less important for public speakers to rehearse their talks."

Based on the above passage, which one of the following is TRUE?

(a) The author is of the opinion that rehearsing is important for musicians, actors and public speakers.

(b) The author is of the opinion that rehearsing is less important for public speakers than for musicians and actors.

(c) The author is of the opinion that rehearsing is more important only for musicians than public speakers.

(d) The author is of the opinion that rehearsal is more important for actors than musicians.

[GATE 2021]

118. 1. Some football players play cricket.
2. All cricket players play hockey.

Among the options given below, the statement that logically follows from the two statements 1 and 2 above, is:

(a) No football player plays hockey.

(b) Some football players play hockey.

(c) All football players play hockey.

(d) All hockey players play football.

[GATE 2021]

119.

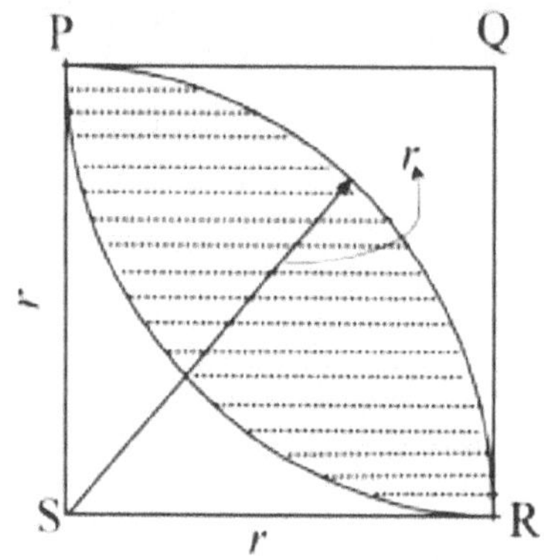

In the figure shown above, PQRS is a square. The shaded portion is formed by the intersection of sectors of circles with radius equal to the side of the square and centers at S and Q.

The probability that any point picked randomly within the square falls in the shaded area is ______

(a) $4 - \frac{\pi}{2}$ (b) $\frac{1}{2}$

(c) $\frac{\pi}{2} - 1$ (d) $\frac{\pi}{4}$

[GATE 2021]

120. In an equilateral triangle PQR, side PQ is divided into four equal parts, side QR is divided into six equal parts and side PR is divided into eight equal parts. The length of each subdivided part in cm is an integer. The minimum area of the triangle PQR possible, in cm^2, is

(a) 18 (b) 24

(c) $48\sqrt{3}$ (d) $144\sqrt{3}$

[GATE 2021]

ANSWER KEY

1. (*b*)	**2.** (*a*)	**3.** (*d*)	**4.** (*c*)	**5.** (*d*)	**6.** (*c*)	**7.** (*c*)	**8.** (*d*)	**9.** (*b*)	**10.** (*b*)
11. (*d*)	**12.** (*b*)	**13.** (*a*)	**14.** (*c*)	**15.** (*c*)	**16.** (*b*)	**17.** (*c*)	**18.** (*a*)	**19.** (*b*)	**20.** (*d*)
21. (*c*)	**22.** (*c*)	**23.** (*d*)	**24.** (*a*)	**25.** (*c*)	**26.** (*a*)	**27.** (*b*)	**28.** (*a*)	**29.** (*c*)	**30.** (*a*)
31. (*b*)	**32.** (*c*)	**33.** (*c*)	**34.** (*d*)	**35.** (*b*)	**36.** (*c*)	**37.** (*c*)	**38.** (*c*)	**39.** (*b*)	**40.** (*c*)
41. (*a*)	**42.** (*b*)	**43.** (*d*)	**44.** (*c*)	**45.** 1300	**46.** (*d*)	**47.** (*b*)	**48.** 180	**49.** (*d*)	**50.** (*b*)
51. (*a*)	**52.** (*c*)	**53.** (*b*)	**54.** (*b*)	**55.** (*b*)	**56.** (*c*)	**57.** (*c*)	**58.** 280	**59.** (*c*)	**60.** (*b*)
61. (*c*)	**62.** (*b*)	**63.** (*b*)	**64.** (*b*)	**65.** (*c*)	**66.** 120	**67.** (*c*)	**68.** (*c*)	**69.** (*a*)	**70.** 7
71. (*c*)	**72.** (*a*)	**73.** (*b*)	**74.** (*b*)	**75.** (*d*)	**76.** (*a*)	**77.** (*d*)	**78.** (*c*)	**79.** (*c*)	**80.** (*b*)
81. (*d*)	**82.** (*a*)	**83.** (*b*)	**84.** (*a*)	**85.** (*c*)	**86.** (*c*)	**87.** (*c*)	**89.** (*d*)	**90.** (*b*)	**91.** (c)
92. (c)	**93.** (c)	**94.** (c)	**95.** (a)	**96.** (c)	**97.** (a)	**98.** (c)	**99.** (b)	**100.** (d)	**101.** (*a*)
102. (*b*)	**103.** (*b*)	**104.** (*c*)	**105.** (*a*)	**106.** (*a*)	**107.** (*b*)	**108.** (*c*)	**109.** (*c*)	**110.** (*b*)	**111.** (b)
112. (b)	**113.** (d)	**114.** (d)	**115.** (c)	**116.** (c)	**117.** (a)	**118.** (b)	**119.** (c)	**120.** (d)	

EXPLANATIONS

1. Circuitous is marked by obliqueness or indirection in speech or conduct.

2. Fallow means left unploughed or unseeded during a growing season.

3. Conserve the resources.

4. Betrayed: reveal unintentionally.

5. No. of person in room = 25

Play hockey = 15

Play football =17

Play both = 10

Therefore, No. of persons playing neither hockey or football is

$25-[(17+15)-10]=3$

6. As per the passage the appropriate sentence is

Use of chemical agents in warfare would be undesirable.

7. These are base-eight numbers.

8^0 place: $1+2=3$

8^1 place: $3+7=10$ (base 10)

= 1 eight + 2 remainder base 8

= 12 (base 8)

8^2 place: carry the $1+7+6=14$ (base 10)

= 1 eight + 6 remainder

= 16 (base 8)

Therefore, $731+672=1623$ (base 8)

8. 5 skilled workers can build a wall in 20 days.

1 skilled worker can build a wall in $20\times5=100$ days

1 skilled worker can build in 1 day $=\frac{1}{100}$ wall

Similarly,

1 semi-skilled worker can build in 1 day $=\frac{1}{200}$ wall

1 unskilled worker can build in 1 day $=\frac{1}{300}$ wall

So, $2\times\frac{1}{100}+6\times\frac{1}{200}+5\times\frac{1}{300}$ is the portion of wall made in 1 day by 2 skilled, 6 semi-skilled and 5 unskilled workers which is $\left(\frac{1}{15}\text{ wall}\right)$.

Therefore, they will take 15 days together.

9. Case-1

1st digit is 3.

Rest three digits can be filled in 3×3×3 ways but excluding two numbers 3222 and 3333 because they exceed the limit of available options.

Therefore, $27-2=25$

Case-2

1st digit is 4.

Rest of the digits again can be filled in 3×3×3 ways but 4222 is not possible.

Thus, $27-1=26$

Therefore, total of $25+26=51$ numbers are possible.

10. There are no twins and age difference between Gita and Saira is 1 year, they must be born one after another. This deduction rules out option (A)

Option (C) and (D) are ruled out by condition 1 which says

$H + G > I + S$

Therefore, only the answer is SGHI

11. Deference means - polite submission and respect.

So, the near opposite to the given word is contempt which means - disrespect, disregard, slighting, neglect; contumacy

12. We lost confidence in him because he never delivered on the grandiose promises he had made.

(B) Delivered on is most appropriate.

(A) Delivered this is not correct because of "The" in next part. Does not sound appropriate.

(C) forgot: Does not fit properly

(D) reneged on: This does not seem appropriate, As he never go back on a promise then we should not loose confidence.

13. <u>To survive</u> in the frozen wastes of Arctic takes special equipment.

14. One scholarship can be awarded to 4 applicants in 4 ways. Three scholarship can be awarded to 4 applicants in $4^3 = 64$ ways.

All scholarship are independent, any student can get any no of scholarships between 0 and 3

15. The <u>preponderance</u> of evidence was on the side of the plaintiff since all but one witness testified that his story was correct.

Preponderance: the quality or fact of being greater in number, quantity, or importance.

"the preponderance of women among older people" C is correct option.

Here, plaintiff means a person who brings a case against another in a court of law.

"the plaintiff commenced an action for damages".

16. Given, $\frac{(2y+1)}{(y+2)} < 1$

$2y + 1 < y + 2$

$y < 1$ upper limit

Also $y = -2$ cannot be used as it makes it infinite.

Therefore, $-2 < y < 1$

17. If equation is $ax^2 + bx + c = 0$ and roots of equation are x_1 and x_2 respectively.

then, $x_1 + x_2 = \frac{-b}{a}$

and $x_1x_2 = \frac{c}{a}$

We find that, $x_1 + x_2 = 4 + 3 = 7$ in first attempt

c is incorrect as here $\frac{-b}{a}$ is correct

Again, $x_1x_2 = 3 \times 2 = 6$ in second attempt b is incorrect as here $\frac{c}{a}$ is correct.

Now solving above equation,

$x_2 = \frac{6}{x_1}$

Putting x_2 in $x_1 + x_2 = 7$

we get,

$x_1 + \frac{6}{x_1} = 7$

$\Rightarrow x_1^2 - 7x_1 + 6 = 0$

$\Rightarrow x_1 = 1$ or $x_1 = 6$

18.

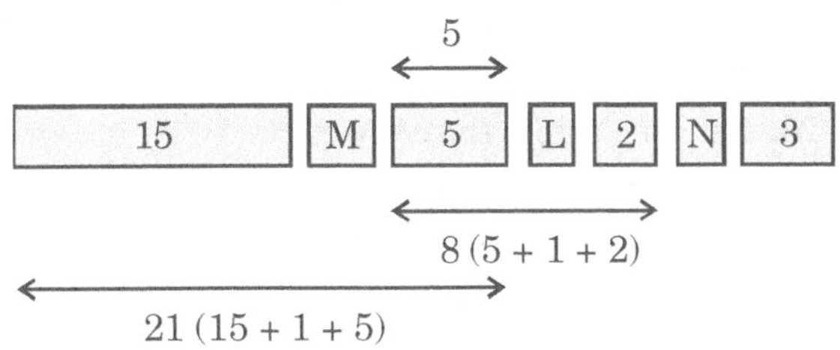

Total number of children = 21 + 1 + 2 + 1 + 3 = 28.

19. Given,

Player's receives 10 points for hitting the bull's eye.

Player's receives 5 points for hitting the within the inner circle.

Player's receives 1 points for hitting the within the outer circle.

P = 1 + 5 + 1 + 10 + 1 + 10 + 5 = 32

Q = 5 + 10 + 1 + 10 + 5 + 5 + 10 = 46

R = 1 + 10 + 1 + 1 + 5 + 5 + 10 = 20

S = 10 + 1 + 5 + 1 + 10 + 1 + 1 = 29

Thus, the analysis gives most accurate player is Q and most consistent player is R with most consistent hitting of the outer circle.

20. Option (*a*) is wrong because nowhere 'fog' is referenced to any type of cloud in paragraph. Option (*b*) is wrong because it is easy to predict whether from cloud but it is not referenced from given paragraph.

Option (*c*) is wrong because paragraph talks about the colour of all clouds but size of only 'stratus' cloud. So, whether size of all clouds are same or different, we can't conclude.

But one thing is sure that all clouds are different type because at least we can differentiate them based on their colours.

21. I would like to have bought a diamond ring.

22. Food price have been rising again this month.

23. The administrators went on to implement yet another unreasonable measure, arguing that the measures were already unpopular and one more would hardly make a difference.

24. To those of us who had always thought him timid, his intrepidity came as a surprise.

Explanation: Timid means fearful and intrepid means fearless which is the most suitable contrast.

25. It is 50 obviously.

Total sum of 5 natural numbers = $12 \times 5 = 60$

If 60 is one natural number, then the other 4 numbers must be 0. As 0 is not a natural number (it is a whole number), 60 is not right, so next option remaining is 50. Also in question they have told as different natural numbers. So 50, 1, 2, 3, 4 are the numbers.

26. Let X be A hits the convict

And Y be B hits the convict

Let, Z be convict is injured

Z' is convict is not injured

Given,

P(Z') = 0.5

$\Rightarrow P(Z) = 1 - P(Z') = 1 - 0.5 = 0.5$

Now,

$P(Z) = P(X) \times P(Y') + P(X') \times P(Y) + P(X) \times P(Y)$

Let P(Y) = t

P(X) = 3t

P(Y') = 1 – t

P(X') = 1 – 3t

Substituting in above equation,

$0.5 = (3t \times (1-t)) + ((1-3t) \times t) + (t \times 3t)$

$\Rightarrow 3t - 3t^2 + t - 3t^2 + 3t^2 = 0.5$

$\Rightarrow 3t^2 - 4t + 0.5 = 0$

$\Rightarrow 6t^2 - 8t + 1 = 0$

We get,

t = 1.193,0.1396

Therefore, $P(Y) = t = 0.1396 \approx 0.14$

27. Total number of runs

P = 1810

q = 1677

r = 1048

s = 1202

Now % increase is defined as

$$= \left(\frac{\text{final value} - \text{initial value}}{\text{initial value}}\right) \times 100$$

$$p = \frac{206}{802} \times 100 = 25.68\%$$

$$q = \frac{147}{756} \times 100 = 19.21$$

$$r = \frac{190}{429} \times 100 = 44.28$$

$$s = \frac{200}{501} \times 100 = 39.42$$

So, the lowest increase is for q.

28. For getting the prime number as per given question the format is

Prime Number = 4n + 1

Let us take a prime number from above expression

$4 \times 1 + 1 = 5$

$4 \times 3 + 1 = 13$

$4 \times 4 + 1 = 17$

Now as per question both the numbers can be given as

$1^2 + 2^2 = 1 + 4 = 5$

$2^2 + 3^2 = 4 + 9 = 13$

$1^2 + 4^2 = 1 + 16 = 17$

Thus, it proves that the option A satisfies the condition i.e. sum of squares of two natural numbers.

29. For two points (x_1, y_1) and (x_2, y_2) on a line, Slope of the line $= \frac{y_2 - y_1}{x_2 - x_1}$

So, $\frac{q-p}{0-4} = \frac{3}{4}$

$\therefore (p - q) = 3$

30. Option (*a*): Did not question that progress was a fact. True they took progress for granted.

Option (*b*): Did not approve of Biology: False. No mention regarding this in paragraph. Its mention that it was not inspired from biology and that is a different thing.

Option (*c*): Framed the laws of progress: False They attempted to discover its laws. Frame and discover are different things.

Option (*d*): Clearly false.

31. $\frac{X}{Y}$ is minimum for the given range when X is minimum possible and Y is maximum possible

i.e., $\frac{3}{11}$

$\frac{X}{Y}$ is maximum for the given range when X is maximum possible and Y is minimum possible

i.e., $\frac{5}{8}$

32. The Headmaster want to speak to you

Rule: Subjects and verbs must AGREE with one another in number (singular or plural). Thus, if a subject is singular, its verb must also be singular; if a subject is plural, its verb must also be plural.

Here headmaster is singular subject so we do not use plural verb 'want' with Headmaster.

33. He displayed modesty in his interactions.

Humility - a modest or low view of one's own importance; humbleness.

Synonyms: modesty, humbleness, meekness, diffidence, unassertiveness

34. All engineering students should learn mechanics, mathematics and computation.

or possibly

All engineering students should learn how to do mechanics, mathematics and computation.

35. Water runs through a pipe like electricity runs through a wire.

36. $32 < 80 - 32t < 64$

$-48 < -32t < -16$

$48 > 32t > 16$

$\frac{3}{2} > t > \frac{1}{2}$

37. Let P (Mi) denote the probability that the component is manufactured by machine Mi, and P (def) denote the probability that the component is defective.

$$P\left(M_{2_{def}}\right) = \frac{P\left(M_2 \cap def\right)}{p\left(def\right)}$$

$$= \frac{0.4 \times 0.03}{\left(0.4 \times 0.03\right) \times \left(0.6 \times 0.02\right)} = 0.5$$

38. Total number of tourists who visited India in 2011 = 13500

One third of total tourists =13500 ÷ 3 = 4500

OPTION A: USA and Japan = 4400

OPTION B: USA and Australia = 4300

OPTION D: Japan and Australia = 4700

OPTION C: England and France = 4500 which is one third of total tourists.

39. Given,

$|-2X + 9| = 3$

$\Rightarrow -2X + 9 = 3 \text{ or } -(-2X + 9) = 3$

$\Rightarrow X = 3 \text{ or } X = 6$

$\Rightarrow |-X| - X^2 = |-3| - 3^2 = -6$

$\text{or } |-X| - X^2 = |-6| - 6^2 = -30$

40. If some scientists are professors and all professors are researchers, then that means that some scientists are researchers.

41. Comprehension means understanding.

42. Virtues is related to moral standards while strength is related to physical ability.

43. Ooption (*a*) discarded: Sajan preferred to work independently

Option (*b*) discarded: No expectation so no unhappiness

Option (*c*) discarded: Sajan preferred to work independently

Option (*d*) correct: unhappy as he believed to be working together

44. Given, $y = 5x^2 + 3$

Slope $\frac{dy}{dx} = 10x$

Given point $x = 0, y = 3 : (0,3)$

Now, at (0, 3) we have the slope as

$\frac{dy}{dx} = 10 \times 0 = 0,$ which means line is parallel to X-axis.

Options (*b*) and (*d*) eliminated as slope is 0.

Option (a) is eliminated as given the tangent is at point Equationand we found slope = 0, which means it is parallel to X axis. The equation of line is nothing but y = 3, which never passes through (0,0). Hence (a) too eliminated.

Therefore, Option C is correct.

45. For a daily production of 100 tonnes, the daily cost of foundry would be

Rs 50,000 + (800 × 100) = Rs 130,000.

Hence cost of production in Rs per tonne

$\frac{130000}{100} = 1300$

46. Option (a), ALRVX - only one vowel

Option (b), EPVZB - only one vowel

Option (c), ITZDF - only one vowel

Option (d), OYEIK - three vowels

Therefore, option (d) is odd one

47. Option (a): Bhola's floor number –2 and Eswar's floor number –3

Option (b): Bhola's floor number –6 and Eswar's floor number –3

Option (c): Bhola's floor number –2 and Eswar's floor number –1

Option (d): Bhola's floor number – 4 and Eswar's floor number –3

Here, we can clearly see that in options (a), (c) and (d) the statement is not followed.

Therefore, the answer is option (b).

48. Clearly, the angles of quadrilateral are 60, 80, 100, and 120.

That makes smallest angle of triangle $60 \times \frac{2}{3} = 40$

That makes largest angle of triangle 40 × 2 = 80

So third angle of triangle = 60.

The required value is = 120 + 60 = 180

49. Let the number of people in Y = a

Number of people in Y greater than 6 feet = $\frac{2a}{100}$

Number of people in X = 3a

Number of people in X greater than 6 feet = $\frac{3a}{100}$

Total number of people greater than 6 feet

$= \frac{2a}{100} + \frac{3a}{100} = \frac{5a}{100}$

Total number of people = a + 3a = 4a

Percentage of people in X + Y greater than 6 feet

$= \left(\frac{5a}{100} \div 4a\right) \times 100 = 1.25\ \%$

50. First statement - true since average of July is more than that of December in the Average plot.

Second statement - not true since, we have data until 95% only -- in all data of 50 years, amount of rainfall is below the value presented by graph in 95% cases. Therefore, in the remaining 5% of years, there can be a year where it rains more in January than August.

Third statement - true, since difference between 5 percentile plot and 95?percentile plot indicates that this was range for 90% of years. In February, this range is bigger compared to July.

Fourth statement - not necessarily true. Here, the 5 percentile plot is touching 500 mm level. However, any year in this 5% could have got a lower level of rainfall.

51. Apparent lifelessness harbours dormant life.

Harbours- To give shelter.

52. The boy from the town was a fish out of water in the sleepy village.

Fish out of water is an idiom used to refer to a person who is in unfamiliar, and often uncomfortable, surroundings.

53. **Elusive** – difficult to find, catch or achieve.

Allusive – using or containing suggestion rather than explicit mention.

54. CLIFF<TANYA<ERIC

It clear form question, third statement is false.

55. Number of combination is given by

$${}^5C_2 = \frac{n(n-1)}{2} = \frac{5 \times 4}{2} = 10$$

56. Increase in productivity necessarily reflects greater efforts made by the employees.

57. Every leader is executive and no manager is leader so manager can be other than leader which is also executive so (I) is false and (II) is also false because some executive which is not leader can be manager. Therefore, neither conclusion I nor II follows.

58. Given,

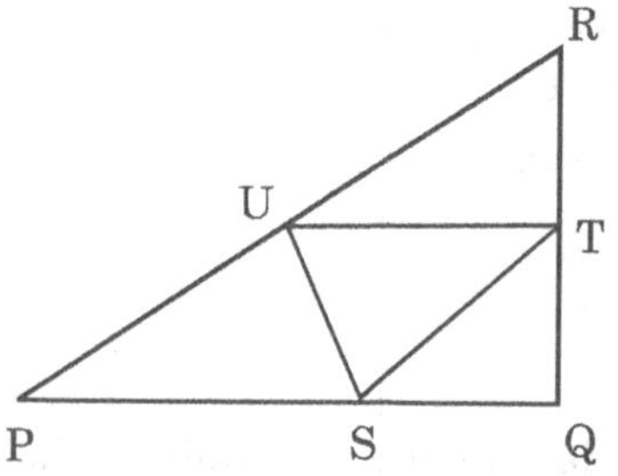

$\frac{RT}{QT} = \frac{5x}{2x}$

$\Rightarrow RQ = 7x$

$$\frac{PS}{QS} = \frac{3y}{1y}$$

$\Rightarrow PQ = 4y$

Area of the triangle QTS $= \frac{1}{2} \times QS \times QT$

$$= \frac{1}{2} \times 1y \times 2x = xy = 20 \text{ cm}^2$$

Area of triangle PQR $= \frac{1}{2} \times PQ \times RQ$

$$= \frac{1}{2} \times 4y \times 7x = 14xy = 14 \times 20 = 280 \text{ cm}^2$$

59. Since, PR is parallel to x-axis and P is right angled therefore, PQ is parallel to y-axis.

Total number of x-values available is 10

Total number of y-values available is 11

At point P

Coordinate of P can have any value of x and y, hence it can be chosen in = $10 \times 11 = 110$ ways

At point Q

PQ is parallel to y-axis, the value of x-coordinate is already fixed due to point P, hence Q can be chosen in = 10 ways.

At point R

PR is parallel to x-axis its y-coordinate is already fixed due to point P, hence point p can be chosen in = 9 ways (not 10 ways as x-coordinate is already occupied by P)

Total number of ways are given by

$= 110 \times 9 \times 10 = 9900.$

60. X and Y are independent if,

$$P(X \cap Y) = P(X)P(Y)$$

Sample space is $2 \times 2 \times 2 = 8$

Y = {HHT, TIT, HTT, THT}

$$P(Y) = \frac{4}{8}$$

Z = {HTT, THT, TTH}

$$P(Z) = \frac{3}{8}$$

$$Y \cap Z = \{HTT, THT\}$$

$$P(Y \cap Z) = \frac{2}{8}$$

To check if Y and Z are independent

$$P(Y)P(Z) = \frac{3}{16}$$

$$P(Y \cap Z) = \frac{2}{8}$$

Hence, we have

$$P(Y)P(Z) \neq P(Y \cap Z)$$

$\therefore$ Y and Z are dependent

61. The chairman requested the aggrieved shareholders to bear with him.

62. Manageable is the right spelling, which means, able to be controlled or dealt with without difficulty.

63. Given,

13, 23, 33, 43, 53

All the numbers are prime except

$33 = 3 \times 11$ is not prime, therefore 33 is the odd one out.

64. Option (*a*) and (*c*), say same thing - R2D2 can repair only Aeroplanes – no other things it can repair. There is no mention in the passage about other tasks that R2D2 can perform. This makes options (*a*) and (*c*) false.

However, the passage says that R2D2 is the only robot, which can repair Aeroplanes.

65. Given,

$$|9y - 6| = 3$$

$$\therefore y = \frac{3+6}{9} = \frac{9}{9} = 1$$

$$\text{Now, } y^2 - \frac{4y}{3} = 1^2 - \frac{4 \times 1}{3}$$

$$= \frac{3-4}{3} = \frac{-1}{3}$$

66. A plant with installed capacity of at least 200 tonnes is called a large plant.

Installed capacity is greater than 200 tonnes for plant number 1, 4, 8, and 9.

The remaining plant numbers i.e. 2, 3, 5, 6, and 7 are small plants.

Thus the difference between the total production of large plants and small plants, in tonnes is

$= (160 + 190 + 230 + 190) - (150 + 160 + 120 + 100 + 120)$

$= 120$ tonnes

67. The sentence in passage "A research study on women with masters or higher degrees in mechanical engineering found that 99% of such women were successful in their professions." directly supports option (*c*). Option (*d*) and option (*a*) are false as there is no relation with passage. Option (*b*) says about men with advanced degree but what is stated in passage is about a group of students, so this option is false.

68. Statement (i) - nothing is mentioned in the passage from where SIT students will get degree. Therefore, it is wrong statement.

Statement (ii) - "to cater to the technological and industrial needs of a developing country." this line supports statement (ii). It means Sourya has distinct national objectives.

Statement (iii) - Nothing is mention about consultation with IIT so this makes statement (iii) false.

Statement (iv) - Again this statement "to cater to the technological and industrial needs of a developing country." directly supports statement (iv).

Therefore, only statements (ii) and (iv) are correct

69. Let p be the probability of success cases and q be the probability of failure cases.

Given, p = 0.6, q = 0.4

The probability of 6 successful throws out of 10 can be given as:

$$= {}^{10}C_6 p^6 q^{10-6}$$

$$= {}^{10}C_6 (0.6)^6 (0.4)^4$$

$$= 210 \times 0.046656 \times 0.0256$$

$$= 0.2508$$

70. $211^{870} + 146^{127} \times 3^{424}$

units digit $= 1^{870} + 6^{127} \times 3^{424}$

units digit $= 1 + 6 \times 1$

$= 7$

71. The ninth and the tenth of this month are Monday and Tuesday <u>respectively</u>.

72. It is <u>easier</u> read this year's textbook <u>than</u> the last year's.

73. In the given question, the rule states that "in order to drink beer the person must be over 18 years of age". Therefore, we will consider the following options according to the rule:

For P it is given that his/her age is 16 years and since he/she is below 18 years of age, they cannot have been so we have to check P's drink to see if the rule is followed.

For Q it is given that his/her age is 25 years and the rule states nothing about what anyone above 18 years cannot drink therefore there is no restriction on him/her. We do not need to check his/her drink.

For R it is given that he/she is having a milkshake. Now for milkshake also the rule puts no restriction therefore there is no need to check his/her age

Thus, finally for S it is given that he/she is having a beer and now since the rule clearly states that if anyone is having beer then he/she must be over 18 years of age therefore we need to check S's age.

Therefore, the correct option is Option (*b*) we need to check P's drink and S's age.

74.

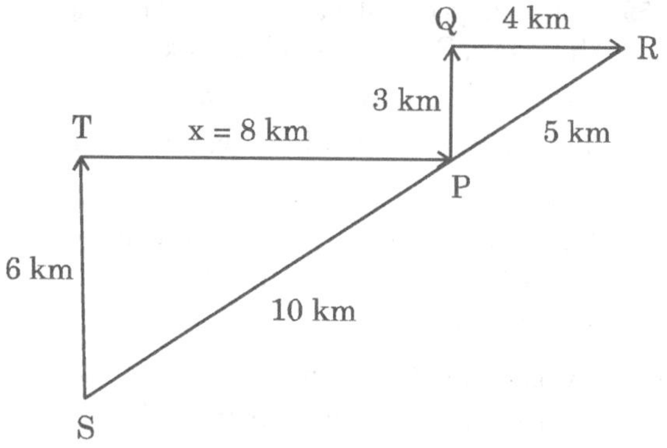

Consider PST as a triangle as given in figure above

As per Pythagoras theorem

$$(10)^2 = (6)^2 + x^2$$

$$\Rightarrow x^2 = 100 - 36 = 64$$

$$\Rightarrow x = 8 \text{ km}$$

75. Given,

The total number of students = 500

Let, Number of student enrolled in Chemistry be 'A' = 329

Number of student enrolled in Physics be 'B' = 186

Number of student enrolled in Mathematics be 'C' = 295

As per formula

$$P(A \cup B \cup C) = P(A) + P(B) + P(C) - P(A \cap B) - P(A \cap C) - P(B \cap C) + P(A \cap B \cap C)$$

$$500 = 329 + 186 + 295 - 83 - 63 - 217 + x$$

$$\Rightarrow x = 500 - 447 = 53$$

76. According to author, the passage reflects an intimate association does not allow for necessary perspective.

77. 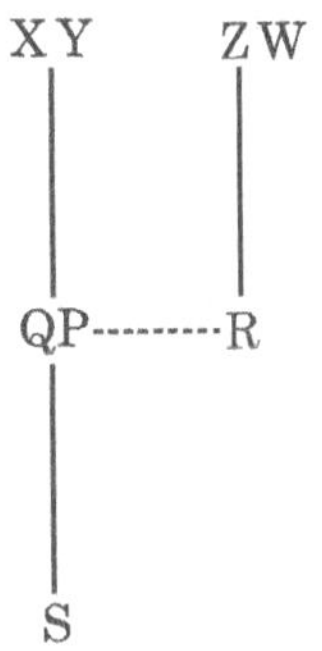

--- line represents that there is a possibility either true or false.

| line represents that it directly inferred from question statements.

Thus, we get that,

P and R should be married to each other. Otherwise, Z cannot be grandfather of S. This might not be true ,as per question " Z and W are married and are parents of R. "there is a possibility that R may have another siblings ... there nothing is mention Z and W have only a single child so this possibility still exists.

From question, it is clear that P is Not a child of Z and W.

P is child of X and Y only, but daughter of P is S and S's grandfather is Z so P is somehow related to Z.

That relation is only possible when Z and W have another child and P married to that child only, so that means R can marry Q.

So in that case this statement "P and R are not married to each other "may not always false, it can be true also.

In addition, "Q cannot be married to R." This statement must necessarily be false.

As there is a possibility, that Q can marry R. Therefore, Q cannot marry R ever in any case is always false statement.

Hence, correct statement is statement D.

78. Let, 1 day task of 1 man = u

1 day task of 1 woman = v

As per equations

$1200u + 500v = \frac{1}{14}$ [As they complete in 14 days

so 1 day work together = $\frac{1}{14}$] ... (1)

$\left(900u + 250v = \frac{1}{21}\right) \times 2$

$\Rightarrow 1800u + 500v = \frac{2}{21}$ ... (2)

Subtracting equation 2 from 1, we get,

$1800u + 500v = \frac{2}{21}$

$\Rightarrow 600u = \frac{2}{21} - \frac{1}{14}$

$\Rightarrow 600u = \frac{1}{42}$

$u = \frac{1}{42 \times 600}$ is 1 man's 1 day work

So time taken by 1 man to complete work

$= 600 \times 42$ days

$= 600 \times 6$ weeks

$= 3600$ weeks

Hence to get the work completed in 1 week:

No of men needed = $\frac{3600}{1} = 3600$

79. We have total 900 three digit numbers from 100-999.

We have numbers in which 1 is immediate right to 2 are 210, 211, 212, 213, 214, 215, 216, 217, 218, 219, 121, 221, 321, 421, 521, 621, 721, 821, 921 = 19 numbers

So, we have numbers in which 1 is never immediate right of 2 = 900 – 19 = 881 numbers.

80. Explanation: Closer lines represents steepest path

Alternate method:

The steepest path will be the path, which is deepest from sea level. Therefore, P to R is the steepest path.

81. The dress complemented her so well that they all immediately complimented her on her appearance.

The word compliment is both a noun and a verb. As a noun, compliment means praise, an expression of admiration, or more simply, something nice you say to someone. As a verb, compliment has the same meaning. To compliment means to praise, to express admiration, or to say something nice to someone. Both the noun and verb forms of the word are equally common.

The word complement is also both a noun and a verb, but unlike compliment, complement is more commonly used as a verb. To complement means to go well with or to be the perfect addition to. The word complement is typically used to speak about clothing and food. If a skirt and t-shirt look very nice together, they complement each other. When two parts of a meal combine to form a taste that is especially delicious, it is said that the two foods complement one another.

82. The judge is standing in the legal community, though shaken by false allegations of wrongdoing, remained undiminished.

Undiminished (adjective)

Meaning- as good, strong, or important as always:

Example-A hundred years after his death, his influence remains undiminished.

83. Given,

BC, FGH, LMNO

Taking first letter B + 4 = F + 6 = L + 8 = T

Therefore, answer is TUVWX

84. Let the perimeter of each of the above geometrical figures be P.

Let us now find the area of circle (Ac) w.r.t its circumference/ perimeter (P), where r is the radius of the circle.

$P = 2\pi r$

$\therefore\ r = \dfrac{P}{2\pi}$

$A_c = \pi r^2$

$A_c = \pi\left(\dfrac{P}{2\pi}\right)^2$

Let us now find the area of square (As) w.r.t its perimeter (P), where l is the side of the square

$P = 4l$

$l = \dfrac{P}{4}$

$A_s = l^2$

$A_s = \dfrac{P^2}{16}$

Let us now find the area of equilateral triangle (Ae) w.r.t its perimeter (P), where s is the side of the triangle

$P = 3s$

$s = \dfrac{P}{3}$

$A_e = \dfrac{\sqrt{3}}{4} \times s^2$

$A_e = \dfrac{\sqrt{3}}{4} \times \dfrac{P^2}{9}$

$A_e = \dfrac{P\sqrt{3}}{36}$

$A_e = \dfrac{P}{12\sqrt{3}}$

Comparing the denominators of equations Ac, As and Ae $A_c > A_s > A_e$,

$4\pi < 16 < 12\sqrt{3}$

As the above numbers are in the denominator,

$A_c > A_s > A_e$

Thus, the circle contains he maximum area and the equilateral triangle contains the minimum area.

85. Given,

$$\frac{1}{1+\log_u vw} + \frac{1}{1+\log_v wu} + \frac{1}{1+\log_w uv}$$

$$\Rightarrow \frac{1}{1+\dfrac{\log vw}{\log u}} + \frac{1}{1+\dfrac{\log wu}{\log v}} + \frac{1}{1+\dfrac{\log uv}{\log w}}$$

$$\Rightarrow \frac{\log u}{\log u + \log vw} + \frac{\log v}{\log v + \log wu} + \frac{\log w}{\log w + \log uv}$$

$$\Rightarrow \frac{\log u}{\log(uvw)} + \frac{\log v}{\log(uvw)} + \frac{\log w}{\log(uvw)} = \frac{\log(uvw)}{\log(uvw)} = 1$$

86. Given,

Total number of students who watched films A, B and C over a week = n(s) = 40.

Also given that, each student watched either only one film or all three.

$n(A \cap B \cap \bar{C}) = n(B \cap C \cap \bar{A}) = n(C \cap A \cap \bar{B}) = 0$

Where, $\bar{A}, \bar{B}, \bar{C}$ are complementary events of A, B, C respectively.

As per question,

n(A) = 13

n(B) = 16

n(c) = 19

Let us consider no. of students who watch all the flims $= n(A \cap B \cap C) = x$

$n(A \cap \bar{B} \cap \bar{C}) = 13 - x$

$n(B \cap \bar{C} \cap \bar{A}) = 16 - x$

$n(C \cap \bar{B} \cap \bar{A}) = 19 - x$

$n(s) = 13 - x + 16 - x + 19 - x + x$

$\Rightarrow 40 = 48 - 2x$

$\Rightarrow 2x = 48 - 40$

$\Rightarrow x = \dfrac{8}{2} = 4$

$\therefore$ No. of students who watched all the three films is 4.

87. Given,

A wire bent over square area of 1936 m^2.

Let us consider that the side of a square is x

Then, $x^2 = 1936$

$\Rightarrow x = 44m$

Therefore, Length of wire = Perimeter of square = 4x = 176 m

Also,

a + b = 176, such that a = 3b

$\Rightarrow 3b + b = 176$

$\Rightarrow b = 44$

$\therefore$ a = 132.

Given, a is bent over square and b bent over circle.

i.e., 4x' = a = 132

$\Rightarrow x' = \frac{132}{4}$

$= 33$ (x' = length of the side of square)

$\Rightarrow x' = 33$

$\because$ Area of square = $33 \times 33 = 1089$

$\therefore 2\pi r = b(\because b'$ bent over circle)

$\Rightarrow r = \frac{44}{2\pi} = 7$

$\therefore$ Area of circle = $\pi(7^2) = 154$

Sum of the area's = $1089 + 154 = 1243\ m^2$

88. Given,

Persons worked = 125

Number of hours each person worked per day = 7

Number of days they worked = 39.

Work completed = $\frac{5}{7}$

Remaining days = 52 – 39 = 13

Remaining Work to be completed = $1 - \frac{5}{7} = \frac{2}{7}$

Let the total number of persons who do the remaining work = x

Number of hours each person needs to be work per day = 8

Amount of work 1 person did in 1 day, working 1 hours a day = $\frac{\left(\frac{5}{7}\right)}{125 \times 39 \times 7}$

Now, the amount of work x person should do in 1 day, working 1 hours a day = $\frac{\left(\frac{2}{7}\right)}{13 \times 8}$

$$\Rightarrow x = \frac{\dfrac{\left(\frac{2}{7}\right)}{13 \times 8}}{\dfrac{\left(\frac{5}{7}\right)}{125 \times 39 \times 7}} = \frac{125 \times 39 \times 7 \times 2}{13 \times 8 \times 5} = 131.25$$

Number of additional persons required = 131.25 – 125 = 6.25 or 7

89. From Statement- i

We get, if house number is multiple of 3, then house number $\in [50,59]$

From Statement- ii, we get

If house number is not a multiple of 4, then house number $\in [60,69]$

From Statement- iii, we get

If house number is not a multiple of 6, then house number $\in [70,79]$

Therefore, the required house number, is 76 among $\{54,65,66,76\}$

Since,

(i) 76 is not a multiple of 3, so house number $\notin [50,59]$

(ii) 76 is not a multiple of 4, so house number $\notin [60,69]$

(iii) 76 is not a multiple of 6, so house number $\notin [70,79]$

Therefore, required house number is 76.

90. It has probability $\frac{1}{2}$ in contrast to the other options that all have probability $\frac{1}{4}$.

Option (A) - TT has probability $\frac{1}{4}$

Option (B) - HT or TH has probability $\frac{1}{4} + \frac{1}{4}$ (summation of two probabilities of mutually exclusive events)

Option (C) - HH has probability $\frac{1}{4}$

Option (D) - HT has probability $\frac{1}{4}$

Essential for this conclusion is the fact that the coin is unbiased.

91. "To whom" is used when we don't have a specific person to correspond. In the given case, the subject being corresponded is "The fisherman", So, option (c) is correct.

92. Some student were not involved in the strike also means that some students were also involved in the strike(due the presence of "Some. Thus, option (c) is correct.

93. Let r be the radius and h be the height of a circular cone. Then

$$R = r + \frac{r \times 10}{100} = \frac{11r}{10}$$

$$H = h + \frac{h \times 10}{100} = \frac{11h}{10}$$

$\therefore$ Increase percent in volume

$$= \left[\frac{\text{New volume} - \text{old volume}}{\text{old volume}} \times 100\right]\%$$

$$= \frac{\frac{1}{3} \times \pi \times \frac{11r}{10} \times \frac{11r}{10} \times \frac{11h}{10} - \frac{1}{3} \times \pi \times r^2 h}{\frac{1}{3} \times \pi \times r^2 \times h}$$

$$= \frac{\frac{1}{3} \times \pi \times r^2 h\left(\frac{11}{10} \times \frac{11}{10} \times \frac{11}{10} - 1\right)}{\frac{1}{3} \times \pi \times r^2 \times h} \times 100$$

$$= \left(\frac{11 \times 11 \times 11}{1000} - 1\right) \times 100$$

$$= \left[\frac{1331 - 1000}{1000} \times 100\right]\% = \frac{331}{10} = 33.1\%$$

$\therefore$ option (c) is correct

94. We have following sequence

10	5	4	7	2

$\therefore$ second number from right is 7.

$\therefore$ option (c) is correct.

95. The sentence is in passive form and we require the use of "had + been + V3". In the given sentence, third form of "defeat" will be used. So, "defeated" is the correct option.

96. Refer to the line, "Since the last one year, after a 125 basis point reduction in repo rate by the Reserve Bank of India, banking institutions have been making a demand to reduce interest rates on small saving schemes.". So, we can infer from the above lines that the government sometimes takes into consideration the demands of banking institutions before reducing the rates. So, option (c) is correct.

97. The required percentage

$$= \frac{147}{294} \times 100 = \frac{1}{2} \times 100 = 50\%$$

$\therefore$ Option (d) is correct.

98. Refer to the lines, "Terms like gayaki and baaf were used to refer to vocal and instrument styles, respectively. So, we can infer from these lines that baaf is a type of vocal and instrument style. Thus baaf is not an institution. Moreover, option (c) is correct.

99. $t_1\ 80 + t_2\ 100 = 540$

$\Rightarrow t_1 = 3,\ t_2 = 2$

$\therefore$ total time = 3 + 2 = 5h.

= 7 + 5 = 11 A.M.

100. Referring to the above paragraph, we can infer that there is a direct relation between prestige of a kingdom and prestige of a head hunter. As the prestige of a kingdom shows his power by collecting taxes, in the same way, head hunter collects the heads of enemies. Thus, option (d) is correct.

101. It is a punctuality checking, there is always 'in' used with accidentand of used with diseases.

102. There is always 'but also' used if a sentence contains 'not only' before in the sentence.

103. We repress the feeling at the point where we cannot express it.

104. In the given section there are 15 minutes used for English announcement and only 5 minutes used for French announcement.

105. Since x > 1 then let $x_1 = 2$ and $x_2 = 3$

Let $f(x) = \frac{1}{x}$, then $f(x_1) = \frac{1}{x_1}$

$$f(2) = \frac{1}{2}$$

Similarly, $$f(3) = \frac{1}{3}$$

$$f(x_1 + x_2) = f(2 + 3) = f(5) = \frac{1}{5} = 0.2$$

$$f(x_1) + f(x_2) = \left(\frac{1}{2}\right) + \left(\frac{1}{3}\right) = \frac{5}{6} = 0.833$$

But for super additive function

$$f(x_1 + x_2) \geq f(x_1) + f(x_2).$$

Let $f(x) = e^x$

Then $f(x_1) = e^2$ and $f(x_2)$

$= e^3$

$f(x_1 + x_2) = f(2 + 3)$

$= f(5)$

$= e^5$

$= 148.413$

$f(x_1) + f(x_2) = f(2) + f(3)$

$= e^2 + e^3$

$= 27.475$

Because $f(x_1 + x_2) > f(x_1) + f(x_2)$

Therefore $f(x) = e^x$ is a super additive function.

106. As we can see in the given section global crisis happened in 2008 while subprime landing crisis happened in USA in 2007. The subprime lending crisis led to the banking crisis in 2008 and subprime lending crisis happened due to excess liquidity followingthe East Asian crisis. So making correct arrangement of crisis we shall get.

East Asian crisis → subprime lending crisis → banking crisis → global financial crisis

107. Hour needle rotates 3600 in 12 hours.

Hour needle rotates $3600 \times \frac{1}{(12 \times 60)}$ in 1 minute

Hour needle rotates $3600 \times \frac{15}{(12 \times 60)}$ in 15 minutes.

The angle between the hour needle and minute needle

$$= \frac{(360 \times 15)}{(12 \times 60)}$$

$$= 7.50$$

108. The area of shaded region can be determine by simple subtracting the area of rectangle from area of circle.

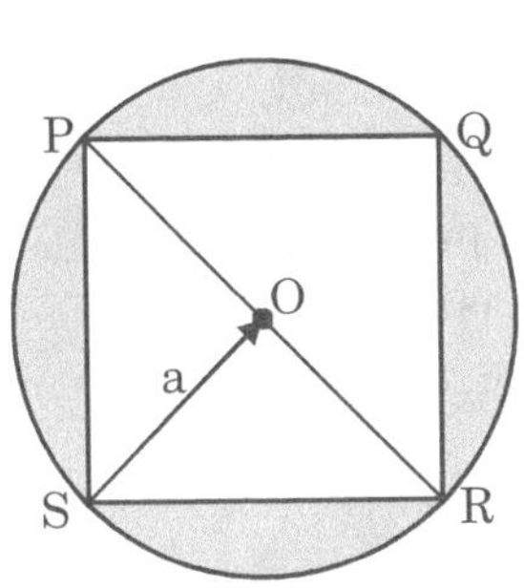

Area of the circle is πa^2 and the area of rectangle will be PQ*PS. Now we have to find out the area of rectangle so that we can determine the area of shaded region. If we draw a line or hypotaneous PR in the rectangle then we see two triangles POS and SOR are similar and their sides are equals to each other respectively.

Using Pythagoras theorem

Now we have PR = 2a and PS = SR

$$PS^2 + SR^2 = PR^2$$

$$PS^2 + PS^2 = (2a)^2$$

$$2PS^2 = 4a^2$$

$PS^2 = 2a^2$ then $PS = \sqrt{2}a$ and $PS = SR = \sqrt{2}a$

The area of shaded region

= area of circle – area of rectangle

$= \pi a^2 - PS \times SR$

$= \pi a^2 \times \sqrt{2}a \times \sqrt{2}a$

$= \pi a^2 - 2a^2$

109. The quadratic equation is $ax^2 - bx + c = 0$.

Let assume the equal roots are β then

Sum of roots $\beta + \beta$ = coefficient of x/coefficient of x^2

$$2\beta = \frac{b}{a}$$

Product of roots β^2 = constant value/coefficient of x^2

$$= \frac{c}{a}$$

Now $2\beta \times \beta^2 = \frac{b}{a} \times \frac{c}{a}$

$$\beta^3 = \frac{bc}{2a^2}$$

110. Total number of students enrolled in school P during 2014-2018 are 23

Average students in School P

$$= \frac{(3+5+5+6+4)}{5} = \frac{23}{5}$$

Average of Difference of students enrolled in school P and school Q are 8

$$\frac{(\text{School P} - \text{School Q})}{5} = \frac{(1+2+3+1+1)}{5}$$

$$= \frac{8}{5}$$

Now the ratio will be

$$= \left(\frac{\text{Average students in school P}}{\text{Average of difference of students Enrolled in school P and school Q}}\right)$$

$$= \frac{\left(\frac{23}{5}\right)}{\left(\frac{8}{5}\right)} = \frac{23}{8} = 23:8$$

111. Option (i) is correct.

Arun and Aparna are two persons, i.e., they are plural. Therefore, are is the appropriate linking verb used here.

Option (iv) is correct

Here arun's denotes a singular person. Therefore, family is the appropriate word used here.

112. As the mirror is in x-axis. The correct option is Option (b).

113. Total number of cases = 6 × 6 = 36

For even number of choices we have,

2, 4, 6 i.e., 3 choices

As they are 2 dices

Thus, even choices are = 3 × 3 = 9

$\therefore$ P (even number is rolled) = $\frac{9}{36} = \frac{1}{4}$

114. Given, $p \odot q = p - q$

$p \oplus q = p \times q$

$\therefore$ we get

$\odot = -$

$\oplus = \times$

then, $(9\odot(6\oplus7))\odot(7\oplus(6\odot5))$

$= (9-(6\times7))-(7\times(6-5)) = -40$

115. Total number of arrangement in which four persons can sit

= 4 × 3 × 2 × 1

= 24

Number of arrangement for which R is seated at second position is

3 × 2 × 1 = 6

$\therefore$ Number of arrangement for which R is not seated at second position is 24 – 6 = 18

116.

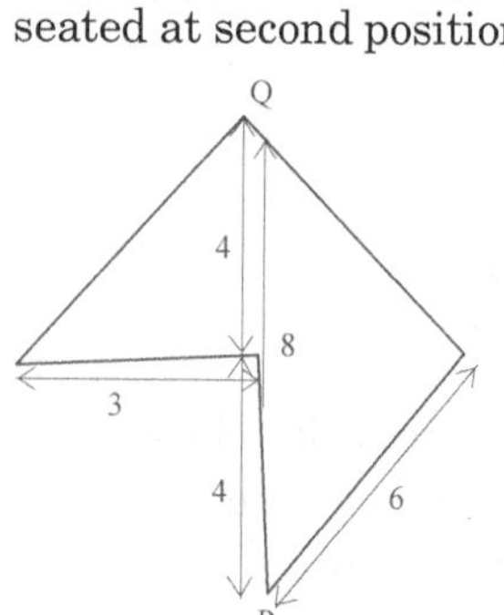

Therefore, distance between OQ is

OQ = $\sqrt{3^2 + 4^2} = 5$ units

117. As per the passage option (a) is correct.

The author is of the opinion that rehearsing is important for musicians, actors and public speakers.

118. Given,

1. Some football players play cricket
2. All cricket players play hockey.

As per above statement it is clear that option (b) gives the correct statement as

Some football players play hockey.

119.

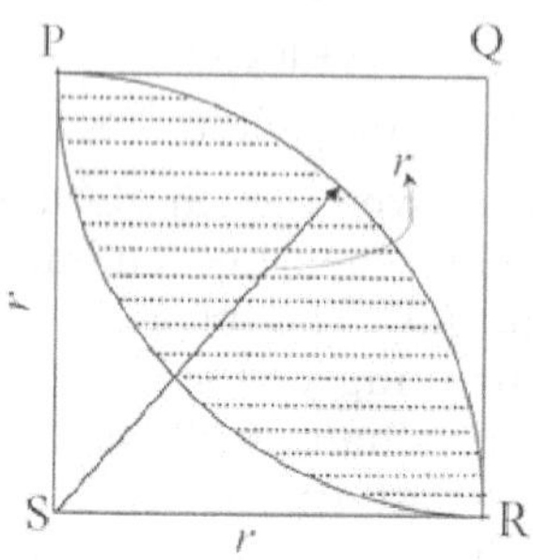

P (points which falls in the shaded area)

$= \frac{\text{Shaded area}}{\text{Total area}}$

Total area = r^2

Shaded area = 2 × (area of sector – area of remaing triangle of square)

$= 2\times\left(\frac{1}{2}\times r^2\times\frac{\pi}{2}-\frac{1}{2}\times r^2\right) = r^2\left(\frac{\pi}{2}-1\right)$

P (points which falls in the shaded area)

$= \frac{r^2\left(\frac{\pi}{2}-1\right)}{r^2} = \left(\frac{\pi}{2}-1\right)$

120. Given,

As equalateral triangle

PQ = QR = PR

$\frac{PQ}{4}, \frac{QR}{6}, \frac{PR}{8}$ and each are integers

Firstly, we find the smallest side length which is computed as

LCM of 4, 6, 8 which is 24

Now,

Area = $\frac{1}{2}ab\sin\theta$

As equalateral triangle

$\therefore$ $\theta = 60°$

Thus, Minimum area = $\frac{1}{2}\times24\times24\times\sin60 = 144\sqrt{3}$

SUGGESTIVE QUESTIONS & ANSWERS (Section-Wise)

1 CHAPTER

ENGINEERING MATHEMATICS

LINEAR ALGEBRA

MCQ Type questions

1. Eigen values of the matrix $\begin{bmatrix} 3 & -1 & -1 \\ -1 & 3 & -1 \\ -1 & -1 & 3 \end{bmatrix}$ are

(a) 1, 1, 1 (b) 1, 1, 2
(c) 1, 4, 4 (d) 1, 2, 4

2. Rank of the following $(n + 1)\times(n + 1)$ matrix, where a is a real number, is

$$\begin{bmatrix} 1 & a & a^2 & . & . & . & a^n \\ a & a & a^2 & . & . & . & a^n \\ . & . & . & . & . & . & . \\ . & . & . & . & . & . & . \\ . & . & . & . & . & . & . \\ 1 & a & a^2 & . & . & . & a^n \end{bmatrix}$$

(a) 1
(b) 2
(c) n
(d) dependent upon the value of A

3. Matrices $\begin{bmatrix} \cos\theta & -\sin\theta \\ \sin\theta & \cos\theta \end{bmatrix}$ and $\begin{bmatrix} a & 0 \\ 0 & b \end{bmatrix}$ commute under multiplication

(a) if $a = b$ or $\theta = n\pi$, where n is an integer
(b) always
(c) never
(d) if $a \cos \theta \neq b \sin \theta$

4. If $A = \begin{bmatrix} 0 & 2 & 3 \\ -2 & 0 & 5 \\ -3 & -5 & 0 \end{bmatrix}$, then

(a) $A' = -A$ (b) $A' = A$
(c) $A' = -\dfrac{1}{A}$ (d) none of these

5. Eigen values of the system represented by

$$\dot{X} = \begin{bmatrix} 0 & 1 & 0 & 0 \\ 0 & 0 & 1 & 0 \\ 0 & 0 & 0 & 1 \\ 0 & 0 & 0 & 1 \end{bmatrix} X \text{ are}$$

(a) 0, 0, 0, 0 (b) 1, 1, 1, 1
(c) 0, 0, 0, –1 (d) 1, 0, 0, 0

6. Matrix, $A = \begin{bmatrix} \cos\theta & \sin\theta & 0 \\ \sin\theta & \cos\theta & 0 \\ 0 & 0 & 1 \end{bmatrix}$ is not

(a) orthogonal (b) non–singular
(c) singular (d) have A^{-1} exists

7. In the matrix equation $Px = q$, which of the following is a necessary condition for the existence of at least one solution for the unknown vector x ?

(a) Augmented matrix [Pq] must have the same rank as matrix P
(b) Vector q must have only non-zero elements
(c) Matrix P must be singular
(d) Matrix P must be square

8. Eigen vectors of a real symmetric matrix corresponding to different eigen values are

(a) orthogonal (b) singular
(c) non-singular (d) none of these

9. An arbitrary vector X is an eigen vector of the matrix $A = \begin{bmatrix} 1 & 0 & 0 \\ 0 & a & 0 \\ 0 & 0 & b \end{bmatrix}$, if (a, b) =

(a) (0, 0) (b) (1, 1)
(c) (0, 1) (d) (1, 2)

10. The system of equations

$$a_1x + a_2y = 0$$
$$b_1x + b_2y = 0$$

where a_1, a_2, b_1, b_2, are real numbers, has a non-trivial solution if

(a) $a_1b_1 = a_2b_2$ (b) $a_1b_2 = b_1a_2$
(c) $a_1a_2 = b_1b_2$ (d) none of these

11. Solution of the given matrix equation is

$$\begin{bmatrix} 2 & 2 & 2 \\ 1 & 5 & -3 \\ -1 & -3 & 1 \end{bmatrix} \begin{bmatrix} x_1 \\ x_2 \\ x_3 \end{bmatrix} = \begin{bmatrix} 0 \\ 0 \\ 0 \end{bmatrix}$$

(a) $x_1 = 0, x_2 = 0, x_3 = 0$
(b) $x_1 = 1, x_2 = 1, x_3 = 0$
(c) $x_1 = 0, x_2 = 1, x_3 = 1$
(d) $x_1 = 2, x_2 = -5, x_4 = -1$

12. If product of matrix

$$A = \begin{bmatrix} \cos^2\theta & \cos\theta\sin\theta \\ \cos\theta\sin\theta & \sin^2\theta \end{bmatrix} \text{ and}$$

$$B = \begin{bmatrix} \cos^2\phi & \cos\phi\sin\phi \\ \cos\phi\sin\phi & \sin^2\phi \end{bmatrix}$$

is a null matrix, then θ and ϕ differ by an

(*a*) odd multiple of π (*b*) even multiple of π

(*c*) odd multiple of $\pi/2$ (*d*) even multiple of $\pi/2$

13. Inverse of the matrix $\begin{Bmatrix} ab & 0 \\ 0 & 1 \end{Bmatrix}$ is

(*a*) $\begin{Bmatrix} a & 0 \\ 0 & a \end{Bmatrix}$ (*b*) $\begin{Bmatrix} b & 0 \\ a & b \end{Bmatrix}$

(*c*) $\begin{pmatrix} 1/a & 0 \\ 0 & 1/b \end{pmatrix}$ (*d*) $\begin{Bmatrix} a & 0 \\ 0 & 1/b \end{Bmatrix}$

14. The system of equations

$$2x + 4y = 10,$$
$$5x + 10y = 25$$

has

(*a*) no unique solution (*b*) only one solution

(*c*) only two solutions (*d*) infinite solutions

15. Sum of the eigen values of the matrix $\begin{pmatrix} 3 & 4 \\ x & 1 \end{pmatrix}$ for real and negative values of x is

(*a*) greater than zero

(*b*) less than zero

(*c*) zero

(*d*) dependent on the value of x

16. The system of equations

$$4x + 6y + 8z = 0$$
$$7x + 8y + 9z = 0$$
$$3x + 2y + z = 0$$

has

(*a*) no solution

(*b*) only one solution

(*c*) two solutions

(*d*) infinite number of solutions

17. Eigen values of $\begin{bmatrix} 1 & 1 & 1 \\ 1 & 1 & 1 \\ 1 & 1 & 1 \end{bmatrix}$ are

(*a*) 0, 0 0 (*b*) 0, 0 1

(*c*) 0, 0, 3 (*d*) 1, 1, 1

18. For following set of simulataneous equations

$$1.5x - 0.5y = 2$$
$$4x + 2y + 3z = 9$$
$$7x + y + 5z = 10$$

(*a*) solution is unique

(*b*) infinitely many solutions exist

(*c*) equations are incompatible

(*d*) finite number of multiple solutions exist

19. If $R = \begin{bmatrix} 1 & 0 & -1 \\ 2 & 1 & -1 \\ 2 & 3 & 2 \end{bmatrix}$, then top row of R^{-1} is

(*a*) [5 6 4] (*b*) [5 –3 1]

(*c*) [2 0 –1] (*d*) [2 –1 1/2]

20. The system of simultaneous equations

$$x + 2y + z = 6$$
$$2x + y + 2z = 6$$
$$x + y + z = 5$$

has

(*a*) unique solution

(*b*) infinite number of solutions

(*c*) no solution

(*d*) exactly two solutions

21. Eigen vectors of the matrix $\begin{bmatrix} 1 & -1 \\ -1 & 1 \end{bmatrix}$ is (are)

(*a*) (1, 0) (*b*) (0, 1)

(*c*) (1, 1) (*d*) (1, –1)

22. Inverse of the matrix $\begin{bmatrix} -3 & 5 \\ 2 & 1 \end{bmatrix}$ is

(*a*) $\begin{bmatrix} \frac{5}{13} & \frac{-1}{13} \\ \frac{2}{13} & \frac{3}{13} \end{bmatrix}$ (*b*) $\begin{bmatrix} \frac{2}{13} & \frac{5}{13} \\ \frac{-1}{13} & \frac{3}{13} \end{bmatrix}$

(*c*) $\begin{bmatrix} \frac{-1}{13} & \frac{5}{13} \\ \frac{2}{13} & \frac{3}{13} \end{bmatrix}$ (*d*) $\begin{bmatrix} \frac{1}{13} & \frac{-5}{13} \\ \frac{2}{13} & \frac{3}{13} \end{bmatrix}$

23. If A and B are square matrices of size n × n, then which of the following statement is not true ?

(*a*) det (AB) = det (A) det (B)

(*b*) det (kA) = k^n det (A)

(*c*) det (A + B) = det (A) + det (B)

(*d*) det (A^T) = 1/det (A^{-1})

24. If matrix A is m × n and B is n× p, then number of multiplication operations and addition operations needed to calculate the matrix AB, respectively are

(a) mn^2p, mpn (b) $mpn, (n-1)$

(c) mpn, mpn (d) $mn^2 p, (m+p)n$

25. If A be an invertible matrix and inverse of 7A is $\begin{bmatrix} -1 & 2 \\ 4 & -7 \end{bmatrix}$, then matrix A is

(a) $\begin{bmatrix} 1 & 2/7 \\ 4/7 & 1/7 \end{bmatrix}$ (b) $\begin{bmatrix} 7 & 2 \\ 4 & 1 \end{bmatrix}$

(c) $\begin{bmatrix} 1 & -4/7 \\ -2/7 & 1/7 \end{bmatrix}$ (d) $\begin{bmatrix} 7 & 4 \\ 2 & 1 \end{bmatrix}$

26. Matrix $B = A^T$, where A is any matrix is

(a) skew symmetric

(b) symmetric about the secondary diagonal

(c) always symmetric

(d) another general matrix

27. If $A^T = A^{-1}$, where A is a real matrix, then A is

(a) normal (b) symmetric

(c) Hermitian (d) orthogonal

28. If A and B are non-zero squarematrices, then AB = I implies

(a) A and B are orthogonal

(b) A and B are singular

(c) B is singular

(d) A is singular

29. Eigen values of a square symmetric matrix are always

(a) positive (b) real and imaginary

(c) negative (d) real

30. If $A = \begin{bmatrix} 1 & 0 \\ 0 & 1 \end{bmatrix}$, then state transition matrix e^{At} is

(a) $\begin{bmatrix} 0 & e^{-t} \\ e^{-t} & 0 \end{bmatrix}$ (b) $\begin{bmatrix} e^{t} & 0 \\ 0 & e^{t} \end{bmatrix}$

(c) $\begin{bmatrix} e^{-t} & 0 \\ 0 & e^{-t} \end{bmatrix}$ (d) $\begin{bmatrix} 0 & e^{t} \\ e^{t} & 0 \end{bmatrix}$

31. For matrix $p = \begin{bmatrix} 3 & -2 & 2 \\ 0 & -2 & 1 \\ 0 & 0 & 1 \end{bmatrix}$, if one of the eigen values is equal to –2, then which of the following is an eigen vector ?

(a) $\begin{bmatrix} 3 \\ -2 \\ 1 \end{bmatrix}$ (b) $\begin{bmatrix} -3 \\ 2 \\ -1 \end{bmatrix}$

(c) $\begin{bmatrix} 1 \\ -2 \\ 3 \end{bmatrix}$ (d) $\begin{bmatrix} 5 \\ 2 \\ 0 \end{bmatrix}$

32. If $A = \begin{bmatrix} 2 & -0.1 \\ 0 & 3 \end{bmatrix}$ and $A^{-1} = \begin{bmatrix} \frac{1}{2} & a \\ 0 & b \end{bmatrix}$, then (a + b) equals

(a) $\frac{7}{20}$ (b) $\frac{3}{20}$

(c) $\frac{19}{60}$ (d) $\frac{11}{20}$

33. Eigen values and corresponding eigen vectors of a 2 × 2 matrix are given by

Eigenvalue	Eigenvector
$\lambda_1 = 4$	$v_1 = \begin{bmatrix} 1 \\ 1 \end{bmatrix}$
$\lambda_2 = 8$	$v_2 = \begin{bmatrix} 1 \\ -1 \end{bmatrix}$

The matrix is

(a) $\begin{bmatrix} 6 & 2 \\ 2 & 6 \end{bmatrix}$ (b) $\begin{bmatrix} 4 & 6 \\ 6 & 4 \end{bmatrix}$

(c) $\begin{bmatrix} 2 & 4 \\ 4 & 2 \end{bmatrix}$ (d) $\begin{bmatrix} 4 & 8 \\ 8 & 4 \end{bmatrix}$

34. A second order system starts with an initial condition of $\begin{bmatrix} 2 \\ 3 \end{bmatrix}$ without any external input. The state transition matrix for the system is given by $\begin{bmatrix} e^{-2t} & 0 \\ 0 & e^{-t} \end{bmatrix}$. The state of the system at the end of 1 second is given by

(a) $\begin{bmatrix} 0.271 \\ 1.100 \end{bmatrix}$ (b) $\begin{bmatrix} 0.135 \\ 0.368 \end{bmatrix}$

(c) $\begin{bmatrix} 0.271 \\ 0.736 \end{bmatrix}$ (d) $\begin{bmatrix} 0.135 \\ 1.100 \end{bmatrix}$

Numerical Type questions

1. Rank of the matrix A = $\begin{bmatrix} 0 & 0 & 0 & 0 \\ 4 & 2 & 3 & 0 \\ 1 & 0 & 0 & 0 \\ 4 & 0 & 3 & 0 \end{bmatrix}$ is ______

2. Rank of the diagonal matrix $\begin{pmatrix} 3 & & & & & & & & \\ & 0 & & & & & & & \\ & & 0 & & & & & & \\ & & & 3 & & & & & \\ & & & & 0 & & & & \\ & & & & & 4 & & & \\ & & & & & & 0 & & \\ & & & & & & & 0 & \\ & & & & & & & & 4 \end{pmatrix}$ is ____

3. Rank of the matrix A = $\begin{bmatrix} 1 & 3 & 5 & 1 \\ 2 & 4 & 8 & 0 \\ 3 & 1 & 7 & 5 \end{bmatrix}$ is______

4. Rank of the matrix $\begin{bmatrix} 1 & 1 \\ 0 & 0 \end{bmatrix}$ is______

5. For which value of k, following system is consistent?

$$2x - 5ky + 6z = 0$$
$$kx + 2y - 2z = 0$$
$$2x + 2y - kz = 0$$

6. The value of λ for which the equations

$$2x + y + 2z = 0$$
$$x + y + 3z = 0$$
$$4x + y + \lambda z = 0$$

have a non-zero solution, is_____

7. The system of linear equations

$$(4d - 1)\,x + y + z = 0$$
$$-y + z = 0$$
$$(4d - 1)\,z = 0$$

has a non-trivial solution, if d equals ______

8. Rank of a 3 × 3 matrix C (= AB), found by multiplying a non-zero column matrix A of size 3 × 1 and a non-zero row matrix B of size 1 × 3, is

______

9. Sum of eigen values of the matrix $\begin{bmatrix} 1 & 1 & 3 \\ 1 & 5 & 1 \\ 3 & 1 & 1 \end{bmatrix}$ is

_____

10. For which value of x, will the matrix $\begin{bmatrix} 8 & x & 0 \\ 4 & 0 & 2 \\ 12 & 6 & 0 \end{bmatrix}$

become singular?

11. Rank of the matrix, A = $\begin{bmatrix} 1 & -2 & -1 \\ -3 & 3 & 0 \\ 2 & 2 & 4 \end{bmatrix}$ is ______

12. Solution(s) to the equations

$$2x + 3y = 1$$
$$x - y = 4$$
$$4x - y = a$$

will exist if a is equal to______

13. For matrix $\begin{bmatrix} 4 & 2 \\ 2 & 4 \end{bmatrix}$, eigen value corresponding

to the eigen vector $\begin{bmatrix} 101 \\ 101 \end{bmatrix}$ is_____

CALCULUS

MCQ Type questions

1. $\lim_{x \to \infty} x^n e^{-x}$ is equal to

(a) ∞ (b) 1

(c) $\underline{|n}$ (d) 0

2. If | x | denotes the greatest integer not greater than 2, then $\lim_{x \to 2} [x]$ is

(a) 0 (b) 1

(c) 2 (d) does not exist

3. If two functions $f(x) = \begin{cases} x \sin \dfrac{1}{x} & , x \neq 0 \\ 0 & , x \neq 0 \end{cases}$ $g(x) = |x|$

are given, then at $x = 0$

(a) both f and g continuous

(b) f is continuous and g is discontinuous

(c) f is discontinuous and g is continuous

(d) both f and g are discontinuous

4. If $f(x) = |x|$, then $f(x)$ is

(a) discontinuous at $x = 0$

(b) continuous only at $x = 0$

(c) continuous at all values of x

(d) discontinuous at $x = 1$

5. The function $f(x) = \begin{cases} ax^2 + b & if \quad x > 2 \\ 2 & if \quad x = 2 \\ 2ax - b & if \quad x < 2 \end{cases}$

is continuous at $x = 2$, if

(a) $a = 0, b = \dfrac{1}{2}$ (b) $a = \dfrac{1}{2}, b = 0$

(c) $a = -\dfrac{1}{2}, b = 0$ (d) $a = 0, b = -\dfrac{1}{2}$

6. If $f(x) = x - [x]$, then $f(x)$ is discontinuous

(a) $\forall\, x \in Z$

(b) $\forall\, x \in Q$

(c) $\forall\, x \in R$

(d) at very real number except integers

7. The interval in which Lagrange's theorem is applicable for the function $f(x) = \frac{1}{x}$ is

(a) $[-3, 3]$

(b) $[-2, 2]$

(c) $[2, 3]$

(d) $[-1, 1]$

8. If $f(x) = 3x^4 - 4x^2 + 5$, then the interval for which $f(x)$ satisfied all the condition of Rolle's theorem is

(a) $[0, 2]$ (b) $[-1, 1]$

(c) $[-1, 0]$ (d) $[1, 2]$

9. If $f(x) = |x|$, then In the interval $[-1, 1]$, $f(x)$ is

(a) satisfied all the conditions of Rolle's theorem

(b) satisfied all the conditions of Mean Value theorem

(c) does not satisfied the conditions of Mean Value theorem

(d) none of these

10. If f and F be both continuous in $[a, b]$ and are derivable in (a, b) and if $f'(x) = F'(x)$ for all x in (a, b), then $f(x)$ and $F(x)$ differ

(a) by 1 in $[a, b]$

(b) by x in $[a, b]$

(c) by constant in $[a, b]$

(d) none of these

11. $x^{1/x}$ is a decreasing function if

(a) $x < e$ (b) $x > e$

(c) $x = e$ (d) $x > \frac{1}{e}$

12. $1 - \frac{x^2}{\underline{|2}} + \frac{x^4}{\underline{|4}} + \ldots + (-1)^n \frac{x^{2n}}{\underline{|2n}} + (-1)^{n+1} \frac{x^{2n+1}}{\underline{|2n+1}} \sin\theta x$

is equal to

(a) $\sin x$ (b) $\cos x$

(c) $\tan x$ (d) $\log x$

13. Taylor's expansion of the function $f(x) = \frac{1}{1+x^2}$ is

(a) $\sum_{n=0}^{\infty} (-1)^n x^{2n}$ where $-1 < x < 1$

(b) $\sum_{n=0}^{\infty} (-1)^n x^{2n}$ for all real x

(c) $\sum_{n=0}^{\infty} x^{2n}$ where $-1 < x < 1$

(d) $\sum_{n=0}^{\infty} (-1)^n x^{n}$ where $-1 < x < 1$

14. If $f(x)$, $\phi(x)$ and $\psi(x)$ have derivatives when $a \le x \le b$, then $\exists$ a value of c of x lying between a and b such that

(a) $\begin{vmatrix} f'(c) & \phi'(c) & \psi'(c) \\ f(b) & \phi(b) & \psi(b) \\ f(a) & \phi(a) & \psi(a) \end{vmatrix} = 0$

(b) $\begin{vmatrix} f(c) & \phi(c) & \psi(c) \\ f'(a) & \phi'(a) & \psi'(a) \\ f'(b) & \phi'(b) & \psi'(b) \end{vmatrix} = 0$

(c) $\begin{vmatrix} f(a) & \phi(a) & \psi(a) \\ f'(b) & \phi'(a) & \psi'(b) \\ f''(c) & \phi''(c) & \psi''(c) \end{vmatrix} = 0$

(d) none of these

15. If $f(x) = ax + b$, $x \in [-1, 1]$, then the point $c \in (-1, 1)$ where $f'(c) = \frac{f(1) - f(-1)}{2}$

(a) does not exist (b) can be any $c \in (-1, 1)$

(c) can be only $\frac{1}{2}$ (d) can be only $\frac{-1}{2}$

16. Let $f(x) = x(x-1)(x-2)$, $a = 0$, $b = \frac{1}{2}$.

The value of 'c' in the Lagrange's mean value theorem is

(a) $\frac{6+\sqrt{11}}{6}$ (b) $\frac{6-\sqrt{11}}{6}$

(c) $\frac{6+\sqrt{-12}}{6}$ (d) $\frac{6-\sqrt{21}}{6}$

17. Maxima and Minima occur

(a) simultaneously (b) once

(c) alternately (d) rarely

18. If $x + y = k$, $x > 0$, $y > 0$, then xy is the maximum when

(a) $x = ky$ (b) $kx = y$

(c) $x = y$ (d) none of these

19. The triangle of maximum area inscribed in a circle of radius r is

(*a*) a right angled triangle with hypotenuse measuring $2r$

(*b*) an equilateral triangle

(*c*) an isosceles triangle of height r

(*d*) does not exist

20. Sum of the perimeters of a circle and a square is l. If the sum of area is least, then

(*a*) side of the square is double the radius of the circle

(*b*) side of the square is half of the radius of the circle

(*c*) side of the square is equal to radius of the circle

(*d*) none of these

21. Which of the following is false?

(*a*) $f(a)$ is an extreme value of $f(x)$ if $f'(a) = 0$

(*b*) If $f(a)$ is an extreme value of $f(x)$, then $f'(a) = 0$

(*c*) If $f'(a) = 0$, then $f(a)$ is an extreme value of $f(x)$

(*d*) None of these

22. Let $f(x) = |x|$, then

(*a*) $f'(0) = 0$

(*b*) $f(x)$ is maximum at $x = 0$

(*c*) $f(x)$ is minimum at $x = 0$

(*d*) none of these

23. The greatest and the least value of $f(x) = x^4 - 8x^3 + 22x^2 - 24x + 1$ in $[0, 2]$ are

(*a*) 0, 8 (*b*) 0, – 8

(*c*) 1, 8 (*d*) 1, – 8

24. The function $f(x) = x^5 - 5x^4 + 5x^3 - 1$ has

(*a*) one minima and two maxima

(*b*) two minima and one maxima

(*c*) two minima and two maxima

(*d*) one minima and one maxima

25. The maximum value of $\frac{\log x}{x}$ in $(0, \infty)$ is

(*a*) e (*b*) $\frac{1}{e}$

(*c*) 1 (*d*) none of these

26. Maximum volume of a cylinder in a cone with semi vertical angle 'α' and height 'b' is

(*a*) $\frac{4}{27}\pi b^3 \tan^2\alpha$ (*b*) $\frac{4}{27}\pi b^2 \tan\alpha$

(*c*) $\frac{4}{27}\pi b \tan^3\alpha$ (*d*) none of these

27. The maximum point on the curve $x = e^x y$ is

(*a*) $(1, e)$ (*b*) $(1, e^{-1})$

(*c*) $(e, 1)$ (*d*) $(e^{-1}, 1)$

28. The value of function $f(x) = x + \frac{1}{x}$ at the points of minimum and maximum are respectively

(*a*) – 2 and 2 (*b*) 2 and – 2

(*c*) – 1 and 1 (*d*) 1 and – 1

29. The value of $\int_0^{5\pi}(2 - \sin x)dx$ is

(*a*) > 0 (*b*) 2

(*c*) 0 (*d*) undefined

30. The unit normal to the plane $2x + y + 2z = 6$ can be expressed in the vector form as

(*a*) $i3 + j2 + k2$ (*b*) $i\frac{2}{3} + j\frac{1}{3} + k\frac{2}{3}$

(*c*) $i\frac{1}{3} + j\frac{1}{2} + k\frac{1}{2}$ (*d*) $i\frac{2}{6} + j\frac{1}{6} + k\frac{2}{6}$

31. The function $f(x) = 3x(x - 2)$ has a

(*a*) minimum at $x = 1$ (*b*) maximum at $x = 1$

(*c*) minimum at $x = 2$ (*d*) maximum at $x = 2$

32. The value of integral $\int_{-2}^{2}\frac{dx}{x^2}$ is

(*a*) 0 (*b*) 0.25

(*c*) 1 (*d*) ∞

33. The value of ξ in the mean value theorem of
$$f(b) - f(a) = (b - a)f'(\xi)$$
for $f(x) = Ax^2 + Bx + C$ in (a, b) is

(*a*) $b + a$ (*b*) $b - a$

(*c*) $\frac{(b+a)}{2}$ (*d*) $\frac{(b-a)}{2}$

34. If $\vec{V}$ is a differentiable vector function and f is a sufficient differentiable scalar function, then curl $\left(f\vec{V}\right)$ is equal to

(*a*) $(\text{grad } f) \times \left(\vec{V}\right) + (f \text{ curl } \vec{V})$

(*b*) $\vec{O}$

(*c*) $f \text{ curl} \left(\vec{V}\right)$

(*d*) $(\text{grad } f) \times \left(\vec{V}\right)$

35. Among the following, the pair of vectors orthogonal to each-other is

(a) [3, 4, 7], [3, 4, 7] (b) [1, 0, 0], [1, 2, 0]

(c) [1, 0, 2], [0, 5, 0] (d) [1, 1, 1], [–1, –1, –1]

36. The function $f(x) = x^3 - 6x^2 + 9x + 25$ has

(a) a maxima at $x = 1$ and a minima at $x = 3$

(b) a maxima at $x = 3$ and a minima at $x = 1$

(c) no maxima, but a minima at $x = 1$

(d) a maxima at $x = 1$, but no minima

37. The temperature field in a body varies according to the equation $T(x, y) = x^3 + 4xy$. The direction of fastest variation in temperature at the point (1, 0) is given by

(a) $3\hat{i} + 8\hat{j}$ (b) $\hat{i}$

(c) $0.6\hat{i} + 0.8\hat{j}$ (d) $0.5\hat{i} + 0.866\hat{j}$

38. If $\Phi(x) = \int_0^{x^2} \sqrt{t}\, dt$, then $\frac{d\Phi}{dx}$ is

(a) $2x^2$ (b) $\sqrt{x}$

(c) 0 (d) 1

39. If velocity vector in a two dimensional flow field is given by $\vec{v} = 2xy\,\vec{i} + (2y^2 - x^2)\vec{j}$, then vorticity vector, curl $\vec{v}$ will be

(a) $2y^2\,\vec{j}$ (b) $6y\,\vec{k}$

(c) zero (d) $-4x\,\vec{k}$

40. Divergence of the vector $\vec{r} = x\,\vec{i} + y\,\vec{i} + z\,\vec{i}$ is

(a) $\vec{i} + \vec{j} + \vec{k}$ (b) 3

(c) 0 (d) 1

41. What is the derivative of $f(x) = |x|$ at $x = 0$?

(a) 1 (b) –1

(c) 0 (d) Does not exist

42. The minimum point of the function $f(x) = (x^3/3) - x$ is at

(a) $x = 1$ (b) $x = -1$

(c) $x = 0$ (d) $x = \frac{1}{\sqrt{3}}$

43. Angle between two unit-magnitude coplanar vectors P(0.866, 0.500, 0) and Q(0.259, 0.966, 0) will be

(a) 0° (b) 30°

(c) 45° (d) 60°

44. Volume of an object expressed in spherical coordinates is given by

$$V = \int_0^{2\pi}\int_0^{\pi/3}\int_0^1 r^2 \sin\phi \, dr\, d\phi\, d\theta .$$

Value of the integral is

(a) $\frac{\pi}{3}$ (b) $\frac{\pi}{6}$

(c) $\frac{2\pi}{3}$ (d) $\frac{\pi}{4}$

45. The function, $y = x^2 + \frac{250}{x}$ $x = 5$ attains

(a) maximum (b) minimum

(c) neither (d) 1

46. Volume generated by revolving the area bounded by the parabola $y^2 = 8x$ and the line $x = 2$ about y-axis is

(a) $\frac{128\pi}{5}$ (b) $\frac{5}{128\pi}$

(c) $\frac{127}{5\pi}$ (d) none of these

47. The integration of $\int \log x \,.\, dx$ has the value

(a) $(x \log x - 1)$ (b) $\log x - x$

(c) $x(\log x - 1)$ (d) None of these

48. The derivative of f (x, y) at point (1, 2) in the direction of vector i + j is $2\sqrt{2}$ and in the direction of the vector –2j is –3. Then derivative of f (x, y) in the direction –i –2j is

(a) $2\sqrt{2} + 3/2$ (b) $-7/\sqrt{5}$

(c) $-2\sqrt{2} - 3/2$ (d) $1/\sqrt{5}$

49. Area bounded by the parabola $2y = x^2$ and the line $x = y - 4$ is equal to

(a) 6

(b) 18

(c) ∞

(d) none of these

50. If a function is continuous at a point, then its first derivative

(a) may or may not exist

(b) exists always

(c) will not exist

(d) has a unique value

51. The integral $\lim_{a\to\infty}\int_1^a x^{-4}dx$

(*a*) diverges (*b*) converges to $\frac{1}{3}$

(*c*) converges to $-\frac{1}{a^3}$ (*d*) converges to 0

Numerical Type questions

1. $\lim_{x\to 0}\frac{e^x-1}{x}$ is equal to ______

2. If $\lim_{x\to 1}\frac{f(x)-2}{f(x)+2}=0$, then $\lim_{x\to 1} f(x)$ is equal to ______

3. $\lim_{x\to 0}\frac{1-\cos x}{2}$ is equal to ______

4. The value (N) of $\lim_{x\to 1}\frac{\sqrt{1+x}-\sqrt{1-x}}{x}$ is given by ______

5. $\lim_{n\to 0} e^{\left(\frac{1}{n\log n}\right)}$ is ______

6. $\lim_{x\to\infty} x\sin\frac{1}{x}$ is ______

7. $\lim_{x\to\infty}\left[\frac{1}{n^2}+\frac{2}{n^2}+\frac{3}{n^2}+\ldots+\frac{1}{n}\right]$ is

8. The 'c' of the mean value theorem for the function $f(x) = x(x-2)$, when $a = 0$, $b = \frac{3}{2}$ is ______

9. The minimum value of $|x^2 - 5x + 2|$ is ______

10. The maximum value of $\frac{1}{\sqrt{2}}(\sin x - \cos x)$ is ______

11. Maximum slope of the curve $-x^3 + 6x^2 + 2x + 1$ is ______

12. A rectangular box with square base in open at the top. The maximum volume of the box made from 300 m^3 wood is ______

13. Given $f(x, y) = x^2 + y^2$. Then $\nabla^2 f$ is ______

14. The directional derivative of

$$f(x, y, z) = x^2 + y^2 + z^2$$

at the point (1, 1, 1) in the direction i – k is ______

15. A triangle ABC consists of vertex points A (0, 0), B(1, 0) and C(0, 1).

Value of integral $\iint 2x\,dx\,dy$ over the triangle is ______

16. The magnitude of the gradient of the function $f = xyz^3$ at (1, 0, 2) is ______

17. For the function, $\phi = ax^2 y - y^3$ to represent the velocity potential of an ideal fluid, $\nabla^2\phi$ should be equal to zero. In that case, the value of 'a' has to be ______

18. Maximum value of the directional derivative of the function $\phi = 2x^2 + 3y^2 + 5z^2$ at a point (1, 1, –1) is ______

19. $\int_0^{\pi/2}\int_0^{\pi/2} \sin(x+y)\,dx\,dy$ is ______

DIFFERENTIAL EQUATIONS

MCQ Type questions

1. The solution for the differential equation

$$\frac{d^2y}{dx^2}+5\frac{dy}{dx}+6y=0 \text{ is}$$

(*a*) $C_1e^{-2t}+C_2e^{3t}$ (*b*) $C_1\sin 2t + C_2\cos 2t$
(*c*) $C_1e^{2t}+C_2e^{-2t}$ (*d*) $C_1e^{-2t}+C_2e^{-3}$

2. Integrating factor for the differential equation

$$\left(\cos^2 x\right)\frac{dy}{dx}+y=\tan x, \text{ is}$$

(*a*) $e^{\tan x}$ (*b*) $\cos 2x$
(*c*) $e^{-\tan x}$ (*d*) $\sin 2x$

3. Find range of values for a constant K to yield a stable system in the following set of time dependent differential equations

$$\frac{dy_1}{dt}=-5y_2+(4-K)y_2;\quad \frac{dy_2}{dt}=y_1-2y_2$$

(*a*) $0 < K < 7$ (*b*) $6.25 < K < 10$
(*c*) $-6 < K \le 6.25$ (*d*) $0 \le K \le 7$

4. Differential equation,

$$\frac{d^2x}{dt^2}+10\frac{dx}{dt}+25x=0$$

will have a solution of the form

(*a*) $(C_1 + C_2 t)e^{-5t}$ (*b*) C_1e^{-2t}
(*c*) $C_1e^{-5t}+C_2e^{5t}$ (*d*) $C_1e^{-5t}+C_2e^{2t}$

where C_1 and C_2 are constants.

5. Differential equation for the variation of the amount of salt x in a tank with time t is given by

$$\frac{dx}{dt} + \frac{x}{20} = 10$$

where x is in kg and t is in minutes.

Assuming that there is no salt in the tank initially, the time (in min.) at which the amount of salt increases to 100 kg is

(*a*) 10 ln 2 (*b*) 20 ln 2

(*c*) 50 ln 2 (*d*) 100 ln 2

6. Solve the differential equation

$$xdx + ydy + 2(x^2 + y^2)\, dx = 0$$

(*a*) $x^2 - y^2 = ce^{-4x}$ (*b*) $x^2 + y^2 = ce^{+4x}$

(*c*) $x^2 + y^2 = ce^{-4x}$ (*d*) $x^2 - y^2 = ce^{4x}$

7. The differential equation

$$\frac{d^2y}{dx^2} + \sin x \frac{dy}{dx} + ye^x = \sinh x$$

is

(*a*) first order and linear

(*b*) first order and non-linear

(*c*) second order and linear

(*d*) second order and non-linear

8. For the differential equation $\frac{dy}{dt} + 5y = 0$ with y(0) = 1, the general solution is

(*a*) e^{5t} (*b*) e^{-5t}

(*c*) $5e^{-5t}$ (*d*) $e^{\sqrt{-5t}}$

9. The necessary and sufficient for the differential equation of the from

$$M(x, y)\, dx + N(x, y)\, dy = 0$$

to be exact is

(*a*) M = N (*b*) $\frac{\partial M}{\partial x} = \frac{\partial N}{\partial y}$

(*c*) $\frac{\partial M}{\partial y} = \frac{\partial N}{\partial x}$ (*d*) $\frac{\partial^2 M}{\partial x^2} = \frac{\partial^2 N}{\partial y^2}$

10. The differential equation

$$y'' + (s^3 \sin x)^5 y' + y = \cos x^3 \text{ is}$$

(*a*) homogeneous

(*b*) nonlinear

(*c*) second order linear

(*d*) non-homogeneous with constant coefficients

11. For the differential equation,

$$f(x, y)\frac{dy}{dx} + g(x, y) = 0$$

to be exact

(*a*) $\frac{\partial f}{\partial y} = \frac{\partial g}{\partial x}$ (*b*) $\frac{\partial f}{\partial x} = \frac{\partial g}{\partial y}$

(*c*) $f = g$ (*d*) $\frac{\partial^2 f}{\partial x^2} = \frac{\partial^2 g}{\partial y^2}$

12. If $f(D)\, y = 0$ be a linear differential equation with constant coefficients, then its auxiliary equation is

(*a*) $f(m^3) = 0$ (*b*) $f(m) = 0$

(*c*) $f(D - m) = 0$ (*d*) $f(x) = 0$

13. The general solution of the linear differential equation $(D^4 - 81)y = 0$ is given by

(*a*) $(C_1 + C_2x)\, e^{3x} + (C_3 + C_4x) \sin 3x$

(*b*) $(C_1 + C_2x)\, e^{3x} + (C_3 + C_4x)\, e^{-3x}$
$+ (C_5 + C_6x) \cos 3x + (C_7 + C_8x) \sin 3x$

(*c*) $C_1e^{3x} + C_2e^{-3x} + C_3 \cos 3x + C_4 \sin 3x$

(*d*) $C_1e^{3x} + C_2e^{-3x} + e^{3x} (C_3 \cos x + C_4 \sin x)$

14. $\frac{1}{D - \alpha}$ Q(x) is equal to

(*a*) $e^{\alpha x} \int Q(x)dx$ (*b*) $e^{-\alpha x} \int e^{\alpha x}Q(x)dx$

(*c*) $e^{-\alpha x} \int Q(x)dx$ (*d*) $e^{\alpha x} \int e^{-\alpha x}Q(x)dx$

15. P.I. of $(D^2 - 2D)\, y = e^x \sin x$ is

(*a*) $-\frac{1}{2}e^x \sin x$ (*b*) $e^x \cos x$

(*c*) $-\frac{1}{2}e^x \cos x$ (*d*) none of these

16. General solution of the differential equation $(D^2 - m^2)\, y = 0$ is

(*a*) $y = (C_1 + C_2x)e^{mx}$

(*b*) $y = C_1 \sin x + C_2 \cos x$

(*c*) $y = C_1 \cosh x + C_2 \sinh x$

(*d*) none of these

17. $\frac{1}{\phi(D)(D - a)^n} e^{ax}$ is equal to

(*a*) $\frac{x^n e^{ax}}{\phi(a)}$ (*b*) $\frac{x^n e^{ax}}{n!\phi(a)}$

(*c*) $\frac{x^n e^{ax}}{\phi(a)}, \phi(a) \neq 0$ (*d*) $\frac{x^n e^{ax}}{n!\phi(a)}, \phi(a) \neq 0$

18. If V be any function of x, then $\frac{1}{f(D)}x V$ is equal to

(a) $\left[x+\frac{f(D)}{f'(D)}\right]f'(D).V$ (b) $\left[x-\frac{f'(D)}{f(D)}\right]f(D).V$

(c) $\left[x-\frac{f'(D)}{f(D)}\right]\frac{1}{f'(D)}.V$ (d) $\left[x-\frac{f'(D)}{f(D)}\right]\frac{1}{f(D)}.V$

19. If $y = 1 + \cos x$ and $y = 1 + \sin x$ solutions of the differential equation $\frac{d^2y}{dx^2}+y=1$, then its solution will be also

(a) $y = 2(1 + \cos x)$ (b) $y = 2 + \cos x + \sin x$
(c) $y = \cos x - \sin x$ (d) $y = 1 + \cos x + \sin x$

20. Solution of the differential equation

$$\frac{d^2y}{dx^2}-\frac{dy}{dx}-2y=3e^{2x},$$

where $y(0) = 0$ and $y'(0) = -2$ is

(a) $y = e^{-x} - e^{2x} + xe^{2x}$ (b) $y = e^{x} - e^{-2x} - xe^{2x}$

(c) $y = e^{-x} + e^{2x} + \frac{x}{2}e^{2x}$ (d) $y = e^{x} - e^{-2x} - \frac{x}{2}e^{2x}$

21. The general solution of the differential equation

$\frac{d^2y}{dx^2}+n^2y=0$ is

(a) $c_1\sqrt{\cos nx}+c_2\sqrt{\sin nx}$
(b) $c_1\cos nx + c_2 \sin nx$
(c) $c_1\cos^2 nx + c_2 \sin^2 nx$
(d) $c_1\cos^3 nx + c_2 \sin^3 nx$

22. The particular integral for the differential equation $\frac{d^2y}{dx^2}-5\frac{dy}{dx}+6y=e^{3x}$ is

(a) xe^{-x} (b) x^2e^{-x}
(c) $(x-1)e^{3x}$ (d) $(x-2)e^{4x}$

23. Solution of the differential equation $\frac{d^2y}{dx^2}=x$ is

(a) $y=\frac{x^3}{6}+c_1x^2+c_2$ (b) $y=\frac{x^3}{6}+c_1x+c_2$

(c) $y=\frac{x^2}{6}+c_1x+c_2$ (d) $y=-\frac{x^2}{6}+c_1x+c_2$

24. Particular integral of $\frac{d^2y}{dx^2}+\frac{dy}{dx}=x^2+2x+4$ is

(a) $\frac{x^2}{3}+4x$ (b) $\frac{x^3}{3}+4x$

(c) $\frac{x^3}{3}+4$ (d) $\frac{x^3}{3}+4x^2$

Numerical Type questions

1. The value of y as $t \to \infty$ for an initial value of $y(1) = 0$, for the differential equation

$$\left(4t^2+1\right)\frac{dy}{dt}+8yt-t=0 \text{ is } \_\_\_\_$$

2. The number of boundary conditions required to solve the following differential equation

$$\frac{\partial^2\phi}{\partial x^2}+\frac{\partial^2\phi}{\partial y^2}=0 \text{ is/are } \_\_\_\_$$

NUMERICAL METHODS

MCQ Type questions

1. The order of error is the Simpson's rule for numerical integration with a step size h is

(a) h (b) h^2
(c) h^3 (d) h^4

2. Four arbitrary points (x_1, y_1), (x_2, y_2), (x_3, y_3), (x_4, y_4) are given in the x, y-plane. Using the method of least squares, if regressing y upon x gives the fitted line $y = ax + b$; and regressing x upon y gives the fitted line $x = cy + d$, then

(a) two fitted lines must coincide
(b) two fitted lines need not coincide
(c) it is possible that $ac = 0$
(d) a must be $1/c$

3. The accuracy of Simpson's rule quadrature for a step size h is

(a) $O(h^2)$ (b) $O(h^3)$
(c) $O(h^4)$ (d) $O(h^5)$

4. The Newton's-Raphson iterative formula for finding $f(x) = x^2 - 1$, is

(a) $x_{i+1}=\frac{x_i^2-1}{2x_i}$ (b) $x_{i+1}=\frac{x_i^2+1}{2x_i}$

(c) $x_{i+1}=\frac{2x_i^2+1}{2x_i}$ (d) $x_{i+1}=\frac{2x_i}{2x_i^2+1}$

5. The convergence of which of the following method is sensitive to starting value?
   (*a*) False position
   (*b*) Gauss seidal method
   (*c*) Newton-Raphson method
   (*d*) All of these
6. Newton-Raphson method of solution of numerical equation is not preferred when
   (*a*) graph of A(B) is vertical
   (*b*) graph of $x(y)$ is not parallel
   (*c*) graph of $f(x)$ is nearly horizontal where it crosses the x-axis.
   (*d*) none of these
7. In which of the following methods, proper choice of initial value is very important ?
   (*a*) Bisection method (*b*) False position
   (*c*) Newton-Raphson (*d*) Bairsto method
8. Newton-Raphson method is applicable to the solution of
   (*a*) both algebraic and trans-cendental equations
   (*b*) both algebraic and trans-cendental and also used when the roots are complex
   (*c*) algebraic equations only
   (*d*) transcendental equations only

Numerical Type questions

1. Solving $x^2 - 2 = 0$ by Newton Raphson technique, when initial guess $x_0 = 1.0$, then subsequent estimate of x (*i.e.* x_1) will be _____
2. Following are the values of a function
   $y(x) : y(-1) = 5, y(0), y(1) = 8$
   $\frac{dy}{dx}$ at x = 0 as per Newton's central difference scheme is _____
3. The root of $x^3 - 2x - 5 = 0$ correct to three decimal places by using Newton-Raphson method is _____
4. Using Newton-Raphson method, a root correct to three decimal places of the equation $e^x = 1 + 2x$ is _____
5. Using Newton-Raphson method, a root correct to three decimal places of the equation $x^3 - 3x - 5 = 0$ is _____
6. Using Newton-Raphson method, a root correct to three decimal places of the equation sin x = 1 – x is _____
7. A root of the equation $x^3 - x - 11 = 0$ correct to four decimals using bisection method, is _____
8. Using Bisection method, the negative of $x^3 - x + 11 = 0$ is _____
9. A real root of equation $x^3 + x^2 - 1 = 0$ by iteration (method of successive approximation) method is _____
10. A real root of equation $\cos x = 3x - 1$ correct to seven decimal places by the method of successive approximation is _____
11. A real root of the equation $x^3 - 2x - 5 = 0$ by the method of false position correct to three decimal places is _____
12. Using Bisection method, the negative root of $x^3 - 4x + 9 = 0$ correct to three decimal places is _____
13. A real root of the equation x – cos x = 0 by the method of false position correct to four decimal places is _____
14. From certain experiment following data has been obtained:

x	1	3	4
y	4	12	19

   Using Langrange's inverse formula, the value of x for which y = 7, will be _____
15. The double (repeated) root of $4x^3 - 8x^2 - 3x + 9 = 0$ by Newton-raphson method will be _____
16. Using trapezoidal rule, and the table given below,
   $\int_4^{5.2} \ln x \, dx$ will be _____

x :	4	4.2	4.4	4.6	4.8	5	5.2
In x:	1.39	1.44	1.48	1.53	1.57	1.61	1.65

17. Using trapezoidal rule, taking 10 equal interval
   $\int_0^{\pi} \sin x \, dx$ will be _____
18. The Lagrange interpolating polynomial of degree 2 approximating the function y = ln x defined by the following table of values. The value of ln 2.7 will be ___

x	2	2.5	3.0
y = ln x	0.69315	0.91629	1.09861

19. If $y_1 = 4$, $y_3 = 12$, $y_4 = 19$ and $y_x = 7$, then x will be _____
20. The function y = sin x is tabulated below

x	0	π/4	π/2
y = sin x	0	0.70711	1.0

   Using Lagrange's interpolation formula, the value of sin (π/6) will be _____

TRANSFORM THEORY

MCQ Type questions

1. The fourier transform of the function sgn(t) defined in the figure is

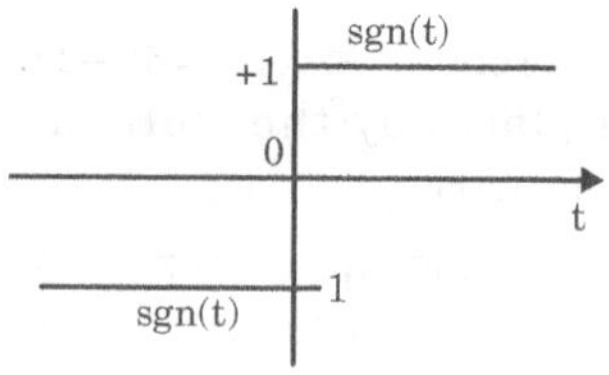

(a) $\frac{1}{j\omega}$ (b) $-\frac{1}{j\omega}$

(c) $\frac{2}{j\omega}$ (d) $-\frac{2}{j\omega}$

2. Which one of the following is the fourier transform of the signal given in figure (b), if the fourier transform of the signal in figure (a) is given by

$$\frac{2\sin\omega T_1}{\omega}$$

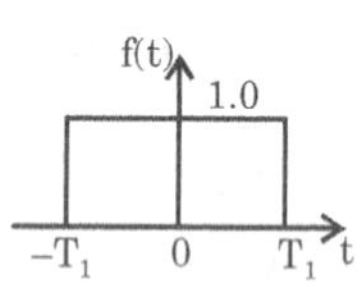

Fig. (a)

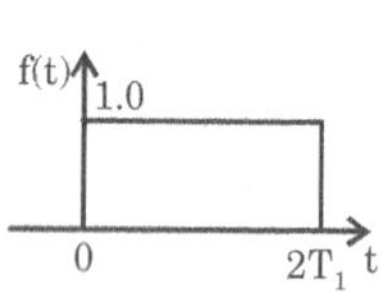

Fig. (b)

(a) $\frac{2\sin\omega T_1}{\omega}e^{j\omega T_1}$ (b) $\frac{2\sin\omega T_1}{\omega}e^{-j\omega T_1}$

(c) $\frac{\sin\omega T_1}{\omega}e^{-j\omega T_1}$ (d) $\frac{\sin\omega T_1}{\omega}e^{j\omega T_1}$

3. Fourier transform of an arbitrary signal has the property.

(a) $F(j\omega) = F(-j\omega)$ (b) $F(j\omega) = -F(-j\omega)$

(c) $F(j\omega) = F^*(-j\omega)$ (d) $F(j\omega) = -F^*(-j\omega)$

4. The two inputs to an analog multiplier are $x(t)$ and $y(t)$ with fourier transforms X(f) and Y(t) respectively. The output $z(t)$ will have a transform Z(f) given by

(a) X(f). Y(f)

(b) X(F) + Y(f)

(c) $\frac{1}{2\pi}\int_{-\lambda}^{\lambda} x(\lambda)y(\lambda - f)d\lambda$

(d) $\int_{-\infty}^{\infty} x(\lambda)y(f-\lambda)d\lambda$

5. The fourier transform of a Gaussian time pulse is

(a) uniform (b) a pair of impulse

(c) gaussian (d) rayleigh

6. The response of a system is $h(t)$ when the input is $x(t)$. The output $y(t)$ will be [H(f) is the fourier transform of $h(t)$]

(a) $h(t)x(t)$ (b) $[h(t)]\, x(t)$

(c) $h(t) * x(t)$ (d) $h(t)$

7. Which one of the following is the correct fourier transform of the unit step signal

$$u(t) = 1 \text{ for } t > 0$$
$$= 0 \text{ for } t \le 0$$

(a) $\pi\delta(\omega)$ (b) $\frac{1}{j\omega}$

(c) $\frac{1}{j\omega} + \pi\delta(\omega)$ (d) $-2\pi\delta(\omega)$

8. The fourier transform of

$v(t) = \cos\omega_0 t$ is given by

(a) $v(t) = \frac{1}{2}\delta(f - f_0)$

(b) $v(t) = \frac{1}{2}\delta(f + f_0)$

(c) $v(t) = \frac{1}{2}[\delta(f - f_0) + \delta(f + f_0)]$

(d) $v(t) = \frac{1}{2}[\delta(f - f_0) - \delta(f + f_0)]$

9. If $g(t) \leftrightarrow G(f)$ represents a fourier transform pair then according to the duality property of fourier transform.

(a) $G(t) \leftrightarrow g(f)$ (b) $G(t) \leftrightarrow g^*(f)$

(c) $G(t) \leftrightarrow g(-f)$ (d) $G(t) \leftrightarrow g^*(-f)$

10. If F(f) be the fourier transform of a function $f(t)$, then F(0) is

(a) $\int_{-\infty}^{\infty} f(t)dt$ (b) $\int_{-\infty}^{\infty} |f(t)|^2 dt$

(c) $\int_{-\infty}^{\infty} |tf(t)|^2 dt$ (d) $\int_{-\infty}^{\infty} tf(t)^2 dt$

11. The fourier transform of $x(f)$ is the periodic delta function

$$x(t) = \sum_{k=-\infty}^{\infty} \partial(t - kT) \text{ is}$$

(a) $T\sum_{k=-\infty}^{\infty} \partial(f - kT)$ (b) $T\sum_{k=-\infty}^{\infty} \partial\left(f - \frac{k}{T}\right)$

(c) $\frac{1}{T}\sum_{k=-\infty}^{\infty} \partial\left(f - \frac{k}{T}\right)$ (d) $\frac{1}{T}\sum_{k=-\infty}^{\infty} \partial(f - kT)$

12. z-transform of $x(n) = u(n)$ and ROC is

(a) $X(z) = \frac{1}{1-z^{-1}}$ $|z| < 1$

(b) $X(z) = \frac{z^{-1}}{1-z^{-1}}$ $|z| > 1$

(c) $X(z) = \frac{1}{1-z^{-1}}$ $|z| > 1$

(d) $X(z) = \frac{z}{1-z}$ $|z| < 1$

13. If $x(n) = -a^n u(-n-1)$, then z-transform and its ROC is

(a) $X(z) = \frac{z}{z-a}$ $|z| < a$

(b) $X(z) = \frac{1}{1+az^{-1}}$ $|z| < a$

(c) $X(z) = \frac{1}{1-a^{-1}z}$ $|z| < a$

(d) $X(z) = \frac{z}{z-a}$ $|z| > a$

14. If $X(z) = \frac{1}{1-az^{-1}}$ then two discrete time signals are

(a) $x(n) = a^n u(n)$ and $x(n) = a^n u(-n)$

(b) $x(n) = a^n u(n)$ and $x(n) = a^n u(-n-1)$

(c) $x(n) = a^n u(n)$ and $x(n) = -a^n u(-n-1)$

(d) none of these

15. If $x(n) = \left[7\left(\frac{1}{3}\right)^n u(n) - 6\left(\frac{1}{2}\right)^n u(n)\right]$, then ROC for the discrete time signal is

(a) $|z| < \frac{1}{3}$ (b) $|z| < \frac{1}{2}$

(c) $|z| > \frac{1}{2}$ (d) $|z| > \frac{1}{3}$

16. If $x(n) = b^{|n|}$, $b>0$, then z-transform and its ROC is

(a) $X(z) = \left(\frac{b^2-1}{b}\right)\frac{z}{(z-b)(z-b^{-1})}$; $b < |z| < \frac{1}{b}$

(b) $X(z) = \left(\frac{b^2-1}{b}\right)\frac{z}{(z-b)(z+b^{-1})}$; $b < |z| < \frac{1}{b}$

(c) $X(z) = \left(\frac{b}{b^2-1}\right)\frac{z}{(z+b)(z+b^{-1})}$; $b < |z| < \frac{1}{b}$

(d) none of these

17. For the z-transform

$X(z) = 4z^2 + 2 + 3z^{-1}$; $0 < |z| < \infty$ $x(n)$ is

(a) $x(n) = 4\delta[n-2] + 2\delta[n] + 3\delta[n+1]$

(b) $x(n) = 4\delta[n+2] + 2 + 3\delta[n-1]$

(c) $x(n) = 4\delta[n+2] + 2\delta[n] + 3\delta[n-1]$

(d) none of these

18. If $X(z) = \log(1 + az^{-1})$, find $X(n)$ if ROC is $|z| > a$.

(a) $x(n) = \frac{-(a)^n}{n}u(n-1)$

(b) $x(n) = \frac{-(-a)^n}{n}u(n+1)$

(c) $x(n) = \frac{-(-a)^n}{n}u(n-1)$

(d) $x(n) = \frac{-(a)^n}{n}u(n+1)$

19. An LTI system is stable if and only if ROC of its system function $H(z)$

(a) inside the unit circle $|z| = 1$

(b) includes the unit circle $|z| = 1$

(c) outside the unit circle $|z| = 1$

(d) none of these

20. If $x(n) = \{1, 2, 5, 7, 0, 1\}$, then ROC of $X(z)$ is entire z-plane except

(a) $z = 0$

(b) $z = \infty$

(c) $z = 0$ and $z = \infty$

(d) none of these

21. z-transform and ROC of $x(n) = [3(2)^n - 4(3)^n]u(n)$ is

(a) $X(z) = \frac{3}{1-2z^{-1}} - \frac{4}{1-3z^{-1}}$; $|z| > 2$

(b) $X(z) = \frac{3}{1-2z^{-1}} - \frac{4}{1-3z^{-1}}$; $|z| > 3$

(c) $X(z) = \frac{3}{1-2z^{-1}} - \frac{4}{1-3z^{-1}}$; $|z| < 2$

(d) $X(z) = \frac{3}{1-2z^{-1}} - \frac{4}{1-3z^{-1}}$; $|z| < 3$

22. If $x(n) = \cos \omega_0 nu(n)$, then X (z) and its ROC is

(a) $X(z) = \frac{1}{2}\left[\frac{1}{1-e^{+j\omega_0}z^{-1}} + \frac{1}{1-e^{-j\omega_0}z^{-1}}\right]; \ |z| < 1$

(b) $X(z) = \frac{1}{2}\left[\frac{1}{1-e^{+j\omega_0}z^{-1}} + \frac{1}{1-e^{-j\omega_0}z^{-1}}\right]; \ |z| > -1$

(c) $X(z) = \frac{1}{2}\left[\frac{1}{1-e^{+j\omega_0}z^{-1}} + \frac{1}{1-e^{-j\omega_0}z^{-1}}\right]; \ |z| > 1$

(d) None of these

23. If $x(n) = u(-n)$, then X(z) and ROC is

(a) $\frac{1}{1-z^{-1}} \ |z| < 1$ (b) $\frac{1}{1-z} \ |z| > 1$

(c) $\frac{z}{1-z} \ |z| < 1$ (d) $\frac{1}{1-z} \ |z| < 1$

24. If $x(n) = n\left(\frac{1}{3}\right)^n u(n)$, then X (z) and its ROC is

(a) $X(z) = \frac{\frac{1}{3}z^{-1}}{\left(1-z^{-1}\right)^2} \ |z| > \frac{1}{3}$

(b) $X(z) = \frac{\frac{1}{3}z^{-1}}{\left(1-\frac{1}{3}z^{-1}\right)^2} |z| > \frac{1}{3}$

(c) $X(z) = \frac{z^{-1}}{\left(1-\frac{1}{3}z^{-1}\right)^2} |z| > \frac{1}{3}$

(d) None of these

25. What should be the ROC of H(z) and $h(n)$ the system is anti-casual if

(a) $h(n) = \left(\frac{1}{2}\right)^n u(n) - 2(3)^n(-n-1) \ \frac{1}{2} < |z| < 3$

(b) $h(n) = -\left[\left(\frac{1}{2}\right)^n u(n) + 2(3)^n u(-n-1)\right] \ |z| < \frac{1}{2}$

(c) $h(n) = \left(\frac{1}{2}\right)^n u(n) + 2(3)^n u(n) \ |z| > 3$

(d) None of these

26. Inverse z-transform of 1 is

(a) $\delta(n+1)$ (b) $\delta(n)$

(c) $\delta(n-1)$ (d) $u(n)$

27. z-transform of $y(n) = n[u(n)]$ is

(a) $\frac{z}{\left(1-z^{-1}\right)^2}$ (b) $\frac{z^{-1}}{(1-z)^2}$

(c) $\frac{z}{(1-z)^2}$ (d) $\frac{z^{-1}}{\left(1-z^{-1}\right)^2}$

28. The region of convergence of X(z) consists of a ring in the z-plane centered about

(a) origin (b) pole

(c) Γ (d) curve

29. If $x(n)$ is anticausal (*i.e.* if it is left sided and equal to 0 for for $n > 0$), then ROC also includes as

(a) $z = 0$ (b) $z = \infty$

(c) $z = -\infty$ (d) $z = 1$ (pole)

30. For z-transform

$$X(z) = \frac{3-\frac{5}{6}z^{-1}}{\left(1-\frac{1}{4}z^{-1}\right)\left(1-\frac{1}{3}z^{-1}\right)}, \quad |z| > \frac{1}{3}$$

if one pole is $z = \frac{1}{3}$, then other pole is

(a) $z = -\frac{1}{3}$ (b) $z = \frac{1}{4}$

(c) $z = -\frac{1}{4}$ (d) $z = 0$

31. What is the value of $z\{\cos(\omega_0 n)\, u(n)\}$ when ROC $|z| > 0$?

(a) $\frac{1-z^{-1}\cos\omega_0}{1-2z^{-1}\cos\omega_0 + z^{-2}}$ (b) $\frac{1-z^{-1}\cos\omega_0}{1+z^{-1}+z^{-2}}$

(c) $\frac{1-z^{-1}}{1-2z^{-1}\cos\omega_0}$ (d) None of these

32. z-transform of $\delta(n-k)$ is

(a) z^{-k} (b) z^{k}

(b) $z^{\frac{1}{k}}$ (d) z^{-k+1}

33. z-transform of $x(n) = [4(2^n) - 5(3^n)]\,u(n)$ is

(a) $\dfrac{4}{1-2z^{-1}} - \dfrac{5}{1-3z^{-1}}$ (b) $\dfrac{4}{1-3z^{-1}} + \dfrac{5}{1-3z^{-1}}$

(c) $\dfrac{5}{1-2z^{-1}} - \dfrac{4}{1-3z^{-1}}$ (d) $\dfrac{6}{1-2z} \times \dfrac{4}{1-3z^{-1}}$

34. Inverse of z-transform of $\dfrac{Az}{(z-A)^2}$ is

(a) $nA^n\,u(n-1)$

(b) $nA^n\,u(n+1)$

(c) $(n-1)A^{(n-1)}u(n)$

(d) $nA^n u(n)$

35. A LTI system is stable if and only if ROC of its system function H(z) includes the unit circle, *i.e.*

(a) $|z| > 0$ (b) $|z| \geq \pm 1$

(c) $|z| = 1$ (d) $|z| = \dfrac{1}{z^{-1}k}$

36. The inverse z-transform of

$$X(z) = \frac{z}{\left(z-\frac{1}{2}\right)\left(z-\frac{1}{4}\right)} \text{ is}$$

(a) $4\left[\left(\frac{1}{2}\right)^n - \left(\frac{1}{4}\right)^n\right]u(n)$

(b) $4\left[\left(\frac{1}{2}\right)^n - \left(\frac{1}{4}\right)^n\right]u(n-1)$

(c) $\left[\left(\frac{1}{2}\right)^n - \left(\frac{1}{4}\right)^n\right]u(n-1)$

(d) $4\left[\left(\frac{1}{2}\right)^n - \left(\frac{1}{4}\right)^n\right]$

37. If $|\alpha| < 1$, then value of $\sum_{n=0}^{n} \alpha^n$ is

(a) $\dfrac{1}{1+\alpha}$ (b) $\dfrac{1}{1+\alpha^2}$

(c) $\dfrac{1}{1-\alpha}$ (d) $\dfrac{1}{\alpha}$

38. The value of

$$\lim_{x\to\infty} X(z) = \lim_{z\to\infty} \frac{1-\frac{3}{2}z^{-1}}{\left(1-\frac{1}{3}z^{-1}\right)\left(1-\frac{1}{2}z^{-1}\right)} \text{ is equal to}$$

(a) 0 (b) 2

(c) -1 (d) 1

39. If z-transform, X(z) of $x(n)$ is rational and if $x(n)$ is right hand sided, then ROC is region of the z-plane outside outermost pole, *i.e* outside the circle of radius equal to the largest magnitude of the poles of

(a) X(n) (b) $x(n)$

(c) $z = 0$ (d) X(z)

40. z-transform of a discrete-time signal is defined as

(a) $X(z) = \sum_{n=-\infty}^{\infty} x(n)z^{-n}$

(b) $X(z) = \sum_{n=-\infty}^{0} x(n)z^{-n}$

(c) $X(z) = \sum_{n=0}^{\infty} x(n)z^{-n}$

(d) $X(z) = \sum_{n=0}^{1} x(n)z^{-n}$

41. If $x(n) \overset{z}{\longleftrightarrow} X(z)\ x(n-k) \overset{z}{\longleftrightarrow} z^{-k}X(z)$, then

ROC of z^{-k} X(z) is same as that of X(z) except for $z = 0$ if K > 0 and $z = \infty$ if

(a) $k = 0$ (b) $k \geq 0$

(c) $k < 0$ (d) $k \leq 0$

42. Unit impulse signal $d(n)$ z-transform of

$$\delta(n) \overset{z}{\longleftrightarrow} \sum_{n=-\infty}^{\infty} \delta(n)z^{-n} \text{ is equal to}$$

(a) 1 (b) 0

(c) ∞ (d) ± 1

43. For an LTI system with input $x(n)$ and unit impulse response $h(x)$ specified as $x(n) = 2^n\,u(-n)$ and

$h(n) = u\,(n)$, the value of $\sum_{h=-\infty}^{0} 2^k$ will be

(a) 1 (b) 2

(c) 3 (d) 4

44. A causal LTI system with rational system function H(z) is stable if and only if all the poles of H(z) lie inside the unit circle *i.e.* they must all have magnitude

(a) greater than 1

(b) smaller than 1

(c) greater than equal to 1

(d) equal to zero

45. A stable and casual system is described by the difference equation

$$y(n)+\frac{1}{4}y(n-1)-\frac{1}{8}y(n-2)=-2x(n)+\frac{5}{4}x(x-1)$$

The system impulse response is

(a) $h(n)=\left(\frac{1}{4}\right)^n u(n)-3\left(-\frac{1}{2}\right)^n u(n)$

(b) $h(n)=\left(\frac{1}{4}\right)^n u(n)-3u(n)$

(c) $h(n)=u(n)$

(d) None of these

46. Consider discrete time singal $x(n)$ defined by

$$x(n)=\begin{cases}1, & -2\le n\le 2\\ 0, & |n|>2\end{cases}$$

then $y(n) = x(3n - 2)$ is equal to

(a) $x(n)=\begin{cases}1, & n=0,1\\ -1, & \text{otherwise}\end{cases}$

(b) $x(n)=\begin{cases}1, & n=1\\ -1, & n=-1\end{cases}$

(c) $x(n)=\begin{cases}1, & n=0,1\\ 0, & \text{otherwise}\end{cases}$

(d) None of these

ANSWER KEY

LINEAR ALGEBRA

MCQ Type Questions

1. (c)	**2.** (a)	**3.** (a)	**4.** (a)	**5.** (d)	**6.** (b,d)	**7.** (a)	**8.** (a)	**9.** (b)	**10.** (b)
11. (a)	**12.** (c)	**13.** (c)	**14.** (d)	**15.** (a)	**16.** (b)	**17.** (c)	**18.** (c)	**19.** (b)	**20.** (c)
21. (c,d)	**22.** (c)	**23.** (c)	**24.** (b)	**25.** (a)	**26.** (c)	**27.** (d)	**28.** (a)	**29.** (d)	**30.** (b)
31. (d)	**32.** (a)	**33.** (a)	**34.** (a)						

Numerical Type Questions

1. 3	**2.** 4	**3.** 3	**4.** 1	**5.** 2	**6.** 8	**7.** $\frac{1}{4}$	**8.** 1	**9.** 7	**10.** 4
11. 2	**12.** 12	**13.** 6							

CALCULUS

MCQ Type Questions

1. (d)	**2.** (d)	**3.** (a)	**4.** (c)	**5.** (b)	**6.** (d)	**7.** (c)	**8.** (c)	**9.** (c)	**10.** (c)
11. (b)	**12.** (b)	**13.** (b)	**14.** (a)	**15.** (b)	**16.** (d)	**17.** (c)	**18.** (d)	**19.** (b)	**20.** (a)
21. (d)	**22.** (c)	**23.** (d)	**24.** (d)	**25.** (b)	**26.** (a)	**27.** (b)	**28.** (d)	**29.** (a)	**30.** (b)
31. (a)	**32.** (d)	**33.** (c)	**34.** (a)	**35.** (c)	**36.** (a)	**37.** (c)	**38.** (a)	**39.** (d)	**40.** (b)
41. (b)	**42.** (a)	**43.** (c)	**44.** (a)	**45.** (b)	**46.** (a)	**47.** (c)	**48.** (b)	**49.** (b)	**50.** (d)
51. (a)									

Numerical Type Questions

1. (1)	**2.** (2)	**3.** (0)	**4.** (1)	**5.** (0)	**6.** (1)	**7.** (1/2)
8. (3/4)	**9.** (0)	**10.** (1)	**11.** (14)	**12.** (500)	**13.** (4)	**14.** (0)
15. (1/3)	**16.** (8)	**17.** (3)	**18.** $(\sqrt{152})$	**19.** (2)		

DIFFERENTIAL EQUATIONS

MCQ Type Questions

1. (*d*) **2.** (*c*) **3.** (*c*) **4.** (*a*) **5.** (*b*) **6.** (*c*) **7.** (*d*) **8.** (*b*) **9.** (*c*) **10.** (*c*)
11. (*b*) **12.** (*b*) **13.** (*c*) **14.** (*d*) **15.** (*a*) **16.** (*c*) **17.** (*d*) **18.** (*d*) **19.** (*d*) **20.** (*a*)
21. (*b*) **22.** (*c*) **23.** (*b*) **24.** (*b*)

Numerical Type Questions

1. (1/8) **2.** (2)

NUMERICAL METHODS

MCQ Type Questions

1. (*b*) **2.** (*d*) **3.** (*d*) **4.** (*b*) **5.** (*c*) **6.** (*c*) **7.** (*c*) **8.** (*a*)

Numerical Type Questions

1. (1.5) **2.** (1.5) **3.** (2.0946) **4.** (1.256) **5.** (2.279) **6.** (0.511) **7.** (2.3736)
8. (–2.3736) **9.** (0.7548776) **10.** (0.6071016) **11.** (2.0944) **12.** (–2.706) **13.** (0.7391) **14.** (1.857)
15. (1.5) **16.** (1.8277) **17.** (1.984) **18.** (0.9941164) **19.** (1.86) **20.** (0.51743)

TRANSFORM THEORY

MCQ Type Questions

1. (*c*) **2.** (*b*) **3.** (*c*) **4.** (*d*) **5.** (*c*) **6.** (*c*) **7.** (*c*) **8.** (*c*) **9.** (*c*) **10.** (*a*)
11. (*d*) **12.** (*c*) **13.** (*a*) **14.** (*c*) **15.** (*c*) **16.** (*a*) **17.** (*b*) **18.** (*c*) **19.** (*c*) **20.** (*a*)
21. (*b*) **22.** (*c*) **23.** (*d*) **24.** (*b*) **25.** (*c*) **26.** (*b*) **27.** (*d*) **28.** (*a*) **29.** (*a*) **30.** (*b*)
31. (*a*) **32.** (*a*) **33.** (*a*) **34.** (*d*) **35.** (*c*) **36.** (*a*) **37.** (*c*) **38.** (*d*) **39.** (*d*) **40.** (*a*)
41. (*c*) **42.** (*a*) **43.** (*b*) **44.** (*b*) **45.** (*b*) **46.** (*c*)

EXPLANATIONS

LINEAR ALGEBRA

MCQ Type Questions

1. Characteristic equation is

$$[A - \lambda I] = 0$$

$$\Rightarrow \begin{bmatrix} 3-\lambda & -1 & -1 \\ -1 & 3-\lambda & -1 \\ -1 & -1 & 3-\lambda \end{bmatrix} = 0$$

$$\Rightarrow (3-\lambda)\{(3-\lambda)^2 - 1\} - 1(3-\lambda) + 1\} - 1\{(3-\lambda)+1\} = 0$$

$$\Rightarrow \{(3-\lambda)+1\}\{(3-\lambda)(3-\lambda-1)\} - 2\} = 0$$

$$\lambda = 1, 4, 4$$

$\therefore$ λ = **1, 4, 4** are eigen values.

4. $$A' = \begin{bmatrix} 0 & -2 & -3 \\ 2 & 0 & -5 \\ 3 & 5 & 0 \end{bmatrix} = -\begin{bmatrix} 0 & 2 & 3 \\ -2 & 0 & 5 \\ -3 & -5 & 0 \end{bmatrix} = -A.$$

This is a skew-symmetric matrix.

6. Determinant A = 1 $(\cos^2\theta - \sin^2\theta)$

Hence A is non-singular and A^{-1} exists.

7. According to *Rouche's theorem,* the system is consistent if and only if coefficient matrix and augmented matrix K are of the same rank, otherwise the system is inconsistent.

8. Let A be a real symmetric matrix, therefore

$$A^T = A$$

Let α_1 and α_2 be different eigen values of the matrix A, and X_1 and X_2 be the corresponding vectors, then

$AX_1 = \alpha_1 X_1$ and $AX_2 = \alpha_2 X_2$

Taking transpose of the second equation

$$(AX_2)^T = (\alpha_2 X_2)^T$$

$\Rightarrow$ $X_2^T A^T = \alpha_2 . X_2^T$

But $A^T = A,$

$\therefore$ $X_2^T A = \alpha_2 . X_2^T$

Post multiply by X_1, we get

$$X^T_2 AX_1 = \alpha_2 X_2^T X_1$$

But $AX_1 = \alpha_1 X_1$

$\therefore$ $X^T_2 \alpha_1 X_1 = \alpha_2 X_2^T X_1$

$\Rightarrow$ $(\alpha_1 - \alpha_2) X_2^T X_1 = 0$

Since $\alpha_1 \neq \alpha_2, \alpha_1 - \alpha_2 \neq 0$

$\therefore$ $X_2^T X_1 = 0$

i.e. X_2 and X_1 are orthogonal.

9. Since matrix is triangular, hence eigen values are α, a, b.

If (X_1, X_2, X_3) is an arbitrary eigen vector, say corresponding to 1, then

$$\begin{bmatrix} 1 & 0 & 0 \\ 0 & a & 0 \\ 0 & 0 & b \end{bmatrix} \begin{bmatrix} X_1 \\ X_2 \\ X_3 \end{bmatrix} = 1 \begin{bmatrix} X_1 \\ X_2 \\ X_3 \end{bmatrix}$$

X_2, X_3 being not zero, we have

$X_1 = X_1 ; aX_2 = X_2$

which gives $a = 1$

and $bX = X_3$

$\Rightarrow$ $b = 1$

$\therefore$ $(a, b) = (1, 1)$.

10. $\frac{x}{y} = -\frac{a_2}{a_1}$ and $\frac{x}{y} = -\frac{b_2}{b_1}$

Equations are consistent only if

$$-\frac{a_2}{a_1} = -\frac{b_2}{b_1} \Rightarrow a_1 b_2 = b_1 a_2$$

11. Let matrics, $A = \begin{bmatrix} 2 & 2 & 2 \\ 1 & 5 & -3 \\ -1 & -3 & 1 \end{bmatrix}$

$$X = \begin{bmatrix} x_1 \\ x_2 \\ x_3 \end{bmatrix} \text{ and } B_1 = \begin{bmatrix} 0 \\ 0 \\ 0 \end{bmatrix}$$

$\therefore AX = B$

Multiplying both sides by A^{-1},

$X = A^{-1} B$

But as $B = 0$; therefore $X = 0$

Hence $\begin{bmatrix} x_1 \\ x_2 \\ x_3 \end{bmatrix} = \begin{bmatrix} 0 \\ 0 \\ 0 \end{bmatrix}$

$\therefore x_1 = 0, x_2 = 0, x_3 = 0.$

12. AB

$$= \begin{bmatrix} \cos\theta.\cos\phi\cos(\theta-\phi) & \cos\theta.\sin\phi.\cos(\theta-\phi) \\ \cos\phi\sin\theta.\cos(\theta-\phi) & \sin\theta.\sin\phi\cos(\theta-\phi) \end{bmatrix}$$

= A null matrix, when $\cos(\theta - \phi) = 0$

i.e. if $(\theta - \phi)$ is an odd multiple of $\left(\frac{\pi}{2}\right)$.

15. Eigen values are given by the solution of equation,

$$\begin{vmatrix} 3-\lambda & 4 \\ x & 1-\lambda \end{vmatrix} = 0$$

Since x is real and negative, put $x = -k$, where k is positive constant

$\therefore$ $(3-\lambda)(1-\lambda) + 4k = 0$

$\Rightarrow$ $\lambda^2 - 4\lambda + 3 + 4k = 0$

If λ_1 and λ_2 be the solutions of above equation, then λ_1 and λ_2 are eigen values.

Now, Sum of eigen values

= Sum of roots of above equation

i.e. $\lambda_1 + \lambda_2 = \frac{-(-4)}{1} = 4 (> 0)$

16. For given system of equations,

$$\Delta = \begin{vmatrix} 4 & 6 & 8 \\ 7 & 8 & 9 \\ 3 & 2 & 1 \end{vmatrix} = 4(8-18) - 6(7-27) + 8(14-24)$$

$$= -40 + 120 - 80 = 0$$

Since $\Delta = 0$, hence given system of equations has unique solution, *i.e.* only one solution.

17. Given matrix is $\begin{bmatrix} 1 & 1 & 1 \\ 1 & 1 & 1 \\ 1 & 1 & 1 \end{bmatrix}$

Now $\begin{bmatrix} (1-\lambda) & 1 & 1 \\ 1 & (1-\lambda) & 1 \\ 1 & 1 & (1-\lambda) \end{bmatrix} = 0$

$R_1 \to R_1 + R_2 + R_3$

$$\Rightarrow \begin{bmatrix} (3-\lambda) & (3-\lambda) & (3-\lambda) \\ 1 & (1-\lambda) & 1 \\ 1 & 1 & (1-\lambda) \end{bmatrix} = 0$$

$$\Rightarrow (3-\lambda) \begin{vmatrix} 1 & 1 & 1 \\ 1 & (1-\lambda) & 1 \\ 1 & 1 & (1-\lambda) \end{vmatrix} = 0$$

$C_2 \to C_2 - C_1$
$C_3 \to C_3 - C_1$

$$\Rightarrow \quad (3-\lambda)\begin{vmatrix} 1 & 0 & 0 \\ 1 & -\lambda & 1 \\ 1 & 0 & -\lambda \end{vmatrix} = 0$$

Hence, eigen values are 0, 0, 3.

19. $$\text{Adj R} = \begin{bmatrix} 5 & -6 & 4 \\ -3 & 4 & -3 \\ 1 & -1 & 1 \end{bmatrix}^T = \begin{bmatrix} 5 & -3 & 1 \\ -6 & 4 & -1 \\ 4 & -3 & 1 \end{bmatrix}$$

Since, $|R| = 1$, therefore top row $\to$ [5 –3 1]

20. Given equations are :

$$x + 2y + z = 6$$
$$2x + y + 2z = 6$$
$$x + y + z = 5$$

Given system can be written as

$$\begin{bmatrix} 1 & 2 & 1 \\ 2 & 1 & 2 \\ 1 & 1 & 1 \end{bmatrix} \cdot \begin{bmatrix} x \\ y \\ z \end{bmatrix} = \begin{bmatrix} 6 \\ 6 \\ 5 \end{bmatrix}$$

Applying row operation

$R_2 = R_2 - 2R_1$, $R_3 = R_3 - R_1$, we get

$$\begin{bmatrix} 1 & 2 & 1 \\ 0 & -3 & 0 \\ 0 & -1 & 0 \end{bmatrix} \cdot \begin{bmatrix} x \\ y \\ z \end{bmatrix} = \begin{bmatrix} 6 \\ -6 \\ -1 \end{bmatrix}$$

Applying $R_3 = 3R_3 - R_2$, we get

$$\begin{bmatrix} 1 & 2 & 1 \\ 0 & -3 & 0 \\ 0 & 0 & 0 \end{bmatrix} \cdot \begin{bmatrix} x \\ y \\ z \end{bmatrix} = \begin{bmatrix} 6 \\ -6 \\ 3 \end{bmatrix}$$

Since rank of co-efficient matrix is 2 and rank of argument matrix is 3, which is not equal. Hence system has no solution.

21. Eigen value are roots of the equation

$$\begin{bmatrix} 1-\alpha & -1 \\ -1 & 1-\alpha \end{bmatrix} = 0$$

i.e. $$(1-\alpha)^2 - 1 = 0$$

$$\Rightarrow \quad \alpha^2 - 2\alpha = 0$$

$$\Rightarrow \quad \alpha = 0 \text{ or } \alpha = 2$$

When $\alpha = 0$, $$\begin{bmatrix} 1 & -1 \\ -1 & 1 \end{bmatrix} \begin{bmatrix} x \\ y \end{bmatrix} = 0$$

$$\Rightarrow \quad \begin{bmatrix} x \\ y \end{bmatrix} = \begin{bmatrix} 0 \\ 0 \end{bmatrix}$$

This gives, $x = y$

When $\alpha = 2$, $$\begin{bmatrix} 1 & -1 \\ -1 & 1 \end{bmatrix} \begin{bmatrix} x \\ y \end{bmatrix} = 2$$

$$\Rightarrow \quad \begin{bmatrix} x \\ y \end{bmatrix} = \begin{bmatrix} 2x \\ 2y \end{bmatrix}.$$

This gives, $x = -y$

Hence, eigen vectors corresponding to 0 and 2

are $\begin{bmatrix} x \\ x \end{bmatrix}$ and $\begin{bmatrix} x \\ -x \end{bmatrix}$, or $\begin{bmatrix} 1 \\ 1 \end{bmatrix}$ and $\begin{bmatrix} 1 \\ -1 \end{bmatrix}$

22. For matrix $A = \begin{bmatrix} -3 & 5 \\ 2 & 1 \end{bmatrix}$

cofactors are :

$$a_{11} = 1, a_{21} = -5, \ a_{12} = -2, a_{22} = -3$$

$$\therefore \text{ Matrix by cofactors} = \begin{bmatrix} 1 & -2 \\ -5 & -3 \end{bmatrix}$$

$$\text{Adjoint A} = \begin{bmatrix} 1 & -5 \\ -2 & -3 \end{bmatrix}$$

$$|A| = -3(1) - 5(2)$$
$$= -3 - 10 = -13$$

$$\therefore \ A^{-1} = \frac{\text{Adjoint A}}{|A|} \begin{bmatrix} -\frac{1}{13} & \frac{5}{13} \\ \frac{2}{13} & \frac{3}{13} \end{bmatrix}$$

25. $$\text{Inverse of 7A} = \begin{bmatrix} -1 & 2 \\ 4 & -7 \end{bmatrix}$$

$$\det |\Delta| = \begin{vmatrix} -1 & 2 \\ 4 & -7 \end{vmatrix} = 7 - 8 = -1$$

$$\therefore \text{ Matrices of cofactor} = \begin{bmatrix} -7 & -4 \\ -2 & -1 \end{bmatrix},$$

and it's transpose, $$A^T = \begin{bmatrix} -7 & -2 \\ -4 & -1 \end{bmatrix}$$

$$\therefore \text{ Inverse} = \frac{A^T}{|\Delta|} = \begin{bmatrix} 7 & 2 \\ 4 & 1 \end{bmatrix}$$

$$\therefore \ A = \frac{1}{7}\left[\frac{A^T}{|\Delta|}\right] = \begin{bmatrix} 1 & \frac{2}{7} \\ \frac{4}{7} & \frac{1}{7} \end{bmatrix}$$

30. Here $[s\,I - A]^{-1} = \begin{bmatrix} s-1 & 0 \\ 0 & s-1 \end{bmatrix}^{-1}$

Then, $2^{-1}[sI - A]^{-1} = 2^{-1}\left\{\begin{bmatrix} \frac{1}{s-1} & 0 \\ 0 & \frac{1}{s-1} \end{bmatrix}\right\}$

$$= \begin{bmatrix} e^t & 0 \\ 0 & e^t \end{bmatrix}$$

31. $|P - \lambda I| = \begin{vmatrix} 3-\lambda & -2 & 2 \\ 0 & -2-\lambda & 1 \\ 0 & 0 & 1-\lambda \end{vmatrix} = 0$

$\therefore\quad \lambda = -2, 3, 1$

Putting $\lambda = -2$, we get

$5x - 2y + 2z = 0$

$z = 0$

So, $5x = 2y$ eigen vector $= \begin{bmatrix} 5 \\ 2 \\ 0 \end{bmatrix}$

32. We know, $AA^{-1} = I$

$\therefore\quad \begin{bmatrix} 2 & -0.1 \\ 0 & 3 \end{bmatrix}\begin{bmatrix} \frac{1}{2} & a \\ 0 & b \end{bmatrix} = \begin{bmatrix} 1 & 2a - 0.1b \\ 0 & 3b \end{bmatrix}$

$\Rightarrow\quad 3b = 1 \quad\Rightarrow\quad b = \frac{1}{3}$

or $\quad 2a - 0.1b = 0$

$\Rightarrow\quad 2a = \frac{b}{10} \quad\Rightarrow\quad a = \frac{1}{60}$

$\therefore\quad a + b = \frac{1}{3} + \frac{1}{60} = \frac{7}{20}$

33. Given matrix : $A = \begin{bmatrix} 6 & 2 \\ 2 & 6 \end{bmatrix}$

We know $|\lambda I - A| = 0$

$\Rightarrow\quad \begin{vmatrix} \lambda-6 & -2 \\ -2 & \lambda-6 \end{vmatrix} = 0$

$\Rightarrow\quad \lambda^2 - 12\lambda + 32 = 0$

$\Rightarrow\quad \lambda = 4, 8$ (eigen values)

For $\lambda_1 = 4$,

$$(\lambda_1 I - A) = \begin{bmatrix} -2 & -2 \\ -2 & -2 \end{bmatrix},\ v_1 = \begin{bmatrix} 1 \\ 1 \end{bmatrix}$$

For $\lambda_1 = 8$,

$$(\lambda_2 I - A) = \begin{bmatrix} 2 & -6 \\ -6 & 2 \end{bmatrix},\ v_2 = \begin{bmatrix} 1 \\ -1 \end{bmatrix}$$

34. State of the system at time t is,

$$X(t) = [sI - A]^{-1}X(0) = \phi(t)X(0)$$

$$= \begin{bmatrix} e^{-2t} & 0 \\ 0 & e^{-t} \end{bmatrix}\begin{bmatrix} 2 \\ 3 \end{bmatrix} = \begin{bmatrix} 2e^{-2t} \\ 3e^{-t} \end{bmatrix}$$

At $t = 1$, $X(1) = \begin{bmatrix} 2e^{-2} \\ 3e^{-1} \end{bmatrix} = \begin{bmatrix} 0.271 \\ 1.100 \end{bmatrix}$

Numerical Type Questions

1. Determinant A = 0

$\Rightarrow\quad \begin{vmatrix} 4 & 2 & 3 \\ 1 & 0 & 0 \\ 4 & 0 & 3 \end{vmatrix} = (-1) \times 6 = -6 \neq 0$

Hence rank of A = 3

2. Since the number of non-zero elements on this diagonal matrix is four, hence the rank is four.

3. Given matrix A possesses a minor of order 3

viz. $\begin{vmatrix} 1 & 3 & 1 \\ 2 & 4 & 0 \\ 3 & 1 & 5 \end{vmatrix} = \begin{vmatrix} 0 & 0 & 1 \\ 2 & 4 & 0 \\ -2 & -14 & 5 \end{vmatrix}$

Replacing C_1 and C_2 by $C_1 - C_3$ and $C_2 - 3C_3$.

$= \begin{vmatrix} 2 & 4 \\ -2 & -14 \end{vmatrix}$ expanding with respect to R_1

$= 2(-14) - (4)(-2) = -28 + 8 \neq 0$

$\therefore\quad p(A) \geq 3 \quad\quad ...(i)$

Also A does not possess any minor of order 4, *i.e.* 3 + 1

$\therefore\quad p(A) \leq 3 \quad\quad ...(ii)$

From equations (*i*) and (*ii*), we get

$p(A) = 3$ *i.e.* rank of A is 3.

5. $A = \begin{bmatrix} 2 & -5k & 6 \\ k & 2 & -2 \\ 2 & 2 & -k \end{bmatrix}$

Equations are consistent, if rank of A and that of k are equal. But in this case it is always true. Hence equations will have a trivial solution if $|A| \neq 0$ Therefore only non-trivial solution will exist if $|A| = 0$

i.e. $\begin{bmatrix} 2 & -5k & 6 \\ k & 2 & -2 \\ 2 & 2 & -k \end{bmatrix} = 0$

$\Rightarrow\ 2(-2k+4)+5k(-k^2+4)+6(2k-4)$

$\Rightarrow\ -5k^3+20k-4k+8+12k-24=0$

$\Rightarrow\ 5k^3-28k+16=0$

$\Rightarrow\ 5k^3-10k^2+10k^2-20k-8k+16=0$

$\Rightarrow\ (5k^2+10k-8)(k-2)=0$

$$k = \frac{-10 \pm \sqrt{100+160}}{2} \text{ or } 2$$

$$= -1 \pm \sqrt{13/5} \text{ or } 2$$

6. Equivalent matrix equation is

$$\begin{bmatrix} 2 & 1 & 2 \\ 1 & 1 & 3 \\ 4 & 3 & \lambda \end{bmatrix}\begin{bmatrix} x \\ y \\ z \end{bmatrix} = 0$$

In order that given system of equations may have non-zero solution, rank of A should be less than 3. This requires that

$$\begin{bmatrix} 2 & 1 & 2 \\ 1 & 1 & 3 \\ 4 & 3 & \lambda \end{bmatrix} = 0$$

Interchanging R_1 and R_2,

$$\begin{bmatrix} 1 & 1 & 3 \\ 2 & 1 & 2 \\ 4 & 3 & \lambda \end{bmatrix} = 0$$

By $(R_2 - 2R_1)$ and $(R_3 - 4R_1)$,

$$\begin{bmatrix} 1 & 1 & 3 \\ 1 & -1 & -4 \\ 0 & -1 & \lambda-12 \end{bmatrix} = 0$$

$$\Rightarrow \begin{bmatrix} -1 & -4 \\ -1 & \lambda-12 \end{bmatrix} \Rightarrow -1(\lambda-12)-(-1)(4)=0$$

$\Rightarrow\ \lambda = \mathbf{8}$

7. The system of homogeneous linear equations has a non-trivial solution if

$$\begin{bmatrix} 4d-1 & 1 & 1 \\ 0 & -1 & 1 \\ 0 & 0 & 4d-1 \end{bmatrix} = 0$$

$$\Rightarrow -(4d-1)^2 = 0 \quad \Rightarrow d = \frac{1}{4}$$

8. Let $A = \begin{bmatrix} a_1 \\ b_1 \\ c_1 \end{bmatrix}$; $B = [a_2\ b_2\ c_2]$

$$\therefore \quad C = \begin{bmatrix} a_1a_2 & a_1b_2 & a_1c_2 \\ b_1a_2 & b_1b_2 & b_1c_2 \\ c_1a_2 & c_2b_2 & c_1c_2 \end{bmatrix}$$

$$= a_1\, a_2\, [b_1\, b_2\, c_1\, c_2 - b_1 b_2\, c_1 c_2] = 0$$

$$\text{Cofactor} = \begin{vmatrix} a_1a_2 & a_1b_2 \\ b_1a_2 & b_1a_b \end{vmatrix} = 0$$

So rank = 1

9. Sum of eigen values of given matrix
= Sum of element of diagonal of given matrix
= 1 + 5 + 1 = 7

10. For singular matrix,

$$\begin{vmatrix} 8 & x & 0 \\ 4 & 0 & 2 \\ 12 & 6 & 0 \end{vmatrix} = 0$$

$\Rightarrow\ 8[0-12]-x[0-2\times 12]=0$

$\Rightarrow\ -96+24x=0$

$\Rightarrow\ x=4$

11. Since, $det\ A = \begin{vmatrix} 1 & -2 & -1 \\ -3 & 3 & 0 \\ 2 & 2 & 4 \end{vmatrix}$

$= 12 - 24 + 12 = 0$

and $M_{11} = 12 \neq 0$. Hence rank is 2.

12. Equations are consistent when

$$\begin{vmatrix} 2 & 3 & -1 \\ 1 & -1 & -4 \\ 4 & -1 & -a \end{vmatrix} = 0$$

$\Rightarrow\ 2(a-4)-3(-a+16)-1(-1+4)=0$

$5a-8-48-3=0$

$5a=59$

$$a = \frac{59}{5} \approx 12$$

13. Given : Matrix $M = \begin{bmatrix} 4 & 2 \\ 2 & 4 \end{bmatrix}$

$$\therefore \quad M - \lambda I = \begin{bmatrix} 4-\lambda & 2 \\ 2 & 4-\lambda \end{bmatrix}$$

For the eigen vector $\begin{bmatrix} 101 \\ 101 \end{bmatrix}$

$(4-\lambda)(101)+2(101)=0$

$\Rightarrow\ 4-\lambda+2=0$

$\Rightarrow\ \lambda = 6$ (eigen value)

CALCULUS

MCQ Type Questions

1. $\lim_{x\to\infty} x^n e^{-x} = \lim_{x\to\infty} \frac{x^n}{e^x}$ $\left[\text{form } \frac{\infty}{\infty}\right]$

By L' Hospital Rule Differentiating denominator and numerator n times, we get

$$\lim_{x\to\infty} \frac{x^n}{e^x} = \lim_{x\to\infty} \frac{\underline{|n}}{e^x} = 0$$

2. $\text{L.}\lim_{x\to 2}[x] = \lim_{h\to 0}[2-h] = \lim_{h\to 0} 1 = 1$

$\text{R.}\lim_{x\to 2}[x] = \lim_{h\to 0}[2+h] = \lim_{h\to 0} 2 = 2$

$\therefore\ \text{L.}\lim_{x\to 2}[x] \neq \text{R.}\lim_{h\to 2}[x]$

3. $\text{L.}\lim_{x\to 0} f(x) = \lim_{h\to 0} f(0-h) = \lim_{h\to 0} -h\sin\left(-\frac{1}{h}\right)$

$= \lim_{h\to 0} +h\sin\frac{1}{h} = 0$

$\text{R.}\lim_{x\to 0} f(x) = \lim_{h\to 0} f(0+h)$

$= \lim_{h\to 0} -h\sin\left(\frac{1}{h}\right) = 0$

Also $f(0) = 0$

Since $\text{L.}\lim_{x\to 0} f(x) = \text{R.}\lim_{x\to 0} f(x) = f(0) = 0$

Hence $f(x)$ is continuous at $x = 0$

Now, $\text{L.}\lim_{x\to 0} g(x) = \lim_{h\to 0} g(0-h)$

$= \lim_{h\to 0}|-h| = \lim_{h\to 0} h = 0$

$\text{R.}\lim_{x\to 0} g(x) = \lim_{h\to 0} g(0+h)$

$= \lim_{h\to 0}|h| = \lim_{h\to 0} h = 0$

$g(0) = |0| = 0$

$\therefore\ \text{L.}\lim_{x\to 0} g(x) = \text{R.}\lim_{x\to 0} g(x) = g(0) = 0$

Hence $g(x)$ is also continuous at $x = 0$

4. Let $a \in$ R (real numbers)

$\text{L.}\lim_{x\to a} f(x) = \lim_{h\to 0}|a-h| = \lim_{h\to 0}(a-h) = a$

$\text{R.}\lim_{x\to a} f(x) = \lim_{h\to 0}|a+h| = \lim_{h\to 0} a+h = a$

and $f(a) = |a| = a$

$\therefore\ \lim_{x\to a}|x| = a = f(a)$

Hence $\lim_{x\to a}|x|$ continuous for all values of x.

5. $\text{L.}\lim_{x\to 2} f(x) = \lim_{h\to 0} 2a(2-h) - b = 4a - b$

$\text{R.}\lim_{x\to 2} f(x) = \lim_{h\to 0} a(2+h)^2 + b = 4a + b$

The function will be continuous if

$\text{L.}\lim_{x\to 2} f(x) = \text{R.}\lim_{x\to 2} f(x) = f(2)$

$\therefore\ 4a - b = 2$

$4a + b = 2$

Solving we get, $a = \frac{1}{2}$, $b = 0$

6. Given function is, $f(x) = x - [x]$

Let a is a real number except integer and h be a very small real number, then

$[a-h] = b$ (an integer less than to a)

$[a+h] = b$ (an integer greater than to a)

Also $[a] = b$

$\therefore\ \text{L.}\lim_{x\to 0} f(x) = \lim_{h\to 0}(a-h) - [a-h]$

$= \lim_{h\to 0}(a-h) - b = a - b$

Also $f(a) = (a) - [a] = a - b$

$\therefore\ \text{L.}\lim_{x\to a} f(x) = \text{R.}\lim_{x\to a} f(x) = f(a) = a - b$

$\therefore\ f(x) = x - [x]$

is continuous all real except integers.

7. Since $f(x) = \frac{1}{x}$ is not continuous in [– 3, 3] [– 2, 2] or, [– 1, 1], hence the point of discontinuity is '0'. Only in [2, 3] the function is continuous, and differentiable. Hence mean value theorem is applicable in [2, 3].

8. Given function is

$$f(x) = 3x^4 - 4x^2 + 5$$

(*i*) $f(x)$ is continuous in every real interval

(*ii*) $f'(x)$ exist in any real interval

(*iii*) $f(-1) = 3(-1)^4 - 4(-1)^2 + 5 = 4$

$f(1) = 3(1)^4 - 4(1)^2 + 5 = 4$

$f(-1) = f(1)$

Also $f'(c) = 12c^3 - 8c = 0$

$$\Rightarrow \qquad c = 0, c = \pm\sqrt{\frac{2}{3}}$$

$$\therefore \qquad c \in [-1, 1]$$

Hence all the condition of Rolle's Theorem are satisfied in the interval [–1, 1]

9. Since $f(x) = |x|$ is continuous in [–1, 1] but it is not differentiable at $x = 0 \in (-1, 1)$

10. Since f and F are two continuous function in [a, b] such that

$$f'(x) = F'(x)$$

Let $\phi(x) = f(x) - F(x)$

$\therefore \quad \phi'(x) = f'(x) - F'(x)$

$= 0$ [from equation]

$\therefore \quad \phi(x) = \text{constant}$

11. Let $f(x) = x^{1/x}$

$$\therefore \qquad f'(x) = x^{1/x}\left(\frac{1-\log_e x}{x^2}\right) < 0$$

$f(x)$ is decreasing if $f'(x) < 0$

$$i.e., \ x^{1/x}\left(\frac{1-\log x}{x^2}\right) < 0$$

$$\Rightarrow 1 - \log_e x < 0 \quad \Rightarrow \quad \log_e x > 1 \quad \Rightarrow \quad x > e$$

12. Let $f(x) = \cos x$

$$f''(x) = \cos\left(x + \frac{n\pi}{2}\right)$$

$$\therefore \qquad f'(0) = 0$$
$$f''(0) = -1$$
$$f'''(0) = 0$$
$$f''''(0) = 1$$

....

....

and so on

By Maclaurin's theorem, we get

$$\cos(x) = f(0) + xf'(0) + \frac{x^2}{\underline{|2}} f''(0) + \frac{x^2}{\underline{|3}} f'''(0)$$

$$+ \ldots + \frac{x^{2n}}{\underline{|2n}} f^{2n}(0) + \frac{x^{2n+1}}{\underline{|2n+1}} f^{2n+1}(\theta x)$$

$$= 1 + 0 - \frac{x^2}{\underline{|2}} + 0 + \frac{1}{\underline{|4}} \ldots + (-1)^{n+1} \frac{x^{2n+1}}{\underline{|2n+1}} \sin\theta x$$

$$= 1 - \frac{x^2}{\underline{|2}} + \frac{1}{\underline{|4}} x^4 + \ldots + (-1)^{n+1} \frac{x^{2n+1}}{\underline{|2n+1}} \sin\theta x$$

13. Given: $f(x) = \dfrac{1}{1+x^2}$

$\therefore \quad (1 + x^2) f(x) = 1;\ f(0) = 1$

$(1 + x^2) f'(x) + 2x f(x) = 0;\ f'(0) = 0$

$(1 + x^2) f''(x) + 4x f'(x) + 2f(x) = 0;\ f''(0) = -2$

$(1 + x^2) f'''(x) + 6x f''(x) + 6f'(x) = 0;\ f'''(0) = 0$

$(1 + x^2) f^{iv}(x) + 8x f'''(x) + 12 f''(x) = 0;\ f^{iv}(0) = 24$

and so on

$$\therefore \quad f(x) = f(0) + x f'(0) + \frac{x^2}{\underline{|2}} f'(0) + \ldots$$

$$\therefore \quad \frac{1}{1+x^2} = 1 + x.\ 0 + \frac{x^2}{\underline{|2}}(-2) + \frac{x^2}{\underline{|3}}.0 + \frac{x^4}{\underline{|4}}.24$$

$$+ \ldots + (-1)^n . \frac{x^{2n}}{\underline{|2n}} . \underline{|2n}$$

$$= 1 - x^2 + x^4 + \ldots + (-1)^n . x^{2n}$$

$$= \sum_{n=0}^{\infty} (-1)^n . x^{2n} \quad \text{for all real } x.$$

19. Let ABC be a triangle inscribed in the circle with centre O and radius r.

If area of this triangle is maximum, then vertex C should be at a maximum distance from the base AB i.e., CD must be perpendicular to AB. Hence ABC is an isosceles triangle.

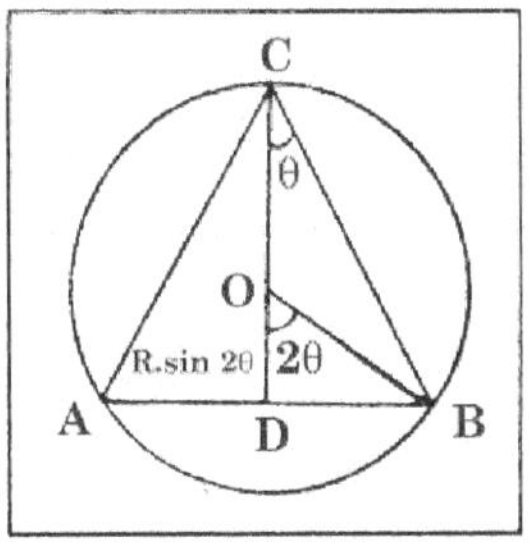

If $\angle BCD = \theta$,

where D is the mid-point of AB, then

$\angle BOD = 2\theta$

$\therefore \quad AB = 2BD = 2r \sin 2\theta$

$CD = CO + OD = r + r \cos 2\theta$

If S be the area of the triangle ABC, then

$$S = \frac{1}{2} AB \times CD = \frac{1}{2} \times 2r \sin 2\theta\ (r + r \cos 2\theta)$$

$$\therefore \quad \frac{ds}{d\theta} = r^2[\sin 2\theta\,(-2 \sin 2\theta] + (1 + \cos 2\theta)(2 \cos 2\theta)]$$

For maximum and minimum,

$$\frac{dS}{d\theta} = 0$$

$\therefore$ $\cos 4\theta + \cos 2\theta = 0$

$\Rightarrow$ $2 \cos 3\theta \cos \theta = 0$

Hence either $\cos 3\theta = 0$, or $\cos \theta = 0$

If $\cos \theta = 0$, then $\theta = \frac{\pi}{2}$

If $\cos 3\theta = 0$, then $3\theta = \frac{\pi}{2}$

$\Rightarrow$ $\theta = \frac{\pi}{6}$

$\left(\frac{d^2S}{d\theta^2}\right)_{\theta=\pi/6}$ is negative.

Hence S is maximum for $\theta = \frac{1}{6}\pi$

$$\angle ACB = 2\theta = 2\left(\frac{\pi}{6}\right) = \frac{\pi}{3}$$

$$= \angle ABC = \angle BAC$$

Hence ΔABC is an equalateral triangle.

20. Let radius of circle is r and side of a sqaure is x.

$\therefore$ Perimeter $= 2\pi r + 4x = l$ ...(*i*)

Let area be A

$\therefore$ $A = \pi r^2 + x^2$

$$= \pi r^2 + \left(\frac{l - 2\pi r}{4}\right)^2 \quad \text{[from equation (i)]}$$

$$\frac{dA}{dr} = 2\pi r + \frac{2}{16}(1 - 2\pi r).(-2\pi)$$

and $\frac{d^2A}{dr^2} = 2\pi + \frac{\pi^2}{2}$

For maximum and minimum,

$$\frac{dA}{dr} = 0$$

$\Rightarrow$ $2\pi r - \frac{\pi}{4}(l - 2\pi r) = 0$

$\Rightarrow$ $l = 8r + 2\pi r$ ...(*ii*)

$\frac{d^2A}{dr^2} > 0$, hence area is least

From equations (*i*) and (*ii*),

$$2\pi r + 4x = 8r + 2\pi r$$

$\Rightarrow 4x = 8r \quad \Rightarrow \quad x = 2r$

Hence side of a square is double the radius of circle.

21. Since statements (*a*), (*b*), (*c*) are all correct, hence the statement (*d*) is false.

23. $f(x) = x^4 - 8x^3 + 22x^2 - 24x + 1$

$f(0) = 1$

$f(2) = 2^4 - 8.2^3 + 22.2^2 - 24.2 + 1 = -7$

Now $f'(x) = 4x^3 - 24x^2 + 44x - 24$

For maximum and minimum,

$f'(x) = 0$

$\Rightarrow$ $4x^3 - 24x^2 + 44x - 24 = 0$

$\Rightarrow$ $4(x - 1)(x - 2)(x - 3) = 0$

$\therefore$ $x = 1, 2, 3$

Since $x = 3$ does not lie in $[0, 2]$

Hence, consider only $x = 1$ and $x = 2$.

We have

$f(1) = 1^4 - 8.1^3 + 22.1^2 - 24.1 + 1 = -8$

Greatest of $f(x) =$ largest of $\{1, -7, -8\} = 1$

Least of $f(x) =$ smallest of $\{1, -7, -8\} = -8$

24. $f(x) = x^5 - 5x^4 + 5x^3 - 1$

$f'(x) = 5x^4 - 20x^3 + 15x^2$

$f''(x) = 20x^3 - 60x^2 + 30x$

$f'''(x) = 60x^2 - 120x + 30$

For maximum and minimum

$f'(x) = 0$

$\therefore$ $x^4 - 4x^3 + 3x^2 = 0$

$\Rightarrow$ $x^2(x - 1)(x - 3) = 0$

$\therefore$ $x = 0, 1, 3$

$f''(0) = 0$

$f'''(0) = 30 \neq 0$

$\therefore f(x)$ is neither maximum nor minimum at $x = 0$

$f''(1) = 20 - 60 + 30$

$= -10 < 0$

$\therefore f(x)$ is maximum at $x = 1$

$f''(3) = 540 - 540 + 90$

$= 90 > 0$

$\therefore f(x)$ is maximum at $x = 3$

Hence there is one maximum and one minimum for $f(x)$.

25. Given function is

$$f(x) = \frac{\log x}{x}$$

$$\therefore \quad f'(x) = \frac{x.\frac{1}{x} - \log x.1}{x^2} = \frac{1-\log x}{x^2}$$

$$\text{and} \quad f''(x) = \frac{x^2.\left(-\frac{1}{x}\right) - (1-\log x).2x}{x^4}$$

$$= \frac{2\log x - 3}{x^3}$$

For maximum and minimum of $f(x)$,

$f'(x) = 0 \Rightarrow 1 - \log x = 0 \quad \Rightarrow \quad x = e$

$$\therefore \quad f''(e) = \frac{2-3}{e^3} = -\frac{1}{e^3} < 0$$

Hence $f(x)$ is maximum at $x = e$

$$\therefore \text{ Maximum value, } f(e) = \frac{\log e}{e} = \frac{1}{e}$$

29. $\int_0^{5\pi}(2-\sin x)dx = [2x - \cos x]_0^{5\pi}$

$= 10\pi - 1 - 1 = 10\pi - 2 > 0$

31. $f(x) = 3x(x-2) = 3x^2 - 6x$

Differentiating, we get

$$\frac{df(x)}{dx} = f'(x) = 6x - 6$$

Again differentiating,

$$\frac{d^2f(x)}{dx} = f''(x) = 6 > 0$$

For maxima or minima,

$f'(x) = 0$

$\therefore \quad 6x - 6 = 0,$

$\Rightarrow \quad x = 1$

As $f''(x) = 6 > 0$, $f(x)$ is minimum at $x = 1$

32. $$\int_{-2}^{2}\frac{dx}{x^2} = \int_{-2}^{2}\frac{1}{x^2}dx$$

$f(x) = \frac{1}{x^2}$ = even function

For even function,

$$\int_{-a}^{a} f(x)dx = 2\int_0^a f(x)dx = 2\int_0^2 \frac{1}{x^2}dx = 2\left[-1\frac{1}{x}\right]_0^2 = \infty$$

$$\therefore \quad \int_{-2}^{2}\frac{1}{x^2} = \infty$$

33. Given : $f(x) = Ax^2 + Bx + C$

$$\Rightarrow \quad f'(x) = 2Ax + B \quad \text{and} \quad \frac{f(b)-f(a)}{b-a} = f'(\xi)$$

$$\Rightarrow 2A\xi + B = \frac{(Ab^2 + Bb + C) - (Aa^2 + Ba + C)}{b-a}$$

$$= \frac{A(b^2 - a^2) + B(b-a)}{b-a} = A(b+a) + B$$

$$\therefore \quad \xi = \frac{b+a}{2}$$

34. $$\text{curl}(f\vec{V}) = i \times \frac{\partial}{\partial x}\left(f\vec{V}\right) + j \times \frac{\partial}{\partial y}\left(f\vec{V}\right) + k \times \frac{\partial}{\partial z}\left(f\vec{V}\right)$$

$$= i \times \left[f\frac{\partial\vec{V}}{\partial x} + \frac{\partial f}{\partial x}\vec{V}\right] + \text{two similar terms}$$

$$= i \times \left[f\frac{\partial\vec{V}}{\partial x} + j \times \partial\vec{V} + k \times \frac{\partial\vec{V}}{\partial z}\right]$$

$$+ \left[i \times \frac{\partial f}{\partial x} + j\frac{\partial f}{\partial y} + k\frac{\partial f}{\partial x}\right] \times \vec{V}$$

$= f \text{ curl } \vec{V} + (\text{grad} f) \times \vec{V}$

35. (a_1, b_1, c_1) and (a_2, b_2, c_2) are orthogonal, only if

$$a_1a_2 + b_1b_2 + c_1c_2 = 0$$

Only C fits, i.e. $[1, 0, 2] \times [0, 5, 0] = 0 + 0 + 0 = 0$

36. $f'(x) = 3(x^2 - 4x + 3)$, $f''(x) = 6x - 12$

At $x = 1$; $f''(1) = -ve$; and, at $x = 3$, $f''(3) = +ve$

37. Given $\quad T = x^3 + 4xy$

$$\Rightarrow \quad \frac{\partial T}{\partial x} = 3x^2 + 4y$$

$$\therefore \quad \left(\frac{\partial T}{\partial y}\right)_{(1,0)} = 3, \quad \Rightarrow \quad \left(\frac{\partial T}{\partial y}\right)_{(1,0)} = 4$$

Hence direction of fastest variation in temperature at (1, 0) is given by $3\hat{i} + 4\hat{j}$ or $0.6\hat{i} + 0.8\hat{j}$

38. $$\phi(x) = \int_0^{x^2} \sqrt{t}\, dt$$

$$= \left|\frac{2}{3}t^{3/2}\right|_0^{x^2} = \frac{2}{3}x^3$$

$$\therefore \quad \frac{d\phi}{dx} = \frac{2}{3} \times 3x^2 = 2x^2$$

39. Given: $\vec{v} = 2xy\vec{i} + (2y^2 - x^2)\vec{j}$

$$\therefore \text{Curl } \vec{v} = \begin{vmatrix} i & j & k \\ \frac{\partial}{\partial x} & \frac{\partial}{\partial k} & \frac{\partial}{\partial z} \\ 2xy & 2y^2 - x^2 & 0 \end{vmatrix}$$

$$= i(0) - j(0) + k \left[\frac{\partial}{\partial x}(2y^2 - x^2) - \frac{\partial}{\partial y}(2xy)\right]$$

$$= (-2x - 2x)\, k$$

$$= -4xk.$$

40. div. $\vec{r} = \vec{\Delta}.\vec{r}$

$$= \left(\frac{\partial}{\partial x}\hat{i} + \frac{\partial}{\partial y}j + \frac{\partial}{\partial z}\hat{k}\right)\left(x\hat{i} + y\hat{j} + z\hat{k}\right)$$

$$= 1 + 1 + 1 = 3$$

41. $f(x) = |x|$, and $f(x) = x$ for $x > 0$

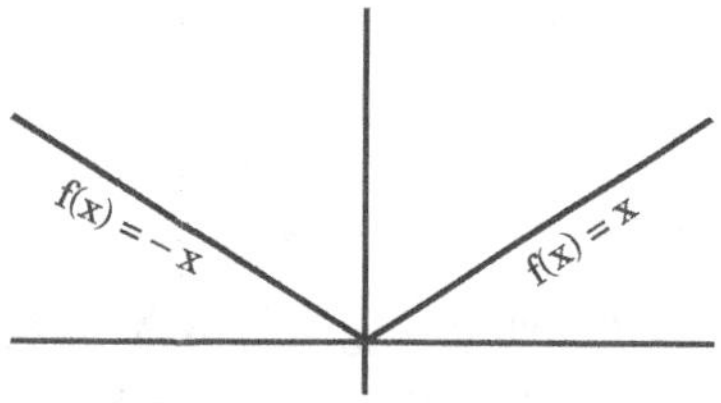

$\Rightarrow \quad f(x) = -x$ for $x < 0$

Hence from graph it is clear that derivative does not exist for $x = 0$.

42. $$f(x) = \frac{x^3}{3} - x$$

$\Rightarrow \; f'(x) = x^2 - 1$

For maximum and minimum value, $f'(x) = 0$

$\therefore \quad x^2 - 1 = 0$

$\Rightarrow \quad x = \pm 1$

Again $\quad f''(x) = 2x$

At $x = 1$, $\quad f''(x) = 2 > 0$

$x = 1$, $\quad f''(x) = -2 < 0$

Hence minimum at $x = 1$

43. $\vec{P} = 0.866i + 0.500j + 0k$, $\vec{Q} = 0.259i + 0.966j + 0k$

$$\therefore \quad \vec{P}\cdot\vec{Q} = |P|\,|Q|\cos\theta$$

$$= (0.866i + 0.5j + 0k)\;(0.259i + 0.966j + 0k)$$

$$= \sqrt{(0.866)^2 + (0.5)^2} \times \sqrt{(0.259)^2 + (0.966)^2}\cos\theta$$

$$\therefore \quad \cos\theta = \frac{0.866 \times 0.259 + 0.5 \times 0.966}{\sqrt{0.9999} \times \sqrt{1.001}} = 0.707$$

$$\Rightarrow \quad \theta = 45°$$

44. $$V = \int_0^{2\pi}\int_0^{\pi/3}\int_0^1 r^2 \sin\phi\, dr\, d\phi\, d\theta$$

$$= \int_0^{2\pi}\int_0^{\pi/3} \left[\frac{r^3}{3}\right]_0^1 \sin\phi\, d\phi\, d\theta$$

$$= \int_0^{2\pi}\int_0^{\pi/3} \frac{1}{3}\sin\phi\, d\phi\, d\theta$$

$$= \frac{1}{3}\int_0^{2\pi} [-\cos\phi\,]_0^{\pi/3}\, d\theta$$

$$= \frac{1}{3}\int_0^{2\pi} \frac{1}{2}\, d\theta = \frac{1}{6} \times 2\pi = \frac{\pi}{3}.$$

45. Given: $$y = x^2 + \frac{250}{x}$$

$$\therefore \quad \frac{dy}{dx} = 2x - \frac{250}{x^2},$$

and $$\frac{dy^2}{dx^2} = 2 + \frac{250 \times 2}{x^3}$$

When $x = 5$, $\frac{dy}{dx} = 10 - 10 = 0$

and $\frac{d^2y}{dx^2}$ is positive

Therefore function is minimum at $x = 5$.

46. Given, $y^2 = 8x$ and $P(x, y)$ is any point

Volume of the element

$$= \pi[2^2 - x^2]\,dy = \pi\left[4 - \frac{y^4}{64}\right]dy$$

$\therefore$ Required volume $= 2\pi \int_0^4 \left[4 - \frac{y^4}{64}\right] dy$

$$= 2\pi \left[4y - \frac{y^5}{320}\right]_{y=0}^4 = 2\pi \left[16 - \frac{4^5}{320}\right] = \frac{128\,\pi}{5}$$

47. $\int \log x \,.\, dx = \log x \,.\, x - \int x \,.\, \frac{d}{dx}(\log x)\, dx$

$$= x \log x - \int 1 \,.\, dx = x \log x - x = x\,(\log x - 1)$$

48. $$\frac{1}{\sqrt{2}}\frac{\partial f}{\partial x} + \frac{1}{\sqrt{2}}\frac{\partial f}{\partial y} = 2\sqrt{2},$$

$\Rightarrow \quad \frac{\partial f}{\partial x} + \frac{\partial f}{\partial y} = 4$ and $\frac{\partial f}{\partial y} = -3$

$\Rightarrow \quad \frac{\partial f}{\partial y} = 3.$

Hence $\frac{\partial f}{\partial x} = 1$ and $\frac{\partial f}{\partial y} = 3$

$\therefore$ Derivative in the direction $-i - 2j$

$$= \left(\frac{-1}{\sqrt{5}}\right)(1) + \left(\frac{-2}{\sqrt{5}}\right)3 = \frac{-7}{\sqrt{5}}$$

49. Given : $\quad 2y = x^2$

$\Rightarrow \quad y = \frac{x^2}{2}$

and $\quad y = x + 4$

$\therefore \quad x^2 - 2x - 8 = 0$

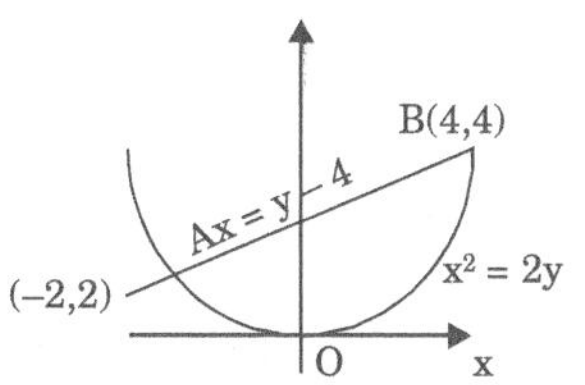

Given curve and the line intersect at A(–2, 2) and B (4, 8)

$\therefore$ Area $= \int_{-2}^{4} (y_1 - y_2)\, dx = \int_{-2}^{4} \left\{(x+4) - \frac{x^2}{2}\right\} dx$

$$= \left[\frac{x^2}{2} + 4x - \frac{x^3}{6}\right]_{-2}^{4}$$

$$= \left(8 + 16 - \frac{32}{3}\right) - \left(2 - 8 - \frac{4}{3}\right) = 18$$

51. $\lim_{a\to\infty} \int_1^a x^{-4} dx = \lim_{a\to\infty}\left[-\frac{1}{3}\left(x^{-3}\right)_1^a\right]$

$$= \lim_{a\to\infty} -\frac{1}{3}\left[a^{-3} - 1\right]$$

$$= -\frac{1}{3}[0 - 1] = \frac{1}{3}$$

Numerical Type Questions

1. $$\lim_{x\to 0}\frac{e^x - 1}{x} = \lim_{x\to 0}\frac{\left(1 + x + \frac{x^2}{\lfloor 2} + \frac{x^3}{\lfloor 3} + \ldots\right) - 1}{x}$$

$$= \lim_{x\to 0}\left(1 + \frac{x}{\lfloor 2} + \frac{x^2}{\lfloor 3} + \ldots\right) = 1$$

By L' Hospital's Rule

$$\lim_{x\to 0}\frac{e^x - 1}{x}\left[\text{form}\,\frac{0}{0}\right] = \lim_{x\to 0}\frac{e^x}{1} = 1$$

4. $\lim_{x\to 0}\frac{\sqrt{1+x} - \sqrt{1-x}}{x}$

$$= \lim_{x\to 0}\frac{\sqrt{1+x} - \sqrt{1-x}}{x} \times \left(\frac{\sqrt{1+x} + \sqrt{1-x}}{\sqrt{1+x} + \sqrt{1-x}}\right)$$

$$= \lim_{x\to 0}\frac{1 + x - 1 + x}{x(\sqrt{1+x} + \sqrt{1+x})}$$

$$= \lim_{x\to 0}\frac{2x}{x(\sqrt{1+x} + \sqrt{1-x})}$$

$$= \frac{2}{\sqrt{1+0} + \sqrt{1-0}} = \frac{2}{1+1} = 1$$

5. Let A $= \lim_{n\to 0} e^{\frac{1}{n \log n}}$

$\therefore \quad \log_e A = \lim_{n\to 0}\frac{1}{n \log n}.$

$\Rightarrow \log_e A = \lim_{n\to 0}\frac{1/n}{\log n} \qquad \left[\text{form}\,\frac{\infty}{\infty}\right]$

$$= \lim_{n\to 0}\frac{-1/n^2}{1/n} = \lim_{n\to 0} -\frac{1}{n}$$

$\log_e A = -\infty$

$$A = e^{-\infty} = 0$$

6. $\lim_{x\to\infty} x\sin\frac{1}{x}$ = Let $\frac{1}{x} = n$

So, $x \to \infty,\ n \to 0$

$$\lim_{n\to 0}\frac{\sin n}{n} = 1$$

7. $\lim_{n\to\infty}\left[\frac{1}{n^2}+\frac{2}{n^2}+\frac{3}{n^2}+\ldots+\frac{1}{n}\right]$

$$= \lim_{n\to\infty}\left[\frac{1+2+3+\ldots+n}{n^2}\right] = \lim_{n\to\infty}\frac{n(n+1)/2}{n^2}$$

$$= \lim_{n\to\infty}\frac{1}{2}\left(1+\frac{1}{n}\right) = \frac{1}{2}(1+0) = \frac{1}{2}$$

8. Given function is

$$f(x) = x(x-2) = x^2 - 2x$$

$f(x)$ is continuous function in $\left[0,\frac{3}{2}\right]$ and differentiable in $\left(0,\frac{3}{2}\right)$. Therefore, mean value thorem is applicable in $\left[0,\frac{3}{2}\right]$

$$f\left(\frac{3}{2}\right) = \left(\frac{3}{2}\right)^2 - 2\left(\frac{3}{2}\right) = \frac{-3}{4}$$

$$f(0) = 0$$

By Lagrange's mean value theorem, there exist $c \in \left(0,\frac{3}{2}\right)$ such that

$$\therefore \quad f'(c) = \frac{f\left(\frac{3}{2}\right) - f(0)}{\frac{3}{2} - 0}$$

$$\Rightarrow \quad 2c - 2 = \frac{-\frac{3}{4} - 0}{\frac{3}{2}} \text{ or } -\frac{1}{2}$$

$$\Rightarrow \quad c = +\frac{3}{4} \in \left(0,\frac{3}{2}\right)$$

9. Value of a mod function cannot be less than zero, therefore

$f(x) = |x^2 - 5x + 2|$ is zero.

10. Let $f(x) = \frac{1}{\sqrt{2}}(\sin x - \cos x)$

$$= \sin x \cos\frac{\pi}{4} - \cos x \sin\frac{\pi}{4} = \sin\left(x - \frac{\pi}{4}\right)$$

Since maximum value of sin θ is 1, hence maximum value of $f(x) = 1$

11. Let $f(x) = -x^3 + 6x^2 + 2x + 1$

Slope of the function is,

$$f'(x) = -3x^2 + 12x + 2 = F(x) \text{ (say)}$$

Now to find minimum or maximum of $F(x)$.

$\therefore \quad F'(x) = -6x + 12$

and $\quad F''(x) = -6$

For maxima and minima

$$F'(x) = 0$$

$\Rightarrow \quad x = 2$

$\therefore \quad F''(2) = -6 < 0$

Hence $F(x)$ (slope) is maximum at $x = 2$

$\therefore$ Maximum slope $= F(2) = -3(2)^2 + 12(2) + 2 = 14$

12. Volume is maximum at $x = 10$

At $x = 10$, $\quad y = \frac{300-100}{40} = 5$

$\therefore$ Maximum volume $= x^2y = (10)^2 \times 5 = 500$

13. $f(x, y) = x^2 + y^2$

$$\nabla^2 f = \vec{\nabla}.\vec{\nabla} f$$

$$= z\left(\vec{i}\frac{\partial}{\partial x} + \vec{j}\frac{\partial}{\partial y}\right)\left(\vec{i}\frac{\partial}{\partial x} + \vec{j}\frac{\partial}{\partial y}\right)\left[x^2 + y^2\right]$$

$$= \frac{\partial^2}{\partial x^2}(x^2) + \frac{\partial^2}{\partial y^2}(y^2) = 4$$

14. $f(x, y, z) = x^2 + y^2 + z^2$

$$\Delta f(x, y, z) = 2x\,\vec{i} + 2y\,\vec{j} + 2z\,\vec{k} = (\vec{A}) \text{ say}$$

Now, $\quad \vec{B} = \vec{i} - \vec{k}\,;\ \hat{B} = \frac{1}{\sqrt{2}}(\vec{i} - \vec{k})$

$$\therefore \quad \vec{A}.\hat{B} = \frac{2}{\sqrt{2}}(x - z)$$

$$\Rightarrow \quad \vec{A}.\hat{B}\Big|_{(1,1,1)} = \frac{2}{\sqrt{2}}(1-1) = 0$$

15. Equation of the straight line joining (1, 0) and (0, 1) is

$$x + y = 1, \quad y = 1 - x$$

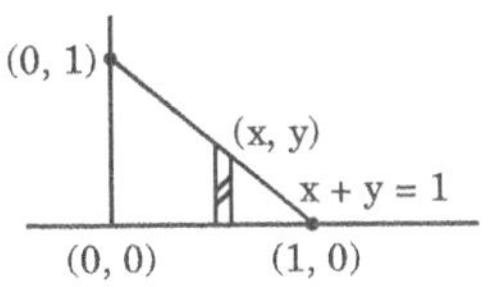

Consider $\iint 2x\,dx\,dy = \int 2x\,dx \int_0^{1-x} dy$

$= \int 2x\,dx\,(y)_0^{1-x} = \int 2x\,dx\,(1-x)$

$= \int (2x - 2x^2)\,dx$

Since, limits of the integral are now between 0 and 1

$\therefore \iint 2x\,dx\,dy = \int_0^1 (2x - 2x^2)\,dx$

$= \left[x^2 - \frac{2x^3}{3}\right]_0^1 = 1 - \frac{2}{3} = \frac{1}{3}$

16. $\nabla . f = i\frac{\partial}{\partial x} + j\frac{\partial}{\partial y} + k\frac{\partial}{\partial z}$

$= iyz^3 + jxz^3 + k \times 3xyz^2$

$= i \times 0 + j \times 1 \times 2^3 + +k \times 0 = j \times 8$

$\therefore \quad |\nabla . f| = 8$

17. Given: $\phi = ax^2y - y^3$

$\therefore \quad \frac{\partial \phi}{\partial y} = ax^2 - 3y^2$; and $\frac{\partial^2 \phi}{\partial y^2} = -6y$

$\frac{\partial \phi}{\partial x} = 2\,axy$; and $\frac{\partial^2 \phi}{\partial x^2} = 2\,ay$

Now $\frac{\partial^2 \phi}{\partial x^2} + \frac{\partial^2 \phi}{\partial y^2} = 0$

$\therefore \quad 2ay - 6y = 0$

$\Rightarrow \quad a = 3$

18. $\phi = 2x^2 + 3y^2 + 5z^2$

At point (1, 1, –1), $\nabla\phi = i.\,4x + j.\,6y + k.\,10\,z$

Given : $x = 1, y = 1, z = -1$

$\therefore \quad |\nabla\phi| = \sqrt{(4)^2 + (6)^2 + (-10)^2} = \sqrt{152}$

19. $\int_0^{\pi/2}\int_0^{\pi/2} \sin(x+y)\,dx.dy = [\cos(x+y)]_0^{\pi/2}\,dy$

$= \left[\sin y - \sin\left(\frac{\pi}{2} + y\right)\right]_0^{\pi/2}$

$= (1 - 0) - (0 - 1) = 2$

DIFFERENTIAL EQUATIONS

MCQ Type Questions

3. Given : $\frac{dy_1}{dt} = -5y_2 + (4-k)\,y_2 \ldots (i)$

and $\frac{dy_2}{dt} = y_1 - 2y_2 \quad \ldots (ii)$

Putting $\frac{d}{dt}$ = D, we have from equation (*i*),

$Dy_1 + 5y_2 - (4 - K)\,y_2 = 0$

$\Rightarrow \quad (D + 5)y_1 - (4 - K)\,y_2 = 0 \quad \ldots (iii)$

From equation (*ii*)

$Dy_2 = y_1 - 2y_2$

$\Rightarrow \quad (D + 2)y_2 - y_1 = 0 \quad \ldots (iv)$

$\Rightarrow \quad y_1 = (D + 2)\,y_2$

Putting in equation (*iii*), we get

$(D + 5)(D + 2)y_2 - (4 - K)y_2 = 0$

$\Rightarrow \quad [D^2 + 7D + (6 + K)]y_2 = 0$

$\Rightarrow \quad D^2 + 7D + 6K = 0$

$\Rightarrow \quad D = \frac{-7 \pm \sqrt{49 - 4(6 + K)}}{2 \times 2}$

For real value of D,

$49 - 4(6 + K) \geq 0 \Rightarrow 49 - 24 - 4K \geq 0 \Rightarrow 4K \leq 25$

$\Rightarrow K \leq \frac{25}{4} \Rightarrow K \leq 6.25$

4. $\frac{d^2x}{dt^2} + 10\frac{dx}{dt} + 25x = 0$

$\Rightarrow (D^2 + 10D + 25)\,x = 0 \quad \Rightarrow (D+5)^2 = 0$

$\Rightarrow D = -5, -5$

Hence soloution is $(C_1 + C_2 t)\,e^{-5t}$

5. Given differential equation is,

$$\frac{dx}{dt} + \frac{x}{20} = 10$$

$$\Rightarrow \quad \frac{dx}{dt} = 10 - \frac{x}{20} = \frac{200 - x}{20}$$

Integrating,

$$20\int_0^{100}\frac{dx}{200-x} = \int_0^t dt$$

$$\therefore \quad t = 20\left[\ln\frac{1}{200-x}\right]_0^{100} = 20\ln$$

6. Let $\frac{dy}{dx} = p$, therefore

$$\frac{d^2y}{dx^2} + \frac{dp}{dx} = \frac{dp}{dy}\cdot\frac{dy}{dx} = p\frac{dp}{dy}$$

$$\therefore \quad p^2 + yp\frac{dp}{dy} = 0$$

$$\Rightarrow \quad p = -y\frac{dp}{dy}$$

$$\Rightarrow \quad \frac{dp}{p} = -\frac{dy}{y}$$

7. Given: $\frac{d^2y}{dx^2} + \sin x\frac{dy}{dx} + ye^x = \sinh x$

The above equation is second order and non linear.

8. Given : $\frac{dy}{dt} + 5y = 0$

$\Rightarrow \quad \frac{dy}{y} = -5\,dt$

Integrating, we get $\log_e y = c - 5t$

When $t = 0$, $y = 1$.

$\therefore \quad \log_e 1 = c - 5 \times 0$

$\Rightarrow \quad c = 0$

$\therefore \quad \log_e y = -5t$

$\Rightarrow \quad y = e^{-5t}$

10. y'' indicates that it is of second order. Degree of y'', y' and y is one. Hence given differential equation is second order linear equation.

11. For the differential equation $M dx + N dy = 0$,

to be exact, $\frac{\partial M}{\partial y} = \frac{\partial N}{\partial x}$

Hence for the differential equation

$f(x, y)\frac{dy}{dx} + g(x, y) = 0$ to be exact $\frac{\partial f}{\partial x} = \frac{\partial g}{\partial y}$

13. $(D^4 - 81)y = 0$

Auxiliary equation is

$m^4 - 81 = 0 \Rightarrow (m+3)(m-3)(m+3i)(m-3i) = 0$

Hence Solution is

$$y = C_1e^{3x} + C_2e^{-3x} + C_3\cos 3x + C_4\sin 3x$$

15. $\text{P.I.} = \frac{1}{D^2 - 2D}e^x\sin x = e^x\frac{1}{(D+1)^2 - 2(D+1)}\sin x$

$$= e^x\frac{1}{D^2-1}\sin x = e^x\frac{1}{(-1^2)-1}\sin x$$

$$= -\frac{1}{2}e^x\sin x$$

16. Auxiliary equation of the given equation is,

$D^2 - m^2 = 0$

$\Rightarrow D = \pm m$

Hence C.F. is $y = C_1e^x + C_2e^{-x} = C_1\cosh x + C_2\sinh x$

17. $\frac{1}{\phi(D)(D-a)^n}e^{ax} = \frac{1}{\phi(a)}\cdot\frac{1}{(D-a)^n}e^{ax}$

$$= \frac{1}{\phi(a)}x.\frac{1}{\frac{d}{dD}(D-a)^n}e^{ax} \quad (\because f(a) = 0)$$

$$= \frac{1}{\phi(a)}x.\frac{1}{n(D-a)^{n-1}}e^{ax}$$

$$= \frac{1}{\phi(a)}x^2.\frac{1}{n\frac{d}{dD}(D-a)^{n-1}}e^{ax}$$

$(\because f'(a) = 0)$

$$= \frac{1}{\phi(a)}x^2.\frac{1}{n(n-1)(D-a)^{n-2}}e^{ax}$$

$$= \frac{1}{\phi(a)}x^n.\frac{1}{n(n-1)(n-2)...2.1}e^{ax}$$

$$= \frac{1}{\phi(a)}x.\frac{x^ne^{ax}}{n!}$$

19. $y = 1 + \cos x + \sin x$

$\therefore \quad \frac{dy}{dx} = -\sin x + \cos x$ and $\frac{d^2y}{dx^2} = -\cos x - \sin x$

$\therefore \quad y + \frac{d^2y}{dx^2} = 1$

Hence $y = 1 + \cos x + \sin x$ is solution of, $\frac{d^2y}{dx^2} + y = 1$

20. Auxiliary equation is,

$$m^2 - m - 2 = 0$$

$$\therefore \quad m = -1, 2$$

$$CF = c_1e^{-x} + c_2e^{2x}$$

$$P.I. = \frac{1}{D^2 - D - 2}.3.e^{2x}$$

$$= 3.\frac{1}{D^2 - D - 2}e^{2x}$$

$$= 3x\frac{1}{2D - 1}e^{2x} = 3x\frac{1}{3}.e^{2x} = xe^{2x}$$

$$\therefore \quad y = c_1e^{-x} + c_2e^{2x} + xe^{2x}$$

$$\frac{dy}{dx} = -c_1e^{-x} + c_2e^{2x} + 2xe^{2x} + e^{2x}$$

$$y(0) = c_1 + c_2 = 0 \text{ (given)}$$

$$y'(0) = -c_2 + 2c_2 + 1 = -2 \text{ (given)}$$

$$\therefore \quad c_1 + c_2 = 0 \quad \quad ...(i)$$

$$c_1 - 2c_2 = 3 \quad \quad ...(ii)$$

Solving equations (*i*) and (*ii*), we get

$$c_1 = 1, c_2 = -1$$

Hence required solution is

$$y = e^{-x} - e^{2x} + xe^{2x}$$

22. Given equation is,

$$(D^2 - 5D + 6)y = e^{3x}$$

$$\therefore \quad P.I. = \frac{1}{D^2 - 5D + 6}e^{3x} = \frac{1}{(D-3)(D-2)}e^{3x}$$

$$= \frac{1}{D-3}e^{3x} - \frac{1}{D-2}e^{3x}$$

$$= x.e^{3x} - \frac{1}{3-2}e^{3x} = (x-1)e^{3x}$$

24. Given differential equation is, $D^2 + D = x^2 + 2x + 4$

$$\therefore \quad P.I. = \frac{1}{D(D+1)}(x^2 + 2x + 4)$$

$$= \frac{1}{D}(D+1)^{-1}(x^2 + 2x + 4)$$

$$= \frac{1}{D}(1 - D + D^2)(x^2 + 2x + 4)$$

$$= \frac{1}{D}(x^2 + 2x + 4 - 2x - 2 + 2) = \frac{x^3}{3} + 4x$$

Numerical Type Questions

1. Given: $(4t^2 + 1)\frac{dy}{dt} + 8yt - t = 0$

$$\Rightarrow \quad \frac{dy}{dt} + \frac{8t}{4t^2+1}y = \frac{t}{4t^2+1} \quad \quad ...(i)$$

$$\therefore \quad I.F. = e^{\int .8t/(4t^2+1)} = (4t^2 + 1)$$

Hence solution of equation (*i*) is

$$y(4t^2 + 1) = \int \frac{t}{(4t^2+1)}.(4t^2+1)\,dt + C = \frac{t^2}{2} + C$$

$$\Rightarrow \quad y = \frac{t^2}{2(4t^2+1)} + \frac{C}{(4t^2+1)} \quad \quad ...(ii)$$

But $y(1) = 0$,

$$\therefore \quad 0 = \frac{1}{10} + \frac{C}{5}$$

$$\Rightarrow \quad C = -\frac{1}{2}$$

From equation (*ii*), we have

$$y = \frac{t^2}{2(4t^2+1)} - \frac{1}{2(4t^2+1)}$$

$$\therefore \quad y\big|_{t\to\infty} = \underset{t\to\infty}{\text{Lt}} \frac{1}{2(4+1/t^2)} - \frac{1}{2(4t^2+1)}$$

$$= \frac{1}{2(4+0)} - 0$$

$$= \frac{1}{8}$$

NUMERICAL METHODS

MCQ Type Questions

2.

$$y = ax + b \quad \quad ....(i)$$

$$x = cy + d$$

$$\Rightarrow \quad x - d = cy$$

$$\Rightarrow \quad y = \frac{1}{c}x - d$$

From equatons (*i*),

$$a = \frac{1}{c} \text{ and } b = -d$$

Numerical Type Questions

1. $x^2 - 2 = 0$

$$x_1 = x_0 - \frac{f(x_0)}{f'(x_0)} = 1 + \frac{1}{2} = 1.5$$

$$f(x_0) = 1^2 - 2 = -1$$
$$f(x_0) = 2x$$
$$f(x_0) = 2x_0 = 2(1) = 2$$

2. $$\left(\frac{dy}{dx}\right)_{at\ x=0} = \frac{y_2 - y_1}{x_2 - x_1} = \frac{y(1)-y(-1)}{1-(-1)} = \frac{8-5}{2} = 1.5$$

18. $$L_0(x) = \frac{(x-2.5)(x-3.0)}{(-0.5)(-1.0)} = 2x^2 - 11x + 15$$

Similarly,

$$L_1(x) = -(4x^2 - 20x + 24)$$
$$L_2(x) = 2x^2 - 9x + 10$$

$$\therefore \quad L_2(x) = (2x^2 - 11x + 15)(0.69315) - (4x^2 - 20x + 24)(0.091629) + (2x^2 - 9x + 10)(1.09861)$$
$$= -0.08164x^2 + 0.81366x - 0.60761$$

which is the required quadratic polynomial.

Putting $x =$ 2.7 in the above polynomial, we get

$\ln 2.7 \approx L_2(2.7)$

$$= -0.08164(2.7)^3 + 0.81366(2.7) - 0.60761$$
$$= 0.9941164$$

19. $$x = \frac{(-5)(-12)}{(-8)(-15)}(1) + \frac{(3)(-12)}{(8)(-7)}(3) + \frac{(3)(-5)}{(15)(7)}(4)$$

$$= \frac{1}{2} + \frac{27}{14} - \frac{4}{7} = \mathbf{1.86}$$

20. $$\sin\frac{\pi}{6} \approx \frac{\left(\frac{\pi}{6}-0\right)\left(\frac{\pi}{6}-\frac{\pi}{2}\right)}{\left(\frac{\pi}{4}-0\right)\left(\frac{\pi}{4}-\frac{\pi}{2}\right)}(0.70711) + \frac{\left(\frac{\pi}{6}-0\right)\left(\frac{\pi}{6}-\frac{\pi}{4}\right)}{\left(\frac{\pi}{2}-0\right)\left(\frac{\pi}{2}-\frac{\pi}{4}\right)}(1)$$

$$= \frac{8}{9}(0.70711) - \frac{1}{9}$$

$$= \mathbf{0.51743}$$

TRANSFORM THEORY

MCQ Type Questions

1. We know that, $\text{sgn}(t) = 2u(t) - 1$ ...(*i*)

Let, $\text{sgn}(t) = s(t)$

$$\frac{ds(t)}{dt} \leftrightarrow j\omega S(\omega)$$

Now differentiate equation (*i*)

$$\frac{ds(t)}{dt} = 2\delta(t) - 0$$

Fourier transform of above equation is

$$F\left[\frac{ds(t)}{dt}\right] = 2$$

Therefore, $2 = j\omega S(\omega)$

$$S(\omega) = \frac{2}{j\omega}$$

2. If $f(t) \leftrightarrow F(\omega)$

then by $f(t-b) \leftrightarrow e^{-j\omega b} F(\omega)$ time shifting property

$$\frac{2\sin\omega T_1}{\omega} e^{-jwT_1}$$

4. Frequency Convolution Theorem : Multiplication of two signals in time domain is equivalent of their convolution in frequency domain.

i.e. $x(t).y(t) \leftrightarrow X(f) * 2\pi Y(f)$

5. Let $f(t)$ denote a function of time and $F(f)$ its fourier transform. Then differentiating the fourier transform $F(f)$ with respect to f, we get,

$$-j2\pi t f(f) \leftrightarrow \frac{d}{df}F(f)$$

Similarly differentiating $f(t)$ w.r.t. we get,

$$\frac{d}{dt}f(t) = -j2\pi f(f)$$

then, $$\frac{d}{df}F(f) = -2\pi f F(f)$$

This shows that the gaussian pulse and its fourier transform are the same function.

6. Using convolution theorem,

$$y(t) = h(t) * x(t)$$

7. $$u(t) = \frac{1}{2}[1 + \text{sgn}(t)]$$

Taking the fourier transform both sides

$$F[u(t)] = \frac{1}{2}\left[2\pi\delta(\omega) + \frac{2}{j\omega}\right]$$

$$F[u(t)] = \pi\delta(\omega) + \frac{1}{j\omega}$$

8. $$v(t) = \cos \omega_0 t$$

$$v(t) = \frac{1}{2}\left[e^{j\omega_0 t} + e^{-j\omega_0 t}\right]$$

Taking the fourier transform,

$$V(\omega) = \frac{1}{2}[2\pi\delta(\omega - \omega_0) + 2\pi\delta(\omega + \omega_0)]$$

$$V(\omega) = \pi[\delta(\omega - \omega_0) + \delta(\omega + \omega_0)]$$

or $$V(f) = \frac{1}{2}[\delta(f - f_0) + \delta(f + f_0)]$$

9. If, $g(t) \leftrightarrow G(f)$
then from the Duality property
$G(t) \leftrightarrow g(-f)$

10. Area under the curve is

$$\int_{-\infty}^{\infty} f(t)d(t) = F(0)$$

11. Fourier transform of periodic delta function $x(t)$ is also a periodic delta function with same period.

14. $$x(n) = -a^n u(-n-1)$$

$$x(z) = \sum_{-\infty}^{\infty} -a^n u(-n-1)z^{-n}$$

$$= -\sum_{-\infty}^{-1} a^n z^{-n} = -\sum_{1}^{\infty} a^{-n} z^n = -\sum_{-\infty}^{-1} (a^{-1}z)^n$$

$$= -[a^{-1}z + (a^{-1}z)^2 + (a^{-1}z)^3 + ...]$$

$$= -a^{-1}z[1 + (a^{-1}z) + (a^{-1}z)^2 + ...]$$

$$= -a^{-1}z \times \frac{1}{1 - a^{-1}z} = \frac{1}{1 - az^{-1}} \quad |a^{-1}z| < a$$

or $|z| < a$

15. $z > \frac{1}{3}, z > \frac{1}{2}$

24. $$x(n) = a^n u(n) \leftrightarrow \frac{1}{1 - az^{-1}}$$

$$nx(n) \leftrightarrow -z\frac{dX(z)}{dz}$$

$$n\left(\frac{1}{3}\right)^n u(n) \leftrightarrow -z\frac{(-az^{-2})}{(1 - az^{-1})^2} = \frac{az^{-1}}{(1 - az^{-1})^2}$$

$$n\left(\frac{1}{3}\right)^n u(n) \leftrightarrow \frac{\frac{1}{3}z^{-1}}{\left(1 - \frac{1}{3}z^{-1}\right)^2}$$

ROC : $|z| > \frac{1}{3}$

2

CHAPTER

FLIGHT MECHANICS

1. The slope of $C_L - \alpha$ curve ----- with increase is aspect ratio of wing
   (a) Remains same
   (b) Decrease
   (c) Increases
   (d) Depends on the shape of Aerofoil
2. A tangent is drawn from origin to $C_L - C_D$ curve of Aerofoil. What does the point of tangency represent.
   (a) Maximum value of $\frac{C_L}{C_D}$
   (b) Maximum value of $\frac{C_L}{C_D}$
   (c) Maximum value of C_L
   (d) $\frac{C_L}{C_D}$ of Cruise
3. Which of the following correctly represents range of aeroplane from height h above ground
   (a) $h\left(\frac{L}{D}\right)^2_{max}$ (b) $h\left(\frac{L}{D}\right)_{max}$
   (c) $h\left(\frac{D}{L}\right)_{max}$ (d) $h\left(\frac{D}{L}\right)_{min}$
4. A positive rudder deflection in a connection at Aircraft produces
   (a) A negative Yawing and a positive rolling moment
   (b) A negative rolling moment positive yawing moment
   (c) Negative yawing moment
   (d) Positive yawing moment
5. An aircraft in straight and level flight starts raising at double its normal speed. What happens to induced drag?
   (a) one fourth (b) one-half
   (c) become double (d) becomes four times
6. Which of the following quantities do not vary during a phugoid oscillation?
   (a) Aircraft (b) Aircraft - Altitude
   (c) Angle of attack (d) Aircraft pitch angle
7. Lateral directional characteristics of airplane gave following sets of roots $\lambda_1 = -0.000479$, $\lambda_2 = -0.118$, $\lambda_{3,4} = 0.00724 \pm 10.068$.
   What conclusion can be made about the stability of airplane?
   (a) Spiral mode : Divergence, roll damping: convergence, dutch roll mode.
   (b) Spiral mode: convergence, roll damping: convergence, dutch roll mode: convergence
   (c) Spiral mode: convergence, roll damping: divergence, dutch roll mode: divergence
   (d) None of these
8. Data for jet powered Aircraft given below:-
   Weight W = 28 KN, Wing Plan form Area S = 20 m^2, maximum lift coefficient $C_{L\,max} = 2.2$ and drag coefficient $C_D = 0.021 + 0.26C_L^2$. (Assume density = 1.23 kg/n^3) at what velocity is should fly to maximise it range?
   (a) 66.5/sec (b) 46.1 m/sec
   (c) 33 m/sec (d) 51 m/sec
9. Common data for lateral - directional characteristics equation for airplane gave the following jet of roots: $\lambda_1 = -0.55$, $\lambda_2 = -0.0054$, $\lambda_{3,4} = -0.06 \pm 1.8\,i$
   the root corresponding to roll made is
   (a) –0.55 (b) –0.0034
   (c) –0.06 (d) –0.06 – 1.8
10. Damping ratio for dutch roll mode is
   (a) 0.033 (b) 0.018
   (c) 0.025 (d) 0.009
11. Damped frequency for dutch roll mode
   (a) 0.018 rad/sec (b) 1.8 rad/sec
   (c) 2.5 rad/sec (d) 2.8 rad/sec
12. In which of the following cases, trimming of aircraft is possible?
   (a) When CG is ahead to stick free neutral point
   (b) When CG is behind to stick free neutral point
   (c) When CG is behind the wing-body aerodynamic centre.
   (d) None of the above

13. Which of the following features does not contribute in lateral stability?

(*a*) Lower position of CG
(*b*) Dihedral
(*c*) Wing sweep
(*d*) Wing Taper

14. Which of the true for gauge flap for an aircraft?

(*a*) It increases the wing area
(*b*) It increases the chord length
(*c*) It does not have a slot effects
(*d*) It affects train.

15. Slat is leading edge device used to minimising landing and take-off distances a slat

(*a*) increases α_{0L}, decreases $C_{L\alpha}$
(*b*) increases α_{0L}, increase $C_{L\alpha}$
(*c*) increases $C_{L\alpha}$
(*d*) increases $C_{L\,max}$

16. In the lanchester model of phugoid mode, the approximate phugoid frequency is given by the expression $W = g\frac{k}{v}$, when the value of k constant k is equal to

(*a*) $\sqrt{2}$ (*b*) $\frac{1}{\sqrt{2}}$
(*c*) 2 (*d*) $\frac{1}{2}$

17. For lateral and directional stability of the aircraft. Which of the following conditions is true?

(*a*) $\ell_\beta < 0, N_\beta < 0$ (*b*) $\ell_\beta > 0, N_\beta > 0$
(*c*) $\ell_\beta < 0, N_\beta > 0$ (*d*) $\ell_\beta > 0, N_\beta < 0$

18. A Rocket in flying at air speed at 1000 m/sec angle of attack 30° and side flip angle is 20°. Then the axial, side flip and normal velocity components are

(*a*) 813.80 m/sec, 342 m/sec, 469 m/sec
(*b*) 820 m/sec, 342 m/sec, 470 m/sec
(*c*) 825 m/sec, 350 m/sec, 880 m/sec
(*d*) None of the above

19. An airplane which in flying is steady level condition, if the zero lift drag condition is 0.014 aspect ratio in 8, Ostwald's efficiency factor is 0.9 then what is maximum $\frac{L}{D}$ ratio aircraft can maintain under such conditions?

(*a*) 18 (*b*) 25
(*c*) 192 (*d*) 51

20. An aircraft is flying having lift to drag ratio 20 the damping ratio of phugoid mode will be _____.

21. Determine maximum range for steady level flight of the airplane,

Airplant design

$$\frac{V}{D_{min}} = 170\ m/sec$$

$$\left(\frac{L}{D}\right)_{max} = 23$$

Propulsion design

$I_{sp} = 2550$ sec

Structural design

$$\frac{W_i}{W_f} = 1.4$$

(*a*) 3150 km (*b*) 3105 km
(*c*) 3065 km (*d*) 3605 km

22. For spiral mode eigen value are approximated as

(*a*) $N_r - \frac{L_y \ell\beta}{L_p^2} N_\beta$ (*b*) $N_\beta - \frac{\ell_\beta}{L_y} N_\beta$

(*c*) $N_y - \frac{l_y}{L_\beta} N_\beta$ (*d*) $N_y + \frac{l_\beta}{L_y} N_\beta$

also

$$N_y = \frac{\partial N}{I_{zz}\partial y} . N_\beta = \frac{\partial N}{I_{zz}\,\partial\beta} . L_y = \frac{\partial L}{I_{zz} d_y}$$

$$L_\beta = \frac{\partial L}{I_{\alpha\alpha} \times \partial\beta}$$

23. A Biplane of given wing area as of monoplane wing will have

(*a*) More drag and more lift
(*b*) Less drag and more lift
(*c*) More drag and less lift
(*d*) None of above

24. An whose CG on the neutral point then

(*a*) Airplane has maximum possible stability
(*b*) Airplane is neutrally stable
(*c*) Static stability is minimum
(*d*) None of the above

25. Dutch roll instability of an airplane is

(a) Highly damped oscillatory motion with high frequency

(b) Highly damped oscillation motion will low frequency.

(c) Lightly damped oscillation motion with low frequency.

(d) None of the above.

26. Characteristics of short-term mode of longitudinal instability are

(a) Constant speed, highly damped

(b) Constant angle of attack highly damped

(c) Constant speed, lightly damped

(d) Constant speed, constant angle of attack.

27. $C_{L\alpha}$ of wing is less than $C_{L\alpha}$ of the Aerofoil because of

(a) Wing is bigger

(b) Wing tip vortices

(c) Unsteadiness of flow association with it

(d) downwash

28. Consider an aircraft flying of an altitude of 30000 ft at a velocity 540 mi/h calculate the energy height

(a) 39,740 ft (b) 55000 ft

(c) 60000 ft (d) 10000 ft

29. Which of the following can increase the range of propeller powered aircraft&

i. Increasing propeller efficiency

ii. Decreasing specific fuel consumption

iii. Flying at the altitude near to mean sea level

iv. Increasing the value of $\dfrac{C_L^{\frac{1}{2}}}{C_D}$ maximum

(a) only i (b) i and ii

(c) i, ii and iii (d) i, ii, iii and iv

30. An aircraft is flying vertically upward the load factor on the aircraft is

(a) 1 (b) 0

(c) less than 1 (d) more than

31. An aircraft is designed to sustain maximum load factor of 8, if the weight of aircraft is 20000 N, how much lift it must produce cruise on the steady level flight path

(a) 18000 N (b) 20000 N

(c) 25000 N (d) 3000 N

32. In the above question, how much maximum lift the aircraft can produce?

(a) 20000 (b) 18000

(c) 160000 (d) 20000

33. For shorten the take-off distance of an airplane. Which of the following is true?

i. Increasing thrust to weight ratio $\left(\dfrac{T}{W}\right)$

ii. Increasing wing loading $\left(\dfrac{W}{S}\right)$.

iii. Take off from the altitude nearer to mean sea level.

iv. Decreasing wing loading $\left(\dfrac{W}{S}\right)$

v. Increasing lift coefficient $C_{L\,max}$

(a) i, iii, iv, v (b) i, ii, iii, iv

(c) i, iii, iv (d) i, ii, iii, iv

34. Match the following pairs.

(i) Maximum range for jet aircraft	(a) maximum $\dfrac{C_L}{C_D}$
(ii) Maximum range for propeller driven aircraft	(b) maximum $\dfrac{C_L^{\frac{3}{2}}}{C_D}$
(iii) Maximum endurance for jet aircraft	(c) maximum $\dfrac{C_L^{\frac{1}{2}}}{C_D}$
(iv) Maximum endurance for propeller driven aircraft	

(a) (i)-(c), (iii)-(a), (iv)-(b), (ii)-(a)

(b) (i)-(b), (iii)-(b), (iv)-(c), (ii)-(c)

(c) (i)-(b), (ii)-(c), (iii)-(c), (iv)-(b)

(d) (i)-(b), (ii)-(c), (iii)-(c), (ii)-(a)

35. Minimum velocity required for aircraft take off is

(a) Significantly more than stalling velocity

(b) Less than stalling velocity

(c) Equal to stalling velocity

(d) Depending open the aircraft-weight

36. Neutral point represents

(i) Aerodynamic centre of full aircraft

(ii) Most aft location of CG for stable aircraft

(iii) Location of centre of gravity such that $C_{m,cg}$ does not changes with angle of attack

(iv) Most forward location of CG for stable aircraft

(*a*) only iv (*b*) iv, iii, ii

(*c*) i, ii, iii (*d*) i, ii, iii, iv

37. An aircraft flying at the altitude where density is 0.4135 kg/m at the speed of 250 m/sec requires power of 4000 hp. What is the power required for the aircraft to fly at the same lift coefficient at mean sea level conditions?

(*a*) 2324 hp (*b*) 2700 hp

(*c*) 3500 hp (*d*) 4000 hp

38. Rate of climb of an aircraft varies with altitude is given as $\frac{R}{C} = -1.89 \times 10^{-3}\ h + 28.52$. Where $\frac{R}{C}$ is in m/sec and h is altitude which is in meter. How much time the aircraft-would take to climb 11.5 km altitude from mean sea level condition.

(*a*) 20 min (*b*) 12 min 6 sec

(*c*) 25 min 10 sec (*d*) none of these

39. Consider an airplane with instantaneous acceleration of 8 ft/sec² at an instantaneous velocity of 800 ft/sec. At the existing flight condition the specific excess power is 300 ft/sec. Calculate the instantaneous maximum rate of climb that can be obtained as these alleviated flight conditions.

(*a*) 101 ft/sec (*b*) 100 ft/sec

(*c*) 540 ft/sec (*d*) None of these

40. For stalling speed of an aircraft which of following is true?

(i) Increasing with altitude

(ii) Increasing with increasing wing loading (W/S).

(iii) Increasing with decreasing $C_{L\,max}$

(iv) Increasing with increasing $C_{L\,max}$

(*a*) (i), (ii), (iii) (*b*) (i), (iii), (iv)

(*c*) (i), (ii), (iv) (*d*) (i), (ii), (iii), (iv)

41. Total take-off distance for a jet airplane at the sea-level (ρ = 1.225 kg/m³) is 800 m. What will be take off distance at the altitude of 1200 m where density of air is 1.10 kg/m³.

(*a*) 200 m (*b*) 1500 m

(*c*) 1000 m (*d*) None

Common data question

Figure shows the V-η diagram for an aircraft whose data is given below

Weight of aircraft = 55000 Kg

Wing area = 12 m².

Altitude = 10000 m above mean sea level air density = 0.413 kg/m³

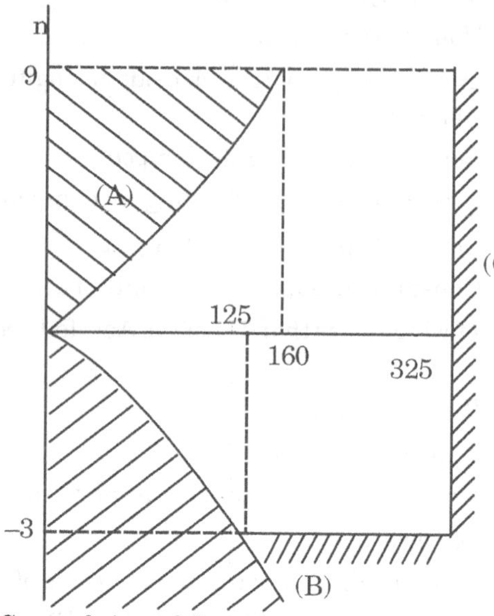

42. $C_{L\,max}$ of aircraft is

(*a*) 9 (*b*) 4.5

(*c*) 7.8 (*d*) 10

43. Minimum cruise lift coefficient for aircraft in the above question?

(*a*) 1.7 (*b*) 2.5

(*c*) 1.25 (*d*) 0.87

44. Minimum instantaneous radius of vertical upward loop at the given altitude?

(*a*) 500 m (*b*) 400 m

(*c*) 1292 m (*d*) None of these

45. Region (A), (B), (C) corresponds to

(*a*) Stall, structural, propulsion limit

(*b*) Stall, structural, Aerodynamic limit, respectively

(*c*) Stall, propulsion, structural limit.

(*d*) None of the above

46. Corner velocity at this altitude?

(*a*) 125 m/sec (*b*) 160 m.sec

(*c*) 325 m.sec (*d*) None

Common data question

Wing and tail of an aircraft have same air foil section, the distance between two aerodynamic centre is equal to wing semi-span (b/2) in additional, following data is given $C_{L\alpha}$ = 0.08/degree, $C_{L\delta c\ tail}$ = 0.04/degree, $C_{M_{ac,wb}} = -0.0218$, b = 29.2, C = 2.8 m. For tail ξ = 0, ξ_a = 0.3, η_t = 0.9, i_t = 2.1° s_t = 0.35 the symbols have their usual aerodynamics interpretation.

47. What is value of C_{mo} for aircraft?

(*a*) 0.2410 (*b*) 0.2147

(*c*) –0.2410 (*d*) –0.2147

48. Maximum distance by which centre of gravity can be behind Aerodynamic centre without destabilizng the aircraft?

(*a*) 1.55 m (*b*) 3.26 m

(*c*) 9.56 m (*d*) 2.73 m

49. In the given problem, what would be the separation between wing-body Aerodynamic center and centre of gravity for aircraft for 10% static margin?

(*a*) 1.98 m (*b*) 0.28 m

(*c*) 3.54 m (*d*) 2.46 m

50. For an aircraft, the wind tunnel data is given by

$S = 1.5\ m^2$, $C = 0.45$ m, $C_{M_{cg}}$ at zero lift

$= -0.002$, $C_{M_{cg}} = 0.004$ at $C_L = 0.3$

What is location of Aerodynamic centre?

(*a*) 9 mm behind CG

(*b*) 9 mm ahead of CG

(*c*) 18 mm behind CG

(*d*) 18 mm ahead of CG.

51. In the above question if centre of gravity is moved ahead of chord by length equal to 10% of chord. What is the value of moment coefficient about centre of gravity at 0.85 lift coefficient.

(*a*) –0.070 (*b*) –0.066

(*c*) 0.100 (*d*) 10.104

52. For a wing body combinations the aerodynamic centre lies 0.05 chord length ahead of cente of gravity moment coefficient about Aerodynamic centre is 0.016. If the tail coefficient is 0.45 the moment coefficient about the centre of gravity is ______.

53. A wing body model is tested in a subsonic wind tuned the lift coefficient in found to be zero at geometric angle of attack $\alpha = -1.5$. At $\alpha = 5°$ the lift coefficient is measured as 0.52. Also at $\alpha = 1.0°$ and 7.88° the moment coefficient about the centre of gravity are measured as –0.01 and 0.05, respectively the centre of gravity is located at 0.356 the location of aerodynamic centre and the value of $C_{M_{ac,wb}}$ is ______

54. The lateral- directional characteristics equation for an airplane gave the following set of roots $\lambda_1 = -0.22$, $\lambda_2 = 0.0035$ $\lambda_{3,4} = -0.091 \pm 2.1i$ the damping ratio correspond to dutch roll mode is ______.

55. The relationships between body fixed angular velocity vector $[p\ q\ r]^T$ and the rate of change of euler angles $[\dot{\phi}\ \dot{\theta}\ \dot{\Psi}]$ is given by

$$\begin{bmatrix} p \\ q \\ r \end{bmatrix} = j^{-1} \begin{bmatrix} \dot{\varphi} \\ \dot{\theta} \\ \dot{\gamma} \end{bmatrix}$$

then j^{-1} equals to

(*a*) $\begin{bmatrix} \cos\Psi & 0 & 0 \\ 0 & \cos\Psi & \sin\theta \\ 0 & \sin\Psi & -\sin\phi\sin\theta \end{bmatrix}$

(*b*) $\begin{bmatrix} 1 & 0 & -\sin\theta \\ 0 & \cos\varphi & \sin\varphi\cos\theta \\ 0 & -\sin\varphi & \cos\varphi\cos\theta \end{bmatrix}$

(*c*) $\begin{bmatrix} 1 & 0 & -\sin\theta \\ 0 & \cos\theta & \sin\varphi\sin\theta \\ 0 & -\sin\varphi & \cos\theta.\cos\varphi \end{bmatrix}$

(*d*) $\begin{bmatrix} 1 & 0 & -\cos\theta \\ 0 & \cos\varphi & \sin\varphi\cos\theta \\ 0 & -\sin\varphi & \cos\theta\cos\varphi \end{bmatrix}$

Common data question

An aeroplane flying straight and level at an speed of 150 m/sec at flight 9 km is disturbed by symmetric. Vertical gust resulting is a phugoid oscillation

56. The approximate frequently of phugoid oscillation is ______ sec.

57. It $\dfrac{X\mu}{m} = -0.045\,\text{sec}^{-1}$ for airplane the damping ratio of phugoid oscillation is ______.

Common data question

The roots of characteristics equation for the longitudinal dynamics of centre air craft are

$\lambda_{1,2} = -0.02 \pm 0.2i$

$\lambda_{3,4} = -2.5 \pm 2.6i$

58. The pair of eigen values correspond to short period mode is ______.

59. No. of cycles from t = 0 to $t\frac{1}{2}$ is ______.

60. The time that magnitude decreases half of its amplitude ______.

ANSWER KEY

1. (*c*)	**2.** (*a*)	**3.** (*b*)	**4.** (*a*)	**5.** (*a*)	**6.** (*c*)	**7.** (*b*)	**8.** (*a*)	**9.** (*a*)	**10.** (*a*)
11. (*b*)	**12.** (*a*)	**13.** (*d*)	**14.** (*d*)	**15.** (*d*)	**16.** (*a*)	**17.** (*c*)	**18.** (*a*)	**19.** (*a*)	**20.** 0.35
21. (*c*)	**22.** (*c*)	**23.** (*a*)	**24.** (*b*)	**25.** (*c*)	**26.** (*a*)	**27.** (*d*)	**28.** (*a*)	**29.** (*b*)	**30.** (*b*)
31. (*b*)	**32.** (*c*)	**33.** (*a*)	**34.** (*a*)	**35.** (*a*)	**36.** (*c*)	**37.** (*a*)	**38.** (*b*)	**39.** (*b*)	**40.** (*a*)
41. (*c*)	**42.** (*c*)	**43.** (*d*)	**44.** (*c*)	**45.** (*a*)	**46.** (*b*)	**47.** (*b*)	**48.** (*d*)	**49.** (*d*)	**50.** (*b*)
51. (*a*)	**52.** = 0.0065		**53.** $\boxed{C_{M\,ae\,wb} = -0.032}$			**54.** 2.1 rad/sec		**55.** (*b*)	
56. 0.09266 sec^{-1}		**57.** 0.24	**58.** 0.1608		59. 0.1608		**60.** 0.28 sec.		

EXPLANATIONS

3. gliding angle

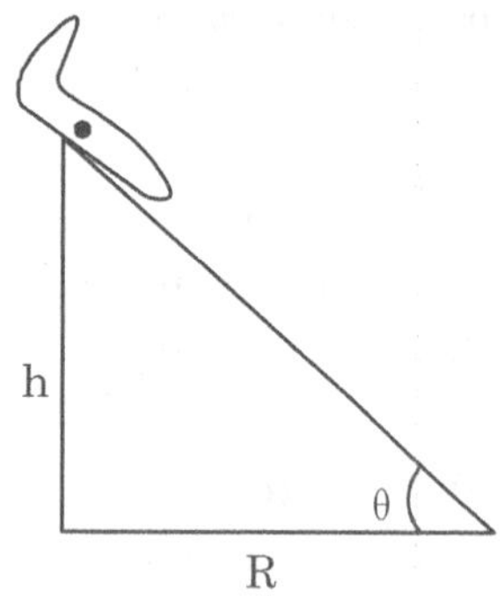

$$\tan\theta = \frac{1}{\left(\frac{L}{D}\right)}$$

$$\frac{h}{R} = \frac{1}{\left(\frac{L}{D}\right)}$$

$$\boxed{R = h\left(L/D\right)}$$

for maximum $R = \left(\frac{L}{D}\right)$ must be maximum.

5. $C_L = \dfrac{2W}{\rho_\infty S v^2}$

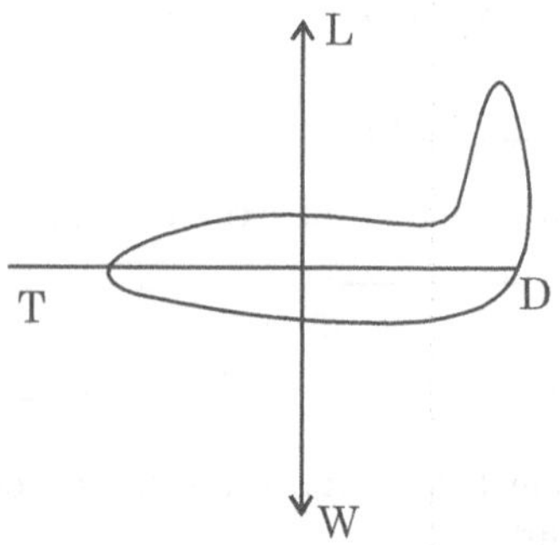

$L = W$

$$C_L \propto \frac{1}{v^2}$$

When $v \to 2v$

$$C_L \propto \frac{1}{4v^2}$$

$$C_{Di} = \frac{{C_L}^2}{\pi\, eAR}$$

$$= C_{Di} \propto C_L^2$$

$$\propto \frac{1}{4v^2}$$

C_{Di} becomes one-fourth.

6. During phugoid oscillation, velocity pitch angle changes in periodic manner, but angle of attack remains constant.

7. In the given roots of lateral dynamics, two are real and two are in complex conjugate form.

$\lambda_1 = -0.000479$ (spiral mode)

$\lambda_2 = -0.118$ (roll mode)

$\lambda_{3,4} = -0.00723 \pm 0.06\ 82j$ (dutch roll mode)

In the real part of the above roots negative is present hence the spiral roll, roll mode, dutch roll mode all are convergence.

8. For jet powered aircraft the condition for maximum range is the aircraft should fly at

$$\left(\frac{C_L^{\frac{1}{2}}}{C_D}\right)_{max}$$

The velocity for $\left(\frac{C_L^{\frac{1}{2}}}{C_D}\right)_{max}$ is

$$V\left(\frac{C_L^{Y_2}}{C_D}\right)_{max} = \left(\frac{2}{e_\infty}\sqrt{\frac{3K}{C_{D0}}} \times \frac{W}{S}\right)^{\frac{1}{2}}$$

$$= \left(\frac{2}{\rho_\infty}\sqrt{\frac{K}{C_{D_0}}} \times \frac{W}{S}\right)^{\frac{1}{2}} \times (\sqrt{3})^{\frac{1}{2}}$$

$$= 1.32\left(\frac{2}{1.23} \times \sqrt{\frac{0.026}{0.021}} \times \frac{28 \times 10^3}{20}\right)^{\frac{1}{2}}$$

= 66.5 m/sec.

9. Lateral dynamics of aircraft motion gives four roots the real larger value correspond to roll mode, complex roots corresponds to dutch roll mode.

10. $\lambda_{3,4} = -0.06 \pm 108i$

$$= -\xi\omega_n + \omega_n\sqrt{1-\xi^2}$$

$$\xi\omega_n = 0.06 \qquad ...(1)$$

$$\omega\sqrt{1-\xi^2} = 1.8 \qquad ...(2)$$

equation (2) and equation (1)

$$\frac{\sqrt{1-\xi^2}}{\xi} = \frac{1.8}{0.06}$$

$$\Rightarrow \frac{1-\xi^2}{\xi^2} = \left(\frac{1.8}{0.06}\right)^2$$

$$1 = 901\xi^2$$

$$\xi^2 = \sqrt{0.001109}$$

= 0.033.

11. damped frequency

$$\omega_d = \omega_n\sqrt{1-3^2}$$

= 1.8 rad/sec.

14. Gauge flap mainly increases the wing area and chord of the wing it is used to reducing take off speed for the given load. If doesnot effect trime condition of aircraft.

15. Slats are high lift devices which is used near leading edge of wing used to increase the maximum lift and hence reducing stalling speed of aircraft

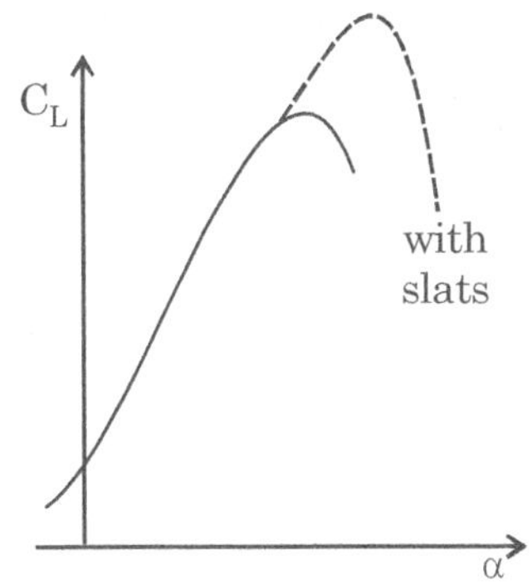

In the above figure we can see slats does not effect zero lift coefficient or $\frac{dC_L}{\partial\alpha}$.

16. The frequency of phugoid mode is given by

$$W = \sqrt{2}\frac{g_0}{u}$$

17. for lateral and directional stability of aircraft

$l_\beta < 0$

$N_\beta > 0$

β = side slip angle, l = rolling moment

N = yawing moment

for the position side ship the aircraft should rate over list an should yaw towards the right.

18. We have the velocity component along. axial, side ship and normal direction are

$V_{xb} = V \cos\alpha \cos\beta$.

= 1000 cos 36° cos 20° = 813 80 m/sec

$V_{yb} = V \sin\beta = 1000 \sin 20° = 342$ m/sec

$V_{zb} = V \sin\alpha \cos\beta = 1000 \sin 36°, \cos 0$

= 46.9 85 m/h

19. Given

ρ = 0.9

$C_{D0} = 0.018$

AR = 8

At steady level flight $\left(\frac{L}{D}\right)_{max}$

$$= \frac{1}{2\sqrt{C_{D0}K}} = \frac{1}{2\sqrt{0.018}} \times \sqrt{3.14 \times 0.9 \times 8}$$

= 17.71 = 18

under steady level flight $\left(\frac{L}{D}\right)_{max}$

$$= \frac{1}{\sqrt[2]{C_{D0}K}} = \frac{1}{\sqrt[2]{0.018}} \times \frac{1}{3.19 \times 0.9 \times 8}$$

= 17.71 = 18

20. Damping ratio for phugoid mode is

$$\xi = \frac{1}{\sqrt{2}} \times \frac{1}{\left(\frac{L}{D}\right)}$$

$$= \frac{1}{\sqrt{2}} \times \frac{1}{20}$$

$= 0.35$

21. Brequest range equation

$$R_{max} = V|_{D\,min} \times \frac{L}{D}\bigg|_{max} \times I_{sp} \times \ln\frac{W_i}{W_f}$$

$= 170 \times 23 \times 2330 \ln 1.4$

$= 3065.36$ km.

22. For spiral mode, roll rate = 0

$$p = \frac{L_y + i_x N_y}{1 - i_x 1_z}\gamma$$

$p = 0$

$$0 = -\frac{L_y + i_x N_y}{L_B + i_x N_\beta}\gamma = \frac{-L_y}{L_v}\gamma$$

$$i_x = \frac{I_{xz}}{I_{zz}}.\ I_z = \frac{I_{xz}}{I_{zz}}$$

Neglecting i_x and i_z and substituting β from equation (1) we get yaw rate equation

$$\gamma = \left(N_\gamma - \frac{L_\gamma N_\beta}{L_\beta}\right)\gamma$$

$$\lambda = N_\gamma - \frac{L_y N_\beta}{L_\beta}$$

25. Dutch roll motion is lateral dynamic which is highly damped motion with low frequency.

26. Short period oscillation is heavily damped where angle of attack pitch rate changes rapidly but velocity remains approximately constant.

27. In case of larger finite wing there in downwash experienced at various wing sections, due to which effective angle of attack gets reduced and only a component of lift acts is direction in to free stream velocity this results in effective loss of lift and the presence of induce drag.

28. Given

$V_\infty = 540$ m/h

$$= 540 \times \frac{88}{60} = 792\text{ft/sec}$$

Energy light

$$He = h + \frac{V_\infty^2}{2g}$$

$$= 30000 + \frac{(792)^2}{2 \times 32.2}$$

$= 39.740$ft

29. Brequest range equation for propeller driven aircraft is

$$R = \frac{n_{pr}}{C}\frac{L}{D}\ell n\frac{W_o}{W_1}$$

$R \propto n_{pr}$

n_{pr} – increases

$R \propto \frac{1}{C}$ R – increases

C – decreases

R – increases

30.

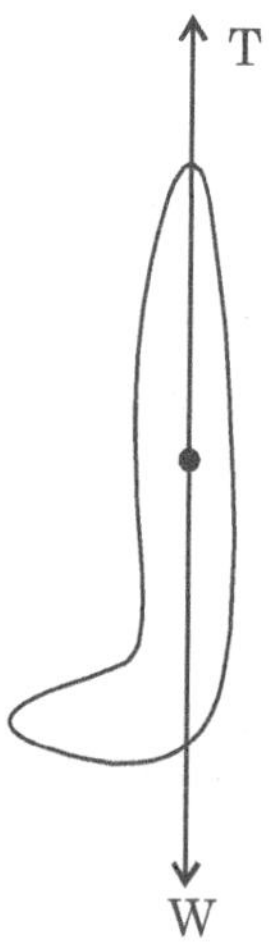

In case of vertically upward,

$L = 0$

load factor

$$n = \frac{L}{W} = 0$$

31. In steady level flight condition

$L = W$

$T = D$

$\boxed{L = W = 20000N}$

32. Load factor $n = \frac{L}{W}$

$L = n \times W$
$= 20000 \times 8$
$= 160000$

33. $S_{LO} = \frac{1.44W^2}{\rho_\infty gs\, C_{L\,max} T}$

To shorter the lift off distance

$\frac{T}{W} \rightarrow$ higher

$\frac{W}{S} \rightarrow$ lower

take off from near sea level because the density is higher at near the level

$S_{L_o} \propto \frac{1}{\rho_\infty^2} \cdot T \propto \rho_\infty$

$C_{L\,max}$ – higer

34. Condition for

$(R)_{max\ jet}$ ---- $\left(\frac{C_L^{\frac{1}{2}}}{C_D}\right)_{max}$

$(R)_{max\ propeller}$ ---- $\left(\frac{C_L}{C_D}\right)_{max}$

$(E)_{max\ jet}$ ---- $\left(\frac{C_L}{C_D}\right)_{max}$

$(E)_{max\ propeller}$ ---- $\left(\frac{C_L^{\frac{3}{2}}}{C_D}\right)_{max}$

37. Power required at mean sea level

$$(P_R)_{sea} = P_{R_a}\left(\frac{\rho}{\rho_s}\right)^{\frac{1}{2}}$$

$$= 400\left(\frac{0.4135}{1.225}\right)^{\frac{1}{2}}$$

$= 2324$ H.P

38. Given,

$$R/C = -1.89 \times 10^{-3} h + 28.52$$

$$R/C = \frac{dh}{dt}$$

$$dt = \frac{dh}{R/C}$$

Now time of climb from the near sea level to 11.5 km altitude

$$\int_0^t dt = \int_0^{11.5\times10^3} \frac{dh}{-1.89 \times 10^{-3} h + 28.82}$$

$$t_m = \left|\ell n\left(1.89 \times 10^{-3} h + 28.52\right)\right|_0^{11.5\times10^3}$$

$$= \frac{1}{1.89 \times 10^{-3}} \frac{\left(-1.89 \times 10^{-3} \times 11.5 \times 10^3 + 28.52\right)}{28.52}$$

$$= \frac{1}{1.89 \times 10^{-3}} \ell n\left(\frac{28.52 - 21.735}{28.52}\right)$$

$$= \frac{1}{1.89 \times 10^{-3}} \ell n\left(\frac{6.78}{28.52}\right)$$

$$= \frac{1.4366}{1.81 \times 10^{-3}}$$

$= 0.7566 \times 10^3$.
$= 756.60$
$= 12$ min 61 sec.

39. Rate of climb $(R/C) = P_5 - \frac{V_\infty}{g} A$

A = Instantaneous acceleration
V_∞ = Instantaneous velocity
P_S = Specific excess power
Given A = 8 ft/sec²
P_s = 300 ft/sec
V_∞ = 800 ft/sec
g = 32.2 ft/sec²

$$\frac{dh}{dt} = 300 - \frac{800}{32.2} \times 8$$

$= 101$ ft/sec

40. Stalling speed of aircraft is given by

$$V_{stall} = \sqrt{\frac{2W}{\rho_\infty SC_{L\,max}}}$$

41. Lift off distance $S_{L_0} = \frac{1.44W^2}{\rho_\infty g\,S\,C_{L\,max}T}$

Where $T \propto \rho_\infty$

$$S_{L_0} \propto \frac{1}{\rho_\infty^2}$$

$$S_{L_0/200} = S_{L_0\,sea}\left(\frac{\rho_s}{\rho_a}\right)^2 = 800\left(\frac{1.225}{1.10}\right)^2$$

$= 800 \times (1.10)^2$

$= 986$ m

42. Expensing for $C_{L\,max}$.

$$C_{L\,max} = \frac{2W \times n}{\rho SV_\infty^2}$$

$$= \frac{2 \times 9 \times 55000}{0.413 \times 12 \times 160^2}$$

$= 7.8$

43. $C_{L\,min} = \frac{2W}{\rho_s V_\infty^2}$

$$= \frac{2 \times 5500}{0.413 \times 12 \times 160^2} = 0.86$$

44. $n = 9$

$V_\infty = 160$ m/sec

$$\text{Radius of turns} = \frac{V_\infty{}^2}{g\sqrt{n^2 - 1}}$$

$$= \frac{160^2}{9.8\sqrt{9^2 - 1}}$$

$= 292.05$ m

45. By Definition.

47. $C_{m.\,cg}$ at zero list condition is know as $C_{m.0}$

$$C_{M.o} = C_{M.ae\,wb} + \frac{S_t\,l_t}{SC}\eta_t\,a_t(i_t - \xi_0)$$

$$= = -0.0218 + \frac{0.3 \times \left(\frac{29.2}{2}\right)}{2.8} \times 0.9 \times 0.08(2.1 - 0)$$

$= 0.2147$

48. Expression for neutral point

$$h_n = h_{acwb} + V_H\frac{a_t}{a} \times \eta_t(1 - \xi_a)$$

$$V_H = \frac{s_t + l_t}{CS} = \frac{0.3 \times \frac{29.2}{2}}{2.8}$$

$= 1.56$

$$\boxed{h_n - h_{acwb} = 0.9855}$$

0.9855 is non-dimension with chord, hence and to multiply with chord length.

$h_n - h_{ac} = 0.9805 \times 2.8$

$= 2.76$ m

49. Static margin $= -\frac{dC_{m\,cg}}{\partial C_L}$

$$\frac{\partial C_{m\,cg}}{\partial C_L} = \left[h - h_{ac} - V_H\frac{a_t}{a}n_t(1 - \xi_a)\right]$$

$$-\frac{\partial C_{m\,cg}}{\partial C_L} = -\left[h - h_{ac} - V_H\frac{a_t}{a}n_t(1 - \xi q)\right] = 0$$

$$-SM = -\left[h - h_{ac} - V_H\frac{a_t}{a}n_t(1 - \xi_a)\right]$$

$$h - h_{ac} = -SM + V_H\frac{a_t}{a}n_t(1 - \xi_a)$$

$$= -0.1 + 1.56 \times \frac{0.08}{0.08} \times 0.9(1 - 0.5)$$

$= 0.8855$

$h - h_{ae} = 0.8855$ in non-dimensional with chord here nead to multiply with C

$h - h_{ae} = 0.8855 \times 2.8$

$= 2.47$ m

50. For an aircraft, considering wing and body coefficient of moment about center of gravity (CG)

$$C_{M,.cg} = C_{M,ac} + C_{Lwb}(h - h_{acwb})$$

$$C_{M,.ac\,wb} = C_{Mo} \text{ at zero lift coordination}$$

$$C_{Mo} = C_{Mac} + C_{Lwb}(h - h_{ac})$$

$0.004 = -0.002 + 0.3\,(h - h_{ae})$

$h - h_{ac} = 0.02$ ....(1)

$h - h_{ac}$ is non-dimensional with chord hence multiplying with C

$h - h_{ac} = 0.02 \times 0.45$

$= 9$ mm

$h_{ac} = h - 9$ m

here aerodynamic centre lies 9 mm ahead of CG.

51. In the above problem if CG is moved ahead by 10% then new value of h will be h – 10%. Now substituting there values we will get

$$C_{meq.\,wb} = C_{M\,ac} + C_{L\,wb}(h - h_{ac})$$

$$C_{meq.\,wb} = C_{M\,ac\,wb} + C_{L\,wb}(h - 0.1 - h_{ac\,wb})$$

from equation (1), putting value of h. h_{ae}.

$$C_{ac.wb} = -0.002 + 0.85(0.02 - 0.1)$$

$= -0.070$

52. $C_{Meq.\,wb} = C_{Mac.\,wb} + C_{Lwb}(h - h_{ac\,wb})$

given

$h - h_{ac\,wb} = 0.05$

$C_{L\,wb} = 0.45$

$C_{M.\,ac\,wb} = -0.016$

$C_{m\,eq\,wb} = -0.016 + 0.45\,(0.05)$

$= 0.0065$

53. Lift slope $= \dfrac{\partial C_L}{\partial \alpha} = \dfrac{(.52 - 0)}{5 - (1 - 1 - 5)} = \dfrac{0.52}{6.5}$

$= 0.08$ per degree

Now

$$C_{m\,eq} = C_{m\,ac\,wb} + C_{L\,wL}(h - h_{ac\,wb})$$

$$= C_{m\,ac\,wb} + a_{wb}\alpha_{wb}(h - h_{ac\,wb})$$

per 1°

$$-0.01 = -0.01 = C_{M\,ac\,wb} + 0.08(1 + 1.5)(h - h_{ac}) \quad \text{...(i)}$$

for 7.88°.

$$0.05 = C_{m\,ac\,wb} + 0.08(7.88 + 1.5)(h - h_{ae}) \quad \text{...(ii)}$$

solving there two equation we get

$-0.06 = 0 - 0.55\,(h - h_{ae})$

$$h - h_{ae\,wb} = \frac{-0.06}{0.55} = 0.11$$

$$h_{ae\,wb} = 0.35 - 0.11 = 0.24$$

from equation (1) we get

$-0.01 = C_{Mac,\,wb} + 0.08\,(1 + 1.\sigma) \times 0.1)$

$\boxed{C_{M\,ac\,wb} = -0.032}$

54. Lateral directional characteristics of an aircraft have from roots have, complex roots corresponds to dutch roll motion,

$\therefore\ \lambda_{3,4} = (-0.09 \pm 2.1\,i)$

$$= -\xi\omega_n \pm i\omega_n\sqrt{1 - \xi^2}$$

Now comparing seal and mapping values are will get

$(\xi\omega_n) = 0.09$ ....(1)

$\omega_n\sqrt{1 - \xi^2} = 2.1$ ....(2)

dividing equation (1) by equation (2), we get

$$\frac{\xi^2}{1 - \xi^2} = \frac{0.09}{2.1}$$

$$\frac{\xi^2}{1 - \xi^2} = \left(\frac{0.09}{21}\right)^2$$

$$\Rightarrow \frac{1.\xi^2}{\xi^2} = 544.44$$

$$\xi^2 = \frac{1}{545.5}$$

$$\xi = \sqrt{\frac{1}{545.5}}$$

$$\xi = \sqrt{0.001833} = 0.042$$

damped frequency $\omega d = \omega_n\sqrt{1 - \xi^2}$

$= 2.1$ rad/sec

Natural frequency

$\omega_n = 49$ rad/sec

$$\frac{2\pi}{T} = 49$$

Time period $T = \frac{2\pi}{49} = 0.1279$ sec

From equation (1)

$$\omega_n = \frac{0.09}{0.042} = 2.14 \text{ rad/sec}$$

56. At 9 km altitude

$$g = g_e\left(\frac{r^2}{R^2}\right)$$

$$= g_e\left(\frac{R+h}{R}\right)^2 = g_e\left(\frac{1+h}{R}\right)^2$$

$$= 9.8\left(1+\frac{9}{6400}\right)^2 = 9.82 \text{ m/sec}^2$$

Frequency of the phugoid oscillation is given by Lanchester is

$$\omega_n = \sqrt{2}\frac{g_0}{u}$$

$$= \sqrt{2} \times \frac{9.82}{150} = 0.09266 \text{ sec}^{-1}$$

57. Damping ratio

$$\xi_{\text{phugoid}} = \left(\frac{-\frac{x_\mu}{m}}{2\omega_n}\right)$$

$$= \frac{-(-0.045)}{2 \times 0.09266} = 0.24$$

58. $\lambda_{3,4}$.
the eigen values corresponds to short period mode in $2.5 \pm 2.6\hat{i}$ because the short period mode heavily damped oscillation.

Time to half $= \frac{0.693}{\lambda}$

$$= \frac{0.693}{2.5} = 0.277$$

= 0.28 sec

$$\lambda_{3,4} = 2.5 \pm i2.6$$

$$= -\xi\omega_n \pm i\omega_n\sqrt{1-\xi^2}$$

$\therefore \quad \xi_n = 2.5 \qquad$ ...(1)

$\omega_n\sqrt{1-\xi^2} = 2.6 \qquad$ ....(2)

solving this we will get

$$\frac{1-\xi^2}{\xi^2} = \left(\frac{2.6}{2.5}\right)^2$$

$$1-\xi^2 = 1.083^2$$

$$\xi^2 = \frac{1}{2.08}$$

$$3 = \sqrt{\frac{1}{2.08}}$$

$$3 = 0.6931$$

$$\omega_n = \frac{2.5}{0.6931}$$

$\omega_n = 3.6$ rad/sec

$\therefore$ Number of cycle

from $t = 0$ to $t\frac{1}{2}$

$$N\frac{1}{2} = \frac{t^{\frac{1}{2}}}{T}$$

$$= \frac{0.28}{\frac{2\pi}{3.6}}$$

= 0.1608

59. Number of cycles from t = 0 to the = 0.1608

60. Time to half = 0.28 sec.

3 CHAPTER

SPACE DYNAMICS

1. The total energy of a Missile launched in space with a velocity less than Earth's escape velocity is
   (*a*) Negative
   (*b*) Positive
   (*c*) Zero
   (*d*) Either positive or negative
2. The geostationary artificial satellite has a period of
   (*a*) 6 hours (*b*) 12 hours
   (*c*) 24 hours (*d*) 365 days
3. The second law of Kepler stating constancy of aerial velocity of a planet is a consequence of the law of Conservation of
   (*a*) angular momentum
   (*b*) energy
   (*c*) linear momentum
   (*d*) none of these
4. Let us consider a planet comes nearer the sun moves
   (*a*) slow
   (*b*) fast
   (*c*) constant at every point
   (*d*) none of the above
5. If a body is to be projected in a direction making an angle 45° to the vertical, then the escape velocity is (Given the escape velocity from earth is 11.2 km per sec.)
   (*a*) $11.2 \times \sqrt{2}$ km/sec (*b*) 11.2 km/sec
   (*c*) 11.2×2 km/sec (*d*) $11.2 \times \frac{1}{\sqrt{2}}$ km / sec
6. The gain in the potential energy of an object of mass m raised from the surface of the earth to a height equal to the radius R of the earth is, where g is the acceleration due to gravity of the earth's surface
   (*a*) 2 mgR (*b*) $\frac{1}{2}$ mgR
   (*c*) mgR (*d*) $\frac{1}{4}$ mgR
7. If the acceleration due to gravity at the surface of the earth is g, an earth's satellite of mass m revolves in a circular orbit at a height h from the surface. The velocity of the satellite in the orbit is given by
   (*a*) gR (*b*) $\frac{gR}{R+h}$
   (*c*) $\frac{gR^2}{R+h}$ (*d*) $\sqrt{\frac{gR^2}{R+h}}$
8. What happens to a satellite that is revolving around the sun in a circular orbit with uniform velocity v? If the gravitational force suddenly disappears, the velocity of the satellite will be
   (*a*) Zero (*b*) v
   (*c*) 2v (*d*) infinity
9. A planet has twice the radius but the mean density is $\frac{1}{4}^{th}$ as compared to earth. What is the radio of the escape velocity from the earth to that from the planet?
   (*a*) 3 : 1 (*b*) 1 : 2
   (*c*) 1 : 1 (*d*) 2 : 1
10. The radii of two planets are in the ratio 1: 2 and their masses are in the ratio 1: 2. The acceleration due to gravity on the planets are in the ratio.
    (*a*) 1 : 2 (*b*) 2 : 1
    (*c*) 3 : 5 (*d*) 5 : 3
11. An artificial satellite has a total (K.E + P.E) energy E_0 is moving around the earth in a circular orbit. Its potential energy is
    (*a*) $2\,E_0$ (*b*) $1.5\,E_0$
    (*c*) E_0 (*d*) $-E_0$
12. A circular thin uniform ring is rolling down an inclined plane of inclination 30° without slipping. Its linear acceleration along the inclined plane will be
    (*a*) $\frac{g}{2}$ (*b*) $\frac{g}{3}$
    (*c*) $\frac{g}{4}$ (*d*) $\frac{2g}{3}$

13. If the radius of planet is half the radius of the earth and the acceleration due to gravity of a planet is half the acceleration due to gravity of the earth's surface, the mass of planet in terms of mass of earth is

(a) $\frac{M_e}{2}$ (b) $\frac{M_e}{4}$

(c) $\frac{M_e}{6}$ (d) $\frac{M_e}{8}$

14. If the acceleration due to gravity on earth's surface is g and radius of earth is R , the mean density of the earth is

(a) $\frac{4\pi G}{2}$ (b) $\frac{3g}{4\pi RG}$

(c) $\frac{3\pi G}{4gR}$ (d) $\frac{\pi Rg}{4G}$

15. Acceleration due to gravity g in terms of mean density of Earth d (where R is radius of earth and G - universal gravitational constant) is

(a) $g = 4\pi R^2 dG$

(b) $g = \frac{4\pi R^2 G}{d}$

(c) $g = \frac{3}{4}\pi RdG$

(d) $g = \frac{4}{3}\pi RdG$

16. Mars orbiter sent by ISRO was parked in its first earth orbit of size 264 km × 23903 km. The orbit was then raised to 264 km × 28825 km. How much velocity increment was required at perigee.

(a) 321.4 m/s (b) 285.6 m/s

(c) 127.2 m/s (d) 96.8 m/s

17. According to kepler's second law of planetary motion, a line joining planet to center of the sun (which is at focus) sweeps out equal interval of time. Mathematically, $\frac{dA}{dt}$ = const. What is the value of this constant for a planet of mass M having L angular momentum in the orbit?

(a) $\frac{L}{2}$ (b) $\frac{L}{2M^2}$

(c) $\frac{L}{2M}$ (d) $\frac{L^2}{2M}$

18. A planet P has 6 times less acceleration due to gravity at its surface than planet Q .How much is the escape velocity at P, if escape velocity at Q is V_e ?

(a) $\frac{V_e}{6}$

(b) $\frac{v_e}{\sqrt{6}}$

(c) $\frac{V_e}{36}$

(d) more data required

19. A satellite is revolving around earth is 400 km × 1200 km elliptical orbit. What is velocity of satellite at its apoapsis?

(a) 7.87 km/s (b) 7.04 km/s

(c) 7.56 km/s (d) 6.82 km/s

20. In above question, what is the velocity of satellite at its periapsis?

(a) 7.87 km/s (b) 7.04 km/s

(c) 7.56 km/s (d) 6.82 km/s

21. In above question, let velocity is represented by V and distance of satellite from center of the earth by R. What is the ratio of product of V and R at apoapsis and periapsis, $\frac{(VR_a)}{(VR_p)}$?

(a) 0.8945 (b) 1.1179

(c) 1.0000 (d) 0.8002

22. What is the time period of the satellite in above question?

(a) 5241 s (b) 5569 s

(c) 5897 s (d) 6077 s

23. A satellite is an elliptical orbit has time period equal to a satellite in Geo- stationary orbit (GEO). What is the semi- major axis of this orbit?

(a) 42226 km (b) 37254 km

(c) 35856 km (d) Data insufficient

24. In above question, what is the minimum height of satellite above Earth's surface.

(a) 42226 km (b) 37254 km

(c) 35856 km (d) Data insufficient

25. If the Earth stops rotating (Neglecting the effect of the revolution), then gravity of pole and equation will

(a) Increase (b) decrease

(c) Remain Same (d) Changes irratically

26. If a man of weight W jumps from a roof of height h carrying a load M. What is the weight experienced by the man under free fall?

(a) 2W (b) W
(c) 0 (d) None of these

27. If F_1 – Magnitude of force on sun by Earth, F_2 – Magnitude of force on Earth by sun, then which of the following is correct?

(a) $F_1 = F_2$ (b) $F_1 < F_2$
(c) $F_1 > F_2$ (d) None of these

28. Acceleration due to gravity at poles, is g_p and acceleration due to gravity at equator is g_e. Which of the following is true, caused due to rotation of the earth.

(a) $g_p > g_e$ (b) $g_p < g_e$
(c) $g_p = g_e$ (d) None of these

29. A man standing in a freely falling lift, if he drops a ball from his pocket, then the ball will have a motion with respect to the man is

(a) Exerts downwards
(b) Exerts upwards
(c) Exerts simple harmonically
(d) Remains stationary

30. Two spheres of same radius r and material ρ has been placed such that they are touching with each other. Then the force of attraction between them will be

(a) $F \propto r^3$ (b) $F \propto r^4$
(c) $F \propto r^2$ (d) $F \propto \frac{1}{r^2}$

31. If the force of attraction is proportional to $F \propto r^{\frac{-5}{2}}$ rather than inverse square as a universal law, then the potential energy will be -

(a) $U \propto r^2$ (b) $U \propto r^3$
(c) $U \propto r^{-2}$ (d) $F \propto r^{\frac{-3}{2}}$

32. Compare the weights of the bodies which are at h m below the surface of Earth (w_1), on the surface of Earth (w_2) and h m above the surface of Earth (w_3) having same mass m?

(a) $w_1 > w_2 > w_3$ (b) $w_3 < w_1 < w_2$
(c) $w_1 < w_2 < w_3$ (d) $w_1 < w_3 > w_2$

33. A satellite is moving around Earth with speed V in circular orbit of radius r. If orbit radius decreases by 1%, then the velocity will be.

(a) Increase (b) Decrease
(c) remain same (d) none

34. In an circular orbit if KE of the satellite is proportional to $(T)^k$. Then the value of k is where T is time period of satellite.

(a) 2 (b) 3
(c) $\frac{3}{2}$ (d) $\frac{2}{3}$

35. Three identical mass m are located at corners of equilateral triangle and revolves in a circular orbit of radius R. Calculate the velocity of each planet in an orbit and total potential of the system?

(a) $\frac{Gmm}{R}$ (b) $\frac{-Gmm}{R}$
(c) $\frac{\sqrt{2}Gmm}{R}$ (d) $\frac{-\sqrt{3}Gmm}{R}$

36. The time period of satellite revolving around Earth in circular orbit is

(a) Independent on the mass of the Earth
(b) Independent on distance from the planet
(c) Independent of the inclination angle
(d) Option a and b are true
(e) Option a and c are true

37. The ratio of two planet's A and B radii and the ratio of two planet's gravity are $k_1 = \frac{R_A}{R_B}$ and $k_2 = \frac{g_A}{g_B}$. Then the ratio of their escape velocity from the surface of their respective planets will be -

(a) $k_1 \times k_2$ (b) $\sqrt{\frac{k_1}{k_2}}$
(c) $\sqrt{k_1 \times k_2}$ (d) $\sqrt{\frac{k_2}{k_1}}$
(e) Option a and c are true

38. A rocket of mass m is launched vertically with initial speed u. Assuming negligible air resistance. Then the maximum height attained by the rocket will be

(a) $h = \frac{R^2}{\frac{2gR}{u} - 1}$ (b) $h = \frac{R^2}{\frac{2gR}{u} + 1}$
(c) $h = \frac{R}{\frac{2gR}{u^2} - 1}$ (d) $h = \frac{R}{\frac{2gR}{u^2} + 1}$

39. The change in P.E. to throw a mass from surface of earth to height h where h = nR is, where R is radius of earth :-

(a) $mgR\left(\frac{n}{n-1}\right)$ (b) mgRn

(c) $mgR\left(\frac{n^2}{n^2-1}\right)$ (d) $mgR\left(\frac{n}{n+1}\right)$

40. The velocity required to throw a mass m from the surface of to the height h will be. Given r = R + h

(a) $\sqrt{2gR}$ (b) $R\left[2g\left(\frac{1}{R}-\frac{1}{R}\right)\right]^{\frac{1}{2}}$

(c) $\sqrt{2g(R+r)}$ (d) $2g\left(\frac{1}{R}-\frac{1}{r}\right)$

41. If a satellite is revolving around Earth in a circular orbit at a distance r from the center of Earth. Find the extra energy that must be provided to the satellite to escape from Earth's gravitational field

(a) $\frac{GMm}{r}$ (b) $\frac{-GMm}{r}$

(c) $\frac{GMm}{2r}$ (d) $\frac{-GMm}{2r}$

42. As per property of ellipse, the distance of point P from foci on ellipse is e times distance from latus rectum. Find r in terms of e.

(a) $r = \frac{ek}{1+e\cos\theta}$ (b) $r = \frac{ek}{1-e\cos\theta}$

(c) $r = \frac{e}{1+e\cos\theta}$ (d) $r = \frac{e}{1-e\cos\theta}$

43. A satellite is launched into Earth orbit where its launch vehicle burns out at an altitude of 250 km. At burnout, the satellite's velocity is 7950 m/s with zenith angle equal to 89 degrees. Calculate the satellite's altitude at perigee and apogee.

(a) 239 km and 971 km

(b) 12995 km and 13728 km

(c) 250 km and 1008 km

(d) 115 km and 654 km

44. Calculate the eccentricity of the orbit for the satellite is launched into Earth orbit where its launch vehicle burns out at an altitude of 250 km. At burnout, the satellite's velocity is 7950 m/s with zenith angle equal to 89 degrees. Calculate the satellite's altitude at perigee and apogee.

(a) 0.03 (b) 0.04

(c) 0.05 (d) 0.06

45. Calculate the semi-major axis of the orbit for the satellite is launched into Earth orbit where its launch vehicle burns out at an altitude of 250 km. At burnout, the satellite's velocity is 7950 m/s with zenith angle equal to 89 degrees. Calculate the satellite's altitude at perigee and apogee.

(a) a = 7,349,944 m

(b) a = 6,983,739 m

(c) a = 6,617,534 m

(d) a = 6,628,140 m

ANSWER KEY

1. (a)	2. (c)	3. (a)	4. (b)	5. (b)	6. (b)	7. (d)	8. (b)	9. (c)	10. (a)
11. (a)	12. (c)	13. (d)	14. (b)	15. (d)	16. (c)	17. (c)	18. (d)	19. (b)	20. (a)
21. (c)	22. (d)	23. (a)	24. (d)	25. (a)	26. (c)	27. (a)	28. (a)	29. (d)	30. (b)
31. (d)	32. (b)	33. (a)	34. (d)	35. (d)	36. (c)	37. (c)	38. (c)	39. (d)	40. (b)
41. (d)	42. (a)	43. (a)	44. (c)	45. (b)					

EXPLANATIONS

1. Kinetic Energy of a missile is given as,

$$K.E. = \frac{1}{2}mV^2$$

For circular orbits it is given as,

$$K.E. = \frac{GMm}{2r}$$

where m is the mass of missile and r = R + h is the distance of missile from the centre if Earth.

The potential energy is given as,

$$U = \frac{-GMm}{r}$$

Therefore, the total Energy is given as,

T. E. = K. E. + P. E.

$$= -\frac{GMm}{2r}$$

2. A geostationary orbit can be achieved only at an altitude very close to 35,786 km and directly above the equator.

Therefore, as per given formula

$$T = \sqrt{\frac{4\pi^2 r^3}{GM}}$$

Here,

Radius of Earth, R = 6400 km

Altitude, a = 35,786 km

$GM = 3.96 \times 10^{14}$.

$$T = \sqrt{\frac{4 \times \pi^2 \times ((6400 + 35786) \times 10^3)^3}{3.96 \times 10^{14}}}$$

= 86513.68s = 23.96 hours

3. According to Kepler's second law, Area swept out by radial vector from the primary planet to satellite in equal interval of time is constant, i.e.

$$\frac{dA}{dt} = \text{constant}$$

$$h = \vec{v} \times \vec{r}$$

$$h = vr\sin\theta$$

$$\therefore \frac{dA}{dt} = \frac{1}{2} \times v \times r\sin\theta = \frac{1}{2}h = \text{constant}$$

where, h is the angular momentum which is conserved.

4. As per Kepler's second law, a line joining a planet and the Sun sweeps out equal areas during equal intervals of time. The orbital radius and angular velocity of the planet in the elliptical orbit will vary. Thus, the planet travels faster when closer to the sun, then slower when farther from the sun.

5. The escape velocity is given by,

$$V_e = \sqrt{2gR} = \sqrt{2 \times 9.81 \times 6378 \times 10^3} = 11.2 \text{ km/s}$$

6. At earth surface, P.E. is given by

$$U_{surface} = \frac{-GMm}{r}$$

At a distance R from the earth's surface, P.E. of system is

$$U_{distance} = \frac{-GMm}{2r}$$

$$\therefore \Delta U = \frac{-GMm}{2r} - \left(-\frac{GMm}{r}\right)$$

$$= \frac{GMm}{2r}$$

$$\because g = \frac{GM}{r^2}$$

$$\Rightarrow GM = gr^2$$

$$\therefore \Delta U = \frac{1}{2}mgr$$

7. As per formula,

$$V = \sqrt{\frac{GM}{(R+h)}} \text{ and}$$

$$g = \frac{GM}{R^2}$$

$$\Rightarrow GM = gR^2$$

$$\therefore V = \sqrt{\frac{gR^2}{(R+h)}}$$

8. Satellite will move tangentially to its circular orbit with the same velocity. As per Newton's first law, the body tends to be in motion unless an external force acts on it, so the satellite will be in motion

due to inertia of motion of the satellite. In absence of gravitational force there will be no centripetal force to keep the satellite in its circular orbit, therefore, it will fly off its orbit with a tangential velocity.

9. Given,

Radius of the planet = 2R

Mean density of planet = $\frac{1}{4}\rho$

As per formula,

$$\Rightarrow GM = gR^2$$

$$M = \rho \times \frac{4}{3}\pi R^3$$

$$\frac{V_{e_{Planet}}}{V_{e_{earth}}} = \sqrt{\frac{2GM_{planet}}{R_{Planet}}} \times \sqrt{\frac{R_{earth}}{2GM_{earth}}}$$

$$= \sqrt{\frac{2 \times G \times \rho \times 4 \times \pi \times (2R)^3}{4 \times 3 \times 2R}} \times \sqrt{\frac{R \times 3}{2 \times G \times \rho \times 4 \times \pi \times R^3}}$$

$$= \sqrt{\frac{8}{8}} \times 1 = 1$$

10. Given,

$$\frac{R_1}{R_2} = 1:2, \quad \frac{M_1}{M_2} = 1:2$$

As per given formula

$$g = \frac{GM}{R^2} = \frac{G \times d \times \frac{4}{3}\pi R^3}{R^2} = \frac{4}{3}\pi RGd$$

$$\therefore \frac{g_1}{g_2} = \frac{\frac{4}{3}\pi R_1 Gd}{\frac{4}{3}\pi R_2 Gd} = \frac{R_1}{R_2} = 1:2$$

11. Given,

The Total Energy

T. E. = K. E. + P. E.

Kinetic Energy is given by,

For circular orbits it is given as,

$$K.E. = \frac{GMm}{2r} = E_0$$

$$\therefore P.E = T.E. - K.E$$

$$= -E_o - E_o$$

$$= -2E_o$$

12. Acceleration on an inclined plane

$$a = \frac{g\sin\theta}{\left(1 + \frac{I}{MR^2}\right)}$$

For circular rings: $I = MR^2$.

so by putting the value in above equation we get,

$$\therefore a = \frac{g\sin\theta}{2} = \frac{g\sin 30^\circ}{2} = \frac{g}{4}$$

13. Given,

$$\frac{R_p}{R_e} = \frac{1}{2}$$

$$\frac{g_p}{g_e} = \frac{1}{2}$$

Also, we know

$$g = \frac{GM}{R^2}$$

$$\therefore M = \frac{gR^2}{G}$$

$$\frac{M_P}{M_e} = \frac{g_p R_p^2}{G} \times \frac{G}{g_e R_e^2} = \frac{1}{2} \times \frac{1}{4} = \frac{1}{8}$$

$$\therefore M_p = \frac{M_e}{8}$$

14. As per question,

$$g = \frac{GM}{R^2}$$

and $g = \frac{4G\pi R^3 d}{3 \times R^2}$, where d is the mean density.

$$\therefore d = \frac{3g}{4G\pi R}$$

15. As per given formula,

$$g = \frac{GM}{R^2}$$

and $g = \frac{4G\pi R^3 d}{3 \times R^2}$, where d is the mean density.

$$\therefore g = \frac{4}{3}G\pi Rd$$

16. Given,

$R_p = 264$ km

$R_{a_1} = 23903$ km

$R_{a_2} = 28825$ km

$R_e = 6378$ km

$r_p = 6378 + 264 = 6642$ km

$r_{a_1} = 6378 + 23903 = 30281$ km

$r_{a_2} = 6378 + 28825 = 35203$ km

Increment in velocity is given as,

$\Delta V = V_2 - V_1$

$$\Delta V = \sqrt{\frac{2GMr_{a_2}}{r_p(r_p + r_{a_2})}} - \sqrt{\frac{2GMr_{a_1}}{r_p(r_p + r_{a_1})}}$$

$$= \sqrt{\frac{2 \times 6.67 \times 10^{-11} \times 5.98 \times 10^{24} \times 35203 \times 10^3}{6642 \times 10^3 \times (6642 + 35203) \times 10^3}} - \sqrt{\frac{2 \times 6.67 \times 10^{-11} \times 5.98 \times 10^{24} \times 30281 \times 10^3}{6642 \times 10^3 \times (6642 + 30281) \times 10^3}}$$

$= 10015.69 - 9888.94$

$= 126.75$ m/s ≈ 127 m/s

17. According to Kepler's second law, Area swept out by radial vector from the primary planet to satellite in equal interval of time is constant i.e.

$$\frac{dA}{dt} = \text{constant}$$

$$r^2\dot{\theta} = \frac{L}{m} = \text{const.}$$

$$\therefore \frac{dA}{dt} = \frac{1}{2}r^2\dot{\theta} = \frac{L}{2m} = \text{const.}$$

18. Given

$$g_P = \frac{g_Q}{6}$$

$$\frac{V_{eP}}{V_{eQ}} = ?$$

$$V_e = \sqrt{2gR}$$

As the value of radius of planet P and Q are not stated in the question. Thus the exact relationship cannot be found. To find the ratio of escape velocity more information is required.

19. Given,

$R_a = 1200$ km

$R_p = 400$ km

Let, Radius of the Earth, r = 6400 km

$G = 6.67 \times 10^{-11}$

$M = 5.98 \times 10^{24}$

Now, $r_a = r + R_a = 6400 + 1200$

$= 7600$ km $= 7600 \times 10^3$ m

$r_p = r + R_p = 6400 + 400$

$= 6800$ km $= 6800 \times 10^3$ m

Velocity at opoapsis V_a is given as

$$V_a = \sqrt{\frac{2GM \times r_p}{r_a \times (r_a + r_p)}}$$

$$= \sqrt{\frac{2 \times 6.67 \times 10^{-11} \times 5.98 \times 10^{24} \times 6800 \times 10^3}{7600 \times 10^3 \times (6800 + 7600) \times 10^3}}$$

$= 7049.3$ m/s

$= 7.04$ km/s

20. $$V_p = \sqrt{\frac{2GM \times r_a}{r_p \times (r_a + r_p)}}$$

$$= \sqrt{\frac{2 \times 6.67 \times 10^{-11} \times 5.98 \times 10^{24} \times 7600 \times 10^3}{6800 \times 10^3 \times (6800 + 7600) \times 10^3}}$$

$= 7868.63$ m/s $= 7.86$ km/s

21. $$\frac{V_a r_a}{V_p r_p} = \frac{7049.3 \times 7600 \times 10^3}{7867.9 \times 6800 \times 10^3} = 1.0001$$

22. As per formula,

$R_p = a(1 - e)$

$$e = \frac{R_a - R_p}{R_a + R_p} = \frac{7600 - 6800}{7600 + 6800} = 0.055$$

$$\therefore a = \frac{R_p}{1 - e} = \frac{6800 \times 10^3}{1 - 0.055} = 7195.76 \times 10^3 \text{ m}$$

$$V = \sqrt{\frac{GM}{a}} = \sqrt{\frac{6.67 \times 10^{-11} \times 5.97 \times 10^{24}}{7195.76 \times 10^3}}$$

$= 7438.95 \text{ m/s}$

$$T = \frac{2\pi a}{V} = \frac{2 \times \pi \times 7195.76 \times 10^3}{7438.95} = 6077 \text{ s}$$

23. Given,

T = 24 hours =86400 s

As per question,

$$T^2 = \frac{4\pi^2 a^3}{GM}$$

$$a = \sqrt[3]{\frac{T^2 \times GM}{4\pi^2}}$$

$$= \sqrt[3]{\frac{(86400)^2 \times 6.67 \times 10^{-11} \times 5.97 \times 10^{24}}{4 \times \pi^2}}$$

= 42226 km

24. Minimum height = R_p

$\therefore R = a(1 - e)$

To know e we need to know either radius at apogee from formula

$$e = \frac{R_a - R_p}{R_a + R_p}$$

Or velocity at perigee as per formula

$$e = \frac{R_p V_p^2}{GM} - 1$$

Thus, data is not sufficient to find minimum height.

25. For v = 0, g_p will remain same but ge will increase as

$g_e = g - R\omega^2$

$$\Rightarrow mg = \frac{GM}{r^2} - \frac{mv^2}{r}$$

As,v = 0 therefore, $mg = \frac{GM}{r^2}$,thus g has increased.

26. Acceleration of both the load and man are same. Therefore, the man will experience No force.

27. $$F = \frac{GM_{earth}M_{sun}}{r^2}$$

$\therefore$ magnitude of forces are same i.e

$F_{earth} = F_{sun}$

whereas, $g_{sun} = \frac{GM_{sum}}{R^2_{sun}}$ and $g_{earth} = \frac{GM_{earth}}{R^2_{earth}}$

$\therefore g_{sun} > g_{earth}$

28. $g' = g - \omega^2 R\cos^2\theta$

For poles, $\theta = 90°$

$\Rightarrow g' = g$

For equator, $\theta = 0$□

$$g' = g\left(1 - \frac{\omega^2 R}{g}\right)$$

Therefore,

$\Rightarrow g_p > g_e$

29. Remains stationary as both are falling freely and velocity doesn't depend on mass.

30. $$F = \frac{GMm}{(2r)^2}$$

$$\Rightarrow = \frac{G\left(\frac{4}{3}\pi r^3\right)^2}{\left(2r^2\right)}$$

$$\Rightarrow F \propto \frac{r^6}{r^2}$$

$$\Rightarrow F \propto r^4$$

31. Potential Energy $= \int f.\,dr$

$$\Rightarrow U = \int \frac{1}{r^{\frac{5}{2}}}$$

$$\Rightarrow U \propto \frac{1}{r^{\frac{3}{2}}}$$

$$\Rightarrow U \propto r^{-\frac{3}{2}}$$

32. On comparing the value of acceleration due to gravity above and below the surface of Earth,

For r from the center of the Earth g is directly proportional to the r

For r is increasing from the center of the Earth g is also increasing

On the other hand, g is sharply decreasing which is proportional to r^{-2} above the surface of the Earth.

Therefore, if the height and depth is same then, the gravity will be more at same depth comparison to the same height, and as weight is mass times gravity, therefore by comparing we can directly compare the weight of the same mass. Hence, $w_1 < w_2 < w_3$

33. $V_c = \sqrt{\frac{GM}{r}}$

Taking natural logarithm on the both sides,

$$\Rightarrow \ln V_c = \frac{1}{2}\ln\frac{GM}{2}(\ln GM - \ln r)$$

Differentiating both sides:

$$\frac{dV_c}{V_c} = \frac{1}{2}\left[\frac{-dr}{r}\right]$$

$$\frac{\Delta V_c}{V_c} \times 100 = \frac{1}{2}\left[\frac{-\Delta r}{r}\right] \times 100$$

$$\frac{\Delta V_c}{V_c} \times 100 = \frac{1}{2}[-1]\% = 0.5\% \text{ will increase}$$

34. $T_2 = \frac{(2\pi r)^2}{V^2} = \frac{4\pi^2 r^3}{GM}$

$$\Rightarrow T^2 \propto r^3$$

$$\Rightarrow r \propto T^{\frac{2}{3}}$$

Now, $V^2 = \frac{GM}{r}$

$$\Rightarrow V^2 \propto \frac{1}{r} \propto \frac{1}{T^{\frac{2}{3}}}$$

$$\Rightarrow \frac{1}{2}mV^2 = KE \propto \frac{1}{T^{\frac{2}{3}}}$$

35. Assuming the distance between two planets or the side of equilateral triangle is a.

Therefore,

$$\frac{a}{2} = R\cos 30^\circ \frac{\sqrt{3}R}{2}$$

$$\Rightarrow a = \sqrt{3}R$$

Now, as per Newton's law of gravitation.

Let us say F is the force between two masses along triangle's side direction, therefore the total force along the center of the orbit is

$$F_{resultant} = F\cos 30^\circ + F\cos 30^\circ = \sqrt{3}F$$

where, $F = \frac{Gm^2}{a^2}$, thus

$$\frac{mV^2}{R} = \sqrt{3}\frac{Gm^2}{a^2}$$

$$\Rightarrow \frac{V^2}{R} = \sqrt{3}\frac{Gm}{R^2}$$

$$\Rightarrow V^2 = \sqrt{\frac{Gm}{\sqrt{3}R}}$$

The potential between any two mass is, $\frac{-Gmm}{a}$

Thus, for three combination $= 3 \times \frac{-Gmm}{a}$

$$P.E. = \frac{3Gmm}{a} = \frac{\sqrt{3}Gmm}{R}$$

36. It is independent of mass and inclination angle. Therefore, option is e.

37. Ratio of escape velocities

$$V_e = \sqrt{2gR}$$

$$\frac{V_{e1}}{V_{e2}} = \frac{\sqrt{2g_1R_1}}{\sqrt{2g_2R_2}}$$

$$\frac{V_{e1}}{V_{e2}} = \sqrt{k_1 \times k_2}$$

38. $\frac{-GMm}{(R+h)} = \frac{-GMm}{R} + \frac{1}{2}mu^2$

$$\frac{-GMm}{(R+h)} = \times\frac{R^2}{R^2} = \frac{-GMm}{R} + \frac{1}{2}mu^2$$

$$\frac{-GMm}{(R+h)} = \times\frac{R^2}{R^2} = \frac{-GM}{R} + \frac{1}{2}u^2$$

$$-g\left[\frac{R^2 - R^2 - Rh}{(R+h)}\right] = \frac{1}{2}u^2$$

$$-g\left[\frac{-Rh}{(R+h)}\right] = \frac{1}{2}u^2$$

$$\frac{4}{(R+h)} = \frac{1}{2gR}u^2$$

$$\frac{(R+h)}{h} = \frac{2gR}{u^2}$$

$$\frac{R}{h}+1 = \frac{2gR}{u_2}$$

$$h = \frac{R}{\frac{2gR}{u^2}-1}$$

39. Potential Energy at height h is

$\frac{GMm}{R_e+h}$ and h = nR

$$P.E. = \frac{-GMm}{(n+1)R}$$

Change in P.E. is

$$\Delta P.E. = \frac{-Gm}{(n+1)R} - \left(\frac{-GMm}{R}\right)$$

$$\Delta P.E. = \frac{-GMm}{R}.\frac{R}{R} - \left[\frac{1}{n+1}-1\right]$$

$$\Delta P.E. = \frac{-GMm}{R}.\frac{R}{R} - \left[\frac{1-n-1}{n+1}\right]$$

$$\Delta P.E. = mgR\left(\frac{n}{n+1}\right)$$

40. Energy = $\frac{-GMm}{r}$

$$\frac{-GMm}{r} = \frac{-GMm}{R}+\frac{1}{2}mv^2$$

$$\frac{-GM}{r}+\frac{GM}{R} = \frac{1}{2}v^2$$

$$\frac{-GM}{r}+\left[\frac{R^2}{R}-R\right] = \frac{1}{2}v^2$$

$$-gR\left[\frac{1}{e}-\frac{1}{R}\right] = \frac{1}{2}v^2$$

$$v = R\left[2g\left(\frac{1}{R}-\frac{1}{r}\right)\right]^{\frac{1}{2}}$$

41. Satellite has total Energy at r is

$$\frac{-GMm}{r}+\frac{1}{2}mv^2$$

$$\frac{-GMm}{r}+\frac{1}{2}m\frac{Gm}{r} = m\frac{GM}{2r}$$

When we give satellite some energy, such that it will escape from earth then total energy will be zero. Hence the required to escape this satellite is $\frac{-GMm}{2r}$

42.

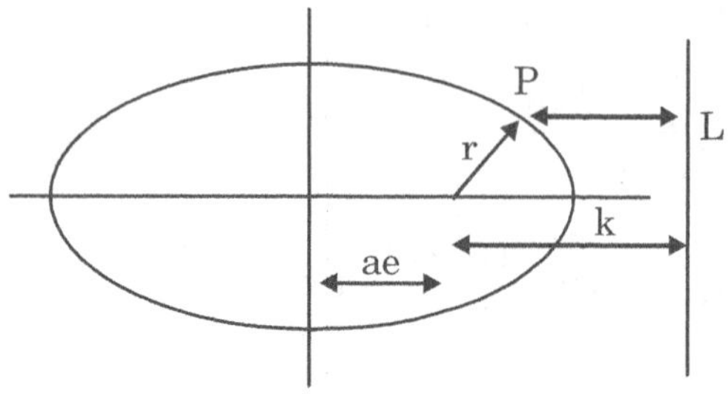

The distance of point P from foci on ellipse is e times the distance from latus rectum.

$|FP| = e|PL|$

$r = e(k - r\cos\theta)$

$r + re\cos\theta = ek$

$$r = \frac{ek}{1+e\cos\theta}$$

43. $r_1 = (6378.14 + 250) \times 1000 = 6{,}628{,}140$ m

$v_1 = 7950$ m/s

$\varphi = 89°$

$$\left(\frac{R_p}{r_1}\right)_{1,2} = \frac{-C \pm \sqrt{C^2+4\times(1-C)\times\sin^2\varphi}}{2\times(1-C)}$$

Where $C = \frac{2\times GM}{r_1 \times v^2_1}$

$$\left(\frac{R_p}{r_1}\right)_{1,2}$$

$$= \frac{-1.903 \pm \sqrt{1.903^2+4\times(1-1.903)\times\sin^2 89°}}{2\times(1-1.903)}$$

$$C = \frac{2\times 3.986005\times 10^{14}}{\left(6{,}628{,}140\times(7950)^2\right)}$$

$$\left(\frac{R_p}{r_1}\right)_{1,2} = 0.9984 \text{ and } 1.1089$$

Perigee Radius, $Rp = R_{p1} = r_1 \times \left(\frac{R_p}{r_1}\right)_1$

$R_p = 6,628,140 \times 0.996019$

$R_p = 6,617,534$ m

Altitude at perigee

$$= \frac{6,617,534}{1000} - 6378.14 = 239.39 \text{ km}$$

Apogee Radius, $Ra = R_{p2} = r_1 \times \left(\frac{R_p}{r_1}\right)_2$

$R_a = 6,628,140 \times 1.1089$

$R_a = 7,349,944$ m

Altitude at apogee

$$= \frac{7,349,944}{1000} - 6378.14 = 971.80446 \text{ km}$$

44. $r_1 = (6378.14 + 250) \times 1000 = 6,628,140$ m

$v_1 = 7950$ m/s

$\varphi = 89°$

$R_p = 6,617,534$ m

$R_a = 7,349,944$ m

$$e = \frac{R_a - R_p}{R_a + R_p}$$

$$e - \frac{7,349,944 - 6,617,534}{7,349,944 - 6,617,534}$$

$e = 0.0524$

45. $r_1 = (6378.14 + 250) \times 1000 = 6,628,140$ m

$\varphi = 89°$

$R_p = 6,617,534$ m

$R_a = 7,349,944$ m

$R_p = a - ae$

$6,617,534 = a(1 - 0.0524)$

$a = 6,983,739$ m

4

CHAPTER

AERODYNAMICS

1. Kelvin circulation theorem is applicable for
   (*a*) Barotropic flows
   (*b*) Isentropic flows
   (*c*) Isothermal and Isochoric flows
   (*d*) None of the above

2. Biot – Savart law is used to calculate the induced velocity by a vortex line which is applicable for states that
   (*a*) Inviscid/irrotational flows and induced velocity is proportional to strength of vortex line and inversely proportional to square of the distance between point and vortex line.
   (*b*) Inviscid/rotational flows and induced velocity is proportional to square of strength of vortex line and inversely proportional to cube of distance between point and vortex line.
   (*c*) Viscous/rotational flows and induced velocity is proportional to square of strength of vortex line and inversely propertional to square of the distance between point and vortex line.
   (*d*) Viscous/rotational flows and induced velocity is proportional to strength of vortex line.

3. The velocity profile given by

   $\overline{V} = A\,xy\hat{i} + By^2\hat{j}$

   where, A = 1 m^{-1} sec^{-1}, B = 0.5 m^{-1} sec^{-1} then equation of streamline is
   (*a*) xy = constant
   (*b*) x^2y = constant
   (*c*) $\frac{x}{y^2}$ = constant
   (*d*) $\frac{y}{x^2}$ = constant

4. The velocity profile given by $\overline{V} = A\hat{i} + Bt\hat{j}$ where A = 2 m sec^{-1} and B = 0.4 m sec^{-2} the equation for path line for the particle passing through the origin at time t = 0
   (*a*) $y = 0.4\,x^2$ (*b*) $y = 0.05\,x^2$
   (*c*) $0.1\,y^2 = x$ (*d*) $0.1\,y = x^2$

5. Induced velocity at point A in given fig. below is, where the strength of vortex line is Γ = 4 m^2/sec?

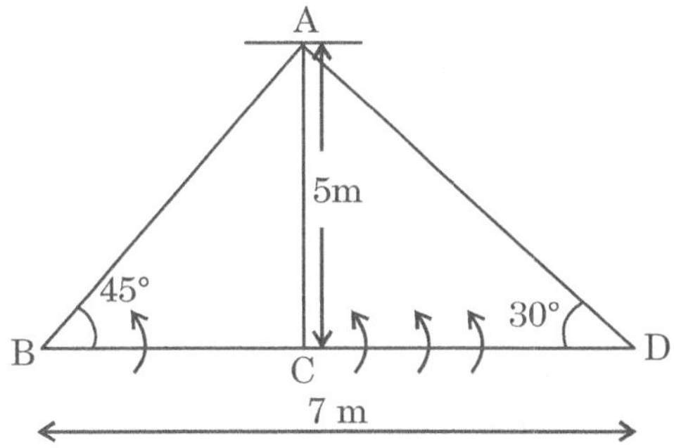

   (*a*) 0.1001 m/sec^2 (*b*) 0.2001 m/sec^2
   (*c*) 0.15001 m/sec^2 (*d*) 0.3001 m/sec^2

6. Laminar flow aerofoils are used to reduce
   (*a*) Skin friction drag (*b*) Wave drag
   (*c*) Trim drag (*d*) Induced drag

7. The white comb area rule is design technique used to reduce aircraft drag in :
   (*a*) Subsonic and transonic
   (*b*) Transonic and supersonic
   (*c*) Supersonic and hypersonic
   (*d*) Transonic and hypersonic

8. Which series of Aerofoil is used in supersonic jets, helicopter blades, shrouds, missile/rocket fins.
   (*a*) 4 series
   (*b*) 5 series
   (*c*) 6 series
   (*d*) 7 series

9. NACA 712A218 has
   (*a*) The area of minimum pressure 10% of chord back the upper surface and 20% of the chord back on the lower surface.
   (*b*) The area of minimum pressure 70% of the chord back on the upper surface and 24% chord back or the lower surface
   (*c*) The area of minimum pressure 10% of chord zero on the upper surface and 30% chord back or the lower surface
   (*d*) None of the above

10. A supercritical aerofoil has its maximum thickness close to

(*a*) Leading edge

(*b*) Trailing edge

(*c*) Middle of the chamber line

(*d*) Quarter chord

11. Centre of pressure which is chord wise location about which :

(*a*) Pitching moment varying with respect to angle of attack

(*b*) Pitching moment could not be determined.

(*c*) Pitching moment is zero

(*d*) Pitching moment is constant

12. The effect on the pattern of the streamlines if $V = 10\ V_\infty$ on :

(A) Rotating cylinder

(B) Stationary cylinder

(*a*) the pattern of stream lines will be same in case (A) and change in case (B)

(*b*) the pattern of stream lines will remains same in case of (A) and (B)

(*c*) We could not predict the pattern of flow from only velocity functions.

(*d*) The pattern of stream line in case of A will change and remains same in case of B

13. Considering stream function as

$$\Psi = Vr\sin\theta\left(1 - \frac{a^2}{r^2}\right) + \frac{\Gamma}{2\pi}\ln\frac{r}{a}$$

where $V = 25$ m/sec, $a = 1$ m, $\Gamma = 50\ \pi m^2/sec$

the stagnation points are at :

(*a*) 45° and 135°

(*b*) 240° and 300°

(*c*) 210° and 330°

(*d*) 60° and 120°

14. The value of u_r, u_θ at $\theta = \pi$ is :

(*a*) $u_r = 0$ m/sec, $u_\theta = -75$ m/sec

(*b*) $u_r = 0$ m/sec, $u_\theta = -125$ m/sc

(*c*) $u_r = 0$ m/sec, $u_\theta = -100$ m/sec

(*d*) $u_r = 0$ m/sec, $u_\theta = -175$ m/sec

15. For what value of θ, $P = P_\infty$ if $\Gamma = 0\ m^2/sec$

(*a*) 0° and 180°

(*b*) 90° and 270°

(*c*) 194° and 346°

(*d*) 210° and 330°

16. Corner flow around two surfaces inclined with an angle of 60°, the values of u_r, u_θ, are :

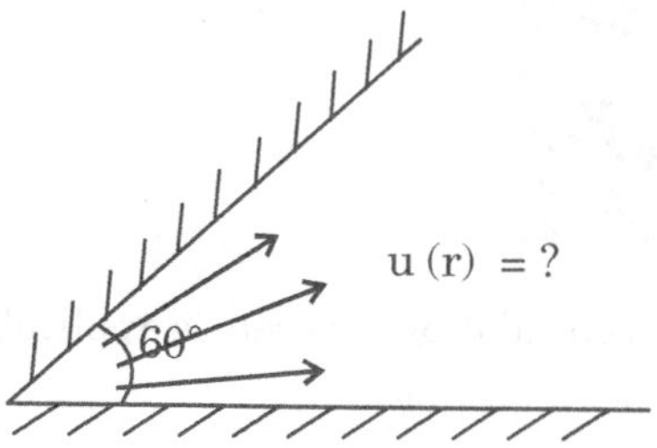

(*a*) $u_r = -3\ Ar^2$, $u_\theta = 0$ (*b*) $u_r = 3Ar^2$, $u_\theta = 5\ Ar^2$

(*c*) $u_r = Ar^2$, $u_\theta = 3\ Ar^2$ (*d*) None of the above

17. What is potential functions for the corner flow the two surfaces inclined at an angle 60° :

(*a*) $\omega = Az$ (*b*) $\omega = Az^2$

(*c*) $\omega = Az^3$ (*d*) $\omega = Az^4$

18. Lift and drag on the stationary cylinder

(*a*) L = 0 and D = 0

(*b*) $L = \rho\Gamma$ and D = 0

(*c*) L = 0 and $D = \rho V\Gamma$

(*d*) $L = \rho V\Gamma$ and $D = 3\rho V\Gamma$

19. Lift and drag on the rotating cylinder

(*a*) $L = \rho V\Gamma$ and D = 0

(*b*) L = 0 and D = 0

(*c*) $L = \rho V\Gamma$ and $D = 0.5\ \rho AV^2$

(*d*) $L = \rho V\Gamma$ and $D = 0.5\ \rho V^2$

20. Navier stokes equations without considering viscocity turn are called

(*a*) Lagrange equations

(*b*) Eulers equations

(*c*) Helmholtz equations

(*d*) Newton's equations

21. Which of the following doesn't appear in Navier stokes momentum equations

(*a*) Pressure

(*b*) Body forces

(*c*) Internal energy

(*d*) Shear stress due to viscocity

22. What happens to temperature, total temperature, pressure, total pressure, density, entropy across a normal shock wave

(A) Increases, remains same, increases, decreases, increases, increases

(B) Increases, decreases, increases, increases, decreases

(C) Decreases, increases, remains same, increase, decreases

23. Aerodynamic lift is component of aerodynamic force

(*a*) opposite to the weight

(*b*) perpendicular to the direction of relative wind

(*c*) perpendicular to the longitudinal axis of the Aircraft

(*d*) Same as the normal

24. According to Prandtl lifting line theory, minimum Induced drag exists for which wing shape.

(*a*) Straight rectangular

(*b*) Elliptical

(*c*) Tapered straight

(*d*) Tapered sweep back

25. The curve for lift coefficient versus angle of attack for an aerofoil is found to intersect y - axis at 0.1 and x – axis at 1.2. The aerofoil may be

(*a*) Symmetrical

(*b*) Positively cambered

(*c*) Negatively cambered

(*d*) Laminar flow

26. A tapered straight aeroplane having root chord and tip chord of 1.1 m and 0.8 m respectively if semi span of the wing is 8.2 m, what is the aspect ratio

(*a*) 20

(*b*) 17.26

(*c*) 25

(*d*) 15

27. Velocity components in x and y directions are given by $u = \frac{2xy^3}{3} - x^2y$ and $v = xy^2 - \frac{2xy^3}{3}$.

The given velocity distribution is :

(*a*) a possible field of flow

(*b*) not a possible field of flow

(*c*) cannot say based on given data

(*d*) none of these

28. A conical pipe diverges uniformly from 100 mm to 200 mm diameter over a length of 1 m. then local and connective acceleration are ________

29. Given that $u = -4ax(x^2 - 3y^2)$, $v = 4ay(3x^2 - y^2)$, examine whether these velocity components represents a physically possible 2-D flow, if so whether the flow is rotational or irrotational?

30. For two-dimensional flow, the velocity components are $u = \frac{x}{(x^2+y^2)}, v = \frac{y}{x^2+y^2}$, then acceleration components a_x and a_y are

(*a*) $\frac{-x}{(x^2+y^2)^2}, \frac{-y}{(x^2+y^2)^2}$

(*b*) $\frac{x^2-y^2}{(x^2+y^2)^2}, \frac{x^2-y^2}{(x^2+y^2)^2}$

(*c*) $\frac{x}{(x^2+y^2)^2}, \frac{y}{(x^2+y^2)^2}$

(*d*) None of these

31. Velocity for a 2-D flow field is given by

$$V = (3 + 2xy + 4t^2)\hat{i} + (xy^2 + 3t)\hat{j},$$

then accelerations at a point (1, 2) after 2 sec will be ___________

32. Stream lines and equipotential lines intersect each other at all points

(*a*) Orthogonally

(*b*) Tangentially

(*c*) Parallelly

(*d*) None of these

33. A fluid flow

$$V_r = \left(1 - \frac{a}{r^2}\right)\cos\theta, V_\theta = -\left(1 + \frac{a}{r^2}\right)\sin\theta$$

represents a

(*a*) Physically possible flow

(*b*) Physically impossible flow

(*c*) Can not say on the above information

(*d*) None of these

34. A fluid flow

$$V_r = \left(1 - \frac{a}{r^2}\right)\cos\theta, V_\theta = -\left(1 + \frac{a^2}{r^2}\right)\sin\theta$$

represents

(*a*) rotational flow

(*b*) irrotational flow

(*c*) may be rotational or irrotational cannot say depends upon nature of fluid

(*d*) None of these

35. In a two-dimensional flow the velocity components are u = Cy, v = 0 (where C = constant). The circulation about the circle $x^2 + y^2 - 2ay = 0$ Situated in a flow (where a is radius of circle) is

____

36. Match the following pairs

P.	Reynolds Number (Re)	L.	$\frac{Fi}{Fv}$
Q.	Froude Number (Fr)	M.	$\sqrt{\frac{Fi}{Fs}}$
R.	Weber Number (We)	N.	$\sqrt{\frac{Fi}{Fg}}$
S.	Euler's Number (Eu)	O.	$\sqrt{\frac{Fi}{F_P}}$

where Fi = inertial force

Fv = viscous force

Fs = surface tension

Fg = gravity force

F_P = pressure force

(*a*) P – L, Q – N, R – M, S – O

(*b*) P – L, Q – N, R – O, S – M

(*c*) P – N, Q – L, R – M, S – O

(*d*) P – M, Q – L, R – N, S – O

37. The velocity distribution in the boundary layer is given by $\frac{u}{v} = \frac{y}{\delta}$, where u is the velocity at a distance y from the plate and u = v at y = δ.δ being boundary layer thickness, then the value of $\frac{\delta^*}{\theta}$ is _______________

38. Match the following pairs

(P) Nominal thickness	L.	$\int_0^\delta \frac{u}{v}\left(1-\frac{u}{v}\right)dy$
(Q) Displacement thickness	M.	$y = \delta$ for $u = 0.99V_\infty$
(R) Momentum thickness	N.	$\int_0^\delta \frac{u}{v}\left(1-\frac{u^2}{v^2}\right)dy$
(S) Energy thickness	O.	$\int_0^\delta \left(1-\frac{u}{v}\right)dy$

(*a*) P – M, Q – O, R – L, S – N

(*b*) P – O, Q – M, R – N, S – L

(*c*) P – M, Q – O, R – N, S – L

(*d*) P – L, Q – O, R – M, S – N

39. The ratio of frictional drag on the front half and rear, half of the flat plate kept on zero incident angle in a stream of uniform velocity, if the boundary layer is laminar over the whole plate

____

40. Boundary layer separation occurs when

(*a*) $\frac{dP}{dx} < 0$ (*b*) $\left(\frac{\partial u}{\partial y}\right)_{y=0} = 0$

(*c*) $\left(\frac{\partial u}{\partial y}\right)_{y=0} > 0$ (*d*) None of the above

41. Which of the following condition is detached flow condition

(*a*) $\left(\frac{\partial u}{\partial y}\right)_{y=0} = 0$ (*b*) $\left(\frac{\partial u}{\partial y}\right)_{y=0} > 0$

(*c*) $\left(\frac{\partial u}{\partial y}\right)_{y=0} < 0$ (*d*) None of the above

42. Von-Karman momentum integral equation $\left(\frac{\Gamma_0}{\rho V^2} = \frac{\partial \theta}{\partial x}\right)$ is applicable to

(*a*) Laminar boundary layer flow only

(*b*) Turbulent boundary layer flow only

(*c*) Transition boundary layer flow only

(*d*) By providing the rotating cylinder near leading edge.

(*e*) Laminar, transition & turbulent boundary layer flows

43. Separation of boundary layer can be prevented by

(*a*) Streamlining the body shape

(*b*) Sucking the retarded flow

(*c*) Providing the slot near leading edge

(*d*) all of the above

44. The boundary layer thickness at a distance of 1 m from the leading edge of flate plate kept over the zero angle of incidence to the flow direction is 1 mm. The velocity outside of the boundary layer is 20 m/sec. the boundary layer thickness at the distance of 4 m is _______________ (consider laminar boundary layer)

45. For compressible flow area – velocity relationship

(*a*) $\frac{dA}{A} = \frac{dV}{V}\left(1 - M^2\right)$ (*b*) $\frac{dA}{A} = \frac{dV}{V}\left(C^2 - 1\right)$

(*c*) $\frac{dA}{A} = \frac{dV}{V}\left(M^2 - 1\right)$ (*d*) $\frac{dA}{A} = \frac{dV}{V}\left(V^2 - 1\right)$

46. Which of the following expression does not represent the speed of sound is a medium?

(*a*) $\sqrt{\frac{k}{e}}$ (*b*) $\sqrt{\gamma RT}$

(*c*) $\sqrt{\frac{KP}{\rho}}$ (*d*) $\sqrt{\frac{dP}{d\rho}}$

47. Which of the following is basic governing equation of compressible fluid flow?

(*a*) Continuity equation

(*b*) Momentum equation

(*c*) Energy equation

(*d*) Equation of states

(v) All of the above

48. Kutta-Joukowshi theorem relates the lift and circulation over a body for

(*a*) 2 - D body

(*b*) 3 - D body

(*c*) Both (*a*) & (*b*)

(*d*) None of the above

49. Mathematical form of Kelvin's circulation theorem

(*a*) $\frac{\Delta T}{\Delta t} = 0$ (*b*) $\frac{\Delta T}{\Delta t} < 0$

(*c*) $\frac{\Delta T}{\Delta t} > 0$ (*d*) $\frac{\Delta T}{\Delta t} = 1$

50. C_p for cylinder and sphere for potential flow is given by

(*a*) $1 - 4\sin^2\theta, 1 - \frac{9}{4}\sin^2\theta$

(*b*) $1 - 4\sin^2\theta, 1 - \frac{9}{2}\sin^2\theta$

(*c*) $1 - \frac{9}{4}\sin^2\theta, 1 - 4\sin^2\theta$

(*d*) None of these

51. For Cusped aerofoil,

(*a*) trailing edge is always stagnation point

(*b*) trailing edge may or may not be stagnation point

(*c*) flow velocity above and below the trailing edge have different magnitudes

(*d*) flow velocity above and below the trailing edge have different directions.

52. If the mach number at the cross-section of nozzle is M = 0.8 and ratio of specific heat is $\gamma = 1.4$ then

$\frac{A}{A^*}$ is __________ Where A* is cross-sectional area at throat.

53. If the stagnation temperature T_0 at compressor inlet is increased 1.44 times and stagnation pressure P_0 is decreased to 0.6 times in values, then by what factor should the mass flow rate be decreased in the compressor to achieve same pressure ratio?

(*a*) 1.63 (*b*) 2.11

(*c*) 2 (*d*) 1.84

Drag and lift on the cylinder with circulation in potential flow is

(ρ = Density, u = free stream velocity)

(*a*) Drag = 0 Lift = $\rho V\Gamma$

(*b*) Drag > 0, Lift > $\rho V\Gamma$

(*c*) Drag < 0, Lift < $\rho V\Gamma$

(*d*) Drag > 0, Lift < $\rho V\Gamma$

54. Pick the wrong statement about the turbulent Boundary layer

(*a*) The Turbulent boundary layer produces more skin frictional drag than laminar boundary layer

(*b*) The turbulent boundary layer is more stable than laminar boundary layer

(*c*) Turbulent boundary layer does not contribute lift on the aerofoil

(*d*) Both (*b*) & (*c*)

55. There are two geometrically similar aircrafts however, due to proportional requirements the aircrafts have different sizes. If the surface area of the larger aircraft is twice the smaller one and mass is 2.4 times of smaller one by what kinetic energy percentage is the stall speed (at steady level flight) of the larger aircraft greater than smaller aircraft (at same flight condition)?

(*a*) 10.2% (*b*) 27%

(*c*) 3.3% (*d*) 9.5%

56. Consider the flow at mach 0.8. The ratio of kinetic energy per unit mass to internal energy per unit mass of the fluid element moving along a stream line is (Take $\gamma = 1.4$) ________

57. The pressure coefficient on the point surface of aerofoil at incoming mach number 0.4 is 0.5. The value of pressure coefficient at same point at free stream mach number 0.7 would be ________.

58. For the preliminary design of a mach 2 supersonic wind tunnel, the ratio of diffuser throat area to nozzle throat area is ________

59. In case of formation of normal shock wave, ρ_1 and ρ_2 are densities before and after of shock wave, if $M_1 \rightarrow \infty$ then M_2 = ________

	Normal shock
M_1	M_2
ρ_1	ρ_2

60. If a point in the airflow the pressure, temperature and velocity are 1 atm, 320 K, 100 m/sec. Total temperature and total pressure at this point will be ________

61. Just upstream of the shock wave, the air temperature and pressure 288 K and 1 atm respectively. Just downstream of the wave, the air temperature and pressure are 690 K and 8.6 atm respectively. The change in entropy across the wave is ________ J/kg-K.

62. Lift and Drag force coefficients for flate plate at an 8° angle of attack at mach 4 is ________

63. Shadograph, schlieren, Interferometer used to measure :

(*a*) $\dfrac{\partial^2 \rho}{\partial y^2}, \dfrac{\partial \rho}{\partial y}, \rho$ respectively

(*b*) $\dfrac{\partial \rho}{\partial y}, \dfrac{\partial^2 \rho}{\partial y^2}, \rho$ respectively

(*c*) $\rho, \dfrac{\partial \rho}{\partial y}, \dfrac{\partial^2 \rho}{\partial y^2}$ respectively

(*d*) None of the above

64. Oblique shock wave is formed at leading edge of 10° half wedge. If the shock wave angle is 30° and downstream velocity is 900 m/sec, the upstream velocity is ________ m/sec

65. In the above problem, if T_1 = 310 K, P_1 = 1 atm then P_2 is ________ atm

66. A supersonics flow with M_1 = 1.5, P_1 = 1 atm, T_1 = 288 K, expanded along the sharp corner as shown in figure through deflection angle of 15°. The angle of forward and rearward mach line is ________

67. In the above problem P_2 is ________

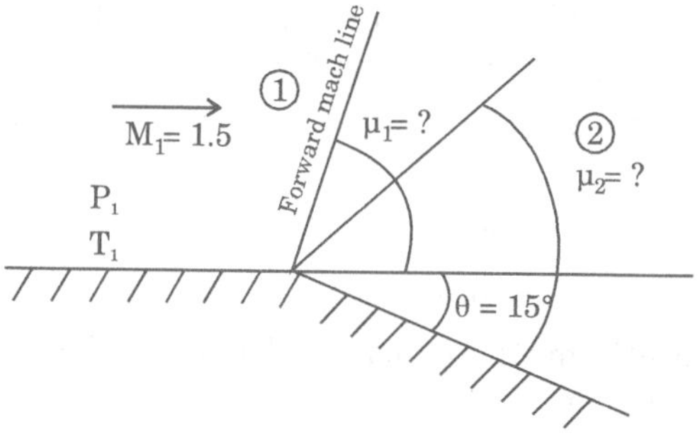

ANSWER KEY

1. (*a*)	**2.** (*a*)	**3.** (*c*)	**4.** (*b*)	**5.** (*a*)	**6.** (*a*)	**7.** (*b*)	**8.** (*a*)	**9.** (*a*)	**10.** (*a*)
11. (*c*)	**12.** (*d*)	**13.** (*c*)	**14.** (*a*)	**15.** (*d*)	**16.** (*a*)	**17.** (*b*)	**18.** (*a*)	**19.** (*a*)	**20.** (*b*)
21. (*c*)	**22.** (*a*)	**23.** (*a*)	**24.** (*a*)	**25.** (*c*)	**26.** (*b*)	**27.** (*b*)	**30.** (*a*)	**31.** 186.03	**32.** (*a*)
33. (*a*)	**34.** (*b*)	**36.** (*a*)	**37.** (*c*)	**38.** (*a*)	**39.** (*a*)	**40.** (*b*)	**41.** (*c*)	**42.** (*d*)	**43.** (*e*)
44. 2 mm	**45.** (*c*)	**46.** (*c*)	**47.** (*d*)	**48.** (*a*)	**49.** (*a*)	**50.** (*a*)	**51.** (*b*)	**52.** 1.04	**53.** (*c*)
54. (*c*)	**55.** (*d*)	**56.** 0.1792	**57.** 0.32	**58.** 1.387	**60.** 817.8K, 26.7 atm		**61.** 258.2 J/kg-K		**62.** 0.02
63. (*a*)	**64.** 976.55	**65.** 2.05 atm		**66.** 41.81°, 30°		**67.** 0.4691 atm			

EXPLANATIONS

1. Kelvin circulation theorem is applicable for barotropic flows, Isentropic and isothermal flows are barotropic is nature

$P = f(e)$

2. According to Biot-savart law, induced velocity at any point P by a vortex line is given by

$$d\vec{V} = \frac{\Gamma}{4\pi} \times \frac{d\vec{l} \times \vec{r}}{|r^3|}$$

$$d\vec{V} \propto \Gamma$$

$$d\vec{V} \propto \frac{1}{r^2}$$

3. Equation for stream line is given by

$$\frac{dx}{u} = \frac{dy}{v}$$

$$\frac{dx}{Axy} = \frac{dy}{By^2}$$

$$\frac{dx}{Ax} = \frac{dy}{By}$$

Putting value of A and B, and integrating we get

$$\int \frac{dx}{1 \times x} = \int \frac{dy}{0.5\,y}$$

$$\Rightarrow \ln x = 2 \ln y + \ln c$$

$$\Rightarrow \ln x = \ln y^2 + \ln c$$

$$\Rightarrow \ln x - \ln y^2 = \ln c$$

$\ln(x/y^2) = \ln(c)$

$$\boxed{\frac{x}{y^2} = C}$$

4. Equation of path line is given by

$$\frac{dx}{dt} = u, \frac{dy}{dt} = v$$

$u = A, v = Bt$

$$\frac{dx}{dt} = 2, \frac{dy}{dt} = 0.4t$$

$$\Rightarrow dx = 2dt$$

$$x = 2t + C \quad ....(1)$$

$$dy = 0.4\frac{t^2}{2} + C \quad ....(2)$$

Eqn. (1) & eqn. (2) passing through origin

$C = 0$

$x = 2t$

$$t = \frac{x}{2}$$

$y = 0.2t^2$

By putting value of t in terms of x, we get

$$y = 0.2\left(\frac{x}{2}\right)^2$$

$$y = 0.2\left(\frac{x^2}{4}\right)$$

$$\Rightarrow 4y = 0.2x^2$$

$$\Rightarrow y = \frac{2}{40}x^2$$

$$\Rightarrow 20y = x^2$$

$$\boxed{y = 0.05\,x^2}$$

5. Induced velocity at point A is given by Biot-Savart law

$$d\vec{V} = \frac{\Gamma}{4\pi h}[\cos\alpha + \cos\beta]$$

$$= \frac{4}{4\pi \times 5}[\cos 30° + \cos 45°]$$

$= 0.1001$ m/sec

9. 712A218, 7 digit series Aerofoil where 1st digit → series (indicating series)

2nd digit → Distance of minimum pressure on the upper surface in tens of percentage of chord.

3rd digit → Distance of minimum pressure on the lower surface in tens of percentage of chord.

Alphabet refers to a standard profile from the earlier NACA series. Next digit → H describes the lift coefficient is tenths. Last two digits describing the maximum thickness as percentage of chord.

Here minimum pressure 10% of chord back on the upper surface and 20% chord back on the lower surface.

11. By definition of centre of pressure.

12. We know equation of stream lines in Cartesian co-ordinate and polar co-ordinates are

$$\frac{dy}{dx} = \frac{v}{u}, \frac{1}{r}\frac{\partial r}{\partial \theta} = \frac{u_r}{u_\theta}$$

For rotating cylinder

$$\frac{u_r}{u_\theta} = \frac{\left(1-\frac{a^2}{r^2}\right)\cos\theta}{-\left(\frac{1+a^2}{r^2}\right)\sin\theta - \frac{\Gamma}{2\pi rv}}$$

$\frac{u_r}{u_\theta}$ is dependent on V_∞, hence the pattern will change.

For stationary cylinder

$$\frac{u_r}{u_\theta} = \frac{\left(1-\frac{a^2}{r^2}\right)\cos\theta}{-\left(1+\frac{a^2}{r^2}\right)\sin\theta}$$

$\frac{u_r}{u_\theta}$ is independent of V_∞, hence pattern will not change.

13. At stagnation points $u_r = 0$, $u_\theta = 0$, $r = 1$

$$u_r = -V\left(1+\frac{a^2}{r^2}\right)\sin\theta - \frac{\Gamma}{2\pi r}$$

$$0 = -25\left(1+\frac{1}{1}\right)\sin\theta - \frac{50\pi}{2\pi \times 1}$$

$$0 = -50\sin\theta - 25$$

$$\Rightarrow \sin\theta = \left(-\frac{1}{2}\right)$$

$$\theta = \sin^{-1}\left(-\frac{1}{2}\right) = 210° \;\&\; 330°$$

$$\theta = 210°, 330°$$

14. $u_r = v\left(1-\frac{a^2}{r^2}\right)\cos\theta$

$$= 25\left(1-\frac{1}{1}\right)\cos\pi = 0$$

$$u_\theta = -V\left(1+\frac{a^2}{r^2}\right)\sin\theta - \frac{\Gamma}{2\pi r}$$

$$= -25(1+1)\sin\pi - \frac{50\pi}{2\pi\times 1}$$

$$= -50 - 25 = -75$$

$u_\theta = 75$ m/sec

15. Pressure coefficient

$$C_P = \frac{P - P_\infty}{a_\infty}, P = P_\infty$$

$$C_P = 0$$
$$C_P = 1 - 4\sin^2\theta$$
$$0 = 1 - 4\sin^2\theta$$
$$4\sin^2\theta = 1$$

$$\sin\theta = \pm\frac{1}{2}$$

$$\theta = \sin^{-1}\left(\pm\frac{1}{2}\right) = 30°, 150°, 210°, 330°$$

16. $\theta = \frac{\pi}{3}$

$$n = \frac{\pi}{\theta} = 3$$

We have

$$w = AZ^n$$
$$= Ar^n \cos n\theta + i\,Ar^n \sin n\theta$$
$$= \phi + i\Psi$$
$$\phi = Ar^n \cos n\,\theta$$

$$u_r = \frac{d\phi}{dr} = nAr^{n-1}\cos n\theta$$
$$= 3A\,r^{3-1}\cos 3\theta$$
$$= 3A\,r^2\cos 3\theta$$

$$u_\theta = \frac{1}{r}\frac{\partial\phi}{\partial\theta} = -nAr^{n-1} \times \sin n\theta$$
$$= -3\,Ar^2 \sin 3\theta$$

Putting value of θ, we get

$$u_r = 3Ar^2\cos\left(3\times\frac{\pi}{3}\right) = -3Ar^2$$

$$u_\theta = -3Ar^2\sin 3\,\theta$$

$$= -3Ar^2 \sin 3 \times \frac{\pi}{3}$$

$$= 0$$

17. We have, potential function for corner flow

$$\omega = Az^n$$

$$n = \frac{\pi}{\theta} = \frac{\pi}{\frac{\pi}{3}} = 3$$

$$\boxed{\omega = Az^2}$$

18. Lift and drag on stationary cylinder is zero

$L = 0, D = 0$

19. Lift and drag on rotating cylinder is $L = \rho v \Gamma$, $D = 0$

20. Without considering viscosity term

N – V equation is written as

$dP = -\rho v dv$ (which is eulers' equation)

21. Navier stokes equation can be written as

$$P_x - \frac{1}{\rho}\frac{\partial P}{\partial x} = \frac{\partial u}{\partial t} - \mu\left(\frac{\partial^2 u}{\partial x^2} + \frac{\partial^2 u}{\partial y^2} + \frac{\partial^2 u}{\partial z^2}\right)$$

$$P_y - \frac{1}{\rho}\frac{\partial P}{\partial y} = \frac{\partial v}{\partial t} - \mu\left(\frac{\partial^2 v}{\partial x^2} + \frac{\partial^2 v}{\partial y^2} + \frac{\partial^2 v}{\partial z^2}\right)$$

$$P_z - \frac{1}{\rho}\frac{\partial P}{\partial y} = \frac{\partial w}{\partial t} - \mu\left(\frac{\partial^2 w}{\partial x^2} + \frac{\partial^2 w}{\partial y^2} + \frac{\partial^2 w}{\partial z^2}\right)$$

N – V in vector form, may be written as

$$\frac{\Delta V}{\Delta t} = R - \frac{1}{e}\text{grad } P + V\overline{V}^2 V$$

$\overline{V}^2 V$ laplace operator

$$= \left[\frac{\partial^2}{\partial x^2} + \frac{\partial^2}{\partial y^2} + \frac{\partial^2}{\partial z^2}\right]$$

Hence the pressure force, Body force (gravity) viscosity present in N – V equation.

22.

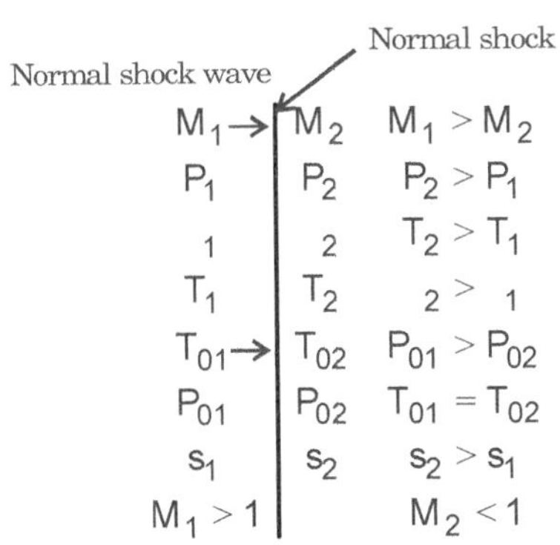

23. By definition of lift force

24. Induce drag coefficient

$$C_{di} = \frac{C_L^2}{\pi AR}(1+\delta) = \frac{C_L^2}{\pi e AR}$$

$$e = \frac{1}{1+\delta}$$

e = Span efficiency factor.

$\delta >, 0$

$\delta = 0$, $e = 1$ (elliptical wing)

$\delta > 0$ for other type of wing hence $e > 1$ for other wing.

25. Negativity cambered aerofoil

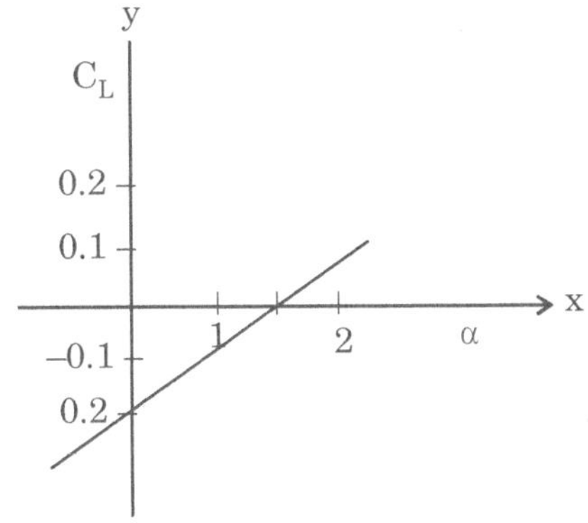

In C_L Vs. α curve, for positive cambered aerofoil, airfoil is found to be intercept y-axis (positive) and x-axis negative for negative cambered case just opposite to positive cambered aerofoil.

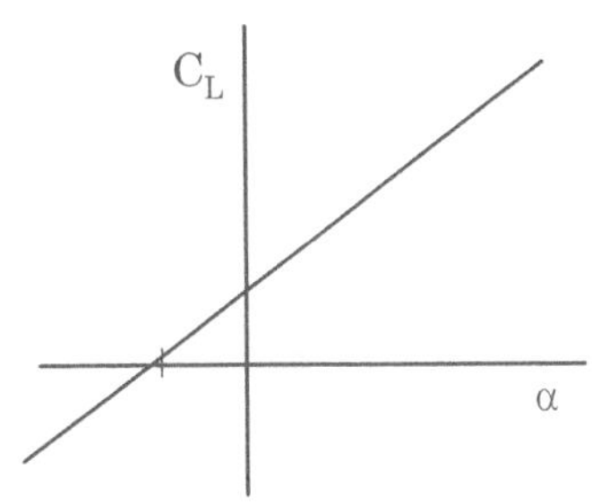

26.

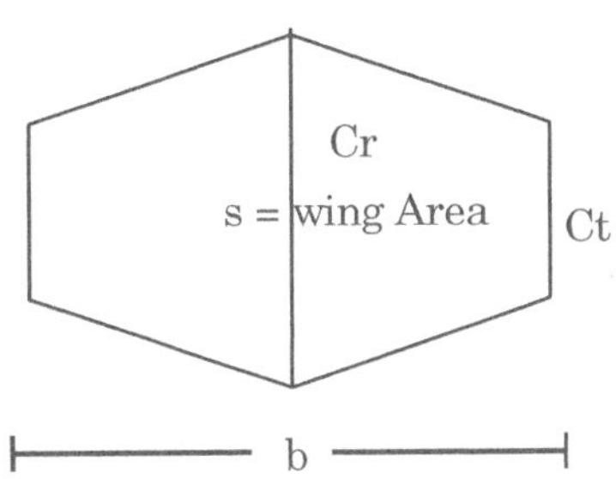

C_t = tip chord

C_r = root chord

$$\text{Chord of wing} = \frac{C_t + C_r}{2}$$

$$= \frac{1.1 + 0.8}{2} = 0.95$$

$$AR = \frac{b^2}{S}$$

$$= \frac{b^2}{b \times c} = \frac{b}{c}$$

$$= \frac{8.2 \times 2}{0.95} = 17.26$$

27. $u = \frac{2xy^3}{3} - x^2 y,\ v = xy^2 - \frac{2yx^3}{3}$

A possible flow field (2-D) must satisfy the continuity equation.

$$\frac{\partial u}{\partial x}+\frac{\partial v}{\partial y}=0$$

$$\frac{2y^3}{3}-\cancel{2xy}+\cancel{2xy}-\frac{2x^3}{3}=\frac{2}{3}\left(y^3-x^3\right)$$

Since continuity equation is not satisfied the given flow field, velocity component do not represent a possible flow field.

28.

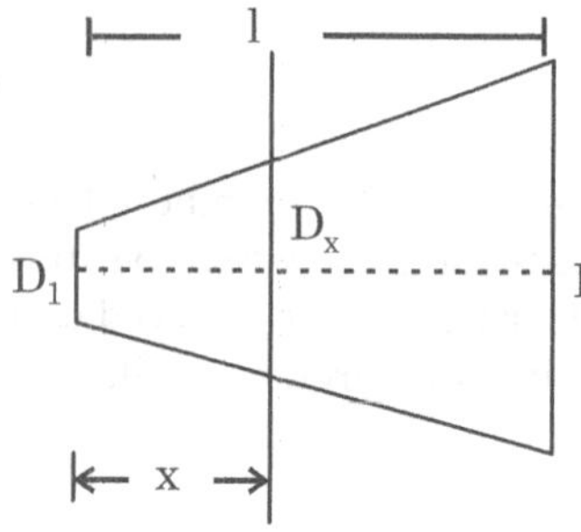

D_2 = 200 mm = 0.2 m

D_1 = 100 m = 0.1 m

l = 1 m

Distance at any distance x from inlet

$$Dx=\left(\frac{D_2-D_1}{l}\times x\right)+D_1$$

$$=0.1+\frac{(0.2-0.1)}{1}\times x$$

= 0.1 + 0.1 x = 0.1 (1 + x)

∴ Cross-sectional area

$$A_x=\frac{\pi}{4}D_x^{\;2}=\frac{\pi}{4}(0.1(1+x))^2$$

$$=\frac{\pi}{4}0.1^2(1+x)^2$$

= 0.00785 $(1+x)^2$

velocity of flow $u_x = u = \dfrac{Q}{A}$

$$=\frac{Q}{0.00785(1+x)^2}$$

Velocity gradient $\dfrac{\partial u}{\partial x}=\dfrac{\partial}{\partial x}\left(\dfrac{Q}{0.00785}(1+x)^2\right)$

$$=\frac{-2Q}{0.00785(1+x)^3}$$

Local acceleration $=\dfrac{\partial v}{\partial t}=0$

Since flow is steady

Convective acceleration

$$a_x=u\frac{\partial u}{\partial x}=\frac{Q}{0.00785(1+x)^2}\times\frac{-2Q}{0.0785(1+x)^3}$$

$$=\frac{-2Q^2}{(0.00785)^2(1+x)^5}$$

29. $u = -4ax(x^2-3y^2)$

$v = 4ay(3x^2-y^2)$

A possible flow field (2-D) flow must satisfying continuity equation.

$$\frac{\partial u}{\partial x}+\frac{\partial v}{\partial y}=0$$

$$\frac{\partial}{\partial x}\left(-4ax\left(x^2-3y^2\right)\right)+\frac{\partial}{\partial y}\left(4ay\left(3x^3-y^2\right)\right)$$

$$\cancel{-12ax^2}+\cancel{12ay^2}+\cancel{12ax^2}-\cancel{12ay^2}=0$$

Hence the given velocity represents the physical possible flow in 2 - D case.

$\dfrac{\partial v}{\partial x}-\dfrac{\partial u}{\partial y}=0$ flow in irrotation

$$\Rightarrow\ \cancel{24axy}-\cancel{24ayx}=0$$

Hence is irrotational.

30. $u=\dfrac{x}{x^2+y^2},\ v=\dfrac{y}{x^2+y^2}$

We know that

$$a_x=u\frac{\partial u}{\partial x}+v\frac{\partial u}{\partial y}$$

$$a_y=u\frac{\partial v}{\partial x}+v\frac{\partial v}{\partial y}$$

$$\frac{\partial u}{\partial x}=\frac{\partial}{\partial x}\left(\frac{x}{x^2+y^2}\right)=\frac{\left(x^2+y^2\right)-(2x)x}{\left(x^2+y^2\right)^2}$$

$$=\frac{y^2-x^2}{\left(x^2+y^2\right)^2}$$

$$\frac{\partial u}{\partial y}=\frac{\partial}{\partial y}\frac{(x)}{x^2+y^2}=\frac{\partial}{\partial y}\left(x\left(x^2+y^2\right)^{-1}\right)$$

$$=x\times-\left(x^2+y^2\right)^{-2}\times xy$$

$$=\frac{-2xy}{\left(x^2+y^2\right)^2}$$

$$\frac{\partial v}{\partial x} = \frac{\partial}{\partial x}\left(\frac{y}{x - y^2}\right) = \frac{\partial}{\partial x}\left(y\left(x^2 + y^2\right)^{-1}\right)$$

$$= y - \left(x^2 + y^2\right)^{-2} \times 2x = \frac{-2xy}{\left(x^2 + y^2\right)^2}$$

$$\frac{\partial v}{\partial y} = \frac{\partial}{\partial y}\left(\frac{y}{x^2 + y^2}\right) = \frac{\left(x^2 + y^2\right) \times 1 - y(xy)}{\left(x^2 + y^2\right)^2}$$

$$= \frac{x^2 - y^2}{\left(x^2 + y^2\right)^2}$$

Substituting all the values we get

$$a_x = \frac{x}{\left(x^2 + y^2\right)} \times \frac{\left(y^2 - x^2\right)}{\left(x^2 + y^2\right)^2} - \frac{y}{\left(x^2 + y^2\right)} \times \frac{2xy}{\left(x^2 + y^2\right)^2}$$

$$= \frac{xy^2 - x^3}{\left(x^2 + y^2\right)^3} - \frac{2xy^2}{\left(x^2 + y^2\right)^3} = \frac{-x^3 - xy^2}{\left(x^2 + y^2\right)^3}$$

$$= \frac{-x\left(x^2 + y^2\right)}{\left(x^2 + y^2\right)^3} = \frac{-x}{\left(x^2 + y^2\right)^2}$$

$$a_x = \frac{-x}{\left(x^2 + y^2\right)^2}$$

$$a_y = \frac{-y}{\left(x^2 + y^2\right)^2}$$

31. $V = \left(3 + 2xy + 4t^2\right)\hat{j} + \left(xy^2 + 3t\right)\hat{j}$

$a_{(1,2)}$ after 2 sec = ?

We know $a = \frac{dv}{dt}$

$$= u\frac{\partial v}{\partial t} + v\frac{\partial v}{\partial y} + \frac{\partial v}{\partial t}$$

$$\frac{\partial v}{\partial x} = 2y\hat{i} + y^2\hat{j}$$

$$\frac{\partial v}{\partial y} = 2x\hat{i} + 2xy\hat{j}$$

$$\frac{\partial v}{\partial t} = 8t\,\hat{i} + 3\hat{j}$$

$$a = \left(3 + 2xy + 4t^2\right)\left(2y\hat{j} + y^2\hat{j}\right)$$

$$+\left(xy^2 + 3t\right)\left(2x\hat{i} + 2xy\hat{j}\right) + \left(8t\hat{i} + 3\hat{j}\right)$$

Putting the value of x, y & t we get

$$23\left(4\hat{j} + 4\hat{j}\right) + (4 + 6)\left(2\hat{j} + 4\hat{j}\right) + \left(16\hat{j} + 3\hat{j}\right)$$

$$= 92\hat{i} + 92\hat{j} + 20\hat{i} + 40\hat{j} + 16\hat{i} + 3\hat{j}$$

$$= 128\hat{i} + 135\hat{j}$$

$$a_{(1,2)} = \sqrt{128^2 + 135^2}$$

= 186.03 unit

32. Stream line and equipotential lines intersects each other orthogonally ,

For curve ϕ = constant

$$\text{Slope of } \phi = \frac{\partial y}{\partial x} = \frac{\frac{\partial \phi}{\partial x}}{\frac{\partial \phi}{\partial y}}$$

$$= \frac{-u}{-u} = \frac{u}{v}$$

For curve ψ = constant

$$\text{Slope} = \frac{\partial y}{\partial x} = \frac{\frac{\partial \psi}{\partial x}}{\frac{\partial \psi}{\partial y}} = \frac{-u}{u}$$

$$= \frac{-v}{u}$$

product of slope

$$\frac{u}{v} \times \frac{-v}{u} = -1$$

$\boxed{m_1 \times m_2 = -1}$ is condition for perpendicularity.

33. $V_r = \left(1 - \frac{a}{r^2}\right)\cos\theta$

$$V_\theta = \left(1 + \frac{a}{r^2}\right)\sin\theta$$

To check the possible flow field it must be satisfy continuity equation.

$$\frac{1}{r}(eVr)+\frac{\partial}{\partial r}(eVr)+\frac{1}{r}\frac{\partial}{\partial\theta}(eV\theta)=0 \quad \ldots (1)$$

$$\frac{V_r}{r}+\frac{\partial V_r}{\partial r}+\frac{\partial V_\theta}{r\partial\theta}=0$$

$$\frac{\partial}{\partial_\gamma}(V_r)=\frac{\partial}{\partial r}\left[\left(1-\frac{a}{r^2}\right)\cos\theta\right]$$

$$=-a\cos\theta(-2).\frac{1}{r^3}=\frac{2a\cos\theta}{r^3}$$

$$\frac{\partial}{\partial\theta}(V_\theta)=\frac{\partial}{\partial\theta}\left[-\left(1+\frac{a}{r^2}\right)\sin\theta\right]$$

$$=\left[-\left(1+\frac{a}{r^2}\right)\cos\theta\right]$$

Substituting these values in equation (1) we get

$$\frac{1}{r}\left(1-\frac{a}{r^2}\right)\cos\theta+\left[\frac{2a\cos\theta}{r^3}\right]+\frac{1}{r}\left[-\left(1+\frac{a}{r^2}\right)\cos\theta\right]$$

$$=\frac{\cos\theta}{r}-\frac{a\cos\theta}{r^3}+\frac{2a\cos\theta}{r^2}-\frac{\cos\theta}{r}-\frac{a\cos\theta}{r^3}=0$$

Hence continuity equation is satisfying a possible flow field.

34. For checking rotationality, vorticity is given by –

$$\Omega=\frac{\partial V_\theta}{\partial r}+\frac{V_\theta}{r}+\frac{1}{r}\frac{\partial V_r}{\partial\theta}$$

$$\frac{\partial V_\theta}{\partial_r}=-a\sin\theta(-2)\frac{1}{r^3}$$

$$\frac{\partial v_r}{\partial_\theta}=\left(1-\frac{a}{r^2}\right)\times(-\sin\theta)$$

$$\Omega=\left[-a\sin(-2)\frac{1}{r^3}\right]+\frac{1}{r}\left[-\left(1+\frac{a}{r^2}\right)\sin\theta\right]-\frac{1}{r}\left[\left(1-\frac{a}{r^2}\right)\times-\sin\theta\right]$$

$$=\frac{2a\sin\theta}{r^3}-\frac{\sin\theta}{r}-\frac{a\sin\theta}{r^3}+\frac{\sin\theta}{r}-\frac{a\sin\theta}{r^3}=0$$

Flow is irrotational.

35. Given u = Cy, v = 0

Equation of circle $x^2-y^2-2ay=0$

We have $\Gamma=\Omega A$

Ω = Vorticity, A = Area

$$\Omega=\frac{\partial v}{\partial x}-\frac{\partial u}{\partial y}=0-\frac{\partial}{\partial y}(Cy)=-C$$

$$A=\pi a^2$$

$$\therefore\quad \Gamma=-C\times\pi a^2$$

36. Reynolds number $(Re)=\frac{F_i}{F_v}$

Froude Number $(Fr)=\sqrt{\frac{F_i}{F_g}}$

Weber Number $(We)=\sqrt{\frac{F_i}{F_s}}$

Evler's number $(Eu)=\sqrt{\frac{F_i}{F_p}}$

37. Displacement thickness

$$\delta^*=\int_0^\delta\left(1-\frac{u}{v}\right)dy$$

$$=\int_0^\delta\left(1-\frac{y}{\delta}\right)dy$$

$$=\left[y-\frac{y^2}{2\delta}\right]_0^\delta=\frac{\delta}{2}$$

Momentum thickness

$$\theta=\int_0^\delta\frac{u}{v}\left(1-\frac{u}{v}\right)dy$$

$$=\int_0^\delta\frac{y}{\delta}\left(1-\frac{y}{\delta}\right)dy$$

$$=\int_0^\delta\frac{y}{\delta}-\int_0^\delta\frac{y^2}{\delta^2}dy$$

$$\theta=\left[\frac{y^2}{2\delta}-\frac{y^3}{3\delta^2}\right]_0^\delta=\frac{\delta^2}{2\delta}-\frac{\delta^3}{3\delta}=\frac{\delta}{6}$$

$$\frac{\delta^*}{\theta}=\frac{\frac{\delta}{2}}{\frac{\delta}{6}}=3$$

38. Nominal thickness y = δ, u = 0.99 V_∞

Displacement thickness $(\delta^*)=\int_0^\delta\left(1-\frac{u}{v}\right)dy$

Momentum thickness $(\theta)=\int_0^\delta\frac{u}{v}\left(1-\frac{u}{v}\right)dy$

Energy thickness $=\int_0^\delta\frac{u}{v}\left(1-\frac{u^2}{v^2}\right)dy$

40. Boundary layer separation occurs when pressure increases in direction of flow $\left(\frac{\partial P}{\partial x} > 0\right)$, the pressure force acts in opposite direction of flow further increase the retarding effect of viscous forces, subsequently the thickness of boundary layer increases rapidly in the direction of flow. These force acts over a long stretched, the boundary layer get separated from the surface and moves in the main stream. This phenomena in called boundary layer separation.

$\frac{\partial u}{\partial y} > 0$ attached flow

$\frac{\partial u}{\partial y} = 0$ flow verge to separate

$\frac{\partial u}{\partial y} < 0$ separated flow

42. Von-Karman momentum integral equation is applicable for laminar transition and turbulent boundary layer.

44. Boundary layer thickness in case of laminar flow is

$$\delta = \frac{5x}{\sqrt{Re_x}}$$

$$\delta \propto \sqrt{x}$$

$$\frac{\delta_1}{\delta_2} = \sqrt{\frac{1}{4}}$$

$$\Rightarrow \frac{1}{\delta_2} = \sqrt{\frac{1}{4}}$$

$$\boxed{\delta_2 = 2\text{ mm}}$$

45. $\frac{dA}{A} = \frac{dV}{V}\left(M^2 - 1\right)$

46. Sonic velocity is given by

$$C = \sqrt{\frac{dP}{de}} = \sqrt{\frac{K}{e}} \text{ is term of bulk modules.}$$

$$C = \sqrt{\frac{P}{e}} = \sqrt{RT} \text{ for Isothermal process}$$

$$C = \sqrt{\frac{\gamma P}{e}} = \sqrt{\gamma RT} \text{ for adiabatic process}$$

47. For compressible flow basic governing equation are :

→ Continuity equation

→ Momentum equation

→ Energy equation

→ Equation of states

48. Kutta joukowshi theorem states that lift per unit span on a 2-D body is directly proportional to circulation around the body.

$$L = e_\infty V_\infty \Gamma$$

Γ = circulation

L = Lift

V_∞ = Free stream velocity

49. Kelvin circulation theorem states that time rate of change of circulation around a close curve consisting of same fluid element is zero.

$$\frac{\Delta T}{\Delta t} = 0$$

$$\Gamma_1 = \Gamma_2$$

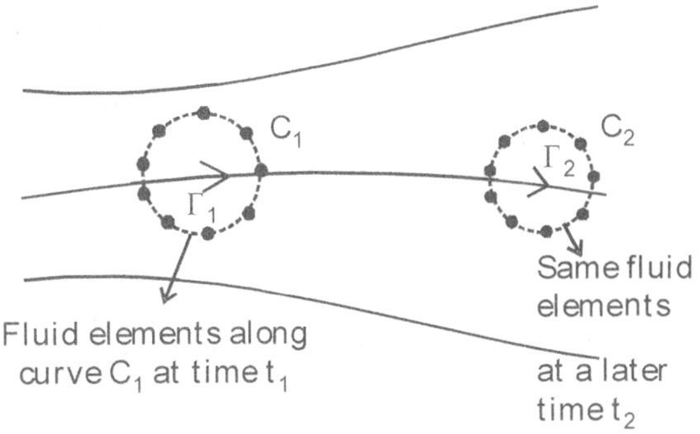

50. $C_P = 1 - 4\sin^2\theta \rightarrow$ for cylinder

$$C_P = 1 - \frac{9}{4}\sin^2\theta \rightarrow \text{for sphere}$$

51. For Cusped aerofoil trailing edge angle is zero, at a point trailing edge $v_1 = v_2 \neq 0$, this comes from Kutta condition.

Velocities leaving the top and bottom surfaces of trailing edge are finite and equal in magnitude and direction.

52. $\frac{A}{A^*} = \frac{1}{M}\left[\frac{2}{\gamma+1}\left(1 + \frac{\gamma-1}{2}M^2\right)\right]^{\frac{\gamma+1}{2(\gamma-1)}}$

$\gamma = 1.4$, M = 0.8

$$= \frac{1}{0.8}\left[\frac{2}{2-4}\left(1 + 0.2 \times 0.8^2\right)\right]^{\frac{1.4+1}{2(1.4-1)}}$$

$$= 1.038$$

$$\frac{A}{A^*} = 1.038 \simeq 1.04 \text{ (Approx)}$$

53. $\dot{m} = \dfrac{P_0 A^*}{\sqrt{T_0}} \sqrt{\dfrac{\gamma}{R}\left(\dfrac{2}{\gamma+1}\right)^{\frac{\gamma+1}{\gamma-1}}}$

$$\dot{m} \propto \frac{P_0}{\sqrt{T_0}}$$

When pressure decreased by 0.6 times and temperature increased by 1.44 times,

$$\dot{m}_{new} \propto \frac{0.6\ P_0}{\sqrt{1.44\ T_0}}$$

$$\frac{\dot{m}}{\dot{m}_{new}} = \frac{\dfrac{P_0}{\sqrt{T_0}}}{\dfrac{0.6\ P_0}{\sqrt{1.44\ T_0}}} = \frac{\sqrt{1.44}}{0.6} = 2$$

$$\dot{m}_{new} = \frac{\dot{m}}{2}$$

Hence the mass flow rate should be decreased by factor 2 to keep the pressure ratio constant.

54. Turbulent boundary layer more stable than laminar boundary layer and less tends to flow separation. To achieve the turbulent boundary layer golf ball has dots, so that flow remains attached and have drag reduces and it can travel long distance also if the flow attached over the sample, it contributes to the lift on the aerofoil produces.

55. In steady level flight minimum air speed at which airplace can flown in steady level flight is called the stall speed.

$$V_{stall} = \sqrt{\frac{2gm}{\rho S C_{L_{max}}}}$$

As the aircraft is geometrically similar the non-dimensional lift coefficiency C_L will be same.

$$\therefore \frac{V_{stall\ larger}}{V_{stall\ smaller}} = \sqrt{\frac{W_l}{S_l} \times \frac{S_s}{W_s}} = \sqrt{\frac{W_l}{W_s} \times \frac{S_s}{S_l}}$$

$$= \sqrt{\frac{2.4}{2}} = \sqrt{1.2} = 1.095$$

% increase in stall speed

$$= \frac{1.095 - 1}{1} \times 100 = 9.5\%\ \text{(Approx)}$$

56. $\dfrac{v^2}{\dfrac{2}{e}} = \gamma \dfrac{(\gamma - 1)}{2} M^2 = \dfrac{1.4 \times 0.4 \times 0.8^2}{2}$

$= 0.1792$

57. $M_1 = 0.5$

$C_{P_1} = 0.4$

$M_2 = 0.7$

$C_{P_2} = ?$

$$\frac{C_{P_2}}{C_{P_1}} = \frac{\dfrac{C_{P_0}}{\sqrt{1 - M_2^2}}}{\dfrac{C_{P_0}}{\sqrt{1 - M_1^2}}} = \sqrt{\frac{1 - M_1^2}{1 - M_2^2}}$$

$$\frac{C_{P_2}}{C_{P_1}} = \sqrt{\frac{1 - 0.7^2}{1 - 0.5^2}}$$

$= 0.82$

$C_{P_2} = C_{P_1} \times 0.82$

$= 0.4 \times 0.82 = 0.32$

58. $M_1 = 2$,

we have

$$\frac{At_2}{At_1} = \frac{P_{01}}{P_{02}}$$

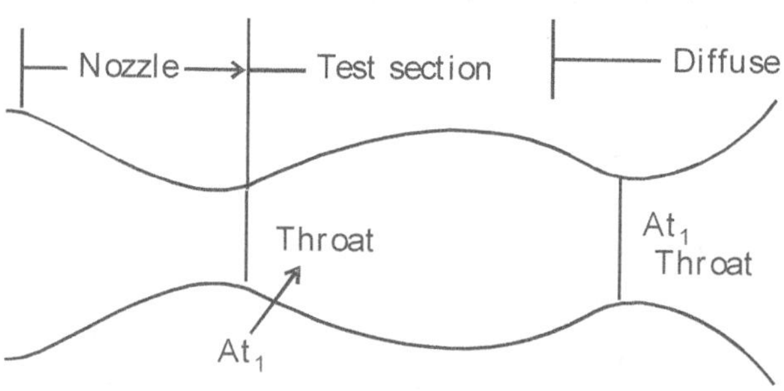

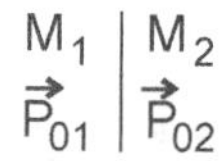

Normal shock wave

$$M_2 = \sqrt{\frac{1 + \dfrac{\gamma - 1}{2} M_1^2}{\gamma M_1^2 - \dfrac{\gamma - 1}{2}}} \qquad M_2 = \sqrt{\frac{1.8}{0.33}} = 0.57$$

$$\frac{P_{01}}{P_{02}} = \frac{P_{01}}{P_1} \times \frac{P_1}{P_2} \times \frac{P_2}{P_{02}}$$

$$\left(1 + \frac{\gamma - 1}{2} M_1^2\right)^{\frac{\gamma}{\gamma-1}} \times \left(1 + \frac{2\gamma}{\gamma - 1}\left(M_1^2 - 1\right)\right)^{-1}$$

$$\times \left(\frac{1}{\left(1 + \dfrac{(\gamma - 1)/2}{2} M_2^2\right)^{\frac{\gamma}{\gamma-1}}}\right)$$

$$= \left(\frac{1+\frac{\gamma-1}{2}M_1^2}{1+\frac{\gamma-1}{2}M_2^2}\right)^{\frac{\gamma}{\gamma-1}} \times \left(1+\frac{2\gamma}{\gamma-1}\left(M_1^2-1\right)\right)^{-1}$$

$$\frac{P_{01}}{P_{02}} = \left(\frac{1+\frac{\gamma-1}{2}M_1^2}{1+\frac{\gamma-1}{2}M_2^2}\right)^{\frac{\gamma}{\gamma-1}} \times \left(\frac{1}{1+\frac{2\gamma}{\gamma-1}\left(M_1^2-1\right)}\right)$$

$$= \left(\frac{1+0.2\times 2^2}{1+0.2\times 0.37^2}\right)^{3.5} \times \frac{1}{1+\frac{2.8}{2.4}\times\left(2^2-1\right)}$$

$= 1.387$

$$\boxed{\frac{At_2}{At_1} = 1.387}$$

59. 0.378

$$\frac{e_2}{e_1} = \frac{(\gamma+1)M_1^2}{2+(\gamma-1)M_1^2}$$

when

$M_1 \to \infty$

$$\frac{e_2}{e_1} = \frac{\gamma+1}{\gamma-1} = 6$$

$$M_2^2 = \frac{1+\frac{\gamma-1}{2}M_1^2}{\gamma M_1^2 - \frac{\gamma-1}{2}} = \frac{\gamma-1}{2\gamma}$$

$$M_2 = \sqrt{\frac{\gamma-1}{2\gamma}} = 0.378$$

60. In steady adiabatic flow, total enthalphy

$$h_0 = h + \frac{v^2}{2}$$

$$T_0 = T + \frac{v^2}{2\,C_P}$$

$$= T + \frac{V^2}{\frac{2\gamma R}{\gamma-1}} = T + \frac{V^2}{2\gamma R}(\gamma-1)$$

$$= 320 + \frac{1000^2\times 0.4}{2\times 1.4\times 287} = 320 + 497.8$$

$T_0 = 817.8$ K

$$\boxed{T_0 = 817.8\ \text{K}}$$

Total pressure, P_0

$$\frac{P_0}{P} = \left(\frac{T_0}{T}\right)^{\frac{\gamma}{\gamma-1}} = \left(\frac{817.8}{320}\right)^{3.5}$$

$$\boxed{P_0 = 26.7\ \text{atm}}$$

61. 258.2 J/kg-K

$$\text{change in entropy } \Delta S = C_P \ln\frac{T_2}{T_1} - R\ln\frac{P_2}{P_1}$$

$$= \frac{\gamma R}{\gamma-1}\ln\frac{690}{288} - 287\ln\frac{8.606}{1}$$

$$= \frac{1.4\times 287}{0.4}\ln\frac{690}{288} - 287\ln 8.606$$

$= 258.2$ J/kg-K

62. For supersonic flow

$$Cl = \frac{4\alpha}{\sqrt{M_\infty^2-1}} = \frac{4\times 0.139}{\sqrt{4^2-1}} = 0.143$$

$$C_d = \frac{4\alpha^2}{\sqrt{M_\infty^2-1}} = \frac{4\times(0.139)^2}{\sqrt{4^2-1}}$$

$= 0.0199$

64.

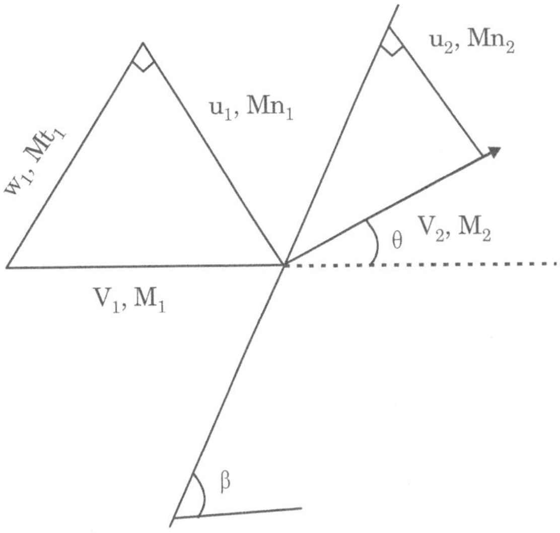

$\theta = 10°$

$\beta = 30°$

V= 900 m/sec

As we know, the tangential component of velocity remains same.

$u_1 \cos \beta = u_2 \cos (\beta - \theta)$

$$u_1 = \frac{u_2 \cos(\beta - \theta)}{\cos \beta}$$

$$= \frac{900 \cos(30 - 10)}{\cos 30°} = 976.55 \text{ m/sec}$$

65. Pressure ratio across oblique shock wave

$$\frac{P_2}{P_1} = 1 + \frac{2\gamma}{\gamma + 1}\left(M_\eta^2 - 1\right)$$

$$M\eta_1 = M_1 \sin \beta$$

$$= \frac{V_1}{\sqrt{\gamma R T_1}} \sin \beta$$

$$= \frac{977 \sin 30°}{\sqrt{1.4 \times 287 \times 310}}$$

$= 2.76 \sin 30°$

$= 1.38$

$$\therefore \quad \frac{P_2}{P_1} = 1 + \frac{2 \times 1.4}{1.4 + 1}\left(1.38^2 - 1\right)$$

$$= 1 + \frac{2.8}{2.4}\left(1.38^2 - 1\right)$$

$= 2.05$

$P_2 = 2.00 \times 1 = 2.05$ atm

66. $M_1 = 1.5$

$P_1 = 1$ atm

$T_1 = 288$ K

Angle of forward mach line

$$\mu_1 = \sin^{-1} \frac{1}{M_1} \qquad \mu_2 = \sin^{-1} \frac{1}{\mu_2}$$

$$= \sin^{-1} \frac{1}{1.5}$$

$= 41.81°$

For finding the angle of rearward mach line, we must have to find mach number downstream of expansion wave.

$\nu(M_2) - \nu(M_1) = \theta$

$\nu(M_2) = \theta + \nu(M_1)$

$$\nu(M_1) = \sqrt{\frac{\gamma + 1}{\gamma - 1}} \tan^{-1} \sqrt{\frac{\gamma - 1}{\gamma + 1}}\left(M^2 - 1\right) - \tan^{-1} \sqrt{M^2 - 1}$$

$\gamma = 1.4$

$$\nu(M_1) = \sqrt{6} \tan^{-1} \sqrt{\frac{1.5^2 - 1}{6}} - \tan^{-1} \sqrt{1.5^2 - 1}$$

$\nu(M_1) = 12°$

$\nu(M_2) = 12 + 15 = 27°$

$$\Rightarrow \quad \sqrt{\frac{\gamma + 1}{\gamma - 1}} \tan^{-1} \sqrt{\left(M_2^2 - 1\right)\frac{\gamma - 1}{\gamma + 1}} - \tan^{-1} \sqrt{M_2^2 - 1}$$

$= 27°$

$$\Rightarrow \quad \sqrt{6} \tan^{-1} \sqrt{\left(M_2^2 - 1\right) \times \frac{1}{6}} - \tan^{-1} \sqrt{M_2^2 - 1}$$

$= 27°$

$\therefore \quad M_2 = 3$ (rounding nearest values)

Angle of rearward mach line

$$\therefore \mu_2 = \sin^{-1} \frac{1}{M_2}$$

$$= \sin^{-1} \frac{1}{2}$$

$$\boxed{\mu_2 = 30°}$$

67. Across the expansion wave, Ratio of pressure

$$\frac{P_2}{P_1} = \left(\frac{1 - \frac{\gamma - 1}{2} M_1^2}{1 + \frac{\gamma - 1}{2} M_2^2}\right)^{\frac{\gamma}{\gamma - 1}}$$

$$= \left(\frac{1 + 0.2 \times 1.5^2}{1 + 0.2 \times 2^2}\right)^{3.5}$$

$$= \left(\frac{1.45}{1.8}\right)^{3.5}$$

$$\frac{P_2}{P_1} = 0.4691$$

$P_2 = 0.4691 \times P_1$

$= 0.4691 \times 1$

$= 0.4691$ atm

5 CHAPTER AIRCRAFT STRUCTURE

1. Prevention of buckling of skin of wing is done by?
2. Which part of the wing takes bending mainly?
3. Give representation of Bending stiffners.
4. What is margin of safety (MOS) in terms of factor of (FOS) safety?
5. How many independent elastic constant are there for orthotropic elastic material ?
6. Give relation between shear modulus and Young's modulus.
7. Give relation between Bulk modulus and Young's Modulus
8. Define Torsional Rigidity.
9. In monocoque structure, what is the major load bearing member?
10. Find no. of degree of freedom of a hinged-hinged beam?
11. Find no. of degree of freedom of figure given below.

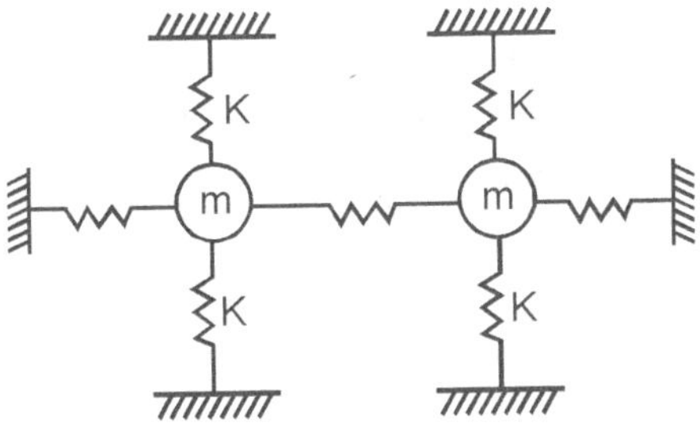

12. What is degree of freedom of a system?
13. What does natural frequencies of a system are equal to?
14. What does damped natural frequency of a system is equal to?
15. What does the relation between damped and undamped natural frequency of single degree of freedom system gives?
16. What is the damping ratio of critically damped system?
17. Give general formula for logarithmic decrement.
18. Find the quality factor for SDOF's.
19. Find no. of natural frequencies of a cantilever beam.
20. Find no. of compatibility equations needed to be satisfied for a 3-D problem of elasticity.
21. Find stresses in term of airy stress function.
22. Find stiffness of cantilever beam with end point load.
23. A steel bar is 900 mm long with variable diameter is given below. Bar is subjected to axial tension of 15 KN. What is total extension of bar. (E = 200 GPa)

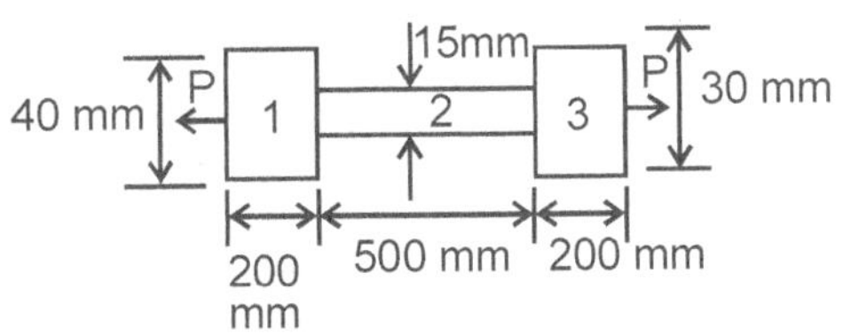

24. A steel rod 15 m long is at a temperature of 15°C. Find free expansion of length when temperature is raised to 65°C. [E = 200 GPa, $\alpha = 12 \times 10^{-6}$/°C]
25. Find temperature stress from the data given in Q.24, when
    (i) Expansion of rod is prevented
    (ii) Rod is permitted to expand by 6 mm
26. Principal stresses in the wall of a container are 40 MN/m^2 and 80 MN/m^2. Determine normal, shear and resultant stress in magnitude and direction in a plane, the normal of which makes an angle of 30° with the direction of maximum principle stress

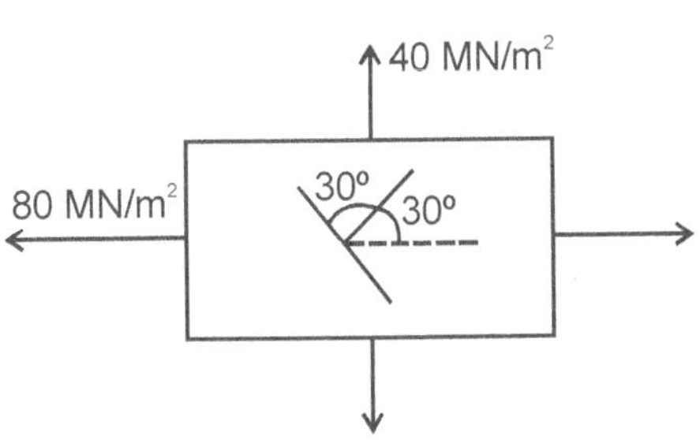

27. Calculate moment of inertia about horizontal and vertical gravity axes (I_{xx} & I_{yy}) of section shown below

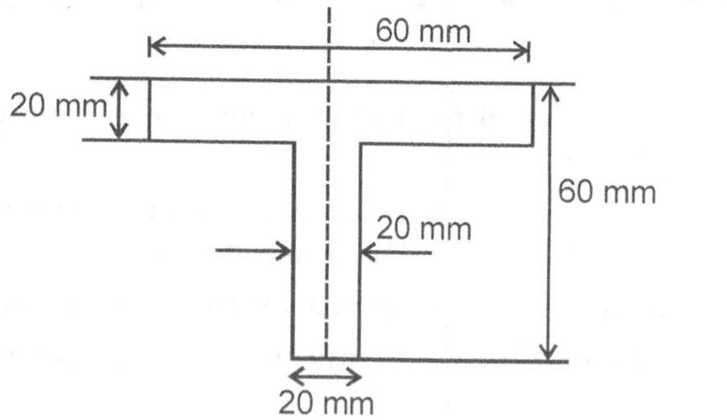

28. Draw Shear force diagram for the given beam.

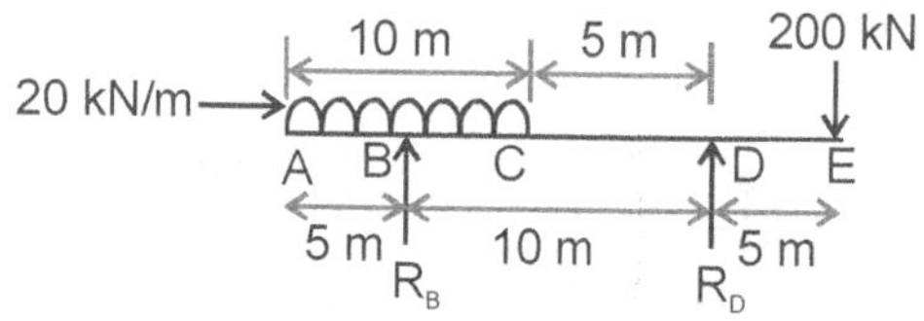

29. A square crossection beam given below is loaded in such a way that maximum bending stress is 20 MPa and maximum moment is 25 KNm. Beam is made up of material of shear modulus G = 100 GPa and poisson ratio υ = 0.2. What is the dimension of cross-section of beam.

30. Calculate hoop stress produced in spherical shell of outer diameter 10 mm and inner diameter 8 mm with internal pressure of 200 MPa. If 450 MPa is the maximum hoop stress it can take. Then what is the factor of safety:

31. What can be the length of 10 mm diameter wire which can be twisted through angle of 360° without exceeding shear stress of 50 MPa.

[Shear modulus G = 20 GPa]

32. What is the angle of twist at point C in stepped shaft given below :

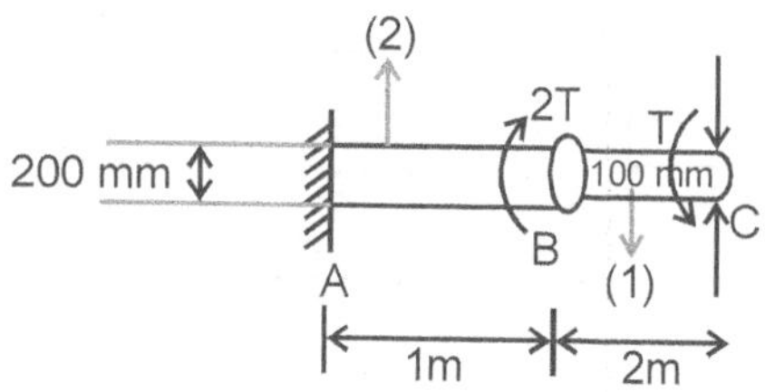

(G = 20 GPa) (Max. shear stress = 50 MPa)

33. Consider a three stringer thin walled beam with cross-section given below. Consider shear flow is constant through the walls :

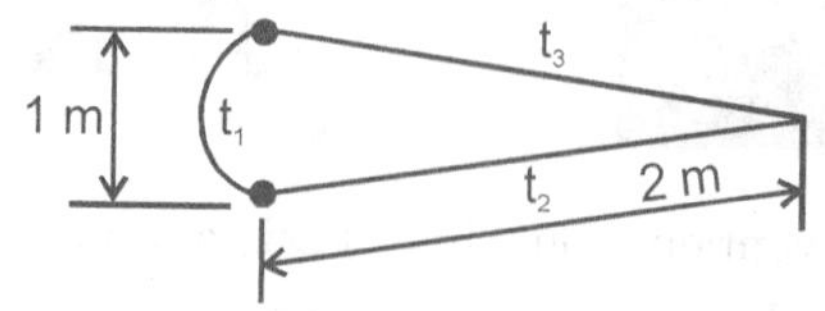

If allowable shear stress is 300 MPa, what is the maximum torque structure can take ?

34. For multicell section given below subjected to a torque T = 20 MNm. Determine the shear flows and angle of twist per unit length. [G = 80 GPa]

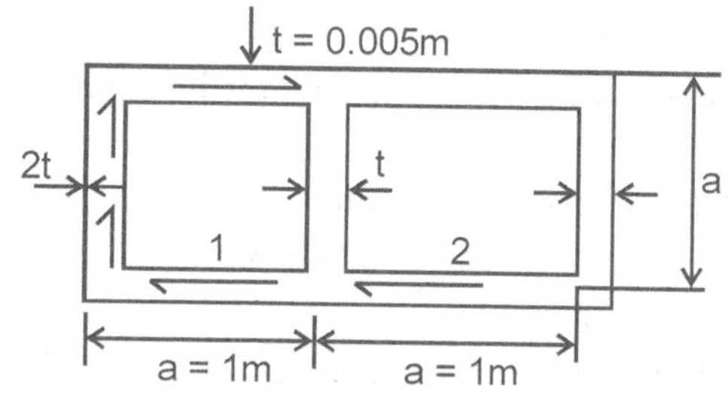

35. A strut made of steel, 2m long with 50 mm diameter is subjected to axial thrust of 20 kN. If compressive-strength of strut is 350 MN/m^2, [strut is simply-supported] E = 200 GPa

Check whether it fails in buckling or compression?

36. Find critical buckling load for both end-fixed column.

37. Give equation of Von Mises theory for 2-D case of failure.

38. Material subjected to principle stress given by:

σ_1 = 150 MPa (2D-case)

σ_2 = 100 MPa

Yield strength of material is given by σ_y = 350 MPa. Check whether material fails under this type of loading.

39. Determine longitudinal stiffness of bar

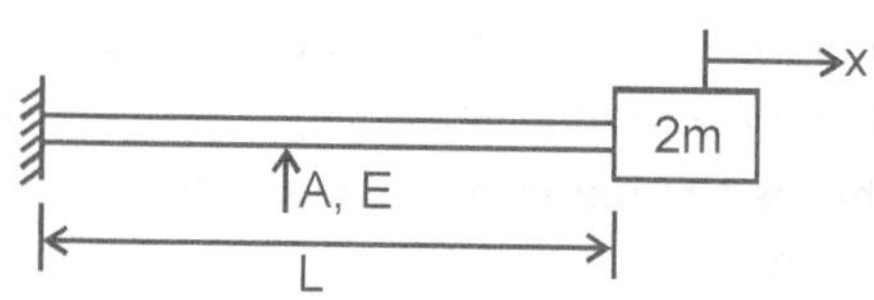

40. Write equation of motion of system

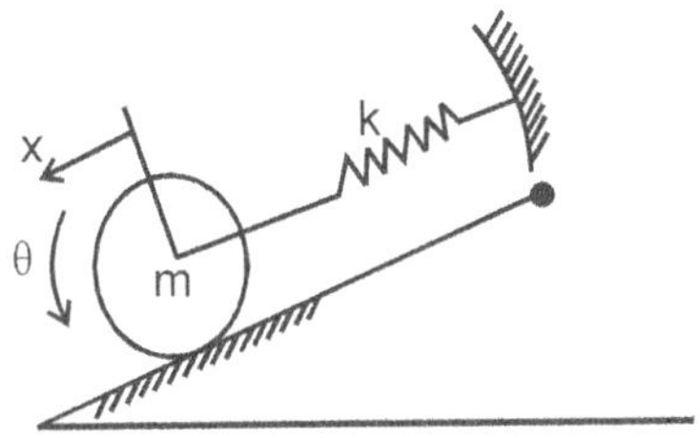

41. What is the natural frequency in above question.

42. Find natural frequency for pendulum given below.

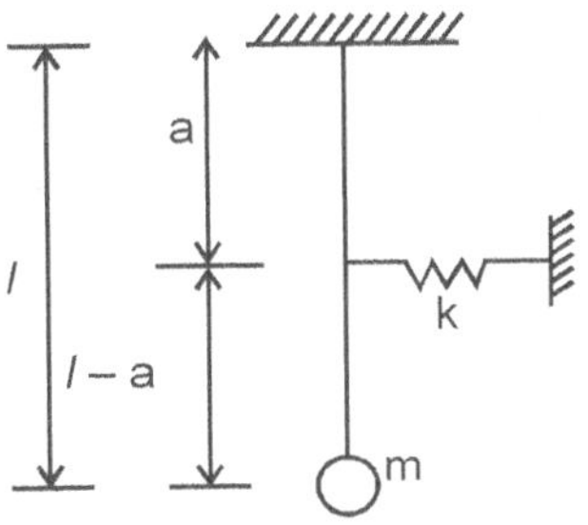

43. For given single degree of freedom system with k = 10,000 N/m

and m = 40 kg

and $\xi = 0.3$

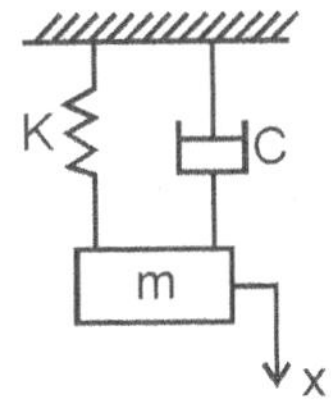

What is the damping factor C and damped natural frequency.

44. For given 2 DOF system

Find out the mass and salttner matrix

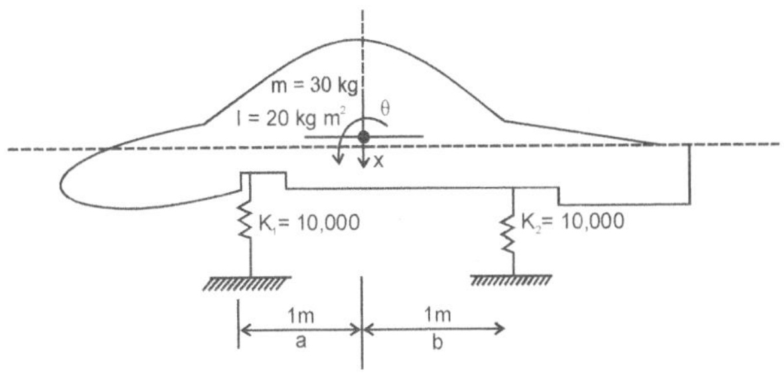

45. Find out natural frequency using data of 46th question.

46. Describe/discuss torsional vibration of Shaft.

47. Briefly discuss Lateral vibration of beam.

EXPLANATIONS

1. Ribs (stiffners). By reducing effective length of skin.
2. Spars
3. EI where

   E = Young's modulus

   I = Moment of inertia [2nd moment of Area]
4. MOS = FOS – 1

   $$FOS = \frac{\text{Ultimate strength or yield strength}}{\text{working stress}}$$
5. 9
6. $G = \frac{E}{2(1+\upsilon)}$ υ = Poissons ratio
7. $k = \frac{E}{3(1-2v)}$
8. GJ, J = polar moment of inertia
9. Thick skin
10. Infinite [For any continuous system, no. of degree of freedom are infinite]
11. There are 4 degree of freedoms.

    1 horizontal and 1 vertical for each mass.
12. No. of independent coordinate needs to explain system.
13. Equal to no. of degree of freedoms.
14. Less than undamped natural frequency.
15. $\omega_d = \omega_n\sqrt{1-\xi^2}$

    Where, ξ = damping Ratio

    ω_d = damped natural frequency

    ω_n = undamped natural frequency.
16. equal to 1

    i.e., $\xi = 1$

    under damped $\xi < 1$

    over damped $\xi > 1$
17. $$\boxed{\delta = \ln\frac{x_1}{x_2} = \frac{2\pi\xi}{\sqrt{1-\xi^2}}}$$ ® first 2 peaks or any 2 consecutive peaks.

    In general,

    $$\boxed{\delta = \ln\left(\frac{A_1}{A_n}\right) = (n-1)\frac{2\pi\xi}{\sqrt{1-\xi^2}}}$$
18. $$\boxed{Q = \frac{1}{2\xi} = \frac{\omega_n}{\omega_2 - \omega_1}}$$

    ξ = damping ratio

    ω_n = natural frequency

    ω_2 and ω_1 are frequency ahead and behind of natural frequency.
19. Infinite
20. 6
21. $\sigma_{xx} = \frac{\partial^2\phi}{\partial y^2}$

    $\sigma_{yy} = \frac{\partial^2\phi}{\partial x^2}$ ϕ = Airy stress functions.

    $\tau_{xy} = -\frac{\partial^2\phi}{\partial x\partial y}$

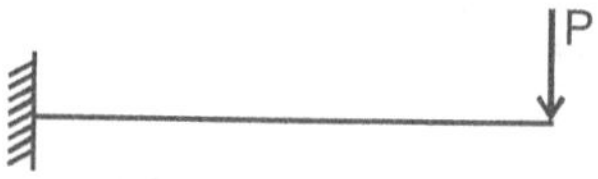

22. $$\boxed{k = \frac{3EI}{l^3}}$$

    l = length of beam

    I = Moment of inertia

    E = Young's modulus
23. P = 15 KN = 15 × 10³ N

    Areas $A_1 = \frac{\pi}{4} \times (40)^2 = 1256.6 \text{ mm}^2$

    $= 0.001256 \text{ m}^2$

    Similary $A_2 = 0.0001767 \text{ m}^2$

    $A_3 = 0.0007068 \text{ m}^2$

    Length l_1 = 200 mm = 0.2 m

    l_2 = 0.5 m

    l_3 = 0.2 m

    Total extension of bar $\delta l = \delta l_1 + \delta l_2 + \delta l_3$

    $$\left[\begin{array}{l}\text{Tensil stress } \sigma = Ee \\ \Rightarrow \frac{P}{A} = E\frac{\delta l}{l} \\ \Rightarrow \delta l = \frac{Pl}{AE}\end{array}\right]$$

    $$\therefore \quad \delta l = \frac{Pl_1}{A_1E} + \frac{Pl_2}{A_2E} + \frac{Pl_3}{A_3E}$$

    $$= \frac{P}{E}\left[\frac{l_1}{A_1} + \frac{l_2}{A_2} + \frac{l_3}{A_3}\right]$$

$$= \frac{15\times10^3}{200\times10^9}\left[\frac{0.2}{0.001256}+\frac{0.5}{0.0001767}+\frac{0.2}{0.0007067}\right]$$

Total extension = 0.0002454 m = 0.2454 mm

24. Free expansion of rod

$\delta l = l\,\alpha(t_2 - t_1)$

$= 15 \times 12 \times 10^{-6} \times [65 - 15]$

= 0.009 m = 9 mm

25. (i) Expansion of rod is prevented

Thermal stress = $\alpha(t_2 - t_1)E$

$= 12 \times 10^{-6} \times (65 - 15) \times 200 \times 10^9$

$= 120 \times 10^6$ Pa

= 120 MPa

(ii) When rod is permitted to expand by 6 mm

Expansion prevented = 9 – 6 = 3 mm

$$\text{Strain} = \frac{\text{Expansion prevented}}{\text{Original length}}$$

$$= \frac{3}{15\times1000} = 0.0002$$

∴ Thermal stress = $e \times E = 0.0002 \times 200 \times 10^9$

$= 40 \times 10^6$ Pa

= 40 MPa

26. $\sigma_x = 80$ MN/m^2

$\sigma_y = 40$ MN/m^2

$\theta = 30°$

Normal stress

$$\sigma_n = \frac{\sigma_x+\sigma_y}{2}+\frac{\sigma_x+\sigma_y}{2}\cos 2\theta$$

$$= \frac{80+40}{2}+\frac{80-40}{2}\cos 60°$$

= 70 MN/m^2

Shear stress

$$\tau = \frac{\sigma_x-\sigma_y}{2}\sin 2\theta = \frac{80-40}{2}\sin 60°$$

= 17.32 MN/m^2

Resultant stress, σ_r:

$$\sigma_r = \sqrt{\sigma_n^2+\tau^2}$$

$$= \sqrt{70^2+17.32^2}$$

= 72.11 MN/m^2

Angle that resultant makes with normal to be plane.

$$\theta = \frac{\tau}{\sigma_n} = \frac{17.32}{70} = 0.2474 \text{ radian}$$

= 13°54'

27. Since section is symmetric about y-axis, therefore we shall find out $\bar{y}$ only

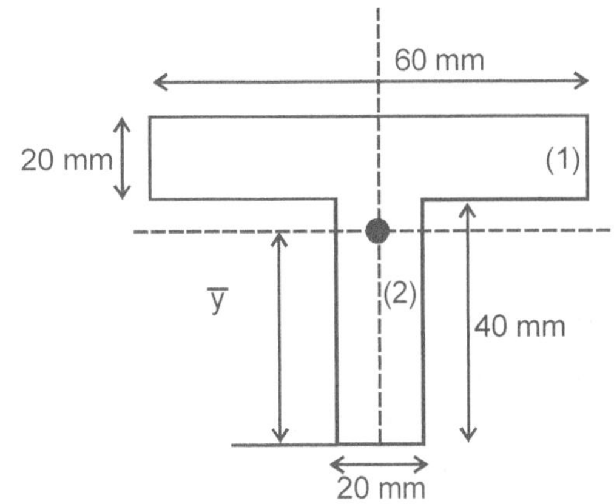

	Area (mm^2)	Centroidal distance y(mm)	ay (mm^3)
(i)	60 × 20 = 1200	$40+\frac{20}{2} = 50$	60000
(ii)	40 × 20 = 800	$\frac{40}{2} = 20$	16000
	Σa = 2000		Σay = 76000

$$\therefore \quad \bar{y} = \frac{\Sigma ay}{\Sigma a} = \frac{76000}{2000} = 38 \text{ mm}$$

Using Theory of parallel axes for I_{xx} and I_{yy}

$$I_{xx} = I_{xx_1} + I_{xx_2} = \left(\frac{60\times20^3}{12}+60\times20\times[50-38]^2\right) + \left[\frac{20\times40^3}{12}+40\times20\times(38-20)^2\right]$$

= 212800 + 365865

= 578666 mm^4

$$\text{similarly } I_{yy} = \left[\frac{20\times60^3}{12}\right]+\left[\frac{40\times20^3}{12}\right]$$

= 360000 + 26666

= 386666 mm^4

28. To determine reactions, taking moment about A

$$R_D\times10+20\times5\times\frac{5}{2} = 20\times5\times\frac{5}{2}+200\times15$$

$R_D \times 10 + 250 = 250 + 3000$

$R_D = 300$ kN

Taking force equilibrium condition.

$R_B + R_D = 20 \times 10 + 200 = 400$ kN

Shear force :

$S_{E-D} = -200$ kN (–ve for direction sign connection)

$S_{D-C} = -200 + 300 = 100$ kN

$S_D = 100 - 20 \times 5 + 100 = 100$ kN

$S_{C-B} = 100 - 20 \times x$

here x = 0 – 5 m

at x= 5

$S_{C-B} = 100 - 100 = 0$

$S_{B-A} = 100 - 20 \times x$

$S_{B-A} = 100 - 20 \times 5 = 0$

∴ Shear force diagram :

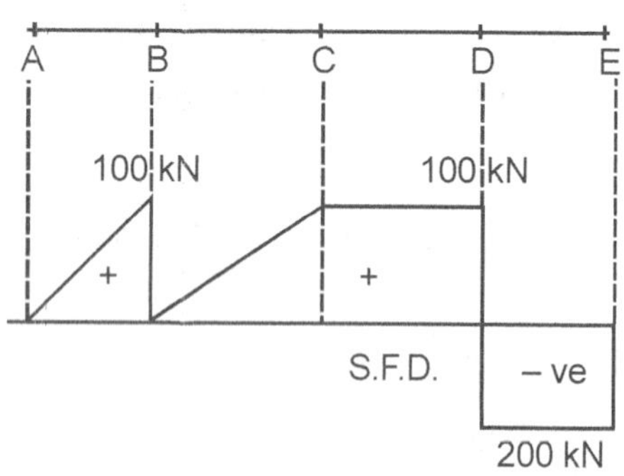

29. Relation between shear modulus and Young's modulus is given by :

$$G = \frac{E}{2(1+\upsilon)}$$

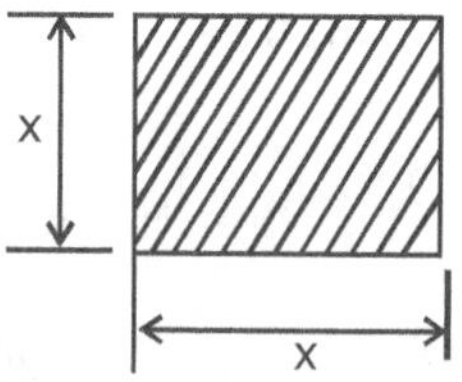

$E = 100 \times 2 \times [1 + 0.2]$

$= 100 \times 2.4 = 240$ GPa $= 240 \times 10^9$ Pa

Maximum bending stress is given by :

$$\sigma_{max} = \frac{M_{max} \times \frac{x}{2}}{I}$$

$$= \frac{M_{max} \times \frac{x}{2}}{\frac{x.x^3}{12}}$$

$$x^3 = \frac{M_{max} \times 6}{\sigma_{max}}$$

$$\Rightarrow \quad x = \left[\frac{25 \times 10^3 \times 6}{20 \times 10^6}\right]^{\frac{1}{3}}$$

$$\Rightarrow \quad \boxed{x = 0.195 \text{ m}}$$

30.

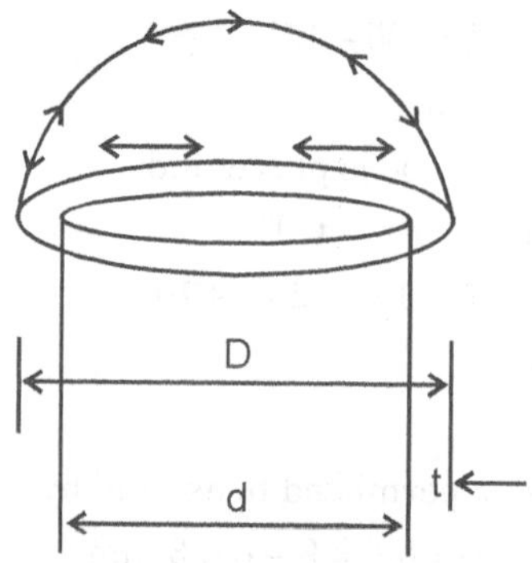

Hoop stress for spherical stress

$$\sigma_c = \frac{Pd}{4t}$$

d = 8 mm = 0.008 m

$$\text{Thickness } t = \frac{D-d}{2} = \frac{10-8}{2} = 1 \text{ mm} = 0.001 \text{ m}$$

$$\therefore \quad \sigma_c = \frac{200 \times 0.008}{4 \times 0.001} = 400 \text{ MPa}$$

$$\text{Factor of safety} = \frac{\text{Max. strength in hoop}}{\text{working load in hoop}}$$

$$= \frac{450}{400} = 1.125$$

$$\boxed{\therefore \text{FOS} = 1.125}$$

31. For torsion theory :

$$\text{Torque } T = \frac{\tau}{R} \times I_P$$

$$= \frac{50 \times 10^6}{\left(\frac{D}{2}\right)} \times \left(\frac{\pi}{32}\right) \times (D^4)$$

$$= 50 \times 10^6 \times \frac{3.14}{6} \times (0.01^3)$$

$= 9.8$ Nm

$$\text{Also, } \frac{T}{I_P} = \frac{G\theta}{l} \qquad (\theta = 360° = 2\pi \text{ rad/sec})$$

$$\Rightarrow \quad l = \frac{G\theta I_P}{T} = \frac{20\times10^9\times2\times\pi\times\frac{\pi}{32}\times(0.01)^4}{9.8}$$

$\Rightarrow$ length $\boxed{l = 12.76 \text{ m}}$

32. For length BC

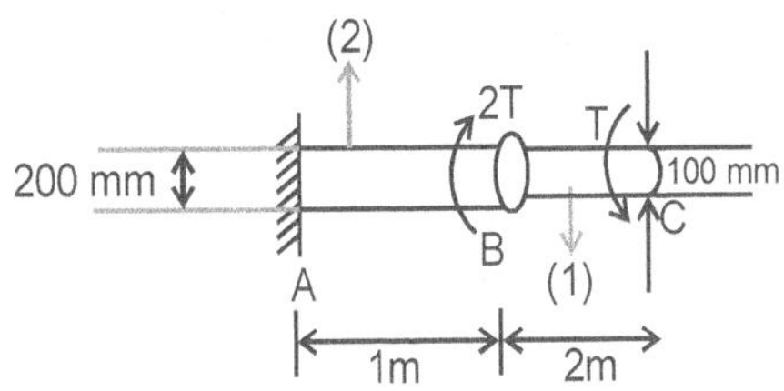

Torque T = T, l = 2m

$$I_P = \left(\frac{\pi}{32}\right)\times(0.1)^4 \text{ m}^4$$

Angle of twist relative to B

$$\theta_{C_1} = \frac{Tl}{GI_P} = \frac{T\times2}{20\times10^9\times\left(\frac{\pi}{32}\right)\times(0.1)^4}$$

Similarly at length AB

$$I_P = \left(\frac{\pi}{32}\right)(0.2)^4$$

Angle of point of B relative to A

$$\therefore \quad \theta_B = \frac{T\times1}{20\times10^9\times\left(\frac{\pi}{32}\right)(0.2)^4}$$

As θ_C & θ_B are in opposite directions

$$\theta_C = \theta_{C_1} - \theta_B$$

$$\text{Torque } T = \frac{\tau}{R}\times I_P$$

$$= 50\times10^6\times\frac{\pi}{16}\times(0.01)^3$$

= 9812.5 Nm

$$\therefore \quad \theta_C = \frac{9812.5\times2}{20\times10^9\times\left(\frac{\pi}{32}\right)\times(0.1)^4} - \frac{9812.5}{20\times10^9\times\left(\frac{\pi}{32}\right)\times(0.2)^4}$$

$= 5\times10^{-6}[20000-625]$

= 0.0968 radius

= 5.549 degree

33. Shear flow is given by :

$$T = 2\bar{A}q$$

$$\bar{A} = \frac{1}{2}\pi(0.5)^2 + \frac{1}{2}[2\times1]$$

$= 1.39 \text{ m}^2$

$$\text{Thus, } q = \frac{T}{2\bar{A}} = \frac{T}{2.785} \qquad \text{... (A)}$$

Because of small thickness shear stress will be maximum in curved walls which is equal to

$$\tau = \frac{q}{t_1} = \frac{T}{0.001\times2.785}$$

$\therefore$ Maximum torque can be given by

$$T_{max} = \tau_{\text{all walls}}\times0.001\times2.785$$

= 300 × 10⁶ × 0.001 × 2.785

= 836700 Nm

34. For cell (1)

$$\oint\frac{ds}{t} = \frac{a}{t}+\frac{a}{2t}+\frac{a}{t}+\frac{a}{t} = \frac{7a}{2t} = a_1$$

Cell (2)

$$\oint\frac{ds}{t} = \frac{a}{t}+\frac{a}{t}+\frac{a}{t}+\frac{2a}{t} = \frac{5a}{t} = a_2$$

$$\text{for web } \oint\frac{dx}{t} = \frac{a}{t} = a_{12}$$

For cell one (1)

$$2G\theta = \frac{1}{A_1}(a_1q_1 - a_{12}q_2)$$

$$= \frac{1}{a^2}\left[\frac{7a}{2t}q_1 - \frac{a}{t}q_2\right]$$

$$= \frac{1}{at}\left[\frac{7}{2}q_1 - q_2\right] \qquad \text{... (i)}$$

Cell (2)

$$2G\theta = \frac{1}{A_2}[a_2q_2 - a_{12}q_1]$$

$$= \frac{1}{a^2}\left[\frac{5a}{t}q_2 - \frac{a}{t}q_1\right]$$

$$= \frac{1}{at}[5q_2 - q_1] \qquad \text{... (ii)}$$

Equating (i) and (ii)

$$\frac{7}{2}q_1 - q_2 = 5q_2 - q_1$$

$$\boxed{q_2 = \frac{3}{4}q_1}$$

$\therefore$ Torque $T = 2q_1A_1 + 2q_2A_2$

$$= 2a^2\left[q_1 + \frac{3}{4}q_1\right]$$

$$= \frac{7}{2}a^2q_1$$

$$\therefore \boxed{q_1 = \frac{2T}{7a^2},\ q_2 - \frac{3T}{14a^2}}$$

$$\boxed{q_1 = \frac{2\times 20\times 10^3}{7\times 1^2} = 5714.28\ \text{N/m}}$$

and $\boxed{q_2 = \dfrac{3\times 20\times 10^3}{14\times 1^2} = 4285.714\ \text{N/m}}$

$$2G\theta = \frac{1}{at}[5q_2 - q_1]$$

$$= \frac{11}{4at}\times q_1$$

$$\Rightarrow \boxed{\theta = \frac{11}{28}\frac{T}{a^3tG} = \frac{11}{25}\times\frac{20\times 10^3\times\left(\frac{180}{\pi}\right)}{1^3\times 0.005\times 80\times 10^9} = 0.00112\ \text{degree}}$$

35. For simply supported case buckling load is given by :

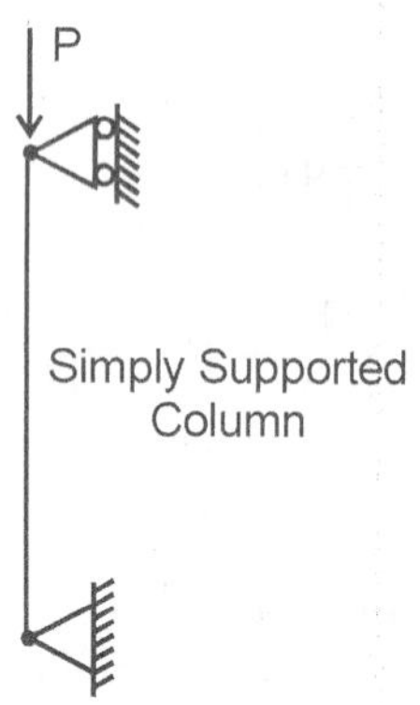

$$P = \frac{\pi^2 EI}{l^2}$$

$$I = \frac{\pi}{64}\times d^4 = \frac{\pi}{64}\times 0.05^4 = 3.066\times 10^{-7}\,\text{m}^4$$

$E = 200 \times 10^9$ Pa

$$P_{cr.} = \frac{3.14^2\times 200\times 10^9\times 3.066\times 10^{-7}}{2^2}$$

= 151147.668 N/m^2

= 0.151147 MPa

Compressive strength of strut is 350 MPa. But as seen from critical buckling load 0.151 MPa, which is very less than the compressive strength. Thus strut will fail in buckling, well before compressive stength reaches.

36. $\boxed{P_{cr} = \dfrac{\pi^2 EI}{l^2}\times 4}$

37. $\boxed{\sigma_1^2 + \sigma_2^2 - \sigma_1\sigma_2 \le \sigma^2}$ can be field strength

38. $\sigma_1^2 + \sigma_2^2 - \sigma_1\sigma_2 \le \sigma_y^2$

$$\sqrt{\sigma_1^2 + \sigma_2^2 - \sigma_1\sigma_2} \le \sigma_y$$

L.H.S. = $\sqrt{(150)^2 + (100)^2 - (150\times 100)}$

= 132.287 MPa which is less than yield strength 350 MPa

Thus material does not fail under this loading.

39. We know than stress generated in bar is :

$\sigma = E\epsilon$ $\qquad\qquad \epsilon$ = strain

if F is the force applied then,

$$\frac{F}{A} = E\,\epsilon = E\frac{\delta l}{l}$$

$$\therefore\quad F = \frac{EA\delta l}{l}$$

$$\boxed{\therefore \text{Stiffness } k_{eq.} = \frac{\text{Force}}{\text{Deflection}} = \frac{F}{\delta l} = \frac{EA}{l}}$$

40. K.E. = $T = \frac{1}{2}m\dot{x}^2 + \frac{1}{2}I(\dot{\theta})^2$

here $x = r\theta,\ \dot{x} = r\dot{\theta}$

$$= \frac{1}{2}m\dot{x}^2 + \frac{1}{2}I\left(\frac{\dot{x}}{r}\right)^2$$

$$= \frac{1}{2}\left(m + \frac{1}{r^2}\right)\dot{x}^2$$

P.E. = U = $\frac{1}{2}kx^2$

∴ single degree of freedom i.e. only one independent coordinate.

Using Lagrangian's equation

$$\frac{d}{dt}\left(\frac{\partial T}{\partial \dot{x}}\right) - \frac{\partial T}{\partial x} + \frac{\partial U}{\partial x} = F(x)$$

$$\boxed{\left(m + \frac{I}{r^2}\right)\ddot{x} + kx = 0}$$

41. $\boxed{\left(m + \frac{I}{r^2}\right)\ddot{x} + kx = 0}$

$$\Rightarrow \quad \ddot{x} + \frac{k}{\left(m + \frac{I}{r^2}\right)}x = 0$$

Natural frequency ∴ $\boxed{\omega = \sqrt{\frac{k}{m + \frac{I}{r^2}}}}$

42.

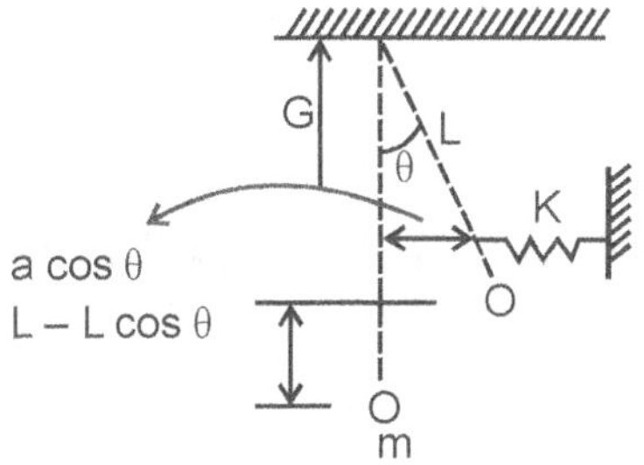

K.E., $T = \frac{1}{2} m l^2 \dot{\theta}^2$

P.E., $U = \frac{1}{2} k (a \sin\theta)^2 + mgl[1 - \cos\theta]$

(Assume θ as small then sin θ = θ, cosθ = 1)

Using lagrangian

$$ml^2\ddot{\theta} + ka^2\theta + mgl\theta = 0$$

∴ Natural frequency.

$$\omega = \sqrt{\frac{ka^2 + mgl}{ml^2}}$$

43. We know that $\frac{C}{C_C} = \xi = 0.3$

$$C = 0.3 \times 2\sqrt{km}$$

$$= 0.6 \times \sqrt{10{,}000 \times 40}$$

= 379.473 Ns/m

Damped natural frequency $\omega_d = \omega_n \times \sqrt{1 - \xi^2}$

$$= \sqrt{\frac{k}{m}} \times \sqrt{1 - \xi^2}$$

$$= \sqrt{\frac{10000}{40}} \times \sqrt{1 - 0.3^2}$$

$$= \sqrt{227.5} \text{ rad/sec}$$

= 15.083 rad/sec

44. If θ small

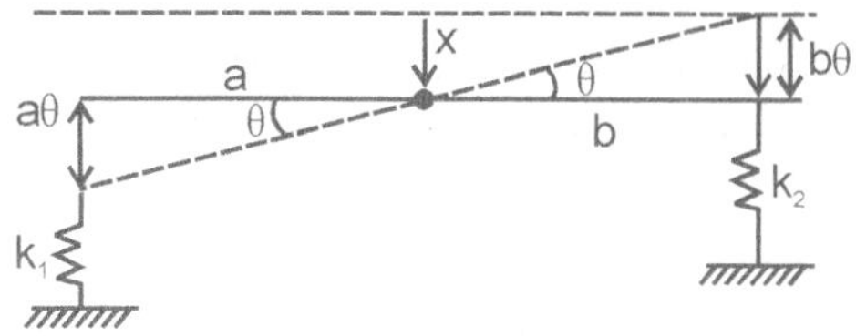

$$\frac{d}{dt}\left(\frac{\partial T}{\partial \dot{x}}\right) \text{ equal by - } \frac{\partial T}{\partial x} + \frac{\partial U}{\partial x} = 0$$

K.E. :

$$T = \frac{1}{2} m\dot{x}^2 + \frac{1}{2} I\dot{\theta}^2$$

P.E. :

$$U = \frac{1}{2} k_1 [x + a\theta]^2 + \frac{1}{2}(k_2)(x - b\theta)^2$$

∴ Using lagrangian on θ and x

$$x \to \quad m\ddot{x} + k_1x + k_1a + k_2x - k_2b\theta = 0$$

$$m\ddot{x} + (k_1 + k_2)x + (k_1a - k_2b)\theta = 0 \qquad \text{... (i)}$$

$$\theta \to I\ddot{\theta} + k_1ax + k_2a^2\theta - k_2bx + k_2b^2\theta = 0$$

$$I\ddot{\theta} + [k_1a - k_2b]x + [k_1a^2 + k_2b^2]\theta = 0 \qquad \text{... (ii)}$$

Therefore,

Mass matrix $M = \begin{bmatrix} m & 0 \\ 0 & I \end{bmatrix}$

Stiffness matrix $K = \begin{bmatrix} (k_1 + k_2) & (k_1a - k_2b) \\ (k_1a - k_2b) & (k_1a^2 + k_2b^2) \end{bmatrix}$

Putting value of m, I, k_1, k_2, a, b these matrices can be calculated.

45. a = 1m, b = 1m

$k_1 = k_2 = 10{,}000$

$$M = \begin{bmatrix} 30 & 0 \\ 0 & 20 \end{bmatrix}$$

$$K = \begin{bmatrix} 20{,}000 & 0 \\ 0 & 20{,}000 \end{bmatrix}$$

Now,

$$A = M^{-1}K = \begin{bmatrix} \frac{1}{30} & 0 \\ 0 & \frac{1}{20} \end{bmatrix} \begin{bmatrix} 20{,}000 & 0 \\ 0 & 20{,}000 \end{bmatrix}$$

$$= \begin{bmatrix} 0.033 & 0 \\ 0 & 0.05 \end{bmatrix} \begin{bmatrix} 20{,}000 & 0 \\ 0 & 20{,}000 \end{bmatrix}$$

$$A = \begin{bmatrix} 0.6667 & 0 \\ 0 & 1.0 \end{bmatrix} \times 10^3$$

For calculating natural frequencies solving eigen-value problem

$|A - \lambda I| = 0$

$$\Rightarrow \ = \left| \begin{bmatrix} 666.7-\lambda & 0 \\ 0 & 1000-\lambda \end{bmatrix} \right| = 0$$

$(666.7 - \lambda)(1000 - \lambda) - 0 = 0$

$666700 - 666.7\lambda - 1000\lambda + \lambda^2 = 0$

$\lambda^2 - 1666.7 + 666700 = 0$

Solving this equation

$\lambda_1 = 666.69$ $\qquad \lambda_2 = 1000.0000001$

Natural frequencies are :

$$\omega_1 = \sqrt{\lambda_1} = 25.82 \text{ rad/sec}$$

$$\omega_2 = \sqrt{\lambda_2} = 31.62 \text{ rad/sec}$$

In Hz, frequencies are

$$f_1 = \frac{\omega_1}{2\pi} \text{Hz}$$

$$f_2 = \frac{\omega_2}{2\pi} \text{Hz}$$

46. Torsional vibration of shaft :

The equation of motion :

Free vibration :

$$C^2 \frac{\partial^2 \theta}{\partial x^2}(x,t) = \frac{\partial \theta^2}{\partial t}(x,t)$$

$$C = \sqrt{\frac{GJ}{I_\theta}}$$

GJ = Torsional rigidity

I_θ = mass moment of inertia

Boundary conditions:

Fixed-free	Boundary	Frequency	Natural frequeies
	$\theta(0, t) = 0$	$\cos \frac{\omega l}{c} = 0$	$\omega_n = \frac{(2n+1)}{2l}\pi c$
	$\frac{\partial \theta}{\partial t}(l, t) = 0$		n = 0, 1, 2, ...
Free free 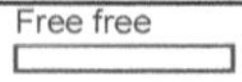	$\frac{\partial \theta}{\partial x}(0, t) = 0$	$\sin \frac{\omega l}{c} = 0$	$\omega_n = \frac{n\pi c}{l}$
	$\frac{\partial \theta}{\partial x}(l, t) = 0$		n = 0, 1, 2, ...
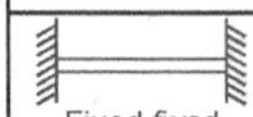 Fixed-fixed	$\theta(0, t) = 0$	$\sin \frac{\omega l}{c} = 0$	$\omega_n = \frac{n\pi c}{l}$
	$\theta(l, t) = 0$		n = 1, 2, 3 ....

47. Lateral vibration of beam :

Governing equation (free-vibration)

$$EI \frac{\partial^4 w}{\partial x^4}(x,t) = -\rho A \frac{\partial^2 w}{\partial t^2}$$

Boundary Condition	**Frequency equation**
Boundary Condition 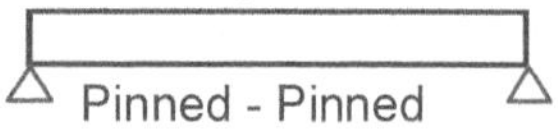 Pinned - Pinned	$\sin \beta_n l = 0 \rightarrow \beta_n l = n\pi$, n = 1, 2, 3, 4, ...
Free-free 	$\cos \beta_n l \cdot \cos \beta_n l = 1$ $\rightarrow \beta_1 l = 4.73$
Fixed-fixed 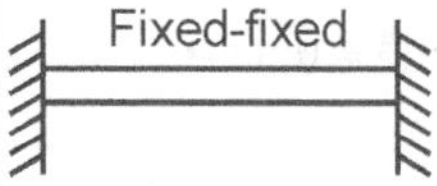	$\cos \beta_n l \cdot \cosh \beta_n l = 1$ $\rightarrow \beta_1 l = 4.73$
Cantilever 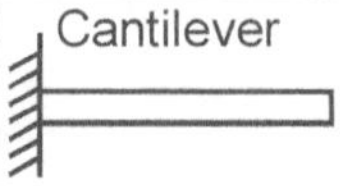 Free Fixed	$\cos \beta nl \cdot \cosh \beta_n l = -1$ $\rightarrow \beta_1 l = 1.875$

6
CHAPTER

JET PROPULSION

1. At high speed and high altitudes which engine has better specific fues consumption:
   (*a*) Turbojet (*b*) Ramjet
   (*c*) Turbofan (*d*) Turboshaft
2. Which engine works nearly identical to otto-cycle?
   (*a*) Pistonprop (*b*) Ramjet
   (*c*) Pulse Jet (*d*) Turbofan
3. Choose the correct option for turbojet engine
   (*a*) for subsonic exhaust, the exhaust pressure will be greater than ambient pressure.
   (*b*) for supersonic exhaust, the exhaust pressure will away be equal to ambient pressure.
   (*c*) for subsonic exhaust, the exhaust pressure will vary from ambient pressure.
   (*d*) for supersonic exhaust, the exhaust pressure may vary from ambient pressure.
4. The working of Jet propulsion is based on:
   (*a*) Reaction principle
   (*b*) Impulse principle
   (*c*) Centrifugal principle
   (*d*) All of these
5. At the condition of maximum propulsive efficiency of turbojet engine:
   (*a*) Thrust developed is maximum.
   (*b*) Thrust developed is zero.
   (*c*) Thrust developed is proportional to twice of velocity ratio.
   (*d*) None of these.
6. The diffusion process in an ideal diffuser is achieved by:
   (*a*) Adiabatic process
   (*b*) Adiabatic and reversible process
   (*c*) Adiabatic and Isobaric process
   (*d*) Adiabatic and irreversible process.
7. For constant velocity ratio of turbojet engine and constant enthalpy rise across the combustion chamber, the thermal efficiency of engine will be greater at combustion efficiency:
   (*a*) 1 (*b*) 0.5
   (*c*) 0.25 (*d*) 0.85
8. The Ram efficiency of diffuser is defined as:
   (*a*) Ratio of isentropic pressure rise to actual pressure rise
   (*b*) Ratio of actual pressure rise to isentropic pressure rise.
   (*c*) Ratio of actual change in enthalpy to isentropic change in enthalpy
   (*d*) None of these
9. In an airbreathing engine turbine inlet temperature is 1100 K. The temperature of air enters to compressor is 300 K and pressure ratio is maintained to 4. What is the temperature of air leaving the turbine when the mechanical efficiency of turbine is 0.98. Value of specific heat at constant pressure is 1 kJ/kg, and ratio of specific heat is 1.33.
10. An aircraft flying at an altitude where pressure and temperature are 0.55 bar and 255 K respectively at mach 0.6 mass flow rate of air is 18.2 kg/s and fuel-air ratio is 0.05. For maximum propulsive efficiency the thrust and thrust power produced by the engine are respectively
    (*a*) 0 N, 0 kW (*b*) 12080 N, 120 kW
    (*c*) 12088 N, 125 kW (*d*) 11000 N, 120 kW
11. A jet propelled plane is flying at an altitude where ambient pressure is 0.55 bar. Pressure and temperature of air entering to nozzle is 1.7 bar 1100 K respectively. The air leaves the nozzle with a Speed, if its efficiency is 92%; value of C_p is $1.147\,\text{kJ}/\text{kg}\cdot\text{k}$ and γ is 1.33.
    (*a*) 785 m/s (*b*) 753 m/s
    (*c*) 783 m/s (*d*) 780 m/s
12. Which is the appropriate pressure rise diagram across a centrifugal compressor.

(*a*)
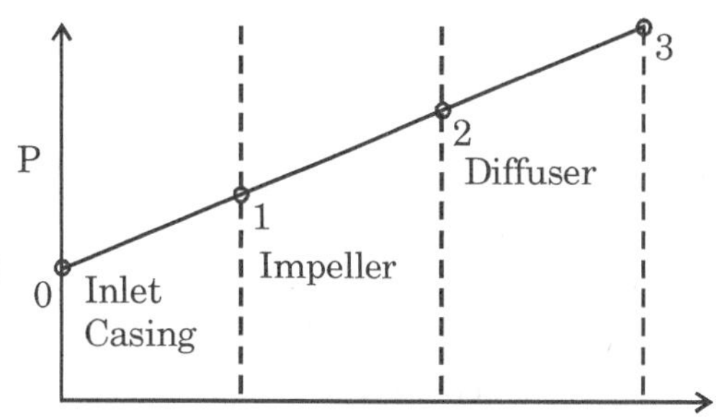

(*b*)

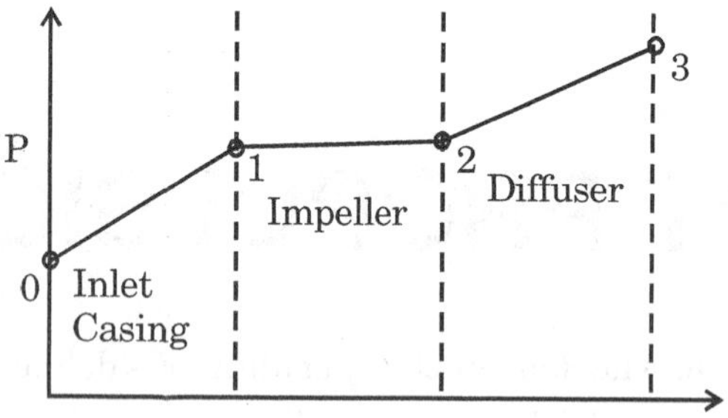

(*c*)

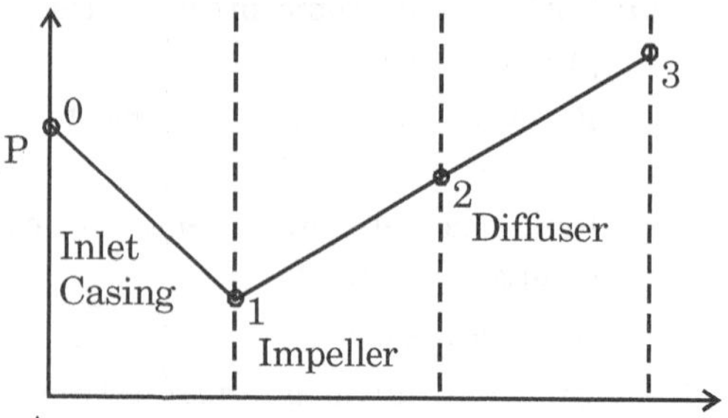

(*d*)

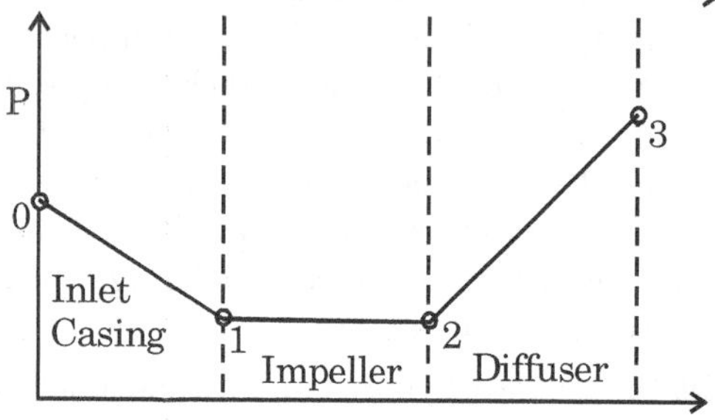

13. It is better to operate a centrifugal compressor with respect to mass flow rate

(*a*) Left side of the surge line.

(*b*) Right side of the surge line.

(*c*) Exctly on the surge line.

(*d*) Any side of surge line.

14. A centrifugal compressor running at 15000 rpm with impeller tip diameter is 600 mm. The temperature rise across the compressor if the power input factor is 1.04 and slip factor is 90%. Take c_p = 1005 J/s

(*a*) 107 K (*b*) 110 K

(*c*) 210 K (*d*) 207 K

15. The incidence losses in a centrifugal compressor with increase in mass flow rate is.

(*a*) Increases linearly

(*b*) Decrease linearly

(*c*) First increase and then decrease

(*d*) First decrease and then increase

16. Which type of impeller blade of centrifugal compressor is most exclusively used in turbojet engine.

(*a*) Forward-curved blades

(*b*) Radial-curved blades

(*c*) Backward-curved blades

(*d*) All of these

17. With increase in mass flow rate the pressure ratio across radial impeller blade of centrifugal compressor.

(*a*) Remains constant

(*b*) Increases linearly

(*c*) Decreases linearly

(*d*) Not depends on mass flow rate.

18. In a high reaction stage axial flow compressor

(*a*) The static pressure rise in diffuser is greater than rotor.

(*b*) The static pressure rise in diffuser is equal to rotor.

(*c*) The static pressure rise in diffuser is less than rotor.

(*d*) It depends on compressor inlet conditions.

19. The euler's energy equation for compressor rotor is

(*a*) Zero (*b*) Tends to infinity

(*c*) Always positive (*d*) Always negative

20. The work absorbing capacity of an axial flow compressor

(*a*) Increases with increase in axial velocity.

(*b*) Increases with increase in blade speed.

(*c*) Both a and b.

(*d*) Neither a nor b.

21. For a 50% reaction axial-flow compressor, the inlet and exit blade angles are 45° and 15° respectively axial velocity is maintained constant to 200 m/s. The tangential force required to produce power of 300 kw, if the inlet pressure and temperature is 1 bar and 300 K.

(*a*) 1183 N (*b*) 3712 N

(*c*) 1085 N (*d*) 3568 N

22. The temperature rise across a axial flow compressor has a loading coefficient 0.88, flow coefficient is 0.9 and axial velocity is 200 m/s The ratio of specific heats is 1.4 and specific heat at constant pressure is 1005 J/kg K.

(*a*) 83.26 K (*b*) 80.26 K

(*c*) 40.24 K (*d*) 43.24 K

23. A multi stage axial compressor operating at an overall isentropic efficiency of 0.85 develops a total pressure ratio of 10. The stagnation temperature at inlet is 300 K and total enthalpy rise across each stage is 32 kJ/kg. No. of stages required are (take $\gamma = 1.4$ and $C_p = 1.005$ kJ/kg k)

(*a*) 9 (*b*) 11

(*c*) 13 (*d*) 7

24. An ideal ramjet engine flying at mach 1.5 at an attitude where ambient temperature is 300 K. The exhaust gases are optimally expanded and leaves the engine at 1100 K. For constant ratio of specific heats what is the value of specific thrust produced by the engine. (R = 287 J/kg K, $\gamma = 1.4$)

(*a*) $476 \dfrac{\text{N-s}}{\text{kg}}$ (*b*) $576 \dfrac{\text{N-s}}{\text{kg}}$

(*c*) $676 \dfrac{\text{N-s}}{\text{kg}}$ (*d*) $376 \dfrac{\text{N-s}}{\text{kg}}$

25. Which of these options are correct for the free vortex design.

(*a*) Stagnation enthalpy remains constant

(*b*) $c_t \times r$ remains constant

(*c*) Entropy remains constant

(*d*) All of the above

26. The blade loading and flow coefficeints of a turbine stage is related as

(*a*) $\phi = \psi(\tan\beta_2 + \tan\beta_3)$ (*b*) $\phi = \psi(\tan\beta_2 - \tan\beta_3)$

(*c*) $\psi = \phi(\tan\beta_2 + \tan\beta_3)$ (*d*) $\psi = \phi(\tan\beta_2 - \tan\beta_3)$

27. Maximum utilization factor of single impulse stage turbine is

(*a*) $\dfrac{1}{2}\sin\alpha_2$ (*b*) $\sin^2\alpha_2$

(*c*) $\dfrac{1}{2}\sin^2\alpha_2$ (*d*) $\sin\alpha_2$

28. In a three-stage velocity compounded impulse turbine the nozzle discharge the fluid on the blade at an angle of $70°$ to the axial direction. The optimum value of blade-to-gas speed ratio of the turbine is.

(*a*) 0.17 (*b*) 0.18

(*c*) 0.16 (*d*) 0.2

29. A single-stage impulse turbine having equal inlet and outlet blades angles has the following data.

Nozzle efflux angle = 65°, Blade speed = 150 m/s

Absolute velocity of fluid leaving the rotor = 300 m/s

Stage exit angle $= 30°$

The blade angles is:

(*a*) $28°$ (*b*) $38°$

(*c*) $48°$ (*d*) $58°$

30. Expansion of the gases in an impulse turbine stage occurs in

(*a*) Rotor blades

(*b*) Stator blades

(*c*) Both rotor and stator blades

(*d*) None of the above

31. Degree of reaction in a reaction turbine stage is given by

(*a*) $R = \dfrac{C_a}{2u}(\tan\beta_3 - \tan\beta_2)$

(*b*) $R = \dfrac{C_a}{2u}(\tan\beta_3 + \tan\beta_2)$

(*c*) $R = \dfrac{C_a}{2u}(\tan\alpha_3 - \tan\alpha_2)$

(*d*) $R = \dfrac{C_a}{2u}(\tan\alpha_3 + \tan\alpha_2)$

32. Isentropic efficiency and polytropic efficiency of an axial comprssor stage are related as:

(*a*) $\eta_{is} = \dfrac{\Pi_c^{\gamma-1/\gamma}}{\Pi_c^{\left(\frac{\gamma-1}{\gamma}\times\frac{1}{\eta_p}\right)} - 1}$ (*b*) $\eta_{is} = \dfrac{\Pi_c^{\left(\frac{\gamma-1}{\gamma}\times\frac{1}{\eta_p}\right)} - 1}{\Pi_c^{\frac{\gamma-1}{\gamma}} - 1}$

(*c*) $\eta_{is} = \dfrac{\Pi_c^{\left(\frac{\gamma-1}{\gamma}\times\frac{1}{\eta_p}\right)} + 1}{\Pi_c^{\frac{\gamma-1}{\gamma}} + 1}$ (*d*) $\eta_{is} = \dfrac{\Pi_c^{\frac{\gamma-1}{\gamma}} - 1}{\Pi_c^{\left(\frac{\gamma-1}{\gamma}\times\frac{1}{\eta_p}\right)} - 1}$

where, η_p is polytropic efficiency, η_{is} is isentropic efficiency and Π_c is compressor pressure ratio.

33. For constant polytropic efficiency of axial flow compressor, The isentropic efficiency ________ with increase of compressor pressure ratio.

(*a*) Remains constant

(*b*) Increases linearly

(*c*) Decreases linearly

(*d*) First decreases than increases linearly

34. The stagnation temperature at compressor inlet is increased by two times and stagnation pressure is reduced by 0.5 times of its value. In order to maintain the same pressure ratio across the compressor, the mass flow rate will be reduced by the factor

(*a*) 0.4 (*b*) 0.35
(*c*) 0.30 (*d*) 0.45

35. The variation of flow area with variation of flow velocity in a compressible flow for subsonic operation is

(*a*) $\frac{dV}{dA} > 0$ (*b*) $\frac{dV}{dA} < 0$
(*c*) $\frac{dV}{dA} = 0$ (*d*) $\frac{dV}{dA} = 1$

36. Combustion efficiency of an aircraft engine

(*a*) remains constant with altitude
(*b*) increases with altitude
(*c*) decreases with altitude
(*d*) first increaes then decreases

37. If the degree of reaction at mid-radius of compressor is 50% and hub-to-tip ratio is 0.8, then the degree of reaction of the compressor at blade tip is:

(*a*) 0.32 (*b*) 0.38
(*c*) 0.62 (*d*) 0.68

38. Air of mass flow rate 80 kg/s enters the combustor of a gas turbine engine at total temperature of 400 K and leaves the combustor at 1300 K. If specific heat at compressor pressure of air and combustion product is 1005 J/kg-K and 1147 J/kg-K, respectively and heating value of fuel is 44 MJ/kg then mass flow rate of fuel (in kg/s) is:

(*a*) 2.05 kg/s (*b*) 2.55 kg/s
(*c*) 3.05 kg/s (*d*) 3.55 kg/s

39. For turbine, which among these is correct. Where, η_p = polytropic efficiency and η_{iso} = isentropic efficiency.

(*a*) $\eta_p > \eta_{ise}$ (*b*) $\eta_p = \eta_{ise}$
(*c*) $\eta_p < \eta_{ise}$ (*d*) None of these

40. A turbofan engine with a bypass ratio of 3.0; the mass flow rate of air going to the cold nozzle, if the total mass flow rate of air is 200 kg/s.

(*a*) 100 kg/s (*b*) 150 kg/s
(*c*) 175 kg/s (*d*) 200 kg/s

ANSWER KEY

1. (*b*)	**2.** (*c*)	**3.** (*d*)	**4.** (*a*)	**5.** (*b*)	**6.** (*b*)	**7.** (*a*)	**8.** (*b*)	**10.** (*a*)	**11.** (*b*)
12. (*c*)	**13.** (*b*)	**14.** (*d*)	**15.** (*d*)	**16.** (*b*)	**17.** (*a*)	**18.** (*c*)	**19.** (*d*)	**20.** (*a*)	**21.** (*a*)
22. (*d*)	**23.** (*b*)	**24.** (*a*)	**25.** (*d*)	**26.** (*c*)	**27.** (*b*)	**28.** (*c*)	**29.** (*d*)	**30.** (*a*)	**31.** (*a*)
32. (*d*)	**33.** (*c*)	**34.** (*b*)	**35.** (*b*)	**36.** (*c*)	**37.** (*d*)	**38.** (*a*)	**39.** (*c*)	**40.** (*b*)	

EXPLANATIONS

2. In pulse-jet engine, combustion process occurs at constant-volume process while in other propulsive device it occurs at constant pressure process.

3. Thrust,

$$T = \dot{m}_a\left[(1+f)V_e - V_i\right] + \left(P_j - P_a\right)Aj - \left(P_i - P_a\right)A_i$$

For subsonic exhaust, the exhaust pressure and inlet pressure is almost equal to ambient pressure and pressure thrust is very small.

For supersonic exhaust, the exhaust pressure may vary from the ambient pressure and there will be contribution of pressure thrust also.

5. Propulsive efficiency $(\eta_p) = \frac{2\alpha}{1+\alpha}$

Where α is velocity ratio

So, for $\alpha = 1$, η_p is 100%

$\alpha = 1$ means $V_e = V_i$

For small fuel air ratio and negligible pressure thrust

$$T = \dot{m}_a\left(V_e - V_i\right)$$

$$T = 0$$

6. The diffusion process in an ideal diffuser is isentropic i.e. adiabatic and reversible.

7. $\eta_{th} = \dfrac{\text{propulsive power (P)}}{\dot{m}_f \times CV}$

for $f \ll 1, p = \dfrac{\dot{m}_a}{2}\left(V_e^2 - V_i^2\right)$ and efficiency of combussion $\left(\eta_{comb}\right) = \dfrac{h_{03} - h_{02}}{f \times CV}$

$$\eta_{th} = \frac{\dot{m}_a\left(V_e^2 - V_i^2\right)}{2\dot{m}_f \times CV}$$

$$\eta_{th} = \frac{\left(V_e^2 - V_i^2\right)}{2 \times CV}$$

$$\eta_{th} = \frac{\eta_{comb} \times \left(V_c^2 - V_i^2\right)}{2\left(h_{o3} - h_{o2}\right)}$$

So, $\eta_{th} \propto \eta_{comb}$

8. $\eta_{ram} = \dfrac{P_{01} - P_a}{P_{01}' - P_a} = \dfrac{\text{Actual pressure rise}}{\text{Isentropic pressure rise}}$

9. 974.328 K

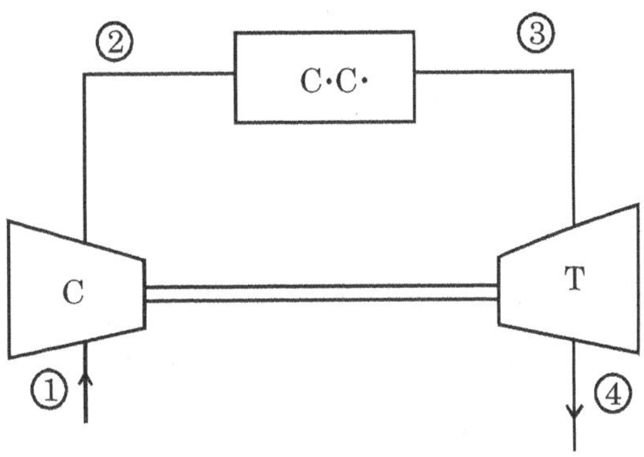

$$\frac{T_1}{P_1^{\frac{\gamma-1}{\gamma}}} = \frac{T_2}{P_2^{\frac{\gamma-1}{\gamma}}}$$

$$T_2 = 300 \times (4)^{\frac{0.33}{1.33}} = 423.159\,\text{K}$$

$$W_c = W_T \times \eta_m$$

$$C_p\left(T_2 - T_1\right) = C_p\left(T_3 - T_4\right) \times \eta_m$$

$$(423.159 - 300) = \left(1100 - T_4\right) \times 0.98$$

$$T_4 = 974.328\,\text{K}$$

10. For maximum propulsive efficiency

$\alpha = 1$ i.e, $V_e = V_i$

Hence thrust produced and thrust power will be zero.

11.

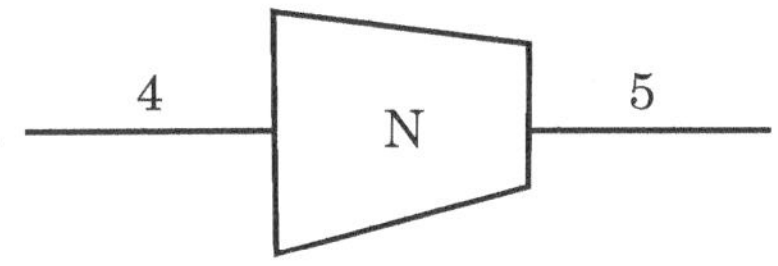

Checking the choking condition of Nozzle.

$$\frac{P_{04}}{P_c} = \frac{1}{\left[1 - \dfrac{1}{\eta_{nozzle}}\left(\dfrac{\gamma_{g-1}}{\gamma_{g+1}}\right)\right]^{\frac{\gamma_g}{\gamma_g - 1}}} = 3.55$$

$$P_c = 0.479\,\text{bar}$$

Since $P_c < P_a \rightarrow$ Nozzle will not choke and taking the exhaust pressure as ambient pressure.

$$T_5' = T_{04}\left(\frac{P_5}{P_{04}}\right)^{\frac{\gamma-1}{\gamma}}$$

$$= 1100 \times \left(\frac{0.55}{1.7}\right)^{\frac{1.33-1}{1.33}}$$

$$= 831.366\,\text{K}$$

$$\eta_{nozz} = \frac{T_{04} - T_5}{T_{04} - T_5'}$$

$$0.92 = \frac{1100 - T_5}{1100 - 831.366} \Rightarrow T_5 = 852.857\,\text{K}$$

$$V_e = \sqrt{2C_p\left(T_{04} - T_5\right)} = \sqrt{2 \times 1147 \times (1100 - 852.857)}$$

$$V_c = 752.958\,\text{m/s}$$

12. The inlet casing with nozzle accelerates the flow to the impeller eye so static pressure reduces here. In impeller, energy transfer takes place which increases the static pressure and the purpose of the diffuser is to convert the high velocity of the air leaving the impeller into pressure.

13.

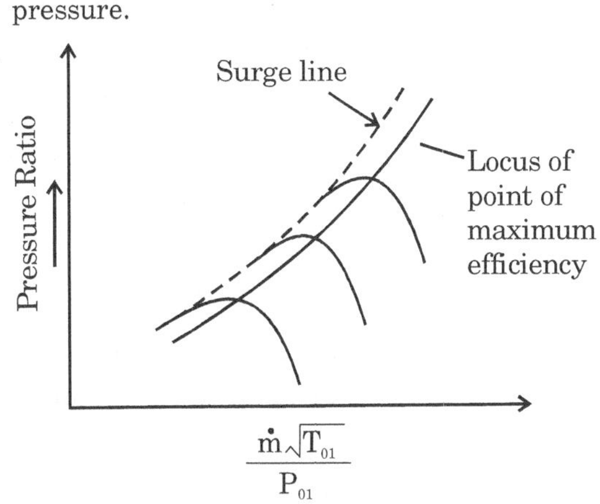

14. $u = \frac{\pi DN}{60} = \frac{\pi \times 0.6 \times 15000}{60} = 471.24\ m/s$

$W = P_{if}\mu u^2$

Where p_{if} is power input factor.

μ is slip factor

$W = 1.04 \times 0.9 \times (471.24)^2$

$C_p\,\Delta T = 207854.8408\ J/kg$

$\Delta T = \frac{207854.8408}{1005} = 206.82K$

15.

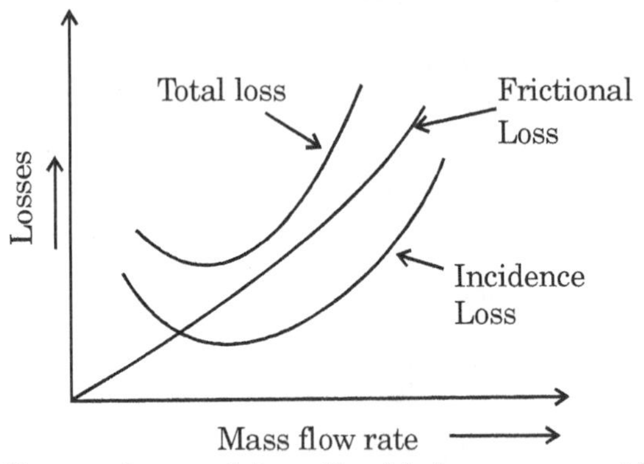

16. Forward-curved impeller blades can provide highest pressure ratio but it is less stable and has a narrow operating range. Also its efficiency is lower than backward and radial blades.

Generally curved blades produce bending moment which reduce the maximum speed of impeller at which it can run. Good performance can be obtained by radial blades.

17.

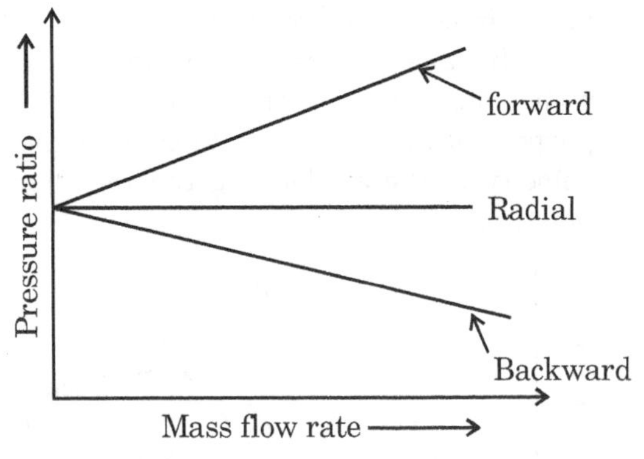

18. $R = \frac{\text{Static pressure rise in rotor}}{\text{Static pressure rise in diffuse}}$

For High Reaction stage R > 1 so, (Static pressure rise in rotor) > (static pressure rise in diffuser)

19. Euler energy equation:

$E = C_{t1}u_1 - C_{t2}u_2$

For compressor $C_{t2}u_2 > C_{t1}u_1$ it means energy is transferred from rotor to fluid in compressor.

20. $W = \Omega u C_a(\tan\beta_1 - \tan\beta_2)$

When, Ω = W.D. factor

Ca = Axial velocity

u = Blade Speed

So, $W \alpha Ca$

21. for 50% Reaction stage: $\alpha_1 = \beta_2 = 15°$

$\beta_1 = \alpha_2 = 45°$

$R = \frac{Ca}{2u}(\tan\beta_1 + \tan\beta_2)$

$\frac{1}{2} = \frac{Ca}{2u}(\tan\beta_1 + \tan\beta_2)$

$\frac{1}{2} = \frac{200}{2 \times u}(\tan 45° + \tan 15°)$

$u = 253.589\ m/s$

$P = \dot{m}w = \dot{m}u(C_{t2} - C_{t1})$

$P = \dot{m}uCa(\tan\alpha_2 - \tan\alpha_1)$

$\dot{m} = \frac{300 \times 10^3}{253.589 \times 200 \times (\tan 45° - \tan 15°)}$

$\dot{m} = 8.08\ kg/s$

Tangential force, $F = \dot{m}(C_{t2} - C_{t1})$

$= \dot{m}C_a(\tan\alpha_2 - \tan\alpha_1)$

$= 8.08 \times 200 \times (\tan 45° - \tan 15°)$

$= 1182.99N$

22. Flow coefficient $(\phi) = \frac{C_a}{u} = 0.9$

$u = \frac{200}{0.9} = 222.22\ m/s$

Loading coefficient $(\Psi) = \frac{w}{u^2} = 0.88$

$$w = 0.88 \times (222.22)^2$$

$$= 43455.921\,J/kg$$

$$\Delta Ts = \frac{W}{C_p} = \frac{43455.921}{1005} = 43.24\,K$$

23. $$\eta_{ison} = \frac{h_{02}' - h_{01}}{h_{o2} - h_{o1}}$$

$$= \frac{C_p\left(T_{02}' - T_{01}\right)}{\left(h_{02} - h_{01}\right)_{overall}}$$

$$\left(h_{02} - h_{01}\right)_{overall} = \frac{C_p T_{01}\left(r_{po}^{\frac{\gamma-1}{\gamma}} - 1\right)}{\eta_{ise}}$$

$$\text{No. of stage} = \frac{\left(h_{02} - h_{01}\right)_{overall}}{\left(h_{02} - h_{01}\right)_{stage}}$$

$$= \frac{C_p T_{01}\left(r_{Po}^{\frac{\gamma-1}{\gamma}} - 1\right)}{32 \times 10^3} = \frac{1005 \times 300\left(\frac{0.4}{1.4}\atop 10 - 1\right)}{32 \times 10^3}$$

$$= 10.312 \approx 11 \text{ stage}$$

24. For ideal ramjet engine, $m_i = m_j$

where m_i is inlet Mach Number

m_j is exhaust Mach Number

$$m_i = \frac{v_i}{\sqrt{\gamma R T_\infty}} \Rightarrow v_i = 1.5 \times \sqrt{1.4 \times 287 \times 300}$$

$$v_i = 520.783 \text{ m/s}$$

$$m_i = m_j$$

$$\frac{v_i}{\sqrt{rRT_i}} = \frac{v_e}{\sqrt{rRT_e}} \Rightarrow v_e = v_i\sqrt{\frac{T_e}{T_i}}$$

$$= 520.783 \times \sqrt{\frac{1100}{300}}$$

$$= 997.224 \text{ m/s}$$

$$\text{specific thrust} = \frac{T}{\dot{m}_a}$$

$$\frac{T}{\dot{m}_a} = V_e - V_i = 997.224 - 520.783$$

$$= 476.441\,N - S/kg$$

26.

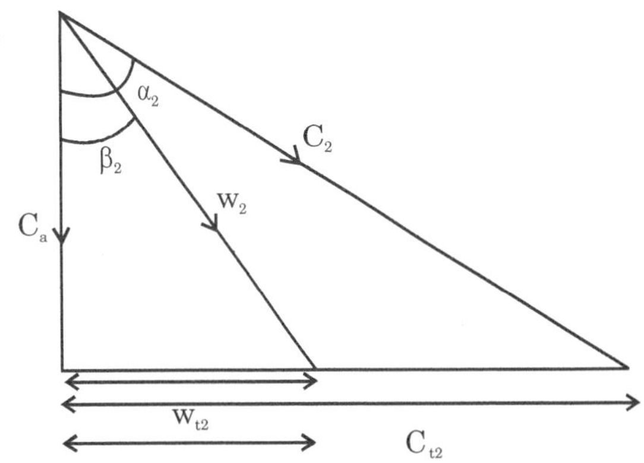

Rotor Inlet

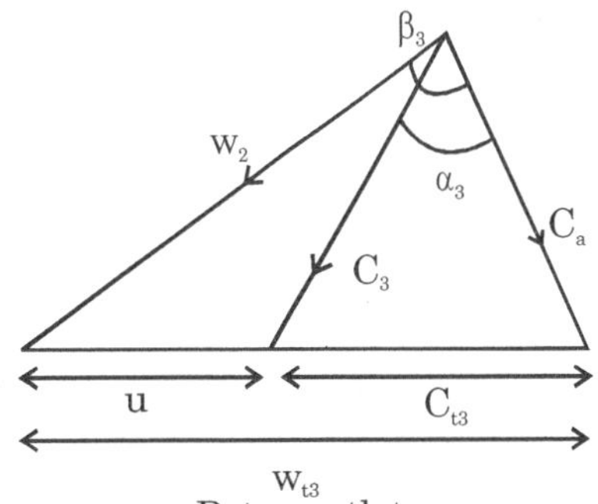

Rotor outlet

work done (w)

$$w = u\left(C_{t2} + C_{t3}\right)$$

$$= u^2\left(\frac{C_{t2}}{u} + \frac{C_{t3}}{u}\right)$$

$$\frac{w}{u^2} = \frac{1}{u}(C_{t2} + C_{t3})$$

$$\frac{w}{u^2} = \frac{C_a}{u}(\tan\alpha_2 + \tan\alpha_3)$$

$$\Psi = \phi\left(\tan\alpha_2 + \tan\alpha_3\right)$$

Also,

$$\tan\alpha_2 + \tan\alpha_3 = \tan\beta_2 + \tan\beta_3$$

$$\Psi = \phi\left(\tan\beta_2 + \tan\beta_3\right)$$

27. visization factor $(\in) = 4\left(\sigma\sin\alpha_2 - \sigma^2\right)$

$$\sigma_{opt} = \frac{u}{C_2} = \frac{1}{2}\sin\alpha_2$$

$$\in_{opt} = 4\left(\frac{\sin^2\alpha_2}{2} - \frac{\sin^2\alpha_2}{4}\right)$$

$$\in_{opt} = \sin^2\alpha_2$$

28. For 'n' velocity stages

$$\sigma_{opt} = \frac{1}{2n}\sin\alpha_2$$

for three-stage, n = 3

$$\sigma_{opt} = \frac{1}{6}\sin\alpha_2$$

$$= \frac{\sin 70°}{6} = 0.16$$

29.

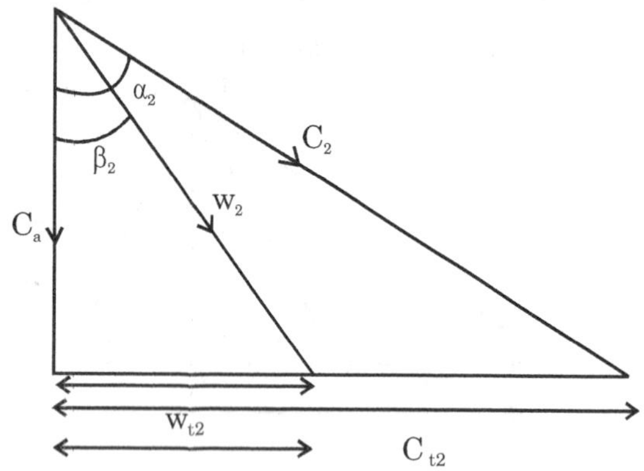

Rotor Inlet

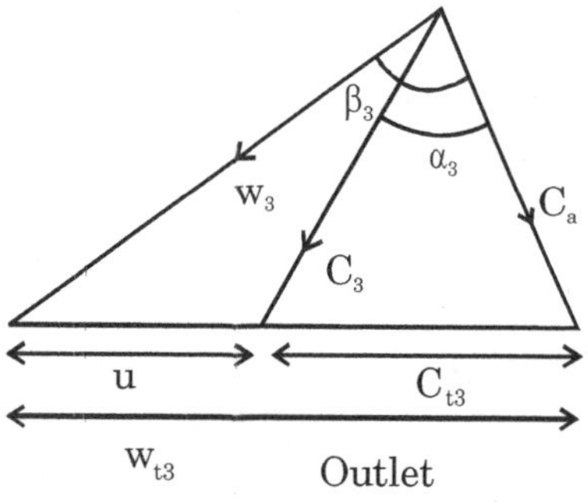

Outlet

$\alpha_2 = 65°, C_3 = 300$ m/s, $\alpha_3 = 30°$

$$\text{Cos}\,\alpha_3 = \frac{C_a}{C_3} \Rightarrow C_a = C_3\,\text{Cos}\,\alpha_3$$

$$= 300 \times \cos 30°$$

$$= 259.808 \text{ m/s}$$

$$W_{t3} = u + C_{t3} = 150 + 259.808$$

$$= 409.808 \text{ m/s}$$

$$\tan\beta_3 = \frac{W_{t3}}{C_a} \Rightarrow \beta_3 = \tan^{-1}\left(\frac{W_{t3}}{C_a}\right)$$

$$= \tan^{-1}\left(\frac{409.808}{259.808}\right)$$

$$\beta_2 = 57.6°$$

32. for compression process, $\frac{n-1}{n} = \frac{\gamma-1}{\gamma} \times \frac{1}{\eta_p}$

$$\eta_{isc} = \frac{W_{ideal}}{W_{actual}} = \frac{h'_{02} - h_{01}}{h_{02} - h_{01}}$$

$$= \frac{C_p\left(T_{02}' - T_{01}\right)}{C_p\left(T_{02} - T_{01}\right)}$$

$$= \frac{\frac{T_{02}'}{T_{01}} - 1}{\frac{T_{02}}{T_{01}} - 1} = \frac{r_p^{\frac{\gamma-1}{\gamma}} - 1}{r_p^{\frac{n-1}{n}} - 1}$$

Writing π_c instead of r_p

$$\eta_{isen} = \frac{\pi_c^{\frac{\gamma-1}{\gamma}} - 1}{\pi_c^{\frac{\gamma-1}{\gamma}\times\frac{1}{\eta_p}} - 1}$$

33.

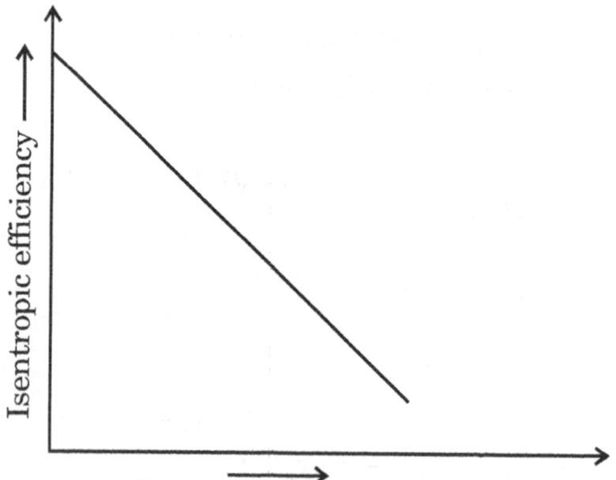

34. The dimensionless parameter for Compressor

$$= \frac{\dot{m}\sqrt{T_{01}}}{p_o}$$

$\dot{m}_f$ = final mass

$\dot{m}_i$ = Initial mass

$$\dot{m}_i \frac{\sqrt{T_{01}}}{p_0} = \frac{\dot{m}_f \times \sqrt{2T_{01}}}{0.5 p_0}$$

$$\dot{m}_f = \dot{m}_i \times \frac{0.5}{\sqrt{2}} = 0.35 \dot{m}_i$$

35. The Area-velocity relationship

$$\frac{dA}{A} = \frac{dV}{V}\left(M^2 - 1\right)$$

for subsonic operation, $M < 1$

$$dA = -dV$$

$$\frac{dV}{dA} = \text{negative term}$$

$$\frac{dV}{dA} < 0$$

36. With increase in altitude, Combustion efficiency of an aircraft engine decreases because pressure reduces with altitude and hence rate of reaction of burning component also reduces, which increase the reaction time.

37. In compressor, the degree of reaction varies with the radius as

$$R = 1 - \frac{K}{r^2}$$

or

$$R = 1 - \frac{k}{\left(r_h / r_{tip}\right)^2}$$

$$0.5 = 1 - \frac{k}{(0.8)^2}$$

$$k = 0.5 \times (0.8)^2 = 0.32$$

at tip, $\frac{r}{r_{tip}} = 1$

$$R = 1 - \frac{0.32}{1} = 0.68$$

38. 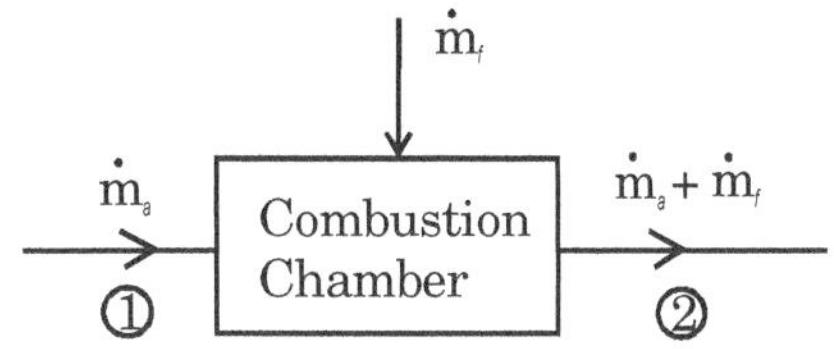

Applying S.F. E. E.

$$\dot{m}_a h_{01} + Q = \left(\dot{m}_a + \dot{m}_f\right) h_{02}$$

$$\dot{m}_a C_{Pa} T_{01} + \dot{m}_f \times HV = \left(\dot{m}_a + \dot{m}_f\right) C_{pg} T_{02}$$

$$\dot{m}_a C_{Pa} T_{01} - \dot{m}_a c_{pg} T_{02} = \dot{m}_f \left(C_{pg} T_{02} HV\right)$$

$$\dot{m}_f = \frac{\dot{m}_a \left(C_{pa} T_{01} - C_{pg} T_{02}\right)}{C_{pg} T_{02} - HV}$$

$$= \frac{80 \times (1005 \times 400 - 1147 \times 1300)}{1147 \times 1300 - 44 \times 10^6}$$

$$= 2.0496 \text{ kg/s}$$

40. $\dot{m}_a = \dot{m}_c + \dot{m}_h = 200$

where, $\dot{m}_c$ is mass flow rate of air through cold nozzle

$\dot{m}_n$ is mass flow rate of air through hot nozzle

$$BPR = \frac{\dot{m}_c}{\dot{m}_h} = 3$$

$$\dot{m}_h = \frac{\dot{m}_c}{3}$$

$$\dot{m}_c + \frac{\dot{m}_c}{3} = 200$$

$$\frac{4\dot{m}_c}{3} = 200$$

$$\dot{m}_c = \frac{\overset{50}{\cancel{200}} \times 3}{\cancel{4}}$$

$$\dot{m}_c = 150 \text{ kg/s}$$

7 CHAPTER ROCKET PROPULSION

1. In a liquid rocket engine, the specific impulse I_{sp} depends on the molecular weight of the propellant $\overline{M}$ as

   (a) $\frac{1}{\overline{M}_e^2}$ (b) $\frac{1}{\overline{M}_e^{\frac{1}{2}}}$

   (c) $\frac{1}{\overline{M}_e}$ (d) $\overline{M}_e$

2. Consider a two-stage rocket with the following design characteristics

   First stage: propellant mass = 8200 kg

   Structural mass = 800 Kg

   Second stage:

   Propellant mass = 5400 Kg

   Structural mass = 500 Kg

   Payload mass = 80 Kg

   If the specific impulse for both the stages is 275 sec, then final burnout velocity is (g = 10 m/s)

   (a) 6416 m/s (b) 7590 m/s

   (c) 8596 m/s (d) 6580 m/s

Common data for Questions 3, 4, 5 & 6

Consider a rocket engine burning hydrogen and oxygen; having the following specification combustion chamber pressure and temperature are 25 atm and 3500 K respectively,

Area of Nozzle throat = 0.2 m². g = 10 m/s².

γ = 1.22, molecular wt. of propellant = 16

The area of nozzle exit is designed in such a way that the exit pressure exactly equals ambient pressure at an altitude of 30000 m.

At this altitude where ambient pressure is 0.01174 atm

3. The value of specific impulse is

   (a) 3887 sec (b) 388.7 sec

   (c) 4887 sec (d) 488.7 sec

4. Specific thrust produced by the rocket is,

   (a) $3887 \frac{N-s}{kg}$ (b) $4887 \frac{N-s}{kg}$

   (c) $3587 \frac{N-s}{kg}$ (d) $4312 \frac{N-s}{kg}$

5. Flow mach number as nozzle exit is

   (a) 3.2 (b) 4.2

   (c) 5.2 (d) 6.2

6. Area of the nozzle exit is

   (a) 30 m² (b) 50 m²

   (c) 17 m² (d) 24 m²

7. In a given rocket engine, mass flow rate of propellant is equals to 40 kg/s. The temperature of the combustion chamber is 3000 K. The values of R and γ for combustion mixture is 400 J/kg K and 1.21 respectively. If the throat area is 0.047 m², then pressure in the combustion chamber is (in KN/m²) ________.

8. The correct relation between thrust coefficient (C_f) and characteristics velocity (C*) of a rocket engine is:

   (a) $C_f = \frac{I_{sp} \times g}{V_e \times C^*}$ (b) $C_f = \frac{I_{sp} \times g}{C^*}$

   (c) $C_f = \frac{V_e \times I_{sp}}{C^*}$ (d) $C_f = \frac{C^* \times I_{sp}}{V_e}$

9. Consider a single stage rocket of propellant mass m_p, structural mass m_s and payload mass is m_L. If σ and λ represents the structural ratio and payload ratio respectively, then equation of burn-out velocity is given by

   (a) $\Delta v_b = v_e \ell n\left[\frac{1}{\sigma + \lambda(1-\sigma)}\right]$

   (b) $\Delta v_b = v_e \ell n[\sigma + \lambda(1-\sigma)]$

   (c) $\Delta v_b = v_e \ell n[\lambda + \sigma(1-\sigma)]$

   (d) $\Delta v_b = v_e \ell n\left[\frac{1}{\lambda + \sigma(1-\lambda)}\right]$

10. A rocket engine is flying at a speed of 1000 m/s having equivalent exhaust velocity of 1900 m/s. What will be the velocity of rocket at a time when its weight becomes the half of the initial weight of rocket.

    (a) 1317 m/s (b) 2317 m/s

    (c) 693 m/s (d) 893 m/s

11. A rocket carrying satellite of initial mass 3000 kg produce specific impulse of 300s. If rocket is fired to give a speed increment of 500 m/s along the direction of motion, then the mass of propellant consumed is ________ (in kg). $g = 10\ m/s^2$.

12. The thrust produced by a rocket motor is dependent of time and given as

$T = 10000\ t^2 - 20t + 1$

If rocket burnt 300 kg of propellant in 3 sec, then the specific impulse of the rocket is : (Take $g = 10\ m/s^2$)

(*a*) 80 sec (*b*) 50 sec

(*c*) 30 sec (*d*) 10 sec

13. Among the following which engine has the highest specific impulse,

(*a*) Solid rocket (*b*) Ramjet

(*c*) Liquid rocket (*d*) Scramjet

Common data for Question 14 and 15:

A liquid rocket engine has a characteristic velocity of 2380 m/s and thrust coefficient of 1.4. If the oxidizer to fuel ratio of 5 : 1 and mass flow rate of propellant is 300 kg/s, then

14. Specific impulse of the engine is (if $g = 9.8\ ms/^2$).

(*a*) 340 sec (*b*) 250 sec

(*c*) 333 sec (*d*) 283 sec

15. Thrust produced by the engine among the following.

(*a*) 2000 kN (*b*) 2020 kN

(*c*) 1000 kN (*d*) 1020 kN

ANSWER KEY

1. (*b*)	**2.** (*c*)	**3.** (*b*)	**4.** (*a*)	**5.** (*c*)	**6.** (*d*)	**8.** (*b*)	**9.** (*a*)	**10.** (*b*)	**12.** (*c*)
13. (*d*)	**14.** (*a*)	**15.** (*c*)							

EXPLANATIONS

1. $I_{SP} = \frac{1}{g}\sqrt{\frac{2\gamma RT_o}{(\gamma-1)}\left[1-\left(\frac{P_e}{P_o}\right)^{\frac{(\gamma-1)}{\gamma}}\right]}, R = \frac{\overline{R}}{M}$

$I_{SP} = \frac{1}{g}\sqrt{\frac{2\gamma \overline{R} T_o}{(\gamma-1)M}\left[1-\left(\frac{P_e}{P_o}\right)^{\frac{(\gamma-1)}{\gamma}}\right]}$

So, $I_{SP} \propto \frac{1}{\sqrt{M_e}}$ or $\frac{1}{M_e^{\frac{1}{2}}}$

2. For first stage

Initial mass,

$m_i = m_L + m_{p1} + m_{s1} + m_{p2} + m_{s2}$

$= 80 + 8200 + 800 + 5400 + 500$

$= 14980$ kg

$m_f = m_L + m_{s1} + m_{p2} + m_{s2}$

$= 80 + 800 + 5400 + 500 = 6780$ kg

so, $(\Delta v_b)_1 = gI_{sp}\ell n\frac{m_i}{m_f}$

$= 10 \times 275 \times \ell n\left(\frac{14980}{6780}\right)$

$= 2180.032$ m/s

for second stage:

Initial mass, $m_i = m_L + m_{p2} + m_{s2}$

$= 80 + 5400 + 500$

$= 5980$ kg

Final mass, $m_f = m_L + m_{s2}$

$= 80 + 500$

$= 580$ Kg.

$(\Delta v_b)_2 = 10 \times 275 \times \ell n\frac{5980}{580}$

$= 6416.156$ m/s

Final burnout velocity $= (\Delta v_b)_1 + (\Delta v_b)_2$

$= 2180.032 + 6416.156$

$= 8596.188$ m/s

3. $R = \frac{\overline{R}}{M} = \frac{8314}{16} = 519.625\ J/kgK$

$I_{sp} = \frac{1}{g}\sqrt{\frac{2\gamma RT_o}{\gamma-1}\left[1-\left(\frac{Pe}{P_o}\right)^{\frac{(\gamma-1)}{\gamma}}\right]}$

$= \frac{1}{10}\sqrt{\frac{2\times1.22\times519.625\times3500}{(1.22-1)}\left[1-\left(\frac{0.01174}{25}\right)^{\frac{0.22}{1.22}}\right]}$

$= 388.67$ S

4. Thrust, $T = \dot{m}_p(V_e) + (P_e - P_a)A_e$

Given, $P_e = P_a$.

$T = \dot{m}_p V_e$

specific thrust, $\frac{T}{\dot{m}_p} = V_e$

$$\frac{T_o}{P_o^{\frac{\gamma-1}{\gamma}}} = \frac{T_e}{P_e^{\frac{\gamma-1}{\gamma}}}$$

$$T_e = T_o \times \left(\frac{P_e}{P_o}\right)^{\frac{\gamma-1}{\gamma}} = 3500 \times \left(\frac{0.01174}{25}\right)^{\frac{0.22}{1.22}}$$

$T_e = 878.796$ K

$$h_o = h_e + \frac{v_e^2}{2}$$

$$V_e = \sqrt{2C_p(T_o - T_e)}$$

$$C_p = \frac{\gamma R}{\gamma - 1} = \frac{1.22 \times 519.625}{0.22}$$

= 2881.557 J/kg-K

$$V_e = \sqrt{2 \times 2881.557 \times (3500 - 878.796)}$$

= 3886.682m/s

Specific thrust = 3886.682 $\frac{Ns}{kg}$

5. $M_e = \frac{V_e}{\sqrt{\gamma R T_e}} = \frac{3886.682}{\sqrt{1.22 \times 519.625 \times 878.796}}$

= 5.2

6. $\rho_o = \frac{P_o}{RT_o} = \frac{25 \times 101325}{519.625 \times 3500} = 1.393\ kg/m^3$

$$\rho^* = \rho_o \times \left(\frac{2}{\gamma+1}\right)^{\frac{1}{\gamma-1}}$$

$$\rho^* = 1.393 \times \left(\frac{2}{2.22}\right)^{\frac{1}{0.22}}$$

= 0.867 kg/m³.

$$T^* = T_o \times \frac{2}{\gamma+1} = 3500 \times \frac{2}{2.22}$$

= 3153. 153 K

$$V^* = \sqrt{\gamma R T^*} = a^*$$

$$\sqrt{1.22 \times 519.625 \times 3153.153}$$

= 1413.83 m/s

'*' shows the condition at throat (M = 1)

$$\rho_e = \frac{P_e}{RTe} = \frac{0.01174 \times 101325}{519.625 \times 878.796}$$

$= 2.604 \times 10^{-3}\ Kg/m^3$

$\rho^* A^* V^* = \rho_e A_e V_e$

$$A_e = \frac{\rho^* A^* V^*}{\rho_e V_e}$$

$$A_e = \frac{0.867 \times 0.2 \times 1413.83}{2.604 \times 10^{-3} \times 3886.682}$$

$A_e = 24\ m^2$

7. 1433 KPa

$$\frac{T_o}{T^*} = \frac{\gamma+1}{2}$$

$$\Rightarrow T^* = T_o \times \frac{2}{\gamma+1}$$

$$= 3000 \times \frac{2}{1.21+1}$$

= 2714.932 K

At throat,

$$V^* = a^* = \sqrt{\gamma R T^*} = \sqrt{1.21 \times 400 \times 2714.932}$$

= 1146.31 m/s

$\dot{m} = \rho^* A^* V^*$

$$\rho^* = \frac{\dot{m}}{A^* V^*}$$

$$\rho^* = \frac{40}{0.047 \times 1146.31}$$

$\rho^* = 0.7424\ Kg/m^3$.

$P^* = \rho^* R T^*$

$= 0.7424 \times 400 \times 2714.932$

= 806226.207 Pa

$$\frac{P_o}{P^*} = \left(\frac{\gamma+1}{2}\right)^{\frac{\gamma}{\gamma-1}}$$

$$P_o = P^* \times \left(\frac{\gamma+1}{2}\right)^{\frac{\gamma}{\gamma-1}}$$

$$= 806226.207 \times \left(\frac{2.21}{2}\right)^{\frac{1.21}{0.21}}$$

= 1433198.111Pa

P_o = 1433 KPa

8. $C_f = \frac{f}{P_o \times A^*}$ and $C^* = \frac{P_o \times A^*}{\dot{m}_p}$

$$C_f = \frac{f}{C^* \times \dot{m}_p} = \frac{\frac{f}{\dot{m}_p}}{C^*}$$

$$I_{sp} = \frac{f}{\dot{m}_p \times g}$$

$$\Rightarrow \frac{f}{\dot{m}_p} = I_{sp} \times g$$

$$\text{So, } C_f = \frac{I_{sp} \times g}{C*}$$

9. We know that,

$$\sigma = \frac{m_S}{m_S + m_P} \text{ and } \lambda = \frac{m_L}{m_S + m_P + m_L}$$

$$\Delta V_b = V_e \ \ell n \frac{m_i}{m_f}$$

$$= V_e \ \ell n \frac{1}{\frac{m_f}{m_i}}$$

$$\frac{m_f}{m_i} = \frac{m_L + m_S}{m_L + m_S + m_P}$$

$$= \frac{(m_L \times m_S) \times (m_S + m_P)}{m_i \times (m_S + m_P)}$$

$$= \frac{m_L m_S + m_L m_P + m_S^2 + m_S m_P}{m_i \times (m_S + m_P)}$$

$$= \frac{m_S(m_L + m_S + m_P)}{m_i \times (m_S + m_P)} + \frac{m_L}{m_i} \times \frac{m_P}{(m_S + m_P)}$$

$$= \frac{m_S}{m_S + m_P} + \left(\frac{m_L}{m_i}\right) \times \left(1 - \frac{m_S}{m_S + m_P}\right)$$

$= \sigma + \lambda(1 - \sigma)$

$$\text{So, } \Delta V_b = V_e \ell n \frac{1}{\sigma + \lambda(1-\sigma)}$$

10. Let initial mass = m_i.

$$\text{Final mass} = \frac{m_i}{2}$$

$$\Delta V_b = V_e \ \ell n \frac{m_i}{m_f}$$

$$V_f - V_i = V_e \ \ell n \frac{m_i}{\frac{m_i}{2}} = V_e \ell n(2)$$

$V_f - 1000 = 1900 \ \ell n(2)$

V_f = 2316.98 m/s

11. 460.56 Kg

$$\Delta V_b \rightarrow g I_{sp} \ell n \frac{m_i}{m_f}$$

$$\ell n \frac{m_i}{m_f} = \frac{\Delta V_b}{g I_{sp}}$$

$$\frac{m_i}{m_f} = e^{\frac{\Delta V_b}{gJ_{sp}}}$$

$$m_f = \frac{m_i}{e^{\Delta V_b / gJ_{sp}}}$$

$$= \frac{3000}{e^{\frac{500}{10 \times 300}}}$$

= 2539.44 Kg

$m_p = m_i - m_f$

= 3000 – 2539.44

= 460.56 Kg

12. $I_{sp} = \frac{\int f\,dt}{m_p \times g}$

$$I_{sp} = \frac{\int_0^3 (10000t^2 - 20t + 1)\,dt}{m_p \times g}$$

$$= \frac{\left[\frac{10000\,t^3}{3} - \frac{20t^2}{2} + t\right]_0^3}{m_p \times g}$$

$$= \frac{10000 \times 9 - 10 \times 9 + 3}{300 \times 10}$$

$= 29.971$ S

14. We know that

$$C_f = \frac{T}{P_c \times A^*} \text{ and } C^* = \frac{P_C \times A^*}{\dot{m}_P}$$

$$C_f = \frac{T}{C^* \times \dot{m}_P} = \frac{\left(\frac{T}{\dot{m}_P}\right)}{C^*}$$

$$I_{sp} = \frac{T}{\dot{m}_P \times g} \Rightarrow \frac{T}{\dot{m}_P} = J_{sp} \times g$$

$$C_f = \frac{I_{sp} \times g}{C^*}$$

$$I_{sp} = \frac{C_f \times C^*}{g} = \frac{1.4 \times 2380}{9.8}$$

$I_{sp} = 340$ sec

15. $I_{sp} = \frac{T}{\dot{m}_p \times g}$

$T = I_{sp} \times \dot{m}_p \times g$

$= 340 \times 300 \times 9.8$

$= 999600$ N

$= 999.6$ KN

8

CHAPTER

GENERAL APTITUDE

ENGLISH GRAMMAR

Directions (Q. 1 – 20)

In this section you are required to spot errors in sentences. Read each sentence to find out whether there is any error in any of the parts. Errors, if any, are only in one of the parts. No sentence has more than one error. When you detect an error, in any of the parts of the sentence, choose the corresponding alphabet.

1. (*a*) The scenery here is not good
(*b*) I have lost my furnitures
(*c*) We have received no information
(*d*) He told his mother this news
(*e*) No error.

2. (*a*) Please excuse the trouble
(*b*) He took pains over his work
(*c*) In India there are many poor
(*d*) Scouts wear shorts
(*e*) No error.

3. (*a*) I spent the holidays with my family
(*b*) Gone him some blotting paper
(*c*) Write this in your note book
(*d*) Gone my kind regards to all
(*e*) No error.

4. (*a*) We had a good play of football
(*b*) We saw a play
(*c*) We like acting
(*d*) He is a tall man
(*e*) No error.

5. (*a*) The boy was wearing a new suit
(*b*) He took offence at this
(*c*) Please put your sign here
(*d*) Please put your signature here
(*e*) No error.

6. (*a*) He is my cousin brother
(*b*) We are all brothers
(*c*) I spent the middle of the day working
(*d*) He had a large amount of money
(*e*) No error.

7. (*a*) He got into bad company
(*b*) One of my servant tells me
(*c*) Some of my servants tell me
(*d*) The front of the house
(*e*) No error.

8. (*a*) I had a bathe in the sea
(*b*) I have hurt a toe
(*c*) Each of these days play games
(*d*) None of us went
(*e*) No error.

9. (*a*) A man should work hard
(*b*) A man should not waste his time
(*c*) A boy should not waste his time
(*d*) Here is my cup; please fill it
(*e*) No error.

10. (*a*) Have you a pencil ? I have not got one
(*b*) Is he coming ? Yes I think so
(*c*) He enjoyed during the holidays
(*d*) I asked for my pencil but he did not give it to me
(*e*) No error.

11. (*a*) The boy who does best he will get a prize
(*b*) Whoever does best will get a prize
(*c*) Who did this ? I
(*d*) He and I are brothers
(*e*) No error.

12. (*a*) I went for a walk with some friends
(*b*) He is wiser than I
(*c*) The master tested the boy if he could read English
(*d*) I shall see whether the brakes work well
(*e*) No error.

13. (*a*) Everyone is frightened when he sees a tiger
(*b*) None of us has seen him
(*c*) People strave when they have no money
(*d*) The size of the shoe should be the same as this shoe
(*e*) No error.

14. (*a*) Everyone knows this
(*b*) Every man knows this
(*c*) These all mangoes are ripe
(*d*) He hold the book in both hands
(*e*) No error.

15. (*a*) I have no any friends
(*b*) Neither man has come
(*c*) All idle man should do some work or other
(*d*) Shakespeare is greater than any other poet
(*e*) No error.

16. (*a*) Open your book at page six
(*b*) He is in class ninth
(*c*) This article costs ten rupees
(*d*) He came a second time
(*e*) No error.

17. (*a*) King George VI
(*b*) He is older than I
(*c*) He is better than I
(*d*) Raipur is Hotter thanSimla
(*e*) No error.

18. (*a*) He is worse than I
(*b*) A horse is more useful than a car
(*c*) He gets a small salary
(*d*) I our library the number of books is less
(*e*) No error.

19. (*a*) He is the more clever of the two
(*b*) He is the cleverer of the two
(*c*) From the three he is more clever
(*d*) Of the two plans this is the better
(*e*) No error.

20. (*a*) He is becoming strong
(*b*) There is a very good teacher in that class
(*c*) He will spend the rest of his life here
(*d*) This is a sight worth seeing
(*e*) No error.

Directions (Q. 21 – 30)

Each question below has a sentence, from the choices provided, identify the one which best restates the given sentence and mark its number as the answer.

21. Uneasy lies the head that wears a crown.
(*a*) Uneasily lies the head that wears a crown.
(*b*) Uneasy is the head wearing a crown.
(*c*) The head which wears the crown lies uneasy.
(*d*) The head which is wearing the crown lies uneasy.

22. Had I been a little early, I would not have missed the train.
(*a*) Had I been a little early, I would not have missed the train.
(*b*) Had I been a little early than now, I would not have missed the train.
(*c*) Had I been a little earlier, I would not have missed the train.
(*d*) Had I only been a little early than now, I would not have missed the train.

23. She is as intelligent as her sister if not more.
(*a*) She is more intelligent than her sister.
(*b*) She is as intelligent as her sister if not more intelligent.
(*c*) She and her sister are equally intelligent.
(*d*) She is less intelligent than her sister.

24. You will not succeed unless you are optimistic.
(*a*) You will not be successfull unless you are optimistic.
(*b*) You will not succeed until you are an optimist.
(*c*) You will not succeed unless you don't be optimistic.
(*d*) You will not succeed unless you don't be an optimistic.

25. The students were advised to follow the instructions of the examiner.
(*a*) The instructions of the examiner were followed by the students.
(*b*) The students followed the examiner's instructions.
(*c*) The examiner's instructions were advised to be followed by the students..
(*d*) It has been advised to the students that they should follow the examiner's instructions.

26. Despite being ill, he attended the classes.
(*a*) Despite of being ill, he attended the classes.
(*b*) Despite his illness, he attended the classes.
(*c*) Inspite of his being ill, he attended classes.
(*d*) The classes were attended by him inspite of being ill.

27. That man is aggressive by nature is a hard fact of life and no one can deny it.
(*a*) That man is aggressive by nature is a hard fact of life which none can deny.
(*b*) That man is aggressive by nature is a hard fact of life and no one can deny it.

(*c*) That man is aggressive by nature is a hard fact of life and not one can deny.

(*d*) That man is aggressive by nature is hard for anyone to deny.

28. He is as old as I.

(*a*) He is as old as me.

(*b*) He and I are equally old.

(*c*) He and I are equally older.

(*d*) Both he and I are of the same age.

29. A man becomes wiser with age and experience.

(*a*) A man as he is aged and experienced becomes wiser.

(*b*) A man with age and experience becomes wiser.

(*c*) A man as he is aged and experienced becomes the wiser.

(*d*) With age and experience, a man becomes wiser.

30. Hazards of life cannot be negated but they can be quite effortlessly evaded.

(*a*) Hazards in life can be quite effortlessly evaded and completely nullified.

(*b*) Hazards of lilfe cannot be avoided but they can be made ineffective.

(*c*) To be made ineffective, hazards of life must be avoided.

(*d*) It may not be possible to nullify the hazards of life but they can quite easily be avoided.

Directions(Q. 31 –40)

Each sentence is broken into four parts a, b, c, d.

Mark the part which has an error. Ignore errors of punctuation.

31. (*a*) Every man, woman (*b*) and child (*c*) in the house on fire (*d*) have been saved.

32. (*a*) One of the (*b*) best lawyers in town (*c*) have been (*d*) hired.

33. (*a*) I request you (*b*) kindly to (*c*) come to me (*d*) immediately.

34. (*a*) My friend's mother (*b*) is the principal (*c*) of a (*d*) girl's college.

35. (*a*) To succeed in these tests

(*b*) it is absolutely necessary

(*c*) for us

(*d*) to aim for speed and accuracy.

36. (*a*) The aeroplane is a

(*b*) powerful instrument of war

(*c*) and their peacetime role

(*d*) is just as important.

37. (*a*) The depletion of ozone in the atmosphere

(*b*) is a reminder to all of us

(*c*) both in the developed and developing world

(*d*) that they cannot continue to use contraptions which give out noxious gases.

38. (*a*) Scotsmen are notorious

(*b*) for their thrift

(*c*) and he is the butt

(*d*) of many exaggerated jokes.

39. (*a*) A mixed economy

(*b*) is it in which

(*c*) the public sector

(*d*) and the private sector co - exist.

40. (*a*) Although the beaver's hind feet

(*b*) are webbed for swimming

(*c*) their front feet

(*d*) are small and handlike.

Directions(Q. 41 – 45)

Given below are five sentences, each of which may or may not have errors.

Mark (a), if there is only one error.

Mark (b), if there are two errors.

Mark(c), if there are more than two errors.

Mark (d), if there is no error.

41. "This feels like one big dream," says a weary American woman to her husband, as they sit in a plane flying home.

42. The worst mistake a non - profit organization can make is to take all the money it is given and be beholden to doing things on someone else's terms.

43. Although most hotels in Bhutan has a distinctly local flavour, some pander with the perceived tastes of Western tour groups - and can be rather bland.

44. The crustaceans find sanctuary in the jords, often clinging to the walls in large, gregarious groups.

45. In these tropical lowlands the Maya built an agrarian civilization that eventually supported one of the highest population densities in the pre - industrial America.

Directions (Q. 46–50)

Each of the following questions has two sentences A and B.

Mark (a), if you think sentence A has an error.

Mark (b), if you think sentence B has an error.

Mark (c), if you think both sentences A and B have errors.

Mark (d), if you think neither sentence has an error.

46. A. The boss himself or his secretary answer the phone on Saturday.
    B. Neither the quality nor the prices have changed.
47. A. The members re-commended that all delinquents be fined.
    B. A stone lying in one position for a long time may gather moss.
48. A. Until I received that letter, I was hoping to have had a visit from Krishnan.
    B. Follow the main road for a mile; then you need to take the next road at the left.
49. A. The people to watch closely are the ones ruling behind the political scene.
    B. Give the tools to whoever can use them best.
50. A. The magazine has been published continuous since 1951, and it has the funniest cartoons you can possibly imagine.
    B. I feel glumly when I read his nightmarish tales.

SENTENCE COMPLETION

Directions for Q. 1 to 41 : *In each of the following sentences, a part/parts of the sentence is left unfinished. Beneath each sentence, four different ways of completing the sentence are indicated. Choose the best alternative among the four.*

1. For some of our ruling politicians, social justice has apparently come to mean that __________.
   (*a*) they have a right to plunder public money from the treasury
   (*b*) they have a right to the plunder of public money from the treasury
   (*c*) theirs is the right to plunder public money from the treasury
   (*d*) the right is theirs to plunder public money from the treasury
2. Some critics believe that Satyajit Ray never quite came back to the great beginning he made in his path-breaking film Pather Panchali. __________ have endured decades of well-travelled bad prints to become a signpost in cinematic history.
   (*a*) The bizarre history of its misty origins
   (*b*) Its haunting images
   (*c*) Its compelling munificence
   (*d*) The breathtaking awe it inspires
3. __________ , the more they remain the same.
   (*a*) The less the dynamism
   (*b*) The more things change
   (*c*) The more pronounced the transformation
   (*d*) The more the merrier
4. Revenues are likely to register a dramatic increase if __________ on the hundred million people who are said to comprise the risingIndian middle class.
   (*a*) a flat tax is collected
   (*b*) a flat tax is placed
   (*c*) taxes at a flat rate are charged
   (*d*) tax is imposed at a flat rate
5. I am an entertainer. __________ I have to keep smiling because, deep in my heart, laughter and sorrow have an affinity.
   (*a*) Even if I have tears in me
   (*b*) Despite conditions of extreme adversity
   (*c*) Inspite of misery around me
   (*d*) Although I have yet to make it big
6. The stock market is probably __________ And the way the markets are plunging says a lot about investor confidence.
   (*a*) an ideal indication of the health of public sentiment
   (*b*) the least imperfect mechanism for judging the quantity of the sentiment of the public.
   (*c*) the best indicator of public sentiment
   (*d*) the best barometer to assess the sentiment of the public
7. Home is __________ one has yet to improve.
   (*a*) a discovery over which
   (*b*) a certain over which
   (*c*) an invention on which
   (*d*) an institution upon which
8. The highest reward for a man's toil is not what he gets for it but what __________.
   (*a*) he makes out of it.
   (*b*) he gets for others.
   (*c*) he has overcome through it.
   (*d*) he becomes by it.
9. An act of justice closes the book on a misdeed; an act of vengeance __________.
   (*a*) reopens the first chapter
   (*b*) writes an epoilogue
   (*c*) writes one of its own
   (*d*) opens new books

10. The trail is the thing , not the end of the trail. Travel too fast and you miss _________.

(*a*) all you are travelling for.

(*b*) all the sights you are supposed to see.

(*c*) the very excitement of your travel.

(*d*) all the enjoyments of travel.

11. ______________ that in this apparent mess, two things not be interfered with.

(*a*) It is important.

(*b*) it is of cardinal important.

(*c*) It should be urgently under-stood.

(*d*) It cannot be emphasised.

12. A country's wealth is its people. But instead of drawing out the strengths of the people, instead of drawing out their talent, this use of religion debases, degrades and depresses than ____________.

(*a*) in greater and greater measure

(*b*) further

(*c*) beyond reasonable limit

(*d*) more and more

13. Wines that yielded a good commercial profit ___________ in the same limited areas of France as of now.

(*a*) seem to have been produced

(*b*) appear to have a remarkable semblance

(*c*) bear a significant similarity in terms of production to those grown

(*d*) appear to have been similarly produced

14. This is about ___________ a sociological analysis can penetrate.

(*a*) the other limits that

(*b*) just how far into the subject that

(*c*) the relative distance that

(*d*) as far as

15. The safest general characterisation of the European philosophical tradition as it has developed up to now, with all its diverse proponents, is that it consists of a________Plato.

(*a*) series of footnotes to

(*b*) set of prologues to

(*c*) collection of chapters on

(*d*) string of commentatries to

16. The ideas that these companies used seem so simple with __________ that their competitors will now __________ themselves for not thinking of them first.

(*a*) advanced technology, hit

(*b*) analysis , abuse

(*c*) passage of time, curse

(*d*) hindsight, kick

17. No doubt, it was our own government but it was being run on borrowed ideas, using ___________solutions.

(*a*) worn out (*b*) second hand

(*c*) impractical (*d*) appropriate

18. Overall, all the recent policy changes by the government only amount to a _________ in the sugar industry.

(*a*) superficial attemp at liberalisation

(*b*) drop in the ocean

(*c*) large change

(*d*) small regulating authority

19. In pusuance of their decision to resist what they saw as anti-labour policies, the company employees, union launched agitation to ___________.

(*a*) show their virility

(*b*) reaffirm their commitment to the company.

(*c*) bring down the government

(*d*) demonstrate their strength

20. Scientists have found ________ evidence that Neanderthals, now an extinct species of humans ___________ Europe, were cannibals.

(*a*) grisly, living in

(*b*) incontrovertible, a cross

(*c*) chilling inhabiting

(*d*) proper, in

21. The telephone symbolises that awkard __________ in all communication technologies ; while it ____________ to bring us together, it keeps us apart.

(*a*) paradox, needs (*b*) irony, intends

(*c*) paradox , tries (*d*) irony wishes

22. The world is fast moving into a period of 'civilisational clash' in which the primary identification of people and countries will not be ____________ as during the Cold War, but ____________.

(*a*) ideological, technological

(*b*) cultural, ethnic

(*c*) political, economic

(*d*) ideological, cultural

23. In our country, the challenges are to raise ______ incomes to reduce poverty, and to ______ inefficient public sector enterprises.

(a) farm, liberalise

(b) middle-class, privatise

(c) workers, take over

(d) rural, restructure

24. The interest generated by the Soccer World Cup is ______ compared to the way cricket ______ the nation.

(a) milder, fascinates

(b) lukewarm, electrifies

(c) tepid, inspires

(d) unusual, grips

25. Though one eye is kept firmly on the ______, the company now also promotes ______ contemporary art.

(a) present, experimental

(b) future, popular

(c) present, popular

(d) market, popular

26. The law prohibits a person from felling a sandalwood tree, even if it grows on one's own land, without prior permission from the government. As poor people cannot deal with the government, this legal provision leads to a rip-roaring business for ______, who care neither for the ______, nor for the trees.

(a) middlemen, rich

(b) the government, poor

(c) touts, rich

(d) touts, poor

27. It will take some time for many South Koreans to ______ the conflicting images of North Korea, let alone to ______ what to make of their northern cousins.

(a) reconcile, decide

(b) understand, clarify

(c) make out, decide

(d) reconcile, understand

28. In these bleak and depressing times of ______ prices, non-performing governments and ______ crime rates, Sourav Ganguly has given us, Indians, a lot to cheer about.

(a) escalating, increasing

(b) spiralling, booming

(c) spiralling, soaring

(d) ascending, debilitating

29. The manners and ______ of the nouveau riche is a recurrent ______ in the literature.

(a) style, motif (b) morals, story

(c) wealth, theme (d) morals, theme

30. Many people suggest ______ and still others would like to convince people not to buy pirated cassettes.

(a) to bring down audio-cassette prices to reduce the incidence of music piracy, others advocate strong legal action against the offenders,

(b) bringing down audio-cassette prices to reduce the incidents of music piracy, others are advocating strong legal action against offenders,

(c) bringing down audio-cassette prices to reduce the incidence of music piracy, others advocate strong legal action against offenders,

(d) audio-cassette prices to be brought down to reduce incidence of music piracy, others advocate that strong legal action must be taken against offenders,

Directions for Q. 31 to 36.

Fill the gaps in the passages below with the most appropriate word from the options given for each gap. The right words are the ones used by the author. Be guided by the author's overall style and meaning when you choose the answers.

Von Nuemann and Morgenstern assume a decision framework in which all options are thoroughly considered, each option being independent of others, with a numerical value derived for the utility of each possible outcome (these outcomes reflecting, in turn, all possible combinations of choices). The decision is then made to maximize the expected utility.

(31) ______, such a model reflects major simplifications of the way decisions are made in the real world. Humans are not able to process information as quickly and effectively as the model assumes; they tend not to think (32) ______ as easily as the model calls for; they often deal with a particular option without really assessing its (33) ______, and when they do assess alternatives, they may be extremely nebulous about their criteria of evaluation.

31. (a) Regrettably (b) Firstly

(c) Obviously (d) Apparently

32. (a) quantitatively (b) systematically

(c) scientifically (d) analytically

33. (*a*) implications (*b*) disadvantages
(*c*) utility (*d*) alternatives

In a large company, (34) _____ people is about as common as using a gun or a switch-blade to (35) _____ an argument. As a result, most managers have little or no experience of firing people, and they find it emotionally traumatic; as result, they often delay the act interminably, much as an unhappy spouse will prolong a bad marriage. And when the firing is done, it's often done clumsily, with far worse side effects than are necessary.

Do the world-class softward organizations have a different way of firing people ? No, but they do the deed swiftly, humanely, and professionally.

The key point here is to view the fired employees as a "failed product" and to ask how the process (36) _____ such a phenomenon in the first place.

34. (*a*) dismissing (*b*) punishing
(*c*) firing (*d*) admonishing

35. (*a*) resolve (*b*) thwart
(*c*) defeat (*d*) close

36. (*a*) derived (*b*) engineered
(*c*) produced (*d*) allowed

SYNONYMS

Directions: Each of the questions below consists of a word followed by four lettered words or phrases. Choose the lettered word or phrase that is most nearly same in meaning to the word.

1. **Abase**
(*a*) incur (*b*) tax
(*c*) estimate (*d*) humiliate

2. **Aberration**
(*a*) deviation (*b*) abhorrence
(*c*) dislike (*d*) absence

3. **Abet**
(*a*) conceive (*b*) wager
(*c*) encourage (*d*) evade

4. **Aabeyance**
(*a*) obedience (*b*) discussion
(*c*) excitement (*d*) suspended action

5. **Abjure**
(*a*) discuss (*b*) renounce
(*c*) run off secretly (*d*) perjure

6. **Ablution**
(*a*) censure (*b*) forgiveness
(*c*) mutiny (*d*) washing

7. **Abnegation**
(*a*) blackness (*b*) self-denial
(*c*) selfishness (*d*) cause

8. **Aborigine**
(*a*) first design (*b*) absolution
(*c*) finale (*d*) primitive inhabitant

9. **Abortive**
(*a*) unsuccessful (*b*) consuming
(*c*) financing (*d*) familiar

10. **Abstinence**
(*a*) restrained (*b*) vulgar display
(*c*) deportment (*d*) reluctance

11. **Abstruse**
(*a*) profound (*b*) irrespective
(*c*) suspended (*d*) protesting

12. **Abut**
(*a*) stimulate (*b*) grasp
(*c*) oppose (*d*) adjoin

13. **Abysmal**
(*a*) bottomless (*b*) eternal
(*c*) meteric (*d*) diabolic

14. **Accede**
(*a*) fail (*b*) compromise
(*c*) correct (*d*) consent

15. **Acclivity**
(*a*) index (*b*) report
(*c*) upslope of a hill (*d*) negotiator

16. **Astute**
(*a*) sheer (*b*) noisy
(*c*) astral (*d*) clever

17. **Atrocity**
(*a*) endurance (*b*) fortitude
(*c*) session (*d*) heinous act

18. **Atrophy**
(*a*) capture (*b*) waste away
(*c*) govern (*d*) award prize

19. **Attenuate**
(*a*) appear (*b*) be absent
(*c*) weaken (*d*) testify

20. **Atypical**
(*a*) superfluous (*b*) fortitude
(*c*) unusual (*d*) clashing

21. **Audacity**
(*a*) boldness (*b*) asperity
(*c*) strength (*d*) stature

22. Augment
(*a*) make noble (*b*) anoint
(*c*) increase (*d*) harvest

23. Auxiliary
(*a*) righteous (*b*) prospective
(*c*) assistant (*d*) archaic

24. Avarice
(*a*) easiness (*b*) greed
(*c*) statement (*d*) invoice

25. Avert
(*a*) entertain (*b*) transform
(*c*) turn away (*d*) lead toward

26. Awry
(*a*) recommended (*b*) commiserating
(*c*) started (*d*) crooked

27. Baleful
(*a*) doubtful (*b*) virtual
(*c*) deadly (*d*) conventional

28. Balmy
(*a*) venturesome (*b*) dedicated
(*c*) mild (*d*) fanatic

29. Banal
(*a*) philosophical (*b*) trite
(*c*) dramatic (*d*) heedless

30. Baneful
(*a*) intellectual (*b*) thankful
(*c*) decisive (*d*) poisonous

31. Boisterous
(*a*) conflicting (*b*) noisy
(*c*) testimonial (*d*) grateful

32. Bombastic
(*a*) sensitive (*b*) pompous
(*c*) rapid (*d*) sufficient

33. Bouillon
(*a*) insight (*b*) chowder
(*c*) gold (*d*) clear soup

34. Brackish
(*a*) careful (*b*) salty
(*c*) chosen (*d*) tough

35. Braggadocio
(*a*) weaponry (*b*) boasting
(*c*) skirmish (*d*) encounter

36. Brazen
(*a*) shameless (*b*) quick
(*c*) modest (*d*) pleasant

37. Brindled
(*a*) equine (*b*) pathetic
(*c*) hasty (*d*) spotted

38. Brochure
(*a*) opening (*b*) pamphlet
(*c*) censor (*d*) bureau

39. bucolic
(*a*) diseased (*b*) replusive
(*c*) rustic (*d*) twinkling

40. Buxom
(*a*) voluminous (*b*) indecisive
(*c*) convincing (*d*) plump

41. Cache
(*a*) lock (*b*) hiding place
(*c*) tide (*d*) automobile

42. Cacophony
(*a*) discord (*b*) dance
(*c*) applause (*d*) type of telephone

43. Callow
(*a*) youthful (*b*) holy
(*c*) mild (*d*) colored

44. Candid
(*a*) vague (*b*) outspoken
(*c*) experienced (*d*) anxious

45. Chaste
(*a*) loyal (*b*) timid
(*c*) curt (*d*) pure

46. Chide
(*a*) unite (*b*) fear
(*c*) record (*d*) scold

47. Chimerical
(*a*) developing (*b*) brief
(*c*) distant (*d*) fantastic

48. Choleric
(*a*) musical (*b*) episodic
(*c*) hot headed (*d*) global

49. Churlish
(*a*) marine (b) economical
(*c*) impolite (*d*) compact

50. Ciliated
(*a*) foolish (*b*) swift
(*c*) early (*d*) hairy

51. Circuitous
(*a*) indirect (*b*) complete
(*c*) obvious (*d*) aware

52. City
(*a*) galvanize (*b*) visualize
(*c*) locate (*d*) quote

53. Clandestine
(*a*) abortive (*b*) secret
(*c*) tangible (*d*) doomed

54. Claustrophobia
(*a*) lack of confidence
(*b*) fear of spiders
(*c*) love of books
(*d*) fear of closed places

55. Cleft
(*a*) split (*b*) waterfall
(*c*) assembly (*d*) adherence

56. Cliche
(*a*) increase (*b*) vehicle
(*c*) morale (*d*) platitude

57. Coerce
(*a*) recover (*b*) begin
(*c*) force (*d*) license

58. Cognizance
(*a*) policy (*b*) knowledge
(*c*) advance (*d*) omission

59. Cohere
(*a*) hold together
(*b*) occur simultaneously
(*c*) recollect
(*d*) materialize

60. Conjecture
(*a*) magic (*b*) guess
(*c*) position (*d*) form

61. Connoisseur
(*a*) gourmand (*b*) lover of art
(*c*) humidor (*d*) delinquent

62. Consanguinity
(*a*) kinship (*b*) friendship
(*c*) bloodletting (*d*) relief

63. Consensus
(*a*) general agreement (*b*) project
(*c*) insignificance (*d*) sheaf

64. Construe
(*a*) explain (*b*) promote
(*c*) reserve (*d*) erect

65. Contaminate
(*a*) arrest (*b*) prepare
(*c*) pollute (*d*) beam

66. Contentious
(*a*) squealing (*b*) surprising
(*c*) quarrelsome (*d*) smug

67. Continence
(*a*) humanity (*b*) research
(*c*) embryology (*d*) self-restraint

68. Contraband
(*a*) purpose (*b*) rogue
(*c*) rascality (*d*) difficulty

69. Contrite
(*a*) smart (*b*) penitent
(*c*) restful (*d*) recognized.

70. Controvert
(*a*) turnover (*b*) contradict
(*c*) mind (*d*) expolain

71. Convene
(*a*) propose (*b*) restore
(*c*) question (*d*) gather

72. Conversant
(*a*) ignorant (*b*) speaking
(*c*) incorporated (*d*) familiar

73. Copious
(*a*) plentiful (*b*) cheating
(*c*) dishonorable (*d*) adventurous

74. Corpulent
(*a*) regenerate (*b*) obese
(*c*) different (*d*) hungry

75. Decimate
(*a*) kill (*b*) disgrace
(*c*) search (*d*) collide

76. Declivite
(*a*) trap (*b*) quadrangle
(*c*) quarter (*d*) downward slope

77. Decollete

(*a*) flavored (*b*) demure
(*c*) flowery (*d*) low-necked

78. Decorous

(*a*) momentary (*b*) emotional
(*c*) suppressed (*d*) proper

79. Decrepitude

(*a*) feebleness (*b*) disease
(*c*) coolness (*d*) melee

80. Default

(*a*) failure to act (*b*) tendency to err
(*c*) desire to remedy (*d*) debt

81. Defection

(*a*) determination (*b*) desertion
(*c*) invitation (*d*) affection

82. Defile

(*a*) maincure (*b*) ride
(*c*) pollute (*d*) assemble

83. Degraded

(*a*) surpprised (*b*) lowered
(*c*) ascended (*d*) learned

84. Deleterious

(*a*) delaying (*b*) experimental
(*c*) harmful (*d*) graduating

85. Deluge

(*a*) confusion (*b*) deception
(*c*) flood (*d*) mountain

86. Denigrate

(*a*) refuse (*b*) blacken
(*c*) terrify (*d*) admit

87. Denouement

(*a*) action (*b*) scenery
(*c*) resort (*d*) solution

88. Depravity

(*a*) wickedness (*b*) sadness
(*c*) heaviness (*d*) tidiness

89. Deranged

(*a*) insane (*b*) systematic
(*c*) neighborly (*d*) alphabetical

90. Dwindle

(*a*) blow (*b*) inhabit
(*c*) spin (*d*) lessen

91. Ecstasy

(*a*) joy (*b*) speed
(*c*) treasure (*d*) warmth

92. Edify

(*a*) mystify (*b*) suffice
(*c*) improve (*d*) erect

93. Efface

(*a*) countenance (*b*) encourage
(*c*) recognize (*d*) rub out

94. Effigy

(*a*) requisition (*b*) organ
(*c*) charge (*d*) dummy

95. Egregious

(*a*) pious (*b*) shocking
(*c*) anxious (*d*) sociable

96. Egress

(*a*) entrence (*b*) bird
(*c*) exit (*d*) double

97. Elated

(*a*) debased (*b*) respectful
(*c*) drooping (*d*) gay

98. Elusive

(*a*) deadly (*b*) eloping
(*c*) evasive (*d*) simple

99. Emaciated

(*a*) garrulous (*b*) primeval
(*c*) vigorous (*d*) thin

100. Emancipate

(*a*) set free (*b*) take back
(*c*) make worse (*d*) embolden

101. Embellish

(*a*) doff (*b*) don
(*c*) abscond (*d*) adon

102. Embroil

(*a*) cherish (*b*) overtheat
(*c*) entangle (*d*) assure

103. Emendation

(*a*) correction (*b*) interpretation
(*c*) exhumation (*d*) inquiry

104. Eminent

(*a*) purposeful (*b*) high
(*c*) delectable (*d*) currious

105. Flored
(*a*) ruddy (*b*) rusty
(*c*) ruined (*d*) patient

106. Foil
(*a*) bury (*b*) frustrate
(*c*) shield (*d*) desire

107. Foment
(*a*) spoil (*b*) instigate
(*c*) interrogate (*d*) spray

108. Foolhardy
(*a*) strong (*b*) unwise
(*c*) brave (*d*) futile

109. Foppish
(*a*) scanty (*b*) radical
(*c*) orthodox (*d*) dandyish

110. Foray
(*a*) exursion (*b*) contest
(*c*) ranger (*d*) intuition

111. Formidable
(*a*) dangerous (*b*) outlandish
(*c*) grandiloquent (*d*) impenetrable

112. Foster
(*a*) accelerate (*b*) fondle
(*c*) become infected (*d*) raise

113. Franchise
(*a*) subway (*b*) discount
(*c*) license (*d*) reason

114. Fritter
(*a*) sour (*b*) chafe
(*c*) dissipate (*d*) cancel

115. Frugality
(*a*) foolishness (*b*) extremity
(*c*) indifference (*d*) economy

116. Fulminate
(*a*) fulfill (*b*) contemplate
(*c*) talk nonsense (*d*) protest

117. Furor
(*a*) excitement (*b*) worry
(*c*) flux (*d*) anteroom

118. Furtive
(*a*) underhanded (*b*) coy
(*c*) brilliant (*d*) quick

119. Gadfly
(*a*) humorist (*b*) nuisance
(*c*) scholar (*d*) bum

120. Galleon
(*a*) liquid measure (*b*) ship
(*c*) armada (*d*) company

121. Garisth
(*a*) sordid (*b*) flashy
(*c*) prominent (*d*) lusty

122. Garner
(*a*) prevent (*b*) assist
(*c*) collect (*d*) compute

123. Garnish
(*a*) paint (*b*) gamer
(*c*) adorn (*d*) abuse

124. Garrulity
(*a*) credulity (*b*) senility
(*c*) loquaciousness (*d*) speciousness

125. Garrulous
(*a*) arid (*b*) hasty
(*c*) sociable (*d*) talkative

126. Gauche
(*a*) rigid (*b*) swift
(*c*) awkward (*d*) taciturn

127. Gaudy
(*a*) holy (*b*) showy
(*c*) sentimental (*d*) mild

128. Gaunt
(*a*) victorious (*b*) tiny
(*c*) stylish (*d*) haggard

129. Genuflect
(*a*) falsify (*b*) trick
(*c*) project (*d*) bend the knee

130. Germane
(*a*) bacteriological (*b*) middle european
(*c*) prominent (*d*) relevant

131. Germtinal
(*a*) creative (*b*) excused
(*c*) sterilized (*d*) primitive

132. Gist
(*a*) chaff (*b*) summary
(*c*) expostulation (*d*) expiation

133. Glib

(*a*) slippery (*b*) fashionable
(*c*) antiquated (*d*) articulate

134. Gnome

(*a*) fury (*b*) giant
(*c*) dwarf (*d*) native

135. Inclement

(*a*) unfavorable (*b*) abandoned
(*c*) kindly (*d*) selfish

136. Incompatible

(*a*) capable (*b*) reasonable
(*c*) faulty (*d*) alienated

137. Inconsequential

(*a*) disorderly (*b*) insignificant
(*c*) subsequent (*d*) insufficient

138. Incontinent

(*a*) insular (*b*) complaisant
(*c*) crass (*d*) wanton

139. Incorrigible

(*a*) narrow (*b*) straight
(*c*) inconceivable (*d*) unreformable

140. Incriminate

(*a*) exacerbate (*b*) involve
(*c*) intimidate (*d*) lacerate

141. Inculcate

(*a*) exculpate (*b*) educate
(*c*) exonerate (*d*) prepare

142. Indigent

(*a*) lazy (*b*) pusillanimous
(*c*) penurious (*d*) affluent

143. Indigniy

(*a*) pomposity (*b*) bombast
(*c*) obeisance (*d*) insult

144. Indolence

(*a*) sloth (*b*) poverty
(*c*) latitude (*d*) aptitude

145. Indubitably

(*a*) flagrantly (*b*) doubtfully
(*c*) carefully (*d*) cortainly

146. Inebriety

(*a*) revelation (*b*) drunkenness
(*c*) felony (*d*) starvation

147. Inept

(*a*) outward (*b*) spiritual
(*c*) foolish (*d*) clumsy

148. Infallible

(*a*) final (*b*) unbelievable
(*c*) perfect (*d*) inaccurate

149. Infirmity

(*a*) disability (*b*) age
(*c*) inoculation (*d*) hospital

150. Ingenuous

(*a*) clever (*b*) stimulating
(*c*) naive (*d*) worried

151. Inimical

(*a*) antagonistic (*b*) anonymous
(*c*) fanciful (*d*) accurate

152. Innocuous

(*a*) not capable (*b*) not dangerous
(*c*) not eager (*d*) not frank

153. Insinuate

(*a*) resist (*b*) suggest
(*c*) report (*d*) rectify

154. Insipid

(*a*) witty (*b*) flat
(*c*) wily (*d*) talkative

155. Integrate

(*a*) tolerate (*b*) unite
(*c*) flow (*d*) copy

156. Inter

(*a*) bury (*b*) amuse
(*c*) relate (*d*) frequent

157. Interdict

(*a*) acclain (*b*) dispute
(*c*) prohibit (*d*) decide

158. Intermittent

(*a*) heavy (*b*) fleet
(*c*) occasional (*d*) fearless

159. Intractable

(*a*) culpable (*b*) flexible
(*c*) unruly (*d*) efficient

160. Intransigence

(*a*) lack of training (*b*) stubbornness
(*c*) novelty (*d*) timidity

161. Intrepid
(*a*) cold (*b*) hot
(*c*) understood (*d*) courageous

162. Intrinsic
(*a*) extrinsic (*b*) abnormal
(*c*) above (*d*) basic

163. Inundate
(*a*) abuse (*b*) deny
(*c*) swallow (*d*) flood

164. Inveigh
(*a*) speak violently (*b*) orate
(*c*) disturb (*d*) apply

165. Lethargy
(*a*) dynamism (*b*) orate
(*c*) disturb (*d*) apply

166. Loquacious
(*a*) verbose (*b*) tacturn
(*c*) rational (*d*) alluring

167. Loath
(*a*) tired (*b*) unwilling
(*c*) sickly (*d*) spirited

168. Mawkish
(*a*) sentimental (*b*) true
(*c*) certain (*d*) devious

169. Mediocre
(*a*) average (*b*) bitter
(*c*) medieval (*d*) industrial

170. Melee
(*a*) heat (*b*) brawl
(*c*) attempt (*d*) weapon

171. Mellifluous
(*a*) porous (*b*) honeycombed
(*c*) strong (*d*) smooth

172. Menial
(*a*) intellectual (*b*) clairvoyant
(*c*) servile (*d*) arrogant

173. Mentor
(*a*) guide (*b*) genius
(*c*) talker (*d*) philosopher

174. Mesmerize
(*a*) remember (*b*) hypnotize
(*c*) delay (*d*) bore

175. Meticulous
(*a*) steadfast (*b*) recent
(*c*) quaint (*d*) painstaking

176. Miasma
(*a*) dream (*b*) noxious fumes
(*c*) scenario (*d*) quantity

177. Militant
(*a*) combative (*b*) dramatic
(*c*) religious (*d*) quaintity

178. Minon
(*a*) monster (*b*) quorum
(*c*) majority (*d*) dependent

179. Mirage
(*a*) dessert (*b*) illusion
(*c*) water (*d*) mirror

180. Misanthrope
(*a*) benefactor (*b*) philanderer
(*c*) man-hater (*d*) aesthete

181. Mischance
(*a*) gamble (*b*) ordinance
(*c*) aperture (*d*) anecdote

182. Misdemeanor
(*a*) felony (*b*) peccadillo
(*c*) indignity (*d*) flat

183. Modish
(*a*) sentimental (*b*) stylish
(*c*) vacillating (*d*) contrary

184. Mollify
(*a*) avenge (*b*) attenuate
(*c*) attibute (*d*) appease

185. Monetary
(*a*) boring (*b*) fascinating
(*c*) fiscal (*d*) stationary

186. Moratorium
(*a*) burial (*b*) gathering
(*c*) delay (*d*) refusal

187. Mordant
(*a*) dying (*b*) trenchant
(*c*) fabricating (*d*) controlling

188. Moribund
(*a*) dying (*b*) trenchant
(*c*) fabricating (*d*) controlling

189. Motley
(*a*) active (*b*) disguised
(*c*) variegated (*d*) somber

190. Muggy
(*a*) attacking (*b*) fascinating
(*c*) humid (*d*) characteristic

191. Mulct
(*a*) swindle (*b*) hold
(*c*) record (*d*) pring

192. Multilingual
(*a*) variegated (*b*) polyglot
(*c*) multilateral (*d*) polyandrous

193. Mundane
(*a*) global (*b*) futile
(*c*) spatial (*d*) worldly

194. Munificent
(*a*) grandiose (*b*) puny
(*c*) philanthropic (*d*) poor

195. Musty
(*a*) flat (*b*) necessary
(*c*) indifferent (*d*) nonchalant

196. Myopic
(*a*) visionary (*b*) nearsighted
(*c*) moral (*d*) glassy

197. Nascent
(*a*) incipient (*b*) ignorant
(*c*) loyal (*d*) treacherous

198. Phlegmatic
(*a*) calm (*b*) cryptic
(*c*) practical (*d*) salivary

199. Physiognomy
(*a*) posture (*b*) head
(*c*) physique (*d*) face

200. Piebald
(*a*) motley (*b*) coltish
(*c*) hairless (*d*) thoroughbred

201. Pillage
(*a*) hoard (*b*) plunder
(*c*) versify (*d*) denigrate

202. Pinion
(*a*) express (*b*) report
(*c*) reveal (*d*) restrain

203. Pinnacle
(*a*) foothills (*b*) card game
(*c*) pass (*d*) peak

204. Pious
(*a*) historic (*b*) devout
(*c*) multiple (*d*) fortunate

205. Pique
(*a*) pyramid (*b*) revolt
(*c*) resentment (*d*) struggle

206. Placate
(*a*) determine (*b*) transmit
(*c*) pacify (*d*) allow

207. Plagiarism
(*a*) theft of funds (*b*) theft of ideas
(*c*) belief in god (*d*) arson

208. Plaintive
(*a*) moumful (*b*) senseless
(*c*) persistent (*d*) rural

209. Platitude
(*a*) fatness (*b*) bravery
(*c*) dimension (*d*) trite remark

210. Polemic
(*a*) blackness (*b*) lighting
(*c*) magnetism (*d*) controversy

211. Precarious
(*a*) priceless (*b*) premature
(*c*) primitive (*d*) hazardous

212. Refractory
(*a*) articulate (*b*) sinkable
(*c*) vaunted (*d*) unmanageable

213. Regal
(*a*) oppressive (*b*) royal
(*c*) major (*d*) basic

214. Reiterate
(*a*) gainsay (*b*) revive
(*c*) revenge (*d*) repeat

215. Relish
(*a*) desire (*b*) nibble
(*c*) savor (*d*) violating

216. Remiss
(*a*) lax (*b*) lost
(*c*) foolish (*d*) violating

217. Remonstrate
(*a*) display (*b*) restate
(*c*) protest (*d*) resign

218. Repartee
(*a*) witty retort (*b*) willful departure
(*c*) spectator (*d*) monologue

219. Repllent
(*a*) propulsive (*b*) unattractive
(*c*) porous (*d*) stiff

220. Rpercussion
(*a*) reaction (*b*) restitution
(*c*) resistance (*d*) magnificence

221. Replenish
(*a*) polish (*b*) repeat
(*c*) reinstate (*d*) refill

222. Replica
(*a*) museum piece (*b*) famous site
(*c*) battle emblem (*d*) facsimile

223. Reprisal
(*a*) reevaluation (*b*) assessment
(*c*) loss (*d*) retaliation

224. Reprove
(*a*) prevail (*b*) rebuke
(*c*) ascertain (*d*) prove false

225. Repudiate
(*a*) besmirch (*b*) appropriate
(*c*) annoy (*d*) reject

226. Repugnance
(*a*) belligerence (*b*) tanacity
(*c*) renewal (*d*) loathing

227. Tarry
(*a*) polish (*b*) restrain
(*c*) surpass (*d*) linger

228. Tautology
(*a*) memory (*b*) repetition
(*c*) tension (*d*) smile

229. Tawdry
(*a*) orderly (*b*) meretricious
(*c*) reclaimed (*d*) filtered

230. Temerity
(*a*) timidity (*b*) resourcefulness
(*c*) boldness (*d*) tremulousness

231. Temporal
(*a*) priestly (*b*) scholarly
(*c*) secular (*d*) sleeply

232. Tenacious
(*a*) fast running (*b*) intertional
(*c*) obnoxious (*d*) holding fast

233. Tenacity
(*a*) splendor (*b*) perseverance
(*c*) tendency (*d*) ingratitide

234. Tendentious
(*a*) biased (*b*) likely
(*c*) absurd (*d*) festive

235. Tentative
(*a*) prevalent (*b*) portable
(*c*) mocking (*d*) wry

236. Tenuous
(*a*) vital (*b*) thin
(*c*) careful (*d*) dangerous

237. Tepid
(*a*) boiling (*b*) luckewarm
(*c*) freezing (*d*) gaseous

ANTONYMS

Directions : *Each question below consists of a word printed in capital lettes, followed by four lettered words of phrases. Choosed the lettered word or phrase that is most nearly opposite in meaning to the word in bold letters.*

1. Abrogate
(*a*) transgress (*b*) signify
(*c*) allevate (*d*) ratify

2. Admonition
(*a*) premonit ion (*b*) hallucination
(*c*) escape (*d*) commendation

3. Adulation
(*a*) youth (*b*) purity
(*c*) brightness (*d*) criticism

4. Advocate
(*a*) define (*b*) oppose
(*c*) remove (*d*) inspect

5. Affable
(*a*) rude (*b*) ruddy
(*c*) needy (*d*) useless

6. Affecfted
(*a*) weary (*b*) unfriendly
(*c*) divine (*d*) unfeigned

7. Affluence
(*a*) poverty (*b*) fear
(*c*) persuasion (*d*) consideration

8. Agility
(*a*) awkwardness (*b*) solidity
(*c*) temper (*d*) harmony

9. Alacrity
(*a*) slowness (*b*) plenty
(*c*) fifth (*d*) courtesy

10. Alleviate
(*a*) endure (*b*) worsen
(*c*) enlighten (*d*) maneuver

11. Allure
(*a*) hinder (*b*) repel
(*c*) ignore (*d*) leave

12. Aloof
(*a*) triangular (*b*) gregarious
(*c*) comparable (*d*) honorable

13. Amalgamate
(*a*) equip (*b*) separate
(*c*) generate (*d*) materialize

14. Ambiguous
(*a*) salvageable (*b*) corresponding
(*c*) responsible (*d*) clear

15. Amble
(*a*) befriend (*b*) hasten
(*c*) steal (*d*) browse

16. Ambulatory
(*a*) convalescent (*b*) valedictory
(*c*) bedridden (*d*) emergency

17. Ameliorate
(*a*) make slow (*b*) make sure
(*c*) make young (*d*) make worse

18. Amicable
(*a*) penetrating (*b*) compensating
(*c*) unfriendly (*d*) zigzag

19. Amoral
(*a*) unusal (*b*) unfriendly
(*c*) ethical (*d*) suave

20. Amorphous
(*a*) nauseous (*b*) obscene
(*c*) providential (*d*) definite

21. Amplify
(*a*) distract (*b*) infer
(*c*) publicize (*d*) decrease

22. Analogous
(*a*) not comparable (*b*) not capable
(*c*) not culpable (*d*) not corporeal

23. Anthematize
(*a*) locate (*b*) deceive
(*c*) regulate (*d*) bless

24. Anemic
(*a*) pailid (*b*) cruel
(*c*) red-blooded (*d*) ventilating

25. Animated
(*a*) worthy (*b*) dull
(*c*) humorous (*d*) lengthy

26. Animus
(*a*) pterodactyl (*b*) bastion
(*c*) giraffe (*d*) favor

27. Anomaly
(*a*) desperation (*b*) requisition
(*c*) registry (*d*) regularity

28. Anonymous
(*a*) desperate (*b*) signed
(*c*) defined (*d*) expert

29. Antediluvian
(*a*) transported (*b*) subtle
(*c*) isolated (*d*) modern

30. Antipathy
(*a*) profundity (*b*) objection
(*c*) willingness (*d*) foundness

31. Antithesis
(*a*) velocity (*b*) maxim
(*c*) similarity (*d*) acceleration

32. Aphasia
(*a*) volubility (*b*) necessity
(*c*) pain (*d*) crack

33. Baroque
(*a*) polished (*b*) constant
(*c*) transformed (*d*) simple

34. **Beatific**
(*a*) glorius (*b*) dreadful
(*c*) theatrical (*d*) crooked

35. **Belittle**
(*a*) disobey (*b*) forget
(*c*) magnify (*d*) extol

36. **Bellicose**
(*a*) peaceful (*b*) naval
(*c*) amusing (*d*) piecemeal

37. **Benign**
(*a*) tenfold (*b*) peaceful
(*c*) blessed (*d*) malignant

38. **Benison**
(*a*) curse (*b*) bachelor
(*c*) wedding (*d*) orgy

39. **Berate**
(*a*) grant (*b*) praise
(*c*) refer (*d*) purchase

40. **Bestial**
(*a*) animated (*b*) noble
(*c*) zoological (*d*) clear

41. **Bigotry**
(*a*) arrogance (*b*) approval
(*c*) mourning (*d*) tolerance

42. **Bizarre**
(*a*) roomy (*b*) veiled
(*c*) subduced (*d*) normal

43. **Blanch**
(*a*) bleach (*b*) scatter
(*c*) darken (*d*) analyze

44. **Bland**
(*a*) caustic (*b*) meager
(*c*) soft (*d*) uncooked

45. **Blase**
(*a*) fiery (*b*) clever
(*c*) intriguing (*d*) ardent

46. **Bleak**
(*a*) pale (*b*) sudden
(*c*) dry (*d*) cheerful

47. **Blithe**
(*a*) spiritual (*b*) profuse
(*c*) cheerless (*d*) hybrid

48. **Capacious**
(*a*) warlike (*b*) cordial
(*c*) curious (*d*) not spacious

49. **Capricious**
(*a*) satisfied (*b*) insured
(*c*) photographic (*d*) steadfast

50. **Captious**
(*a*) tolerant (*b*) capable
(*c*) frivolous (*d*) winning

51. **Carnal**
(*a*) impressive (*b*) minute
(*c*) spiritual (*d*) actual

52. **Carnivorous**
(*a*) gloomy (*b*) tangential
(*c*) productive (*d*) vegetarian

53. **Carping**
(*a*) rapid (*b*) uncritical
(*c*) unintellectual (*d*) illegal

54. **Castigation**
(*a*) commendation (*b*) patience
(*c*) generosity (*d*) understatement

55. **Categorical**
(*a*) negative (*b*) ironic
(*c*) impartial (*d*) qualified

56. **Catholic**
(*a*) religious (*b*) pacific
(*c*) narrow (*d*) weighty

57. **Celerity**
(*a*) assurance (*b*) state
(*c*) acerbity (*d*) delay

58. **Celibate**
(*a*) investing (*b*) married
(*c*) retired (*d*) commodious

59. **Censure**
(*a*) process (*b*) enclose
(*c*) interest (*d*) praise

60. **Centrifugal**
(*a*) centripetal (*b*) ephemeral
(*c*) lasting (*d*) barometric

61. **Cessation**
(*a*) premium (*b*) gravity
(*c*) beginning (*d*) composition

62. Chaffing
(*a*) achieving (*b*) serious
(*c*) caitalistic (*d*) sneezing

63. Coy
(*a*) weak (*b*) airy
(*c*) brazen (*d*) old

64. Cozen
(*a*) amuse (*b*) treat honestly
(*c*) prate (*d*) shackle

65. Craven
(*a*) desirous (*b*) direct
(*c*) bold (*d*) civilized

66. Crux
(*a*) affiliction (*b*) spark
(*c*) events (*d*) trivial point

67. Cryptic
(*a*) tomblike (*b*) futile
(*c*) famous (*d*) candid

68. Cupidity
(*a*) anxiety (*b*) tragedy
(*c*) generosity (*d*) entertainment

69. Curtail
(*a*) mutter (*b*) lengthen
(*c*) express (*d*) burden

70. Cynical
(*a*) trusting (*b*) effortless
(*c*) conclusive (*d*) gallant

71. Dank
(*a*) dry (*b*) guiltless
(*c*) warm (*d*) babbling

72. Dapper
(*a*) unintelligent (*b*) untidy
(*c*) uncertain (*d*) ungrateful

73. Dauntless
(*a*) stolid (*b*) cowardly
(*c*) irrelevant (*d*) peculiar

74. Dearth
(*a*) life (*b*) abundance
(*c*) brightness (*d*) terror

75. Debacle
(*a*) progress (*b*) refusal
(*c*) masque (*d*) cowardice

76. Debilitate
(*a*) bedevil (*b*) repress
(*c*) strengthen (*d*) animate

77. Debonair
(*a*) awkward (*b*) windy
(*c*) balmy (*d*) strong

78. Derogatory
(*a*) roguish (*b*) immediate
(*c*) openionated (*d*) praising

79. Desecrate
(*a*) desist (*b*) integrate
(*c*) confuse (*d*) consecrate

80. Despicable
(*a*) streeing (*b*) worthy of esteem
(*c*) inevitable (*d*) featureless

81. Destitute
(*a*) affluent (*b*) dazzling
(*c*) stationary (*d*) characteristic

82. Devoid
(*a*) latent (*b*) eschewed
(*c*) full of (*d*) suspecting

83. Devout
(*a*) quiet (*b*) dual
(*c*) impious (*d*) straightforward

84. Diabolical
(*a*) mischievous (*b*) lavish
(*c*) seraphic (*d*) redolent

85. Diatribe
(*a*) mass (*b*) range
(*c*) eulogy (*d*) elegy

86. Diffidence
(*a*) sharapness (*b*) boldness
(*c*) malcontent (*d*) dialogue

87. Dilate
(*a*) procrastinate (*b*) contract
(*c*) conclude (*d*) participate

88. Dilatory
(*a*) narrowing (*b*) prompt
(*c*) enlarging (*d*) portentous

89. Diminution
(*a*) expectation (*b*) context
(*c*) validity (*d*) appreciation

90. Din
(*a*) lightness (*b*) safety
(*c*) silence (*d*) hunger

91. Disabuse
(*a*) crash (*b*) violate
(*c*) renege (*d*) deceive

92. Disconsolate
(*a*) unprejudiced (*b*) thankful
(*c*) threatrical (*d*) joyous

93. Enervate
(*a*) strengthen (*b*) sputter
(*c*) arrange (*d*) scrutinize

94. Enhance
(*a*) degrade (*b*) doubt
(*c*) scuff (*d*) gasp

95. Ennui
(*a*) hate (*b*) excitement
(*c*) seriousness (*d*) humility

96. Enunciate
(*a*) pray (*b*) request
(*c*) deliver (*d*) mumble

97. Ephemeral
(*a*) sensuous (*b*) passing
(*c*) popular (*d*) eternal

98. Equable
(*a*) flat (*b*) decisive
(*c*) stormy (*d*) dishonest

99. Equanimity
(*a*) agitation (*b*) stirring
(*c*) volume (*d*) identity

100. Equilibrium
(*a*) imbalance (*b*) peace
(*c*) inequity (*d*) directness

101. Equitable
(*a*) able to leave (*b*) able to learn
(*c*) unfair (*d*) preferable

102. Equivocal
(*a*) mistaken (*b*) quaint
(*c*) azure (*d*) clear

103. Erratic
(*a*) unromantic (*b*) free
(*c*) popular (*d*) steady

104. Erroneous
(*a*) accurate (*b*) dignified
(*c*) curious (*d*) abrupt

105. Erudite
(*a*) professorial (*b*) stately
(*c*) short (*d*) ignorant

106. Ethereal
(*a*) long-lasting (*b*) earthy
(*c*) ill (*d*) critical

107. Eulogistic
(*a*) pretty (*b*) critical
(*c*) brief (*d*) stern

108. Euphonious
(*a*) strident (*b*) lethargic
(*c*) literary (*d*) significant

109. Evasive
(*a*) frank (*b*) correct
(*c*) empty (*d*) fertile

110. Exasperate
(*a*) confide (*b*) formalize
(*c*) placate (*d*) betray

111. Excoriate
(*a*) scandalize (*b*) encourage
(*c*) avoid (*d*) praise

112. Exculpate
(*a*) blame (*b*) prevail
(*c*) acquire (*d*) ravish

113. Execrable
(*a*) innumerable (*b*) philosophic
(*c*) physical (*d*) excellent

114. Excrate
(*a*) decipher (*b*) sadden
(*c*) integrate (*d*) admit

115. Exhume
(*a*) decipher (*b*) sadden
(*c*) integrate (*d*) inter

116. Exodus
(*a*) neglect (*b*) consent
(*c*) entry (*d*) gain

117. Exonerate
(*a*) forge (*b*) accuse
(*c*) record (*d*) doctor

118. Exorbitant

(a) moderate (b) partisan
(c) military (d) barbaric

119. Extemporaneous

(a) rehearsed (b) hybrid
(c) humilating (d) satetesmanlike

120. Extraneous

(a) modern (b) decisive
(c) essential (d) effective

121. Extrinsic

(a) reputable (b) inherent
(c) swift (d) ambitious

122. Extrovert

(a) clown (b) hero
(c) ectomorph (d) introvert

123. Grandiose

(a) false (b) ideal
(c) proud (d) simple

124. Gratuitous

(a) warranted (b) frank
(c) ingenuous (d) frugal

125. Gregarious

(a) antisocial (b) anticipatory
(c) glorious (d) horrendous

126. Grisly

(a) suggestive (b) doubtful
(c) untidy (d) pleasant

127. Gullible

(a) incredulous (b) fickle
(c) tantamount (d) easy

128. Gusto

(a) noise (b) panic
(c) atmosphere (d) distaste

129. Ghusty

(a) clam (b) noisy
(c) fragrant (d) routine

130. Hackneyed

(a) carried (b) original
(c) banned (d) timely

131. Haggard

(a) shrewish (b) inspired
(c) plump (d) maidenly

132. Halcyon

(a) wasteful (b) prior
(c) subsequent (d) martial

133. Haphazard

(a) safe (b) indifferent
(c) deliberate (d) tense

134. Hapless

(a) cheerful (b) consistent
(c) fortunate (d) considerate

135. Heed

(a) ignore (b) hope
(c) overtake (d) nurture

136. Heretic

(a) sophist (b) believer
(c) interpreter (d) pacifist

137. Heterogeneous

(a) orthodox (b) pagan
(c) unlikely (d) similar

138. Hibernal

(a) musical (b) summerlike
(c) local (d) seasonal

139. Hilarity

(a) gloom (b) heartiness
(c) weakness (d) casualty

140. Hirsute

(a) scaly (b) bald
(c) erudite (d) quiet

141. Hortatory

(a) inquiring (b) denying
(c) killing (d) dissuading

142. Hovr

(a) commence (b) soothe
(c) leave (d) transform

143. Hunbbub

(a) calm (b) fury
(c) capital (d) axle

144. Hummock

(a) ummusical (b) scorn
(c) wakefulness (d) vale

145. Husbandry

(a) sportsmanship (b) dishonesty
(c) wastefulness (d) friction

146. Hybrid
(*a*) productive (*b*) special
(*c*) purebred (*d*) oafish

147. Hyperbole
(*a*) velocity (*b*) climax
(*c*) curve (*d*) understatement

148. Hypercritical
(*a*) tolerant (*b*) false
(*c*) extreme (*d*) inarticulate

149. Hypocritical
(*a*) sincere (*b*) narrow-minded
(*c*) shameful (*d*) amiable

150. Hypothetial
(*a*) rational (*b*) fantastic
(*c*) wizened (*d*) axiomatic

151. Ignoble
(*a*) produced by fire (*b*) worthy
(*c*) given to questioning (*d*) huge

152. Illusive
(*a*) not deceptive (*b*) not certain
(*c*) not obvious (*d*) not choherent

153. Irksome
(*a*) interesting (*b*) lazy
(*c*) tireless (*d*) devious

154. Irrelevant
(*a*) lacking piety (*b*) fragile
(*c*) congruent (*d*) pertinent

155. Irreparable
(*a*) legible (*b*) correctable
(*c*) proverbial (*d*) concise

156. Irtreverent
(*a*) related (*b*) mischievous
(*c*) respective (*d*) pious

157. Jaded
(*a*) upright (*b*) stimulated
(*c*) aspiring (*d*) applied

158. Jaundiced
(*a*) whitened (*b*) inflamed
(*c*) quickened (*d*) unbiased

159. Jaunty
(*a*) youthful (*b*) ruddy
(*c*) strong (*d*) sedate

160. Jeopardy
(*a*) patience (*b*) courage
(*c*) safety (*d*) willingness

161. Jettison
(*a*) salvage (*b*) submerge
(*c*) descend (*d*) decelerate

162. Jocular
(*a*) arterial (*b*) bloodless
(*c*) verbose (*d*) serious

163. Judicious
(*a*) punitive (*b*) unwise
(*c*) criminal (*d*) licit

164. Kindle
(*a*) dislike (*b*) quench
(*c*) gather (*d*) sparkle

165. Lachrymose
(*a*) cheering (*b*) smooth
(*c*) passionate (*d*) curt

166. Lackadasical
(*a*) monthly (*b*) possessing time
(*c*) ambitious (*d*) pusillanimous

167. Laconic
(*a*) milky (*b*) verbose
(*c*) wicked (*d*) flagrant

168. Lampoon
(*a*) darken (*b*) praise
(*c*) abandon (*d*) sail

169. Languor
(*a*) vitality (*b*) length
(*c*) embarrassment (*d*) wine

170. Latent
(*a*) trim (*b*) forbidding
(*c*) execrable (*d*) obvious

171. Lavish
(*a*) hostile (*b*) unwashed
(*c*) timely (*d*) frugal

172. Laudatory
(*a*) dritying (*b*) disclaiming
(*c*) defamatory (*d*) inflammatory

173. Lax
(*a*) salty (*b*) strick
(*c*) shrill (*d*) boring

174. Lechery
(*a*) trust (*b*) compulsion
(*c*) zeal (*d*) purity

175. Lethargic
(*a*) convalescent (*b*) beautiful
(*c*) enervating (*d*) invigorating

176. Levite
(*a*) bridge (*b*) dam
(*c*) praise (*d*) solemnity

177. Lilliputian
(*a*) destructive (*b*) proper
(*c*) gigantic (*d*) elegant

178. Limpid
(*a*) erect (*b*) turbid
(*c*) tangential (*d*) timid

179. Lithe
(*a*) stiff (*b*) limpid
(*c*) facetious (*d*) insipid

180. Livid
(*a*) alive (*b*) mundane
(*c*) positive (*d*) undiscolored

181. Loath
(*a*) loose (*b*) evident
(*c*) deliberate (*d*) eager

182. Loquacious
(*a*) taciturn (*b*) sentimental
(*c*) soporific (*d*) soothing

183. Moist
(*a*) crisp (*b*) hard
(*c*) dry (*d*) parched

184. Mask
(*a*) hit (*b*) expose
(*c*) benign (*d*) pompous

185. Moisture
(*a*) dryness (*b*) delegate
(*c*) dampness (*d*) fdefect

186. Magnify
(*a*) produce (*b*) support
(*c*) reduce (*d*) destroy

187. Mendacious
(*a*) broken (*b*) efficacious
(*c*) truthful (*d*) destructive

188. Manumit
(*a*) throw (*b*) lock
(*c*) promise (*d*) enslave

189. Nebulous
(*a*) starry (*b*) clear
(*c*) cold (*d*) fundamental

190. Nefarious
(*a*) various (*b*) lacking
(*c*) benign (*d*) pompous

191. Negation
(*a*) postulation (*b*) hypothecation
(*c*) affirmation (*d*) violation

192. Neophyte
(*a*) veteran (*b*) satellite
(*c*) desperado (*d*) handwriting

193. Niggardly
(*a*) protected (*b*) biased
(*c*) prodigal (*d*) bankrupt

194. Nocturnal
(*a*) harsh (*b*) marauding
(*c*) patrolling (*d*) daily

195. Noisome
(*a*) quiet (*b*) dismayed
(*c*) fragrant (*d*) sleepy

196. Notorious
(*a*) fashionable (*b*) renowned
(*c*) inactive (*d*) interepid

197. Obdurate
(*a*) yielding (*b*) fleeting
(*c*) finite (*d*) fascinating

198. Obese
(*a*) skillful (*b*) cadaverous
(*c*) clever (*d*) unpredictable

199. Objective
(*a*) indecisive (*b*) apathetic
(*c*) markedly inferior (*d*) emotionally involved

200. Obligatory
(*a*) demanding (*b*) optional
(*c*) facile (*d*) friendly

201. Obloquy
(*a*) praise (*b*) rectangle
(*c*) circle (*d*) dialogue

202. Obsequious
(*a*) successful (*b*) democratic
(*c*) supercilious (*d*) ambitious

203. Obsession
(*a*) whim (*b*) loss
(*c*) phobia (*d*) delusion

204. Obsolete
(*a*) heated (*b*) desolate
(*c*) renovated (*d*) frightful

205. Obstreperous
(*a*) turbid (*b*) quiet
(*c*) remote (*d*) lucid

206. Obtuse
(*a*) sheer (*b*) transparent
(*c*) tranquil (*d*) shrewd

207. Odious
(*a*) fragrant (*b*) redolent
(*c*) fetid (*d*) delightful

208. Odium
(*a*) noise (*b*) liking
(*c*) dominant (*d*) hasty

209. Omnipotent
(*a*) weak (*b*) democratic
(*c*) despotic (*d*) passionate

210. Omniscient
(*a*) sophisticated (*b*) ignorant
(*c*) essential (*d*) trivial

211. Opiat
(*a*) distress (*b*) sleep
(*c*) stimulant (*d*) laziness

212. Opportune
(*a*) occasional (*b*) fragrant
(*c*) fragile (*d*) awkward

213. Opportunist
(*a*) man of destiny (*b*) man of principle
(*c*) changeling (*d*) adversary

214. Opporbrium
(*a*) delineation (*b*) aptitude
(*c*) majesty (*d*) praise

215. Optimum
(*a*) pessimistic (*b*) knowledgeable
(*c*) worst (*d*) minimum

216. Opulence
(*a*) pessimism (*b*) patrionsm
(*c*) potency (*d*) poverty

217. Ostentatious
(*a*) inactive (*b*) unassuming
(*c*) impolite (*d*) illicit

218. Overweening
(*a*) humble (*b*) impotent
(*c*) avid (*d*) acrimonious

219. Pejorative
(*a*) positive (*b*) legal
(*c*) determining (*d*) delighting

220. Pellucid
(*a*) logistical (*b*) philandering
(*c*) incomperhensible (*d*) vagrant

221. Penchant
(*a*) distance (*b*) imminence
(*c*) dislike (*d*) attitude

222. Penurious
(*a*) imprisoned (*b*) captivated
(*c*) generous (*d*) vacant

223. Perfunctory
(*a*) official (*b*) through
(*c*) insipid (*d*) vicarious

224. Perigee
(*a*) eclipse (*b*) planet
(*c*) apogee (*d*) refugee

225. Peripatetic
(*a*) wordly (*b*) stationary
(*c*) disarming (*d*) seeking

226. Permeable
(*a*) perishable (*b*) effective
(*c*) plodding (*d*) impenetrable

227. Pernicious
(*a*) practical (*b*) comparative
(*c*) harmless (*d*) tangible

228. Perpetual
(*a*) momentary (*b*) standard
(*c*) serious (*d*) industrial

229. Perspicuity
(*a*) grace (*b*) feature
(*c*) review (*d*) vagueness

230. Pert
(*a*) polite (*b*) perishable
(*c*) moral (*d*) deliberate

231. Pertinacious
(*a*) vengeful (*b*) consumptive
(*c*) superficial (*d*) skilled

232. Pertient
(*a*) understood (*b*) living
(*c*) discontented (*d*) irrelevant

233. Petulant
(*a*) angry (*b*) moral
(*c*) declining (*d*) uncomplaining

234. Precipitate
(*a*) dull (*b*) anticipatory
(*c*) cautious (*d*) considerate

235. Prefatory
(*a*) outstanding (*b*) magnificent
(*c*) conclusive (*d*) intelligent

236. Prelude
(*a*) intermezzo (*b*) diva
(*c*) aria (*d*) aftermath

237. Presumption
(*a*) assertion (*b*) activation
(*c*) motivation (*d*) humility

238. Pretentious
(*a*) ominous (*b*) calm
(*c*) unassuming (*d*) futile

239. Prim
(*a*) informal (*b*) prior
(*c*) exterior (*d*) private

240. Pristine
(*a*) cultivated (*b*) condemned
(*c*) irreligious (*d*) cautious

241. Probity
(*a*) aregert (*b*) assumption
(*c*) corruptibility (*d*) extent

242. Prodigal
(*a*) large (*b*) thrifty
(*c*) consistent (*d*) compatible

243. Prodigious
(*a*) infinitesimal (*b*) indignant
(*c*) indifferent (*d*) indisposed

244. Profane
(*a*) sanctify (*b*) desecrate
(*c*) define (*d*) manifest

245. Prolific
(*a*) unworkable (*b*) backward
(*c*) barren (*d*) controversial

246. Prolix
(*a*) stupid (*b*) indifferent
(*c*) redundant (*d*) pithy

247. Prophylactic
(*a*) causing growth (*b*) causing disease
(*c*) antagonistic (*d*) brushing

248. Propinquity
(*a*) remoteness (*b*) uniqueness
(*c*) health (*d*) virtue

249. Propitious
(*a*) rich (*b*) induced
(*c*) promoted (*d*) unfavorable

250. Prosaic
(*a*) pacified (*b*) reprieved
(*c*) pensive (*d*) imaginative

251. Protean
(*a*) amateur (*b*) catholic
(*c*) unchanging (*d*) rapid

252. Protract
(*a*) make circular (*b*) shorten
(*c*) further (*d*) rectrace

253. Provident
(*a*) unholy (*b*) rash
(*c*) miserable (*d*) disabled

254. Provincial
(*a*) wealthy (*b*) crass
(*c*) literary (*d*) sophisticated

255. Psychotic
(*a*) dangerous (*b*) clairvoyant
(*c*) criminal (*d*) sane

256. Puerile
(*a*) fragrant (*b*) adult
(*c*) lonely (*d*) feminine

257. Pugnacious
(*a*) pacific (*b*) feline
(*c*) mature (*d*) angular

258. Puissant
(*a*) pouring (*b*) fashionable
(*c*) articulate (*d*) weak

259. Pulchritude
(*a*) ugliness (*b*) notoriety
(*c*) bestiality (*d*) masculinity

260. Punctilious
(*a*) happy (*b*) active
(*c*) vivid (*d*) careless

261. Punitive
(*a*) large (*b*) humorous
(*c*) rewarding (*d*) restive

262. Pusillanimous
(*a*) poverty-stricken (*b*) chained
(*c*) posthumous (*d*) courageous

263. Putative
(*a*) colonial (*b*) quarrelsome
(*c*) undisputed (*d*) powerful

264. Resilient
(*a*) pungent (*b*) foolish
(*c*) worthy (*d*) unyielding

265. Restive
(*a*) buoyant (*b*) placid
(*c*) remorseful (*d*) resistant

266. Retentive
(*a*) forgetful (*b*) accepting
(*c*) repetitive (*d*) avoiding

267. Reticence
(*a*) fatigue (*b*) fashion
(*c*) treachery (*d*) loquaciousness

268. Retrograde
(*a*) progressing (*b*) inclining
(*c*) evaluating (*d*) concentrating

269. Revere
(*a*) advance (*b*) dishonor
(*c*) age (*d*) precede

270. Rife
(*a*) direct (*b*) scant
(*c*) peaceful (*d*) grim

271. Robust
(*a*) weak (*b*) violent
(*c*) vicious (*d*) villainous

272. Rotundity
(*a*) promenade (*b*) nave
(*c*) grotesqueness (*d*) slimness

273. Rubble
(*a*) artificial facade (*b*) unbroken stone
(*c*) pale complexion (*d*) strong defense

274. Ruddy
(*a*) robust (*b*) witty
(*c*) wan (*d*) exotic

275. Rudimentary
(*a*) pale (*b*) polite
(*c*) asinine (*d*) developed

276. Rueful
(*a*) trite (*b*) content
(*c*) capable (*d*) capital

277. Rustic
(*a*) urban (*b*) slow
(*c*) corroded (*d*) mercenary

278. Ruthless
(*a*) merciful (*b*) majestic
(*c*) mighty (*d*) militant

279. Sadistic
(*a*) happy (*b*) quaint
(*c*) kindhearted (*d*) vacant

280. Sagacious
(*a*) foolish (*b*) bitter
(*c*) voracious (*d*) veracious

281. Sallow
(*a*) salacious (*b*) ruddy
(*c*) colorless (*d*) permitted

282. Salubrious
(*a*) salty (*b*) bloody
(*c*) miasmic (*d*) maudlin

283. Salvage
(*a*) remove (*b*) outfit
(*c*) burn (*d*) lose

284. Sanctimonious
(*a*) hypothetical (*b*) paltry
(*c*) mercenary (*d*) pious

285. Sanguine
(*a*) choleric (*b*) sickening
(*c*) warranted (*d*) pessimistic

286. Satiety
(*a*) emptiness (*b*) warmth
(*c*) erectness (*d*) straightness

287. Scanty
(*a*) collected (*b*) remote
(*c*) invisible (*d*) plentiful

288. Scurrilous
(*a*) savage (*b*) scabby
(*c*) decent (*d*) volatile

289. Secular
(*a*) vivid (*b*) clerical
(*c*) punitive (*d*) positive

290. Sedentary
(*a*) vicarious (*b*) loyal
(*c*) accidental (*d*) active

291. Sedulous
(*a*) indolent (*b*) guileless
(*c*) vindictive (*d*) upright

292. Senlity
(*a*) verility (*b*) loquaciousness
(*c*) forgetfulness (*d*) youth

293. Sententious
(*a*) paragraphed (*b*) positive
(*c*) posthumous (*d*) wordy

294. Tractable
(*a*) unmanageable (*b*) irreligious
(*c*) mortal (*d*) incapable

295. Traduce
(*a*) exhume (*b*) increase
(*c*) purchase (*d*) extol

296. Tranquillity
(*a*) lack of sleep (*b*) lack of calm
(*c*) emptiness (*d*) renewal

297. Transient
(*a*) carried (*b*) close
(*c*) permanent (*d*) removed

298. Tremulous
(*a*) steady (*b*) obese
(*c*) young (*d*) healthy

299. Trenchant
(*a*) lacking bite (*b*) imperious
(*c*) inessential (*d*) unafraid

300. Trepidation
(*a*) slowness (*b*) amputation
(*c*) fearlessness (*d*) adroitness

301. Trite
(*a*) correct (*b*) original
(*c*) distinguished (*d*) premature

302. Truculent
(*a*) juicy (*b*) overflowing
(*c*) peaceful (*d*) determined

303. Tumult
(*a*) scarcity (*b*) defeat
(*c*) cooleness (*d*) secrenity

304. Turbid
(*a*) clear (*b*) improbable
(*c*) invariable (*d*) honest

305. Turbulence
(*a*) reaction (*b*) approach
(*c*) impropriety (*d*) calm

306. Turgid
(*a*) rancid (*b*) shrunken
(*c*) cool (*d*) explosive

307. Turpitude
(*a*) amplitude (*b*) heat
(*c*) wealth (*d*) virtue

308. Tyro
(*a*) infant (*b*) rubber
(*c*) personnel (*d*) expert

309. Unearth
(*a*) conceal (*b*) gnaw
(*c*) clean (*d*) fling

310. Unfeigned
(*a*) pretended (*b*) fashionable
(*c*) wary (*d*) switched

311. Ungainly
(*a*) ignorant (*b*) graceful
(*c*) detailed (*d*) dancing

312. Unimpeachable
(*a*) fruitful (*b*) rampaging
(*c*) faulty (*d*) pensive

313. Unkempt
(*a*) bombed (*b*) washed
(*c*) neat (*d*) tawdry

314. Unruly

(*a*) chatting (*b*) obedient
(*c*) definite (*d*) lined

315. Unseemly

(*a*) effortless (*b*) proper
(*c*) conducive (*d*) pointed

316. Unsullied

(*a*) tranished (*b*) countless
(*c*) soggy (*d*) papered

317. Untenable

(*a*) supportable (*b*) tender
(*c*) sheepish (*d*) tremulous

318. Unwitting

(*a*) clever (*b*) intense
(*c*) sensitive (*d*) intentional

319. Vacillation

(*a*) remorse (*b*) relief
(*c*) respect (*d*) steadfastness

320. Valedictory

(*a*) sad (*b*) collegiate
(*c*) derivative (*d*) salutatory

321. Valor

(*a*) admonition (*b*) injustice
(*c*) cowardice (*d*) generosity

322. Vanguard

(*a*) regiment (*b*) rear
(*c*) echelon (*d*) protection

323. Vaunted

(*a*) unvanquished (*b*) fell
(*c*) belittled (*d*) believed

REASONING ABILITY

MCQ TYPE QUESTIONS

1. Given two numbers x and y, define

Addition A(x, y) = x + y
Subtraction S (x, y) = x – y
Multiplication M (x, y) = x, y
Division D (x, y) = x/y

S[M(D(A(a, b), 2), D(A(a, b), 2)), M(D(S(a, b), 2), D(S(a, b),2))] is equal to

(*a*) $(a - b)^2$ (*b*) ab
(*c*) $(a + b)^2$ (*d*) None of these

Directions Q. 2 – 5 :

Kamal Babu came home just after judging a beauty contest where there were four semi-finalists, Ms. Andhra Pradesh, Ms. Uttar Pradesh, Ms. Maharashtra and Ms. West Bengal. His wife was very keen on knowing who the winner was and Kamal Babu replied immediately that it was the one wearing the yellow saree. When his wife asked for more details, he gave the following information:

* The four girls were wearing sarees of different colours (yellow, red, green, white) and the runner-up was wearing green.
* The four girls were sitting in a row, and Ms. West Bengal was not sitting at either end.
* There was only one runner-up and she was sitting next to Ms. Maharashtra.
* The girls wearing yellow and white sarees occupied the seats at either end.
* Ms. West Bengal was nither the winner nor the runner-up.
* Ms. Maharashtra was wearing white.
* The winner and the runner-up were not sitting next to each other.
* The girl wearing the green saree was not Ms. Andhra Pradesh.

Answer the following questions based on the above information:

2. What was the colour of the saree that Ms. Andhra Pradesh was wearing?

(*a*) White
(*b*) Yellow
(*c*) Red
(*d*) Cannot be determined

3. Between which two was Ms.West Bengal sitting?

(*a*) Ms. Andhra Pradesh and Ms. Maharashtra
(*b*) Ms. Andhra Pradesh and Ms. Uttar Pradesh
(*c*) Ms. Uttar Pradesh and Maharashtra
(*d*) Cannot be deternmined

4. Who was wearing the red saree?

(*a*) Ms. Andhra Pradesh
(*b*) Ms. Uttar Pardesh
(*c*) Ms. West Bengal
(*d*) Cannot be determined

5. What was the colour of the saree that Ms. Uttar Pradesh was wearing?

(*a*) White (*b*) Green
(*c*) Red (*d*) Yellow

Directions Q. 6 – 7 : A, B, C, D, E and F are a group of friends from a club. There are two housewives, one lecturer, one architect, one accountant and one lawyer in the group. There are two married couples in the group. The lawyer is married to D who is a housewife. No lady in the group is either an architect or an accountant C. The accountant, is married to F who is a lecturer. A is married to D and E is not a housewife.

6. What is E ?
   (*a*) Lawyer
   (*b*) Architect
   (*c*) Lecturer
   (*d*) Accountant
7. How many members of the group are males?
   (*a*) 2 (*b*) 3
   (*c*) 4 (*d*) none of these

Directions Q. 8 – 9 :

Seven university cricket players are to be honoured at a special luncheon. the players will be seated on the dais along one side of a single rectangular table.

A and G have to leave for the lunch early and must be seated at the extreme right end of the table which is closest to the exit.

B will receive the Man of the Match award and must be in the center chair.

C and D, who are bitter rivals for the position of wicket keeper, dislike each other and should be seated as far apart as possible.

E and F are best friends and want to sit together.

8. Which of the following may not be seated at either end of the table ?
   (*a*) C (*b*) D
   (*c*) G (*d*) F
9. Which of the following pairs may not be seated together ?
   (*a*) E & A (*b*) B & D
   (*c*) C & F (*d*) G & D

Directions Q. 10 –13 :

A study was conducted to ascertain the relative importance that employees in five different countries assigned to five different traits in their Chief Executive Officers. The traits were compassion (C), decisiveness (D), negotiation skills (N), public visibility (P), and vision (V). The level of dissimilarity between two countries is the maximum difference in the ranks allotted by the two countries to any of the five traits. The following table indicates the rank order of the five traits for each country.

COUNTRY

RANK	INDIA	CHINA	JAPAN	MALAYSIA	THAILAND
1	C	N	D	V	V
2	P	C	N	D	C
3	N	P	C	P	N
4	V	D	V	C	P
5	D	V	P	N	D

10. Which of the following pairs of countries are most dissimilar?
    (*a*) China and Japan
    (*b*) India and China
    (*c*) Malaysia and Japan
    (*d*) Thailand and Japan
11. Which of the following countries is least dissimilar to India?
    (*a*) China (*b*) Japan
    (*c*) Malaysia (*d*) Thailand
12. Which amongst the following countries is most dissimilar to India?
    (*a*) China (*b*) Japan
    (*c*) Malaysia (*d*) Thailand
13. Three of the following four pairs of countries have identical levels of dissimilarity. Which pair is the odd one out?
    (*a*) Malaysia and China
    (*b*) China and Thailand
    (*c*) Thailand and Japan
    (*d*) Japan and Malaysia

Directions Q. 14 – 17 :

Four families decided to attend the marriage ceremony of one of their colleagues. One family has no kids, while others have at least one kid each. Each family with kids has at least one kid attending the marriage.

Given below is some information about the families, and who reached when to attend the marriage.

The family with 2 kids came just before the family with no kids.

Shanthi who does not have any kids reached just before Sridevi's family.

Sunil and his wife reached last with their only kid.

Anil is not the husband of Joya.

Anil and Raj are fathers.

Sridevi's and Anita's daughters go to the same school.

Joya came before Shanthi and met Anita when she reached the venue.

Raman stays the farthest from the venue.

Raj said his son could not come because of his exams.

14. Which woman arrived third ?
(*a*) Shanthi (*b*) Sridevi
(*c*) Anita (*d*) Joya

15. Name the correct pair of husband and wife ?
(*a*) Raj and Shanthi (*b*) Sunil and Sridevi
(*c*) Anil and Sridevi (*d*) Raj and Anita

16. Of the following pairs, whose daughters go to the same school ?
(*a*) Anil and Raman (*b*) Sunil and Raman
(*c*) Sunil and Anil (*d*) Raj and Anil

17. Whose family is known to have more than one kid for certain ?
(*a*) Raman's (*b*) Raj's
(*c*) Anil's (*d*) Sunil's

Directions Q. 18 – 20 :

Answer the following questions based on the statements given below:

(*i*) There are three houses on each side of the road.
(*ii*) These six houses are labeled as P, Q, R, S, T and U.
(*iii*) The houses are of different colours, namely, Red, Blue, Green, Orange, Yellow and White.
(*iv*) The houses are of different heights.
(*v*) T, the tallest house, is exactly opposite to the Red coloured house.
(*vi*) The shortest house is exactly opposite to the Green coloured house.
(*vii*) U, the Orange coloured house, is located between P and S.
(*viii*) R, the Yellow coloured house, is exactly opposite to P.
(*ix*) Q, the Green coloured house, is exactly opposite to U.
(*x*) P, the White coloured house, is taller than R, but shorter than S and Q.

18. What is the colour of the house diagonally opposite to the Yellow coloured house?
(*a*) White (*b*) Blue
(*c*) Green (*d*) Red

19. Which is the second tallest house?
(*a*) P (*b*) S
(*c*) Q (*d*) cannot be determined

20. What is the colour of the tallest house?
(*a*) Red (*b*) Blue
(*c*) Green (*d*) Yellow

NUMERICAL TYPE QUESTIONS

1. Given two numbers x and y, define

Addition	A(x, y) = x + y
Subtraction	S (x, y) = x – y
Multiplication	M (x, y) = x, y
Division	D (x, y) = x/y

If x = 2, y = 3 and z = 5, then value of M[A(S (x, y), M (y, z)), S (A(x, z), S (z, y))] is ______

Directions Q. 2 – 3 :

These questions are based on the following data

A, B, C, D are standing in a queue. It is known that C and D are not standing adjacent to each other and that B is not in the third place.

2. If A is not in the second place and B is in the first place, then which is the place for D is ______
3. If A is not standing adjacent to B and D is the last in the queue, then the place for C is ______

Directions Q. 4 – 5 :

A boy is asked to put in a basket one mango when ordered 'One', one orange when ordered 'Two', one apple when ordered 'Three' and is asked to take out from the basket one mango and an orange when ordered 'Four'. A sequence of orders is given as
1 2 3 3 2 1 4 2 3 1 4 2 2 3 3 1 4 1 1 3 2 3 4

4. The ______total organges were in the basket at the end of the above sequence.
5. The ______total fruits will be in the basket at the end of the above order sequence.

Directions Q. 6 – 7

Each of the 11 letters A, H, I, M, O, T, U, V, W, X and Z appears same when looked at in a mirror. They are called symmetric letters. Other letters in the alphabet are asymmetric letters.

6. The______ number of four-letter computer passwords can be formed using the symmetric letters (no repetition allowed)
7. The ______ number of three-letter computer passwords can be formed (no repetition allowed) with at least one symmetric letter.

Directions Q. 8 – 9 :

A string of three English letters is formed as per the following rules:

(*a*) The first letter is any vowel
(*b*) The second letter is m, n or p.
(*c*) If the second letter is m then the third letter is any vowel which is different from the first letter
(*d*) If the second letter is n then the third letter is *e* or *u*
(*e*) If the second letter is p then the third letter is the same as the first letter.

8. The ______ strings of letters can possible be formed using the above rules.

9. The ______ strings of letters can possibly be formed using the above rules such that the third letter of the string is e.

NUMBERS AND ALGEBRA

1. If x = b + c, y = c – a, z = a – b, then $x^2 + y^2 + z^2 - 2xy - 2xz + 2yz$ is equal to

(a) a + b + c (b) $4b^2$

(c) abc (d) $a^2 + b^2$

2. HCF of $y^3 - 4y$ and $4y(y^3 + 8)$ is

(a) y(y + 2) (b) y – 2

(c) 2(y + 2) (d) $(y + 2)(y^2 - 2y + 4)$

3. A tiled floor of a room has dimensions m × m sq.m. Dimensions of the tile used are n × n sq.m. All tiles used are green tiles except diagonal tiles which are red. After some years some green tiles are replaced by red tiles to form an alternate red and green tile pattern. How many green tiles are removed?

(m ≠ n and total number of tiles are odd).

(a) $\frac{m^2 - 4mn + 2n^2}{2n^2}$ (b) $\frac{(m - 2n)^2 - n^2}{2n^2}$

(c) $\frac{m^2 - 4mn - n^2}{2n^2}$ (d) $\frac{m^2 - 4mn - 2n^2}{2n^2}$

4. If $\frac{x-4}{-x^2-6} \le \frac{2}{-x^2-6}$ then

(a) $x \le 2$ (b) $x \ge 2$

(c) $x \le 6$ (d) $x \ge 6$

5. Which of the following is true?

(a) $\sqrt{2} + \sqrt{5} = \sqrt{7}$

(b) $\sqrt{2} + \sqrt{5} \le \sqrt{7}$

(c) $\sqrt{2} + \sqrt{5} < \sqrt{7}$

(d) $\sqrt{2} + \sqrt{5} > \sqrt{7}$

6. If $6 \ge x \ge -2$ and $4 \ge y \ge -4$, then limits for $\frac{y}{x}$, where x and y are non zero integers, is

(a) $\frac{y}{x} \ge 2, \frac{y}{x} \le \frac{2}{3}$ (b) $\frac{y}{x} \ge \frac{-2}{3}, \frac{y}{x} \le 2$

(c) $\frac{y}{x} \ge \frac{-2}{3}, \frac{y}{x} \le \frac{1}{4}$ (d) $\frac{y}{x} \ge -4, \frac{y}{x} \le 4$

7. What are the limits of $\frac{2^n(n+1)^n}{n^n}$; where n is a positive integer?

(a) $2 \le \frac{2^n(n+1)^n}{n^n} \le 3$

(b) $2^{n+1} \le \frac{2^n(n+1)^n}{n^n} \le 3.2^n$

(c) $2^n \le \frac{2^n(n+1)^n}{n^n} \le 3.2^n$

(d) None of these

8. If $g = 10^{100}$ and $H = 10^g$, then in which interval does $g! = 1.2.3 .... 10^{100}$ lie?

(a) 10H < g! < H

(b) $H < g! < 10^H$

(c) $10H < g! < 10^H$

(d) $10^H < g!$

9. Which of the following is true?

(a) Sum of four consecutive even numbers is always divisible by 8.

(b) Sum of four consecutive odd numbers is always divisible by 8.

(c) Product of any n consecutive natural numbers may not be divisible by n!.

(d) Product of 4 consecutive odd numbers is always divisible by 15.

10. If a and b are prime numbers, which of the following is true?

I. a^2 has three positive integer factors.

II. ab has four positive integer factors.

III. a^3 has four positive integer factors.

Codes :

(a) I and II only

(b) II and III only

(c) All of these

(d) None of these

11. If 7x + 6y = 420, x and y are natural numbers, then what can be said about x?

(a) x is always odd

(b) x is always even.

(c) x is even only if y is odd.

(d) x is odd if y is even.

12. If a, b, c, d, p and q are non-zero, unequal integers and $\frac{a+bi}{c+di}=\frac{p}{q}$, then $\frac{a^2+b^2}{c^2+d^2}$ equals

(a) $\frac{p}{q}$ (b) $\frac{p^2}{q^2}$

(c) 1 (d) None of these

13. An Egyptian fraction has a numerator equal to 1, and its denominator is a positive integer. What is the maximum number of different Egyptian fraction such that their sum is equal to 1, and their denominators are equal to 10 or less?

(a) 3 (b) 5

(c) 7 (d) 9

14. What is the least number which must be subtracted from 1936 so that the remainder when divided by 9, 10, 15 will leave in each case the same remainder 7?

(a) 32 (b) 53

(c) 46 (d) 39

15. What is the greatest number consisting of six digits which on being divided by 6, 7, 8, 9, 10 leaves 4, 5, 6, 7, 8 as remainders respectively?

(a) 997920 (b) 997918

(c) 999999 (d) 997922

16. What is the least number which on being divided by 5, 6, 8, 9, 12 leaves in each case a remainder 1 but when divided by 13 leaves no remainder?

(a) 2987 (b) 3601

(c) 3600 (d) 2986

17. A group of four numbers has only one prime number amongst them. Which of the following must be true about the group?

I. HCF of the four numbers of the group is either 1 or equal to that prime number.

II. LCM of the four numbers of the group is same as product of the prime number and LCM of the remaining three numbers.

III. Product of four numbers is equal to product of the prime number × HCF of the group × LCM of the group.

(a) I only (b) II only

(c) I and II only (d) All of these

18. If (a, n)! is defined as product of n consecutive numbers starting from a, where a and n are both natural numbers, and if H is the HCF of (a, n)! and n!, then what can be said about H?

(a) H = a! (b) H = n!

(c) $H \geq n!$ (d) $H \geq a \times n$

19. HCF and LCM of two numbers is given. It is possible to find out the two numbers uniquely if

I. either sum or difference between the two numbers is known.

II. HCF of two numbers = LCM of two numbers.

III. $\frac{\text{LCM}}{\text{HCF}}$ = Prime number.

(a) I and II only

(b) II only

(c) II and III only

(d) I, II and III

20. A ball is dropped from a height of 12 m and it rebounds 1/2 of the distance it falls. If it continues to fall and rebound in this way, how far will it travel before coming to rest?

(a) 36 m (b) 30 m

(c) 48 m (d) 60 m

21. There are N questions in an exam.

For i = 1, 2, ..., N, there are 2^{N-1} students who answered 1 or more questions wrongly. If total number of wrong answers is 8191, then N will be

(a) 12 (b) 11

(c) 10 (d) 13

22. Two new charity organizations C_1 and C_2 were formed, with x members each, on January 1, 2003. On first day of each, subsequent month, in C_1, number of members increases by a certain number a, while in C_2, number of members increases in such a way that ratio of the number of members in a month to the preceding month bear a ratio equal to b. On May 1, 2003, both organizations had the same number of members. If a = 20x, then b will be

(a) 2 (b) 3

(c) 2.5 (d) 3.5

23. What is the ratio of common differences d_1 and d_2 of two arithmetic progressions if respective n^{th} terms are in the ratio of 2n + 3 : n – 11?

(a) 1 : 2 (b) 2 : 3

(c) 2 : 1 (d) 1 : 3

24. What is the sum upto 20 terms of the series $\frac{2}{3}+\frac{8}{9}+\frac{26}{27}+\frac{80}{81}+......?$

(a) $\frac{19.3^{20}+1}{3^{20}}$ (b) $\frac{19.3^{20}+1}{2.3^{20}}$

(c) $\frac{19.3^{20}-1}{2.3^{20}}$ (d) $\frac{39.3^{20}+1}{2.3^{20}}$

25. What is the sum to infinity of the series, $3 + 6x^2 + 9x^4 + 12x^6 + \ldots$ given $|x| < 1$?

(*a*) $\frac{3}{(1+x^2)}$ (*b*) $\frac{3}{(1-x^2)^2}$

(*c*) $\frac{3}{(1+x^2)^2}$ (*d*) $\frac{3}{(1-x^2)}$

26. A school has 5 divisions in a class IX having 60, 50, 55, 62 and 58 students. Mean marks obtained in a History test were 56, 64, 72, 63 and 50 by each division respectively. What is overall average of the marks per students?

(*a*) 56.8 (*b*) 58.2

(*c*) 62.4 (*d*) 60.8

27. Average age of a committee of seven trustees is the same as it was five years ago, a younger man having been substituted for one of them. How much younger was he then trustee whose place he took?

(*a*) 32 years (*b*) 35 years

(*c*) 33 years (*d*) 34 years

28. Six men A, B, C, D, E, F agree with a seventh man G to provide a sum of money among them. A, B, C, D, E, F are to subscribe ₹10 each, and G is to pay ₹3 more than the average of the seven. What is the whole sum to be provided?

(*a*) ₹73.50 (*b*) ₹ 74

(*c*) ₹ 73 (*d*) ₹ 72.50

29. Which of the following is/are true?

I. When Arithmetic Mean of a set of values and number of values in the set are known, then Geometric Mean of the set can be found out.

II. When Arithmetic Mean of a set of values and number of values in the set are known, sum of all the elements in the set can be found out.

III. When Geometric Mean of a set of values and number of values in the set are known, then Arithmetic Mean of the set can be found out.

Codes :

(*a*) Only I (*b*) Only II

(*c*) Only I and II (*d*) All except II

30. Which of the following statement/s is/are true?

I. Average of a set of values will always lie between lowest and the largest of these values.

II. If each of the values in a set is increased by a constant k, then new average of the set is increased by k.

III. If each of the values in a set is multiplied by a constant m, then new average will also be 'm' times the old average.

Codes :

(*a*) Only I and II (*b*) Only II and III

(*c*) Only I and III (*d*) I, II and III

31. Three people A, B and C weigh themselves in a particular order. First A, B, C weigh themselves individually and then AB, BC CA and ABC together respectively. Recorded weight for the last measured is 180 kgs. Then average of the 7 measures is

(*a*) 320 kgs (*b*) $\frac{360}{7}$ kgs

(*c*) $\frac{720}{7}$ kgs (*d*) Cannot be determined

32. Average age of a team of 15 employees is 36. Youngest of them is 20 years old and eldest is 36 years old. Two of them with average age 28 leave the team. If one of the two comes back on the condition that he will be made the team leader, then which of the following can possibly be average age of the new team so formed?

(*a*) 35 (*b*) 36

(*c*) 38 (*d*) 39

PERCENTAGE AND ITS APPLICATIONS

1. The Manager (M) of the company along with his Assistant Manager (AM) and a Business Analyst (BA) invest ₹ 55000, ₹ 45000 and ₹ 60000 respectively in a partnership firm for six months. The Manager being the highest authority among the three gets one-tenth of the total profit and Assistant Manager being second highest authority gets $\frac{1}{20}$th of the remaining and the remaining profit is divided among the three in the ratio of their investments. If the total profit was ₹ 40000, who gets the minimum share and how much?

(*a*) BA, ₹ 11400 (*b*) AM, ₹ 11419

(*c*) AM, ₹ 12500 (*d*) None of these

2. In a business partnership of three people A, B and C, B's share in total profit is ₹ 250 per ₹ 2000. At this profit his return on investment is 15%. If B earns total of ₹ 900. what is the total investment made by A and C in business? Assume that each of A, B and C have invested their money for the entire year and profits are shared in the ratio of their investments.

(*a*) ₹ 48000 (*b*) ₹ 42000

(*c*) ₹ 6000 (*d*) Data insufficient

3. Abhay and Bimal started a business in partnership. Abhay invested a capital of ₹ 50000. Bimal agrees to pay Abhay an interest on half the capital at 8% per annum. Also Bimal receives a salary of ₹ 500 per month for managing the business. After six months Abhay withdraws ₹ 20000 and Bimal pitches in an equal amount. Bimal continues to pay Abhay interest at 8% per annum for the balance amount. At the end of the year they divide the profit equally and found that total earning of Bimal in the year was exactly thrice that of Abhay. The total profit ?

(*a*) ₹ 7200 (*b*) ₹ 6800

(*c*) ₹ 5400 (*d*) ₹ 4800

4. How much stock must be bought in 3 percent $89\frac{1}{8}$ in order that by selling it at $91\frac{3}{8}$ a profit of ₹ 45 may be made?

(*a*) 20 (*b*) 30

(*c*) 40 (*d*) 50

5. Which is the better investment (the security being equal), 3 percents perpetually at par, or 5 percents at 124, the latter stock being redeemable in 20 years at par? (Simple interest to be reckoned).

(*a*) 3% at par

(*b*) 5% at 124

(*c*) Both are same

(*d*) Cannot be determined

6. What rate percent per annum does a man get by buying ₹100 share (₹ 70 paid-up) at ₹ 60 when dividend is 5% per annum?

(*a*) 4.28% (*b*) 7.14%

(*c*) $8\frac{1}{3}$ % (*d*) $5\frac{5}{6}$ %

7. A rejects 0.08% of the meters as defective .How many will he examine to reject 2?

(*a*) 2500 (*b*) 1250

(*c*) 625 (*d*) 5000

8. (a% of b) + (b% of a) is equal to

(*a*) a% of b (*b*) 2% of ab

(*c*) 20% of ab (*d*) 2% of 100 ab

9. Ryan's stock of goods is worth \$9462, which is 15% more than Harland's, and 15% less than Lambert's. What is the value of stock carried by Harland and Lambert respectively?

(*a*) \$ 8218, \$ 10032 (*b*) \$ 1419, \$ 10881

(*c*) \$ 8042, \$ 10881 (*d*) \$ 8228, \$ 11132

10. Numerator of a fraction increases by 25%. What can be said about the resultant fraction?

(*a*) It is 67% less than original.

(*b*) It is equal to the original fraction.

(*c*) It is 67% more than the original fraction.

(*d*) The change cannot be determine.

11. Which of the following statement/s is/are true?

I. If two numbers are such that one is 25% more than the second, then sum of the two numbers is nine times that of the difference of the two numbers.

II. If 56% of a number A is 24% of another number B then 49% of A is 21% of B.

Codes:

(*a*) Only I (*b*) Only II

(*c*) Only I and II (*d*) None of these

12. In the Annual General Meeting of the Rotary club a chairman is to be selected for which A and B is the only two candidates. After the first round, each of them gave a speech on their policy and work plan for the coming year after which another round of election was held. The number of votes not cast for A increased by 25% in the second round over those not cast for him in the first round and A lost second round to B by twice as many votes as that by which he had won in the first round. If total 260 people voted each time what was the percent decrease in votes received by A in the second round assuming that no vote was void in any of the rounds?

(*a*) 18% (*b*) 25%

(*c*) 21.43% (*d*) 30%

13. A watermelon weighs 5000 gm. 99% of its weight is water. It is kept in a drying room and after some time it turns out that it is only 98% water by weight. What is its weight now?

(*a*) 2500 gm (*b*) 4500 gm

(*c*) 4950 gm (*d*) None of these

14. A sum of ₹ 5000 is divided into two parts A and B such that simple interests at the rate of 20% p.a. on A and B after 2 and 3 years respectively are equal. Which of the following is false about A and B?

(*a*) Ratio of A to B is 2 : 3

(*b*) Ratio of A to B is 3 : 2

(*c*) A exceeds B by ₹ 1000

(*d*) Total interest earned on A and B is ₹ 2400

15. Which amongst the following two offers is the better one? Consider with respect to interest accrued.

I. Investing an amount compounded annually at 1% per annum for 100 years.

II. Investing the amount compounded annually at 100% per annum for 1 year.

Codes:

(*a*) First offer

(*b*) Second offer

(*c*) Both are same

(*d*) Cannot be determined

16. The population of rate in a locality x increases by 20% in one year. Observing this, the pest control committee decided to use a special kind of pesticide 'xyz' which effectively kills 160 rats in 3 months. Just after 2 years, what is the net increase or decrease in the population of rats if, initially the population of rats is 3200 and pesticide is used effectively?

(*a*) Increase of 128 rats.

(*b*) Decrease of 128 rats.

(*c*) Neither an increase nor a decrease in the population.

(*d*) None of these.

17. A man travels from A to B to buy goods which he can get 10% cheaper in B than in A. If expenses of the journey are ₹ 15 and he makes a clear saving of ₹ 10, how much does he pay for the goods?

(*a*) ₹ 225

(*b*) ₹ 200

(*c*) ₹ 150

(*d*) Cannot be determined

18. By selling 5 dozen mangoes for ₹ 156, it was found that $\frac{3}{10}$th of the outlay was gained. What should the retail price per mango be in order to gain 60%?

(*a*) ₹ 4 (*b*) ₹ 2

(*c*) ₹ 3.2 (*d*) ₹ 4.2

19. A man sells sugar to a tradesman at a profit of 20% but the tradesman becoming bankrupt pays only 80 paise in the rupee. How much percentage does the man gain or lose by his sale?

(*a*) 2.5% (*b*) 3%

(*c*) 4% (*d*) 5.2%

20. A trader allows a discount of 5 percent to his customers. What price should he mark on a article the cost price of which is ₹ 800 so as to make clear profit of 25 percent on his outlay?

(*a*) ₹ 1000 (*b*) ₹ 1053

(*c*) ₹ 1200 (*d*) ₹ 1123

21. A camera costing ₹550 is marked to be sold at a price, which gives a profit of 30%. What will be its selling price in a sale when 20% is taken off the marked price?

(*a*) ₹ 600 (*b*) ₹ 572

(*c*) ₹ 635 (*d*) ₹ 605

22. Buy two and get one free. Which of the following is true?

(*a*) The person is selling at a loss.

(*b*) The person is selling at a profit.

(*c*) The loss or gain depends on the profit from sale of two articles.

(*d*) The person is selling at the cost price.

Directions (Q.23–25): *Refer to the data below and answer the questions that follow.*

A profit-maximizing producer has the option of discriminating between markets A and B (i.e., charging different prices in markets A and B for the same product) for a product X. Demand for the product in each of the markets A and B is given by

$$Q_A = 42 - 0.2\,P_A$$
$$Q_B = 100 - 0.8\,P_B$$

Total cost in any market is given by

$C = 1000 + 5Q$, where Q is quantity sold in that market.

Consider two cases:

***Case 1** in which she discriminates between the markets and*

***Case 2** in which she does not discriminate.*

23. What is the approximate ratio of the price in market A to price in market B in Case 1?

(*a*) 1.65 (*b*) 2.00

(*c*) 2.25 (*d*) 2.50

24. What is the ratio of the total quantity of product X sold in Case 1 to the total quantity sold in Case 2?

(*a*) 0.50 (*b*) 1

(*c*) 1.10 (*d*) 1.05

25. What is the difference in total profit in Case 1 and the total profit in Case 2?

(*a*) ₹ 289 (*b*) ₹ 829

(*c*) ₹ – 289 (*d*) ₹ – 829

TIME AND WORK

MCQ TYPE QUESTIONS

1. A barrel full of beer has 2 taps, one midway which draws a litre in 6 minutes and the other at the bottom which draws a litre in 4 minutes. The lower tap is normally used after the level of beer in the barrel is lower than midway. The capacity of the barrel is 36 litres. A new assistant opens the lower tap when the barrel is full and draws out some beer. As a result the lower tap has been used 24 minutes before the usual time. For how long was the beer drawn out by the new assistant?

   (*a*) 10 mins. (*b*) 8 mins.
   (*c*) 16 mins. (*d*) 32 mins.

2. Pipe A can fill a cistern in 36 minutes and B in 48 minutes. If both the pipes are opened together, when should pipe B be closed so that the cistern may be just full in 24 minutes?

   (*a*) 8 mins. (*b*) 9 mins.
   (*c*) 12 mins. (*d*) 16 mins.

3. Pipes P, Q and R are attached to a tank and each can act as either an inlet or outlet pipe. Pipes P, Q and R respectively take 8, 10 and 12 hours to fill the empty tank or empty the full tank. In the first hour, pipes P and R work as inlet and Q work as outlet. In the second hour, pipes P and Q work as inlet and pipe R as outlet. In the third hour pipes Q and R work as inlet and pipe P as outlet and the process goes on like this. When will the cistern be filled?

   (*a*) In the 8th hour. (*b*) In the 9th hour.
   (*c*) In the 10th hour. (*d*) In the 11th hour.

4. There are 2 inlet pipes and an outlet pipe. The efficiency of one of the inlet pipes is double than that of the other. Also, the efficiency of the outlet pipe is half that of the lesser efficient inlet pipe. The empty tank gets filled in 16 hours when all the pipes are opened. How many hours will be taken to fill the empty tank when the 'lesser' efficiency inlet pipe is plugged and the rest kept opened?

   (*a*) $33\frac{1}{3}$ hours (*b*) $26\frac{2}{3}$ hours
   (*c*) 100 hours (*d*) 80 hours

5. A train overtakes 2 persons walking at 2 km/hr and 4 km/hr respectively in the same direction and completely passes them in 9 seconds and 10 seconds. Length of the train and its speed in km/hr respectively are

   (*a*) 75 m, 18 km/hr
   (*b*) 80 m, 21 km/hr
   (*c*) 60 m, 20 km/hr
   (*d*) 50 m, 22 km/hr

6. A policeman goes after a thief who has a 100 m start. The policeman runs 1 km. in 6 mins. and the thief 1 km in 10 mins. How far did the thief go before he was overtaken?

   (*a*) 100 m (*b*) 125 m
   (*c*) 150 m (*d*) 250 m

7. Everyday Sachin exercises for $2\frac{1}{2}$ hours. He exercises by walking, jogging and running. His speeds while walking, jogging and running in the ratio 1 : 2 : 4. Which of the following statements are true?

   I. If he spends equal time on the three activities, then he runs twice the distance that he jogs.
   II. If he runs at 8 km per hour and distance covered by running and jogging together is 12 km, then he walks a distance of 1 km in 1 hour.
   III. If the time for which he walks, jogs and runs is in the ratio 2 : 2 : 1 and he covers a total distance of 10 km, then his speeds while walking, jogging and running are 2 km/hr, 4 km/hr and 8 km/hr res-pectively.

   Codes:

   (*a*) I and II (*b*) II and III
   (*c*) I and III (*d*) I, II and III

8. A monkey climbing up a greased pole ascends 10 metres and slips down 2 metres in alternate minutes. If the pole is 64 metres high, how long will it take him to reach the top?

   (*a*) 16 min
   (*b*) 14 min and 48 sec
   (*c*) 12 min
   (*d*) 14 min

9. A train travelling at 10 m/sec from A to B at 7 a.m. meets a train leaving B at 7:20 a.m. and coming to A at a speed $\frac{1}{3}$ times faster than the first train. If the distance from A to B is 68 km. then, at what distance from A will the two trains meet?

   (*a*) 72 km (*b*) 36 km
   (*c*) 60 km (*d*) 50 km

10. Two trains A and B start from stations X and Y towards each other. B leaves station Y half an hour after train A leaves station X. Two hours after train A has started, the distance between trains A and B is $\frac{19}{30}$th of the distance between stations X and Y. How much time would it take each train (A and B) to cover the distance X to Y, if train A reaches half an hour later to its destination as compared to B?

(a) 8 hrs, 6 hrs (b) 5 hrs, 4 hrs
(c) 10 hrs, 9 hrs (d) 9 hrs, 8 hrs

11. Two men, A and B, run a 4 km race on a circular course of $\frac{1}{4}$ km. If their speeds are in the ratio of 5 : 4, how often does the winner pass the other?

(a) Once (b) Twice
(c) Thrice (d) Four times

12. A and B start from the same point to run in opposite directions round a circular path 550 yards in length, A giving B a start of 100 yards. They pass each other when A has run 250 yards. Who will come first to the starting point and at what distance will they be apart?

(a) A, 20 yards (b) B, 10 yards
(c) B, 20 yards (d) A, 10 yards

13. Two runners run in the same direction along a circular track 2 km long. The faster runner overtakes the slower one every 30 minutes. What are their speeds, if faster one completes a round one minute sooner than the other?

(a) 24 km/hr and 20 km/hr
(b) 44 km/hr and 28 km/hr
(c) 20 km/hr and 18 km/hr
(d) 20 km/hr and 16 km/hr

14. To do certain work alone 'A' takes 4 hrs, 'B' takes 5 hrs and 'C' takes 6 hrs. How long would A and B together take to do a work which C can do in $4\frac{1}{2}$ hrs.?

(a) 2 hrs 10 mins. (b) 2 hrs 30 mins.
(c) 1 hr 20 mins. (d) 1 hr 40 mins.

15. A can do piece of work in 8 days which B and C can do in 12 and 16 days respectively. If they work on alternate days, then which of the following is true?

(a) If A started the work on the first day followed by B and c, then the work will be finished on the 10th day.
(b) If B started the work on the first day followed by C and A, then the work will be finished on the 11th day.
(c) If C started the work on the first day followed by A and B, then the work will be finished on the 11th day.
(d) None of these

NUMERICAL TYPE QUESTIONS

1. Meera paints walls for a living. She sometimes has Ashok to work with her. Meera takes 2 days to paint a 10 ft by 10 ft wall. Ashok takes 3 days to do the same job. Meera has to pay Ashok for his work on a per day basis, so she'd rather do without him. Meera has recently got a contract to paint the walls and ceiling of a room of dimensions 10 ft by 15 ft and a height of 10 ft. She had promised to do the job within 9 days. The ______ number of days does she need to employ Ashok?

2. 3 pipes can fill a reservoir in 10, 15 and 20 hours respectively. If the three taps are opened one after another in the given order, with a certain fixed time gap between them, the reservoir fills in 5 hours. The time gap will be _____ hr.

3. Two pipes A and B can fill a tank in 8 hours. If only pipe A is open, then it would take 4 hours longer to fill the tank. The _______ hours it would take if only pipe B is open.

Directions (Q.4 – 5): *Refer to the data below and answer the questions that follow.*

The boiler tank in a chemical factory holds 10^5 litres. 5 tanks each having one-fifth the capacity of the boiler tank fill in 'hard water' at same rates in the boiler tank in 2 hours. The outlet of the two of smaller tanks work as inlet pipes and other two work as outlet and the fifth tank fill in the main 'boiler' at half its efficiency.

4. The number of hours taken to fill in the main boiler when all the pipes are opened at once is ________

5. Only three pipes are working, two at their full efficiency and the third one with half of its efficiency and all three are acting as an inlet. If all the three are opened alternately with the latter pipe being first to be opened to fill the boiler such that each smaller tank fills 'hard water' in the main boiler for equal time, then number of hours taken to fill the boiler to half its capacity is______

6. A contract is to be completed in 46 days and 117 men were set to work, each working 8 hours a day. After 33 days, 4/7 of the work is completed. The _______ number of additional men may be employed so that the work may be completed in time, each man now working 9 hours a day?

7. A train travelling at 42 km/hr. passes a cyclist going in the same direction in 9 secs.; if cyclist had been going in the opposite direction, the train would have passed him in 5 secs. Length of the train is______ metres

8. A man rows upstream 13 km. and downstream 28 km. taking 5 hrs. each time. Velocity of the current in km/hr_______ is

Directions (Q. 9 – 11): *Refer to the data below and answer the questions that follow.*

The variation in the speed of a car on a particular day at the respective times is shown in the table below:

s(km/hr)	t(hr)
0	11.00 am
40	11.30 am
50	1.00 pm
85	1.30 pm
10	3.30 pm
10	4.30 pm

9. The distance travelled by the car from 11 a.m. to 1 p.m. is _______km

10. If car maintains the speed it has at 4.30 p.m. then at _______ time will the car cover the same distance as it had covered from 11.00 a.m. to 1.00 p.m.?

11. Average speed of the car from 11.00 a.m. to 4.30 p.m. is ________ kmph

12. A cheetah chase a deer which is 100 m ahead. The time in which the deer takes 10 leaps the cheetah takes only 6 leaps. In one leap, the deer covers 1 m while the cheetah covers 2 m. In ______ leaps would the cheetah catch up the deer.

13. A hare pursued by a hound who is 50 of her own leaps before him. When the hare takes 4 leaps, the hound takes 3. In one leap, the hare goes $1\frac{3}{4}$ metres and the hound $2\frac{3}{4}$ metres. In how _______ leaps will the hound overtake the hare.

Directions (Q. 14 – 15): *Refer to the graph below and answer the questions that follow.*

The engine of the new car Palio brought by Fiat in the market follows certain mileage characteristics as shown in the diagram below:

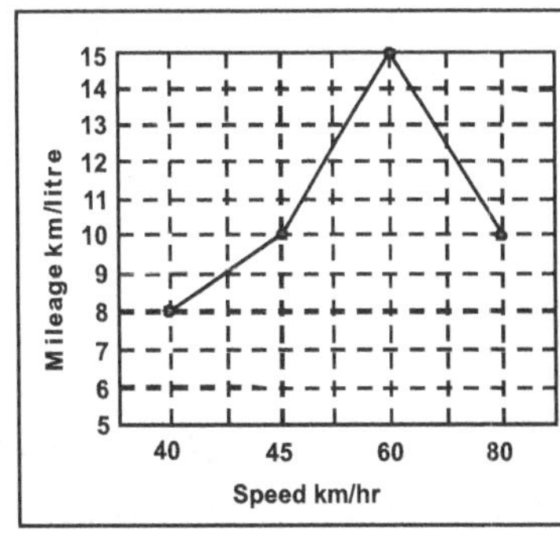

Vijay bought the new car Palio and filled exactly 10 litres in his car. He goes at a speed of 40 km/hr for first 20 km and then at a speed of 60 km/hr for next 60 km and driving at a constant speed reaches his destination which was 115 km away from his starting point.

14. The time in hours taken by Vijay if he reaches his destination in minimum possible time and finishes all the fuel.

15. In the above question, if Vijay had travelled at a constant average speed rather than changing his speed, taking the same minimum time, then the approximate saving in the fuel consumed is______ litres

RATIO, PROPORTION AND MIXTURES

MCQ TYPE QUESTIONS

1. A cask of wine when fully filled holds 10 litres. 2 litres of wine is removed and filled with water. Then 4 litres in the solution is replaced with water. Then, 6 and 8 litres respectively. At the end of the 4th operation, the ratio of wine to water is

(*a*) $\frac{4!}{(5)^4}$ (*b*) $\frac{8!}{(10)^4}$

(*c*) $\frac{4!}{5^4 - 4!}$ (*d*) $\frac{8!}{(10)^4 - 8!}$

2. A 20 litre vessel is filled with alcohol. Some of the alcohol is poured out into another vessel of an equal capacity, which is then completely filled by adding water. Then mixture obtained is poured into the first vessel to capacity. Then $6\frac{2}{3}$ litres is poured from the first vessel into second. Both vessels now contain an equal amount of alcohol. How much alcohol was originally poured from the first vessel into the second?

(*a*) 9 litres (*b*) 10 litres

(*c*) 12 litres (*d*) 12.5 litres

3. What is the ratio compounded of x – y : x + y and $y^2 + xy : x^2 - xy$?

(*a*) $y^2 : x^2$ (*b*) xy : 1

(*c*) y : x (*d*) $\sqrt{x} : \sqrt{y}$

4. If ratio of two natural numbers x and y is 'a' and that of y and x is 'b', then value of 'a + b' is

(*a*) greater than 2 (*b*) lesser than 1

(*c*) greater than 1 (*d*) data insufficient.

5. The charges of a goldsmith is partly fixed and partly variable with the amount in grams of gold. The charge is ₹ 300 for 20 grams of gold and ₹ 550 for 45 grams of gold. Find the charge of work for 100 grams of gold.

(*a*) ₹ 900 (*b*) ₹ 1000

(*c*) ₹ 1100 (*d*) ₹ 9000

Directions (Q.6–7) : *Refer to the data below and answer the questions that follow.*

The ratio of white collar to blue collar employees in a firm is 8 : 3. Also ratio of male employees to female employees is 7 : 4. It is observed that 60% of the white collar employees are males.

6. What is the ratio of female white collar employees to male blue collar employees?

(*a*) 16 : 11 (*b*) 24 : 11

(*c*) 3 : 1 (*d*) Data insufficient

7. If there are 48 female blue collar workers, what is the difference between male white collar employees and female blue collar employees?

(*a*) 288 (*b*) 240

(*c*) 220 (*d*) Data insufficient

8. In a chemical experiment, two NaOH solution bottles are used. Bottle A contains salt and water in the ratio 7 : 3 and bottle B contains salt and water in the ratio 4 : 3. In what proportion should the quantities be taken from A and B to give the 2 : 1 NaOH solution?

(*a*) 2 : 1 (*b*) 10 : 7

(*c*) 20 : 7 (*d*) 1 : 2

9. A triswitch mixer grinder operates in three stages of speed in revolutions per min i.e., rpm. The speed of the motor with some load in the second stage varies as a square of the speed in stage 1 and that in stage 3 varies as a cube of that in stage 2. If mixer motor makes 50 revolutions in 3 minutes in stage 1, its speed in stage 2 and 3 with the same load is 2500 and 12500 rpm respectively. What will be the speed of motor (in rpm) in stage 3, if at the same load motors speed in rpm is 25 revolutions in 6 minutes in stage 1?

(*a*) $\frac{3125}{16}$ rpm (*b*) $\frac{5^5}{2^{10}}$ rpm

(*c*) $\frac{5^{10}}{2^5}$ rpm (*d*) None of these

10. Two bags have certain number of mangoes. If half the mangoes from bag 1 are transferred to bag 2, the ratio of mangoes in bag 1 to bag 2 gets reversed. Now one-third and two-third of the mangoes from the first and the second bag are transferred to the second and first bag respectively. The ratio of mangoes in bag 1 to bag 2 now is

(*a*) 2 : 1 (*b*) 2 : 2

(*c*) 4 : 5 (*d*) 5 : 4

NUMERICAL TYPE QUESTIONS

1. Five litres are drawn from a cask full of wine and it is then filled with water. Five litres of the mixture are drawn and the cask is again filled with water. The quantity of wine now left in the cask to that of the water in it is in the ratio 361 : 39. The cask hold ______ litres.

2. In a mixture of wheat and barley, wheat is 60%. To 400 kg of the mixture a quantity of barley is added and then the wheat is $53\frac{1}{3}\%$. The_____kg of barley are added?

3. Two jars of capacity 3 and 5 litres are filled with mixtures of alcohol and water. In the smaller jar 25% of the mixture is alcohol and in the larger 25% of the mixture is water. The jars are emptied into a 9 litre cask and remaining volume is filled up with water. Then percentage of alcohol in the cask will be ________

4. A man buys milk at ₹ 5 a litre and after adding water, sells it at ₹ 6 a litre, thereby making a profit of $33\frac{1}{3}\%$. The proportion of water to milk in the mixture is_______

5. The percentage of alcohol in two wine tanks is 20 and 40 respectively and rest is grape juice. If a wine glass is filled to its half by the tank 1 and 2 in the ratio 2 : 3 and the rest is filled again by pure alcohol, then the ratio of the grape juice to alcohol in the wine glass is _______

6. In ______ ratio must a grocer mix two varieties of tea worth ₹ 60 a kg and ₹ 65 a kg so that by selling the mixture at ₹ 68.20 a kg he may gain 10%?
7. A container contains 40 lit of milk. From this container 4. lit of milk was taken out and replaced by water. This process was repeated further two times. The ______litres of milk is now contained by the container?
8. A merchant has 1000kg of sugar,part of which he sells at 8% profit and the rest at 18% profit. He gains 14% on the whole. The quantity sold at 18% profit is_____ kg
9. Volumes in the proportion 3 : 4 : 7 of three different substances are mixed together. The densities of equal volume of the substances are in the ratio 5 : 2 : 6 respectively. The weight of the third substance contained in 130 lbs. of the mixture is _______ lbs
10. In a bag, there are 50 paise coins, 25 paise coins and rupee coins, proportional to the numbers 3, 4 and $2\frac{1}{2}$. If they amount to ₹ 210, then number of rupee coins is______

PERMUTATIONS AND COMBINATIONS & PROBABILITY

MCQ TYPE QUESTIONS

Directions (Q. 1 – 2) :

If A and B are two possible events of an experiment such that P (A ∩ B) = 0.6 and P(A) = 0.3 then find P (B) given that

1. A and B are mutually exclusive events

(a) $\frac{7}{10}$ (b) $\frac{4}{10}$

(c) $\frac{3}{10}$ (d) $\frac{1}{2}$

2. A and B are independent events

(a) $\frac{4}{7}$ (b) $\frac{9}{10}$

(c) $\frac{3}{10}$ (d) $\frac{3}{7}$

3. Raju throws a fair dice. He is promised an amount (in rupees), which is twice the number showing up if that number is odd and an amount thrice the number showing up, if it is even. What is the maximum amount Raju would be willing to pay each time to throw the dice, if in the long run he wants to make an average profit of ₹ 7 per throw?

(a) ₹ 6 (b) ₹ 7

(c) ₹ 3 (d) ₹ 2

Directions (Q. 4 – 5) :

7 letters are to be placed in seven addressed envelops. If the letters are placed at random into the envelopes, the probability that

4. All of them are placed in the corresponding envelops is

(a) 1 (b) $\frac{1}{6!}$

(c) $\frac{1}{7!}$ (d) None of these

5. Exactly six letters are placed in their corresponding envelopes is

(a) $1-\frac{1}{7!}$ (b) $\frac{1}{7!}$

(c) 1 (d) 0

Directions (Q. 6 – 8) :

If three cards are drawn at random, from a well shuffled pack of cards, then what is the probability that

6. All of them are from the same suit?

(a) $\frac{4\left({}^{13}C_3\right)}{{}^{52}C_3}$ (b) $\frac{\left({}^{13}C_3\right)^3}{{}^{52}C_3}$

(c) $\frac{4\left({}^{13}C_3\right)^2}{{}^{52}C_3}$ (d) $\frac{{}^{13}C_3}{{}^{52}C_3}$

7. All of them are from different suits?

(a) $\frac{13}{{}^{52}C_3}$ (b) $\frac{4(13)^3}{{}^{52}C_3}$

(c) $\frac{{}^{39}C_3}{{}^{52}C_3}$ (d) $\frac{(13)^4}{{}^{52}C_3}$

8. Two of them are number cards from the same suit and the remaining number card is from a different suit?

(a) $\frac{{}^9C_2\,{}^9C_1}{{}^{52}C_3}$ (b) $\frac{{}^9C_2\,{}^{27}C_1}{{}^{52}C_3}$

(c) $\frac{4\left({}^9C_2\,{}^{27}C_1\right)}{{}^{52}C_3}$ (d) $\frac{4\left({}^9C_2\,{}^9C_1\right)}{{}^{52}C_3}$

9. A bag contains 5 white balls and 7 red balls. If 5 balls are drawn at random, then probability that 3 balls are red and the rest are white is

(*a*) $\frac{175}{198}$ (*b*) $\frac{175}{396}$

(*c*) $\frac{38}{132}$ (*d*) $\frac{221}{396}$

10. A bag contains 6 five rupee coins, 5 two rupee coins and 4 one rupee coins. If 5 coins are selected at random from the bag, then odds in favour of the draw yielding minimum possible amount are

(*a*) 1:3002 (*b*) 5:2998

(*c*) 2:3003 (*d*) 5:3003

11. In how many ways can 6 girls and 6 boys sit around a circular table so that no two boys sit together?

(*a*) $(5!)^2$ (*b*) $(6!)^2$

(*c*) 5! 6! (*d*) 11!

12. How many ten-digit numbers can be formed using all the digits of 1324642119 such that the even digits appear only in odd places?

(*a*) $(5!)^2$ (*b*) $\frac{(5!)^2}{3!}$

(*c*) 5! 3! 2! (*d*) $\frac{(5!)^2}{(2!)^2\,3!}$

Directions (Q. 13 – 14) :

A man has 12 friends whom he wants to invite for lunch. In how many ways can he invite

13. At least one of them?

(*a*) 4096 (*b*) 4095

(*c*) 2047 (*d*) 2048

14. At least 10 of them?

(*a*) 66 (*b*) 140

(*c*) 153 (*d*) None of these

15. In how many ways can 6 prizes be distributed among 3 boys, if each boy is eligible to receive one or more prizes?

(*a*) 3^6 (*b*) 6^3

(*c*) ${}^6C_3$ (*d*) ${}^6P_3$

Directions (Q. 16 – 17) :

Find the number of ways of dividing 16 different books equally

16. Among 4 boys

(*a*) $\frac{16!}{(4!)^4}$ (*b*) $\frac{16!}{(4!)^3}$

(*c*) $\frac{16!}{(4!)^5}$ (*d*) $(4!)^4$

17. Into 4 parcels (Parcels are not distinct)

(*a*) $\frac{16!}{(4!)^4}$ (*b*) $\frac{16!}{(4!)^3}$

(*c*) $\frac{16!}{(4!)^5}$ (*d*) $(4!)^4$

18. How many linear arrangements can be made using all the letters of the word ILLUSTRATE which begin with S and end with a vowel?

(*a*) $\frac{8!}{(2!)^2}$ (*b*) 8!

(*c*) $\frac{8!}{2!}$ (*d*) $\frac{8!}{(2!)^3}$

19. From a group of 10 professors and 6 assistant professors, a management institute desires to send to delegation of 8 persons consisting of 5 professors and 3 assistant professors to the IIMs annual meet. If Prof. Balamurali, a science professor refuses to be in the delegation if assistant Prof. Sheshdari, an arts professor is included in the delegation, then in how many ways can the delegation be formed?

(*a*) ${}^9C_4\,{}^4C_3$

(*b*) ${}^9C_5\,{}^4C_2$

(*c*) ${}^{10}C_5\,{}^6C_3 - {}^9C_4\,{}^5C_2$

(*d*) ${}^9C_4\,{}^4C_3 + {}^9C_5\,{}^4C_2$

20. Two red pencils, three black pencils and two white pencils are to be arranged in a row such that

1. No two adjacent pencils are of the same colour and
2. Pencils at the two ends of the row are of same colour.

(*a*) 12 (*b*) 8

(*c*) 9 (*d*) 10

ANSWER KEY

ENGLISH GRAMMAR

1. (b)	**2.** (a)	**3.** (e)	**4.** (a)	**5.** (c)	**6.** (a)	**7.** (b)	**8.** (c)	**9.** (e)	**10.** (c)
11. (a)	**12.** (c)	**13.** (d)	**14.** (c)	**15.** (a)	**16.** (b)	**17.** (d)	**18.** (d)	**19.** (c)	**20.** (a)
21. (c)	**22.** (a)	**23.** (c)	**24.** (a)	**25.** (d)	**26.** (b)	**27.** (a)	**28.** (d)	**29.** (d)	**30.** (d)
31. (d)	**32.** (c)	**33.** (b)	**34.** (d)	**35.** (d)	**36.** (c)	**37.** (d)	**38.** (c)	**39.** (b)	**40.** (c)
41. (a)	**42.** (d)	**43.** (b)	**44.** (d)	**45.** (d)	**46.** (a)	**47.** (d)	**48.** (c)	**49.** (d)	**50.** (c)

SENTENCE COMPLETION

1. (a)	**2.** (b)	**3.** (b)	**4.** (d)	**5.** (b)	**6.** (c)	**7.** (d)	**8.** (d)	**9.** (c)	**10.** (c)
11. (a)	**12.** (d)	**13.** (a)	**14.** (d)	**15.** (a)	**16.** (d)	**17.** (b)	**18.** (a)	**19.** (d)	**20.** (c)
21. (3)	**22.** (c)	**23.** (d)	**24.** (b)	**25.** (b)	**26.** (d)	**27.** (a)	**28.** (c)	**29.** (d)	**30.** (c)
31. (d)	**32.** (d)	**33.** (d)	**34.** (c)	**35.** (a)	**36.** (d)				

SYNONYMS

1. (d)	**2.** (a)	**3.** (c)	**4.** (d)	**5.** (b)	**6.** (d)	**7.** (b)	**8.** (d)	**9.** (a)	**10.** (a)
11. (a)	**12.** (d)	**13.** (a)	**14.** (d)	**15.** (d)	**16.** (d)	**17.** (d)	**18.** (b)	**19.** (c)	**20.** (c)
21. (a)	**22.** (c)	**23.** (c)	**24.** (b)	**25.** (c)	**26.** (d)	**27.** (c)	**28.** (c)	**29.** (b)	**30.** (d)
31. (b)	**32.** (b)	**33.** (d)	**34.** (b)	**35.** (b)	**36.** (a)	**37.** (d)	**38.** (b)	**39.** (c)	**40.** (d)
41. (b)	**42.** (a)	**43.** (a)	**44.** (c)	**45.** (d)	**46.** (d)	**47.** (d)	**48.** (c)	**49.** (c)	**50.** (d)
51. (a)	**52.** (d)	**53.** (b)	**54.** (d)	**55.** (a)	**56.** (d)	**57.** (c)	**58.** (b)	**59.** (a)	**60.** (b)
61. (b)	**62.** (a)	**63.** (a)	**64.** (a)	**65.** (c)	**66.** (c)	**67.** (d)	**68.** (d)	**69.** (b)	**70.** (b)
71. (d)	**72.** (d)	**73.** (a)	**74.** (b)	**75.** (a)	**76.** (d)	**77.** (d)	**78.** (d)	**79.** (a)	**80.** (a)
81. (b)	**82.** (c)	**83.** (b)	**84.** (c)	**85.** (c)	**86.** (b)	**87.** (d)	**88.** (a)	**89.** (a)	**90.** (d)
91. (a)	**92.** (c)	**93.** (d)	**94.** (d)	**95.** (b)	**96.** (c)	**97.** (d)	**98.** (c)	**99.** (d)	**100.** (a)
101. (d)	**102.** (c)	**103.** (a)	**104.** (b)	**105.** (a)	**106.** (b)	**107.** (b)	**108.** (b)	**109.** (d)	**110.** (a)
111. (a)	**112.** (c)	**113.** (c)	**114.** (c)	**115.** (d)	**116.** (d)	**117.** (a)	**118.** (a)	**119.** (b)	**120.** (b)
121. (b)	**122.** (c)	**123.** (c)	**124.** (c)	**125.** (d)	**126.** (c)	**127.** (b)	**128.** (d)	**129.** (d)	**130.** (d)
131. (a)	**132.** (b)	**133.** (d)	**134.** (c)	**135.** (a)	**136.** (d)	**137.** (b)	**138.** (d)	**139.** (d)	**140.** (b)
141. (b)	**142.** (c)	**143.** (d)	**144.** (a)	**145.** (d)	**146.** (b)	**147.** (d)	**148.** (c)	**149.** (a)	**150.** (c)
151. (a)	**152.** (b)	**153.** (b)	**154.** (b)	**155.** (a)	**156.** (a)	**157.** (c)	**158.** (c)	**159.** (c)	**160.** (b)
161. (d)	**162.** (d)	**163.** (d)	**164.** (a)	**165.** (b)	**166.** (a)	**167.** (b)	**168.** (d)	**169.** (a)	**170.** (b)
171. (d)	**172.** (c)	**173.** (a)	**174.** (b)	**175.** (d)	**176.** (b)	**177.** (a)	**178.** (d)	**179.** (b)	**180.** (c)
181. (d)	**182.** (b)	**183.** (b)	**184.** (d)	**185.** (c)	**186.** (c)	**187.** (b)	**188.** (a)	**189.** (c)	**190.** (c)
191. (a)	**192.** (b)	**193.** (d)	**194.** (c)	**195.** (a)	**196.** (b)	**197.** (a)	**198.** (a)	**199.** (d)	**200.** (a)
201. (b)	**202.** (d)	**203.** (d)	**204.** (b)	**205.** (c)	**206.** (c)	**207.** (b)	**208.** (a)	**209.** (d)	**210.** (d)
211. (d)	**212.** (d)	**213.** (b)	**214.** (d)	**215.** (c)	**216.** (a)	**217.** (c)	**218.** (a)	**219.** (b)	**220.** (a)

221. (*d*) **222.** (*d*) **223.** (*d*) **224.** (*b*) **225.** (*d*) **226.** (*d*) **227.** (*d*) **228.** (*b*) **229.** (*b*) **230.** (*c*)
231. (*c*) **232.** (*d*) **233.** (*b*) **234.** (*a*) **235.** (*d*) **236.** (*b*) **237.** (*b*)

ANTONYMS

1. (*d*) **2.** (*b*) **3.** (*d*) **4.** (*b*) **5.** (*a*) **6.** (*d*) **7.** (*a*) **8.** (*a*) **9.** (*a*) **10.** (*b*)
11. (*b*) **12.** (*b*) **13.** (*b*) **14.** (*d*) **15.** (*b*) **16.** (*c*) **17.** (*d*) **18.** (*c*) **19.** (*c*) **20.** (*d*)
21. (*d*) **22.** (*a*) **23.** (*d*) **24.** (*c*) **25.** (*b*) **26.** (*d*) **27.** (*b*) **28.** (*b*) **29.** (*d*) **30.** (*d*)
31. (*c*) **32.** (*a*) **33.** (*d*) **34.** (*b*) **35.** (*d*) **36.** (*a*) **37.** (*d*) **38.** (*a*) **39.** (*b*) **40.** (*b*)
41. (*d*) **42.** (*d*) **43.** (*c*) **44.** (*a*) **45.** (*d*) **46.** (*d*) **47.** (*c*) **48.** (*d*) **49.** (*d*) **50.** (*a*)
51. (*c*) **52.** (*d*) **53.** (*b*) **54.** (*a*) **55.** (*d*) **56.** (*c*) **57.** (*b*) **58.** (*a*) **59.** (*b*) **60.** (*a*)
61. (*c*) **62.** (*b*) **63.** (*c*) **64.** (*b*) **65.** (*c*) **66.** (*d*) **67.** (*d*) **68.** (*c*) **69.** (*b*) **70.** (*a*)
71. (*a*) **72.** (*b*) **73.** (*b*) **74.** (*b*) **75.** (*a*) **76.** (*c*) **77.** (*a*) **78.** (*d*) **79.** (*d*) **80.** (*b*)
81. (*a*) **82.** (*c*) **83.** (*c*) **84.** (*d*) **85.** (*c*) **86.** (*b*) **87.** (*b*) **88.** (*b*) **89.** (*d*) **90.** (*c*)
91. (*d*) **92.** (*d*) **93.** (*a*) **94.** (*a*) **95.** (*b*) **96.** (*d*) **97.** (*d*) **98.** (*c*) **99.** (*a*) **100.** (*a*)
101. (*c*) **102.** (*b*) **103.** (*d*) **104.** (*a*) **105.** (*d*) **106.** (*b*) **107.** (*b*) **108.** (*a*) **109.** (*a*) **110.** (*c*)
111. (*d*) **112.** (*a*) **113.** (*d*) **114.** (*b*) **115.** (*d*) **116.** (*c*) **117.** (*b*) **118.** (*a*) **119.** (*b*) **120.** (*c*)
121. (*b*) **122.** (*d*) **123.** (*d*) **124.** (*a*) **125.** (*a*) **126.** (*d*) **127.** (*a*) **128.** (*d*) **129.** (*d*) **130.** (*b*)
131. (*c*) **132.** (*d*) **133.** (*c*) **134.** (*c*) **135.** (*a*) **136.** (*b*) **137.** (*d*) **138.** (*b*) **139.** (*a*) **140.** (*b*)
141. (*d*) **142.** (*c*) **143.** (*a*) **144.** (*d*) **145.** (*c*) **146.** (*c*) **147.** (*d*) **148.** (*a*) **149.** (*a*) **150.** (*d*)
151. (*b*) **152.** (*a*) **153.** (*a*) **154.** (*d*) **155.** (*b*) **156.** (*d*) **157.** (*b*) **158.** (*d*) **159.** (*d*) **160.** (*c*)
161. (*a*) **162.** (*d*) **163.** (*a*) **164.** (*b*) **165.** (*a*) **166.** (*c*) **167.** (*d*) **168.** (*b*) **169.** (*a*) **170.** (*d*)
171. (*d*) **172.** (*c*) **173.** (*b*) **174.** (*d*) **175.** (*d*) **176.** (*d*) **177.** (*c*) **178.** (*b*) **179.** (*a*) **180.** (*d*)
181. (*d*) **182.** (*a*) **183.** (*c*) **184.** (*b*) **185.** (*a*) **186.** (*c*) **187.** (*c*) **188.** (*d*) **189.** (*b*) **190.** (*c*)
191. (*c*) **192.** (*a*) **193.** (*c*) **194.** (*d*) **195.** (*c*) **196.** (*b*) **197.** (*d*) **198.** (*d*) **199.** (*d*) **200.** (*b*)
201. (*a*) **202.** (*c*) **203.** (*a*) **204.** (*c*) **205.** (*b*) **206.** (*d*) **207.** (*d*) **208.** (*b*) **209.** (*a*) **210.** (*a*)
211. (*c*) **212.** (*d*) **213.** (*b*) **214.** (*d*) **215.** (*c*) **216.** (*d*) **217.** (*b*) **218.** (*a*) **219.** (*a*) **220.** (*c*)
221. (*c*) **222.** (*c*) **223.** (*b*) **224.** (*c*) **225.** (*b*) **226.** (*d*) **227.** (*c*) **228.** (*a*) **229.** (*d*) **230.** (*a*)
231. (*c*) **232.** (*d*) **233.** (*d*) **234.** (*c*) **235.** (*c*) **236.** (*d*) **237.** (*d*) **238.** (*c*) **239.** (*a*) **240.** (*a*)
241. (*c*) **242.** (*b*) **243.** (*a*) **244.** (*a*) **245.** (*c*) **246.** (*d*) **247.** (*b*) **248.** (*a*) **249.** (*d*) **250.** (*d*)
251. (*c*) **252.** (*b*) **253.** (*b*) **254.** (*d*) **255.** (*d*) **256.** (*b*) **257.** (*a*) **258.** (*d*) **259.** (*a*) **260.** (*b*)
261. (*c*) **262.** (*d*) **263.** (*c*) **264.** (*b*) **265.** (*a*) **266.** (*d*) **267.** (*a*) **268.** (*b*) **269.** (*b*) **270.** (*b*)
271. (*a*) **272.** (*d*) **273.** (*b*) **274.** (*c*) **275.** (*d*) **276.** (*b*) **277.** (*a*) **278.** (*a*) **279.** (*c*) **280.** (*a*)
281. (*b*) **282.** (*c*) **283.** (*d*) **284.** (*d*) **285.** (*d*) **286.** (*d*) **287.** (*d*) **288.** (*c*) **289.** (*b*) **290.** (*d*)
291. (*a*) **292.** (*d*) **293.** (*d*) **294.** (*a*) **295.** (*d*) **296.** (*b*) **297.** (*c*) **298.** (*a*) **299.** (*a*) **300.** (*c*)
301. (*b*) **302.** (*c*) **303.** (*d*) **304.** (*a*) **305.** (*d*) **306.** (*b*) **307.** (*d*) **308.** (*d*) **309.** (*a*) **310.** (*a*)
311. (*b*) **312.** (*c*) **313.** (*d*) **314.** (*b*) **315.** (*b*) **316.** (*a*) **317.** (*a*) **318.** (*d*) **319.** (*d*) **320.** (*d*)
321. (*c*) **322.** (*b*) **323.** (*d*)

REASONING ABILITY

MCQ Type Questions

1. (*b*) **2.** (*b*) **3.** (*b*) **4.** (*c*) **5.** (*b*) **6.** (*b*) **7.** (*b*) **8.** (*d*) **9.** (*a*) **10.** (*a*)

11. (*b*) **12.** (*d*) **13.** (*d*) **14.** (*a*) **15.** (*b*) **16.** (*c*) **17.** (*b*) **18.** (*d*) **19.** (*d*) **20.** (*b*)

Numerical Type Questions

1. 70 **2.** 105 **3.** 2 **4.** 2 **5.** 11 **6.** 7920 **7.** 12870 **8.** 35 **9.** 10

NUMBERS AND ALGEBRA

1. (*b*) **2.** (*a*) **3.** (*b*) **4.** (*d*) **5.** (*d*) **6.** (*d*) **7.** (*b*) **8.** (*c*) **9.** (*b*) **10.** (*c*)

11. (*b*) **12.** (*b*) **13.** (*a*) **14.** (*d*) **15.** (*b*) **16.** (*b*) **17.** (*a*) **18.** (*b*) **19.** (*d*) **20.** (*a*)

21. (*d*) **22.** (*b*) **23.** (*c*) **24.** (*d*) **25.** (*b*) **26.** (*d*) **27.** (*b*) **28.** (*a*) **29.** (*b*) **30.** (*d*)

31. (*c*) **32.** (*b*)

PERCENTAGE AND ITS APPLICATIONS

1. (*b*) **2.** (*b*) **3.** (*a*) **4.** (*a*) **5.** (*b*) **6.** (*d*) **7.** (*a*) **8.** (*b*) **9.** (*d*) **10.** (*c*)

11. (*c*) **12.** (*c*) **13.** (*a*) **14.** (*a*) **15.** (*a*) **16.** (*c*) **17.** (*a*) **18.** (*c*) **19.** (*c*) **20.** (*b*)

21. (*b*) **22.** (*c*) **23.** (*a*) **24.** (*b*) **25.** (*a*)

TIME AND WORK

MCQ Type Questions

1. (*c*) **2.** (*d*) **3.** (*c*) **4.** (*b*) **5.** (*d*) **6.** (*c*) **7.** (*c*) **8.** (*b*) **9.** (*b*) **10.** (*c*)

11. (*c*) **12.** (*d*) **13.** (*a*) **14.** (*d*) **15.** (*c*)

Numerical Type Questions

1. 6 **2.** $\frac{1}{2}$ **3.** 16 **4.** 20 **5.** 6 **6.** 81 **7.** 75 **8.** 1.5 **9.** 77.5

10. 12.15 a.m. **11.** 39.3 **12.** 300 **13.** 210 **14.** 2 **15.** $2\frac{1}{3}$

RATIO, PROPORTION AND MIXTURES

MCQ Type Questions

1. (*c*) **2.** (*b*) **3.** (*c*) **4.** (*c*) **5.** (*c*) **6.** (*a*) **7.** (*b*) **8.** (*c*) **9.** (*b*) **10.** (*a*)

Numerical Type Questions

1. 100 **2.** 50 **3.** 50 **4.** 1 : 9 **5.** 17 : 33 **6.** 3 : 2 **7.** 29.16 **8.** 600 **9.** 84 **10.** 105

PERMUTATIONS AND COMBINATIONS & PROBABILITY

1. (*c*) **2.** (*d*) **3.** (*d*) **4.** (*c*) **5.** (*d*) **6.** (*a*) **7.** (*b*) **8.** (*c*) **9.** (*b*) **10.** (*b*)

11. (*c*) **12.** (*d*) **13.** (*b*) **14.** (*d*) **15.** (*a*) **16.** (*a*) **17.** (*c*) **18.** (*b*) **19.** (*c*) **20.** (*d*)

EXPLANATIONS

ENGLISH GRAMMAR

21. Uneasily does not denote a state, which is required in the context. 'Wearing a crown' denotes that the head is doing it, 'is wearing' denotes a continuous action, which is impossible, as the head cannot do it. Hence choice (*c*) restates the given sentence correctly.

22. All the statements have the correct combination of tenses. So, we need not check it. Let us find the errors in other statements. 'Early than now' and 'earlier' suggest that I reached early; in that case, I did not miss the train. So, choices (*b*), (*c*), and (*d*) do not state the same idea as in the question statement. Hence correct choice is (*a*).

23. The given statement states that both she and her sister are equally intelligent. This is given in statement. Hence correct choice is (*c*).

24. 'Unless' denotes condition, 'until' denotes time. Choice (*b*) has adverb clause of time.With unless, another negative is not used. 'Unless ..... don't' is wrong. So, Choices (*c*) and (*d*) are incorrect. Hence correct choice is (*a*).

25. Choices (*a*) and (*b*) state the same idea [(*a*) – passive voice; (*b*) Active voice] that the students followed the instructions, which differs from the question statement. Instructions cannot be advised. So, choice (*c*) is meaningless. Choice (*d*) states the context of the given statement aptly.

26. Despite means inspite of. Despite of is a wrong usage. 'His' need not be used in choice (*c*), as the sentence is about one person. Choice (*d*) means that the classes were ill. Hence correct choice is (*b*).

27. Choice (*a*) best restates the given sentence, though choice (*b*) is the repetition of the given statement. 'Not one can deny' in choice (*c*) and for any one to deny' in choice (*d*) make them incorrect. Hence correct choice is (*a*).

28. In this type of comparison, the pronouns are in subjective case 'Me', which is in objective case is wrongly used in choice (*a*). Equally is not used with 'old'. Of the same age is the correct expression. Hence correct choice is (*d*).

29. Choice (*a*) and (*b*) mean that the prerequisite for wisdom is age and experience, which is incorrect. Just by growing old and working for a long period, a person cannot improve his wisdom. 'The wiser' is used only when we compare wisdom with some other quality. Choice (*d*) is the correct restatement of the given one. Hence correct choice is (*d*).

30. The given statement says that one cannot completely nullify the hazards of life but they can be evaded quite easily. Choice (*a*) which states that hazards cannot be avoided is wrong (*b*) states that hazards can be made ineffective and hence this is also wrong (*c*) does not convey the meaning as stated and hence it too is incorrect. (*d*) correctly restates the idea and is our answer.

31. Every man refers to a singular noun, therefore.... has been saved.

32. has

33. drop 'kindly'

34. Girl's implies possessive noun, whereas the college is meant for girls.

35. aim at, aim for is wrong usage.

36. delete their - add : its

37. delete they - add : we

38. delete he is - add : they are.

39. delete it - add : that

40. delete their - add : its.

41. One sits 'on' a plane.

42. There are no errors.

43. Hotels 'have' and some pander 'to....'

44. There are no errors in this sentence.

45. There are no errors.

48. (A) hoping to have a visit;
(B) .... then take the next road to

50. (A) continuously; (B) glum.

SENTENCE COMPLETION

1. They in choice (*a*) means 'some of our ruling politicians' (the subject). Also grammatically 'they have a right to plunder ....' is correct wheras'...... right to the punder ......' is incorrect.
2. 'Bizarre' means 'weird'. Hence, choice (*a*) is ruled out. "Its haunting images have endured ' fits syntactically. choice (*c*) can also be ruled out as 'munificence' means generosity and this is not intended in the sentence. Also, the word 'prints' in the given sentence gets with the word 'images' of choice (*b*).
3. 'They' in the part after the comma must refer to somebody or some-thing. Hence, only choice (*b*) will fit as 'they' in that case refers to 'things'.
4. The right word to precede the proposition 'on' is imposed. Tax is not collected on .... people. Tax is not placed on people, Tax is not charged on people. Hence, choices (*a*), (*b*) and (*c*) are not the answers. Tax is imposed ..... on ..... .

5. Choice (*d*) is ruled out as it does not logically fit. Choice (*a*) is also ruled out because of the expression 'tears in me'.

6. The key words are 'plunging' and 'investor confidence'. If markets are plunging, investor confidence is low. This aspect is brought out best by choice (*c*).

7. (*a*) 'Discovery'refers to something that existed previously and came to knowledge later. Home is not a discovery.

(*b*) Home is not a certain. It is a set up which comes into being as a result of a certain composition of various factors.

(*c*) Home is not a new invention of anyone at any stage of development.

So choice (*d*) is the best answer. Home is an institution just like 'marriage. Home is a reality only when its members accept certain principles.

8. (*a*) A man's toil may help to make out wealth, but it is not the ultimate aim.

(*b*) A man's toil may secure a lot of things for his children or friends . That too is not something great.

(*c*) A man's toil may help him to overcome many difficulties, but that does not give him much pleasure.

So, choice (*d*) is the best answer. What one becomes by one's toil is greater than what one gets.

9. (*a*) 'Reopens the first chapter' means begins from the very start.

(*b*) 'Writes an epilogue' means ends the book.

(*c*) 'Opens new books' means creates new problems.

The above three are only partly correct. An act of justice pardons for ever an act of vengeance creates new problems. It is a saying. "An act of justice closes the book on a misdeed; an act of vengeance writes one of its own."

10. Here, the trail is the important thing and hence the excitement of the trail is what you will be looking for.

11. (*b*) The expression must be "it is of cardinal importance".

(*c*) There is no urgency in understanding, "It should be understood" is enough.

(*d*) Instead of "it cannot be,' the expression should be it must be'.

So choice (*a*) is the best, It is important that the two things are not to be interfered with.

13. (*b*) It is not mentioned with what do wines bear semblance.

(*c*) The 'similarity' is only in terms of production. The similarity in production does not achieve commercial profit.

(*d*) For the same commercial profit wines need not be produced in the same of similar manner.

So, choice (*a*) is the best. It does not mention that the wines of similar profit are similarily produced or have semblance with one another.

14. (*a*) A sociological analysis does not stop at the outer limits of the subjects.

(*b*) No one can indicate how far into the subject can a sociological analysis go.

(*c*) 'Relative distance' refers to the distance of something else which is not mentioned.

So, choice (*d*) is the best. This indicates the probable distance to which the sociological analysis can go to.

30. Choices (*a*) and (*d*) are ruled out because 'suggest' does not go with 'to+ verb' form. Choice (2) cannot fit in the blank because of the error in parallelism. The appropriate structure is "Many people suggest.....and others advocate ....".

42. Going by options, 'regrettably' doesn't fit as there is nothing to regret. 'Firstly' is no choice and nothing 'secondly' is mentioned. Paragraph has nothing aggresive tone so 'obviously' does not suit. 'Apparently' is only and most suitable option.

43. All options are suitable for the space but to choose one, the best, is required here 'Analytically' appears to be best. It expresses to analyse something using statistically proven methods to reach a decision.

44. It is a very obvious choice determined from the paragraph giving it an access.

45. 'Firing' comes as best suited option as paragraph explains it as a subject of discussion.

46. An argument is always 'resolved', it is never thwart, defeat or closed.

47. Going by options makes choice easier. Only "how the process allowed......." suits logically.

REASONING ABILITY

MCQ Type Questions

1. S[M(D(A(*a*, *b*)2, D(A(*a*, *b*)2)), M(D(S(*a*,*b*)2)) D (*s*(*a*,*b*)2)]

$$= S\left[\left|M\frac{a+b}{2},\frac{a+b}{2}\right|\left|M\frac{a-b}{2};\frac{a-b}{2}\right|\right]$$

$$= \left(\frac{a+b}{2}\right)^2 - \left(\frac{a-b}{2}\right)^2 = ab$$

6 – 7.

A → Lawyer → Male
B → Housewife → Female
C → Accountant → Male
D → Housewife → Female
E → Architect → Male
F → Lecturer → Female
Married Couple : AD & CF

8—9. The seating arrangement is shown below:

Left ... Right

7	6	5	4	3	2	1
D/C	E/F	F/E	B	C/D	G/A	A/G

10 – 13.

Level of Dissimilarity is equal to Maximum difference in ranks alloted to any two countries on any of the five traits.

For example, dissimilarity between India and China.

Trait	India's Rank	China's Rank	Difference
C	1	2	1
P	2	3	1
N	3	1	2
V	4	5	1
D	5	4	1

From the above table, maximum difference is 2.
So dissimilarity level between India and China is 2.

10. Calculating the level of dissimilarity for India Vs remaining countries.
China : For the parameter N (3 – 1) = 2
Japan : For the parameter D (5 – 1) = 4
Malaysia : For the parameter C (4 – 1) = 3
Thailand : For the parameter V (4 – 1) = 3
Hence least level of dissimilarity is 2 which is for China.

11. From above calculations, highest level of dissimilarity is 4 which is for Japan.

12. *The level of dissimilarity between*
China and Japan is for the parameter D (4 – 1) = 3
India and China is for the parameter N (3 – 1) = 2
Malaysia and Japan is for the parameter V (4 – 1) = 3
Thailand and Japan is for the parameter D (5 – 1) = 4

13. *The level of dissimilarity between*
Malaysia and China is for the parameter V (5 – 1) = 4
China and Thailand is for the parameter V (5 – 1) = 4
Thailand and Japan is for the parameter D (5 – 1) = 4
Japan and Malaysia is for the parameter V (4 – 1) = 3

14 to 17 :

The given set of questions is based on distribution and sequencing.

Firstly, let us put the names of women and men,
Women : Shanthi, Sridevi, Joya, Anita
Men : Sunil, Anil, Raj, Raman

It is given that(2 kids family) $\xleftarrow{\text{came before}}$ (No kid family)

Also Shanti (0 kids) $\xrightarrow{\text{came before}}$ Sridevi

As Sunil and his wife came with their only kid, hence Sunil cannot be the husband of Shanthi.

As Anil and Raj are fathers, hence neither of them is the husband of Shanthi. Hence <u>Raman is the husband of Shanthi.</u>

Also, Joya came before Shanthi and met Anita on the venue, this gives the following order of their reaching the venue. [Anita, Joya, Shanthi, Sridevi]

As Sunil reached last, <u>Sunil is the husband of Sridevi</u>. This means that <u>Raj is Joya's husband and Anil is Anita's husband.</u>

Following table gives the desired details

Order	Women	Men	No. of Kids
1	Anita	Anil	at least 1
2	Joya	Raj	2
3	Shanthi	Raman	0
4	Sridevi	Sunil	1

14. Shanthi arrived third.

15. Out of the given choices. Sunil and Sridevi is the correct pair.

16. The daughters of Anita and Sridevi go to the same school. Anita's husband is Anil and Sridevi's husband is Sunil. Hence, Sunil and Anil is the correct answer.

17. Raj's family consists of two kids for certain (Anil has at least one kid, hence could be having exactly one kid also).

18 to 20 :

From statements (vii), (viii) and (ix), we get

P U S
Orange
R Q
Yellow Green

Then, T is opposite S and the colour of S is red.

From (x), the colour of house P is white, hence the colour of house T is blue.

P	U	S
White	Orange	Red
R	Q	T
Yellow	Green	Blue

Now, from (v) and (x), we get

T > S, Q > P > R

From (vi), U is the shortest.

i.e., the order of houses in the descending order of their heights is T, S/Q, Q/S, P, R, U.

18. R is yellow coloured house and the house diagonally opposite R is S. And the colour of house S is Red.

19. The second tallest house is either S or Q.

20. The tallest house is T and its colour is blue.

Numerical Type Questions

1. $M[A(S(x, y), M(y, z)), S(A(x, z), S(z, y))]$

$= M[A((2-3), (3, \times 5)), S((2+5), (5-3))]$

$= M[A(-1, 15), S(7, 2)]$

$= M[14, 5] = 70.$

2. Since B is in 1, and C and D are not adjacent, the only places possible for them are 2 and 4.

3. Since B is not in the third place and A is not adjacent to B, and D is last in the queue. The position of A will be beside D and B will be at the other corner. C will be between B and A. Hence the positions will be BCAD. Place of C is 2, hence choice (*b*)

The sitting position is as follows.

White	Green	Red	Yellow
Maharashtra	U.P.	West Bengal	A.P.

From the above table,

Ms. Andhra Pradesh ⟶ Yellow

4. (1) (2) 3 3 2 1 4 (2) 3 (1) 4 (2) (2) 3 3 (1) 4 1 (1) (3) (2) 3 4

Total no. oranges were in basket = 2

5. Total fruits in the basket = 11

6. Total number of four letter password formed

$= {}^{11}C_4 \times \angle 4 = 11 \times 10 \times 9 \times 8 = 7920$

7. Required number of passwords

$= {}^{26}C_3 \times \angle 3 - {}^{15}C_3 \times \angle 3$

$= 26 \times 25 \times 24 - 15 \times 14 \times 13 = 12870$

8.

1st Letter	2nd Letter	3rd Letter

There are 5 vowels, *a*, *e*, *i*, *o*,, *u*,.

hence 1st letter can be any of the 5 vowels ;

∴ 5 different methods. ...(*i*)

The second letter can be any of the *m*, *n*, or *p*. But each has its own rule for the third letter. Hence, each case to be considered separately.

1 Vowel	m

In the above case; third letter is one of the remaining 4 vowels

Hence there are 5 × 4 = 20 ways of forming such a string. ...(*ii*)

If the second letter is *n*, the third is either *e* or *u*, *i.e.* 2 different ways.

∴ total of 5 × 2 ways = 10 ways ...(*iii*)

When 2nd position is filled by P, the third letter is same as the first letter. As the 1st letter is already selected, third can be filled only one way.

Hence, there are 5× 1 = 5 ways ....(*iv*)

From equations (*ii*), (*iii*) and (*iv*),

Total number of strings that can be formed

= 20 + 10 + 5 = 35

9. Third letter find is 'e'.

Once again considering each case of the second position being filled by *m*, *n* or *p*, we have

1st position	n	e
	2nd position	3rd position

The 3rd position shall be different form vowel in first position, *i.e.* vowel for 1st position can be selected out of the vowels *a*, *i*, *o*, *u* only.

Hence 4 different ways ....(*i*)

When 2nd possible is filled by *n*, occupying 2nd position, does not create any restriction on the vowel is position 1.

e	n	
3rd	2nd	1st

Hence, 1st position can be occupied by any of the 5 vowels.

Hence, 5 different ways ....(*ii*)

When 2nd position is filled by *p*,

e	p	
3rd	2nd	1st

In this case, letters in 3rd position out 1st position shall be the same.

Hence, 1st position shall be filled by 'e'.

i.e. only one way of filling ....(*iii*)

Total number of possible strings = 4 + 5 + 1 = 10

NUMBERS AND ALGEBRA

1. $x^2 + y^2 + z^2 - 2xy - 2xz + 2yz = (x - y - z)^2$

$= (b + c - c + a - a + b)^2 = 4b^2$

2. $y^3 - 4y = y(y^2 - 4)$

$= y(y-2)(y+2)$

$4y(y^3 + 8) = 4y(y+2)(y^2 - 2y + 4)$

∴ HCF $= y(y + 2).$

3. Total red tiles initially $= 2\left(\frac{m}{n}\right) - 1$

Total number of tiles $= \frac{m^2}{n^2}$

As number of red tiles > number of green tiles (since room has odd number of tiles and corner tiles are red)

Hence total red tiles required to form an alternate pattern

$$= \frac{\left(\frac{m}{n}\right)^2 + 1}{2} = \frac{m^2 + n^2}{2n^2}$$

∴ Number of green tiles removed

$$= \frac{m^2 + n^2}{2n^2} - 2\left(\frac{m}{n}\right) + 1$$

$$= \frac{m^2 + n^2 - 4mn + 2n^2}{2n^2}$$

$$= \frac{(m - 2n)^2 - n^2}{2n^2}$$

4. $$\frac{x - 4}{-x^2 - 6} \le \frac{2}{-x^2 - 6}$$

i.e. $$\frac{x - 4}{-(x^2 + 6)} \le \frac{2}{-(x^2 + 6)}$$

i.e. $x - 4 \ge 2$ ...(cancelling a negative term reverses sign of the inequality)

i.e. $x \ge 6$.

5. $$\left(\sqrt{2} + \sqrt{5}\right)^2 = 2 + 2\sqrt{10} + 5$$

$$= 7 + 2\sqrt{10}$$

∴ $\sqrt{2} + \sqrt{5} > \sqrt{7}$

7. $$\frac{2^n (n+1)^n}{n^n} = 2^n\left(\frac{n+1}{n}\right)^n = 2^n\left(1 + \frac{1}{n}\right)^n$$

Now, $$2 \le \left(1 + \frac{1}{n}\right)^n \le 3$$

∴ $$2.2^n \le 2^n\left(1 + \frac{1}{n}\right)^n \le 3.2^n$$

i.e., $$2^{n+1} \le 2^n\left(1 + \frac{1}{n}\right)^n \le 3.2^n$$

8. Choose k, so that

$$10^H = g^{\lambda} = 10^{100k},$$

$$k = \frac{H}{100}$$

∴ $$10^H = g^{\frac{H}{100}} > g^g > g!$$

On the other hand,

$$10H = 10 \times 10^g = 10^{g+1} < g!$$

(most factors are much bigger than 10)

$$10H < g! < 10^H$$

9. (*a*) Four consecutive even numbers can be written as 2n, 2n + 2, 2n + 4 and 2n + 6, where n is any natural number.

∴ Sum = 2n + 2n + 2 + 2n + 4 + 2n + 6

= 8n + 12

= 4(2n + 3) not always divisible by 8.

Thus, (*a*) is not true.

(*b*) Four consecutive odd numbers can be written as 2n + 1, 2n + 3, 2n + 5, 2n +7

where n is a natural number.

Sum = 2n + 1 + 2n + 3 + 2n + 5 + 2n + 7 = 8n + 16 divisible by 8

Thus, (*b*) is true

(*c*) In product of n consecutive natural numbers atleast one is divisible by n, atleast one by n – 1 ... till 1.

Thus product is atleast divisible by

$n \times (n - 1) \times (n - 2) .... 1 = n!$.

Thus, (*c*) is not true.

(*d*) Take four consecutive odd numbers as 7 × 9 × 11 × 13 which is not divisible by 15.

Thus, (*d*) is not true.

10. Factors of a^2 are 1, a and a^2.

Factors of ab are 1, a, b and ab.

Factors of a^3 are 1, a, a^2 and a^3.

11. $$7x + 6y = 420$$

Equation is of the form:

7x + even number = even number.

∴ 7x has to be even

Hence x has to be even.

12. $$\frac{a + bi}{c + di} = \frac{p}{q}$$

∴ qa + qbi = pc + pdi

Equating real and imaginary parts.

qa = pc and qb = pd.

⇒ $$a = \frac{p}{q}c \text{ and } b = \frac{p}{q}d$$

∴ $$\frac{a^2 + b^2}{c^2 + d^2} = \frac{\frac{p^2}{q^2}c^2 + \frac{p^2}{q^2}d^2}{c^2 + d^2} = \frac{p^2}{q^2}$$

13. We ignore $\frac{1}{7}$ and $\frac{1}{9}$ because no sum of other numbers is going to give 7th or 9th in the denominator.

Also, $\frac{1}{5}$ and $\frac{1}{10}$ are not enough to add up to anything $\left(\frac{1}{10}, \frac{2}{10}, \text{ and } \frac{3}{10}\right.$ are going to leave tenths left over no matter what else you add$\left.\right)$.

What's left is $\frac{1}{2}, \frac{1}{3}, \frac{1}{4}, \frac{1}{6}, \frac{1}{8}$.

Sum total of these is $\frac{11}{8}$. So we need all of them except $\frac{1}{8}$, which means $\frac{1}{2}+\frac{1}{3}+\frac{1}{6}$. Which is the only way to do this with egyptian fractions whose denominators are 10 or less. Hence maximum number of Egyptian fractions needed is 3.

14. L.C.M. of 9, 10 and 15 = 90

$\therefore$ 1936 = 90 × 21 + 46 (Remainder).

But a part of this remainder is 7.

so 46 –7 = 39 should be subtracted from 1936.

15. 6 – 4 = 2, 7 – 5 = 2, 8 – 6 = 2, 9 – 7 = 2, 10 – 8 = 2

L.C.M. of 6, 7, 8, 9, 10 = 2520;

Greatest number of 6 digits = 999999

2520 × 396 + 2079 = 999999

$\therefore$ Reminder = 2079.

Subtract 2079 from 999999, then we get

999999 – 2079 = 997920.

Subtract 2 from this number to get required number, which is 997918 and which will give the remainders 4, 5, 6, 7, 8 when divided by 5, 6, 7, 8, 9 respectively.

16. L.C.M. of 5, 6, 8, 9 and 12 is 360

Required number = 360 K + 1

= (13 × 27 + 9) K + 1

= (13 × 27) K + (9K + 1)

Now this number must be divisible by 13

$\therefore$ K = 10 and required number = 3601.

17. I. In case all the remaining three numbers are multiples of the prime number, HCF of the group is equal to that prime number. otherwise HCF of the group is 1. Thus, I is true.

II. **Case 1:** Let numbers be 2, 4, 6 and 8, where 2 is only prime number, then LCM = 24.

But LCM of 4, 6 and 8 = 24.

LCM = 24 & 2 = 24

Case 2 : Let numbers be 2, 9, 81 and 27 Then LCM = 162 and LCM of 9, 81 and 27 = 81.

Thus, LCM = 81 × 2 = 162

Hence II is not always true.

III. Let numbers be 2, 4, 6 and 8

$\therefore$ LCM = 24 and HCF = 2

Product of four numbers

= 2 × 4 × 6 × 8 = 384

Also, 384 ≠ 24 × 2 × 2.

Thus, III is not true.

18. (a. n)! = product of n consecutive natural number starting from 'a' which is atleast divisible by n!.
(n)! = product of n consecutive natural numbers.

For n = 2 :

(a. n)! = a(a + 1) and n! = 2

a(a + 1) is divisible by 2!.

For n = 3 :

(a n)! = a(a + 1)(a + 2) and n! = 6.

One of the factors of a(a + 1)(a + 2) is divisible by 3 and other by 2.

Thus, proceeding in this manner, (a. n)! and n! have HCF = n!

$\therefore$ H = n!.

19. Let HCF be h and LCM be l.

I. Let numbers be ah and bh.

Then abh = l and (a + b)h = m

$\Rightarrow$ (a – b)h = n

Using these ah and bh can be uniquely determined. Thus, I is true

II. If HCF = LCM, then two numbers are equal and same as HCF or LCM.

Thus, II is true.

III. $\frac{\text{LCM}}{\text{HCF}}$ = a prime i.e., $\frac{\ell}{h} = P$.

Then one of the numbers is equal to h and other is equal to ℓ. Thus, III is true.

20. Total distance travelled by the ball before coming to rest

= 12 + 2 (6 + 3 + 1.5 + ... to ?)

$= 12 + 2.\left(\frac{6}{1-1/2}\right)$

= 12 + 2(12) = 36m

21. Number of students who answered 1 or more questions wrongly = 2^{N-1}.

Number of students who answered 2 or more questions wrongly = 2^{N-2}

Hence number of students who answered 1 question wrongly

$$= 2^{N-1} - 2^{N-2} = 2^{N-2}$$

Similarly it can be shown that number of students who answered 2 questions wrongly

$$= 2^{N-2} - 2^{N-3} = 2^{N-3}$$

Similarly we can find number of students who answered K questions wrongly where $K \geq 3$

Hence total number of questions attempted wrongly

$$s = 2^{N-2}+2(2^{N-3})+3(2^{N-4})+\ldots + (N-1)(2^{\circ})+N(1) \ldots(i)$$

$$\therefore \quad \frac{s}{2} = 2^{N-3} + 2(2^{N-4}) + \ldots + (N-2)(2^{\circ}) + \frac{N-1}{2} + \frac{N}{2} \ldots(ii)$$

Substracting equations (*i*) from (*ii*)

$$\frac{s}{2} = 2^{N-2} + 2^{N-3} + \ldots + 2^{\circ} + \frac{1}{2}$$

$$\Rightarrow \quad s = 2^{N-1} + 2^{N-2} + 2^{N-3} + \ldots + 1$$

$$= 2^{N} - 1 = 8191$$

$$\Rightarrow \quad N = 13$$

22. Number of members in C_1 on May 1, 2003 = $x + 4a$

Number of members in C_2 on May 1, 2003 = xb^4

$$x + 4a = xb^4 \text{ and } a = 20x$$

$$\therefore \quad x(b^4 - 81) = 0$$

As $\quad x \neq 0, \quad b^4 - 81 = 0$

$$\therefore \quad b = 3$$

23. n^{th} terms are in the ratio of $\frac{2n+3}{n-11}$.

If $n = 1$ in $2n + 3$ series, then first term is

$T_1 = 2(1) + 3 = 5$

$n = 2$, then second term is $2(2) + 3 = 7$

$T_2 = 7$

$d = T_2 - T_1 = 7 - 5 = 2$

$\therefore \quad d_1 = 2$

If $n = 1$ in $n - 11$ series, then first term is $1 - 11 = -10$

If $n = 2$, then second term is $2 - 11 = -9$

$\therefore \; d_2 = T_2 - T_1 = -9 - (-10) = 1$

$$\therefore \quad \frac{d_1}{d_2} = \frac{2}{1} \text{ or } 2 : 1$$

24. $\frac{2}{3} + \frac{8}{9} + \frac{26}{27} + \frac{80}{81} + \ldots$ up to 20 terms

$$= \left(1-\frac{1}{3}\right)+\left(1-\frac{1}{9}\right)+\left(1-\frac{1}{27}\right)+\left(1-\frac{1}{81}\right)+ \ldots \text{ up to 20 terms}$$

$$= 20 - \left(\frac{1}{3}+\frac{1}{3^2}+\frac{1}{3^3}+\frac{1}{3^4}+\ldots+\frac{1}{3^{20}}\right)$$

$$= 20 - \frac{1}{3}\left(\frac{1-\left(\frac{1}{3}\right)^{20}}{1-\frac{1}{3}}\right)$$

$$= 20 - \frac{3^{20}-1}{2.3^{20}}$$

$$= \frac{40.3^{20} - 3^{20} + 1}{2.3^{20}} = \frac{39.3^{20}+1}{2.3^{20}}$$

25. Let $\quad s = 3 + 6x^2 + 9x^4 + 12x^6 + \ldots$

$$\therefore \quad x^2 s = 3x^2 + 6x^4 + 9x^6 + 12x^6 + \ldots$$

$$\Rightarrow \quad s - x^2 s = 3 + 3x^2 + 3x^4 + 3x^6 + \ldots$$

$$\Rightarrow \quad s(1-x^2) = 3\,[1 + x^2 + x^4 + x^6 + \ldots]$$

$$\Rightarrow \quad s(1-x^2) = 3\left[\frac{1}{1-x^2}\right]\left[\because |x| < 1\right]$$

$$\therefore \quad s = \frac{3}{\left(1-x^2\right)^2}$$

26. Average

$$= \frac{60\times56+50\times64+55\times72+62\times63+58\times50}{60+50+55+62+58}$$

$$= \frac{3360+3200+3960+3906+2900}{285} = 60.8$$

27. Increase after 5 years should have been

$7 \times 5 = 35$ years.

But since average remains same, difference in their age is 35 years.

28. Let average of the seven men be x.

$$\therefore \quad x = \frac{10\times6+(x+3)}{7}$$

$$\Rightarrow \quad 60 + x + 3 = 7x$$

$$\Rightarrow \quad x = \frac{63}{6} = 10.5$$

$\therefore$ Total amount = $10.5 \times 7 = 73.5$

29. Consider a set of three numbers a, b and c.

I. $AM = \frac{a+b+c}{3}$

Thus, a + b + c = 3AM.

But GM = $\sqrt[3]{abc}$ which cannot be found out.

Thus, I is false,

II. $AM = \frac{a+b+c}{3}$

∴ Sum of the numbers = a + b + c = 3AM.

Thus, II is true.

III. GM = $\sqrt[3]{abc}$.

But AM = $\frac{a+b+c}{3}$, which cannot be found out.

Thus, III is false.

30. I. Average of a set of numbers is greater than smallest and smaller than the greatest number of the set. Thus, I is true.

II. Consider five numbers a, b, c, d and e whose average is $\frac{a+b+c+d+e}{5}$

Now if each of them is increased by k, then we have average

$$= \frac{a+k+b+k+c+k+d+k+e+k}{5}$$

$$= \frac{a+b+c+d+e}{5} + \frac{5k}{5} = \text{old average} + k.$$

Thus, II is true.

III. Consider three numbers x, y and z whose average is $\frac{x+y+z}{3}$

Now if each of them is multiplied by m then

new average = $\frac{xm+ym+zm}{3}$

$$= m\left(\frac{x+y+z}{3}\right) = m \times \text{odd average}.$$

Thus, III is true.

31. Order of measures is

A, B, C, A + B, B + C, C + A, A + B + C.

Given: A + B + C = 180

Hence, average of the 7 measures

$$= \left[\frac{[(A)+(B)+(C)+(A+B)+(B+C)+(C+A)+(A+B+C)]}{7}\right]$$

$$= \frac{4}{7}(A+B+C) = \frac{4}{7} \times 180 = \frac{720}{7} \text{ kgs.}$$

32. Total age = 15 × 36

After two left, to tal age = 15 × 36 – 2 × 28

= 540 – 56

= 484 years.

Let the age of the person who returns be x years.

Then, new average = $\frac{484 + x}{14}$

Now, x lies between 20 and 36 (both inclusive)

∴ New average min = $\frac{484+20}{14}$ = 36 years

And New average max = $\frac{484+36}{14}$

= 37.14. years.

PERCENTAGE AND ITS APPLICATIONS

1. The ratio of their investments is 11 : 9 : 12.

So, Manager gets $\frac{1}{10}$(40000) = 4000 + his share

Remaining amount = 36000.

Asstt.Manager gets $\frac{1}{20}$(36000) = 1800 + his share

Remaining amount = 36000 – 1800 = 34200

∴ M's share = $\frac{11}{32} \times 34200 + 4000$

= 11756.25 + 4000

= ₹ 15756.25

≅ ₹ 15756

AM's share = $\frac{9}{32} \times 34200 + 1800$

= 9618.75 + 1800

= ₹ 11418.75 ≈ ₹ 11419

BA's share = $\frac{12}{32} \times 34200$ = ₹ 12825.

2. B's share in the profit = 250 per 2000

If B gets ₹ 900, then

total profit = $2000 \times \frac{18}{5} = 7200$

B's return on investment = 15%

∴ $\frac{900}{\text{B's investment}} \times 100 = 15$

∴ B's investment = 6000

Hence profits are shared in the ratio of their investments.

Total investments = $6000 \times \frac{7200}{900} = 48000$

∴ Investment made by A and C = 48000 – 6000

= 42000.

3. Interest paid by Bimal to Abhay for first six months

$$= \frac{50000}{2} \times \frac{8}{100} \times \frac{1}{2} = ₹\,1000$$

For last six months

$$= \frac{30000 - 20000}{2} \times \frac{8}{100} \times \frac{1}{2}$$

$$= ₹\,200$$

∴ Total salary paid to Bimal = 12 × 500 = ₹ 6000

Let total profit be ₹ x

∴ Abhay's earing = $1200 + \frac{x - 6000}{2}$

Bimal's earning = $6000 + \frac{x - 6000}{2} - 1200$

$$\therefore \quad 3\left(1200 + \frac{x - 6000}{2}\right) = 4800 + \frac{x - 6000}{2}$$

$$\Rightarrow \quad 3600 + \frac{3x}{2} - 9000 = 4800 + \frac{x}{2} - 3000$$

$$\Rightarrow \quad \frac{3x}{2} - \frac{x}{2} = 7200;$$

$$\therefore \quad x = ₹\,7200$$

4. Profit per stocks = $91\frac{3}{8} - 89\frac{1}{8} = ₹\,2\frac{2}{8}$

$$\therefore \quad \frac{45}{2\frac{1}{4}} = 20 \text{ stocks}$$

5. Income from a stock at 3% in 20 years at par

$$= \frac{100}{100} \times 3 \times 20 = 60$$

Income from a stock at 5% at 124 in 20 years

$$= \frac{100}{124} \times 5 \times 20 = 80.64.$$

6. He invests ₹ 60 and gets ₹ 100 share

Income = ₹ 5

Required % = $\frac{5}{60} \times 100(0.70) = 5.83\% = 5\frac{5}{6}\ \%$

7. Let number of meters to be examined be x.

Then 0.08% of x = 2.

$$\Rightarrow \quad \frac{0.08}{100} \times x = 2$$

$$x = \frac{2 \times 100}{0.08} = \frac{2 \times 100 \times 100}{8} = 2500$$

8. $\left(\frac{a}{100} \times b\right) + \left(\frac{b}{100} \times a\right) = \frac{ab}{100} + \frac{ab}{100} = \frac{2ab}{100}$

$$= \frac{2}{100} \times ab = 2\% \text{ of ab.}$$

9. Ryan's stock = 1.15 × Harland's stock

∴ Harland's stock = $\frac{9462}{1.15}$ = \$8228

Ryan's stock = 0.85 × Lambert's stock

∴ Lambert's stock = $\frac{9462}{0.85}$ = \$11132.

10. Let the fraction be $\frac{x}{y}$.

Let $x_1 = x + 0.25x = 1.25x$

$y_1 = y - 0.25x = 0.75x$

$$\therefore \quad \frac{x_1}{y_1} = \frac{1.25x}{0.75x} = \frac{5}{3}\frac{x}{y}$$

$$\therefore \text{ Increase} = \frac{\frac{5}{3}\frac{x}{y} - \frac{x}{y}}{\frac{x}{y}} \times 100 = \frac{2}{3} \times 100 = 66.67\%$$

Thus, resultant fraction is more than the original fraction by 67%. Thus III is true.

11. I. Let first number be x.

∴ Second number = 1.25x

Sum = x + 1.25x = 2.25x

Difference = 1.25x – x = 0.25x

Also, 9 × 0.25x = 2.25x

Thus I is true.

II. $\frac{56}{100} \times A = \frac{24}{100} \times B$

Multiplying by $\frac{49}{56}$ on both the sides.

$$\therefore \quad \frac{49}{56} \times \frac{56}{100} \times A = \frac{49}{56} \times \frac{24}{100} \times B$$

$$\Rightarrow \quad \frac{49}{100}A = \frac{21}{100}B$$

⇒ 49% of A = 21% of B.

Thus, II is true.

12. In the first round, let A get 'a' votes and B get 'b' votes.

Then, $a + b = 260$

i.e., $a = 260 - b$ ... (*i*)

In second round, B gets

$$b \times \frac{125}{100} = \frac{5b}{4}$$

and A gets $260 - \frac{5b}{4}$

Also, $\frac{5b}{4} - \left(260 - \frac{5b}{4}\right) = 2(a - b)$

$$\frac{10b}{4} - 260 = 2(260 - b - b)$$

$$\frac{10b}{4} - 260 = 520 - 4b$$

$$4b + \frac{10b}{4} = 520 + 260$$

$$b = 120$$

Hence first round, A gets 140 and B gets 120.

In second round A gets 110 and B gets 150.

$\therefore$ % decrease in votes received by A

$$= \frac{140 - 110}{140} \times 100 = 21.43\%$$

13. Non-water matter is 1% of 5000 = 50 gm.

After drying, this matter becomes 2% of the weight of the water-melon. Hence weight of the watermelon now is 2500 gm.

14. Let A = ₹ x and B = ₹ (5000 – x)

Then, $\frac{x \times 2 \times 20}{100} = \frac{(5000 - x) \times 3 \times 20}{100}$

$\Rightarrow$ $2x = 15000 - 3x$

$\Rightarrow$ $5x = 15000$

$\Rightarrow$ $x = 3000$

$A = x =$ ₹ 3000

and $B = 5000 - x$

$= 5000 - 3000$

$=$ ₹ 2000

$\therefore$ $\frac{A}{B} = \frac{3000}{2000} = \frac{3}{2}$

Thus, (*a*) is not true and (*b*) is true.

$A - B = 3000 - 2000 =$ ₹ 1000.

Thus, (*c*) is true.

(*d*) The total interest

$$= \frac{3000 \times 2 \times 20}{100} + \frac{2000 \times 3 \times 20}{100}$$

$=$ ₹ 2400.

Thus, (*d*) is true.

15. First investment gives a return

$$P\left(1 + \frac{1}{100}\right)^{100} - P$$

First term is $P\left(1 + \frac{1}{100}\right)^{100} = X$,

such that $2P < X \le 3P$

Hence return X – P lies between P and 2P

i.e., $2P - P < X - P \le 3P - P.$

i.e., $P < X - P \le 2P.$

Whereas second offer.

$$= \text{Return} = P\left(1 + \frac{100}{100}\right)^{1} - P$$

$$= 2P - P = P.$$

Alternatively

Option of investing at 1% for 100 years is obviously better than investing at 100% for 1 year, because total compound interest for 100 years will be atleast 100% of the total amount (1% for each year), plus whatever the compounding effect will have.

16. Growth rate of rat population in 3 months

$$= 20 \times \frac{3}{12} = 5\%.$$

Increase in first 3 months = $3200 \times 1.05 = 3360$

Also, net decrease in 3 months = 160

$\therefore$ Rat population $= 3360 - 160 = 3200$

In the same way, after every 3 months, rat population remains the same.

Hence, even after 3×8 months i.e., 2 years, the population is maintained

17. Let CP in A be ₹ x

$\therefore$ CP in B = ₹ 0.9x

$\therefore$ $0.9x + 15 + 10 = x$

$\Rightarrow$ $0.1x = 25$

$\Rightarrow$ $x = 250$

$\therefore$ Price in B $= 250 \times 0.9 =$ ₹ 225

18. Let CP of 5 dozen mangoes be ₹ x.

SP = ₹ 156 and Gain = 0.3x

$\therefore \quad 156 - x = 0.3x$

$\Rightarrow \quad x = 120$

$\therefore$ SP of 60 mangoes = $120 \times 1.6 = 192$

$\therefore$ SP per mango = ₹ 3.2

19. Let CP be ₹ x ; SP = 1.2x

But he gets only $1.2x \times 0.8 = 0.96x$

$\therefore \quad$ Loss = 0.04x

$\therefore \quad$ Loss = 4 %

20. Selling price = 800×1.25 = ₹ 1000

$\therefore$ Marked price = ₹ $\frac{1000}{0.95} = 1053$

21. CP = ₹ 550

$\therefore$ Marked price = 1.3×550 = ₹ 715

$\therefore$ Sales price = 0.8×715 = ₹ 572.

22. Profit or loss of the sale depends upon how much profit the person is earning on the sale of two articles.

If SP > CP of 1 article, there is a profit.

If SP < CP of 1 article, there is a loss.

If SP = CP of 1 article, there is not profit no loss.

Thus, only (*c*) is true.

23. *Following two key points need to be noted :*

(A) Producer is profit maximizing which implies she will set prices that maximize the profit function.

(B) Quantity demanded in market A is independent of quantity demanded in market B.

CASE 1:

In this case, producer charges different prices in markets A and B. This implies that she will maximize profits individually.

Equations can be re-written as

$$P_A = 210 - 5Q_A$$

$$P_B = 125 - 1.25\,Q_B$$

Profit in Market A = π_A

$= \text{Revenue} - \text{Cost}$

$= Q_A \times P_A - C_A$

$= Q_A \times (210 - 5Q_A) - (1000 + 5Q_A)$

At the maximum profit level, differential π_A with respect to Q_A is 0

or

$\therefore \quad 210 - 10Q_A - 5 = 0$

$\Rightarrow \quad Q_A = 20.5, P_A = 107.5$

and total revenue in market A

$= 20.5 \times 107.5$ = ₹ 2203.75

Similarly

Profit in Market B = π_B

= Revenue Cost

$= Q_B \times P_B - C_B$

$= Q_B \times (125 - 1.25\,Q_B) - (1000 + 5Q_B)$

At maximum profit level, differ-ential of π_B with respect to Q_B is 0

or $\quad 125 - 2.5Q_B - 5 = 0$

$\Rightarrow \quad Q_B = 48, P_B = 65$

and total revenue in market B

$= 48 \times 65$ = ₹ 3120

Using the above, we get

$$\frac{P_A}{P_B} = \frac{107.5}{65} \sim 1.65.$$

24. Data for 1 is available above. Consider case 2. In this case, prices across the two markets are the same. Hence, quantity equations can be added to give

$$Q = 142 - P$$

where Q is total quantity and P is price in both markets

$\Rightarrow \quad P = 142 - Q$

$\therefore \quad \pi = \text{Profit} = Q \times (142 - Q) - (1000 + 5Q)$

At the maximum profit level, differential of π with respect to Q is 0

$\therefore \quad Q = 68.5, \quad P = 73.5$

and $\quad$ Total Revenue = 5034.75

Given this, we note that quantities sold in both Case 1 and Case 2 are the same. Hence, the ratio is exactly 1.

25. Profit Difference between case 1 and case 2

= Revenue Difference (since quantity sold and hence, cost is the same)

$\therefore$ Profit difference = $2203.75 + 3120 - 5034.75$

$= 289$

TIME AND WORK

MCQ Type Questions

1. Top tap is operational till 18 litres is drawn out.

$\therefore$ Time after which the lower tap is usually open

$= 18 \times 6 = 108$ minutes.

$\therefore$ Time after which it is open now

$= 108 - 24 = 84$ minutes.

$\therefore \quad$ Litres drawn $= \frac{84}{6} = 14$ litres

$\therefore$ $18 - 14 = 4$ litres

were drawn by the new assistant.

$\therefore$ Time $= 4 \times 4 = 16$ minutes.

Alternatively

Upper tap was not used for 24 minutes.

$\therefore$ Litres drawn $= \dfrac{24}{6} = 4$ litres.

$\therefore$ Beer was drawn for $4 \times 4 = 16$ minutes.

2. Let pipe B be turned-off after x minutes.

Part filled in x minutes $= x\left(\dfrac{1}{36}+\dfrac{1}{48}\right) = \dfrac{7x}{144}$.

Part filled in 24 – x minutes $= \dfrac{24-x}{36}$

$$\therefore \quad \frac{7x}{144}+\frac{24-x}{36} = 1$$

$$\Rightarrow \quad \frac{7x+96-4x}{144} = 1$$

$$\Rightarrow \quad x = 16 \text{ minutes}$$

3. In a cycle of 3 hours : pipes P, Q and R are working as inlet pipes for 2 hours each and they are working as outlet pipes for an hour each. So part of tank filled in 3 hours.

$$= 2x\left(\frac{1}{8}+\frac{1}{10}+\frac{1}{12}\right)-\left(\frac{1}{8}+\frac{1}{10}+\frac{1}{12}\right)$$

$$= \frac{1}{8}+\frac{1}{10}+\frac{1}{12} = \frac{15+12+10}{120} = \frac{37}{120}$$

Hence they will take $\dfrac{120\times 3}{37} = 9\dfrac{27}{37}$ hours.

Thus, tank will be filled in 10th hour.

4. Let number of hours taken by the outlet and the 2 inlet pipes be 4x, 2x and x respectively. So in 1 hour the part of empty tank filled is

$$\frac{1}{x}+\frac{1}{2x}-\frac{1}{4x} = \frac{1}{16}$$

$$\frac{5}{4x} = \frac{1}{16}$$

$$\Rightarrow \quad x = 20$$

Hence, inlet pipe with higher efficiency fills the tank in 20 hours.

So in one hour, the desired pipes fill

$$\frac{1}{x}-\frac{1}{4x} = \frac{3}{4x} \text{ th of the tank.}$$

Tank gets filled in

$$\frac{4x}{3} = \frac{80}{3} \text{ hours} = 26\frac{2}{3} \text{ hours.}$$

5. In each case, train has to travel its own length, to pass each man.

If x km/hr is speed of the train, then

$$\text{Length} = \text{Relative speed} \times \text{Time}$$

$$= (x-2)\times\frac{9}{60\times 60}$$

$$= (x-4)\times\frac{10}{60\times 60}$$

$$\therefore \quad 9x - 18 = 10x - 40$$

$$\Rightarrow \quad x = 22 \text{ km/hr.}$$

$$\therefore \quad \text{Length} = (22-2)\times\frac{9}{3600}$$

$$= \frac{20\times 9}{3600} \text{ km} = 50 \text{ m.}$$

6. Policeman gains $= \dfrac{1}{6}-\dfrac{1}{10} = \dfrac{4}{60}$ km/min

$$= 66\frac{2}{3} \text{ m/min.}$$

Hence to gain100 m, time required

$$= \frac{100}{66\frac{2}{3}} = 1\frac{1}{2} \text{ min.}$$

Hence thief has gone ahead by

$$1\frac{1}{2}\times\frac{1}{10}\times 1000 \text{ metres} = 150 \text{ m.}$$

7. I. Ratio of speed = 1 : 2 : 4

If equal time is spent on each of these, then ratio of distance = 1 : 2 : 4

So, Sachin runs twice the distance that he jogs.

Thus, I is true.

II. He runs at 8 km/hr and therefore walks at 2 km/hr.

Hence II is not true.

III. Ratio of speed= 1 : 2 : 4

Ratio of time = 2 : 2 : 1

Ratio of distance

$= 2\times 1 : 2\times 2 : 4\times 1 = 2 : 4 : 4$

If Sachin covers a total distance of 10 km, then he

Walks for $\dfrac{2}{2+4+4}\times 10 = 2$ km,

Joggs for $\frac{4}{2+4+4} \times 10 = 4$ km.

Runs for $\frac{4}{2+4+4} \times 10 = 4$ km.

Total time taken is $2\frac{1}{2}$ hours $= \frac{5}{2}$ hours.

Thus, he walks for $\frac{2}{2+2+1} \times \frac{5}{2} = 1$ hour.

jogs for $\frac{2}{2+2+1} \times \frac{5}{2} = 1$ hour.

runs for $\frac{1}{2+2+1} \times \frac{5}{2} = \frac{1}{2}$ hour.

So speed while,

Walking $= \frac{2}{1} = 2$ km/hr

Jogging $= \frac{4}{1} = 4$ km/hr

Running $= \frac{4}{\frac{1}{2}} = 8$ km/hr.

Thus, III is true.

8. In 1 minute, monkey ascends 10 metres but he takes 1 minute to slip down 2 metres. Thus, at the end of 2 minutes, net ascending of the monkey is $= 10 - 2 = 8$ metres.

Thus, to have a net ascending of 8 metres, process of ascending and then slipping happens once. So, to cover 64 metres, above process is repeated $\frac{64}{8}$ or 8 times.

It is clear that in 8 such happenings, the monkey will slip 7 times, because 8th time, he will ascend to the top. Thus, in climbing 7 times and slipping 7 times, he covers (7×8) or 56 metres.

Time taken to cover 56 metres

$$= \frac{56 \times 2}{8} = 14 \text{ minutes}$$

Remaining distance $= 64 - 56 = 8$ metres

Time taken to ascend 8 metres

$$= \frac{8}{10} \text{ min} = \frac{4}{5} \text{ min.}$$

Total time taken $= 14$ minutes $+ \frac{4}{5}$ min

= 14 mins. 48 sec.

9. If they meet t hrs after 7 a.m,

$$10 \times \frac{18}{5} \times t + 10 \times \frac{18}{5} \times \frac{4}{3}\left(t - \frac{1}{3}\right) = 68$$

$\Rightarrow \quad 36t + 48t - 16 = 68$

$\Rightarrow \quad t = 1$ hr.

They meet at a distance of

$10 \times 60 \times 60 = 36000$ m

i.e., 36 km from A.

10. Two equations are

$$T_1 - T_2 = \frac{1}{2} + \frac{1}{2} = 1 \quad ...(i)$$

where T_1 and T_2 are time taken by trains A and B to cover the whole distance

and $$\frac{2}{T_1} + \frac{1.5}{T_2} = \frac{11}{30} \quad ...(ii)$$

Solving equations (*i*) and (*ii*), we get

$T_1 = 10$ hrs. and $T_2 = 9$ hrs.

11. When A runs 5 rounds, B runs 4 rounds (ratio of speeds)

∴ A passes B each time A has run 5 rounds or

$$5 \times \frac{1}{4} = \frac{5}{4} \text{ km.} = 1\frac{1}{4} \text{ km.}$$

$1\frac{1}{4}$ km is contained in 4 km.

3 times.

Hence A passes B thrice.

12. When A has run 250 yards, B should have run

$550 - 250 = 300$ yards.

∴ B runs $300 - 100 = 200$ yards in the same time taken by A to run 250 yards.

∴ Ratio of the speeds of A and B

$= 250 : 200 = 5 : 4$

If speed of A is 5x yards per minute, then speed at B = 4x yards/min.

Time taken by B to reach the starting point

$$= \frac{450}{4x} = \frac{112.5}{x} \text{ min.}$$

Time taken by A to reach the starting point

$$= \frac{550}{5x} = \frac{110}{x} \text{ min.}$$

∴ A wins by $\frac{2.5}{x}$ min

$\therefore$ Distance travelled by B in $\frac{2.5}{x}$ min.

$$= \left(\frac{2.5}{x}\right) \times (4x) = 10 \text{ yards}$$

Hence B is 10 yards behind A.

13. Let n_1 and n_2 be the speeds in km/hr and let n_1 be the greater speed i.e., $n_1 > n_2$.

$$\therefore \quad \frac{2}{n_2} = \frac{2}{n_1} + \frac{1}{60}.$$

$$\Rightarrow \quad n_1n_2 = 120(n_1 - n_2)$$

Faster runner covers $(n_1 - n_2)$ km. more than the slower runner in 1 hour.

Hence to cover one round of 2 km, he will take

$\frac{2}{n_1 - n_2}$ or $\frac{1}{2}$ hour.

$$\therefore \quad \frac{2}{n_1 - n_2} = \frac{1}{2}$$

$$\Rightarrow \quad n_1 - n_2 = 4$$

$$\Rightarrow \quad n_1 = 4 + n_2$$

$$\Rightarrow \quad n_2(4 + n_2) = 120 \times 4$$

$$\Rightarrow \quad n_2^2 + 4n_2 - 480 = 0$$

$$\Rightarrow \quad n_2 = 20 \text{ or } -24 \text{ (rejected)}$$

$$\therefore \quad n_1 = 4 + n_2 = 24 \text{ km/hr}$$

Hence speeds are 24 km/hr and 20 km/hr

Alternatively

Options may also be substituted to find the result.

14. Work done by A and B in 1 hr

$$= \frac{1}{4} + \frac{1}{5} = \frac{9}{20}^{\text{th}} \text{ part.}$$

$\therefore$ Ratio of work done by (A + B)

and $\quad C = \frac{9}{20} : \frac{1}{6}$

$\therefore$ Ratio of time taken

$$= \frac{1}{6} : \frac{9}{20} = 10 : 27$$

If C takes 27 hrs. (A + B) will take 10 hrs.

If C takes $4\frac{1}{2}$ hrs. (A + B) will take

$$\frac{10 \times 9}{2 \times 27} \cong 1 \text{ hr. } 40 \text{ mins.}$$

15. I. Work done by A, B and C together in 3 days

$$\frac{1}{8} + \frac{1}{12} + \frac{1}{16} = \frac{13}{48} \text{th of the work.}$$

Thus, in 9 days, total work done will be

$$3 \times \frac{13}{48} = \frac{39}{48} \text{th of the work}$$

$\therefore$ Work remaining $= 1 - \frac{39}{48} = \frac{9}{48}$th of the work is left of which A will finish $\frac{6}{48}$th of work on 10th day and B will finish remaining work on 11th day.

Thus, I is false

II. Work done by A, B, C together in 9 days $= \frac{39}{48}$

$$\text{Work left} = 1 - \frac{39}{48}$$

$$= \frac{9}{48} \text{ th of the whole.}$$

On the 10th day B will finish $\frac{1}{12}$th of the whole.

$$\therefore \quad \text{Work left} = \frac{9}{48} - \frac{1}{12}$$

$$= \frac{5}{48} \text{th of the whole.}$$

On 11th day C will finish $\frac{1}{16}$th of the work

$$\therefore \quad \text{Work left} = \frac{5}{48} - \frac{1}{16}$$

$$= \frac{2}{48} \text{th of the whole}$$

which will be done by A on the 12th day. Thus, work will be finished on the 12th day. Thus, II is false.

III. Work done by A, B and C in 9 days $= \frac{39}{48}$

Work left after 9 days

$$= 1 - \frac{39}{48} = \frac{9}{48} \text{th of the whole.}$$

On the tenth day $\frac{1}{16}$th will be finished by C.

$\therefore$ Work left $= \frac{9}{48} - \frac{1}{16}$

$= \frac{6}{48}$ th of the whole,

which will be finished by A on the eleventh day. Thus, III is true.

Numerical Type Questions

1. Total area to be painted

$= 10 \times 15$ (ceiling) $+ 2 \times 10 \times 10$ (wall) $+ 2 \times 10 \times 15$ (wall)

$= 650$ sq.ft

In one day Meera can paint

$$10 \times \frac{10}{2} = 50 \text{ sq.ft}$$

Ashok can paint

$$10 \times \frac{10}{3} = \frac{100}{3} \text{ sq.ft},$$

So, together they can paint $\frac{250}{3}$ sq.ft per day.

Let number of days Ashok works be x.

Work done in 9 days $= 50(9 - x) + \frac{250}{3}x = 650$

$$\therefore \quad 450 - 50x + \frac{250x}{3} = 650$$

$$\Rightarrow \quad \frac{100}{3}x = 200$$

$$\Rightarrow \quad x = 6 \text{ days.}$$

2. Let fixed interval be x

$$\therefore \quad \frac{5}{10} + \frac{5-x}{15} + \frac{5-2x}{20} = 1$$

$$\Rightarrow \quad (5)6 + (5 - x)4 + (5 - 2x)3 = 60$$

$$\Rightarrow \quad 30 + 20 - 4x + 15 - 6x = 60$$

$$\Rightarrow \quad x = \frac{1}{2} \text{ hour.}$$

3. Let pipe B alone take b hours longer than if both A and B were open.

$$8 = \sqrt{4 \times x}$$

$$\Rightarrow \quad 64 = 4 \times x$$

$$\Rightarrow \quad x = 16 \text{ hours.}$$

Alternatively

Part filled if both A and B are open $= \frac{1}{8}$

Part filled if only A is open $= \frac{1}{12}$

$\therefore$ Part filled if only B is open $= \frac{1}{8} - \frac{1}{12} = \frac{1}{24}$

$\therefore$ B alone would take 24 hours, i.e., B alone would take $24 - 8 = 16$ hours longer to fill the tank.

4. Since 5 tanks of same efficiency fill the boiler in 2 hours each tank individually takes

$2 \times 5 = 10$ hours to fill it.

Tank with half the efficiency will take

$10 \times 2 = 20$ hours.

In one hour, fraction of boiler getting filled when 4 fully efficient pipes with two as inlet and two as outlet and 5th acts at 50% efficiency as an inlet is

$$\frac{1}{10} + \frac{1}{10} + \frac{1}{20} - \frac{1}{10} - \frac{1}{10} = \frac{1}{20}$$

Hence it will take 20 hours to fill the boiler tank.

5. Here fraction of tank filled in 1 hour

$$\frac{1}{10} + \frac{1}{10} + \frac{1}{20} = \frac{5}{20} = \frac{1}{4} \text{th}$$

of the tank.If they are opened alternately, then tank gets filled in $4 \times 3 = 12$ hours

Hence half of the tank gets filled in $\frac{12}{2} = 6$ hours.

6. 4/7 of work is completed.

Remaining work $= 1 - \frac{4}{7} = \frac{3}{7}$

Remaining period $= 46 - 33 = 13$ days

Less work, less men(direct proportion)

Less days, more men(Indirect proportion)

More hours/day, less men(Indirect proportion)

work 4/7 : 3/7

Days 13 : 33 :: 117 : x

hrs/day 9 : 8

$$\frac{4}{7} \times 13 \times 9 \times x = \left(\frac{3}{7}\right) 33 \times 8 \times 117$$

$x = (3 \times 33 \times 8 \times 117) / (4 \times 13 \times 9)$

$= 198$ men

So, additional men to be employed

$= 198 - 117 = 81$

7. Let L is length of train in km and x is speed of the cyclist, Then

(*i*) $\frac{3600}{9}$ L = 42 – x = 400L

(*ii*) $\frac{3600}{5}$ L = 42 + x = 720 L

On adding, we get

$$1120 L = 84$$

$$\Rightarrow \quad L = \frac{84 \times 1000}{1120} = 75\text{m}$$

8. $$\frac{13}{\text{Speed upstream}} = 5$$

and $$\frac{28}{\text{Speed downstream}} = 5$$

$\therefore$ Speed upstream = $\frac{13}{5}$ km/hr.

and Speed downstream = $\frac{28}{5}$ km/hr.

$\therefore$ Speed of river $= \frac{1}{2}\left(\frac{28}{5} - \frac{13}{5}\right)$

$$= \frac{1}{2} \times \frac{15}{5} = 1.5 \text{ km/hr}$$

9. Graph of speed vs time is plotted as shown. Area under the graph and time axis gives the distance.

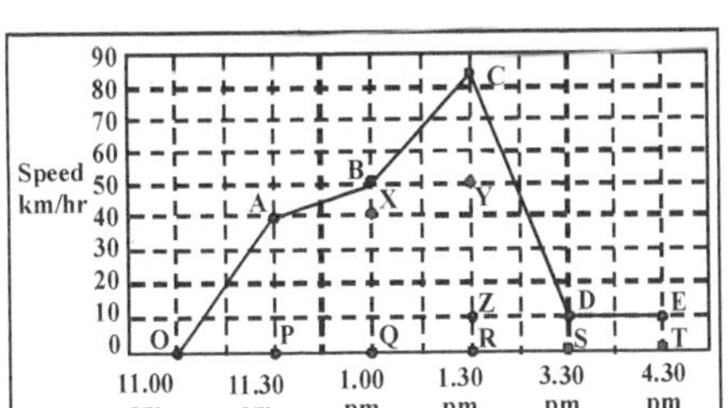

Required distance

= A(Δ OAP) + A(Δ BXA) + A (□ APQX)

$$= \frac{1}{2} \times \left(\frac{1}{2}\right) \times 40 + \frac{1}{2} \times 10 \times \left(\frac{3}{2}\right) + \left(\frac{3}{2}\right) \times 40$$

$$= 10 + \frac{30}{4} + 60 = 77.5 \text{ km.}$$

10. Distance covered are the same.

Speed in the latter is maintained at 10 kmph.

$\therefore$ 77.5 = 10 × x

$\Rightarrow$ x = 7.75 hours.

Hence, clock time will be 4.30 + 7 hours 45 mins = 12.15 a.m.

11. Average speed = $\frac{\text{total distance travelled}}{\text{total time taken}}$

Distance travelled from 1.00 pm to 4.30 p.m.

= A(□BQRY) + A(ΔBYC) + A(ΔCZD) + A(□ZRTE)

$$= 50 \times \left(\frac{1}{2}\right) + \left(\frac{1}{2}\right) \times \left(\frac{1}{2}\right) \times 35 + \frac{1}{2} \times (2) \times 75 + 3 \times 10$$

$$= \frac{50}{2} + \frac{35}{4} + 75 + 30$$

= 130 + 8.75

= 138.75 km.

Total distance travelled = 77.5 + 138.75 = 216.25 km.

$\therefore$ Average speed = $\frac{216.25}{5.5}$ = 39.3 kmph.

12. While deer makes 10 leaps, cheetah makes 6 leaps.

$\therefore$ Distance covered in 10 leaps by deer = 10 × 1 = 10m

Distance covered in 6 leaps by cheetah = 6 × 2 = 12 m

Hence in six leaps, cheetah gains 12 – 10 = 2 m over the deer.

Cheetah has to gain 100 m over the deer.

So, total leaps required by cheetah

$$= \frac{6}{2} \times 100 = 300.$$

13. 50 leaps of the hare $50 \times 1\frac{3}{4} = \frac{175}{2} = 87.5$ m.

So, hound should gain $\frac{175}{2}$ m over the hare.

When hound travels $3 \times 2\frac{3}{4}$ m, then

hare travels $4 \times 1\frac{3}{4}$ m.

Hence in 3 leaps of the hound the hound gains

$$\frac{33}{4} - \frac{28}{4} = \frac{5}{4} \text{ metres}$$

$\therefore$ Number of leaps required $= \dfrac{\frac{175}{2}}{\frac{5}{4}} \times 3$

$= \dfrac{175}{2} \times \dfrac{4}{5} \times 3 = 210$ leaps.

14. Fuel required for first 20 km $= \dfrac{20}{8} = 2.5$ litres.

Fuel required for next 60 km $= \dfrac{60}{15} = 4$ litres.

$\therefore$ Petrol left $= 10 - 6.5 = 3.5$ litres.

Now, he has to travel $= 115 - (60 + 20)$

$= 35$ km in 3.5 litres.

So he can go at a speed of either 45 km/hr or 80 km/hr as mileage is 10 km/litre in both cases. But since he reaches in minimum time, he travels at 80 km/hr

$\therefore$ Time taken $= \dfrac{20}{40} + \dfrac{60}{60} + \dfrac{35}{80} = \dfrac{155}{80}$

$= 1 + \dfrac{75}{80}$ hours

= 1 hours 56 minutes and 25 seconds.

15. Total distance travelled = 115 km

Total time taken $= \dfrac{155}{80}$ hours

$\therefore$ Average speed $= \dfrac{115}{\frac{155}{80}} = 60$ km/hr

$\therefore$ Fuel consumption $= \dfrac{115}{15} = 7\dfrac{2}{3}$ litres.

So, fuel saved $= 10 - 7\dfrac{2}{3} = 2\dfrac{1}{3}$ litres.

RATIO, PROPORTION AND MIXTURES

MCQ Type Questions

1. Stage $\dfrac{\text{Wine left}}{\text{Original quantity}}$

0 $\quad \dfrac{10}{10}$

1 $\quad \dfrac{8}{10}$

2 $\quad \dfrac{8}{10} \times \dfrac{6}{10}$

3 $\quad \dfrac{8}{10} \times \dfrac{6}{10} \times \dfrac{4}{10}$

4 $\quad \dfrac{8}{10} \times \dfrac{6}{10} \times \dfrac{4}{10} \times \dfrac{2}{10} \Rightarrow \dfrac{4 \times 3 \times 2 \times 1}{5^4} = \dfrac{4!}{5^4}$

$\therefore \quad \dfrac{\text{Wine}}{\text{Water}} = \dfrac{\frac{4!}{5^4}}{1 - \frac{4!}{5^4}} = \dfrac{4!}{5^4 - 4!}$

2.

(1) (2)

	C_1	C_2
Operation 1	20 litres alcohol	Empty
Operation 2	$(20 - x)$ alcohol	x litres alcohol $(20 - x)$ litres water
Operation 3	$(20 - x)$ litres alcohol + xlitres mixture i.e., $\left(\dfrac{x^2}{20}\text{litres alcohol}\right)$	$(20 - x)$ litres mixture

Now $\dfrac{1}{3}$rd of C_1 poured into C_2

Quantity of alcohol in $C_1 = C_2 = 10$ litres

In C_1 : $\quad \dfrac{2}{3}\left[20 - x + (x)\left(\dfrac{x}{20}\right)\right] = 10$

Solving we get, $\quad x = 10$ litres

Hence 10 litres of alcohol was initially poured from C_1 to C_2.

3. Required ratio $= \dfrac{x - y}{x + y} \times \dfrac{y^2 + xy}{x^2 - xy} = \dfrac{y}{x} = y : x$

4. *Given* : $\dfrac{x}{y} = a$ and $\dfrac{y}{x} = b$

If $\dfrac{x}{y} > 1$, then $\dfrac{y}{x} < 1$,

$\Rightarrow \quad \dfrac{x}{y} + \dfrac{y}{x} > 1$.

Hence option (*c*) is correct.

5. Total charge = fixed + variable

$$T_C = m + kx$$

where m and k are constants.

$\therefore$ $300 = m + k \times 20$

Also, $550 = m + k \times 45$

Solving for m and k; we get

$$m = 100; k = 10.$$

Hence, for 100 grams,

total charge = 100 + 10 × 100 = ₹ 1100.

Q. 6 and 7.

Using given data, we can draw following table

	Male	Female	
White Collar	4.8x	3.2x	8x
Blue Collar	2.2x	0.8x	3x
	7x	4x	11x

Now, all the questions can be answered.

6. $\dfrac{\text{Female white collar employess}}{\text{Male blue collar employees}} = \dfrac{3.2x}{2.2x} = \dfrac{16}{11}$

7. Since $0.8x = 48$

$\Rightarrow$ $x = 60$

$\therefore$ Required difference = $4.8x - 0.8x$

$= 4.8 \times 60 - 0.8x = 240$

8. Let 10x be taken from bottle A and 7y be taken from bottle B.

$$\text{Total salt taken} = \left(\frac{7}{10} \times 10x\right) + \left(\frac{4}{7} \times 7y\right)$$

$$= 7x + 4y$$

Total water taken = 3x + 3y.

$\therefore$ $\dfrac{7x+4y}{3x+3y} = \dfrac{2}{1}$

$\Rightarrow$ $7x + 4y = 6x + 6y$

$\Rightarrow$ $x = 2y$

$\therefore$ $\dfrac{x}{y} = \dfrac{2}{1}$

$\therefore$ Ratio of quantities $= \dfrac{10x}{7y}$

$$= \frac{10}{7} \times \frac{2}{1} = \frac{20}{7}$$

9. Let speed of the motor in stage 1 be x rpm. Then its speed in stage 2 be $(x^2) = ax^2$ (say) and in stage 3 be $(ax^2)^3 = b.a^3x^6$ (say)

Given that $x = \dfrac{50}{3}$ ...(i)

$ax^2 = 2500$ ...(ii)

$ba^3x^6 = 12500$ ...(iii)

Substituting value of x in (ii), we get

$$a = \frac{2500}{(50)^2} \times 3^2 = 9$$

Again, substituting values of a and x, we get

$$b = \frac{12500}{(2500)^3} = \frac{1}{125} \times 10^{-4}$$

Given : $x = \dfrac{25}{6}$ rpm

$$\therefore \text{ Speed in stage 3} = \frac{10^{-4}}{125} \times (9)^3 \times \left(\frac{25}{6}\right)^6$$

$$= \frac{10^{-4}}{5 \times 25} \times (3)^6 \times \frac{(25 \times 5) \times (5 \times 25^4)}{(2)^6 \cdot (3)^6}$$

$$= \frac{5 \times 25^4}{2^6 \times 10^4} = \frac{5^5}{2^{10}} \text{ rpm}$$

10. Let bag 1 and 2 have

	x mangoes	y mangoes
Step 1	$\frac{x}{2}$	$y + \frac{x}{2}$

But $\dfrac{x}{2} : y + \dfrac{x}{2} = y : x$

Step 2	$\frac{2}{3}y + \frac{2}{3}x$	$\frac{1}{3}y + \frac{1}{3}x$

$$\therefore \text{ Required ratio} = \frac{\frac{2}{3}y + \frac{2}{3}x}{\frac{1}{3}y + \frac{1}{3}x} = \frac{2\left(\frac{x+y}{3}\right)}{1\left(\frac{x+y}{3}\right)}$$

$$= 2 : 1$$

Numerical Type Questions

1. Let quantity of wine be a, i.e., capacity of the cask be 'a'

$$\text{i.e., } \frac{\text{Quantity of wine}}{\text{Quantity of water}} = \frac{\left(1 - \frac{5}{a}\right)^2}{1 - \left(1 - \frac{5}{a}\right)^2}$$

$$\therefore \quad \frac{361}{39} = \frac{\left(1 - \frac{5}{a}\right)^2}{1 - \left(1 - \frac{5}{a}\right)^2}$$

$$\Rightarrow \quad \frac{361}{39} - \frac{361}{39}\left(1-\frac{5}{a}\right)^2 = \left(1-\frac{5}{a}\right)^2$$

$$\Rightarrow \quad \left(1-\frac{5}{a}\right) = \sqrt{\frac{361}{400}} = \frac{19}{20}$$

$$\Rightarrow \quad \frac{5}{a} = \frac{1}{20}$$

$$\therefore \quad a = 100 \text{ litres}$$

2. Quantity of wheat in 400 kg of mixture

$$= 0.6 \times 400 = 240 \text{ kg}$$

Quantity of barley = 400 – 240 = 160 kg

Let x kg of barley be added to 400 kg of the mixture

$$\therefore \quad \frac{240}{160+x} = \frac{53\frac{1}{3}}{46\frac{2}{3}} = \frac{160}{140} = \frac{8}{7}$$

$$\Rightarrow \quad 240 \times 7 = 8 \times 160 + 8x$$

$$\Rightarrow \quad x = 50 \text{ kg}$$

3. Alcohol content in 1st jar = 0.25 × 3 = 0.75 litre

Alcohol content in 2nd jar = 0.75 × 5 = 3.75 litres

∴ Total alcohol in 9 litre cask = 0.75 + 3.75

= 4.5 litres

∴ Percentage of alcohol in the cask

$$= \frac{4.5}{9} \times 100 = 50\%$$

4. Cost price of milk = ₹ 5/litre

$$\text{Cost price of mixture} = \frac{100}{133\frac{1}{3}} \times 6 = ₹\ 4.5/\text{litre}$$

$$\therefore \quad \frac{\text{Quantity of water}}{\text{Quantity of milk}} = \frac{5-4.5}{4.5-0} = \frac{1}{9}$$

i.e.,

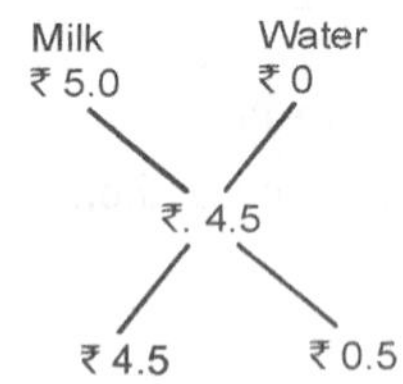

Milk : Water = 9 : 1

5.

	Alcohol	Grape juice	Alcohol : Grape
Tank 1	20%	80%	1 : 4
Tank 2	40%	60%	2 : 3

Ratio of grape juice to alcohol in half filled wine glass

$$= \frac{2\times\frac{4}{5}+3\times\frac{3}{5}}{2\times\frac{1}{5}+3\times\frac{2}{5}} = \frac{17}{8}$$

So, half capacity of glass = 17 + 8 = 25

So, full capacity = 25 × 2 = 50

Now rest 25 parts has to be filled with alcohol.

So, ratio of the grape juice to alcohol

= 17 : (8 + 25) = 17 : 33.

Alternatively,

As the wine glass is filled to half by taking two parts from tank 1 and three parts from tank 2 and second half is equivalent to five parts which is pure alcohol,

percentage concentration of grape juice

$$= \frac{2\times 80+3\times 60+5\times 10}{2+3+5} = 34\%$$

Hence ratio of grape juice to alcohol

= 34 : 66 = 17 : 33.

6. S.P of 1 kg of mixture = ₹ 68.20, Gain =10%

$$\text{S.P of 1 kg of mixture} = ₹\left(\frac{100}{110}\times 68.2\right) = ₹\ 62.$$

Cost of 1 kg tea of 1st kind	Cost of 1 kg tea of 2nd kind
60	65
Mean Price ₹ 62	
3	2

∴ Required ratio = 3 : 2.

7. Amount of milk left after 3 operations

$$= 40\left(1-\frac{4}{40}\right)^3$$

$$= 40 \times \frac{9}{10}\times\frac{9}{10}\times\frac{9}{10}$$

= 29.16 lit

8.

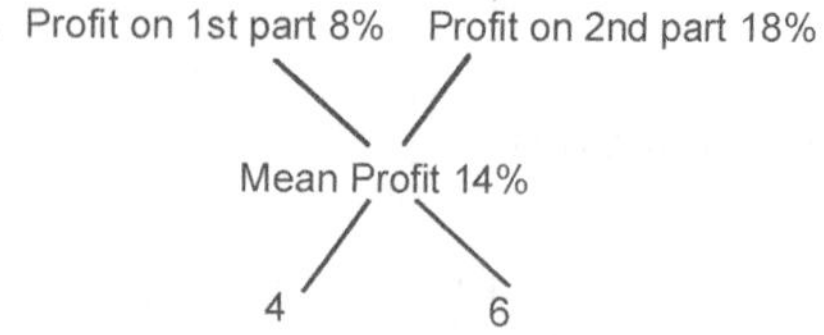

Ratio of 1st and 2nd parts = 4 : 6 = 2 : 3.

$$\text{Quantity of 2nd part} = \frac{3}{5} \times 1000 \text{ kg} = 600 \text{ kg}.$$

9. Ratio of volumes = $\frac{3}{14}:\frac{4}{14}:\frac{7}{14}$

Ratio of densities = $\frac{5}{13}:\frac{2}{13}:\frac{6}{13}$

$\therefore$ Ratio of weights = $\frac{3}{14}\times\frac{5}{13}:\frac{4}{14}\times\frac{2}{13}:\frac{7}{14}\times\frac{6}{13}$

= 15 : 8 : 42

... (Mass = Volume × Density)

Weight of the third substance = $\frac{42}{65}\times 130 = 84\ \ell bs.$

10. Let number of 50p coins be 3x, number of 25p coins be 4x and number of rupee coins be $2\frac{1}{2}$x.

$$\therefore \quad 2\frac{1}{2}x+\frac{1}{2}\times 3x+\frac{1}{4}\times 4x = 210$$

$$\Rightarrow \quad \frac{5}{2}x+\frac{3}{2}x+x = 210$$

$$\Rightarrow \quad 5x = 210$$

$$\Rightarrow \quad x = 42$$

$\therefore$ Number of rupee coins = $\frac{5}{2}\times 42 = 105.$

PERMUTATIONS AND COMBINATIONS & PROBABILITY

1 and 2

Given : $P(A\cup B) = 0.6$

and $P(A) = 0.3$

1. If A and B are mutually exclusive events, then

$$P(A\cap B) = 0$$

$$\therefore \quad P(A\cup B) = P(A) + P(B)$$

$$0.6 = 0.3 + P(B)$$

$$\Rightarrow \quad 0.3 = P(B)$$

Hence $P(B) = 0.3$

2. If A and B are independent events, then

$$P(A\cap B) = P(A).P(B)$$

$$\therefore \quad P(A\cup B) = P(A) + P(B) - P(A\cap B)$$

$$= P(A)+ P(B) - P(A).\ P(B)$$

$$P(A\cap B) = P(A) + P(B)\ [1 - P(A)]$$

$$\Rightarrow \quad 0.6 = 0.3 + P(B)\ (1 - 0.3)$$

$$0.3 = P(B)(0.7)$$

$$\Rightarrow \quad \frac{0.3}{0.7} = P(B)$$

$$\therefore \quad P(B) = \frac{3}{7}$$

3. When a die is rolled, for each number the probability of getting it is $\frac{1}{6}$.

When number on the dice is 1, 3 or 5 Raju receives ₹ 2, ₹ 6 and ₹10 respectively.

When the number on the die is 2, 4 or 6, then Raju receives ₹ 6, ₹ 12 and ₹ 18 respectively.

$\therefore$ Expected value= $\frac{1}{6}$ [2 + 6 + 10 + 6 + 12 + 18]

= $\frac{54}{6}$ = ₹ 9

Hence to make an average profit of ₹ 7 per throw, Raju must pay ₹ 2 (9 – 7) for each time to throw the die.

4. 7 letters can be arranged in 7 addressed envelopes in 7! ways.

$$\therefore \quad n(S) = 7!$$

We can arrange 7 letters into corresponding 7 addressed envelopes in only one way.

$\therefore$ Required probability = $\frac{1}{7!}$.

5. If six letters can be placed in their corresponding addressed envelopes, then seventh letter is also in the correct envelope.

Hence required probability is 0.

6, 7 and 8.

Three cards can be drawn from 52 cards in $^{52}C_3$ ways

$$n(S) = {}^{52}c_3$$

6. Since we have 4 suits i.e., Diamonds, Spades, Clubs and Hearts and each suit contains 13 cards.

Three cards can be selected from 13 cards in $^{13}C_3$ ways.

$\therefore$ Number of favorable outcomes = 4. $^{13}C_3$

Required probability = $4.\frac{^{13}C_3}{^{52}C_3}$

7. From four suits we can select 3 suits in $^4C_3$ ways.

One card can be selected from each suit in $^{13}C_1$. $^{13}C_1$. $^{13}C_1$

$\therefore$ Number of favorable outcomes

$= {}^4C_3.\ {}^{13}C_1.\ {}^{13}C_1.\ {}^{13}C1$

$\therefore$ Required probability = $\frac{4(13)^3}{^{52}C_3}$

8. Each suit contains 9 number cards. 2 number cards can be drawn from 9 cards in ${}^9C_2$ ways..

Third card can be selected from the remaining 27 number cards in ${}^{27}C_1$ ways.

Hence total number of favorable outcomes is $4({}^9C_2)({}^{27}C_1)$

$$\text{Required probability} = \frac{4\left({}^9C_2\right)\left({}^{27}C_1\right)}{{}^{52}C_3}$$

9. Total number of balls in the bag = 5 + 7 = 12.

5 balls can be drawn from 12 balls in ${}^{12}C_5$ ways

Number of ways of drawing 3 red balls and 2 white balls from 5 white balls and 7 red balls is ${}^5C_2 . {}^7C_3$

$$\therefore \text{ Required probability} = \frac{{}^5C_2 . {}^7C_3}{{}^{12}C_5}$$

$$= \frac{10 \times 35}{792} = \frac{175}{396}$$

10. Total number of coins in bag = 6 + 5 + 4 = 15.

5 coins can be drawn from 15 coins in ${}^{15}C_5$ ways.

When 4 coins are one rupee coins and fifth coin is a two rupee coin the amount will be minimum.

$\therefore$ Number of favourable cases = ${}^4C_4 . {}^{45}C_1 = 5$

Probability that the amount will be minimum is

$$P(E) = \frac{5}{{}^{15}C_5} = \frac{5}{3003}$$

$$\therefore \quad P(\bar{E}) = 1 - P(E) = 1 - \frac{5}{3003} = \frac{2998}{3003}$$

$\therefore$ Odds in favour of the required event = 5: 2998

11.

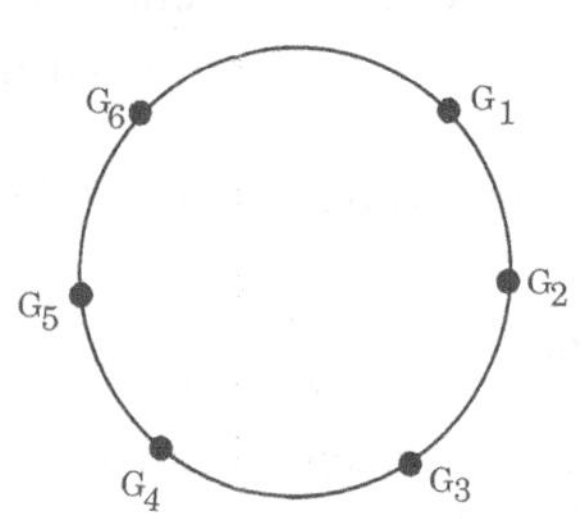

Six girls can be arranged around a circle in 5! ways.

Now, the six boys can be arranged in the six places in between the girls in 6! ways.

Hence, required number of ways = 5 ! 6!

12. Of the 10 places available, there are 5 even places. Digits 2, 4, 6, 4 and 2 can be placed in these 5 even places in $\frac{5!}{2!2!}$ ways.

Remaining digits 1, 3, 1, 1 and 9 can be placed in the remaining 5 places in $\frac{5!}{3!}$ ways.

Hence, required number of ways

$$= \frac{5!}{2!2!}\frac{5!}{3!} = \frac{(5!)^2}{(2!)^2 3!}$$

13. Number of ways of inviting at least one friend is

$$2^{12} - 1 = 4095$$

14. Number of ways of inviting at least 10 friends

$$= {}^{12}C_{10} + {}^{12}C_{11} + {}^{12}C_{12}$$
$$= 66 + 12 + 1 = 79$$

15. Each prize can be distributed in 3 ways, so, 6 prizes can be distributed in 3^6 ways.

16. Number of ways of dividing 16 books equally among 4 boys

$$= \frac{16!}{4!4!4!4!} = \frac{16!}{(4!)^4}$$

17. Number of ways of dividing 16 books equally into 4 parcels

$$= \frac{16!}{4!(4!)^4} = \frac{16!}{(4!)^5}$$

18. There are four vowels viz., U, A, I, E in the word "ILLUSTRATE".

First letter must be S and the last letter can be any one of the vowels U, A, I and E.

Remaining 8 places can be filled with the remaining 8 letters in $\frac{8!}{2!2!}$ ways.

Hence, required number of arrangements is

$$= 4 \times \frac{8!}{2!2!} = 8!$$

19. Prof. Balamurali and Asst. B of Sheshadri cannot be included together in the delegation.

Hence, number of ways of forming delegation

$$= {}^{10}C_5\,{}^6C_3 - {}^9C_4\,{}^5C_2$$

20. Possible ways of arranging the pencils are:

RBWBWBR, WBRBRBW, BWBRWRB, BRBWRWB, BWRBWRB, BWRRBRWB, BRWBRWB, BRWBWRB, BWRWBRB and BRWBBWB

i.e. a total of 10 possible arrangements.